PROJECT PHYSICS

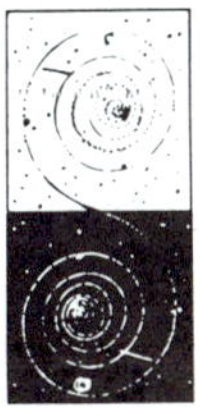

TEXT

Directors of Harvard Project Physics

F. James Rutherford
Department of Science Education, New York University

Gerald Holton
Department of Physics, Harvard University

Fletcher G. Watson
Harvard Graduate School of Education

Special Consultants on the 1981 edition of the Text:
Professor Jay Blake, Harvard University
Dr. Thomas von Foerster, American Institute of Physics
Professor Irving Kaplan, Massachusetts Institute of Technology

Editorial Development: William N. Moore, Roland Cormier, Lorraine Smith-Phelan
Editorial Processing: Margaret M. Byrne, Regina Chilcoat, Holly Massey
Art, Production, and Photo Resources: Vivian Fenster, Fred C. Pusterla, Robin M. Swenson, Annette Sessa, Beverly Silver, Anita Dickhuth, Dorina Virdo
Product Manager: Laura Zuckerman
Advisory Board: John Taggart, Maurice E. Fey, Norman Hughes, David J. Miller, John W. Griffiths, William L. Paul
Consultant: John Matejowsky
Researchers: Pamela Floch, Gerard LeVan

Acknowledgments begin on page 742
Picture Credits begin on page 744

Printed in the United States of America

ISBN 0-03-055141-2

1234-032-98765432

Science is an adventure of the whole human race to learn to live in and perhaps to love the universe in which they are. To be a part of it is to understand, to understand oneself, to begin to feel that there is a capacity within man far beyond what he felt he had, of an infinite extension of human possibilities. . . .

I propose that science be taught at whatever level, from the lowest to the highest, in the humanistic way. It should be taught with a certain historical understanding, with a certain philosophical understanding, with a social understanding and a human understanding in the sense of the biography, the nature of the people who made this construction, the triumphs, the trials, the tribulations.

I. I. Rabi, Nobel Laureate in Physics

Preface

Background

The Project Physics Course is based on the ideas and research of a national curriculum development project that worked for eight years.

Preliminary results led to major grants from the U.S. Office of Education and the National Science Foundation. Invaluable additional financial support was also provided by the Ford Foundation, the Alfred P. Sloan Foundation, the Carnegie Corporation, and Harvard University. A large number of collaborators were brought together from all parts of the nation, and the group worked together intensively for over four years under the title Harvard Project Physics. The instructors serving as field consultants and the students in the trial classes were also of vital importance to the success of Harvard Project Physics. As each successive experimental version of the course was developed, it was tried out in schools throughout the United States and Canada. The instructors and students reported their criticisms and suggestions to the staff in Cambridge. These reports became the basis for the next year's revision. The number of participating instructors during this period grew to over 100. Five thousand students participated in the last year of tryout in a large-scale formal research program to evaluate the results achieved with the course materials. Thereafter, the trial materials were again rewritten. The final version has been revised twice for new editions.

Aims

From the beginning, Harvard Project Physics had three major goals in mind. These were to design a humanistically oriented physics course, to attract more students to the study of introductory physics, and to find out more about the factors that influence the learning of science. The last of these goals involved extensive educational research, and has been reported to the teaching profession in books and journals.

The challenge facing us was to design a humanistic course that would be useful and interesting to students with widely differing skills, backgrounds, and career plans. In practice, this meant designing a new couse that would have the following effects:

1. To help students increase their knowledge of the physical world by concentrating on ideas that characterize physics as a science at its best, rather than concentrating on isolated bits of information.

2. To help students see physics as the wonderfully many-sided human activity that it really is. This meant presenting the subject in historical and cultural perspective,

and showing that the ideas of physics have a tradition as well as ways of evolutionary adaptation and change.

3. To increase the opportunity for each student to have immediately rewarding experiences in science even while gaining the knowledge and skill that will be useful in the long run.

4. To make it possible for instructors to adapt the course to the wide range of interests and abilities of their students.

5. To take into account the importance of the instructor in the educational process, and the vast spectrum of teaching situations that prevail.

Unhappily, it is not feasible to list in detail the contributions of each person who participated in some part of Harvard Project Physics. Previous editions have included a partial list of the contributors. We take particular pleasure in acknowledging the assistance of Dr. Andrew Ahlgren of the University of Minnesota. Dr. Ahlgren was invaluable because of his skill as a physics instructor, his editorial talent, his versatility and energy, and above all, his commitment to the goals of Harvard Project Physics.

We would also especially like to thank Ms. Joan Laws, whose administrative skills, dependability, and thoughtfulness contributed so much to our work. Holt, Rinehart and Winston, Publishers, of New York, provided the coordination, editorial support, and general backing necessary to the large undertaking of preparing the final version of all components of the Project Physics Course. Damon-Educational Division located in Westwood, Massachusetts, worked closely with us to improve the engineering design of the authorized laboratory apparatus and to see that it was properly integrated into the program.

Since their last use in experimental form, all of the instructional materials have been more closely integrated and rewritten in final form. The course now consists of a large variety of coordinated learning materials of which this textbook is only one. With the aid of these materials and the guidance of the instructor, with the student's own interest and effort, every student can look forward to a successful and worthwhile experience.

In the years ahead, the learning materials of the Project Physics Course will be revised as often as is necessary to remove remaining ambiguities, to clarify instructions, and to continue to make the materials more interesting and relevant to the students. We therefore urge all who use this course to send to us (in care of Holt, Rinehart and Winston, Publishers, 383 Madison Avenue, New York, New York 10017) any criticisms or suggestions they may have. And now, welcome to the study of physics!

F. James Rutherford
Gerald Holton
Fletcher G. Watson

Contents

Unit 1 • Concepts of Motion

Unit 2 • Motion in the Heavens

Unit 3 • The Triumph of Mechanics

Unit 4 • Light and Electromagnetism

Unit 5 • Models of the Atom

Unit 6 • The Nucleus

UNIT

Concepts of Motion

CHAPTER 1 **The Language of Motion**
CHAPTER 2 **Free Fall: Galileo Describes Motion**
CHAPTER 3 **The Birth of Dynamics: Newton Explains Motion**
CHAPTER 4 **Understanding Motion**

PROLOGUE

It is January 1934 in the city of Paris. A husband and wife are at work in a university laboratory. They are exposing a piece of ordinary aluminum to a stream of tiny charged bits of matter called alpha particles. Stated so simply, this hardly sounds like an important event. But look more closely, for it is important indeed. Later you will look at the technical details, but for now they will not get in the way of the story.

The story is something of a family affair. The husband and wife are the French physicists Frédéric Joliot and Irène Curie. The alpha particles they are using in their experiment are shooting from a piece of naturally radioactive metal. This metal is polonium, first identified 36 years before by Irène's parents, Pierre and Marie Curie, the discoverers of radium. What Frédéric and Irène have found is that when common aluminum is bombarded by alpha particles, it too becomes radioactive for a short time.

This was a surprise. Until that moment, a familiar, everyday substance becoming artificially radioactive had never been

Physicist Enrico Fermi (1901–1954).

observed. But physicists in the laboratory cannot force new phenomena on nature. They can only show more clearly what nature is like. Scientists know now that this type of radioactivity occurs quite often. It happens, for example, in stars and in the atmosphere when it is bombarded by cosmic rays.

Though it made few, if any, newspaper headlines, the news was exciting to scientists and traveled rapidly. Enrico Fermi, a young physicist at the University of Rome, became intrigued by the possibility of repeating the experiment. But Fermi added an important alteration. The story is told in the book *Atoms in the Family,* written by Enrico Fermi's wife, Laura.

> ... he decided he would try to produce artificial radioactivity with neutrons (instead of alpha particles). Having no electric charge, neutrons are neither attracted by electrons nor repelled by nuclei; their path inside matter is much longer than that of alpha particles; their speed and energy remain higher; their chances of hitting a nucleus with full impact are much greater.

Usually a physicist is guided by some theory in setting up an experiment. This time, no workable theory had yet been developed. Only through actual experiment could one tell whether or not neutrons could trigger artificial radioactivity in the target nuclei. Fermi, already an outstanding theoretical physicist at age 33, decided to design some experiments that could settle the issue. His first task was to obtain instruments suitable for detecting the particles emitted by radioactive materials. The best such laboratory instruments by far were Geiger counters. But in 1934, Geiger counters were still relatively new and not readily available. Therefore, Fermi built his own.

The counters were soon in operation detecting the radiation from radioactive materials. Fermi also needed a source of neutrons. This he made by enclosing beryllium powder and the radioactive gas radon in a glass tube. Alpha particles from the radon, striking the beryllium, caused it to emit neutrons, which passed freely through the glass tube.

All quotations in the Prologue are from Laura Fermi, *Atoms in the Family: My Life With Enrico Fermi,* University of Chicago Press, Chicago, 1954 (available as a paperback book in the Phoenix Books series). Fermi was one of the major physicists of the twentieth century.

> Now Enrico was ready for the first experiments. Being a man of method, he did not start by bombarding substances at random, but proceeded in order, starting from the lightest element, hydrogen, and following the periodic table of elements. Hydrogen gave no results: when he bombarded water with neutrons, nothing happened. He tried lithium next, but again without luck. He went to beryllium, then to boron, to carbon, to nitrogen. None were activated. Enrico wavered, discouraged, and was on the point of giving up his researches, but his stubbornness made him refuse to yield. He would try one more element. That oxygen would not become radioactive he knew already, for his first bombardment had been on water. So he irradiated fluorine. Hurrah! He was rewarded. Fluorine was

strongly activated, and so were other elements that came after fluorine in the periodic table.

This field of investigation appeared so fruitful that Enrico not only enlisted the help of Emilio Segré and of Edoardo Amaldi but felt justified in sending a cable to Rasetti [a colleague then in Morocco], to inform him of the experiments and advise him to come home at once. A short while later a chemist, Oscar D'Agostino, joined the group, and systematic investigation was carried on at a fast pace.

With the help of his co-workers, Fermi pursued his experiments with high spirits, as Laura Fermi's account shows:

Follow the story rather than worry about the techniques of the experiment.

... Irradiated substances were tested for radioactivity with Geiger counters. The radiation emitted by the neutron source would have disturbed the measurements had it reached the counters. Therefore, the room where substances were irradiated and the room with the counters were at the two ends of a long corridor.

Sometimes the radioactivity produced in an element was of short duration, and after less than a minute it could no longer be detected. Then haste was essential, and the time to cover the length of the corridor had to be reduced by swift running. Amaldi and Fermi prided themselves on being the fastest runners, and theirs was the task of speeding short-lived substances from one end of the corridor to the other. They always raced, and Enrico claims that he could run faster than Edoardo....

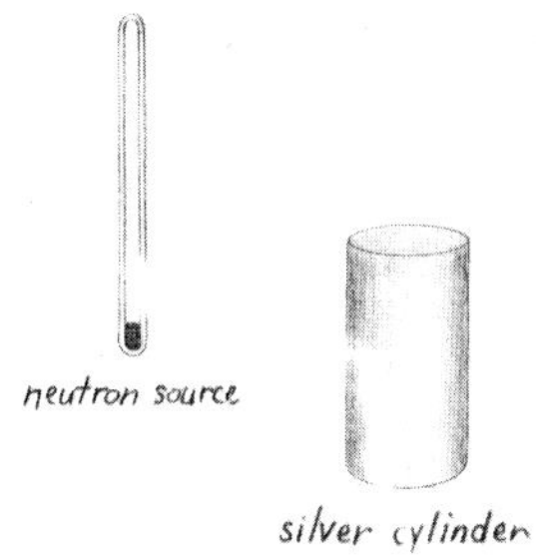

And then, one morning in October 1934, a fateful discovery was made. Two of Fermi's co-workers were irradiating a hollow cylinder of silver to make it artificially radioactive. They were using neutrons from a source placed at the center of the cylinder. They found that the amount of radioactivity induced in the silver depended on other objects that happened to be present in the room!

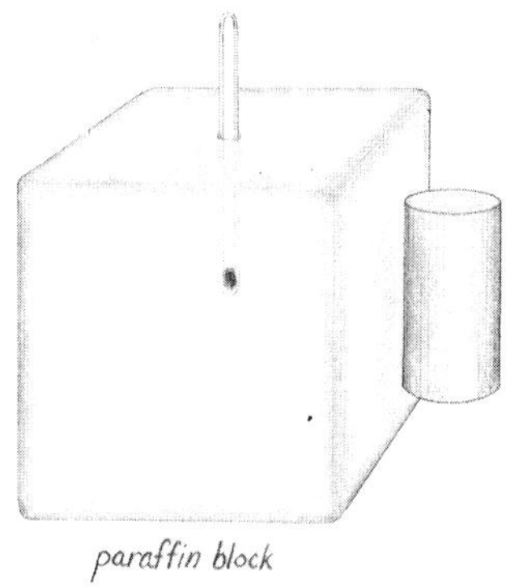

... The objects around the cylinder seemed to influence its activity. If the cylinder had been on a wooden table while being irradiated, its activity was greater than if it had been on a piece of metal.

By now the whole group's interest had been aroused, and everybody was participating in the work. They placed the neutron source outside the cylinder and interposed objects between them. A plate of lead made the activity increase slightly. Lead is a heavy substance. "Let's try a light one next," Fermi said, "for instance, paraffin." The most plentiful element in paraffin is hydrogen. The experiment with paraffin was performed on the morning of October 22.

They took a big block of paraffin, dug a cavity in it, put the neutron source inside the cavity, irradiated the silver cylinder, and brought it to a Geiger counter to measure its activity. The counter clicked madly. The halls of the physics building resounded with loud exclamations: "Fantastic! Incredible! Black Magic!" Paraffin increased the artificially induced radioactivity of silver up to one hundred times.

By the time Fermi came back from lunch, he had already found a theory to account for the strange action of the paraffin.

Paraffin contains a great deal of hydrogen. Hydrogen nuclei are protons, particles having the same mass as neutrons. When the source is enclosed in a paraffin block, the neutrons hit the protons in the paraffin before reaching the silver nuclei. In the collision with a proton, a neutron loses part of its energy, in the same manner as a billiard ball is slowed down when it hits a ball of its same size, whereas it loses little speed if it is reflected off a much heavier ball, or a solid wall. Before emerging from the paraffin, a neutron will have collided with many protons in succession, and its velocity will be greatly reduced. This *slow* neutron will have a much better chance of being captured by a silver nucleus than a fast one, much as a slow golf ball has a better chance of making a hole than one which zooms fast and may bypass it.

Because of Fermi's earlier experiments, they knew the water would not become artificially radioactive. However, they now reasoned that it would slow down neutrons and so allow silver to become more strongly radioactive.

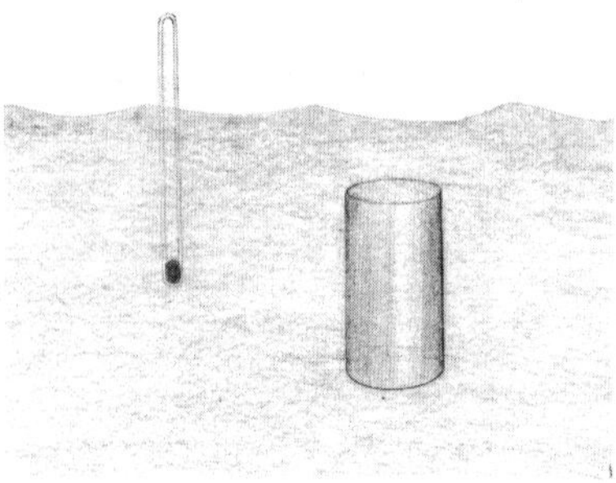

If Enrico's explanations were correct, any other substance containing a large proportion of hydrogen should have the same effect as paraffin. "Let's try and see what a considerable quantity of water does to the silver activity," Enrico said on the same afternoon.

There was no better place to find a "considerable quantity of water" than the goldfish fountain ... in the garden behind the laboratory....

In that fountain the physicists had sailed certain small toy boats that had suddenly invaded the Italian market. Each little craft bore a tiny candle on its deck. When the candles were lighted, the boats sped and puffed on the water like real motorboats. They were delightful. And the young men, who had never been able to resist the charm of a new toy, had spent much time watching them run in the fountain.

It was natural that, when in need of a considerable amount of water, Fermi and his friends should think of that fountain. On that afternoon of October 22, they rushed their source of neutrons and their silver cylinder to that fountain, and they placed both under water. The goldfish, I am sure, retained their calm and dignity, despite the neutron shower, more than did the crowd outside. The men's excitement was fed on the results of this experiment. It confirmed Fermi's theory. Water also increased the artificial radioactivity of silver many times.

Fermi and his co-workers had learned that slowed-down neutrons can produce much stronger effects in making certain atoms radioactive than can fast neutrons. This discovery turned out to be a crucial step toward further discoveries which, years later, led Fermi and others to the controlled production of atomic energy from uranium.

Fermi and his associates did not give up in the face of discouraging results. They showed imagination in the invention of theories and experiments. They remained alert to the appearance of unexpected results and resourceful in using the material resources at hand. Moreover, they found joy in discovering something new and important. These traits are of value in pursuing scientific work no less than elsewhere in life.

Scientists build on what has been found out and reported by other scientists in the past. Yet every advance in science can raise new scientific questions. The work of science is not to produce some day a finished book that can be closed once and for all. Rather, it is to carry investigation and imagination on into fields whose importance and interest have not been realized.

Some work in science depends upon painstaking observation and measurement. The results sometimes stimulate new ideas and sometimes reveal the need to change or even completely discard existing theories. Measurement itself, however, is usually guided by a theory. One does not gather data just for their own sake.

All these characteristics are true of science as a whole and not of physics alone. This being a physics text, you may well ask, "Yes, but just what *is* physics?" The question is fair enough, yet there is no simple answer. Physics can be thought of as an organized body of tested ideas about the physical world. Information about this world is accumulating ever more rapidly. The great achievement of physics has been to find a fairly small number of basic principles which help to organize and to make sense of certain parts of this flood of information.

The Fermi National Accelerator Laboratory is exploring the value of neutron irradiation in the treatment of cancer. A beam of protons from a linear accelerator is directed onto a beryllium target. Neutrons are produced as a result of this collision; these neutrons are used in the cancer therapy research.

Close Up

Our Place in Space and Time

Physics deals with those laws of the universe that apply everywhere, from the largest to the smallest.

	ORDER OF MAGNITUDE
Distance to the furthest observed galaxy	10^{26} meters
Distance to the nearest galaxy	10^{22}
Distance to the nearest star	10^{17}
Distance to the sun	10^{11}
Diameter of the earth	10^{7}
One kilometer	10^{3}
Human height	10^{0}
Finger breadth	10^{-2}
Paper thickness	10^{-4}
Large bacteria	10^{-5}
Small virus	10^{-8}
Diameter of atom	10^{-10}
Diameter of nucleus	10^{-14}

A globular star cluster.
The estimated size of the universe now is of the order of 100 million, million, million, million times a person's height (person's height × 10,000,000,000,000,000,000,000,000). The diameter of the cluster is about 10^{18} m.

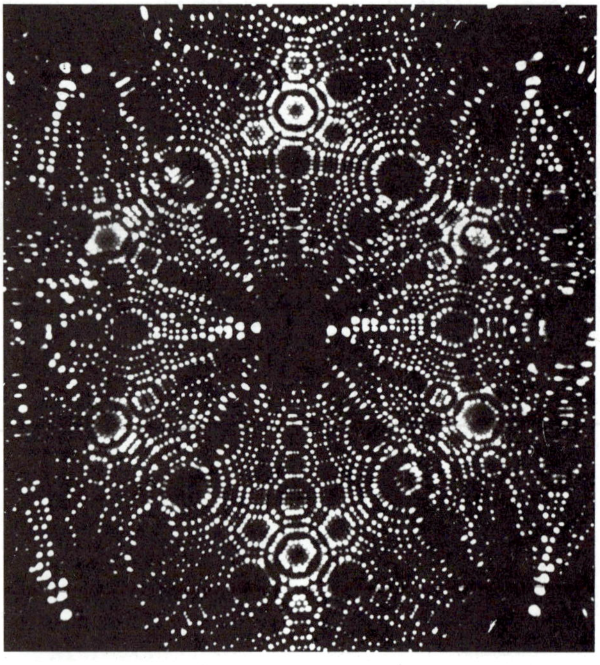

Atomic sites in tungsten.
The smallest known constituent units of the universe are less in size than a hundredth of a millionth of a millionth of a person's height (person's height × 0.000,000,000,000,01). The two bright spots in the middle are produced by atoms about 10^{-9} m apart.

Physicists study phenomena in the extremes of time-space and in the whole region between the longest and shortest.

	ORDER OF MAGNITUDE
Age of universe	10^{17} seconds
Precession of the earth's axis	10^{12}
Human life span	10^{9}
One year	10^{7}
One day	10^{5}
Light from sun to earth	10^{3}
Time between heartbeats	10^{0}
One beat of fly's wings	10^{-3}
Duration of strobe flash	10^{-5}
Short laser pulse	10^{-9}
Time for light to cross an atom	10^{-18}
Shortest-lived subatomic particles	10^{-23}

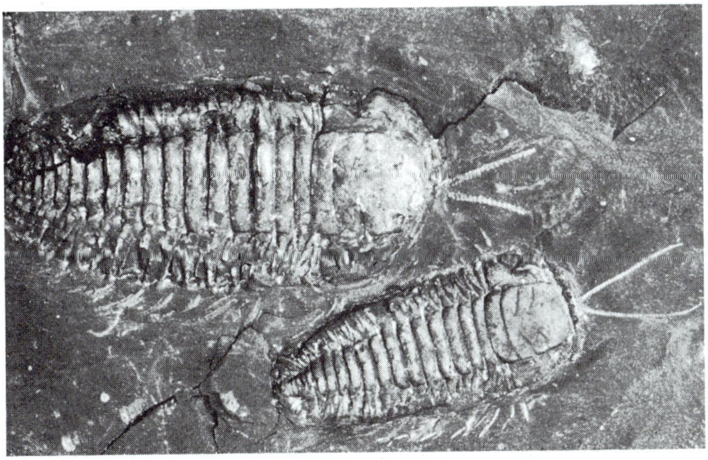

Fossilized trilobites.

These fossils are about 5×10^{8} years old.

The total history of the universe has been traced back as far into the past as 300 million times the length of a human life (human life $\times$ 300,000,000).

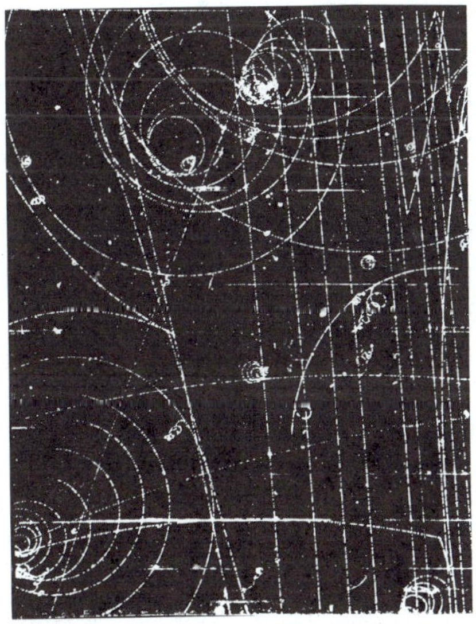

Particle tracks in a bubble chamber.

Events have been recorded that last only a few millionths of a millionth of a millionth of a millionth of a person's heartbeat (person's heartbeat $\times$ 0.000,000,000,000,000,000,000,001). These tracks each took about 10^{-9} sec to make.

Physics started not with these intriguing extremes, but with the human-sized world: the world of horse-drawn chariots, of falling rain, and of flying arrows. It is with the physics of phenomena on that scale that this course begins.

CHAPTER 1

The Language of Motion

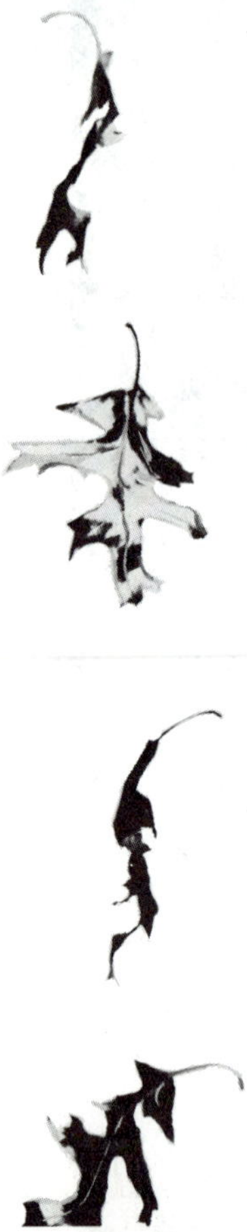

1.1 | The motion of things

The world is filled with things in motion, things as small as dust and as large as galaxies, all continually moving. Your book may seem to be lying quietly on the desk, but each of its atoms is constantly vibrating. The "still" air around you consists of molecules tumbling wild at various speeds, most of them moving as fast as rifle bullets. Light beams dart through the room, covering the distance from wall to wall in about a hundred-millionth of a second and making about 10 million vibrations during that time. The whole earth, our spaceship itself, is moving at about 29 kilometers per second (km/sec) around the sun.

There is an old maxim: "To be ignorant of motion is to be ignorant of nature." So, from this swirling, whirling, vibrating world of ours let us first choose just one moving object for our attention. Then let us describe its motion.

Shall we start with a machine, such as a rocket or a car? Though made and controlled by humans, machines and their parts move in fast and complicated ways. We really ought to start with something simpler, something that our eyes can follow in detail. How about a bird in flight? Or a leaf falling from a tree?

Surely, in all of nature there is no motion more ordinary than that of a leaf fluttering down from a branch. Can you describe how it falls or explain why it falls? You will quickly realize that, while the motion is very "natural," it is also very complicated. The leaf twists and turns, sails right and left, back and forth, as it floats down. Even a motion as ordinary as this may turn out, on closer examination, to be more complicated than the motion of machines. And even if you could describe the motion of a leaf in detail, what would you gain? No two leaves fall in quite the same way. Therefore, each leaf seems to require its own detailed description. This individuality is typical of most events you see occurring in nature.

Study for "Dynamism of a Cyclist" (1913) by Umberto Boccioni. Courtesy Yale University Art Gallery.

And so we face a problem. We want to describe motion, but the motions we encounter under ordinary circumstances appear too complex. Also, from the separate observations we are not likely to find general conclusions that apply to all motions. What shall we do? The answer is that, at least for a while, we must go into the physics laboratory. The laboratory is the place to separate the simple ingredients that make up all complex natural phenomena and to make those phenomena more easily visible to our limited human senses.

1.2 | A motion experiment that does not quite work

Having abandoned the fall of a leaf as the way to start on the physics of motion, we might select a clearly simpler case: a billiard ball, hit squarely in the center and speeding easily across a tabletop in a straight line. An even simpler motion (simpler because there is no rolling) can be obtained. Place a disk of what is called "dry ice" on a smooth floor and give it a gentle push. (Take care not to touch the extremely cold disk with bare hands for more than a brief moment!) The disk will move slowly and with very little friction, supported on its own vapor.

"Dry ice" is really frozen carbon dioxide, at −79°C.

We did this in front of a camera to get a photograph that would record the action for easier measurement later. While the dry-ice disk was moving on the well-leveled surface, the shutter of the camera was kept open; the resulting time-exposure shows the path taken by the disk.

What can you learn about the disk's motion by examining the photograph? As nearly as you can judge by placing a ruler on the photograph, the disk moved in a straight line. This is a very useful result, and you will see later that it is really quite surprising. It shows how simple a situation can be made in the laboratory; the kinds of motion you ordinarily see are almost never that simple. But did the disk move steadily, or did it slow down? From this photograph, we really cannot tell. We must improve the experiment. Before we do so, however, we must decide just how we plan to measure the speed.

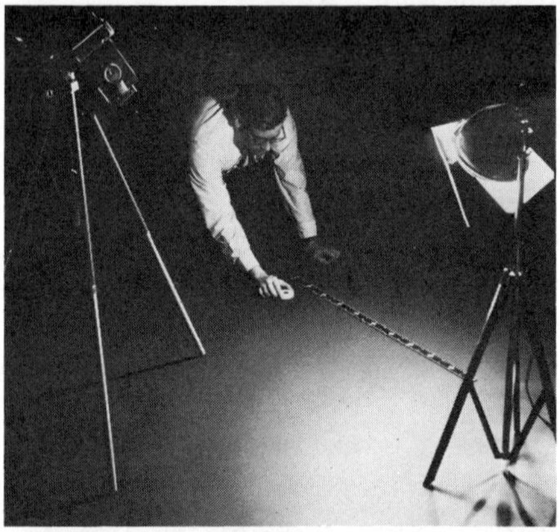

Laboratory setup.

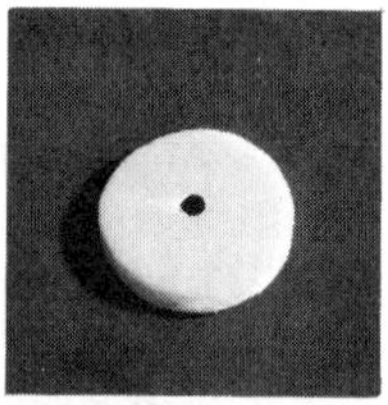

Close-up of a dry-ice disk.

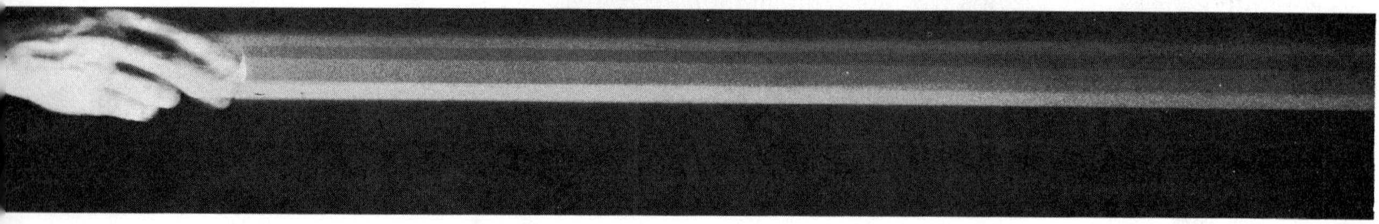

Time exposure of the disk in motion.

It would be nice to use something like an automobile speedometer. A speedometer can tell directly the speed at which a car is moving at any time. We can say, for example, that a car is moving at 100 kilometers per hour (km/hr). This means that if the car continues to move with the same speed it had at the instant the speed reading was taken, the car would move a distance of 100 km in a time interval of 1.0 hr. Or we could say that the car would move 1.7 km in $\frac{1}{60}$ of an hour (1 minute) or 10 km in $\frac{1}{10}$ of an hour. In fact, we could use *any* distance and time intervals for which the ratio of distance to time is 100 km/hr.

Of course, an automobile speedometer cannot be hooked to a disk of dry ice, or to a bullet, or to many other objects. However, there is a rather simple way to measure speeds, at least in most cases that would interest us.

From time to time you will be referred to items in the *Study Guide* found at the end of each chapter. Usually the letters SG plus a number will indicate this. See SG 1 on page 31 for more information on how to study for this course and on the use of the *Study Guide*.

Think of what you could do if the speedometer in your car were broken and you still wanted to know your speed as you moved along a turnpike. You could do one of two things (the result is the same in either case). You could count the number of kilometer markers passed in one hour (or some known fraction of an hour) and find the average speed by computing the ratio of kilometers and hours. Or you could determine the fraction of an hour it takes to go from one kilometer marker to the next (or to another marker a known number of kilometers away) and again find the average speed as a ratio of kilometers to hours.

Either method gives, of course, only the *average* speed for the interval during which speed is measured. That is not the same as *instantaneous* speed. Instantaneous speed is the speed at any given instant as a speedometer might register it. But average speed is good enough for a start. After you understand average speed, you will see a simple way of finding instantaneous speeds.

The speed of an object is, of course, how fast it moves from one place to another. A more formal way to say the same thing is: *Speed is the time rate of change of position*. The term "displacement" is often used to refer to the straight-line distance between the beginning and end points of the change in position of a moving object. We will use this term in connection with vectors in Chapter 3, and more often still with wave motion in Chapter 12.

To find the average speed of an object, measure the distance it moves and the time it takes to move that distance. Then divide the distance by the time. The speed is in kilometers per hour or meters per second, depending upon the units used to measure the distance and time. With this plan of attack, we can return to the experiment with the dry-ice disk. Our task now is to find the speed of the disk as it moves along its straight-line path. If we can do it for the disk, we can do it for many other objects as well.

Note: There will usually be one or more brief questions at the end of each section in the text. Question 1, below, is the first. The end-of-section questions are there for your use, to check on your own progress. Answer the questions *before* continuing to the next section. Check your answers to these end-of-section questions. Whenever you find you did not get the correct answer, study through the section again. Of course, if anything is still unclear after you have tried to study it on your own or together with other students, then ask your instructor.

? 1. *Why is it impossible to determine the speed of the dry-ice disk from the time-exposure photograph on page 10?*

1.3 | A better experiment

To find speed, you need to be able to measure both distance and time. Repeat the experiment with the dry-ice disk. First place a meter stick (100 cm) on the table parallel to the expected path of the disk. This is the photograph obtained:

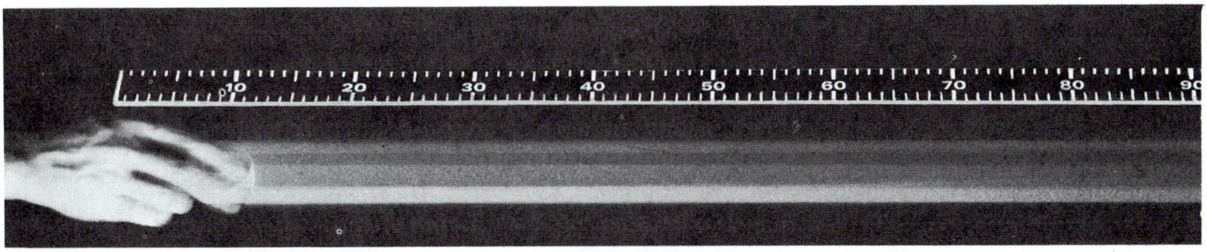

You now have a way of measuring the distance traveled by the disk. But you still need a way to measure the time it takes the disk to travel a given distance.

This can be done in various ways, but there is a fine trick that you can try in the laboratory. The camera shutter is again kept open and everything else is the same as before, except that the only source of light in the darkened room comes from a stroboscopic lamp. This lamp produces bright flashes of light at intervals which we can set as we please. Each pulse or flash of light lasts for only about 10 millionths of a second (10 microseconds). Therefore, the moving disk appears in a series of separate, sharp exposures, rather than as a continuous blur. The photograph below was made by using such a stroboscopic lamp flashing 10 times per second, after the disk had been gently pushed as before.

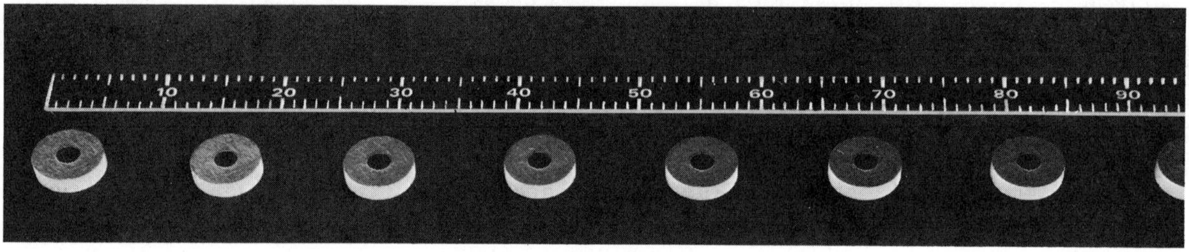

This special setup enables us to record accurately a series of positions of the moving object. The meter stick helps us to measure the distance moved by the front edge of the disk between one light flash and the next. The time interval between images is, of course, equal to the time interval between stroboscopic lamp flashes [0.10 second (sec) in these photos].

You can now determine the speed of the disk at the beginning and end of its photographed path. The front edge of the first clear image of the disk at the left is 19 cm from the zero mark on the meter stick. The front edge of the second image from the left is at the 32-cm position. The distance traveled during that time was the difference between those two positions, or 13 cm. The corresponding time interval was 0.10 sec. Therefore, the speed during the first part of the observation must have been 13 cm/0.10 sec, or 130 cm/sec.

Now look at the two images of the disk farthest to the right in the photograph. Here, too, the distance traveled during 0.10 sec was 13 cm. Thus, the speed at the right end was 13 cm/0.10 sec, or 130 cm/sec.

The disk's motion was not measurably slower at the right end than at the left end. Its speed was 130 cm/sec near the beginning of the path and 130 cm/sec near the end of the path. However, this does not yet prove that the speed was constant all the way.

You might well suspect that it was, and you can easily check to find out. Since the *time* intervals between images are known to be equal, the speeds will be equal if the *distance* intervals are equal to one another. Is the distance between images always 13 cm? Did the speed stay constant, as far as you can tell from the measurement?

When you think about this result, there is something really unusual in it. Cars, planes, and ships do not move in neat, straight lines with precisely constant speed even when using power. Yet this disk did it, coasting along on its own, without any propulsion to keep it moving. You might consider this a rare event which would not happen again. In any case, *you* should try the experiment. The equipment you will need includes cameras, strobe lamps (or mechanical strobes, which work just as well), and low-friction disks of one sort or another. Repeat the experiment with different initial speeds. Then compare your results with those found above.

You may have a serious reservation about the experiment. You might ask: "How do you know that the disk did not slow down an amount too small to be detected by your measurements?" The answer is that we do not know. All measurements involve some uncertainty, though one which can usually be estimated. With a meter stick, you can measure distances reliably to the nearest 0.1 cm. If you had been able to measure to the nearest 0.01 cm or 0.001 cm, you might have detected some slowing down. But if you again found no change in speed, you could still raise the same objection. There is no way out of this dilemma. We must simply acknowledge that *no* physical measurements are ever perfectly precise. The results of any set of measurements are acceptable within its own limits of precision, and you can leave open the question of whether or not measurements made with increased precision could reveal other results.

Uncertainty of measurement is taken up in detail in the *Handbook*, particularly in Experiment 3.

Briefly review the results of this experiment. You devised a way to measure the successive positions of a moving dry-ice disk at known time intervals. From this you calculated first the distance intervals and then the speed between selected positions. You soon discovered that (within the limits of precision of the measurements) the speed did not change. Objects that move in such a manner are said to have *uniform speed*, or *constant speed*. You know now how to measure uniform speed.

But, of course, actual motions are seldom uniform. What about the more usual case of *non*uniform speed? That is our next concern.

Some practice problems dealing with constant speed are given in Study Guide 3 (a, b, c, and d).

?

2. *Suppose the circles on page 14 represent the successive positions of a moving disk as photographed stroboscopically. Did the object move with uniform speed? How do you know?*

3. *Give a general description of uniform speed, without referring to dry-ice disks and strobe photography or to any particular object or technique of measurement.*

1.4 | Leslie's swim and the meaning of average speed

45.7 m/56.1 sec = 0.81 m/sec. That is the equivalent of 2.9 km/hr. No great speed! A sailfish can do over 60 km/hr. But humans are land animals. Even the fastest swimmers are only about twice as fast as Leslie. For short distances, a person can run up to about 30 km/hr. The world's record for a 100-m run is a bit more than 10 sec; the runner thus had an average speed of 36 km/hr.

Consider the situation at a swimming meet. At the end of each race, the name of the winner, the swimmer with the shortest time, is announced. But in any given race, for example, the 100-meter event, every swimmer goes the same distance. Therefore, the swimmer with the shortest time is also the one having the highest average speed while covering the measured distance. The ratio of the distance traveled to the elapsed time is the measure of average speed. This relationship is expressed in the following equation:

$$\text{average speed} = \frac{\text{distance traveled}}{\text{elapsed time of travel}}$$

which, in fact, is the *definition* of average speed. What information does a knowledge of the average speed give you? You can answer this question by studying a real example.

Leslie is not the fastest woman freestyle swimmer in the world, but Olympic speed is not necessary for this purpose. Leslie was timed while swimming two lengths of an old, indoor pool. The pool is 22.5 meters (m) long, which makes it a bit too short for Olympic events but good enough for many sports. It took Leslie 56.1 sec to swim the two lengths (45 m). Thus, her average speed for the 45 m was

$$\frac{45\text{ m}}{56.1\text{ sec}} = 0.80\text{ m/sec}$$

Did Leslie swim the two lengths of the pool at uniform (that is, constant) speed? If not, when was she swimming the fastest? What was her greatest speed? Did she take less time to swim out than to return? How fast was she moving when she was one-quarter or halfway down the pool? The answers to these questions are useful to know when training for a meet.

So far you do not have a way to answer any of these questions, but soon you will. Until then, the average speed over the two lengths (0.80 m/sec) is the best single value you can use to describe the whole of Leslie's swim.

To compare Leslie's speed at different points in the swim, observe the times and distances covered as you did in experimenting with the dry-ice disk. For this purpose, the event was arranged as follows.

Observers stationed at 4.6 m intervals from the zero mark along the length of the pool started their stopwatches when the starting signal was given. Each observer had two watches. The observers stopped one watch as Leslie passed them going down the pool, and they stopped the other as she passed on her return trip. The data are tabulated next to the photograph of this experiment.

d	t
0.0m	0.0 sec
4.6	2.5
9.1	5.5
13.7	11.0
18.3	16.0
22.9	22.0
27.4	26.5
32.0	32.0
36.7	39.5
41.2	47.5
45.7	56.0

From these data, you can, for example, determine Leslie's average speed for the first length (22.5 m) and for the last length separately.

$$\text{average speed for the } \textit{first} \text{ length} = \frac{\text{distance traveled}}{\text{elapsed time}} = \frac{22.5\text{ m}}{22.0\text{ sec}} = 1.02\text{ m/sec}$$

$$\text{average speed for the } \textit{last} \text{ length} = \frac{\text{distance traveled}}{\text{elapsed time}} = \frac{22.5\text{ m}}{56.1\text{ sec} - 22.0\text{ sec}} = \frac{22.5\text{ m}}{34.1\text{ sec}} = 0.66\text{ m/sec}$$

It is now clear that Leslie did not swim with uniform speed. She swam the first length much faster (1.02 m/sec) than the second length (0.66 m/sec). Notice that the overall average speed (0.80 m/sec) does not describe either lap by itself very well.

In a moment we will continue our analysis of the data we have obtained for Leslie's swim. This analysis is important because the concepts we are developing for this everyday type of motion will be needed later to discuss other motions, ranging from that of planets to that of atoms.

The following shorthand notation will simplify the definition of average speed:

$$\text{average speed} = \frac{\text{distance traveled}}{\text{elapsed time}}$$

A more concise statement that says exactly the same thing is

$$v_{av} = \frac{\Delta d}{\Delta t}$$

v_{av} is pronounced "vee av" or "vee sub-av." Δd is pronounced "delta dee."

In this equation, v_{av} is the symbol for the average speed, Δd is the symbol for change in position, and Δt is the symbol for an elapsed interval of time. The symbol Δ is the fourth letter in the Greek alphabet and is called delta. When Δ precedes another symbol, it means "the change in . . . ." Thus, Δd does *not* mean "Δ multiplied by d." Rather, it means "the change in d" or "the distance interval." Likewise, Δt stands for "the change in t" or "the time interval."

You can now go back to the data and compute Leslie's average speed for each 4.5-m interval, from beginning to end. This calculation is easily made, especially if you reorganize the data as shown in the table on page 19. The values of v_{av} calculated at 4.5-m intervals for the first lap are entered in the right-hand column. (The values for the second lap are left for you to complete.)

Much more detail is emerging from the data. Looking at the speed column, you see that Leslie's speed was greatest, as expected, near the start. Her racing jump into the water gave her extra speed at the beginning. In the middle of her first length, she swam at a fairly steady rate, and she slowed down coming into the turn. Use your own calculations to see what happened after the turn.

Practice problems on average speed can be found in Study Guide 4 (e, f, g, and h). Study Guides 6, 8, and 10 offer somewhat more challenging problems. Some suggestions for average speeds to measure are listed in Study Guide 9.

Although you have determined Leslie's speeds at various intervals along the path, you are still dealing with *average* speeds. The intervals during which you determine the average speeds are smaller: 4.5 m rather than the entire 45 m. But you do not know the details of what happened *within* any of the 4.5-m intervals. Thus, you know that Leslie's average speed between the 13.5- and 18-m marks was 0.9 m/sec. You do not know yet how to compute her speed at the very instant and point where she was, say, 16.2 m or 18 m from the start. Even so, the average speed for the 4.5-m interval between the 13.5- and 18-m marks is probably a better estimate of her speed as she went through the 16.2-m mark than is the average speed for the whole 45 m, or for either 22.5-m length. We will come back to this problem of determining "speed at a particular instant and point" in Sec. 1.7.

? *4. A boy on his way to the store about 720 m from his home stopped twice to tie his shoes and once to watch an airplane. What was his average speed for the trip if he took 540 sec to reach the store? In your own words, define the concept "average speed."*

5. Use the formula for average speed to determine how far down the slope a skier will travel in 15 sec if she is moving at a speed of 20 m/sec.

6. If you have not already completed the table on page 19, do so now before going on to the next section.

1.5 | Graphing motion and finding the slope

You can look at the data from Leslie's swim another and more informative way by plotting them on a graph instead of just writing them down in a table. In the first graph on the next page, the distance and time values that were measured for Leslie's swim are shown. (The circles around the points have been put in only to make the points show up more clearly.) Each point on the distance–time graph shows the distance Leslie covered up to that particular time. We know that Leslie was in the pool between our measured points as well, but we do not know precisely where. The usual way to show this is to connect the known points with some kind of line or curve.

In the second graph we have done just that in the simplest way, by drawing straight lines between points. Because we do not know that Leslie was really "on" those lines between the measured points, we have drawn the line segments as broken lines, with dashes, instead of as solid lines.

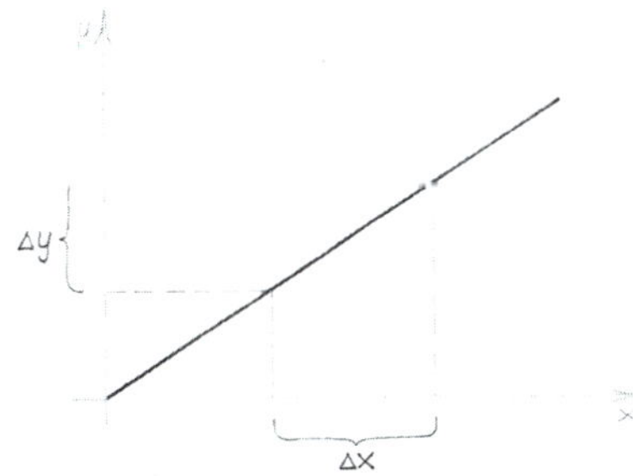

You can get a better approximation of Leslie's actual motion by drawing one continuous "smooth" curve through all the data points. One experimenter's idea of a good curve is shown in the last graph. On first sight, this graph may not look very different from the second graph, but a closer look will show differences in detail. For example, in the second graph there is a rather sharp kink at the point corresponding to 5.5 sec. This would imply that Leslie changed her speed abruptly at that moment. We have no reason to think this happened, and it does not show up on the third graph. This is one of the reasons for preferring the third graph.

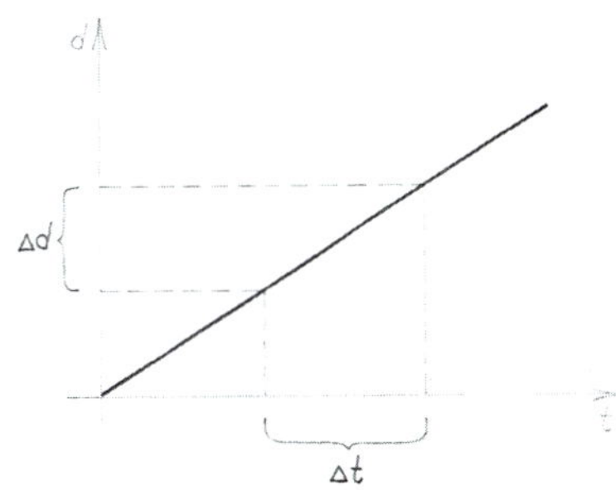

What can you see (or "read") on the last graph? Notice that the line is steepest at the start. This is because there was a comparatively large change in position in the first few seconds. In other words, Leslie got off to a fast start. You can also

If this concept is new to you or if you wish to review it, turn now to Study Guide 11 before continuing here.

determine that from the table. After the third measured point, the graph becomes less steep. The time taken to cover the same distance is longer; therefore, the speed is slower. Again this agrees with what you see in the table of data and computations. *The steepness of the graph line clearly indicates how fast Leslie was moving.* The faster she swam, the steeper is the line on the graph. If you follow the line along, you see that Leslie slowed down coming into the turn, had a brief spurt just after the turn, and then slowed down steadily until the finish.

Looked at in this way, a graph provides you at a glance with a visual representation of motion which can be very useful. But this kind of representation does not tell directly what the actual value of Leslie's speed was at any particular moment. For this, you need a way of measuring the steepness of the graph line. Here we can turn to mathematics for help, as we often shall.

There is an old method in geometry for solving just this problem. The steepness of a graph at any point is related to the change in vertical direction (Δy) and the change in horizontal direction (Δx). By definition, the ratio of these two changes ($\Delta y/\Delta x$) is the slope:

$$\text{slope} = \frac{y}{x}$$

Slope is a widely used mathematical concept. It can be used to indicate the steepness of a line in any graph. In a distance–time graph, like the one for Leslie's swim, the position, or distance from the start, is usually plotted on the vertical axis (d replaces y) and time on the horizontal axis (t replaces x). Thereore, in such a graph, the slope of a straight line is given by

$$\text{slope} = \frac{\Delta d}{\Delta t}$$

This should remind you of the definition of average speed, $v_{av} = \Delta d/\Delta t$. In fact, v_{av} is numerically equal to the slope! In other words, the slope of any straight-line part of a graph of distance versus time gives a measure of the average speed of the object during that interval.

When you measure slope on a graph, you do basically the same thing that highway engineers do when they specify the steepness of a road. They simply measure the rise in the road and divide that rise by the horizontal distance one must go in order to achieve the rise. The only difference between this and what we have done is that the highway engineers are concerned with an actual physical slope. Thus, on a graph of their data, the vertical axis and horizontal axis both show *distance*. We, on the other hand, are using the *mathematical concept* of slope as a way of expressing *distance* measured against *time*.

d	t	Δd	Δt	$\Delta d/\Delta t$ (= v_{av})
0.0 m	0.0 sec	4.6 m	2.5 sec	1.8 m/sec
4.6	2.5	4.6	3.0	1.5
9.1	5.5	4.6	5.5	0.8
13.7	11.0	4.6	5.0	0.9
18.3	16.0	4.6	6.0	0.8
22.9	22.0	4.6	4.5	etc.
27.4	26.5	4.6	5.5	
32.0	32.0	4.6	etc.	
36.7	39.5	4.6		
41.2	47.5	4.6		
45.7	56.0			

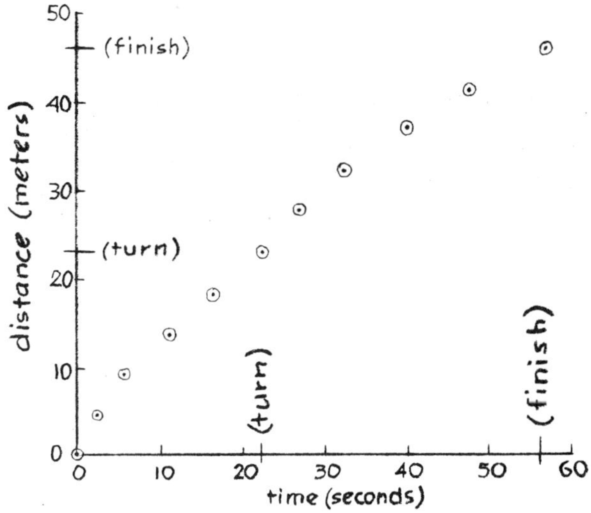

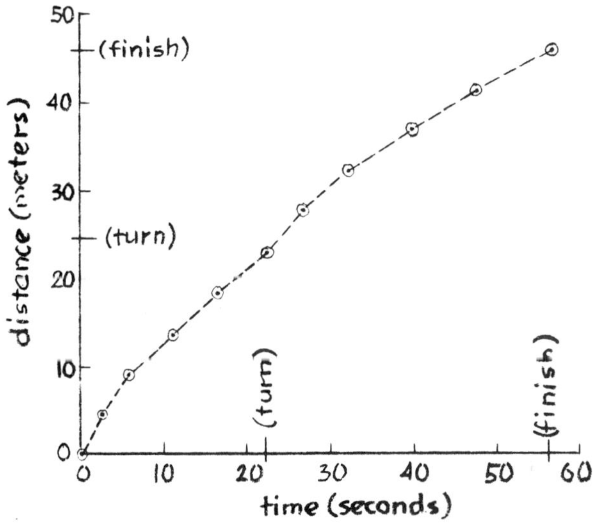

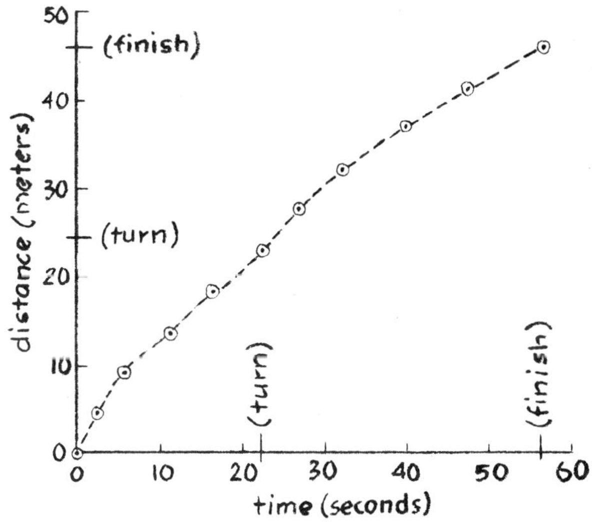

Above are shown four ways of representing Leslie's swim: a table of data and calculations, a plot of the data points, broken straight-line segments that connect the points, and a smooth curve that goes through the points.

The slope, as defined here, is not exactly the same as the steepness of the line on the graph paper. Suppose we had chosen a different scale for either the distance or time axis, making the graph twice as tall or twice as wide. Then the apparent steepness of the entire graph would be different. The slope, however, is measured by the ratio of the distance and time units. A Δd of 10 m in a Δt of 5 sec gives a ratio of 2 m/sec, no matter how much space on the drawing or graph paper is used to represent meters and seconds.

But the graph is more than just a "picture" of the values in the table. You can now ask questions that cannot be answered directly from the original data. For example, what was Leslie's speed 10 sec after the start when she was not even opposite one

The 4-sec value for t is just for convenience; some other value could have been used. Or we could have chosen a value for Δd and then measured the corresponding Δt.

of the data takers? What was her speed as she crossed the 35-m mark? You can answer questions like these by finding the slope of a fairly straight portion of the graph line near the point of interest. Two examples have been worked out on the graph shown below. For each example, Δt was chosen to be a 4-sec interval from 2 sec before the point in question to 2 sec after it. Then the Δd for that Δt was measured.

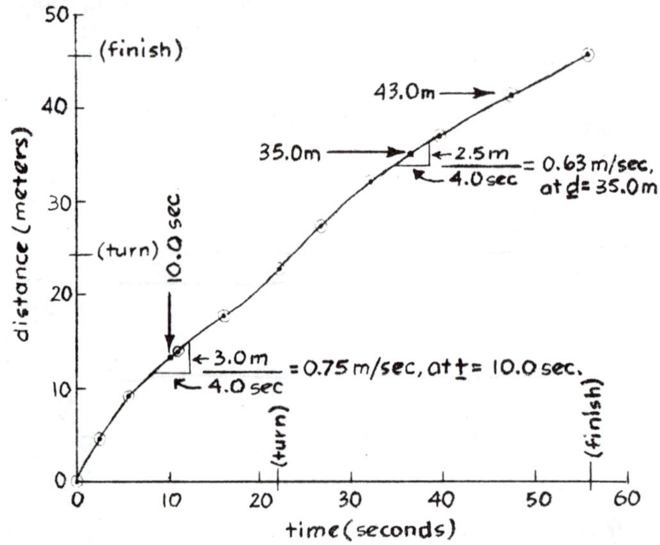

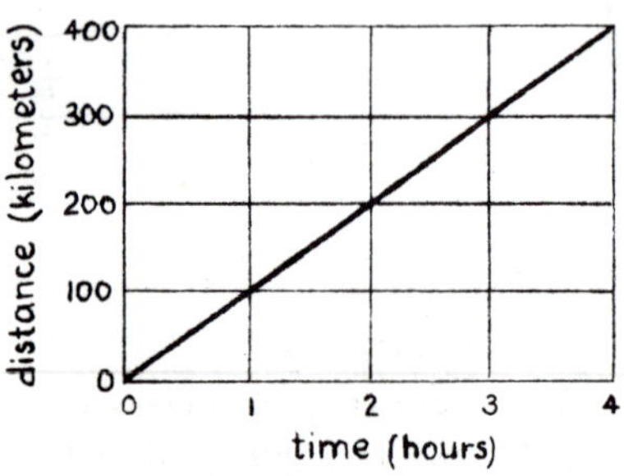

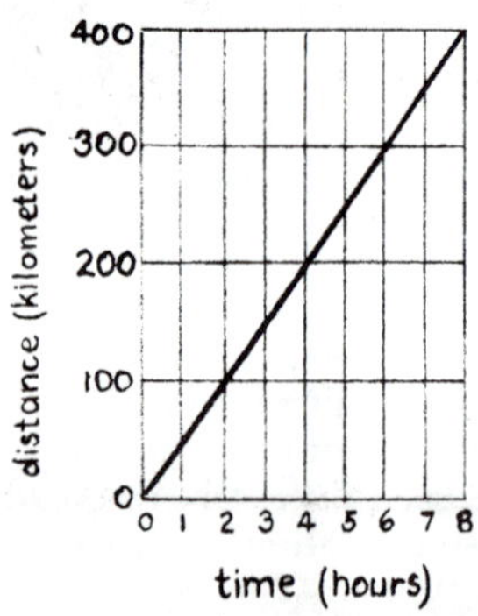

You can check the reasonableness of using the graph in this way by comparing the results with the values listed in the table on page 19. For example, the speed near the 10-sec mark is found from the graph to be about 3.0 m/4.0 sec = 0.75 m/sec. This result is somewhat less than the value of 0.8 m/sec given in the table for the average speed for the time interval between t = 5.5 sec and t = 11 sec. This is just what you would expect, because you can see that the smooth-curve graph does become slightly less steep around the 10-sec point. If the smooth curve really describes Leslie's swimming better than the dashed straight-line graph does, then you can get more information from that last graph than you can get just by looking at the data themselves!

?

7. Turn back to page 12 and draw a distance–time graph for the motion of the dry-ice disk.

8. Which of the two graphs in the margin, for two different objects, has the greater slope?

Close Up

The Language of Motion

A. t = 0 *B. t = 19 min* *C. t = 36 min* *D. t = 63 min*

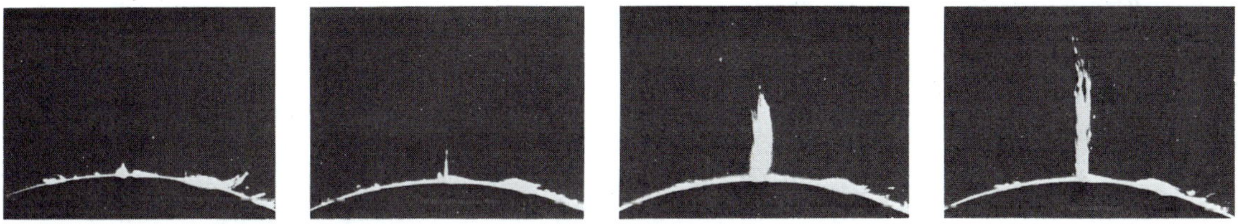

Δt = 19 min Δt = 17 min Δt = 27 min

These photographs show a stormy outburst of incandescent gas at the edge of the sun, a developing chive plant, and a glacier. From these pictures and the time intervals given between pictures, you can determine the average speeds of: (1) the growth of the solar flare with respect to the sun's surface (radius of sun is about 691,200 km), (2) the growth of one of the chive shoots with respect to the graph paper behind it (large squares are 2.5 cm), (3) the moving glacier with respect to its "banks."

A. t = 0 *B. t = 17 hr* *C. t = 50 hr*

A *B*

Δt = 4 years

9. Using the last graph for her swim, where in the pool was Leslie swimming most rapidly? Where was she swimming most slowly?

10. From that graph, find Leslie's speed at the 43.0-m mark. From the table on page 19, calculate her average speed over the last 4.5 m. How do the two values compare?

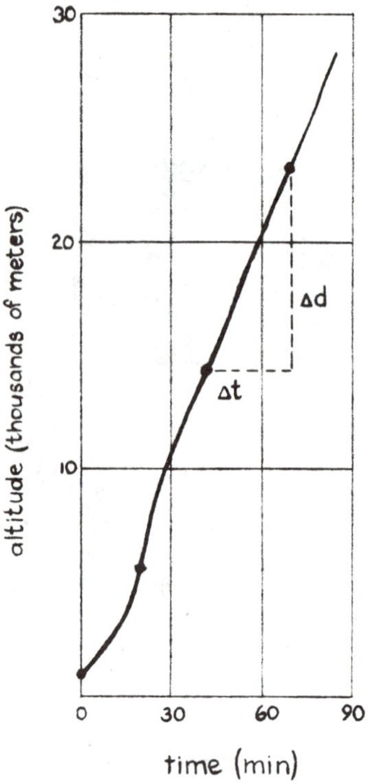

1.6. Time out for a warning

Graphs are useful, but they can also be misleading. You must always be aware of the limitations of any graph you use. The only actual *data* in a graph are the plotted points. There is a limit to the precision with which the points can be plotted, and a limit to how precisely they can be read from the graph.

The placement of a line through a series of data points, as in the graph on page 19, depends on personal judgment and interpretation. The process of estimating values *between* data points is called *interpolation.* That is essentially what you are doing when you draw a line between data points. Even more risky than interpolation is *extrapolation,* where the graph line is extended to provide estimated points *beyond* the known data.

The description of a high-altitude balloon experiment carried out in Lexington, Massachusetts, illustrates the danger of extrapolation. A cluster of helium-filled balloons carried cosmic-ray detectors high above the earth's surface. From time to time, observers measured the altitude of the cluster. The graph on the right shows the data for the first hour and a half. After the first 20 min, the balloons seem to be rising in a cluster with unchanging speed. The average speed can be calculated from the slope: speed of ascent $= \Delta d/\Delta t = 9{,}000$ m/30 min $= 300$ m/min $= 5$ m/sec.

Now, suppose you were asked how high the balloons would be at the very end of the experiment, which came at $t = 500$ min. You might be tempted to extrapolate, either by extending the graph or by computing from the speed. In either case, you would obtain about 500 min × 300 m/min = 150,000 m, which is over 150 km high! Would you be right? (The point is that mathematical aids, including graphs, can be a splendid help, but only within the limits set by physical realities.)

?

11. Look at the pictures of the solar flare on the preceding page. Estimate the height of the flare at (a) 42 min and (b) 72 min. (Draw a graph if necessary.) Explain which estimate required extrapolation and which required interpolation.

12. Which estimate from the graph would you expect to be less accurate: Leslie's speed as she crossed the 30-m mark, or her speed at the end of an additional third lap?

SG 16

1.7 | Instantaneous speed

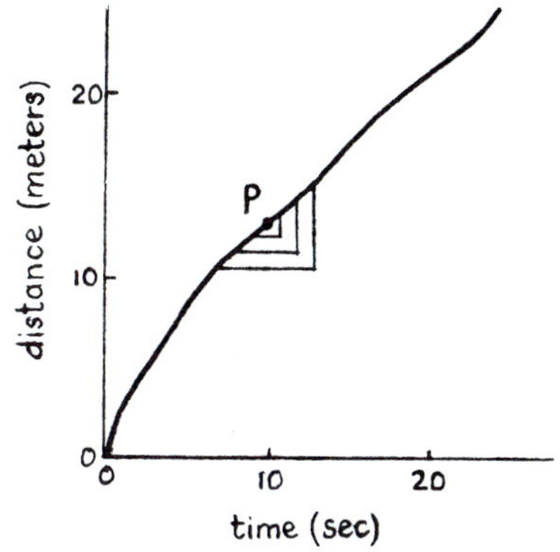

Δt	Δd	$\Delta d/\Delta t$
6.0 sec	4.6 m	0.78 m/sec
4.0 sec	3.0 m	0.75 m/sec
2.0 sec	1.5 m	0.75 m/sec

Now let us summarize the chief lessons of this first chapter. In Sec. 1.5 you saw that distance–time graphs could be very helpful in describing motion. Near the end of the section, specific speeds at particular points along the path ("the 35-m mark") and at particular instants of time ("the instant 10 sec after the start") were mentioned briefly. You may have been bothered by these comments, since the only kind of speed you can actually measure is *average* speed. To find average speed you need a ratio of distance *intervals* and time *intervals*. A particular point on the path, however, does not have any interval. Nevertheless, it does make sense to speak about the *speed at a point*. The following is a summary of the reasons for using "speed" in this way.

You remember that the answer to the question (page 19) "What was Leslie's speed 10 sec after the start?" was 0.75 m/sec. You obtained that answer by finding the slope of a small portion of the curve around the point P when $t = 10$ sec. That section of the curve is reproduced in the margin here. Notice that the part of the curve used appears to be nearly a straight line. As the table under the graph shows, the value of the slope for each interval changes very little as the time interval Δt is decreased. Below $\Delta t = 4$ sec, you keep getting the same value for $\Delta d/\Delta t$. Correspondingly, the chosen segment of the line on which P sits is more and more a straight line. Now imagine that you continued to 700 m, where $t = 10$ sec, until the amount of curve remaining became vanishingly small. Can you safely assume that the slope of that very small part of the curve has the same value as the slope of the short straight-line portion of which it seems to be a part? It seems reasonable. In any case, it is up to you to define what you mean by your concepts. That is why we took the slope of the straight line from $t = 8$ sec to $t = 12$ sec and *called* it the speed *at* the midpoint, $t = 10$ sec. The correct term for this value is the *instantaneous speed* at the instant $t = 10$ sec.

Previously, you estimated Leslie's instantaneous speed at a particular time by actually measuring the average speed over a 4.0-sec interval. This method can be modified so that it can be used in many different contexts in the future. The *instantaneous* speed at a particular instant has the same value as the *average* speed, $\Delta d/\Delta t$, as long as two conditions are met: First, the particular instant must, of course, be included in Δt. Second, the ratio, $\Delta d/\Delta t$ must cover a *small part of the curve*, one that is as nearly as possible a straight-line segment. Under this condition, the ratio $\Delta d/\Delta t$ will not change noticeably when you compute it again over a still smaller time interval.

A second example will help to explain the concept of instantaneous speed. In the oldest known study of its kind, the

French scientist de Montbeillard periodically recorded the height of his son during the years 1759–1777. A graph of height versus age for his son was published and is shown at the lower left.

From the graph, you can compute the average growth rate, or average speed of growth (v_{av}) over the entire 18-year interval or over any shorter time interval within that period. Suppose, however, you wanted to know how fast the boy was growing just as he reached his fifteenth birthday. The answer becomes evident if you enlarge the graph in the vicinity of the fifteenth year. His height at age 15 is indicated at point P, and the other letters indicate instants of time on either side of P. The boy's average growth rate over a 2-year interval is given by the slope of

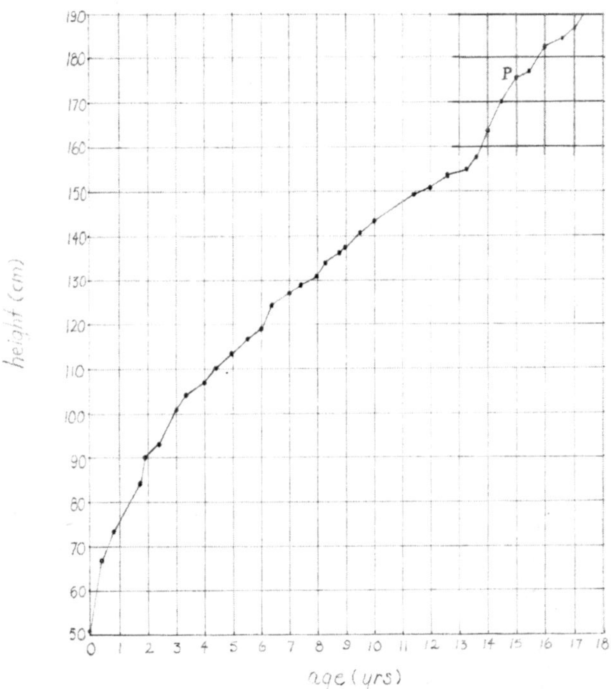

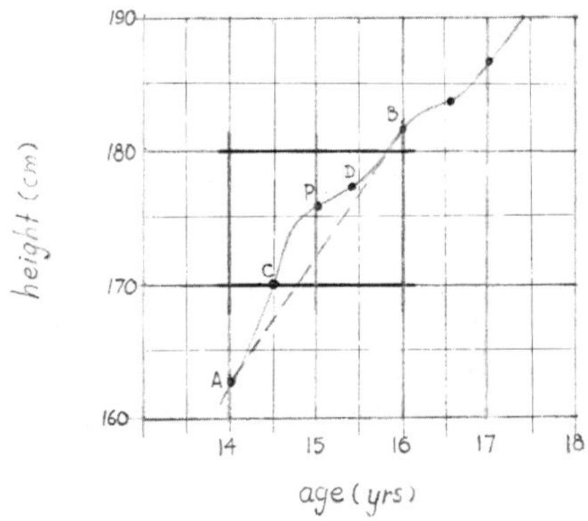

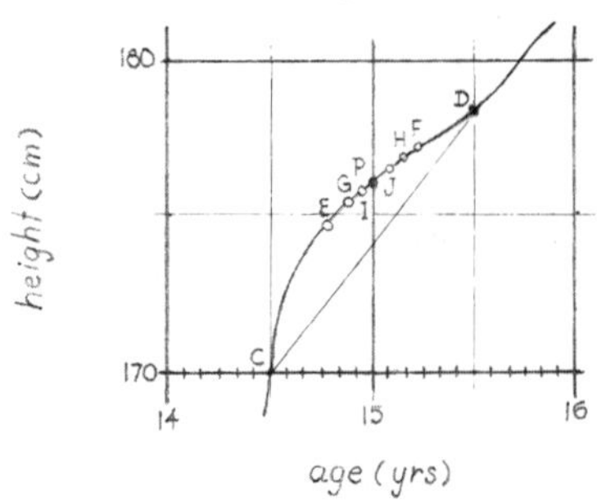

the line AB. Over a 1-year interval, this average growth rate is given by the slope CD. (See the third graph at the middle of this page.) The slope of a straight line drawn from E to F gives the average growth rate over 6 months, etc. The four lines, AB, CD, EF, and GH, are not exactly parallel to each other, and so their slopes are different. However, the difference in slope gets smaller and smaller. It is large when you compare AB and CD, less if you compare CD and EF, and still less between EF and GH. For intervals less than $\Delta t = 1$ year, the line segments do become parallel to each other, as far as you can tell, and gradually merge into the curve. For very small intervals, *you can find the slope by drawing a straight line tangent to this curve at* P. This method involves placing a ruler parallel to line GH at P and extending it on both sides.

The values of the slopes of the straight-line segments in the two right-hand graphs on page 24 have been computed for the corresponding time intervals. These values appear in the table in the margin at the right. Note that values of v_{av} calculated for shorter and shorter time intervals approach closer and closer to 6.0 cm/yr. In fact, for any time interval less than 2 months, v_{av} will be 6.0 cm/yr within the limits of accuracy of measuring height. Thus, you can say that, on his fifteenth birthday, young de Montbeillard was growing at a rate of 6.0 cm/yr. At that instant in his life, $t = 15.0$ yr, this was his instantaneous growth rate. (You might also express it as *instantaneous speed* of his head with respect to his feet when he was lying still!)

Line between points	Δt	Δd	Growth rate $v_{av} = \frac{\Delta d}{\Delta t}$
AB	2 yr	19.0 cm	9.5 cm/year
CD	1 yr	8.0	8.0
EF	6 mo	3.5	7.0
GH	4 mo	2.0	6.0
IJ	2 mo	1.0	6.0

If you have taken a course in calculus, you will recognize that we are discussing "limits," but in a simpler and less rigorous way.

Average speed over a time interval Δt, as mentioned earlier, is by definition the ratio of distance traveled to elapsed time. In symbols,

$$v_{av} = \frac{\Delta d}{\Delta t}$$

SG 17

You now have the *definition of instantaneous speed* at a given instant t: It is the final, limiting value approached by the average speeds when you compute v_{av} for a smaller and smaller range of time intervals that include the desired instant t. In almost all physical situations, such a limiting value can be accurately and quickly estimated by the graphical method used.

From now on, the letter v (without the subscript $_{av}$) will be used to represent instantaneous speed defined in this way. You may wonder why the letter v instead of s was used for speed. The reason is that speed is closely related to velocity. The term "velocity" is used to mean *speed in a specified direction* (such as *50 km/hr to the north*) and will later be represented by the symbol $\vec{v}$. When direction is not specified and only the magnitude (*50 km/hr*) is of interest, remove the arrow and just use the letter v. This symbol represents only the *magnitude* of the velocity, that is, the "speed." This distinction between speed and velocity will be discussed in more detail in later sections. You also will learn why, in physics, velocity is a more important concept than speed.

?

13. Define instantaneous speed in words and in symbols.

14. My average speed during a recent trip by car to Cleveland was 20 m/sec. Estimate my instantaneous speed for the time when I was (a) caught in city traffic, (b) traveling on dry, open highway, and (c) filling up with gas. Explain the difference between average speed and instantaneous speed.

See SG 18, 20, and 24 for problems that check your understanding of the chapter up to this point.

Close Up

Photography, 1839 to the Present

1. *Note the lone figure in the otherwise empty street. He was getting his shoes shined. The other pedestrians did not remain in one place long enough to have their images recorded. With exposure times several minutes long, the outlook for the possibility of portraiture was gloomy.*
2. *However, by 1859, improvements in photographic emulsions and lenses made it possible not only to photograph a person at rest, but also to capture a bustling crowd of people, horses, and carriages. (With a magnifying glass you can see the slight blur of the jaywalker's legs.)*
3. *Today, one can "stop" action with an ordinary camera.*
4. *A new medium—the motion picture. In 1873 a group of California sports enthusiasts called in the photographer Eadweard Muybridge to settle the question, "Does a galloping horse ever have all four feet off the ground at once?" Five years later he answered the question with these photos. The five*

1. Paris street scene, 1839. A daguerreotype made by Louis Daguerre himself.

2. American street scene, 1859.

3. The moving car is seen in focus in the foreground, while the stationary background appears blurred.

4. Muybridge's series, 1878.

pictures were taken with five cameras lined up along the track. Each camera was triggered when the horse broke a string that tripped the shutter. The motion of the horse can be reconstructed by making a flip pad of the pictures.

With the perfection of flexible film, only one camera was needed to take many pictures in rapid succession. By 1895, there were motion picture parlors throughout the United States. Twenty four frames each second were sufficient to give the viewer the illusion of motion.

5. *A light can be flashed successfully at a controlled rate, and a multiple exposure (similar to the strobe photos in this text) can be made. In this photo of a golfer, the light flashed 100 times each second.*

6. *It took another 90 years after the time the crowded street was photographed before a bullet in flight could be "stopped." This remarkable picture was made by Harold Edgerton of MIT, using a brilliant electric spark which lasted for about 1 millionth of a second.*

7. *An interesting offshoot of motion pictures is the high-speed motion picture. In the frames of the milk drop series shown below, 1,000 pictures were taken each second (by Harold Edgerton). The film was whipped past the open camera shutter while the milk was illuminated with a flashing light (similar to the one used in photographing the golfer) synchronized with the film. When the film is projected at the rate of 24 frames each second, action that took place in 1 sec is spread out over 42 sec. Very high-speed photography was used to make several of the* Project Physics *film loops.*

It is clear that the eye alone cannot see the elegant details of this event. This is precisely why photography of various kinds is used in the laboratory.

5. Stroboscopic photo of golfer's swing.

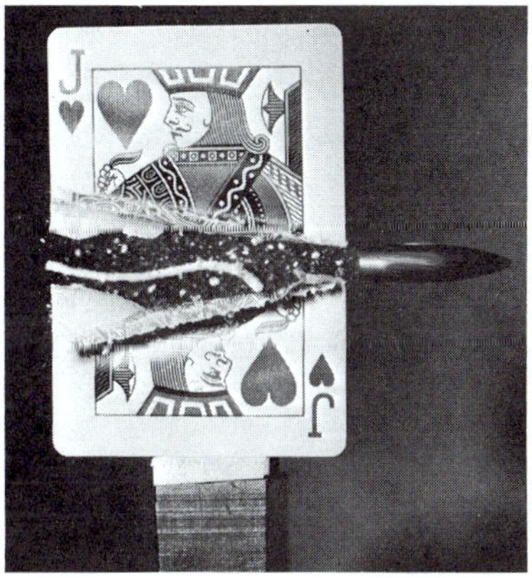

6. Bullet cutting through a playing card.

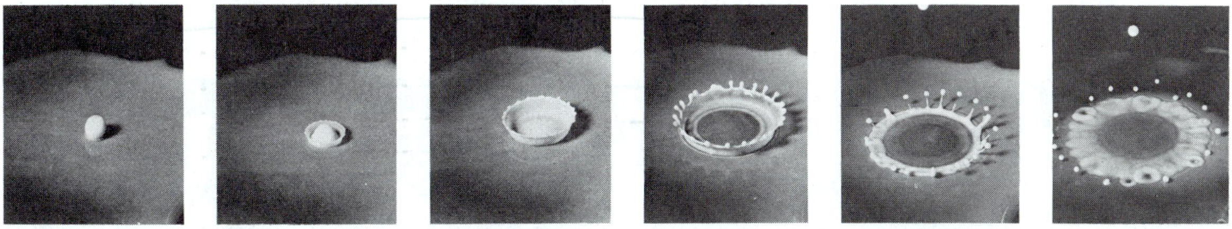

7. Action shown in high-speed film of milk drop.

1.8 | Acceleration by comparison

You can infer from the photograph at the bottom of this page of a baseball rolling on an incline that the ball was changing speed (accelerating). Assuming the time between flashes to have been constant, the increasing distance between the images of the ball give you this information. But how can you tell how *much* acceleration the ball has?

To answer this question, you have to learn the definition of acceleration. The definition itself is simple. The real task is to learn how to *use* it in situations like the one below. For the time being, acceleration can be defined as *rate of change of speed*. Later, this definition will have to be modified somewhat when you encounter motion in which change in *direction* becomes important. For now, you are dealing only with straight-line motion. Therefore, you can equate the rate of change of speed with acceleration.

Unless noted otherwise, "rate of change" will always mean "rate of change with respect to time."

Some of the effects of acceleration are familiar to everyone. It is acceleration, not speed, that you notice when an elevator suddenly starts up or slows down. The flutter in your stomach comes only during the speeding up and slowing down. It is not felt during most of the ride, when the elevator is moving at a steady speed. Likewise, the excitement of the roller coaster and other rides at amusement parks results from their unexpected accelerations. Speed by itself does not cause these sensations. If it did, you would feel them during a smooth plane ride at 900 km/hr, or during the continuous motion of the earth around the sun at 105,000 km/hr.

Simply stated, speed is a relationship between two objects. One object is taken to be the reference object, while the other moves with respect to it. Some examples are the speed of the earth with respect to the sun, the speed of the swimmer with respect to the pool edge, the speed of the top of the growing boy's head with respect to his feet. In a perfectly smooth-riding train, you could tell that you were moving at a high speed only by seeing the scenery speeding by. You would have just the same experience if the train were somehow fixed and the earth, rails, etc., were to speed by in the other direction. If you "lost the reference object" (by pulling down the shades, say), you could not tell whether you were moving or not. In contrast, you "feel"

accelerations. You do not need to look out the train window to realize that the engineer has suddenly started the train or has slammed on the brakes. You might be pushed against the seat, or the luggage might fly from the rack.

For example, if an airplane changes its speed from 800 km/hr to 850 km/hr in 10 min, its average acceleration would be

$$\frac{\Delta v}{\Delta t} = \frac{850\ \text{km/hr} - 800\ \text{km/hr}}{10\ \text{min}}$$

$$= \frac{50\ \text{km/hr}}{10\ \text{min}}$$

$$= \frac{5\ \text{km/hr or } 5\ \text{km/hr/min}}{\text{min}}$$

That is, its speed changed at a rate of 5 km/hr per minute which, expressed in a consistent system of units, is about 0.08 km/min^2 or 300 km/hr^2. (If the speed was decreasing, the value of the acceleration would be negative.)

All this suggests a profound physical difference between motion at constant speed and accelerated motion. It is best to learn about acceleration at first hand (in the laboratory and through the film loops). But the main ideas can be summarized here. For the moment, focus on the similarities between the concepts of speed and acceleration. For motion in a straight line:

The *rate of change of position* is called *speed.*	The *rate of change of speed* is called *acceleration.*

This similarity of form is very helpful. It enables you to use what you have just learned about the concept of speed as a guide for using the concept of acceleration. For example, you have learned that the slope of a line of a *distance*–time graph is a measure of *instantaneous speed.* Similarly, the slope of a *speed*–time graph is a measure of *instantaneous acceleration.*

This section concludes with a list of six statements about motion along a straight line. The list has two purposes: (1) to help you review some of the main ideas about speed presented in this chapter, and (2) to present the corresponding ideas about acceleration. For this reason, each statement about speed is immediately followed by a parallel statement about acceleration.

1. *Speed* is the rate of change of position. *Acceleration* is the rate of change of speed.
2. *Speed* is expressed in units of distance/time. *Acceleration* is expressed in units of speed/time.
3. *Average speed* over any time interval is the ratio of the change of position Δd to the time interval Δt:

$$v_{av} = \frac{\Delta d}{\Delta t}$$

Average acceleration over any time interval is the ratio of the change of speed Δv to the time interval Δt:

$$a_{av} = \frac{\Delta v}{\Delta t}$$

4. *Instantaneous speed* is the value approached by the average speed as Δt is made smaller and smaller. *Instantaneous acceleration* is the value approached by the average acceleration as Δt is made smaller and smaller.
5. On a *distance*–time graph, the *instantaneous speed* at any instant is the slope of the straight line tangent to the curve at the point of interest. On a *speed*–time graph, the *instantaneous*

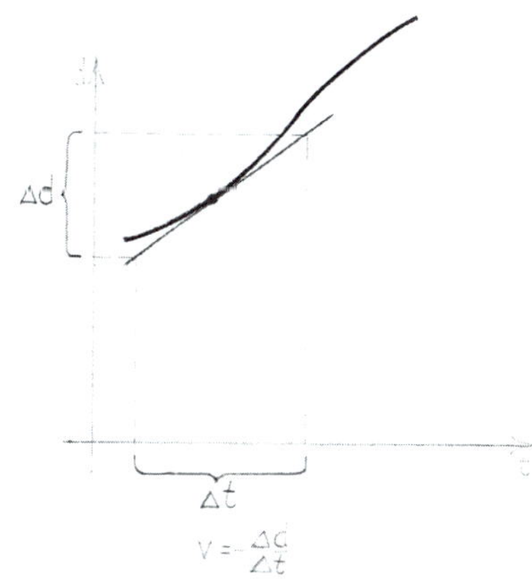

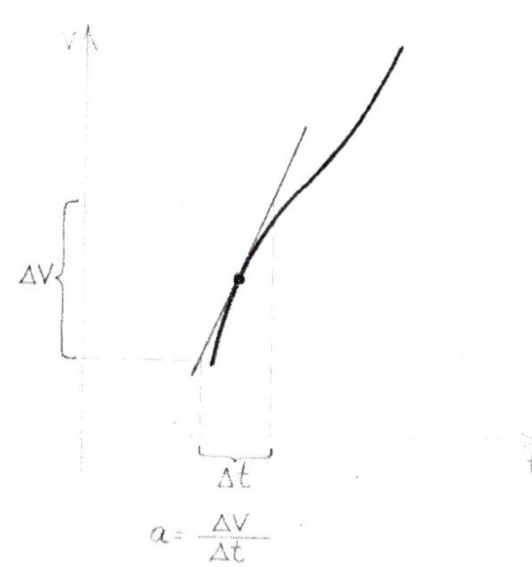

acceleration at any instant is the slope of the straight line tangent to the curve at the point of interest. See the graphs in the margin.

Constant speed and constant acceleration are often called "uniform" speed and "uniform" acceleration. In the rest of this course, we will use the terms interchangeably.

6. For the particular case of *constant speed*, the distance–time graph is a straight line. Therefore, the instantaneous speed has the same value at every point on the graph. Further, this value is equal to the average speed computed for the whole trip. For the particular case of *constant acceleration*, the speed–time graph is a straight line. Therefore, the instantaneous acceleration has the same value at every point on the graph. Further, this value is equal to the average acceleration computed for the whole trip. When speed is constant, its value can be found from any corresponding Δd and Δt. When acceleration is constant, its value can be found from any corresponding Δv and Δt. (This is useful to remember because constant acceleration is the kind of motion you will encounter most often in the following chapters.)

SG 22 provides an opportunity to work with distance–time and speed–time graphs and to see their relationship to one another.

You now have most of the tools needed to get into some real physics problems. The first such problem will involve the accelerated motion of bodies caused by gravitational attraction. It was by studying the motion of falling objects that Galileo, in the early 1600s, first shed light on the nature of accelerated motion. His work remains a wonderful example of how scientific theory, mathematics, and actual measurements can be combined to develop physical concepts. More than that, Galileo's work opened one of the earliest and most important battles of the scientific revolution. The specific ideas he introduced are even today fundamental to the science of *mechanics*, the study of bodies in motion.

SG 23–26 are review problems for this chapter. Some of these will test how thoroughly you grasp the language used for describing straight-line motion.

?

15. What is the average acceleration of an airplane that goes from 0 to 100 km/hr in 5 sec?

16. What is your average acceleration if, while walking, you change your speed from 5 km/hr to 2 km/hr in an interval of 15 min? Is your answer affected by how your change of speed is distributed over the 15 min?

17. Using the formula for average acceleration, determine how long it takes a bird starting from rest to reach a speed of 12 m/sec if it accelerates uniformly at 4 m/sec^2.

18. If the bird in Question 17 continued to accelerate at the same rate, how much speed would it add in another 6 sec?

study guide

1. This book is probably different in many ways from textbooks you have had in other courses. Therefore, it might help to make some suggestions about how to use it.

(**a**) If this is your own personal copy and you intend to retain it after you have completed the course, in short, if you are in a situation that permits you to mark freely in the book, do so. You will note that there are wide margins to record questions or statements as they occur to you. Mark passages that you do not understand so that you can seek help from your instructor.

(**b**) If you may not write in the textbook itself, try keeping a notebook keyed to the text chapters. In this study notebook, jot down the kinds of remarks, questions, and answers that you would otherwise write in the textbook as suggested above. Also, you ought to write down the questions raised by the other learning materials you will use, by the experiments you do, by demonstrations or other observations, and by discussions you may have with fellow students and others with whom you talk physics. Most students find such an informal notebook to be enormously useful when studying, or when seeking help from their instructors (or, for that matter, from advanced students, scientists they may know, or anyone else whose understanding of physics they trust).

(**c**) Always try to answer the end-of-section review questions yourself first, and then check your answers. If your answer agrees with the one in the book, it is a good sign that you understand the main ideas in that section (although it is true that you can sometimes get the right answer for the wrong reason). Also, sometimes there may be other answers as good as (or better than!) those given in the book.

(**d**) There are many different kinds of items in the Study Guide at the end of each chapter. It is not intended that you should do every item. Sometimes material is included in the Study Guide which may especially interest only some students. Notice also that there are several kinds of problems. Some are intended to give practice in the use of a particular concept, while others are designed to help you bring together several related concepts. Still other problems are intended to challenge those students who particularly like to work with numbers.

(**e**) This text is only one of the learning materials of the *Project Physics* course. The course includes several other materials, such as film loops and filmstrips or transparencies. Use them if they are available to you. Be sure to familiarize yourself also with the *Handbook* for students, which describes outside activities and laboratory experiments. Each of these learning aids makes its own contribution to an understanding of physics, and all have been designed to be used together.

Note: The *Project Physics* learning materials particularly appropriate for Chapter 1 include:

Experiments (in the *Handbook*)
Naked Eye Astronomy
Regularity and Time
Variations in Data
Measuring Uniform Motion

2. Define, in words and symbols, the following terms: speed, uniform motion, average speed, slope, instantaneous speed, average acceleration. What does the symbol "Δ" mean?

3. A goalie shoots a puck to his teammate 30 m away. If the puck took 1.5 sec to cover the distance, what was its average speed?

4. Some practice problems:

	Situation	*Find*
a	Speed uniform, distance = 72 cm, time = 12 sec	Speed
b	Speed uniform at 60 km/hr	Distance traveled in 20 min
c	Speed uniform at 36 m/min	Time to move 9 m

d	$d_1 = 0$, $t_1 = 0$ $d_2 = 15$ cm, $t_2 = 5.0$ sec $d_3 = 30$ cm, $t_3 = 10$ sec	Speed and position at 8.0 sec
e	You drive 240 km in 6.0 hr	Average speed
f	Same as **e**	Speed and position after 3.0 hr
g	Average speed is 76 cm/sec, computed over a distance of 418 cm	Time taken
h	Average speed is 44 m/sec, computed over a time interval of 0.20 sec	Distance moved

5. After the parachute opens, a sky diver falls with a roughly uniform speed of 12 m/sec. How long does it take her to fall 228 m? If she continues to fall for another 25 sec, what is the *total* distance she has fallen?

6. What is your average speed in each of these cases?

(**a**) You run 100 m at a speed of 5.0 m/sec and then you walk 100 m at a speed of 1.0 m/sec.

(**b**) You run for 100 sec at a speed of 5.0 m/sec and then you walk for 100 sec at a speed of 1.0 m/sec.

7. A rabbit and a turtle are practicing for their race. The rabbit covers a 30-m course in 5 sec; the turtle covers the same distance in 120 sec. If the race is run on a 96-m course, by how many seconds will the rabbit beat the turtle?

8. A tsunami caused by an earthquake occurring near Alaska in 1946 consisted of several huge waves which were found to travel at the average speed of 790 km/hr. The first of the waves reached Hawaii 4 hr 34 min after the earthquake occurred. From these data, calculate how far the origin of the tsunami was from Hawaii.

9. Design and describe experiments to enable you to make estimates of the average speeds for some of the following objects in motion.

(**a**) A baseball thrown from the outfield to home plate

(**b**) the wind

(**c**) a cloud

(**d**) an ant walking

(**e**) a camera shutter opening and closing

(**f**) an eye blinking

(**g**) a whisker growing

10. Light and radio waves travel through a vacuum in a straight line at a speed of nearly 3×10^8 m/sec.

(**a**) How long is a "light year" (the *distance* light travels in a year)?

(**b**) The nearest star, *Alpha Centauri*, is 4.06×10^{16} m distant from earth. If this star possesses planets on which highly intelligent beings live, how soon, at the earliest, could we expect to receive a reply after sending them a radio or light signal strong enough to be received there?

(**c**) Sound moves very quickly; it is hard to notice the time elapsed between when you *see* somebody say something and when you *hear* the sound.

(**1**) Can you think of situations when you can tell that sound does not reach you virtually instantaneously?

(**2**) Try to design an experiment to measure the speed of sound.

(**3**) The measurements made in many experiments indicate that under ordinary circumstances the speed of sound in air is about 330 m/sec (= 0.33 km/sec). The speed of light is about 300,000 km/sec. Suppose lightning strikes 1 km away. How long does it take before you see the flash? How long before you hear the thunder?

(**4**) Can you use the known speeds of light and sound to find the distance to any lightning stroke?

11. Two cyclists race with nearly uniform speed on a 500-m course. The blue bicycle crosses the finish line 20 sec ahead of the red bicycle. If the red bicycle maintained an average speed of 10 m/sec, what was the average speed of the blue bicycle?

12. After starting from rest, a car reaches a speed of 30 m/sec in 5 sec. What is its average acceleration? If the car accelerates at that rate for an additional 5 sec, what is its final speed?

13. The following graph represents a jogger moving with uniform speed who passes posts at 20 m, 30 m, and 50 m at 4 sec, 6 sec, and 10 sec, respectively. The jogger's average speed over the interval from 4 sec to 10 sec can be calculated as shown on the second graph. (Remember, on a distance–time graph, the slope of the line between two points is the average speed over that interval.)

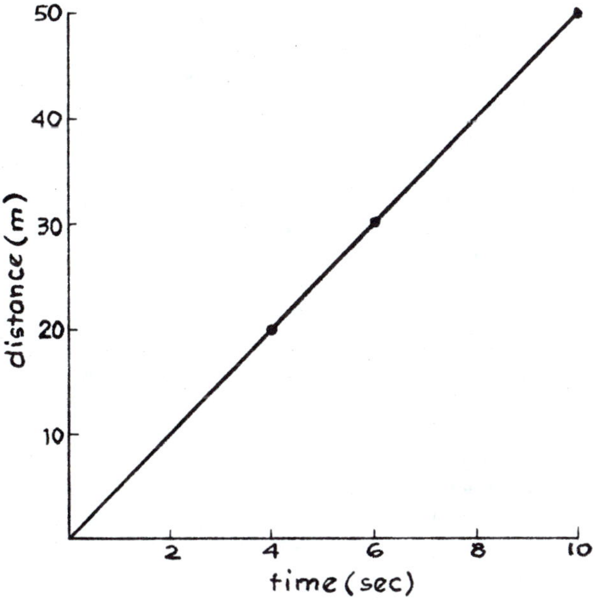

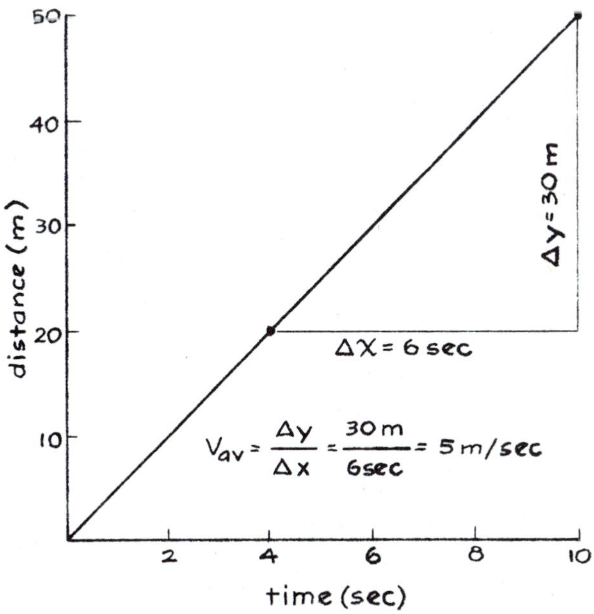

(a) Use the same method to find the jogger's average speed for the following intervals: (1) starting line (0 sec) to 6 sec; (2) starting line to 10 sec; (3) 6 sec to 10 sec; (4) 5 sec to 8 sec.

(b) Your answers to (a) should indicate that no matter how small the interval you choose, you will always find the same speed for a straight line on a distance–time graph. Why?

(c) Now find the instantaneous speed, which is the speed over an extremely small interval. Without measuring the small Δy or Δx, what is the instantaneous speed at the 8-sec mark? How do you know? [*Hint*: see part (b).]

14. The following graph shows the motion of a ball; use it to answer the questions below.

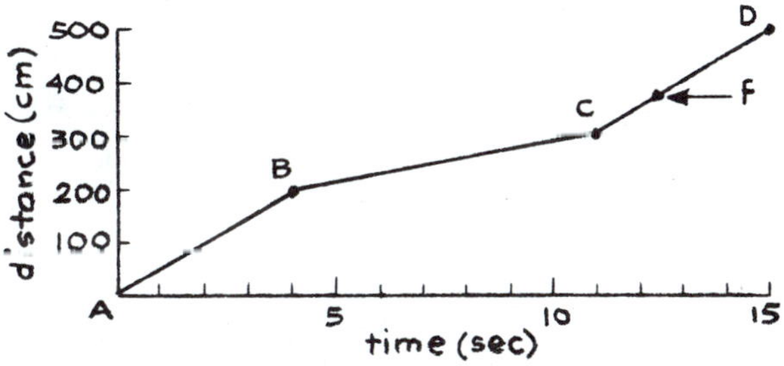

(a) Without doing any calculations, determine in which section the ball was traveling fastest; slowest.

(b) Calculate the average speeds over each interval (AB, BC, CD) and for the entire distance (AD).

(c) Without doing any calculations, determine the instantaneous speed at point f in section CD.

15. The graph on page 34 represents the motion of a jogger who speeds up as she runs. Notice that the graph is not a straight line. Average speed can be calculated from the graph using the same method as for a straight-line graph (see Question 13). However, to find the instantaneous speed at any point on the curved graph, you must draw a line tangent to the curve at that point. This tangent line has the same slope (speed) as the curve at the point where they touch.

(a) What is the instantaneous speed at the 10-sec mark? At the 25-sec mark?

(b) What is the average acceleration between $t = 10$ sec and $t = 25$ sec?

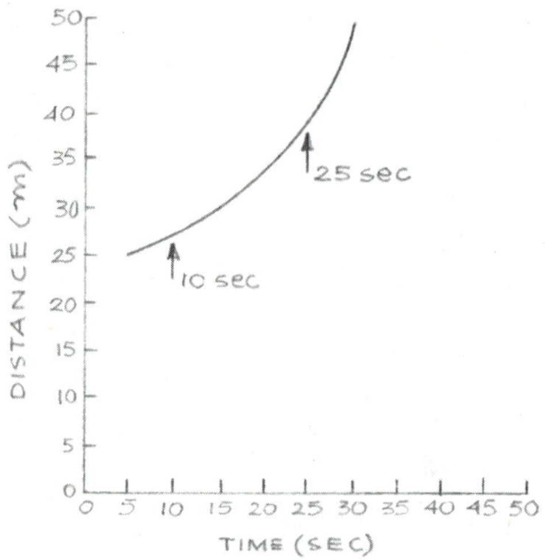

16. World's 400-m swimming record in minutes and seconds for men and women:

1926	4:57.0	Johnny Weissmuller
	5:53.2	Gertrude Ederle
1936	4:46.4	Syozo Makino
	5:28.5	Helene Madison
1946	4:46.4	(1936 record unbroken)
	5:00.1	R. Hveger
1956	4:33.3	Hironoshin Furuhashi
	4:47.2	Lorraine Crapp
1966	4:11.1	Frank Weigand
	4:38.0	Martha Randall
1976	3:15.9	Brian Goodell
	4:09.8	Petra Thumer

By about how many meters would Martha Randall have beaten Johnny Weissmuller if they had been able to race each other? Could you predict the 1986 records for the 400-m race by extrapolating the graphs of world's records vs. dates up to the year 1986?

17. Using the graph on p. 20, find the instantaneous speeds v at several points (0, 10, 20, 30, 40, and 50 sec, and near 0, or at other points of your choice) by finding the slopes of lines tangent to the curve at each of those points. Make a graph of v versus t.

18. Discuss the following quotation from Mark Twain's *Life on the Mississippi* (1875) as an example of extrapolation. "In the space of one hundred and seventy-six years the Lower Mississippi has shortened itself two hundred and forty-two miles. That is an average of a trifle over one mile and a third per year. Therefore, any calm person, who is not blind or idiotic, can see that in the old Colitic Silurian Period, just a million years ago next November, the Lower Mississippi River was upward of one million three hundred thousand miles long, and stuck out over the Gulf of Mexico like a fishing rod. And by the same token any person can see that seven hundred and forty-two years from now the Lower Mississippi River will be only a mile and three-quarters long, and Cairo and New Orleans will have joined their streets together, and be plodding comfortably along under a single mayor and a mutual board of aldermen. There is something fascinating about science. One gets such wholesale returns of conjecture out of such trifling investment of fact."

19. Careful analysis of a stroboscopic photograph of a moving object yielded information that was plotted on the graph below.

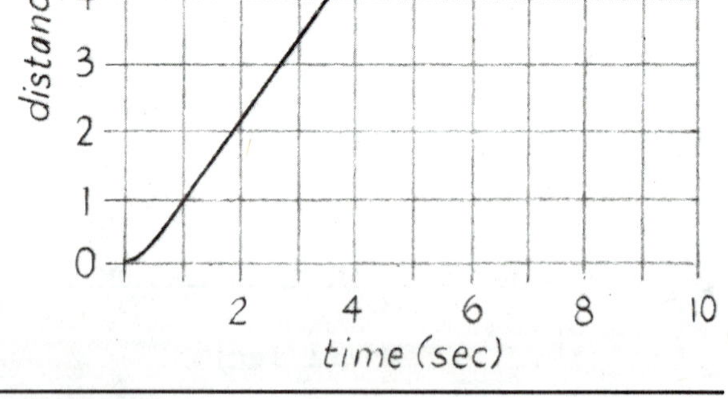

(**a**) At what moment or interval was the speed greatest? What was the speed at that time?

(**b**) At what moment or in which interval was the speed least? What was it at that time?

(**c**) What was the speed at time $t = 5.0$ sec?

(**d**) What was the speed at time $t = 0.5$ sec?

(**e**) How far did the object move from time $t = 7.0$ sec to $t = 9.5$ sec?

20. A band of pioneers left St. Louis in their wagon train. They traveled at a good rate for the first 2.5 weeks (AB), but slowed down as the initial excitement wore off (BC). Then they picked up speed for a week (CD) and really raced through dangerous country (DE). They rested at a watering hole (EF) and finally moved on (FG).

(**a**) By observing the graph, determine which interval was covered fastest; slowest.

(**b**) According to the information given, which interval does *not* look as if it were drawn correctly? What should it look like?

(**c**) Find the average speed over the whole trip.

(**d**) What are the instantaneous speeds at points P and Q?

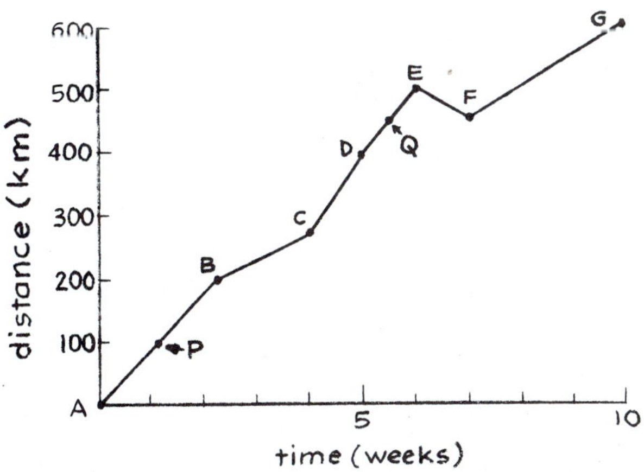

21. The data below show the instantaneous speeds in a test run of a car starting from rest. Plot the speed–time graph, then derive data from it and plot the acceleration–time graph.

(**a**) What is the speed at $t = 2.5$ sec?

(**b**) What is the maximum acceleration?

Time (sec)	*Speed (m/sec)*	*Time (sec)*	*Speed (m/sec)*
0.0	0.0	6.0	27.3
1.0	6.3	7.0	29.5
2.0	11.6	8.0	31.3
3.0	16.5	9.0	33.1
4.0	20.5	10.0	34.9
5.0	24.1		

22. The electron beam in a typical TV set sweeps out a complete picture in 0.03 sec, and each picture is composed of 525 lines. If the width of the screen is 50 cm, what is the speed of that beam horizontally across the screen?

23. Turn back to p. 28. At the bottom of this page there is a multiple-exposure photograph of a baseball rolling to the right. The time interval between successive flashes was 0.20 sec. The distance between marks on the meter stick was 1 cm. You might tabulate your measurements of the ball's progress between flashes and construct a distance–time graph. From the distance–time graph, determine the instantaneous speed at several instants and construct a speed–time graph. You can check your results by referring to the answer page at the end of this book.

24. Suppose you must measure the instantaneous speed of a bullet as it leaves the barrel of a rifle. Explain how you might do this.

25. Discuss the motion of the horse in the following series of photographs by Muybridge. The time interval between exposures is 0.045 sec.

CHAPTER 2

Free Fall
Galileo Describes Motion

A sketch of a medieval world–system.

2.1 | The Aristotelian theory of motion

In this chapter you will follow the development of an important piece of basic research: Galileo's study of freely falling bodies. The phenomenon of free fall is interesting in itself. But the emphasis will be on the way Galileo, one of the first modern scientists, presented his argument. His view of the world, way of thinking, use of mathematics, and reliance upon experimental texts set the style for modern science. These aspects of his work, therefore, are as important as the actual results of his investigation.

SG 1

To understand the nature and importance of Galileo's work, you must first examine the previous system of physical thought which his ideas eventually replaced. Medieval physical science, as Galileo learned it at the University of Pisa, made a sharp distinction between objects on the earth and those in the sky. All *terrestrial* matter, matter on or near the earth, was believed to contain a mixture of four "elements": Earth, Water, Air, and Fire. These elements were not thought of as identical with the natural materials for which they were named. Ordinary water, for example, was thought to be a mixture of all four elements, but mostly the ideal element Water. Each of the four elements was thought to have a natural place in the terrestrial region. The highest place was allotted to Fire. Beneath Fire was Air, then Water, and finally, in the lowest position, Earth. Each was thought to seek its own place. Thus, Fire, if placed below its natural position, would tend to rise through Air. Similarly, Air would tend to rise through Water, whereas Earth would tend to fall through both Air and Water. The movement of any real object depended on its particular mixture of these four elements and on where it was in relation to the natural places of these elements. When water boiled, for example, the element Water would be joined by the element Fire, whose higher natural place would cause the mixture to rise as steam. A stone, on the other hand, was composed mainly of the element Earth. Therefore, a stone would fall when released and would pass through Fire, Air, and Water until it came to rest on the ground, its natural place.

A good deal of common-sense experience supports this natural-place view. See SG 2.

Medieval thinkers also believed that stars, planets, and other *celestial* (heavenly) bodies differed in composition and behavior from objects on or near the earth. Celestial bodies were believed to contain none of the four ordinary elements, but to consist solely of a fifth element, the *quintessence*. The difference in composition required a different physics. Thus, the natural motion of celestial objects was thought to be neither rising nor falling, but an endless revolving in circles around the center of the universe. That center was considered to be identical with the center of the earth. Heavenly bodies, although moving, were at all times in their natural places. In this way, heavenly bodies

From *quinta essentia*, meaning fifth (*quint*) element (*essence*). In earlier Greek writings the term for it was *aether* (also written *ether*).

This painting, entitled "School of Athens," was done by Raphael in the beginning of the sixteenth century. It reflects a central aspect of the Renaissance, the rebirth of interest in classical Greek culture. The central figures are Plato (on the left, pointing to the heavens) and Aristotle (pointing to the ground).

differed from terrestrial objects, which displayed natural motion only as they returned to their natural places from which they had been displaced.

This theory, so widely held in Galileo's time, had originated almost 2,000 years before, in the fourth centry B.C. It is stated clearly in the writings of the Greek philosopher Aristotle. (See the time chart on the opposite page.) This physical science, built on notions of cause, order, class, place, and purpose, seemed to fit well with many everyday observations. Moreover, these ideas about matter and motion were part of an all-embracing universal scheme, or *cosmology*. In this cosmology, Aristotle sought to relate ideas which today are discussed separately under such headings as science, poetry, politics, ethics, and theology.

Not very much is known of Aristotle's physical appearance or life. It is thought that he was born in 384 B.C. in the Greek province of Macedonia. His father was the physician to the King of Macedonia, so Aristotle's early childhood was spent in an environment of court life. He studied in Athens with Plato and later returned to Macedonia to become the private tutor to Alexander the Great. In 335 B.C., Aristotle came back to Athens and founded the Lyceum, a school and center of research.

After the decline of the ancient Greek civilization, Aristotle's writings remained almost unknown in Western Europe for 1,500 years. They were rediscovered in the thirteenth century A.D. and soon began to shape the thinking of Christian scholars and theologians. Aristotle became such a dominant influence in the late Middle Ages that he was referred to simply as "The Philosopher."

Unfortunately, Aristotle's physical theories had serious limitations. (This does not, of course, detract from his great achievements in other fields.) According to Aristotle, the fall of a heavy object toward the center of the earth is an example of "natural" motion. He evidently thought that any object, after release, quickly reaches some final speed of fall which it maintains to the end of its path. What factors determine the final speed of a falling object? It is a common observation that a rock falls faster than a leaf. Therefore, Aristotle reasoned, weight is a factor that governs the speed of fall. This fitted in well with his idea that the *cause* of weight was the presence of the element Earth, whose natural motion was to the center of the earth. Thus, a heavier object, having a greater content of Earth, has a stronger tendency to fall to its natural place. In turn, this stronger tendency creates a greater speed of falling.

Aristotle: Rate of fall is proportional to weight divided by resistance.

The same object falls more slowly in water than in air, so Aristotle reasoned that the resistance of the medium must also affect motion. Other factors, such as the color or temperature of the falling object, also might change the rate of fall. But Aristotle decided that such influences could not be important. He

500 BC
400 BC
384 BC
322 BC
ARISTOTLE
300 BC
200 BC
Historical Events
Persian Wars
Peloponnesian Wars
Sparta Defeats Athens
Rise of Macedonia
The Punic Wars
Ch'in Dynasty in China
Government
XERXES
PERICLES
PTOLEMY I of Egypt
PHILIP II
ALEXANDER
HANNIBAL
Philosophy and Science
CONFUCIUS
ANAXAGORAS
ZENO OF ELEA
PARMENIDES OF ELEA
PROTAGORAS
DEMOCRITUS
SOCRATES
PLATO
DIOGENES
MENCIUS (MENG TZU)
EPICURUS
ZENO OF CITIUM
HSÜN TZU
EUCLID
ARCHIMEDES
ERASTOSTHENES
Literature
AESCHYLUS
PINDAR
SOPHOCLES
HERODOTUS
EURIPIDES
THUCYDIDES
ARISTOPHANES
XENOPHON
DEMOSTHENES
CHUANG TZU
PLAUTUS
Fine Arts
PHIDIAS
MYRON
POLYGNOTUS
POLYCLITUS
ZEUXIS
TIMOTHEUS
SCOPAS
LYSIPPUS
PRAXITELES
APELLES
ARISTOXENUS

concluded that the rate of fall must increase in proportion to the weight of the object and decrease in proportion to the resisting force of the medium. The actual rate of fall in any particular case would be found by dividing the weight by the resistance.

Aristotle also discussed "violent" motion, that is, any motion of an object other than going freely toward its "natural place." Such motion, he argued, must always be caused by a *force*, and the speed of the motion must increase as the force increases. When the force is removed, the motion must stop. This theory agrees with common experience, for example, in pushing a chair or a table across the floor. It does not work quite so well for objects thrown through the air, since they keep moving for a while even after you have stopped exerting a force on them. To explain this kind of motion, Aristotle proposed that the air itself somehow exerts a force that keeps the object moving.

John Philoponus: Rate of fall is proportional to weight *minus* resistance.

Later scientists suggested certain changes in Aristotle's theory of motion. For example, in the fifth century A.D. John Philoponus of Alexandria argued that the speed of an object in natural motion should be found by *subtracting* the resistance of the medium from the weight of the object. (Aristotle, you recall, recommended *dividing* by the resistance.) Philoponus claimed that his experimental work supported his theory, though he did not report the details. He simply said that he dropped two weights, one twice as heavy as the other, and observed that the heavy one did not reach the ground in half the time taken by the light one.

SG 3

There were still other difficulties with Aristotle's theory of motion. However, the knowledge that his teachings had faults did little to lessen their influence in the universities of France and Italy during the fifteenth and sixteenth centuries and during Galileo's lifetime. Aristotle's theory of motion did, after all, fit much of ordinary experience in a general, if qualitative, way. Besides, the study of motion through space was of great interest to only a few scholars, just as it had been only a very small part of Aristotle's own work.

Qualitative refers to quality—the sort of thing that happens. *Quantitative* refers to quantity—the measurement or prediction of *numerical* values. This distinction will be made often in the course.

Two other influences stood in the way of major changes in the theory of motion. First, Aristotle believed that mathematics was of little value in describing terrestrial phenomena. Second, he put great emphasis upon direct, qualitative observation as the basis for forming theories. Simple qualitative observation was very useful in Aristotle's biological studies. But as it turned out, real progress in physics began only when scientists recognized the value of mathematical prediction and detailed measurement.

A number of scholars in the fifteenth and sixteenth centuries took part in this change to a new way of approaching science. But of all these, Galileo was by far the best known and most successful. He showed how to describe mathematically the motions of simple, ordinary objects, such as falling stones and balls rolling on an incline. Galileo's work paved the way for other scholars to describe and explain the motion of everything from pebbles to planets. It also began the intellectual revolution that led to what is now considered modern science.

?

1. Which of the following properties do you believe might affect the rate of fall of an object: color, shape, size, or weight? How could you determine if your answers are correct?

2. Describe two ways in which, according to the Aristotelian view, terrestrial and celestial bodies differ from each other.

3. Which of these statements would be accepted in the fifteenth and sixteenth centuries by persons who believed in the Aristotelian system of thought?

(a) Ideas of motion should fit in with poetry, politics, theology, and other aspects of human thought and activity.

(b) Heavy objects fall faster than light ones.

(c) Except for motion toward their natural location, objects will not move unless acted on violently by a force.

(d) Mathematics and precise measurement are especially important in developing a useful theory of motion.

2.2 | Galileo and his times

Galileo Galilei was born in Pisa in 1564, the year of Michelangelo's death and Shakespeare's birth. Galileo was the son of a noble family from Florence, and he acquired his father's active interest in poetry, music, and the classics. His scientific inventiveness also began to show itself early. For example, as a young medical student, he constructed a simple pendulum-type timing device for the accurate measurement of pulse rates.

Italy about 1600.

After reading the classical Greek philosopher–scientists Euclid and Archimedes, Galileo changed his interest from medicine to physical science. He quickly became known for his unusual scientific ability. At the age of 26, he was appointed Professor of Mathematics at Pisa. There he showed an independence of spirit unmellowed by tact or patience. Soon after his appointment, he began to challenge the opinions of older professors, many of whom became his enemies. He left Pisa before his term was completed, apparently forced out by financial difficulties and by his enraged opponents. Later, at Padua in the Republic of Venice, Galileo began his work in astronomy. He supported the belief that the earth moves around the sun. This belief brought him additional enemies, but it also brought him immortal fame. That part of his work will be covered in Unit 2.

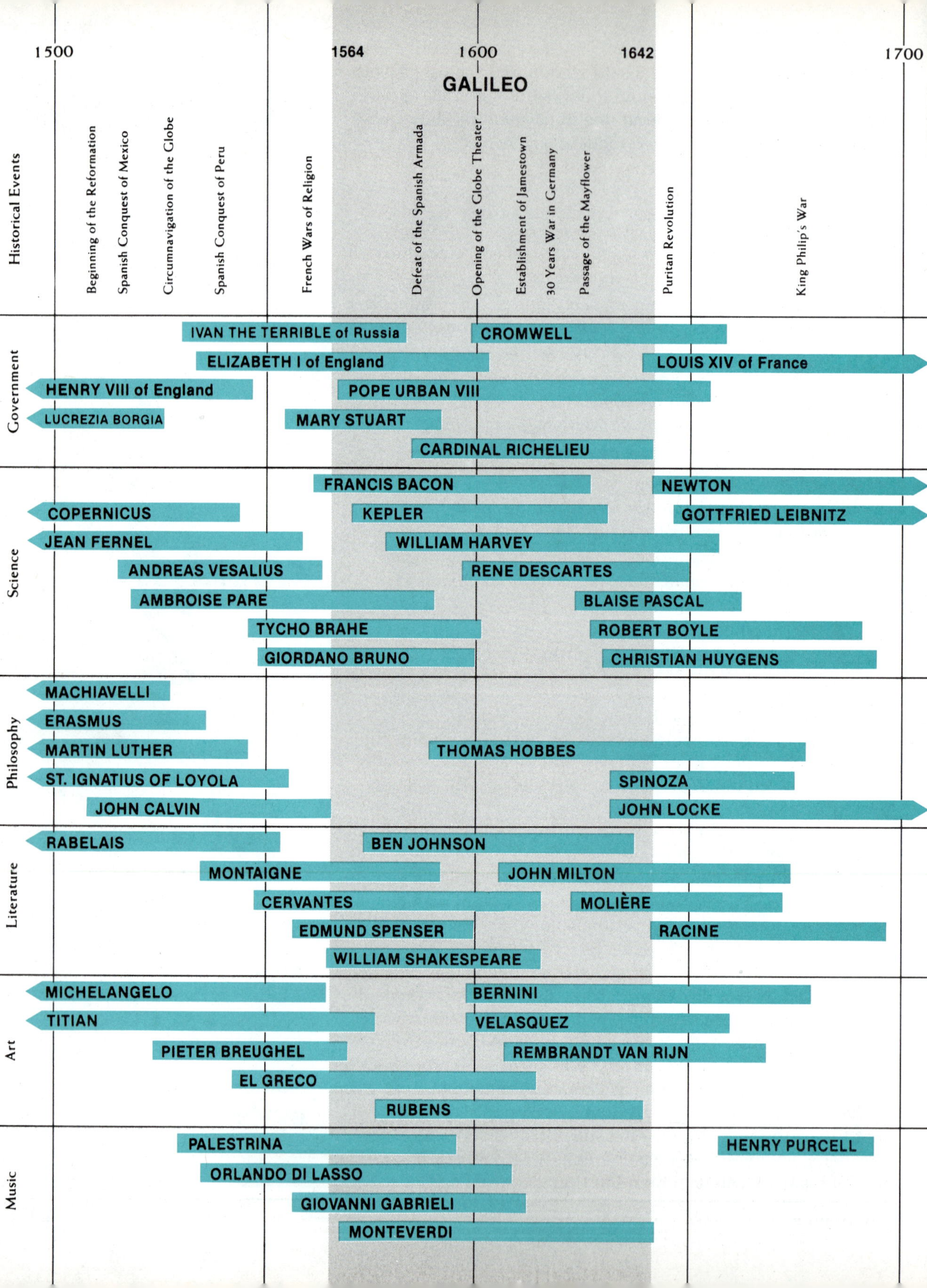

1500
1564
1600
1642
1700
GALILEO
Historical Events
Beginning of the Reformation
Spanish Conquest of Mexico
Circumnavigation of the Globe
Spanish Conquest of Peru
French Wars of Religion
Defeat of the Spanish Armada
Opening of the Globe Theater
Establishment of Jamestown
30 Years War in Germany
Passage of the Mayflower
Puritan Revolution
King Philip's War
Government
IVAN THE TERRIBLE of Russia
CROMWELL
ELIZABETH I of England
LOUIS XIV of France
HENRY VIII of England
POPE URBAN VIII
LUCREZIA BORGIA
MARY STUART
CARDINAL RICHELIEU
Science
FRANCIS BACON
NEWTON
COPERNICUS
KEPLER
GOTTFRIED LEIBNITZ
JEAN FERNEL
WILLIAM HARVEY
ANDREAS VESALIUS
RENE DESCARTES
AMBROISE PARE
BLAISE PASCAL
TYCHO BRAHE
ROBERT BOYLE
GIORDANO BRUNO
CHRISTIAN HUYGENS
Philosophy
MACHIAVELLI
ERASMUS
MARTIN LUTHER
THOMAS HOBBES
ST. IGNATIUS OF LOYOLA
SPINOZA
JOHN CALVIN
JOHN LOCKE
Literature
RABELAIS
BEN JOHNSON
MONTAIGNE
JOHN MILTON
CERVANTES
MOLIÈRE
EDMUND SPENSER
RACINE
WILLIAM SHAKESPEARE
Art
MICHELANGELO
BERNINI
TITIAN
VELASQUEZ
PIETER BREUGHEL
REMBRANDT VAN RIJN
EL GRECO
RUBENS
Music
PALESTRINA
HENRY PURCELL
ORLANDO DI LASSO
GIOVANNI GABRIELI
MONTEVERDI

A generous offer of the Grand Duke drew Galileo back to his native province of Tuscany in 1610. He became Court Mathematician and Philosopher, a title which he chose himself. From then until his death at 78, despite illness, family troubles, occasional poverty, and quarrels with his enemies, Galileo continued his research, teaching, and writing.

Frontispiece of the book Dialogue on Two Great World Systems *(1632).*

2.3 | Galileo's *Two New Sciences*

Mechanics is the study of the behavior of matter under the influence of forces. Galileo's early writings on this subject follow the standard medieval theories of physics, although he was aware of some of the shortcomings of those theories. During his mature years his chief interest was in astronomy. However, his important astronomical book *Dialogue on the Two Great World Systems* (1632) was condemned by the Inquisition. Forbidden to teach the "new" astronomy, Galileo decided to concentrate again on mechanics. This work led to his book *Discourses and Mathematical Demonstrations Concening Two New Sciences Pertaining to Mechanics and Local Motion* (1638), usually referred to as *Two New Sciences*. This book signaled the beginning of the end of the medieval theory of mechanics and of the entire Aristotelian cosmology.

Galileo was old, sick, and nearly blind at the time he wrote *Two New Sciences*. Yet, as in all his writings, his style is lively and delightful. As he had in the *Two Great World Systems*, he presented his ideas in the form of a conversation among three speakers: *Simplicio* competently represents the Aristotelian view; *Salviati* presents the new views of Galileo; and *Sagredo* is a man of good will and open mind, eager to learn. Eventually, of course, Salviati leads his companions to Galileo's views. Listen to Galileo's three speakers as they discuss the problem of free fall:

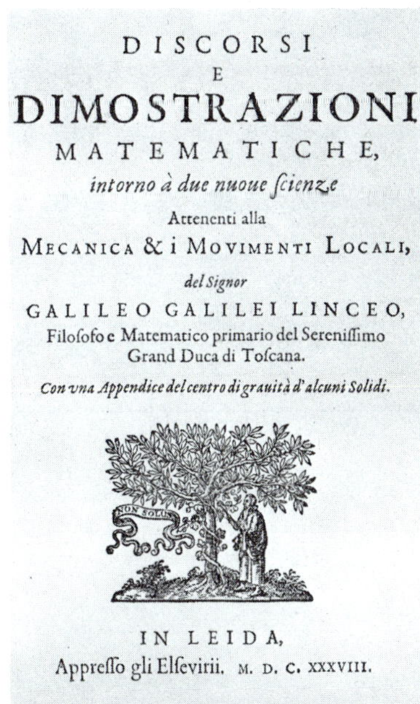
DISCORSI
E
DIMOSTRAZIONI
MATEMATICHE,
intorno à due nuoue ſcienze
Attenenti alla
MECANICA & i MOVIMENTI LOCALI,
del Signor
GALILEO GALILEI LINCEO,
Filoſofo e Matematico primario del Sereniſſimo Grand Duca di Toſcana.
Con vna Appendice del centro di grauità d'alcuni Solidi.

IN LEIDA,
Appreſſo gli Elſevirii. M. D. C. XXXVIII.

Title page of Discourses and Mathematical Demonstrations Concerning Two New Sciences Pertaining to Mechanics and Local Motion *(1638).*

Salviati: I greatly doubt that Aristotle ever tested by experiment whether it is true that two stones, one weighing ten times as much as the other, if allowed to fall at the same instant from a height of, say, 100 cubits, would so differ in speed that when the heavier had reached the ground, the other would not have fallen more than 10 cubits. [A "cubit" is about 50 cm.]

Simplicio: His language would indicate that he had tried the experiment, because he says: *We see the heavier*; now the word *see* shows that he had made the experiment.

Sagredo: But, I, Simplicio, who have made the test can assure you that a cannon ball weighing one or two hundred pounds, or even more, will not reach the ground by as much as a span [hand-breadth] ahead of a musket ball weighing only half a pound, provided both are dropped from a height of 200 cubits.

Here, perhaps, one might expect a detailed report on an experiment done by Galileo or one of his students. Instead, Galileo uses a "thought experiment," that is, an analysis of what would happen in an imaginary experiment, to cast doubt on Aristotle's theory of motion:

SG 4

DEL GALILEO. 63

Vacuo non si farebbe il moto, la posizion del Vacuo assolutamente presa, e non in relazione al moto, non vien destrutta, mà per dire quel che per auuentura potrebber rispondere quegli antichi, acciò meglio si scorga, quanto concluda la dimostrazione d' Aristotele, mi par che si potrebbe andar contro à gli assunti di quello, negandogli amendue. E quanto al primo, io grandemente dubito, che Aristotele non sperimentasse mai quanto sia vero, che due pietre vna più graue dell' altra dieci volte lasciate nel medesimo instante cader da vn altezza, v. gr. di cento braccia fusser talmente differenti nelle lor velocità, che all' arriuo della maggior in terra l'altra si trouasse non hauere nè anco sceso dieci braccia.

Simp. *Si vede pure dalle sue parole, ch' ei mostra d'hauerlo sperimentato, perche ei dice: Veggiamo il più graue: hor quel' vedersi accenna l'hauerne fatta l'esperienza.*

Sagr. *Mà io S. Simp. che n'hò fatto la proua, vi assicuro, che vna palla d'artiglieria, che pesi cento, dugento, e anco più libbre, non anticiperà di vn palmo solamente l'arriuo in terra della palla d'un moschetto, che ne pesi vna mezza, venendo anco dall' altezza di dugento braccia.*

Salu. *Mà senz' altre esperienze con breue, e concludente dimostrazione possiamo chiaramente prouare non esser vero, che vn mobile più graue si muoua più velocemente d'un' altro men graue, intendendo di mobili dell' istessa materia; & in somma di quelli de i quali parla Aristotele. Però ditemi S. Simp. se voi ammettete, che di ciascheduno corpo graue cadente sia vna da natura determinata velocità; si che l'accrescergliela, ò diminuirgliela non si possa se non con vsargli violenza, ò opporgli qualche impedimento.*

Simp. *Non si può dubitare, che l'istesso mobile nell' istesso mezzo habbia vna statuita, e da natura determinata velocità, la quale non se gli possa accrescere se non con nuouo impeto conferitogli, ò diminuirgliela saluo che con qualche impedimento che lo ritardi.*

Salu. *Quando dunque noi hauessimo due mobili, le naturali velo-*

A page from the original Italian edition of Two New Sciences, *showing statements that are translated in this text.*

Salviati: But, even without further experiment, it is possible to prove clearly, by means of a short and conclusive argument, that a heavier body does not move more rapidly than a lighter one provided both bodies are of the same material and in short such as those mentioned by Aristotle. But tell me, Simplicio, whether you admit that each falling body acquires a definite speed fixed by nature, a velocity which cannot be increased or diminished except by the use of violence or resistance?

Simplicio: There can be no doubt but that one and the same body moving in a single medium has a fixed velocity which is determined by nature and which cannot be increased except by the addition of impetus or diminished except by some resistance which retards it.

Salviati: If then we take two bodies whose natural speeds are different, it is clear that on uniting the two, the more rapid one will be partly retarded by the slower, and the slower will be somewhat hastened by the swifter. Do you not agree with me in this opinion?

Simplicio: You are unquestionably right.

Salviati: But if this is true, and if a large stone moves with a speed of, say, eight, while a smaller moves with a speed of four, then when they are united, the system will move with a speed less than eight; but the two stones when tied together make a stone larger than that which before moved with a speed of eight. Hence the heavier body moves with less speed than the lighter one; an effect which is contrary to your supposition. Thus you see how, from your assumption that the heavier body moves more rapidly than the lighter one, I infer that the heavier body moves more slowly.

SG 5

Simplicio: I am all at sea. . . . This is, indeed, quite beyond my comprehension. . . .

Simplicio retreats in confusion when Salviati shows that the Aristotelian theory of fall contradicts itself. But while Simplicio cannot refute Galileo's logic, his own eyes tell him that a heavy object *does* fall faster than a light object:

Simplicio: Your discussion is really admirable; yet I do not find it easy to believe that a birdshot falls as swiftly as a cannon ball.

Salviati: Why not say a grain of sand as rapidly as a grindstone? But, Simplicio, I trust you will not follow the example of many others who divert the discussion from its main intent and hasten upon some statement of mine that lacks a hairsbreadth of the truth, and under this hair hide the fault of another that is as big as a ship's cable. Aristotle says that "an iron ball of one hundred pounds falling from a height of 100 cubits reaches the ground before a one-pound ball has fallen a single cubit." I say that they arrive at the same time. You find, on making the experiment, that the larger outstrips the smaller by two fingerbreadths.... Now you would not hide behind these two fingers the 99 cubits of Aristotle, nor would you mention my small error and at the same time pass over in silence his very large one.

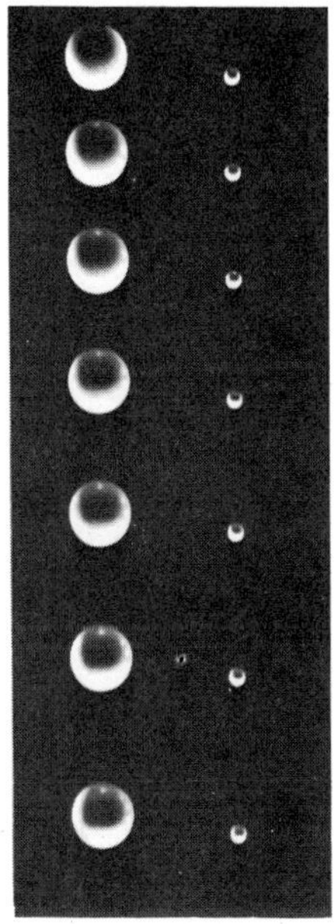

A stroboscopic photograph of two freely falling balls of unequal weight. The picture shows the last part of the total path. The balls had been released simultaneously. The time interval between images is 0.03 sec.

This is a clear statement of an important principle: Even in careful observation of a common natural event, a very minor effect may distract the observer's attention. As a result, a much more important regularity may be overlooked. Different bodies falling in air from the same height, it is true, may *not* reach the ground at exactly the same time. However, the important point is not that the times of arrival are *slightly different*, but that they are *very nearly the same*! Galileo regarded the failure of the bodies to arrive at exactly the same time as a minor effect which could be explained with a better understanding of motion in free fall. He himself correctly attributed the observed results to differences in the effect of air resistance on bodies of different size and weight. A few years after his death, the invention of the vacuum pump allowed others to show that Galileo was right. In one experiment, for example, a feather and a heavy gold coin were dropped from the same height at the same time inside a container pumped almost empty of air. With the effect of air resistance eliminated, the different bodies fell at the same rate and struck the bottom of the container at the same instant. Long after Galileo, scientists learned how to express the laws of air resistance in mathematical form. With this knowledge, one can understand exactly why and by how much a light object will fall more slowly than a heavier one.

The phrase "free fall" as now used in physics generally refers to fall when the only force acting is gravity, that is, when air friction is negligible.

Learning what to ignore has been almost as important in the growth of science as learning what to take into account. In the case of falling bodies, Galileo's explanation depended on his being able to imagine how an object would fall if there were no air resistance. His explanation seems simple today, when we know about vacuum pumps. But in Galileo's time it was difficult to accept. For most people, as for Aristotle, common sense said that air resistance is always present in nature. Thus, a feather and a coin could never fall at the same rate. Why talk about motions in a vacuum, when a vacuum could not be shown to exist? Physics, said Aristotle and his followers, should deal with the world all around us that we can readily observe. It should not bother with imaginary situations which might never be seen or which, like the vacuum, were considered impossible.

One of the arguments against the existence of a vacuum was deduced from Aristotle's theory as follows: If the final speed of fall is proportional to the weight divided by the resistance, then, since the resistance in an assumed vacuum would be zero, the final speed of fall of all bodies must be infinite in a vacuum. But such a result was regarded as absurd, so the assumption of a vacuum was believed to have been shown to be impossible.

Aristotle's physics had dominated Europe since the thirteenth century. To many scientists, it seemed to offer the most

Portrait of Galileo by Ottavio Leoni, a contemporary of Galileo.

reasonable method for describing natural phenomena. To overthrow such a firmly established doctrine required much more than writing reasonable arguments. It even required more than clear experimental proof, such as dropping heavy and light objects from a tall building. (Galileo is often said to have done this from the top of the Leaning Tower of Pisa, but probably did not.) It demanded Galileo's unusual combination of mathematical talent, experimental skill, literary style, and tireless campaigning to discredit Aristotle's theories and thus to begin the era of modern physics.

?

4. If a nail and a toothpick are dropped at the same time from the same height, they do not reach the ground at exactly the same instant. (Try it with these or similar objects.) How would Aristotelian theory explain this? What was Galileo's explanation?

5. A paper bag containing a rock is dropped from a window. Using Aristotle's theory, explain why the bag and the rock together fall slower than the rock would fall by itself. Use the same theory to explain why the two together fall faster than the rock alone.

2.4 | Why study the motion of freely falling bodies?

"Aristotelian cosmology" refers to the whole interlocking set of ideas about the structure of the physical universe and the behavior of all the objects in it. This was briefly mentioned in Sec. 2.1. Other aspects of it will be presented in Unit 2.

In Galileo's attack on the Aristotelian cosmology, few details were actually new. However, Galileo's approach and his findings together provided the first workable presentation of the science of motion. Galileo realized that understanding free-fall motion is the key to understanding all observable motions of all bodies in nature. To know *which* was the key phenomenon to study was a gift of genius. But in many ways Galileo simply worked as do scientists in general. His approach to the problem of motion makes a good "case" to follow as an introduction to strategies of inquiry that are still used in science.

In fact, more than mere "superficial observations" had been made long before Galileo set to work. For example, Nicolas Oresme and others at the University of Paris had by 1330 discovered a distance–time relationship similar to that which Galileo was to announce for falling bodies in *Two New Sciences*. Some of their reasoning is discussed in SG 7. It is, however, questionable how much of this prior work influenced Galileo in detail, rather than just in spirit.

Several reasons for studying in detail Galileo's attack on the problem of free fall have been mentioned. Galileo himself recognized another reason: The study of motion which he proposed was only the starting phase of a mighty field of discovery.

My purpose is to set forth a very new science dealing with a very ancient subject. There is, in nature, perhaps nothing older than motion, concerning which the books written by philosophers are neither few nor small; nevertheless, I have discovered some properties of it that are worth knowing that have not hitherto been either observed or demonstrated. Some superficial observations have been made, as for instance, that the natural motion of a heavy falling body is continuously accelerated; but to just what extent this acceleration occurs has not yet been announced. . . .

Other facts, not few in number or less worth knowing I have succeeded in proving; and, what I consider more important, there have been opened up to this vast and most excellent science, of which my work is merely the beginning, ways and means by which other minds more acute than mine will explore its remote corners.

2.5 | Galileo chooses a definition of uniform acceleration

Two New Sciences deals directly with the motion of freely falling bodies. In studying the following paragraphs from this work, you must be alert to Galileo's overall plan. First he discusses the mathematics of a possible, simple type of motion. (This motion is now called *uniform acceleration* or *constant acceleration*.) Then he proposes that heavy bodies actually fall with just this kind of motion. Next, on the basis of this proposal, he makes certain predictions about balls rolling down an incline. Finally, he shows that experiments bear out these predictions.

It will help you to have a plan clearly in mind as you progress through the rest of this chapter. As you study each succeeding section, ask yourself whether Galileo is
- presenting a definition
- stating an assumption (or hypothesis)
- deducing predictions from his hypothesis
- experimentally testing the predictions

The first part of Galileo's presentation is a thorough discussion of motion with uniform speed, similar to the discussion in Chapter 1. This leads to the second part, where Salviati, one of Galileo's characters, says:

When, therefore, I observe a stone initially at rest falling from an elevated position and continually acquiring new increments of speed, why should I not believe that such increases take place in a manner which is exceedingly simple and rather obvious to everybody? If now we examine the matter carefully we find no addition or increment more simple than that which repeats itself always in the same manner. This we readily understand when we consider the intimate relationship between time and motion; for just as uniformity of motion is defined by and conceived through equal times and equal spaces (thus we call a motion uniform when equal distances are traversed during equal time-intervals), so also we may, in a similar manner, through equal time-intervals, conceive additions of speed as taking place without complication. . . .

This is sometimes known as the rule of parsimony: Unless forced to do otherwise, assume the simplest possible hypothesis to explain natural events.

Hence the definition of motion which we are about to discuss may be stated as follows:

A motion is said to be uniformaly accelerated when, starting from rest, it acquires during equal time-intervals, equal increments of speed.

We can rephrase Galileo's definitions, using the symbols from Chapter 1, as follows: for uniform motion, the ratio $\Delta d/\Delta t$ is constant; for uniformly *accelerated* motion, the ratio $\Delta v/\Delta t$ is constant. These definitions are equivalent to those given in Secs. 1.3 and 1.8. Other ways of describing uniform acceleration are discussed in SG 8 and 9.

Sagredo then questions whether Galileo's arbitrary definition of acceleration actually corresponds to the way objects fall. Is acceleration, as defined, really useful in describing their observed change of motion? Sagredo wonders about a further point, so far not raised by Galileo:

> From these considerations perhaps we can obtain an answer to a question that has been argued by philosphers, namely, what is the *cause* of the natural motion of heavy bodies....

But Salviati, the spokesman of Galileo, rejects the ancient tendency to investigate phenomena by looking first for their causes. As we would today, Salviati says it is pointless to ask about the cause of any motion until an accurate description of it exists:

> Salviati: The present does not seem to be the proper time to investigate the cause of the acceleration of natural motion concerning which various opinions have been expressed by philosophers, some explaining it by attraction to the center, others by repulsion between the very small parts of the body, while still others attribute it to a certain stress in the surrounding medium which closes in behind the falling body and drives it from one of its positions to another. Now, all these fantasies, and others, too, ought to be examined; but it is not really worth while. At present it is the purpose of our Author merely to investigate and to demonstrate some of the properties of accelerated motion, whatever the cause of this acceleration may be.

Here Salviati refers to the Aristotelian assumption that air propels an object moving through it (see Sec. 2.1).

Galileo has now introduced two distinct propositions: (1) "uniform" acceleration means that equal increases of speed Δv occur in equal time intervals Δt; and (2) things actually fall that way.

First look more closely at Galileo's proposed definition. Is this the only possible way of defining uniform acceleration? Not at all! Galileo says that at one time he thought it would be more useful to define uniform acceleration in terms of speed increase in proportion to distance traveled Δd, rather than to time Δt. Notice that both definitions met Galileo's requirement of simplicity. (In fact, both definitions had been discussed since early in the fourteenth century.) Furthermore, both definitions seem to match the common sense idea of acceleration. To say that a body is "accelerating" seems to imply "the farther it goes, the faster it goes" as well as "the longer time it goes, the faster it goes." How should we choose between these two ways of putting it? Which definition will be more useful in describing nature?

This is where experimentation becomes important. Galileo chose to define uniform acceleration as the motion in which the change in speed Δv is proportional to elapsed time Δt. He then demonstrated that his definition matches the real behavior of moving bodies, in laboratory situations as well as in ordinary "un-arranged" experience. As you will see later, he made the right choice. But he was not able to prove his case by direct or obvious means, as you will also see.

?

6. Describe uniform speed without referring to dry-ice disks and strobe photography or to any particular object or technique of measurement.

7. Express Galileo's definition of uniformly accelerated motion in words and in the form of an equation.

8. Use Galileo's definition of acceleration to calculate the acceleration of a car that speeds up from 22 m/sec to 32 m/sec in 5 sec. At this same acceleration, how much more time would the car take to reach 40 m/sec?

9. What two conditions did Galileo want his definition of uniform acceleration to meet?

2.6 | Galileo cannot test his hypothesis directly

After Galileo defined uniform acceleration in terms that matched the way he *believed* freely falling objects behaved, his next task was to show that his definition actually was useful for describing observed motions.

Suppose you drop a heavy object from several different heights, for example, from windows on different floors of a building. You want to check whether the final speed increases in proportion to the time the object falls; that is, you want to know whether $\Delta v \propto \Delta t$ or, in other words, whether $\Delta v/\Delta t$ is constant. In each trial you must observe the time of fall and the speed just before the object strikes the ground. This presents a problem. Even with modern instruments, it would be very difficult to make a *direct measurement* of the speed reached by an object just before striking the ground. Furthermore, the entire time intervals of fall (less than 3 sec from the top of a 10-story building) is shorter than Galileo could have measured accurately with the clocks available to him. So a direct test of whether $\Delta v/\Delta t$ is constant was not possible for Galileo.

The symbol $\propto$ means "directly proportional to," or "changes with."

?

10. Which of these statements accurately explains why Galileo could not test directly whether or not the final speed reached by a freely falling object is proportional to the time of fall?

(a) His definition was wrong.

(b) He could not measure the speed attained by an object just before it hit the ground.

SG 10

(c) He could not measure times accurately enough.

(d) He could not measure distances accurately enough.

(e) Experimentation was not permitted in Italy.

2.7 Looking for logical consequences of Galileo's hypothesis

Galileo's inability to make *direct* measurements to test his hypothesis that $\Delta v/\Delta t$ is constant in free fall did not stop him. He turned to mathematics to derive from this hypothesis some other relationship that *could* be checked by measurements with equipment available to him. You will see that in a few steps he came much closer to a relationship he could use to check his hypothesis.

Large distances and large time intervals are, of course, easier to measure than the very small values of Δd and Δt needed to find the final speed just before a falling body hits. So Galileo tried to determine, by reasoning, how total distance of fall would increase with total time of fall if objects did fall with uniform acceleration. You already know how to find the total distance from total time for motion at constant *speed.* Now you can derive a new equation that relates total distance of fall to total time of fall for motion at constant *acceleration*. In doing so, you will not follow Galileo's own calculations exactly, but the results will be the same. First, recall the definition of average speed as the distance traveled Δd divided by the elapsed time Δt:

$$v_{av} = \frac{\Delta d}{\Delta t}$$

From this general definition you can compute the average speed from measurement of Δd and Δt, whether Δd and Δt are small or large. You can rewrite the equation as

$$\Delta d = v_{av} \times \Delta t$$

More generally, the average speed would be

$$v_{av} = \frac{v_{initial} + v_{final}}{2}$$

This equation, really a definition of v_{av}, is always true. For the special case of motion at a constant speed v, $v_{av} = v$, and therefore, $\Delta d = v \times \Delta t$. Suppose the value of v is known, for example, when a car is driven with a steady reading of 96 km/hr on the speedometer. Then you can use this equation to figure out how far (Δd) the car would go in any given time interval (Δt). But in the case of uniformly accelerated motion the speed is continually *changing*. Therefore, what value can you use for v_{av}?

Galileo used a geometrical argument. Algebra was not used until more than 100 years later.

The answer involves just a bit of algebra and some reasonable assumptions. Galileo reasoned (as others had before him) that for any quantity that changes uniformly, *the average value is just halfway between the beginning value and the final value.* For uniformly accelerated motion starting from rest, $v_{initial} = 0$. Thus, the rule tells you that the average speed is halfway between 0 and v_{final}; that is, $v_{av} = \frac{1}{2}v_{final}$. If this reasoning is correct, it follows that

$$\Delta d = \frac{1}{2} v_{final} \times \Delta t$$

for uniformly accelerated motion starting from rest.

This relation could not be tested directly either, because the equation still contains a speed factor. What is needed is an equation relating total distance and total time, without any need to measure speed.

SG 11, 12

Now look at Galileo's definition of uniform acceleration: $a = \Delta v/\Delta t$. You can rewrite this relationship in the form $\Delta v = a \times \Delta t$. The value of Δv is just $v_{final} - v_{initial}$; and $v_{initial} = 0$ for motion that begins from rest. Therefore, you can write

$$\begin{aligned} \Delta v &= a \times \Delta t \\ v_{final} - v_{initial} &= a \times \Delta t \\ v_{final} &= a \times \Delta t \end{aligned}$$

Now you can substitute this expression for v_{final} into the equation for Δd above. Thus, *if* the motion starts from rest and *if* it is uniformly accelerated (and *if* the average rule is correct, as you have assumed), you can write

$$\begin{aligned} \Delta d &= \frac{1}{2} v_{final} \times \Delta t \\ &= \frac{1}{2} (a \times \Delta t) \times \Delta t \end{aligned}$$

Or, regrouping terms,

$$\Delta d = \frac{1}{2} a(\Delta t)^2$$

This is the kind of relation Galileo was seeking. It relates total distance Δd to total time Δt, without involving any speed term.

Before finishing, though, you can simplify the symbols in the equation to make it easier to use. If you measure distance and time from the position and the instant that the motion starts, then $d_{initial} = 0$ and $t_{initial} = 0$. Thus, the intervals Δd and Δt have the values given by d_{final} and t_{final}. You then can write the equation above more simply as

SG 13

$$d_{final} = \frac{1}{2} a t^2_{final}$$

Remember that this is a very specialized equation. It gives the total distance fallen as a function of total time of fall, but it does so *only* if the motion starts from rest $(v_{initial} = 0)$, if the acceleration is uniform (a = *constant*), and if time and distance are measured from the start ($t_{initial} = 0$ and $d_{initial} = 0$).

Galileo reached the same conclusion, though he did not use algebraic forms to express it. Since you are dealing only with the special situation in which acceleration *a* is constant, the quantity ½*a* is constant also. Therefore, you can write the conclusion in the form of a proportion: In uniform acceleration from rest, the distance traveled is proportional to the square of the time elapsed, or

$$d_{final} \propto t^2_{final}$$

For example, if a uniformly accelerating car starting from rest moves 10 m in the first second, in *twice* the time it would move *four* times as far, or 40 m during the first two seconds. In the first *three* seconds it would move *nine* times as far, or 90 m.

Another way to express this relation is to say that the ratio d_{final} to t^2_{final} has a constant value, that is,

$$\frac{d_{final}}{t^2_{final}} = \text{constant}$$

Because we will use the expression d_{final}/t^2_{final} many times, it is simpler to write it as d/t^2. It is understood that *d* and *t* mean total distance and time interval of motion, starting from rest.

This equation is a logical result of Galileo's original proposal for defining uniform acceleration. This result can be expressed as follows: If an object accelerates uniformly from rest, the ratio d/t^2 should be constant. Conversely, any motion for which this ratio of *d* and t^2 is constant for different distances and their corresponding times is a case of *uniform acceleration* as defined by Galileo.

Of course, you still must test the hypothesis that freely falling bodies actually *do* exhibit just such motion. You know that you cannot test directly whether $\Delta v/\Delta t$ has a constant value. But a constant value of $\Delta v/\Delta t$ means there will be a constant ratio of d_{final} to t^2_{final}. The values for total time and distance of fall d_{final} are easier to measure than are the values for short intervals Δd and Δt needed to find Δv. However, even measuring the total time of fall presented a difficult task in Galileo's time. So, instead of a *direct* test of his hypothesis, Galileo went one step further and deduced a clever *indirect* test.

Physics texts must be read with pencil in hand. Go over each step in this section, starting with the definition of average speed. Make a list of each simplifying assumption and each new definition used in the text.

?

11. *Why was it more reasonable for Galileo to use the equation* $d = \frac{1}{2}at^2$ *for testing his hypothesis than to use* $a = \Delta v/\Delta t$?

12. *If you simply combined the two equations* $\Delta = v\Delta t$ *and* $\Delta v = a\Delta t$, *it looks as if you might get the results* $\Delta d = a\Delta t^2$. *What is wrong with doing this?*

2.8 | Galileo turns to an indirect test

Realizing that direct measurements involving a rapidly and freely falling body would not be accurate, Galileo decided to test an object that was moving less rapidly. He proposed a new hypothesis: *If a freely falling body has constant acceleration, then a perfectly round ball rolling down a perfectly smooth inclined plane will also have a constant, though smaller, acceleration.* Thus, Galileo claimed that if d/t^2 is constant for a body falling freely from rest, this ratio will also be constant, although smaller, for a ball rolling from rest down a straight inclined plane.

Here is how Salviati described Galileo's own experimental test in *Two New Sciences*:

> A piece of wooden moulding or scantling, about 12 cubits long, half a cubit wide, and three finger-breadths thick, was taken; on its edge was cut a channel a little more than one finger in breadth; having made this groove very straight, smooth, and polished, and having lined it with parchment, also as smooth and polished as possible, we rolled along it a hard, smooth, and very round bronze ball. Having placed this board in a sloping position, by lifting one end some one or two cubits above the other, we rolled the ball, as I was just saying, along the channel, noting, in a manner presently to be described, the time required to make the descent. We repeated this experiment more than once in order to measure the time with an accuracy such that the deviation between two observations never exceeded one-tenth of a pulse beat. Having performed this operation and having assured ourselves of its reliability, we now rolled the ball only one-quarter of the length of the channel; and having measured the time of its descent, we found it precisely one-half of the former. Next we tried other distances, comparing the time for the whole length with that for the half, or with that for two-thirds, or three-fourths, or indeed for any fraction; in such experiments, repeated a full

Note the careful description of the experimental apparatus. Today an experimenter would add to his or her verbal description any detailed drawings, schematic layouts, or photographs needed to make it possible for other competent scientists to duplicate the experiment.

Experiment 1.5 in the *Handbook* is very similar to Galileo's test.

This picture, painted in 1841 by G. Bezzuoli, attempts to reconstruct an experiment Galileo is alleged to have made during his time as lecturer at Pisa. Off to the left and right are men of ill will: the blasé Prince Giovanni de Medici (Galileo had shown a dredging-machine invented by the prince to be unusable) and Galileo's scientific opponents. These were leading men of the universities; they are shown here bending over a book of Aristotle, in which it is written in black and white that bodies of unequal weight fall with different speeds. Galileo, the tallest figure left of center in the picture, is surrounded by a group of students and followers.

hundred times, we always found that the spaces traversed were to each other as the squares of the times, and this was true for all inclinations of the ... channel along which we rolled the ball....

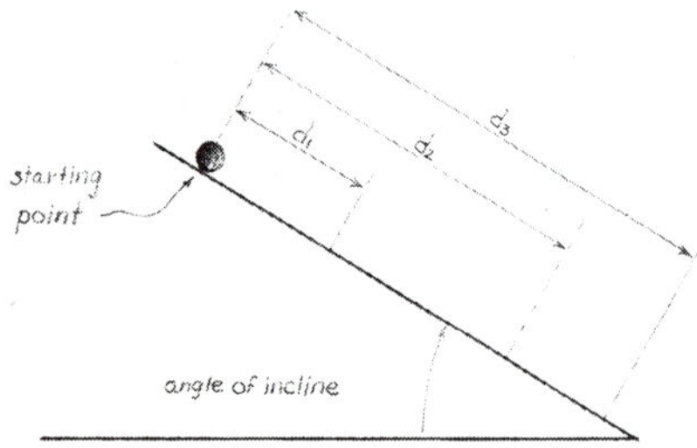

Galileo has packed a great deal of information into these lines. He describes his procedures and apparatus clearly enough to allow others to repeat the experiment for themselves if they wish. He indicates that consistent measurements can be made. Finally, he restates the two chief experimental results which he believes support his free-fall hypothesis. Examine the results carefully.

(a) First, Galileo found that when a ball rolled down the incline, the ratio of the distance covered to the square of the corresponding time was always the same. For example, if d_1, d_2, and d_3 represent distances measured from the same starting point on the incline and t_1, t_2, and t_3 represent the times taken to roll down these distances, then

$$\frac{d_1}{t_1^2} = \frac{d_2}{t_2^2} = \frac{d_3}{t_3^2}$$

In general, for each angle of incline, the value of d/t^2 was constant. Galileo did not present his experimental data in the full detail which since has become the custom. However, others have repeated his experiment and have obtained results which parallel his. (See data in SG 16.) You can perform this experiment yourself with the help of one or two other students.

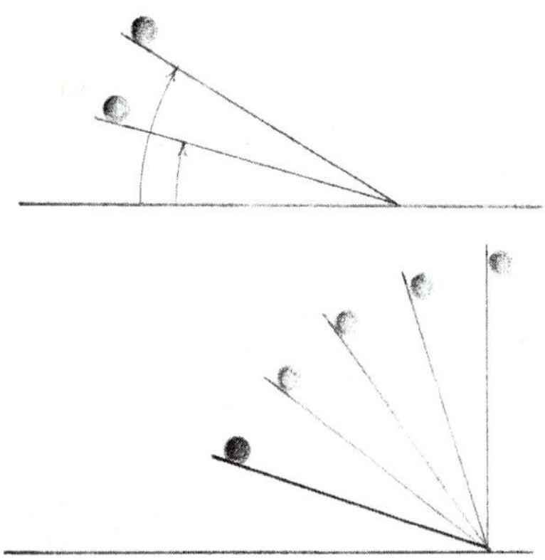

Spheres rolling down planes of increasingly steep inclination. For each angle, the acceleration has its own constant value. At 90° the inclined plane situation looks almost like free fall, except that the ball would still be rolling. Galileo assumed that the difference between free fall and "rolling fall" is not important. (In most real situations, the ball would slide, not roll, down the really steep inclines.)

(b) Galileo's second experimental finding relates to what happened when the angle of inclination of the plane was changed. Whenever the angle changed, the ratio d/t^2 took on a new value, although for any one angle it remained constant regardless of distance of roll. Galileo confirmed this by repeating the experiment "a full hundred times" for each of the many different angles. After finding that the ratio d/t^2 was constant for each angle at which t could be measured conveniently, Galileo was willing to extrapolate. He concluded that the ratio d/t^2 is a constant even for steeper angles, where the ball moves too fast for accurate measurement of t. Now, finally, Galileo was ready to solve the problem that had started the whole argument: He reasoned that when the angle of inclination became 90°, the ball would move straight down as a *freely falling object*. By his reasoning, d/t^2 would still be constant even in that extreme case, although he couldn't say *what* the numerical value was.

Galileo already had deduced that a constant value of d/t^2 was characteristic of uniform acceleration. By extrapolation, he could conclude at last that free fall was uniformly accelerated motion.

? **13.** *In testing his hypothesis that free-fall motion is uniformly accelerated, Galileo made the unproved* assumption *that* *(choose one or more):*

(a) d/t^2 *is constant.*

(b) the acceleration has the same value for all angles of inclination of the plane.

(c) the results for small angles of inclination can be extrapolated to large angles.

(d) the speed of the ball is constant as it rolls.

(e) the acceleration of the rolling ball is constant if the acceleration in free fall is constant, though the value of the two constants is not the same.

14. Which of the following statements best summarizes the work of Galileo on free fall when air friction is not a factor? (Be prepared to defend your choice.) Galileo:

(a) proved that all objects fall at exactly the same speed regardless of their weight.

(b) proved that for any freely falling object the ratio d/t^2 *is constant for any distance of fall.*

(c) proved that an object rolling down a smooth incline accelerates in the same way, although more slowly than, the same object falling freely.

(d) supported indirectly his assertion that the speed of an object falling freely from rest is proportional to the elapsed time.

(e) made it clear that until a vacuum could be produced, it would not be possible to settle the free-fall question once and for all.

Galileo's technique for measuring time is discussed in the next section.

For problems that will check and extend your understanding of uniform acceleration, see SG 14 through 21.

2.9 | Doubts about Galileo's procedure

This whole process of reasoning and experiment appears long and involved on first reading, and you may have some doubts concerning it. For example, was Galileo's measurement of time precise enough to establish the constancy of d/t^2 even for a slowly rolling object? In his book, Galileo tries to reassure possible critics by providing a detailed description of his experimental arrangement:

For the measurement of time, we employed a large vessel of water placed in an elevated position; to the bottom of this vessel was soldered a pipe of small diameter giving a thin jet of water, which we collected in a small cup during the time of each descent, whether for the whole length of the channel or for a part of its length; the water thus collected was weighed on a very accurate balance; the differences and ratios of these weights gave us the differences and ratios of the time intervals, and this with such accuracy that, although the operation was repeated many, many times, there was no appreciable discrepancy in the results.

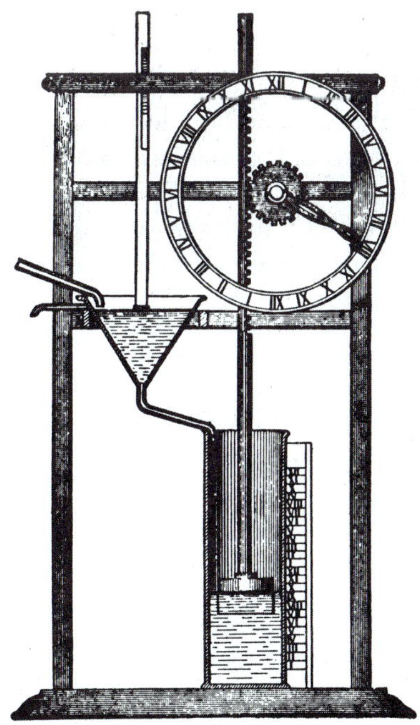

Early water clock.

The water clock described by Galileo was not invented by him. Indeed, water clocks existed in China as early as the sixth century B.C. and probably were used in Babylonia and India even earlier. In the early sixteenth century a good water clock was the most accurate instrument available for measuring short time intervals. It remained so until shortly after Galileo's death, when the work of Christian Huygens and others led to practical pendulum clocks. When better clocks became available, Galileo's results on inclined-plane motion were confirmed.

Another reason for questioning Galileo's results involved the great difference between free-fall and rolling motion on a slight incline. Galileo does not report what angles he used in his experiment. However, as you may have found out from doing a similar experiment, the angles must be kept rather small. As the angle increases, the speed of the ball soon becomes so great that it is difficult to measure the times involved. The largest usable angle reported in a recent repetition of Galileo's experiment was only 6°. It is not likely that Galileo worked with much larger angles. This means that his extrapolation to free fall (90° incline) was bold. A cautious person, or one not already convinced of Galileo's argument, might well doubt it.

There is still another reason for questioning Galileo's results. As the angle of incline is increased, there comes a point where the ball starts to slide as well as roll. This change in behavior could mean that the motion is very different at large angles. Galileo does not discuss these cases. If he had been able to use frictionless blocks that slid down the plane instead of rolling, he would have found that for sliding motion the ratio d/t^2 is also a constant, although having a different numerical value than for rolling at the same angle.

SG 22

? *15. Which of the following statements could be regarded as major reasons for doubting the value of Galileo's procedure?*

(a) His measurement of time was not accurate enough.

(b) He used too large an angle of inclination in his experiment.

(c) It is not clear that his results apply when the ball can slide as well as roll.

(d) In Galileo's experiment the ball was rolling, and therefore he could not extrapolate to the case of free fall where the ball did not roll.

(e) d/t^2 *was not constant for a sliding object.*

2.10 | Consequences of Galileo's work on motion

Galileo seems to have understood that one cannot get the correct numerical value for the acceleration of a body in free fall simply by extrapolating through increasingly large angles of incline. He did not attempt to calculate a numerical value for the acceleration of freely falling bodies. For his purposes it was enough that he could support the hypothesis that the acceleration is *constant* for any given body, whether rolling or falling. This is the first of Galileo's findings, and it has been fully borne out by all following tests.

Scientists now know by measurement that the magnitude of the acceleration of gravity, symbol a_g, is about 9.8 m/sec per sec at the earth's surface. The *Project Physics Handbook* contains five different experiments for finding a value of a_g. (For many problems, the approximate value 10 m/sec/sec is satisfactory.)

Second, spheres of different weights allowed to roll down an inclined plane set at a given angle have the same acceleration. We do not know how much experimental evidence Galileo himself had for this conclusion, but it agrees with his observations for freely falling objects. It also agrees with the "thought experiment" by which he argued that bodies of different weights fall at the same rate (aside from the effects of air resistance). His results clearly contradicted what one would have expected on the basis of Aristotle's theory of motion.

SG 23

Third, Galileo developed a mathematical theory of accelerated motion from which other predictions about motion could be derived. Just one example is mentioned here: it will be very useful in Unit 3. Recall that Galileo chose to define acceleration as the rate at which the speed changes with time. He then found by experiment that falling bodies actually do experience equal changes of speed in equal times, and *not* in equal distances. Still, the idea of something changing by equal amounts in equal distances has an appealing simplicity. You might ask if there is not some quantity that *does* change in that way during uniform acceleration. In fact, there is. If follows without any new assumptions that, during uniform acceleration from rest, the *square* of the speed changes by equal amounts in equal distances. There is a mathematical equation that expresses this result: If $v_{\text{initial}} = 0$ and $a = \textit{constant}$, then

$$v_{\text{final}}^2 = 2ad_{\text{final}}$$

You can derive this equation. (See SG 24.)

In other words: If an object moves from rest with uniform acceleration, the square of its speed at any point is equal to twice the product of its acceleration and the distance it has moved. (You will see the importance of this relation in Unit 3.)

SG 25, 26

These results of Galileo's work were most important to the development of physics. But they could scarcely have brought about a revolution in science by themselves. No sensible scholar in the seventeenth century would have given up a belief in Aristotelian cosmology only because some of its predictions had been disproved. Still, Galileo's work on free-fall motion helped to prepare the way for a new kind of physics, and indeed a new cosmology, by planting the seeds of doubt about the basic assumptions of Aristotelian science. For example, when it was recognized that all bodies fall with equal acceleration if air friction is minor, the whole Aristotelian explanation of falling motion (Sec. 2.1) broke down.

The most disputed scientific problem during Galileo's lifetime was not in mechanics but in astronomy. A central question in cosmology was whether the earth or the sun was the center of the universe. Galileo supported the view that the earth and other planets revolved around the sun, a view entirely contrary to Aristotelian cosmology. But to support such a view required a physical theory of why and how the earth itself moved. Galileo's work on free fall and other motions turned out to be just what was needed to begin constructing such a theory. His work did not have its full effect, however, until it had been combined with the studies of forces and motion by the English scientist Isaac Newton. But as Newton acknowledged, Galileo was the pioneer. (In the next chapter, you will consider Newton's work on force and motion. In Chapter 8, you will see its application to the motions in the heavens as well as the revolution it caused in science.)

Galileo's work on motion introduced a new and important method of doing scientific research. This method is as effective today as when Galileo demonstrated it. The basis of this procedure is a cycle, repeated as often as necessary, entirely or in part, until a satisfactory theory has emerged. The cycle roughly follows this form: general observation → hypothesis → mathematical analysis or deduction from hypothesis → experimental test of deduction → revision of hypothesis in light of test, and so forth.

SG 27

While the mathematical steps are determined mainly by "cold logic," this is not so for the other parts of the process. A variety of paths of thought can lead to a hypothesis in the first place. A new hypothesis might come from an inspired hunch based on general knowledge of the experimental facts. Or it might come from a desire for mathematically simple statements, or from modifying a previous hypothesis that failed. Moreover, there are no general rules about exactly how well experimental data must agree with predictions based on theory. In some areas of science, a theory is expected to be accurate to better than 0.001%. In other areas, or at an early stage of any new work, one might be delighted with an error of only 50%. Finally, note that while experiment has an important place in this process, it is not the only or even the main element. On the contrary, experiments are worthwhile only in combination with the other steps in the process.

The general cycle of observation, hypothesis, deduction, test, revision, etc., so skillfully demonstrated by Galileo in the seventeenth century, commonly appears in the work of scientists today. Though there is no such thing as *the* scientific method, some form of this cycle is almost always present in scientific research. It is used not out of respect for Galileo as a towering figure in the history of science, but because it works so well so much of the time. What is too frequently underplayed is the sheer creativity that enters into each of these phases. There are no fixed rules for doing any one of them or for how to move from one to the next.

Galileo himself was aware of the value of both the results and the methods of his pioneering work. He concluded his treatment of accelerated motion by putting the following words into the mouths of the characters in his book:

> *Salviati:* . . . we may say the door is now opened, for the first time, to a new method fraught with numerous and wonderful results which in future years will command the attention of other minds.
>
> *Sagredo:* I really believe that . . . the principles which are set forth in this little treatise will, when taken up by speculative minds, lead to another more remarkable result; and it is to be believed that it will be so on account of the nobility of the subject, which is superior to any other in nature.

?

16. Which one of the following was not *a result of Galileo's work on motion?*

(a) Determination of the correct numerical value of the acceleration in free fall, obtained by extrapolating the results for larger and larger angles of inclination.

(b) If an object starts from rest and moves with uniform acceleration a *through a distance* d, *then the square of its speed will be proportional to* d.

(c) Bodies rolling on a smooth inclined plane are uniformly accelerated.

Many details of physics, mathematics, and history have appeared in this chapter. For a review of the most important ideas, see SG 28 and 29.

study guide

1. Note that at the beginning of each chapter in this book there is a list of the section titles. This is a sort of road map you can refer to from time to time as you study the chapter. It is important, especially in a chapter such as this one, to know how the part you are studying relates to what preceded it and to have some idea of where it is leading. For this same reason, you will find it very helpful at first to skim through the entire chapter, reading it rapidly and not stopping to puzzle out parts that you do not quickly understand. Then you should return to the beginning of the chapter and work your way through it carefully, section by section. Remember also to use the end-of-section questions to check your progress.

The *Project Physics* learning materials particularly appropriate for Chapter 2 include:

Experiments

A Seventeenth-Century Experiment
Twentieth Century Version of Galileo's Experiment
Measuring the Acceleration of Gravity, a_g

Film Loops

Acceleration Caused by Gravity. Method I
Acceleration Caused by Gravity. Method II

2. Aristotle's theory of motion seems to be supported to a great extent by common sense experience. For example, water bubbles up through earth at springs. When sufficient fire is added to water by heating it, the resulting mixture of elements (what we call steam) rises through the air. Can you think of other examples?

3. Compare Aristotle's hypothesis about falling rate (weight divided by resistance) with Philoponus' (weight minus resistance) for some extreme cases: a very heavy body with no resistance, a very light body with great resistance. Do the two hypotheses suggest very different results?

4. Consider Aristotle's statement "A given weight moves [falls] a given distance in a given time; a weight which is as great and more moves the same distances in less time, the times being in inverse proportion to the weights. For instance, if one weight is twice another, it will take half as long over a given movement." (*De Caelo*)

Indicate what Simplicio and Salviati each would predict for the falling motion in these cases:

(a) A 1-kg rock falls from a cliff and, while dropping, breaks into two equal pieces.

(b) A 5-kg rock is dropped at the same time as a 4.5-kg piece of the same type of rock.

(c) A hundred 4.5-kg pieces of rock, falling from a height, drop into a draw-string sack which closes, pulls loose, and falls.

5. Tie two objects of greatly different weight (like a book and a pencil) together with a piece of string (see below). Drop the combination with different orientations of objects. Watch the string. In a few sentences summarize your results.

6. (a) A bicyclist starting from rest accelerated uniformly at 2m/sec^2 for 6 sec. What distance did he cover in that time? Calculate the average speed for that time (6 sec) by finding the average of the initial speed and final speed. What distance would the bicyclist cover in 6 sec at this average speed? This problem illustrates the Merton theorem (see question 7).

(b) Only in the special case of uniform acceleration does a simple arithmetic average of speeds give the correct average speed. Explain why this is so.

7. A good deal of work on the topic of motion preceded that of Galileo. In the period 1280–1340,

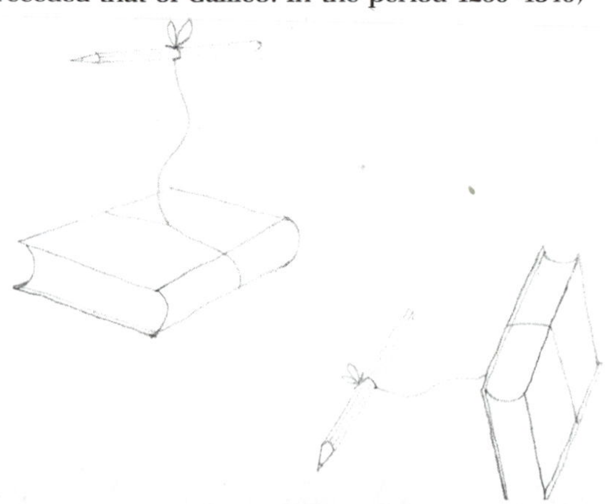

mathematicians at Merton College, Oxford, carefully considered different quantities that change with the passage of time. One result that had profound influence was a general theorem known as the Merton theorem or mean speed rule.

This theorem might be restated in our language and applied to uniform acceleration as follows: The distance an object goes during some time while its speed is changing uniformly is the same distance it would go if it went at the average speed the whole time.

(a) First show that the total distance traveled at a constant speed can be expressed as the area under the graph line on a speed–time graph. ("Area" must be found in speed units × time units.)

(b) Assume that this area represents the total distance even when the speed is not constant. Draw a speed–time graph for uniformly increasing speed and shade in the area under the graph line.

(c) Prove the Merton theorem by showing that the area is equal to the area under a constant-speed line at the average speed.

8. According to Galileo, uniform acceleration means equal Δv's in equal Δt's. Which of the following are other ways of expressing the same idea?

(a) Δv is proportional to Δt

(b) $\Delta v/\Delta t$ = constant

(c) the speed–time graph is a straight line

(d) v is proportional to t

9. In his discussion of uniformly accelerated motion, Galileo introduced another relationship that can also be put to experimental test. Galileo found that "the distances traversed by a body falling from rest during successive intervals of equal times will be in the ratios of the odd integers, 1:3:5:7. . . ."

Show that this experimentally testable result is in accord with our definitions for v_{av} and a for uniformly accelerated motion. (Hint: One way is to proceed as follows. For equal time intervals (Δt), the final speed reached is successively $a\Delta t$, $2a\Delta t$, ... *During* each of these time intervals, the average speeds are $\frac{1}{2}(a\Delta t)$, $\frac{1}{2}(3a\Delta t)$, . . . , and the corresponding distances covered are $\frac{1}{2}(a\Delta t) \cdot (\Delta t)$, $\frac{1}{2}(3a\Delta t) \cdot (\Delta t)$, ...

(*Note:* You can also deduce this result from a speed–time graph. Since the distance traversed is just speed times time, that is, $\Delta d = v\Delta t$, the area of any slice of the v versus t graph that has a height of v and a width Δt is just the distance traversed during Δt. Using this idea, show that the distances, Δd, traversed during equal Δt's obey Galileo's rule, quoted above.)

10. Using whatever modern equipment you wish, describe how you could find an accurate value for the speed of a falling object just before it strikes the ground.

11. Show that the expression

$$v_{av} = \frac{v_{initial} + v_{final}}{2}$$

is equivalent to the Merton theorem discussed in SG 7.

12. For any quantity that *changes uniformly*, the average is the sum of the initial and final values divided by two. Try it out for any quantity you may choose. For example: What is the average age in a group of five people having individually the ages of 15, 16, 17, 18, and 19 years? What is your average earning power over 5 years if it grows steadily from $8,000 per year at the start to $12,000 per year at the end?

13. Lt. Col. John L. Stapp achieved a speed of 284 m/sec in an experimental rocket sled at the Holloman Air Base Development Center, Alamogordo, New Mexico. Running on rails and propelled by nine rockets, the sled reached its top speed within 5 sec. Stapp survived a maximum acceleration of 22 g's in slowing to rest during a time interval of 1.5 sec. (One g is an acceleration equal in magnitude to that due to gravity; 22 g's means 22 × a_g.)

(a) Find the average acceleration in reaching maximum speed.

(b) How far did the sled travel before attaining maximum speed?

(c) Find the *average* acceleration while stopping.

14. Indicate whether the following statements are true or false when applied to the strobe photo below (you may assume that the strobe was flashing at a constant rate):

(**a**) The speed of the ball is greater at the bottom than at the top.

(**b**) This could be a freely falling object. (Make measurements on photograph.)

(**c**) This could be a ball thrown straight upward.

(**d**) If (**b**) is true, the speed increases with time because of the acceleration due to gravity.

(**e**) If (**c**) is true, the speed decreases with time because of the effect of gravity; this effect could still be called acceleration due to gravity.

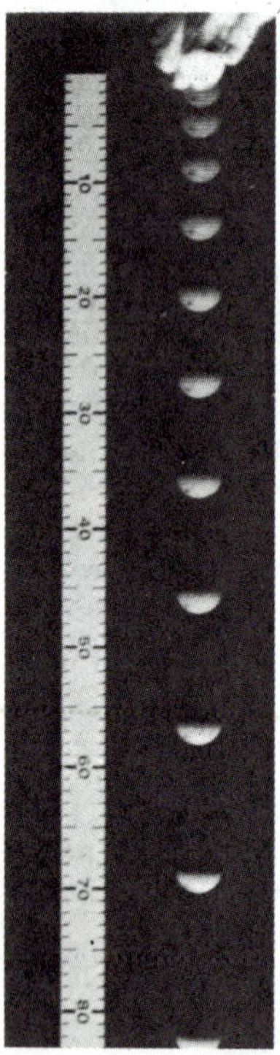

15. The photograph above shows a ball falling next to a vertical meter stick. The time interval between strobe flashes was 0.035 sec. Use this information to make graphs of *d* versus *t* and *v* versus *t*, and find the acceleration of the ball. (*Note:* The bottom of the ball, just on release, was next to the zero point of the meter stick. During the first few flashes, the images of the falling ball may have been so superposed as to be difficult to resolve. But for the purposes of this problem, we can neglect the earliest part of the fall, say, to 10 cm.)

16. The photograph in the figure on page 63 is of a ball thrown upward. The acceleration due to gravity increases the speed of the ball as it descends from its highest point (like any free-falling object) if air friction is negligible. But the acceleration due to gravity, which does not change, acts also while the ball is still on its way up, and for that portion of the path causes the ball to slow down as it ascends to the top point, C.

When there is both up and down motion, it will help to adopt a sign convention, an arbitrary but consistent set of rules, similar to designating the height of a place with respect to sea level. To identify distances measured *above* the point of initial release, give them *positive* values; for example, the distance at B or at D, measured from the release level, is about +60 cm and +37 cm, respectively. If measured *below* the release level, give them *negative* values; for example, E is at −23 cm. Also, assign a positive value to the speed of an object on its way up to the top (about +3 m/sec at A and a negative value to a speed a body has *on the way down* after reaching the top (about −2 m/sec at D and −6 m/sec at E).

(**a**) Fill in the table with + and − signs.

AT POSITION	SIGN GIVEN TO VALUE OF *d*	*v*
A		
B		
C		
D		
E		

(**b**) Show that it follows from this convention and from the definition of $a = \Delta v/\Delta t$ that the value or sign given to the acceleration due to gravity is *negative*, and for both parts of the path.

(**c**) What would the sign of acceleration due to gravity be in each case if we had chosen the + and − sign conventions just the other way, that is, associating − with up, + with down?

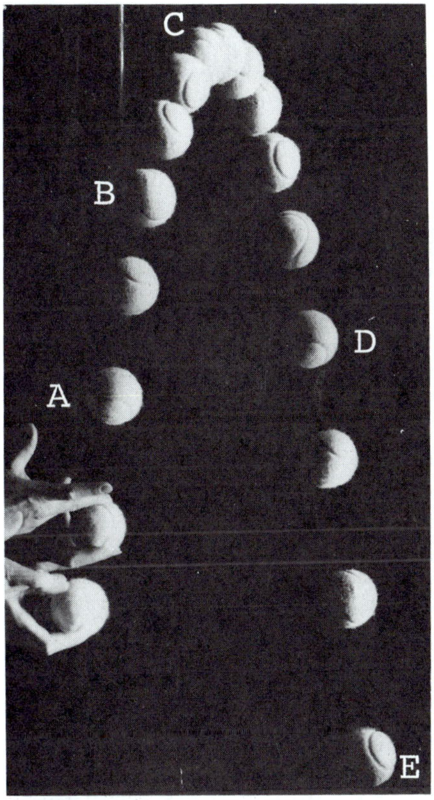

Stroboscopic photograph of a ball thrown into the air.

17. Draw a set of points (as they would appear in a strobe photo) to show the successive positions of an object that by our convention in SG 16 had a positive acceleration, that is, "upward." Can you think of any way to produce such an event physically?

18. Memorizing equations will not save you from having to think your way through a problem. You must decide if, when, and how to use equations. This means analyzing the problem to make certain you understand *what information is given* and *what is to be found*. Test yourself on the following problem. Assume that the acceleration due to gravity is equal to 10 m/sec/sec.

Problem: A stone is dropped from rest from the top of a high cliff.

(a) How far has it fallen after 1 sec?

(b) What is the stone's speed after 1 sec of fall?

(c) How far does the stone fall during the second second (that is, from the end of the first second to the end of the second second)?

19. From the definition for a, show it follows directly that $v_{final} = v_{initial} + at$ for motion with constant acceleration. Using this relation and the sign convention in SG 16, answer the questions below. (Assume a_g = 10 m/sec/sec.) An object is thrown upward with an initial speed of 20 m/sec.

(a) What is its speed after 1.0 sec?

(b) How far did it go in this first second?

(c) How long did the object take to reach its maximum height?

(d) How high is this maximum height?

(e) When it descends, what is its final speed as it passes the throwing point?

If you have no trouble with this, you may wish to try problems SG 20 and 21.

20. A batter hits a pop fly that travels straight upward. The ball leaves the bat with an initial speed of 40 m/sec. (Assume a_g = 10 m/sec/sec.)

(a) What is the speed of the ball at the end of 2 sec?

(b) What is its speed at the end of 6 sec?

(c) When does the ball reach its highest point?

(d) How high is this highest point?

(e) What is the speed of the ball at the end of 10 sec? (Graph this series of speeds.)

(f) What is its speed just before it is caught by the catcher?

21. A ball starts up an inclined plane with a speed of 4 m/sec, and comes to a halt after 2 sec.

(a) What acceleration does the ball experience?

(b) What is the average speed of the ball during this interval?

(c) What is the ball's speed after 1 sec?

(d) How far up the slope will the ball travel?

(e) What will be the speed of the ball 3 sec after starting up the slope?

(f) What is the total time for a round trip to the top and back to the start?

22. As Director of Research in your class, you receive the following research proposals from physics students wishing to improve upon Galileo's free-fall experiment. Would you recommend support for any of them? If you reject a proposal, you should make it clear why you do so.

(a) "Historians believe that Galileo never dropped objects from the Leaning Tower of Pisa. But such an experiment is more direct and more fun than inclined plane experiments, and of course, now that accurate stopwatches are available, it can be carried out much better than in Galileo's time. The experiment involves dropping, one by one, different size spheres made of copper, steel, and glass from the top of the Leaning Tower and finding how long it takes each one to reach the ground. Knowing d (the height of the tower) and time of fall t, I will substitute in the equation $d = ½at^2$ to see if the acceleration a has the same value for each sphere."

(b) "An iron shot will be dropped from the roof of a 4-story building. As the shot falls, it passes a window at each story. At each window there will be a student who starts a stopwatch upon hearing a signal that the shot has been released, and stops the watch as the shot passes the window. Also, each student records the speed of the shot as it passes. From these data, each student will compute the ratio v/t. I expect that all four students will obtain the same numerical value of the ratio."

(c) "Galileo's inclined planes dilute motion all right, but the trouble is that there is no reason to suppose that a ball rolling down a board is behaving like a ball falling straight downward. A better way to accomplish this is to use light, fluffy, cotton balls. These will not fall as rapidly as metal spheres, and therefore it would be possible to measure the time of the fall t for different distances. The ratio d/t^2 could be determined for different distances to see if it remained constant. The compactness of the cotton ball could then be changed to see if a different value was obtained for the ratio."

23. A student on the planet Arret in another solar system dropped an object in order to determine the acceleration due to gravity at that place. The following data are recorded (in local units):

TIME (in surgs)	*DISTANCE (in welfs)*	*TIME (in surgs)*	*DISTANCE (in welfs)*
0.0	0.0	2.2	10.41
0.5	0.54	2.4	12.39
1.0	2.15	2.6	14.54
1.5	4.84	2.8	16.86
2.0	8.60	3.0	19.33

(a) What is the acceleration due to gravity on the planet Arret, expressed in welfs/surg^2?

(b) A visitor from earth finds that one welf is equal to about 6.33 cm and that one surg is equivalent to 0.167 sec. What would this tell us about Arret?

24. **(a)** Derive the relation $v^2 = 2ad$ from the equations $d = \frac{1}{2}at^2$ and $v = at$. What special conditions must be satisfied for the relation to be true?

(b) Show that if a ball is thrown straight upward with an initial speed v, it will rise to a height

$$h = \frac{v^2}{2a_g}$$

25. Sometimes it is helpful to have a special equation relating certain variables. For example, for constant acceleration a, the final speed v_f is related to initial speed v_i and distance traveled d by

$$v_f^2 = v_i^2 + 2ad$$

Try to derive this equation from some other equations you are familiar with.

26. Use a graph like the one sketched below and the idea that the area under the graph line in a speed–time graph gives a value for the distance traveled, to derive the equation

$$d = v_i t + ½at^2$$

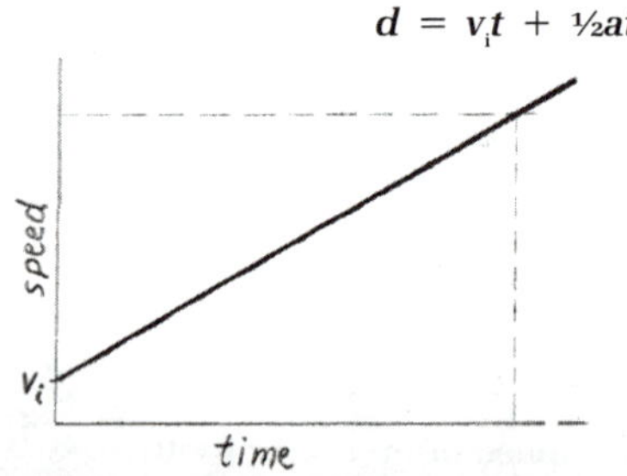

27. List the steps by which Galileo progressed from his first definition of uniformly accelerated motion to his final confirmation that this definition is useful in describing the motion of a freely falling body. Identify each step as a hypothesis, deduction, observation, computation, etc. What limitations and idealizations appear in the argument?

28. In these first two chapters we have been concerned with motion in a straight line. We have dealt with distance, time, speed, and acceleration, and with the relationships among them. Surprisingly, most of the results of our discussion can be summarized in the following three equations.

$$v_{av} = \frac{\Delta d}{\Delta t} \qquad a_{av} = \frac{\Delta v}{\Delta t} \qquad d = \tfrac{1}{2}at^2$$

Because these three equations are so useful, they are worth remembering.

(**a**) State each of the three equations in words, and state explicitly any limitations on when they apply.

(**b**) Make up a *simple* problem to demonstrate the use of each equation. (For example: How long will it take a jet plane to travel 3,200 km if it averages 1,000 km/hr?)

(**c**) Work out the solutions just to be sure the problems can be solved.

29. What is wrong with the following common statements? "The Aristotelians did not observe nature. They took their knowledge out of old books which were mostly wrong. Galileo showed it was wrong to trust authority in science. He did experiments and showed everyone directly that the old ideas on free fall motion were in error. He thereby started science and also gave us the scientific method."

30. (a) What is the acceleration of a car that accelerates uniformly from 5 m/sec to 30 m/sec in 10 sec?

(**b**) How tall is a building if it takes an object 9.0 sec to hit the ground after falling from the roof?

(**c**) A block slides down an inclined plane with a constant acceleration of 2 m/sec^2. How long will it take the block to slide 20 m? How fast will the block be moving at the end of that time?

(**d**) A particle with a velocity of 8 m/sec north starts accelerating. It accelerates uniformly at 5 m/sec^2 north for 10 sec. How far does the particle travel in 10 sec? What is the speed of the particle after 10 sec?

(**e**) What is the final speed of a model train that accelerates uniformly at 2 m/sec^2 for a distance of 4 m?

(**f**) A particle moving with a speed of 6 m/sec enters a region 2 m long where it is uniformly accelerated at 1 m/sec^2. What is the speed of the particle at the end of that region?

(**g**) What is the acceleration of a motor boat that accelerates uniformly from 5 m/sec to 55 m/sec over a distance of 100 m?

(**h**) A ball is dropped from the roof of a 125-m building. At the same time, a second ball is thrown straight up to collide with the first ball. What is the initial speed of the *second* ball if the balls collide 45 m from the ground 4 sec after they were released? (*Hint*: Some of this information is unnecessary.)

31. (a) Using the methods you learned in Chapter 1, calculate the average speed of the object represented by the graph shown below in sections AB and CD.

(**b**) Using the information from (**a**), calculate the average speed and average acceleration in section BC.

(**c**) Discuss your results for (**a**) and (**b**).

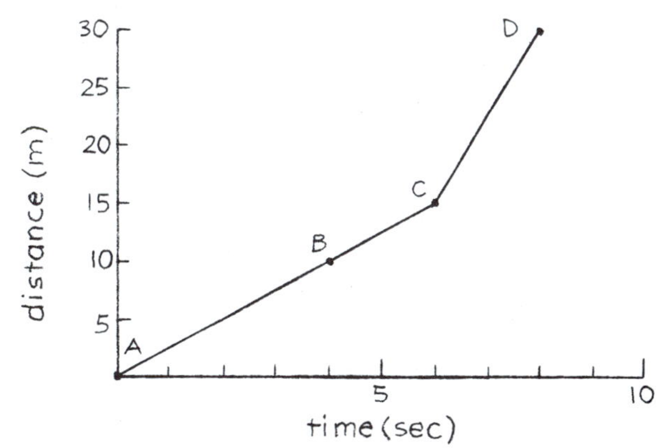

32. You have probably noticed that uniform motion is represented by a straight line on a distance 1/N time graph while accelerated motion is represented by a curved line.

(**a**) Describe the motion represented by each of the graphs below.

(**b**) Identify the direction of motion in each graph.

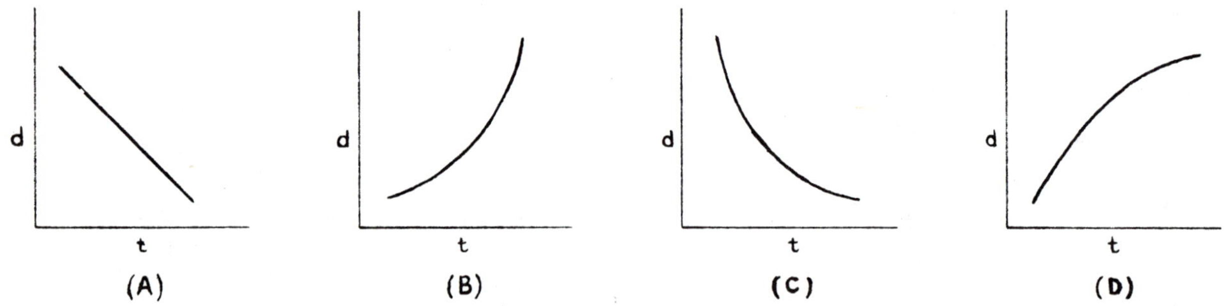

CHAPTER 3

The Birth of Dynamics
Newton Explains Motion

3.1 | "Explanation" and the laws of motion

SG 1

Kinematics is the study of *how* objects move, but not of *why* they move. Galileo investigated many topics in kinematics with insight, originality, and energy. The most valuable part of that work dealt with special types of motion, such as free fall. In a clear and consistent way, he showed how to describe the motion of objects with the aid of mathematical ideas.

Galileo had written that "the present does not seem to be the proper time to investigate the cause of the acceleration of natural motion. . . ." When Isaac Newton began his studies of motion in the second half of the seventeenth century, that statement was no longer appropriate. Indeed, because Galileo had been so effective in describing motion, Newton could turn his attention to *dynamics*. Dynamics is the study of *why* an object moves the way it does, for example, why it starts to move instead of remaining at rest, why it speeds up or moves on a curved path, and why it comes to a stop.

Some kinematics concepts: position, time, speed, acceleration. Some dynamics concepts: mass, force, momentum (Ch.9), energy (Ch. 10).

How does dynamics differ from kinematics? As mentioned above, kinematics treats the description of motion, while dynamics treats the causes of motion, tracing these causes back to the play of forces. Each, of course, depends on the other in order to describe a motion in a way that will make its explanation as simple as possible. Conversely, given an idea for an explanation, that idea can be used to suggest better methods of description.

The study of kinematics in Chapters 1 and 2 revealed that an object may: (a) remain at rest, (b) move uniformly in a straight line, (c) speed up during straight-line motion, (d) slow down during straight-line motion. Because the last two situations are examples of acceleration, the list could actually be reduced to: (a) rest, (b) uniform motion, and (c) acceleration.

In Chapter 4 we will take up motion along *curved* paths.

Rest, uniform motion, and acceleration are therefore the phenomena to explain. The word "explain" must be used with care. To the physicist, an event is "explained" when it is shown to be a logical consequence of a law the physicist has reason to believe is true. In other words, a physicist with faith in a general law "explains" an event by showing that it is consistent (in agreement) with the law. An infinite number of separate, different-looking events occur constantly all around you and within you. In a sense, the physicist's job is to show how each of these events results necessarily from certain general rules that describe the way the world operates. This approach to "explanation" is made possible by the fact that the number of general laws of physics is surprisingly small. This chapter will discuss three such laws. Together with the mathematical schemes of Chapters 1 and 2 for describing motion, they will enable you to understand practically all motions that you can easily observe. Adding one more law, the law of universal gravitation (Unit 2), you can explain the motions of stars, planets, comets, and satellites. In fact, throughout physics one sees again and again that nature has a marvelous simplicity.

Newton's First Law: Every object continues in its state of rest or of uniform motion in a straight line unless acted upon by an unbalanced force.

Newton's Second Law: The acceleration of an object is directly propotional to, and in the same direction as, the unbalanced force acting on it, and inversely proportional to the mass of the object.

Newton's Third Law: To every action there is always opposed an equal reaction; or, mutual actions of two bodies upon each other are always equal and in opposite directions.

To explain rest, uniform motion, and acceleration of any object, you must be able to answer such questions as these: Why does a vase placed on a table remain stationary? If a dry-ice disk resting on a smooth, level surface is given a brief push, why does it move with uniform speed in a straight line? Why does it neither slow down quickly nor curve to the right or left? These and almost all other specific questions about motion can be answered either directly or indirectly from Isaac Newton's three general "Laws of Motion." These laws appear in his famous book,

Philosophiae Naturalis Principia Mathematica (*Mathematical Principles of Natural Philosophy,* 1687), usually referred to simply as the *Principia*. They remain among the most basic laws in physics today.

This chapter will examine Newton's three laws of motion one by one. If your Latin is fairly good, try to translate them from the original, reproduced below. A modern, English version of Newton's text of these laws appears in the margin on page 68.

Before taking up Newton's ideas, it is helpful to see how other scientists of Newton's time, or earlier, might have answered questions about motion. One reason for doing this now is that many people who have not studied physics still tend to think a bit like pre-Newtonians!

Pages 12 and 13 of the original (Latin) edition of Newton's Philosophiae Naturalis Principia *(The Mathematical Principles of Natural Philosophy). These pages contain the three laws of motion and the parallelogram rule for the addition of forces (see Secs. 3.3 and 3.4).*

[12]

AXIOMATA
SIVE
LEGES MOTUS

Lex. I.

Corpus omne perseverare in statu suo quiescendi vel movendi uniformiter in directum, nisi quatenus a viribus impressis cogitur statum illum mutare.

Projectilia perseverant in motibus suis nisi quatenus a resistentia aeris retardantur & vi gravitatis impelluntur deorsum. Trochus, cujus partes cohærendo perpetuo retrahunt sese a motibus rectilineis, non cessat rotari nisi quatenus ab aere retardatur. Majora autem Planetarum & Cometarum corpora motus suos & progressivos & circulares in spatiis minus resistentibus factos conservant diutius.

Lex. II.

Mutationem motus proportionalem esse vi motrici impressæ, & fieri secundum lineam rectam qua vis illa imprimitur.

Si vis aliqua motum quemvis generet, dupla duplum, tripla triplum generabit, sive simul & semel, sive gradatim & successive impressa fuerit. Et hic motus quoniam in eandem semper plagam cum vi generatrice determinatur, si corpus antea movebatur, motui ejus vel conspiranti additur, vel contrario subducitur, vel obliquo oblique adjicitur, & cum eo secundum utriusq; determinationem componitur.

Lex. III.

[13]

Lex. III.

Actioni contrariam semper & æqualem esse reactionem: sive corporum duorum actiones in se mutuo semper esse æquales & in partes contrarias dirigi.

Quicquid premit vel trahit alterum, tantundem ab eo premitur vel trahitur. Siquis lapidem digito premit, premitur & hujus digitus a lapide. Si equus lapidem funi allegatum trahit, retrahetur etiam & equus æqualiter in lapidem: nam funis utrinq; distentus eodem relaxandi se conatu urgebit Equum versus lapidem, ac lapidem versus equum, tantumq; impediet progressum unius quantum promovet progressum alterius. Si corpus aliquod in corpus aliud impingens, motum ejus vi sua quomodocunq: mutaverit, idem quoque vicissim in motu proprio eandem mutationem in partem contrariam vi alterius (ob æqualitatem pressionis mutuæ) subibit. His actionibus æquales fiunt mutationes non velocitatum sed motuum, (scilicet in corporibus non aliunde impeditis:) Mutationes enim velocitatum, in contrarias itidem partes factæ, quia motus æqualiter mutantur, sunt corporibus reciproce proportionales.

Corol. I.

Corpus viribus conjunctis diagonalem parallelogrammi eodem tempore describere, quo latera separatis.

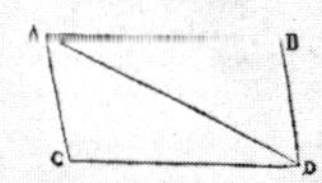

Si corpus dato tempore, vi sola *M*, ferretur ab *A* ad *B*, & vi sola *N*, ab *A* ad *C*, compleatur parallelogrammum *ABDC*, & vi utraq; feretur id eodem tempore ab *A* ad *D*. Nam quoniam vis *N* agit secundum lineam *AC* ipsi *B D* parallelam, hæc vis nihil mutabit velocitatem accedendi ad lineam illam *B D* a vi altera genitam. Accedet igitur corpus eodem tempore ad lineam *B D* sive vis *N* imprimatur, sive non, atq; adeo in fine illius temporis reperietur alicubi in linea

illa

?

1. A baseball is thrown straight upward. Which of these questions about the baseball's motion are kinematic and which dynamic?

(a) How high will the ball go before coming to a stop and starting downward?

(b) How long will it take to reach that highest point?

(c) What would be the effect of throwing it twice as hard?

(d) Which takes longer, the trip up or the trip down?

(e) Why does the acceleration remain the same whether the ball is moving up or down?

3.2 | The Aristotelian explanation of motion

The idea of force played a central role in the dynamics of Aristotle 20 centuries before Newton. You will recall from Chapter 2 that in Aristotle's physics there were two types of motion: "natural" motion and "violent" motion. For example, a falling stone was thought to be in "natural" motion (towards its natural place). On the other hand, a stone being steadily lifted was thought to be in "violent" motion (away from its natural place). To maintain this uniform violent motion, a force had to be continuously applied. Anyone lifting a large stone is very much aware of this force while straining to hoist the stone higher.

Keeping an object in motion at uniform speed.

The Aristotelian ideas agreed with many common-sense observations. But there were also difficulties. Take as a specific example an arrow shot into the air. It cannot be in violent motion without a mover, or something pushing it. Aristotelian physics required that the arrow be constantly propelled by a force. If this propelling force were removed, the arrow should immediately stop its flight and fall directly to the ground in "natural" motion.

But, of course, an arrow does not fall to the ground as soon as it loses direct contact with the bowstring. What, then, is the force that propels the arrow? Here, the Aristotelians offered a clever suggestion: The motion of the arrow through the air is maintained by the air itself! As the arrow starts to move, the air in front of it is pushed aside. More air rushes in to fill the space being vacated by the arrow. This rush of air around the arrow keeps it in flight.

Other ideas to explain motion were developed before the mid-seventeenth century. But in every case, a force was considered necessary to sustain uniform motion. The explanation of uniform motion depended on finding the force, and that was not always easy. There were also other problems. For example, a falling acorn or stone does *not* move with uniform speed. It accelerates. How is acceleration explained? Some Aristotelians thought that the speeding up of a falling object was connected with its approaching arrival at its natural place, the earth. In other words, a falling object was thought to be like a tired horse that starts to gallop as it nears the barn. Others claimed that when an object falls, the weight of the air above it increases, pushing it harder. Meanwhile, the column of air below it decreases, thus offering less resistance to its fall.

SG 2

When a falling object finally reaches the ground, as close to the center of the earth as it can get, it stops. And there, in its "natual place," it remains. Rest, being regarded as the natural state of objects on earth, required no further explanation. The three phenomena of rest, uniform motion, and acceleration could thus be explained more or less reasonably by an Aristotelian. Now examine the Newtonian explanation of the same phenomena. The key to this approach is a clear understanding of the concept of force.

?

2. According to Aristotle, what is necessary to maintain uniform motion?

3. Give an Aristotelian explanation of a dry-ice disk's uniform motion across a tabletop.

3.3 | Forces in equilibrium

The common-sense idea of force is closely linked with muscular activity. You know that a sustained effort is required to lift and support a heavy stone. When you push a lawn mower, row a boat, split a log, or knead bread dough, your muscles indicate that you are applying a force to some object. Force and motion and muscular activity are naturally associated in our minds. When you think of changing the shape of an object, or moving it, or changing its motion, you automatically think of the muscular sensation of applying a force to the object. You will see that many, but not all, of your everyday, common-sense ideas about force are useful in physics.

You know, without having to think about it, that forces can make things move. Forces can also hold things still. The cable supporting the main span of the Golden Gate Bridge is under the influence of mighty forces, yet it remains at rest. Apparently, more is required to start motion than just any application of forces.

Of course, this is not surprising. You have probably seen children quarreling over a toy. If each child pulls with determination in the opposite direction, the toy may go nowhere. On the other hand, the tide of battle may shift if one of the children suddenly makes an extra effort or if two children cooperate and pull side by side against a third.

Likewise, in a tug-of-war between two teams, large forces are exerted on each side, but the rope remains at rest. We might say that the forces "balance" or "cancel." A physicist would say that the rope was in *equilibrium*. That is, the sum of all forces applied to one side of the rope is just as great, though acting in the opposite direction, as the sum of forces applied to the other side. The physicist might also say the *net force* on the rope is zero. Thus, a body in equilibrium cannot start to move. It starts to move only when a new, "unbalanced" force is added, destroying the equilibrium.

In all these examples, both the magnitude (size or amount) of the forces and their directions are important. The effect of a force depends on the direction in which it is applied. You can represent both the sizes and directions of forces in a sketch by using arrows. The direction in which an arrow points represents the direction in which the force acts. The length of the arrow represents how large the force is. For example, the force exerted by a 10-kg bag of potatoes is shown by an arrow twice as long as that for a 5-kg bag.

This force is called the *net* force, because it is the sum of all the forces in one direction *minus* the sum of all the forces in the opposite direction.

If you know separately each of the forces applied to any object at rest, you can predict whether the object will remain at rest. It is as simple as this: An object acted on by forces is in equilibrium and remains at rest *only* if the arrows representing the forces all total zero.

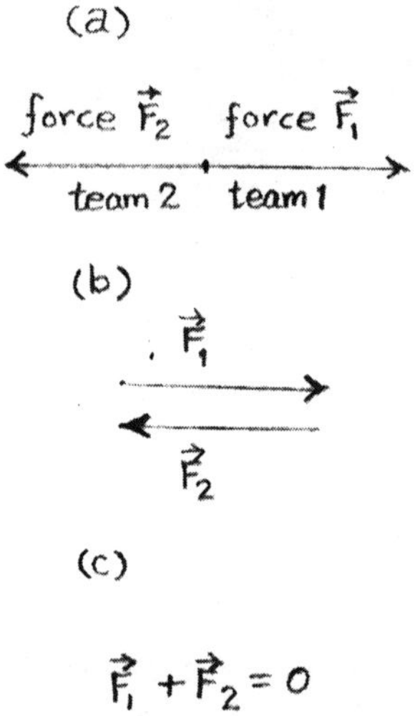

How do you "total" arrows? This can be done by means of a simple technique. Take the tug-of-war as an example. Call the force applied by the team pulling to the right $\vec{F}_1$. (The arrow over the $\vec{F}$ indicates that you are dealing with a quantity for which direction is important.) The force applied to the rope by the second team you can call $\vec{F}_2$. Figure (a) in the margin shows the two arrows corresponding to the two forces, each applied to the same rope, but in opposite directions. Assume that these forces, $\vec{F}_1$ and $\vec{F}_2$, were accurately and separately measured. For example, you might let each team in turn pull on a spring balance as hard as it can. You can then draw the arrows for $\vec{F}_1$ and $\vec{F}_2$ carefully to a chosen scale, such as 1 cm = 100 N. Thus, a 200-N force in either direction would be represented by an arrow 2 cm in length. Next, take the arrows $\vec{F}_1$ and $\vec{F}_2$ and draw them again in the correct directions and to the chosen scale. This time, however, put them "head to tail" as in Figure (b). Thus, you might draw $\vec{F}_1$ first, and then draw $\vec{F}_2$ with the tail of $\vec{F}_2$ starting from the head of $\vec{F}_1$. (Since they would, of course, overlap in this example, they are drawn slightly apart in Figure (b) to show them both clearly.) The technique is this: If the head end of the second arrow falls exactly on the tail end of the first, then you know that the effects of $\vec{F}_1$ and $\vec{F}_2$ balance each other. The two forces, equally large and acting in opposite directions, total zero. If they did not, the excess of one force over the other would be the *net force*, and the rope would accelerate instead of being at rest.

There are several ways of expressing the idea of unbalanced force: *net force, resultant force, total force, vector sum of forces*. All mean the same thing.

To be sure, this was an obvious case. But the "head-to-tail" method, using drawings, also works in cases that are not as simple. For example, apply the same procedure to a boat that is secured by three ropes attached to different moorings. Suppose $\vec{F}_1$, in this case, is a force of 24 N, $\vec{F}_2$ is 22 N, and $\vec{F}_3$ is 19 N, each in the direction shown in the sketch on p. 73. (A good scale for the magnitude of the forces here is 0.1 cm = 1 N of force.) Is the boat in equilibrium when it is acted on by the forces? Yes, if the forces add up to zero. With ruler and protractor you can draw the arrows to scale and in exactly the right directions. Adding $\vec{F}_1$, $\vec{F}_2$, and $\vec{F}_3$ head to tail, as in the diagram to the right of the picture, you see that the head of the last arrow falls on the tail of the first. The sum of the forces, or the *net* force, is indeed zero. The forces are said to cancel, or to be balanced. Therefore, the object (the boat) is in equilibrium. This method tells when an object is in equilibrium, no matter how many different forces are acting on it.

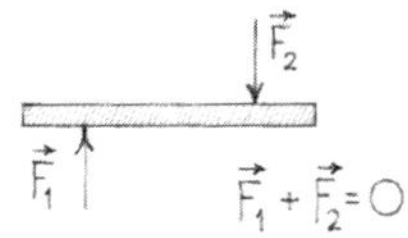

We are defining equilibrium without worrying about whether the object will rotate. For example, the sum of the forces on the plank in the diagram is zero, but it is obvious that the plank will rotate.

SG 5

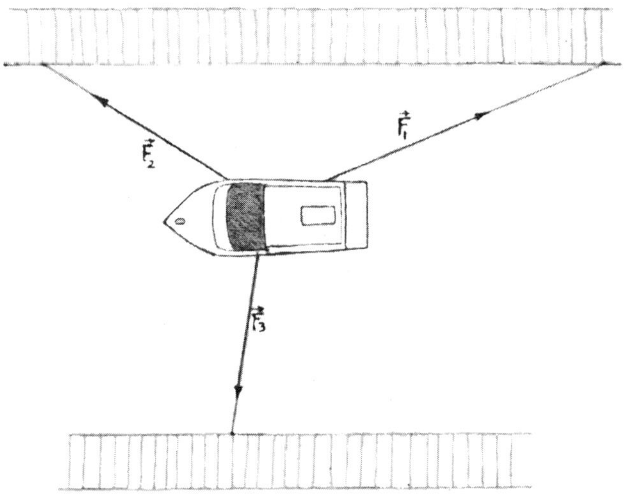

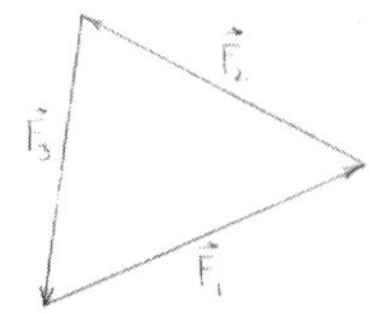

$\vec{F}_1 + \vec{F}_2 + \vec{F}_3 = 0$

We can now summarize our understanding of the state of rest as follows: If an object remains at rest, the sum of all forces acting on it must be zero. Rest is an example of the condition of equilibrium, the state in which all forces on the object are balanced.

An interesting case of equilibrium, different from the tug-of-war, is the last part of the fall of a skydiver. At the beginning, just after the jump, the person is in just the sort of free-fall, accelerated motion discussed in Chapter 2. But the force of air friction on the skydiver increases with speed. Eventually, the upward frictional force becomes large enough to cancel the downward force of gravity (which is the force you experience as your weight). Under these conditions, the skydiver is in equilibrium, going at a constant (terminal) speed, kept from accelerating by friction with the air going by. The net force is zero, just as it is when you are lying still in bed.

The speed at which this equilibrium occurs for a falling person without a parachute is very high. When the parachute is opened, its large area adds greatly to the friction, and therefore equilibrium is established at a much smaller terminal speed.

?

4. A vase is standing at rest on a table. What forces would you say are acting on the vase? Show how each force acts (to some scale) by means of an arrow. Can you show that the sum of the forces is zero?

5. In which of these cases do the forces cancel?

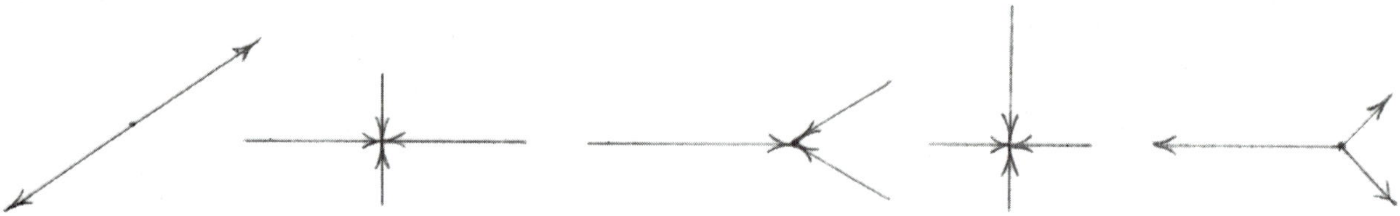

6. What is 3 plus 3? What is 3 north plus 3 south? What is 4 plus 2? What is 4 up plus 2 down? What is 4 in plus 2 in?

7. If an object experiences forces of 5 N right and 3 N left, is it in equilibrium? What is the definition of equilibrium? Does an object in equilibrium have to be at rest?

3.4 | About vectors

The method for representing forces by arrows really can predict whether the forces cancel and will leave the object in equilibrium. It will also show whether any net force is left over, causing the object to accelerate. You can demonstrate for yourself the reliability of the addition rule by doing a few experiments. For example, you could attach three spring scales to a ring. Have three persons pull on the scales with forces that just balance, keeping the ring at rest. While they are pulling, read the magnitudes of the forces on the scales and mark the directions of the pulls. Then make a sketch with arrows representing the forces, using a convenient scale, and see whether they total zero. Many different experiments of this kind ought all to show a net force equal to zero.

It is not obvious that forces should behave like arrows. But arrows drawn on paper happen to be useful for calculating how forces add. (If they were not, we simply would look for other symbols that did work.) Forces belong in a class of concepts called *vector quantities*, or just *vectors* for short. Some characteristics of vectors are easy to represent by arrows. In particular, vector quantities have *magnitude*, which can be represented by the length of an arrow drawn to scale. They also have *direction*, which can be shown by the direction of an arrow. By experiment, we find that vectors can be *added* in such a way that the total effect of two or more can be represented by the head-to-tail addition of arrows. This total effect is called the *vector resultant*, or *vector sum*.

In the example of the tug-of-war, you determined the total effect of equally large, opposing forces. If two forces act in the *same* direction, the resultant force is found in much the same way, as shown below.

Any vector quantity is indicated by a letter with an arrow over it, for example, $\vec{F}$, $\vec{a}$, or $\vec{v}$.

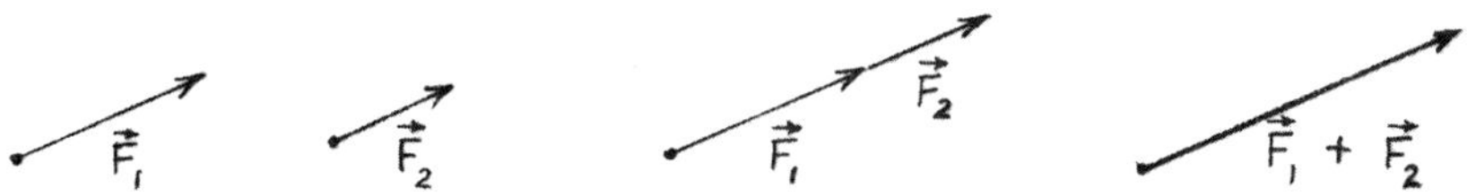

If two forces act as some angle to each other, the same type of sketch is still useful. For example, suppose two forces of equal magnitude are applied to an object at rest but free to move. One force is directed due east and the other due north. The object will accelerate in the northeast direction, the direction of the resultant force. (See sketch in the margin.) The magnitude of the acceleration is proportional to the magnitude of the resultant force, shown by the length of the arrow representing the resultant.

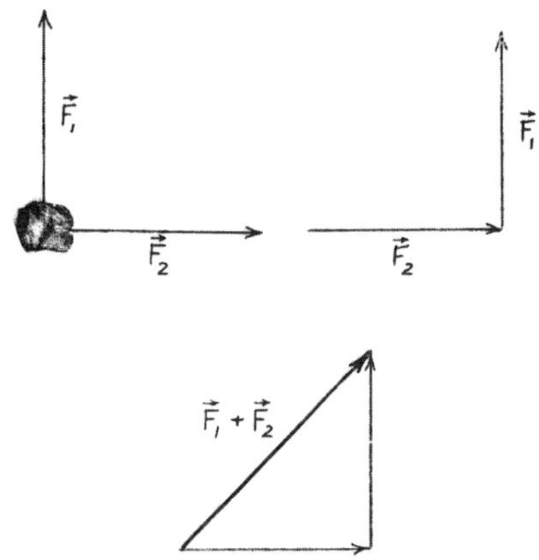

The same adding procedure works for forces of any magnitude and acting at any angle to each other. Suppose one force is directed due east and a somewhat larger force is directed northeast. The resultant vector sum can be found as shown below.

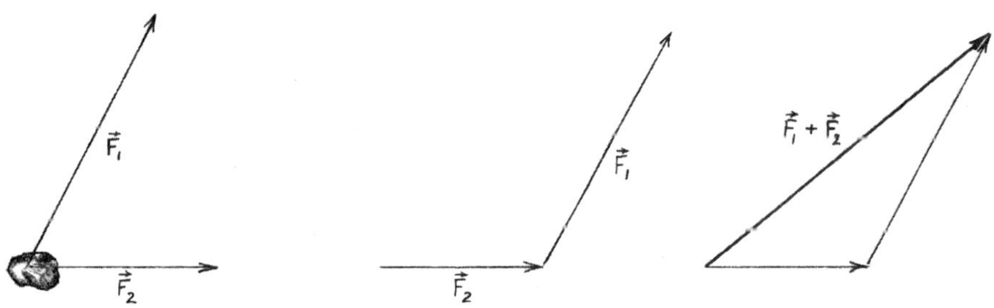

To summarize, a vector quantity has both direction and magnitude. Vectors can be added by constructing a head-to-tail arrangement of vector arrows (graphical method) or by an equivalent technique known as the parallelogram method, which is briefly explained in the marginal note at the right. (Vectors also have other properties which you will study if you take further physics courses.) By this definition, many important concepts in physics are vectors, for example, displacement, velocity, and acceleration. Some other physical concepts, including volume, distance, and speed, do not require a direction, and so are not vectors. Such quantities are called *scalar quantities*. When you add 10 liters (L) of water to 10 L of water, the result is always 20 L; direction has nothing to do with this result. Similarly, the term *speed* has no directional meaning; it is simply the *magnitude* of the velocity vector. Speed is shown by the *length* of the vector arrow, without regard to its direction. By contrast, suppose you

You can use equally well a graphical construction called the "parallelogram method." It looks different from the "head-to-tail" method, but is really exactly the same. In the parallelogram construction, the vectors to be added are represented by arrows joined tail-to-tail instead of head-to-tail, and the resultant is obtained by completing the diagonal of the parallelogram. (See SG 6.)

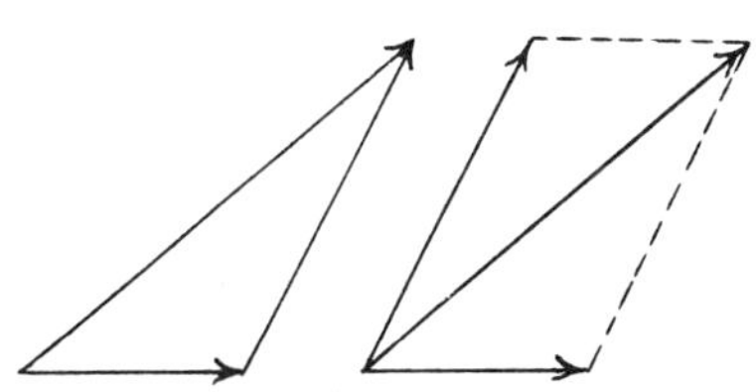

add two forces of 10 N each. The resultant force may be anywhere between 0 and 20 N, depending on the *direction* of the two individual forces.

In Sec. 1.8, acceleration was defined as the rate of change of speed. That was only partly correct because it was incomplete. We must also consider changes in the *direction* of motion as well. Acceleration is best defined as the rate of change of *velocity*, where velocity is a vector having both magnitude and direction. In symbols this definition may be written

$$\vec{a}_{av} = \frac{\Delta\vec{v}}{\Delta t}$$

where $\Delta\vec{v}$ is the change in velocity. Velocity can change in two ways: by changing its magnitude (speed) and by changing its direction. In other words, an object is accelerating when it speeds up, slows down, or changes direction. This definition and its important uses will be explored more fully in later sections.

?

8. Classify each of the following as vectors or scalars: (a) 2 m up; (b) volume; (c) 4 sec; (d) 3 m/sec west; ((e) 2 m/sec^2; (f) velocity.

9. List three properties of vector quantities.

10. How does the new definition of acceleration given above differ from the one used in Chapter 1?

3.5 | Newton's first law of motion

Because constant velocity means both constant speed and constant direction, we can write Newton's first law more concisely:

$$\vec{v} = \text{constant}$$

if and only if

$$\vec{F}_{net} = 0$$

This statement includes the condition of rest, since rest is a special case of unchanging velocity, the case where $\vec{v} = 0$.

You probably were surprised when you first watched a moving dry-ice disk or some other nearly frictionless object. Remember how smoothly it glided along after the slightest shove? How it showed no sign of slowing down or speeding up? From your everyday experience, you automatically think that some force is constantly needed to keep an object moving. But the disk does not act according to common-sense Aristotelian expectations. It is always surprising to see this for the first time.

In fact, the disk is behaving quite naturally. If the forces of friction were absent,, a gentle push would send tables and chairs gliding across the floor like dry-ice disks. Newton's first law directly challenges the Aristotelian idea of what is "natural." It declares that the state of rest and the state of uniform, unaccelerated motion in a straight line are equally natural. Only the existence of some force, friction for example, keeps a moving object from moving *forever*! Newton's first law of motion can be stated in modern language as follows:

Every object continues in its state of rest or of uniform rectilinear (straight-line) motion unless acted upon by an unbalanced force. Conversely, if an object is at rest or in uniform rectilinear motion, the unbalanced force acting upon it must be zero.

In order to understand the motion of an object, you must take into account all the forces acting on it. If *all* forces (including friction) are in balance, the body will be moving at constant $\vec{v}$.

Although Newton was the first to express this idea as a general law, Galileo had made similar statements 50 years before. Of course, neither Galileo nor Newton had dry-ice disks or similar devices. Therefore, they were unable to observe motion in which friction had been reduced so greatly. Instead, Galileo devised a thought experiment in which he imagined the friction to be zero.

This thought experiment was based on an actual observation. If a pendulum bob on the end of a string is pulled back and released from rest, it will swing through an arc and rise to very nearly its starting height. Indeed, as Galileo showed, the pendulum bob will rise almost to its starting level even if a peg is used to change the path as shown in the illustration below.

SG 8

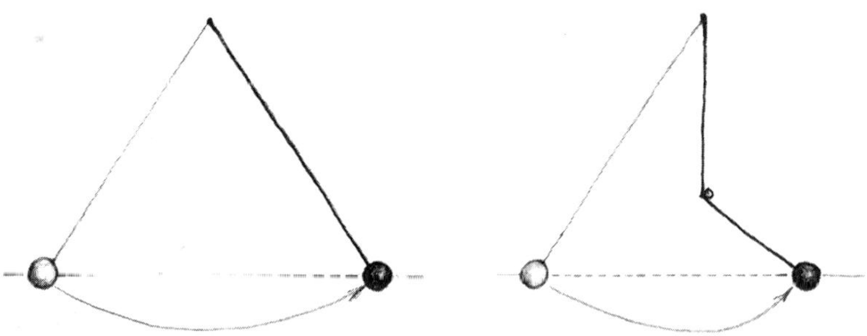

From this observation Galileo went on to his thought experiment. He predicted that a ball released from a height on a frictionless ramp would roll up to the same height on a similar facing ramp. Consider the diagram at the top of the next page. As the ramp on the right is changed from position (a) to (b) and then to (c), the ball must roll farther in each case to reach its original height. It slows down more gradually as the angle of the incline decreases. If the second ramp is exactly *level*, as shown in (d), the ball can *never* reach its original height. Therefore, Galileo believed, the ball on this frictionless surface would roll on in a straight line and at an unchanged speed forever. This could be taken to mean the same as Newton's first law. Indeed, some historians of science do give credit to Galileo for having come up with this law first. Other historians, however, point out that Galileo thought of the "rolling on forever" as "staying at a constant height above the earth." He did not think of it as "moving in a straight line through space."

This tendency of objects to maintain their state of rest or of uniform motion is sometimes called *the principle of inertia*.

Inside the laboratory there is no detectable difference between a straight (horizontal) line and a constant height above the earth. But on a larger scale, Galileo's eternal rolling would become motion in a circle around the earth. Newton made clear what is really important: In the absence of the earth's gravitational pull or other external forces, the ball's undisturbed path would extend straight out into space.

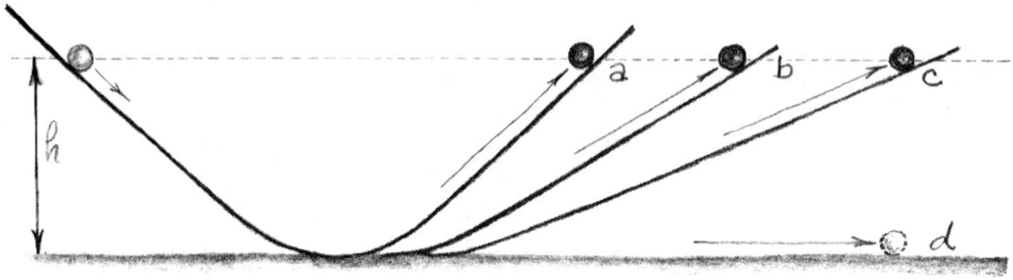

Newton's first law is therefore sometimes referred to as the *law of inertia*. *Inertia* is a property of all objects. Material bodies have, so to speak, a stubborn streak concerning their state of motion. Once in motion, they continue to move with unchanging velocity (unchanging speed and direction) unless acted on by some unbalanced external force. If at rest, they remain at rest. This tendency is what makes seat belts so necessary when a car stops very suddenly. It also explains why a car may not follow an icy road around a turn, but travel a straighter path into a field or fence. The greater the inertia of an object, the greater its resistance to change in its state of motion. Therefore, the greater is the force needed to produce a change in the state of its motion. For example, it is more difficult to start a train or a ship and to bring it up to speed than to keep it going once it is moving at the desired speed. (In the absence of friction, it would keep moving without any applied force at all.) For the same reason it is also difficult to bring it to a stop, and passengers and cargo keep going forward if the vehicle is suddenly braked.

Newton's first law says that if an object is moving with a constant speed in a straight line, the forces acting on it must be balanced; that is, the object is in equilibrium. Does this mean that in Newtonian physics the state of rest and the state of uniform motion are equivalent? It does indeed! If a body is in equilibrium, $\vec{v}$ = constant. Whether the value of this constant is zero or not depends in any case on the frame of reference for measuring the magnitude of $\vec{v}$. You can say whether a body is at rest or is moving with constant $\vec{v}$ larger than zero only by reference to some other body.

Take, for example, a tug-of-war. The two teams are sitting on the deck of a barge that is drifting with uniform velocity down a lazy river. An observer on the same barge and one on the shore report on the incident. Each observes from a particular frame of reference. The observer on the barge reports that the forces on the rope are balanced and that it is at rest. The observer on the shore reports that the forces on the rope are balanced and that it is in uniform motion. Which observer is right? They are both right; Newton's first law of motion applies to both observations. Whether a body is at rest or in uniform motion depends on which frame of reference is used to observe the event. In both cases, the forces on the object involved are balanced.

?

11. How would Aristotle have explained the fact that a bicyclist must keep pedaling in order to move with uniform speed? How would Newton explain the same fact? If Aristotle's explanation is "wrong," why do you think you study it?

12. What is the net force on the body in each of the four cases sketched in the margin of the opposite page?

13. What may have been a difference between Newton's concept of inertia and Galileo's?

3.6 | The significance of the first law

Newton's laws involve many deep philosophical concepts. (See SG 7.) However, the laws are easy to use, and you can see the importance of Newton's first law without going into any complex ideas. For convenience, here is a list of the important insights the first law provides.

Of course, the idea of inertia does not *explain why* bodies resist change in their state of motion. It is simply a term that helps us to talk about this basic, experimentally observed fact of nature. (See SG 10.)

1. It represents a break with Aristotelian physics: The "natural" motion, the motion that needs no further explanation, is not a return to a position of repose at an appropriate place. It is, rather, any motion that takes place with a uniform velocity.

2. It presents the idea of inertia, that is, of the basic tendency of all objects to maintain their state of rest or uniform motion.

3. It says that, from the point of view of physics, a state of rest is equivalent to a state of uniform motion at any speed in a straight line. There is nothing "absolute" or specially distinguished about any one of the states, uniform motion or rest. This raises the need to specify a "frame of reference" for describing motion, since an object that is stationary with respect to one observer, or frame of reference, can be in motion with respect to another.

4. It, like the other physical laws, is a universal law, claiming to be valid for objects anywhere in the universe. That is, the same law applies on the earth, on the moon, throughout the galaxy, and beyond, and the same law applies to the motions of atoms, magnets, tennis balls, stars, and every other thing. (A rather grand claim, but, as far as we can tell, a valid one.)

The correct reference frame to use in our physics turns out to be any reference frame that is at rest or in uniform rectilinear motion with respect to the stars. Therefore, the rotating earth is, strictly speaking, not allowable as a Newtonian reference frame; but for most purposes the earth rotates so little during an experiment that the rotation can be neglected. (See SG 11.)

5. The first law describes the behavior of objects when no net force acts on them. Thus, it sets the stage for the question: Exactly what happens when an unbalanced force *does* act on an object?

SG 12

3.7 | Newton's second law of motion

In Sec. 3.1, we stated that a theory of dynamics must account for rest, uniform motion, and acceleration. So far, we have met two

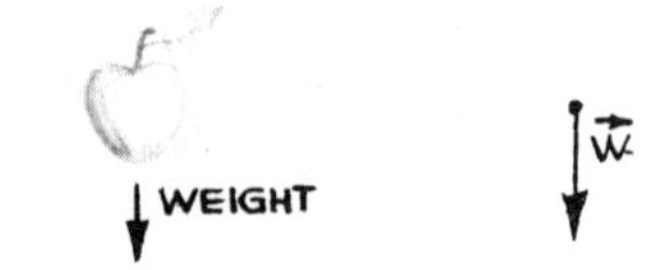

Apple falling: negligible friction

Feather falling at nearly constant speed

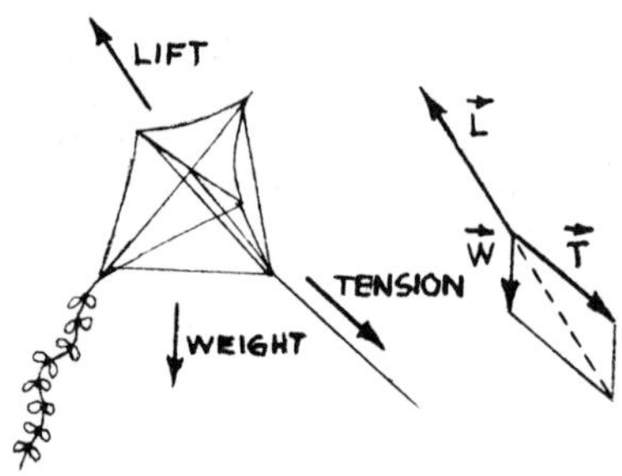

Kite held suspended in the wind

Athlete running against the wind

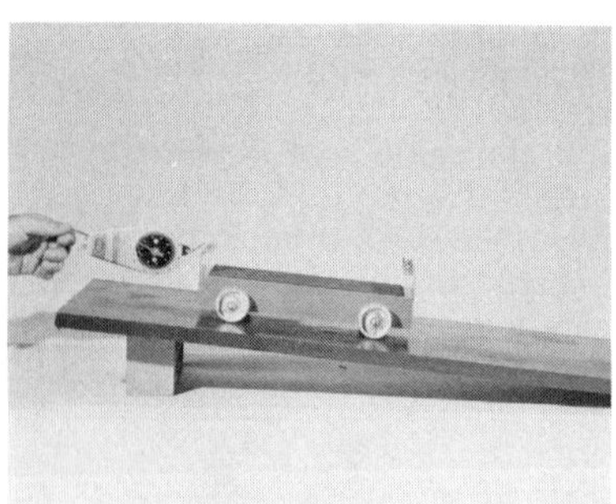

of these three objectives: the explanation of rest and of uniform motion. In terms of the first law, the states of rest and uniform motion are equivalent.

Newton's second law provides an answer to the question: What happens when an unbalanced force acts on an object? Since we think of the force as *causing* the resulting motion, this law is the fundamental law of dynamics.

In qualitative terms, the second law of motion says little more than this: The necessary and sufficient cause for a deviation from "natural" motion (that is, motion with a constant or zero velocity) is that a nonzero net force act on the object. However, the law goes further and provides a simple *quantitative* relation between the change in the state of motion and the net force. In order to be as clear as possible about the meaning of the law, we will first consider a situation in which different forces act on the same object; we will then consider a situation in which the same force acts on different objects; finally, we will combine these results into a general relationship.

Force and acceleration: The acceleration of an object is directly proportional to, and in the same direction as, the net force acting on the object. Note that force and acceleration are both vector quantities. Since acceleration is the rate at which velocity changes, the force is proportional to the *change* in the velocity; the faster or the greater the change, the larger the force must be. If $\vec{a}$ stands for the acceleration of the object and $\vec{F}_{net}$ stands for the net force on it, the relationship is

$$\vec{a} \propto \vec{F}_{net}$$

This relationship is equivalent to the statement above if it is also understood that when *vectors* are proportional, they must point in the same direction as well as have proportional magnitudes.

To say that two quantities are proportional means that if one quantity is doubled (or multiplied by any number), the other quantity is also doubled (or multiplied by the same number). Thus, for example, if a certain force produces a certain acceleration, twice the force (on the same object) will produce twice as great an acceleration in the same direction. In symbols, for the same object,

$$\begin{array}{rrll} \text{if} & \vec{F}_{net} & \text{will cause} & \vec{a}, \text{ then} \\ & 2\vec{F}_{net} & \text{will cause} & 2\vec{a} \\ & \tfrac{1}{2}\vec{F}_{net} & \text{will cause} & \tfrac{1}{2}\vec{a} \\ & 5.2\vec{F}_{net} & \text{will cause} & 5.2\vec{a} \\ & x\vec{F}_{net} & \text{will cause} & x\vec{a} \end{array}$$

It is easy to perform a rough experiment to test this law. Place a dry-ice disk or other nearly frictionless object on a flat table, attach a spring balance, and pull with a steady force so that it accelerates continuously. The pull registered by the balance is the only unbalanced force, so it is the net force on the object. You can determine the acceleration by measuring the time to go a fixed distance ($a \propto \frac{1}{2} t^2$). Repetitions of this experiment should show $\vec{F}_{net} \propto \vec{a}$, to within experimental error; at least, they have shown it whenever such an experiment was done in the past.

Mass and acceleration: The acceleration of an object is inversely proportional to the mass of the object; that is, the larger the mass of an object, the smaller will be its acceleration if a given net force is applied to it. The mass of an object is therefore what determines how large a force is required to change its motion; in other words, the mass of an object is a measure of its inertia. It is sometimes called *inertial mass*, to emphasize that it measures inertia. Mass is a scalar quantity (it has no direction), and it does not affect the direction of the acceleration.

SG 13

What does it mean to say that mass is a scalar quantity?

The relation between acceleration and mass can be written in symbols. Let a stand for the acceleration (the *magnitude* of the acceleration vector $\vec{a}$) and m stand for mass. Then,

See Sec. 3.4.

$$a \propto \frac{1}{m}$$

as long as the same net force is acting. Notice that an object with twice (or three times . . .) the mass of another will experience one half (or one third . . .) the acceleration if subjected to the same net force; that is, for the same net force,

if	m	experiences	a, then
	$2m$	experiences	$\frac{1}{2}a$
	$\frac{1}{3}m$	experiences	$3a$
	xm	experiences	$\frac{1}{x}a$

This law can also be demonstrated to hold by experiment. Can you suggest a way to do that?

The general relationship. Acceleration (for constant m) is proportional to the net force, and (for constant $\vec{F}_{net}$) it is proportional to $1/m$, the reciprocal of the mass. It follows that, in general, acceleration is proportional to the product of the net force and the reciprocal of the mass:

$$\vec{a} \propto \vec{F}_{net} \cdot \frac{1}{m}$$

Are there any other quantities (other than net force and mass) on which the acceleration depends? Newton proposed that the answer is no. Only the *net force on the object* and the *mass of the object* being accelerated affect the acceleration. All experience since then suggests he was right.

Since there are no other factors to be considered, we can make the proportionality into an equality; that is, we can write

Actually, from $a \propto F_{net}/m$ it follows that $a = k(F_{net}/m)$, where k is a constant. But by choosing the units for *a*, *F*, and *m* properly, we have set k = 1.

$$\vec{a} = \frac{\vec{F}_{net}}{m}$$

The same relationship can, of course, also be written as the famous equation

$$\vec{F}_{net} = m\vec{a}$$

In both these equations we have again written $\vec{F}_{net}$ to emphasize that it is the *net* force that determines the acceleration.

This relationship is probably the most basic equation in mechanics and, therefore, in physics. Without symbols we can state it as follows. *Newton's second law: The net force on an object is numerically equal to, and in the same direction as, the acceleration of the object multiplied by its mass.* It does not matter whether the forces that act are magnetic, gravitational, simple pushes and pulls, or any combination; whether the masses are those of electrons, atoms, stars, or cars; whether the acceleration is large or small, in this direction or that. The law applies universally.

Measuring mass and force. We have already mentioned a method of measuring force and have used it in talking about Newton's first law. The method is based on the fact that the extension of a spring (as long as it is not stretched or bent out of shape) is proportional to the force. Therefore, you can use a

SG 15

spring balance to measure forces. A force that is twice as big as another will stretch the same spring twice as far. The spring balance must be calibrated but that can be done after the units of mass have been defined.

Measuring mass (or inertia) is quite different from measuring force. When you think of measuring mass, you might first think of weighing it. But if you take apart a typical scale (the kitchen or bathroom variety, for example) you will find that it is usually just a spring balance. They measure a force and not the inertia of an object. The force they measure, called *weight*, is the gravitational force exerted on the object by the earth. If you stand on your bathroom scale on the moon, it will show a much smaller *weight*, but you will have just the same mass, or inertia.

You will need to calibrate the spring balance, that is, to decide how much stretch of the spring corresponds to *one unit* of force. In modern practice, this decision is made after deciding how to measure masses.

SG 18

A reminder: When we speak of the mass of a body, we mean its inertia, not its weight. The difference will be discussed in more detail in Sec. 3.8 and Chapter 8. For now you can get a quick feeling for the difference between the two by thinking of an experiment in a spacecraft that is moving with constant velocity, far from the earth or other planets. Suppose you are trying to move a puck on a table inside the spacecraft, as in the experiment in the photograph on p. 11. You find that the push needed to bring the puck up to a given speed on its table is just the same as on earth. The mass of an object is a measure of its inertia, its resistance to changes in motion; and that is the same for that object *everywhere*. The weight, on the other hand, depends on its location, for example, on how close the object is to a planet that exerts the gravitational pull called weight. When the spacecraft reaches outer space, the weight of an object becomes negligible, but its mass or inertia remains the same.

The standard kilogram at the U.S. Bureau of Standards.

This fact suggests that the wise thing to do is to choose the unit of mass first, and let the unit of force follow later. The simplest way to define a unit of mass is to choose some convenient object as the *universal standard of mass*, and compare the mass of all other objects with that one. What is selected to serve as the standard object is arbitrary. In Renaissance England the standard used was a grain of barley ("from the middle of the ear"). The original metric commission in France in 1799 proposed the mass of a cubic centimeter of water as a standard. Today, for scientific purposes, the standard mass is a cylinder of platinum–iridium alloy kept at the International Bureau of Weights and Measures near Paris. The mass of this cylinder is *defined* to be 1 kilogram (kg), or 1,000 grams (g). Accurately made copies of this cylinder are used in various standards laboratories throughout the world to calibrate precision equipment. Further copies are made from these for distribution.

The mass of 1 cm^3 of water is 1 gram (g) (approximately). The mass of 1 liter (L) of water is just about 1,000 g, or 1 kg.

The same international agreements that have established the kilogram as the unit of mass also established units of length and time. The meter (m) was originally defined in terms of the circumference of the earth, but modern measurement techniques make it more precise to define the meter in terms of the wavelength of light, generated in a specific way. The second of time (often abbreviated sec, the official symbol is s) was also originally defined with respect to the earth (as a certain fraction of the year), but it, too, is now more precisely defined in terms of light waves emitted by a specific group of atoms. The meter and second together determine the units of speed (m/sec) and acceleration (m/sec^2).

SG 16

1 m/sec^2 is an acceleration of one meter per second per second. For comparison, note that the acceleration in free fall on earth is about 10 m/sec^2. The sec^2 means that division by time units occurs twice.

With these units, you can now go back and calibrate the spring balances used for measuring force. The unit of force, 1 newton (N), is *defined* to be the force required to give an acceleration of 1 m/sec^2 to a mass of 1 kg. Because of Newton's second law ($\vec{F}_{net} = m\vec{a}$),

SG 17, 18

$$1 \text{ N} = 1 \text{ kg} \times 1 \text{ m/sec}^2 = 1 \text{ kg m/sec}^2$$

In this equation we use only the magnitudes; the direction is not part of the definition of the unit of force.

Imagine now, step by step, how you would calibrate a spring balance. To begin, take a 1-kg standard object. Put it on a frictionless horizontal surface. Attach the spring balance, and pull horizontally. The net force on the object, supplied by the spring balance, is the net force. Pull steadily to get an acceleration of 1 m/sec^2. Mark the place to which the pointer of the stretched balance pointed as "1 N." Repeat the procedure with an acceleration of 2 m/sec^2, to get 2 N, etc. Of course, once this is done, other spring balances can be more easily calibrated, for example, by hooking a pair together, one of which is calibrated, and pulling. (Complete this "thought experiment.")

SI stands for *Système Internationale.*

The kilogram, together with the meter and second, are the fundamental units of the "mks" system of measurement; together with units for light and electricity, the mks units form the International System (SI) units. Other systems of units are possible. But since the ratios between related units are more convenient to use in a decimal system, all scientific and technical work, and most industrial work, is now done with SI units in most countries, including the United States.

SG 19–22

?

14. What combination of the three fundamental units is represented by the newton, the unit of force?

15. A net force of 10 N gives an object a constant acceleration of 4 m/sec^2. What is the mass of the object?

16. True or false? Newton's second law holds only when frictional forces are absent.

17. A 2-kg object being pulled across the floor with a speed of 10 m/sec is suddenly released and slides to rest in 5 sec. What is the magnitude of the frictional force producing this deceleration?

18. Newton's second law, $\vec{a} = \vec{F}/m$, *claims that the acceleration of an object depends on three things. What are they?*

19. Complete the table below which lists some accelerations resulting from applying equal forces to objects of different mass.

MASS	*ACCELERATION*
1 kg	30 m/sec^2
2 kg	15 m/sec^2
3 kg	
1/5 kg	
0.5 kg	
45 kg	
	3 m/sec^2
	75 m/sec^2

3.8 | Mass, weight, and free fall

You will now examine some more details concerning the very important topic of mass and weight and how they are related. The idea of force in physics includes much more than muscular pushes and pulls. Whenever you observe an acceleration, you know that there is a force acting. Forces need not be "mechanical" (exerted by contact only). They can also result from gravitational, electric, magnetic, or other actions. Newton's laws hold true for all forces.

The force of gravity acts between objects even without direct contact. Such objects may be separated by only a few meters of air, as is the case with the earth and a falling stone. Or they may be separated by many kilometers of empty space, as are artificial satellites and the earth.

The symbol $\vec{F}_g$ is used for gravitational force. The magnitude of the gravitational pull $\vec{F}_g$ is roughly the same anywhere on the surface of the earth for a given object. When we wish to be very precise, we must take into account the facts that the earth is not exactly spherical and that there are irregularities in the makeup of the earth's crust. These factors cause slight differences (up to 0.5%) in the gravitational force on the same object at different places on the earth. An object having a mass of 1 kg will experience a gravitational force of 9.812 N in London, but only 9.796 N in Denver, Colorado. Geologists make use of these variations in locating oil and other mineral deposits.

The term *weight* is sometimes used in everyday conversation as if it meant the same thing as mass. This is quite wrong, of course. In physics, the weight of an object is defined as the magnitude of the *gravitational force acting on the body*. Your weight is the downward force the planet exerts on you whether you stand or sit, fly or fall, orbit the earth in a space vehicle, or merely stand on a scale to "weigh" yourself. Only in interstellar space, far from planets, would you truly have no weight.

Think for a moment about what a scale does. The spring in it compresses until it exerts on you an upward force strong enough to hold you up. So what the scale registers is really the force with which it pushes up on your feet. When you and the scale stand still and are not accelerating, the scale must be pushing up on your feet with a force equal in magnitude to your weight. That is why you are in equilibrium. The sum of the forces on you is zero.

Now imagine for a moment a ridiculous but instructive thought experiment. As you stand on the scale, the floor (which, while sagging slightly, has been pushing up on the scale) suddenly gives way. You and the scale drop into a deep well in free fall. At every instant, your fall speed and the scale's fall speed will be equal, since you fall with the same acceleration.

Alice falling down the rabbit hole, by Willy Pogany (1929).

In a few physics books, weight is defined as the force needed to support an object. In that case, you would be "weightless" if you were falling freely. We avoid this usage.

Your feet now touch the scale only barely (if at all). You look at the dial and see that the scale registers zero. This does not mean you have lost your weight; that could only happen if the earth suddenly disappeared or if you were suddently removed to deep space. No, $\vec{F}_g$ still acts on you as before, accelerating you downward. But since the scale is accelerating *with* you, you are no longer pushing down on it, nor is it pushing up on you.

You can experience firsthand the difference between the properties of weight and mass by holding a book. First, lay the book on your hand; you feel the weight of the book acting down. Next, grasp the book and shake it back and forth. You still feel the weight downward, but you also feel how hard it is to accelerate the book back and forth. This resistance to acceleration is the book's inertia. You could "cancel" the sensation of the book's *weight* by hanging the book on a string, but the sensation of its *inertia* as you shake it would remain the same. This is only a crude demonstration. More elaborate experiments would show, however, that weight can change without any change of mass. Thus, when an astronaut on the moon's surface uses a large camera, the camera is much easier to hold than on earth. In terms of the moon's gravity, the camera's *weight* is only ⅙ of its weight on earth. But its *mass* or inertia is not less, so it is as hard to swing the camera around suddenly into a new position on the moon as it is on earth.

Consider SG 16 again.

You can now understand more clearly the results of Galileo's experiment on falling objects. Galileo showed that any given object (at a given locality) falls with uniform acceleration, $\vec{a}_g$. What is responsible for this uniform acceleration? Since the object is in free fall, the only force acting on it is $\vec{F}_g$, due to the earth's gravity. Newton's second law allows us to relate this force to the acceleration $\vec{a}_g$ of the object. Applying the equation $\vec{F}_{net} = m\vec{a}$ to this case, where $\vec{F}_{net} = \vec{F}_g$ and $\vec{a}$ and $\vec{a}_g$, we can write

Some books use the symbol g instead of a_g.

$$\vec{F}_g = m\vec{a}_g$$

We can, of course, rewrite this equation as

$$\vec{a}_g = \frac{\vec{F}_g}{m}$$

From Newton's second law, you can now see why the acceleration of a body in free fall is constant. The reason is that, for an object of given mass m, the gravitational force $\vec{F}_g$ over normal distances of fall is nearly constant.

Galileo, however, did more than claim that every object falls with *constant* acceleration. He found that at any one place *all* objects fall with the *same* uniform acceleration. Scientists now know that at the earth's surface this acceleration has the value of 9.8 m/sec^2. Regardless of the mass m or weight $\vec{F}_g$, all bodies in free fall (in the same locality) have the same acceleration $\vec{a}_g$.

Does this agree with the relation $\vec{a}_g = \vec{F}_g/m$? It does so *only* if $\vec{F}_g$ is directly proportional to mass m for every object. In other words, if m is doubled, $\vec{F}_g$ must double; if m is tripled, $\vec{F}_g$ must triple. This is an important result indeed. Weight and mass are entirely different concepts. *Weight* is the gravitational *force* on an object (thus, weight is a vector). *Mass* is a measure of the resistance of an object to a change in its motion, a measure of *inertia* (thus, mass is a scalar). Yet, you have seen that different objects fall freely with the same acceleration in any given locality. Thus, in the same locality, the magnitudes of these two quite different quantities are proportional.

?

20. *What quality of an object is measured by weight; what quality is measured by mass? Using both words and symbols, define weight and mass.*

21. *The force pulling down on a hammer is more than 20 times the force pulling down on a nail. Why, then, do a hammer and a nail fall with nearly equal acceleration?*

22. *If a 30-N force is applied to an object whose mass is 3 kg, what is the resulting acceleration? If the same force is applied to the object on the moon, where the object's weight is one sixth of its weight on earth, what is the acceleration? What is the acceleration in deep space where the object's weight is zero?*

23. *An astronaut has left the earth and is orbiting the earth with a space vehicle. The acceleration due to gravity at that distance is half its value on the surface of the earth. Which of the following are true?*

(a) The astronaut's weight is zero.

(b) The astronaut's mass is zero.

(c) The astronaut's weight is half its original value.

(d) The astronaut's mass is half its original value.

(e) The astronaut's weight remains the same.

(f) The astronaut's mass remains the same.

SG 24–26

Wilma Rudolph at the start of the 200-m sprint in which she set an Olympic record of 23.2 sec.

3.9 | Newton's third law of motion

In his first law, Newton described the behavior of objects when they are in a state of equilibrium, that is, when the net force acting on them is zero. His second law explained how their motion changes when the net force is not zero. Newton's third law added a new and surprising insight into forces.

Consider this problem: In a 100-m dash, an athlete goes from rest to nearly top speed in less than a second. We could measure the runner's mass before the dash, and we could use high-speed photography to measure the initial acceleration. With mass and acceleration known, we could use $\vec{F}_{net} = m\vec{a}$ to find the force acting on the runner during the initial acceleration. But where does the force come from? It must have something to do with the runner herself. Is it possible for her to exert a force on herself as a whole? Can she, for example, ever lift herself by her own bootstraps?

The runner is, to be sure, pushing against the ground, but that is a force acting *on the ground.*

Newton's third law of motion helps explain just such puzzling situations. First, what does the third law claim? In Newton's words:

> To every action there is always opposed an equal reaction: or, mutual actions of two bodies upon each other are always equal and directed to contrary parts.

This is a word-for-word translation from *the Principia*. In modern usage, however, we would use *force* where Newton used the Latin word for action. So we could rewrite this passage as follows: If one object exerts a force on another, then the second also exerts a force on the first; these forces are equal in magnitude and opposite in direction.

SG 27

The most startling idea in this statement is that forces always exist in mirror-twin pairs and act on two different objects. Indeed, the idea of a single force acting without another force acting somewhere else is without any meaning whatsoever.

Now apply this idea to the athlete. You now see that her act of pushing with her feet back against the ground (call it the *action*) also involves a push of the ground forward on her (call it the *reaction*). It is this reaction that propels her forward. In this and all other cases, it really makes no difference which force you call the action and which the reaction, because they occur at exactly the same time. The action does not "cause" the reaction. If the earth could not "push back" on her feet, the athlete could not push on the earth in the first place. Instead, she would slide around as on slippery ice. Action and reaction coexist. You cannot have one without the other. Most important, the two forces are not acting on the same body. In a way, they are like debt and credit. One is impossible without the other; they are equally large but of opposite sign, and they happen to two different objects.

In the collision between the tennis ball and the racket, the force the ball exerts on the racket is equal and opposite to the force the racket exerts on the ball. Both the racket and the ball are deformed by the forces acting on them (see diagram).

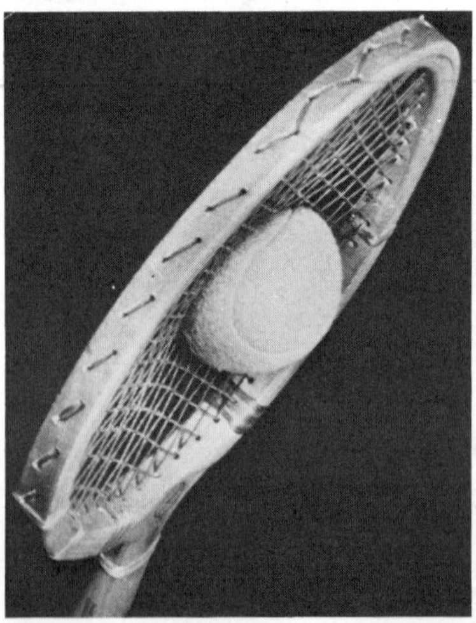

On this point Newton wrote: "Whatever draws or presses another is as much drawn or pressed by that other. If you press a stone with your finger, the finger is also pressed by the stone." This statement suggests that forces always arise as a result of mutual actions ("interactions") between objects. If object A pushes or pulls on B, then at the same time object B pushes or pulls with precisely equal force on A. These paired pulls and pushes are always equal in magnitude, opposite in direction, *and*

on two different objects. Using the efficient shorthand of algebra to express this idea, whenever bodies A and B interact,

$$\vec{F}_{AB} = -\vec{F}_{BA}$$

This equation clearly sums up Newton's third law. A modern way to express it is as follows: *Whenever two bodies interact, the forces they exert on each other are equal in magnitude and opposite in direction.*

Every day you see hundreds of examples of this law at work. A boat is propelled by the water that pushes forward on the oar while the oar pushes back on the water. A car is set in motion by the push of the ground on the tires as they push back on the ground; when friction is not sufficient, the push on the tires cannot start the car forward. While accelerating a bullet forward, a rifle experiences a recoil, or "kick." A balloon shoots forward while the air spurts out from it in the opposite direction. Many such effects are not easily observed. For example, when an apple falls, pulled down by its attraction to the earth, i.e., by its weight, the earth, in turn, accelerates upward slightly, pulled up by the attraction of the earth to the apple.

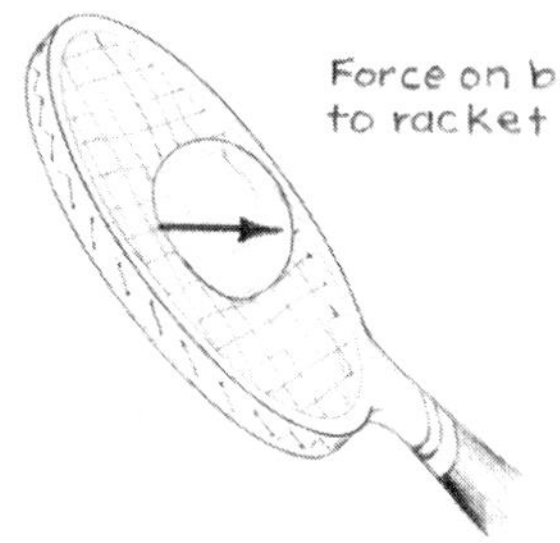

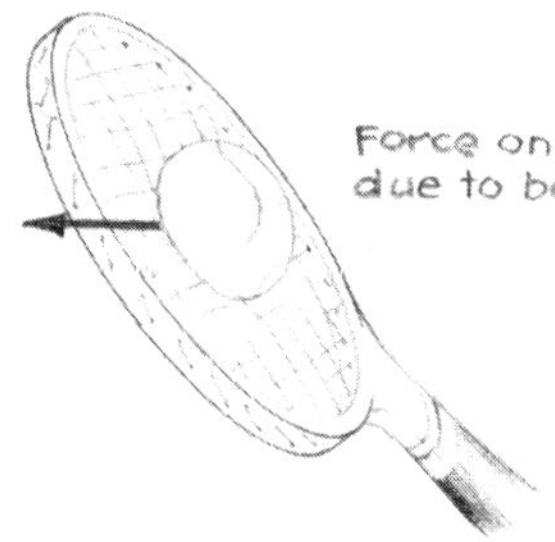

Note what the third law does *not* say. The third law speaks of *forces*, not of the effects these forces produce. Thus, in the last example, the earth accelerates upward as the apple falls down. The force on each is equally large. But the accelerations produced by the forces are quite different. The mass of the earth is enormous, and so the earth's upward acceleration is far too small to notice. The third law also does not describe how the push or pull is applied, whether by contact or by magnetic action or by electrical action. Nor does the law require that the force be either an attraction or repulsion. The third law in fact does not depend on any particular kind of force. It applies equally to resting objects and to moving objects, to accelerating objects as well as to objects in uniform motion. It applies whether or not there is friction present. This universal nature of the third law makes it extremely valuable in physics.

The force on the moon owing to the earth is equal and opposite to the force on the earth owing to the moon.

24. *Two objects are next to one another; one object exerts a force of 3 N to the right on the other object. Describe the three qualities of the second force that is immediately present according to Newton's third law.*

25. *Identify the forces that act according to Newton's third law when a horse accelerates. When a swimmer moves at constant speed.*

26. A piece of fishing line breaks if the force exerted on it is greater than 500 N. Will the line break if two people at opposite ends of the line pull on it, each with a force of 300 N?

27. State Newton's three laws of motion as clearly as you can in your own words.

3.10 | Using Newton's laws of motion

Each of Newton's three laws of motion has been discussed in some detail. The first law emphasizes the modern point of view in the study of motion. It states that what requires explanation is not motion itself, but *change* of motion. The first law stresses that one must account for why an object speeds up or slows down or changes direction. The second law asserts that the rate of change of velocity of an object is related to both the mass of the object and the net force applied to it. In fact, the very meanings of force and mass are shown by the second law to be closely related to each other. The third law describes a relationship between interacting objects.

SG 29–31

Despite their individual importance, Newton's three laws are most powerful when they are used together. The mechanics based on Newton's laws was very successful. Indeed, until the late nineteenth century it seemed that the entire universe must be understood as "matter in motion." Below are two specific examples that illustrate the use of these laws.

Example 1

On September 12, 1966, a dramatic experiment based on Newton's second law was carried out high over the earth. From a previous space flight, a spent piece of an Agena rocket was quietly floating in its orbit around the Earth. In this experiment, the mass of the Agena piece was determined by accelerating it with a push from a Gemini spacecraft. After the Gemini spacecraft made contact with the Agena rocket case, the thrusters on the Gemini were fired for 7.0 sec. These thrusters were set to give an average thrusting force of 890 N. The change in velocity of the spacecraft and Agena was found to be 0.93 m/sec. The mass of the Gemini spacecraft was known to be about 3,400 kg. The question to be answered was: What is the mass of the Agena?

(Actually, the mass of the Agena was known ahead of time, from its construction. But the purpose of the experiment was to develop a method for finding the unknown mass of a foreign satellite in orbit.)

In this case, a known force of magnitude 890 N was acting on two objects in contact, with a total mass of m_{total}, where

$$m_{\text{total}} = m_{\text{Gemini}} + m_{\text{Agena}}$$

$$= 3400 \text{ kg} + m_{\text{Agena}}$$

The magnitude of the average acceleration produced by the thrust is found as follows:

$$a = \frac{\Delta v}{\Delta t}$$

$$= \frac{0.93 \text{ m/sec}}{7.0 \text{ sec}}$$

$$= 0.13 \text{ m/sec}$$

Newton's second law gives the relation

$$F = m_{\text{total}} \times a$$

or

$$F = (m_{\text{Agena}} + 3{,}400 \text{ kg}) \times a$$

Solving for m_{Agena} gives

$$m_{\text{Agena}} = \frac{F}{a} - 3{,}400 \text{ kg} = \frac{890 \text{ N}}{0.13 \text{ m/sec}} - 3{,}400 \text{ kg}$$

$$= 6{,}900 \text{ kg} - 3{,}400 \text{ kg}$$

$$= 3{,}500 \text{ kg}$$

The actual mass of the orbiting portion of the Agena, as previously determined, was about 3,660 kg. The method for finding the mass by pushing the Agena while in orbit therefore gave a result that was accurate to within 5%. This accuracy was well within the margin of error expected in making this measurement.

Example 2

SG 16

Imagine taking a ride on an elevator: (A) At first it is at rest on the ground floor; (B) it accelerates upward uniformly at 2 m/sec^2 for a few seconds; then (C) it continues to go up at a constant speed of 5 m/sec. Suppose a 100-kg man (whose weight would therefore be about 1,000 N) is standing in the elevator. With what force is the elevator floor pushing up on him during (A), (B), and (C)?

Parts (A) and (C) are the same in terms of dynamics. Since the man is not accelerating, the net force on him must be zero. So the floor must be pushing up on him just as hard as gravity is pulling him down. The gravitational force on him (his weight) is 1,000 N. So the floor must be exerting an upward force of 1,000 N.

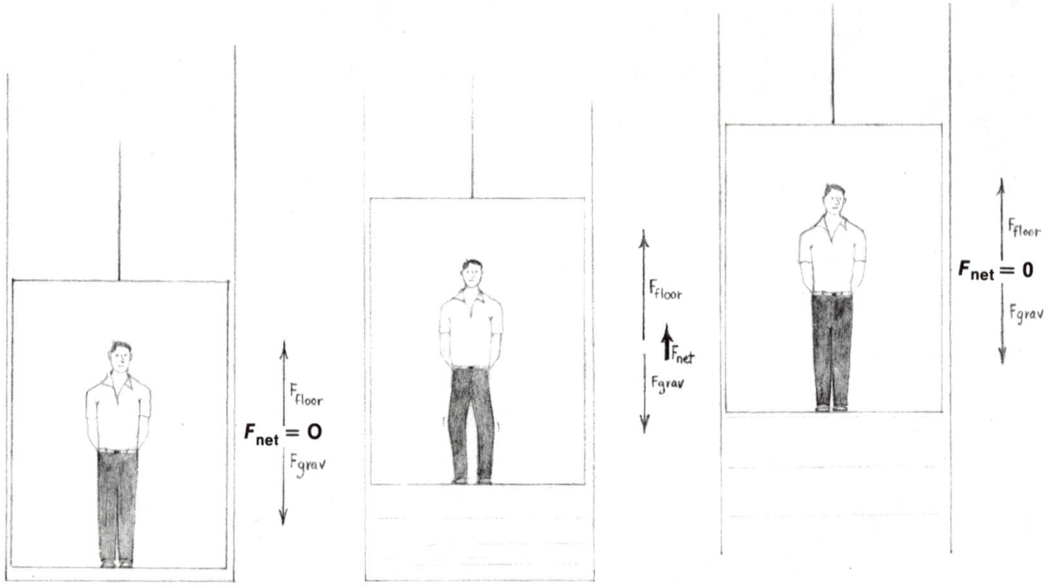

(A) (B) (C)

(A) At rest (B) Accelerating upward (C) Rising at constant speed

In Part (B), since the man is accelerating upward, there must be a net force upward on him. The unbalanced force is

$$F_{net} = ma_{up}$$
$$= 100 \text{ kg} \times 2 \text{ m/sec}^2$$
$$= 200 \text{ N}$$

SG 32 is an elaboration of a similar example. For a difficult, worked-out example, see SG 33.

Clearly, the floor must be pushing up on him with a force 200 N greater than what is required just to balance his weight. Therefore, the total force upward on him is 1,200 N.

?

28. What force must the runner's legs produce in order to accelerate a 70-kg runner at 3 m/sec² (ignore air resistance or other effects)?

29. An object with a mass of 3 kg is pulled down with a force of 29.4 N. What is the acceleration? What is the weight of the object?

3.11 | Nature's basic forces

The study of Newton's laws of motion has increased your understanding of objects at rest, moving uniformly, and accelerating. However, you have learned much more in the process. Newton's first law emphasized the importance of frames of reference. In fact, an understanding of the relationship between decriptions of the same event seen from different frames of reference was the necessary first step toward the theory of relativity.

Newton's second law shows the fundamental importance of the concept of force. It says, in effect, "When you observe acceleration, find the force!" This is how scientists were first made aware of gravitational force as an explanation of Galileo's kinematics. They discovered that, at a given place, $\vec{a}_g$ is constant for all objects. And since $\vec{a} = \vec{F}_g/m$ by Newton's second law, the magnitude of $\vec{F}_g$ is always proportional to m.

But this is only a partial solution. Why is $\vec{F}_g$ proportional to m for all bodies at a given place? How does $\vec{F}_g$ change for a given body as it is moved to places more distant from the earth? Is there a "force law" connecting $\vec{F}_g$, m, and distance? As Unit 2 will show, there is indeed. Knowing that force law, you can understand all gravitational interactions among objects.

Gravitational attraction is not the only basic force by which objects interact. However, there appear to be very few such basic forces. In fact, physicists now believe that everything observed in nature results from only four basic types of interactions. In terms of present understanding, *all* events of nature, from those among subnuclear particles to those among vast galaxies, involve one or more of only these few types of forces. There is, of course, nothing sacred about the number four. New discoveries or insights into present theories might increase or reduce the number. For example, two (or more) of the basic forces might someday be seen as arising from an even more basic force.

The first of the four interactions is the gravitational force. This force becomes important only on a relatively large scale, when tremendous numbers of atoms of matter are involved. Between individual atoms, gravitational force is extremely weak. But it is this very weak force that literally holds the universe together. The second interaction involves electric and magnetic processes and is most important on the atomic and molecular scale. It is chiefly the electromagnetic force that holds together objects in the size range between an atom and a mountain.

Scientists know the force laws governing gravitational and electromagnetic interactions. Therefore, these interactions are fairly well "understood." Considerably less is known about the two remaining basic interactions. They are the subject of much research today. The third interaction (the so-called "strong" interaction) somehow holds the particles of the nucleus together. The fourth interaction (the so-called "weak" interaction) governs certain reactions among subnuclear particles.

There are, of course, other *names* for forces, but each of these forces belongs to one of the basic types. One of the most common is the "frictional" force. Friction is thought to be an electrical interaction; that is, the atoms on the surfaces of the

Einstein spent most of the latter half of his life seeking a theory that would express gravitational and electromagnetic effects in a unified way. A satisfactory "unified field theory" is still being sought.

Recently, however, some successes have come in an entirely different direction. A theory that considers electromagnetism and weak interactions to be aspects of the same fundamental force (in the way that electricity and magnetism are different aspects of the same force) has led to some interesting experimental predictions that have been verified. The details must still be worked out.

objects sliding or rubbing against each other are believed to interact electrically.

You will encounter these ideas again. The gravitational force is covered in Unit 2, the electrical and magnetic forces in Units 4 and 5, and the forces between nuclear particles in Unit 6. In all of these cases, all objects subjected to a force behave in agreement with Newton's laws of motion, no matter what kind of force is involved.

The knowledge that there are so few basic interactions is both surprising and encouraging. It is surprising because, at first glance, the events all around us seem so varied and complex. It is encouraging because it makes the elusive goal of understanding the events of nature look more attainable.

"The Starry Night," by Vincent Van Gogh. The intuitive feeling that all of nature's phenomena are interrelated on a grand scale is shared by scientists as well as artists.

study guide

1. The *Project Physics* learning materials particularly appropriate for Chapter 3 include the following:

Experiments
Newton's Second Law
Mass and Weight

Activities
Checker Snapping
Beaker and hammer
Pulls and Jerks
Experiencing Newton's Second Law
Make One of These Accelerometers

Film Loops
Vector Addition: Velocity of a Boat

2. The Aristotelian explanation of motion should not be dismissed lightly. Great intellects of the Renaissance, such as Leonardo da Vinci, who among other things designed devices for launching projectiles, did not challenge such explanations. One reason for the longevity of these ideas is that they are so closely aligned with our common-sense ideas.

In what ways do your common-sense notions of motion agree with the Aristotelian ones?

3. (a) Explain mechanics, dynamics, and kinematics.

(b) Classify the following values as m, t, d, $\vec{v}$, $\vec{a}$, or $\vec{f}$. Indicate whether they are scalars or vectors.

(1) 5 m

(2) 2 kg · m/sec^2 up

(3) 10 m/sec

(4) −8 N

(5) 5 kg

(6) 5 m west

(7) +10 sec

(8) 6 m/sec^2 left

(9) 4 m/sec down

4. A man walked 3 blocks north, 4 blocks east, 5 blocks south, 1 block west, and 2 blocks north.

(a) Where did he end up?

(b) How far did he walk?

(c) Which part, (a) or (b), is a vector problem? Which is a scalar problem?

5. Three ants are struggling with a breadcrumb. One ant pulls toward the east with a force of 8 units. Another pulls toward the north with a force of 6 units, and the third pulls in a direction 30° south of west with a force of 12 units.

(a) Using the "head-to-tail" construction of arrows, find whether the forces balance, or whether there is a net (unbalanced) force on the crumb.

(b) If there is a net force, you can find its direction and magnitude by measuring the line drawn from the tail of the first arrow to the head of the last arrow. What is its magnitude and direction?

6. Show why the parallelogram method of adding arrows is geometrically equivalent to the head-to-tail method.

7. A parachutist whose weight is 750 N falls with uniform motion. What is the size and direction of the force of air resistance? How do you know?

8. There are many familiar situations in which the net force on a body is zero, and yet the body moves with a constant velocity. One example of such "dynamic equilibrium" is an automobile traveling at constant speed on a straight road. The force the road exerts on the tires is just balanced by the force of air friction. If the gas pedal is depressed further, the tires will push against the road harder, and the road will push against the tires harder. The car will accelerate forward until the air friction builds up enough to balance the greater drive force.

Give another example of a body moving with constant velocity under balanced forces. Specify the source of each force on the body and, as in the automobile example, explain how these forces could be changed to affect the body's motion.

sg

9. Aristotle thought that objects in any kind of motion and objects at rest represented two different dynamical situations and had to be explained separately. Newton claimed that objects in "equilibrium," moving or not, represented one dynamical situation.

(a) What is equilibrium?

(b) What are two possible states of motion for objects in equilibrium?

10. (a) You exert a force on a box, but it does not move. How would you explain this? How might an Aristotelian explain it?

(b) Suppose now that you exert a greater force and the box moves. Explain this from your (Newtonian) point of view and from an Aristotelian point of view.

(c) You stop pushing on the box, and it quickly comes to rest. Explain this from both the Newtonian and the Aristotelian points of view.

11. (a) Assume that the floor of a bus could be made perfectly horizontal and perfectly smooth. A dry-ice puck is placed on the floor and given a small push. Predict the way in which the puck would move. How would this motion differ if the whole bus were moving uniformly during the experiment? How would it differ if the whole bus were accelerating along a straight line? If the puck were seen to move in a curved path along the floor, how would you explain this? From these results, can you argue that uniform motion and rest really represent the same dynamical situation and are different from accelerated motion?

(b) A man gently starts a dry-ice puck in motion while both are on a rotating platform. What will he report to be the motion he observes as the platform keeps rotating? How will he explain what he sees if he believes he can use Newton's first law to understand observations made in a rotating reference frame? Will he be right or wrong?

12. (a) In terms of Newton's first law, explain:

(1) why people in a moving car lurch forward when the car suddenly slows down.

(2) what happens to the passengers of a car that makes a sharp, quick turn.

(b) When a coin is put on a phonograph turntable and the motor is started, does the coin fly off when the turntable reaches a certain speed? Why doesn't it fly off sooner?

13. In an actual experiment on applying the same force to different masses, how would you know it was the "same force"?

14. (a) What three things does Newton's second law tell you about acceleration?

(b) The acceleration of an object is 10 m/sec^2 north. In a second experiment, the force is divided in half and turned to the east, and the mass is reduced to one third. What is the object's acceleration now?

15. Several proportionalities can be combined into an equation only if care is taken with the units in which the factors are expressed. When we wrote $\Delta d = v \times \Delta t$ in Chapter 1, we chose *meters* as units for *d*, *seconds* as units for *t*, and then made sure that the equation came out right by using meters per second as units for *v*. In other words, we let the equation *define* the unit for *v*. If we had already chosen some *other* units for *v*, say miles per hour, then we would have had to write instead something like

$$\Delta d = k \times v\Delta t$$

where *k* is a constant factor that matches up the units of *d*, *t*, and *v*. What value would *k* have if *d* were measured in *miles*, *t* in *seconds*, and *v* in *miles per hour*?

Writing $\vec{a} = \vec{F}_{net}/m$ before we have defined units of *F* and *m* is not the very best mathematical procedure. To be perfectly correct in expressing Newton's law, we would have had to write

$$\vec{a} = k \times \frac{\vec{F}_{net}}{m}$$

where *k* is a universally constant factor that would match up whatever units we choose for *a*, *F*, and *m*. The SI units have been chosen so that $k = 1$.

16. You can confirm the results of Example 2 in Sec. 3.10 by taking a bathroom scale into an elevator. By how much does your registered "weight" seem to be increased when the elevator starts to go up (accelerates upward)? What happens while it slows to a stop? What happens when it goes up or down at constant speed? Does your weight really change? If not, why does the scale show what it does?

How would all these measurements differ if this elevator were in a space vehicle in interstellar space?

17. Ask your instructor for a simple spring balance, and examine how it works. Then describe as a thought experiment how you could calibrate the spring balance in force units. What practical difficulties would you expect if you actually tried to do the experiment?

18. Hooke's law says that the force exerted by a stretched or compressed spring is directly proportional to the amount of the compression or extension. As Robert Hooke put it in announcing his discovery:

> . . . the power of any spring is in the same proportion with the tension thereof: that is, if one power stretch or bend it one space, two will bend it two, three will bend it three, and so forward. Now as the theory is very short, so the way of trying it is very easie.

You can probably think immediately of how to test this law using springs and weights.

(**a**) Try designing such an experiment; then after checking with your instructor, carry it out. What limitations do you find to Hooke's law?

(**b**) How could you use Hooke's law to simplify the calibration procedure asked for in SG 17?

19. Complete this table:

	NET FORCE (N)	MASS (kg)	RESULTING ACCELERATION (m/sec^2)
a	1.0	1.0	1.0 down
b	24.0 west	2.0	12.0 west
c		3.0	8.0 out
d		74.0	0.2 left
e		0.0066	130.0 north
f	72.0 in		8.0 in
g	3.6 right		12.0
h	1.3 up		6.4 up
i	30.0 east	10.0	
j	0.5 left	0.20	
k	120.0 down	48.0	

20. A rocket-sled has a mass of 4,440 kg and is propelled by a solid-propellent rocket motor of 890,000-N thrust which burns for 3.9 sec.

(**a**) What is the sled's average acceleration and maximum speed?

(**b**) This sled has a maximum acceleration of 30 g (= 30 a_g). How can that be, considering the data given?

(**c**) If the sled travels a distance of 1,530 m while attaining a top speed of 860 m/sec, what is its average acceleration? How did it attain *that* high a speed?

21. Describe in detail the steps you would take in an idealized experiment to determine the unknown mass m of a certain object (in kilograms) if you were given nothing but a frictionless horizontal plane, a 1-kg standard, an uncalibrated spring balance, a meter stick, and a stopwatch.

22. A block is dragged with *constant velocity* along a *rough*, horizontal tabletop by means of a spring balance horizontally attached to the block. The balance shows a reading of 0.40 N at this and any other constant velocity. This means that the retarding frictional force between block and table is 0.40 N. Compute the mass of the block.

Now the block is pulled harder and given a constant acceleration of 0.85 m/sec^2; the balance is found to read 2.1 N. Compute the mass of the block.

23. We have claimed that any body in free fall is "weightless" because any weight-measuring device falling with it would read zero. This is not an entirely satisfactory explanation, because you feel a definite sensation during free fall that is exactly the same sensation you would feel if you were truly without weight, for example, deep in space far from any star or planet. (You feel the same sensation on jumping off a roof or a diving board, or when someone pulls a chair out from under you.) Can you explain why your insides react in the same way to lack of weight and to free fall?

24. **(a)** A replica of the standard kilogram is constructed in Paris and then sent to the National Bureau of Standards near Washington, D.C. Assuming that this secondary standard is not damaged in transit, what is

(**1**) its mass in Washington?

(**2**) its weight in Paris and in Washington? (In Paris, $a_g = 9.81\ m/sec^2$; in Washington, $a_g = 9.80\ m/sec^2$.)

(**b**) What is the change in your own weight as you go from Paris to Washington?

25. You have probably seen signs that misleadingly convert people's weights from pounds to kilograms.

(**a**) Since the pound is a unit of force, why are these signs wrong? Explain why they are only valid on earth (and then only approximately).

(**b**) Calculate your mass in kilograms and your weight in newtons.

(**c**) How much force is needed to accelerate you $1 m/sec^2$? How many kilograms can you lift? How many newtons of force must you apply to do this?

26. Why is it often said that astronauts in orbit around a planet or satellite are "weightless"?

27. Quite apart from pushing down on the ground owing to a runner's own weight, the sole of a runner's shoe pushes on the earth in a horizontal direction and the earth pushes with an equal and opposite force on the sole of the shoe. This latter force has an accelerating effect on the runner, but what does the force acting on the earth do to the earth? From Newton's second law, we would conclude that such an unbalanced force would accelerate the earth. The mass of the earth is very great, however, so the acceleration caused by the runner is very small. A reasonable value for the average acceleration of a runner is $5 m/sec^2$, and a reasonable value for the runner's mass would be 60 kg. The mass of the earth is approximately 60×10^{23} kg.

(**a**) What acceleration of the earth would the runner cause?

(**b**) If the acceleration lasts for 2 sec, what speed will the runner have reached?

(**c**) What speed will the earth have reached?

28. A boy of mass 70 kg and a girl of mass 40 kg are on ice skates holding opposite ends of a 10-m rope. The boy pulls on the rope toward himself with a force of 80 N. Assuming that there is virtually no friction between the skates and the ice surface, what is the girl's acceleration? According to Newton's third law, what is the force on the boy? What is his acceleration?

29. In terms of Newton's third law, assess the following statements:

(**a**) You are standing perfectly still on the ground; therefore you and the earth exert equal and opposite forces on each other.

(**b**) The reason that a propeller airplane cannot fly above the atmosphere is that there is no air to push one way while the plane goes the other.

(**c**) Object A rests on object B. The mass of object A is 100 times as great as that of object B, but even so, the force A exerts on B is no greater than the force of B on A.

30. Consider a tractor pulling a heavy log in a straight line. On the basis of Newton's third law, one might argue that the log pulls back on the tractor just as strongly as the tractor pulls the log. But why, then, does the tractor move? (Make a large drawing of the tractor, rope, log, and the earth, and enter all the forces acting on each.)

31. Consider the system consisting of a 1.0-kg ball and the earth. The ball is dropped from a short distance above the ground and falls freely. Assuming that the mass of the earth is 6.0×10^{24} kg,

(**a**) make a vector diagram illustrating the important forces acting on each member of the system.

(**b**) calculate the acceleration of the earth in this interaction.

(c) find the ratio of the magnitude of the ball's acceleration to that of the earth's acceleration (a_b/a_e).

(d) make a vector diagram as in (a) but showing the situation when the ball has come to rest after hitting the ground.

32. (a) A 75-kg person stands in an elevator. What force does the floor exert on the person when the elevator

(1) starts moving upward with an acceleration of 1.5 m/sec^2?

(2) moves upward with a constant speed of 2.0 m/sec?

(3) starts accelerating downward at 1.5 m/sec^2?

(b) If the person were standing on a bathroom (spring) scale during the ride, what readings would the scale have under conditions (1), (2), and (3) above?

(c) It is sometimes said that the "apparent weight" changes when the elevator accelerates. What could this mean? Does the weight really change?

33. Useful hints for solving problems about the motion of an object and the forces acting on it:

(a) Make a light sketch of the physical situation.

(b) In heavy line, indicate the limits of the particular object you are interested in, and draw all the forces acting on that object. (For each force acting on it, it will be exerting an opposite force on something else, but you can ignore these forces.)

(c) Find the vector sum of all these forces, for example, by graphical construction.

(d) Using Newton's second law, set this sum, $\vec{F}_{net}$, equal to $m\vec{a}$.

(e) Solve the equation for the unknown quantity.

(f) Put in the numerical values you know, and calculate the answer.

Example:

A ketchup bottle whose mass is 1.0 kg rests on a table. If the friction force between the table and the bottle is a constant 3 N, what horizontal pull is required to accelerate the bottle from rest to a speed of 6 m/sec in 2 sec?

First, sketch the situation:

Second, draw in arrows to represent all the forces

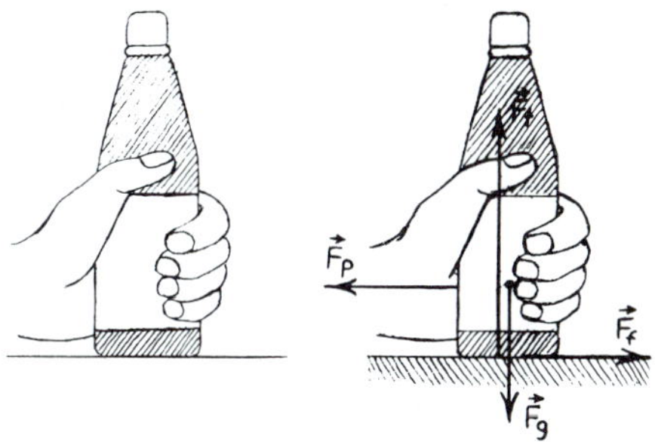

acting on the object of interest. There will be the horizontal pull $\vec{F}_p$, the friction $\vec{F}_f$, the gravitational pull $\vec{F}_g$ (the bottle's weight), and the upward force $\vec{F}_t$ exerted by the table. (There is, of course, also a force acting down on the table, but you are interested only in the forces acting *on the bottle*.)

Next, draw the arrows alone. In this sketch all the forces can be considered to be acting on the center of mass of the object.

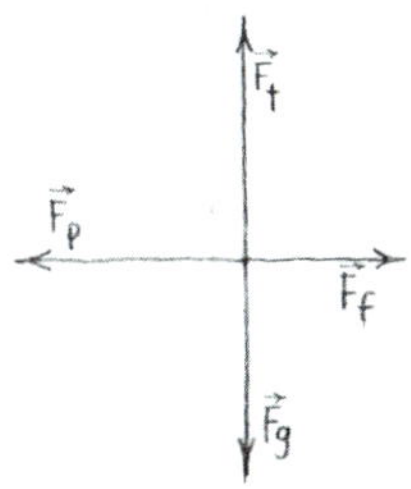

Because the bottle is not accelerating up or down, you know there is no net force up or down. Therefore, $\vec{F}_t$ must just balance $\vec{F}_g$. The net force acting on the bottle is just the vector sum of $\vec{F}_p$ and $\vec{F}_f$. Using the usual tip-to-tail addition:

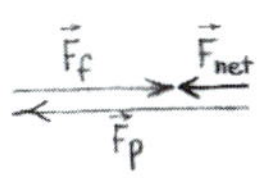

As the last arrow diagram shows, the horizontal pull must be greater than the force required for acceleration by an amount equal to the friction. You already know $\vec{F}_f$. You can find $\vec{F}_{net}$ from Newton's second law if you know the mass and acceleration of the bottle, since $\vec{F}_{net} = m\vec{a}$. The net force required to accelerate the bottle is found from Newton's second law:

$$\vec{F}_{\text{net}} = m\vec{a}$$

The mass m is given as 1.0 kg. The acceleration involved in going from rest to 6.0 m/sec in 2 sec is

$$a = \frac{\Delta v}{\Delta t} = \frac{6.0 \text{ m/sec}}{2 \text{ sec}} = 3.0 \text{ m/sec}^2$$

So the net force required is

$$F_{\text{net}} = 1.0 \text{ kg} \times 3.0 \text{ m/sec}^2$$

$$= 3.0 \text{ kg m/sec}^2$$

$$= 3.0 \text{ N}$$

If you consider toward the right to be the positive direction, F_{net} is 3.0 N, and F_f, which is directed to the left, is -3.0 N.

$$\vec{F}_{\text{net}} = \vec{F}_p + \vec{F}_f$$

$$3.0 \text{ N} = \vec{F}_p + (-3.0 \text{ N})$$

$$\vec{F}_p = 3.0 \text{ N} + 3.0 \text{ N}$$

$$\vec{F}_p = 6.0 \text{ N}$$

If you prefer not to use + and − signs, you can work directly from your final diagram and use only the magnitudes of the forces:

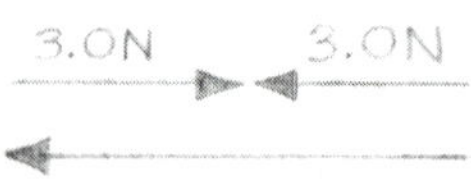

from which the magnitude of F_p is obviously 6.0 N.

34. (a) Two forces act on an object of 5 kg mass. One force is 20 N right and the other is 5 N left. How far will the object move in 10 sec?

(b) A box weighing 500 N can be moved across the floor in uniform motion by a force of 200 N. If the force is suddenly increased to 1,200 N, what will be the speed of the box after 20 sec?

(c) What is the net force on an object of 4 kg mass if its speed is changed from 40 m/sec to 80 m/sec in 10 sec?

(d) A 40-N force acts on an object. This force overcomes a 15-N frictional force and accelerates the object so that its speed increases by 55 m/sec in 11 sec. What is the mass of the object?

(e) A 5-kg object experiences an 80-N force on the earth's surface. What is the acceleration of the object? If the weight of the object is reduced by one half, what is its acceleration given the same 80-N force?

(f) Two wooden blocks resting on a table top have a coiled spring between them. When the spring is released, it exerts a 40-N force on each block. If one block is three times as massive as the other block (whose mass is 3 kg) how long will it take the blocks to move 200 m apart? (The frictional force is estimated at 4 N.)

(g) A 6-kg block is pulled along the floor whose frictional force is estimated at 3 N. Forces of 18 N right and 15 N left are exerted on the block simultaneously. What is the acceleration of the block?

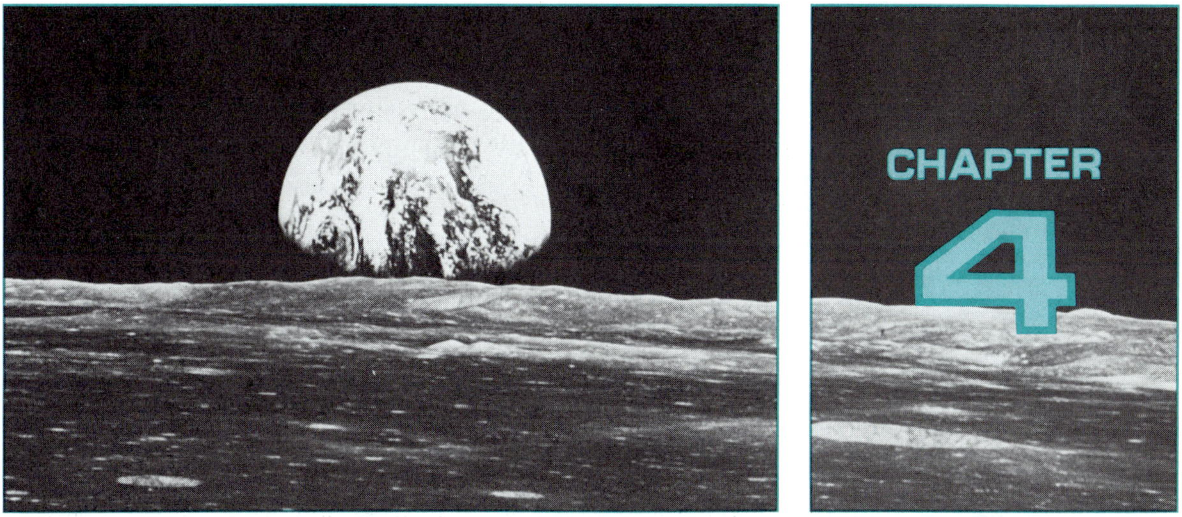

Understanding Motion

4.1 | A trip to the moon

Imagine a Saturn rocket taking off from its launching pad at the Kennedy Space Flight Center on Cape Canaveral. It climbs above the earth, passing through the atmosphere and beyond. Successive stages of the rocket shut off, finally leaving a capsule hurtling through the near-vacuum of space. Approximately 65 hr after takeoff, the capsule reaches its destination 384,000 km away. It circles the moon and descends to its target, the center of the lunar crater Copernicus.

The complexity of such a voyage is enormous. To direct and guide the flight, a great number and variety of factors must be taken into account. The atmospheric drag in the early part of the flight depends upon the rocket's speed and altitude. The engine thrust changes with time. The gravitational pulls of the sun, the earth, and the moon change as the capsule changes its position relative to them. The rocket's mass changes as it burns fuel. Moreover, it is launched from a spinning earth, which in turn is

In his science-fiction novels of more than a hundred years ago, the French author Jules Verne (1828–1905) launched three spacemen to the moon by means of a gigantic charge fixed in a steel pipe deep in the earth.

SG 1

circling the sun. Meanwhile, the moon is moving around the earth at a speed of about 1,000 m/sec relative to the earth.

Yet, like almost any complex motion, the flight can be broken down into small portions, each of which is relatively simple. What you have learned in earlier chapters will be useful in this task.

In simplified form, the earth–moon trip can be divided into eight parts or steps:

Step 1. The rocket accelerates vertically upward from the surface of the earth. The force acting on the rocket is not really constant, and the mass of the rocket decreases as the fuel burns. The value of the acceleration at any instant can be computed by using Newton's second law. The value is given by the ratio of net force (thrust minus weight) at that instant to the mass at that instant.

Step 2. The rocket, still accelerating, follows a curved path as it is "injected" into an orbit about the earth.

Step 3. In an orbit 185 km above the earth's surface, the capsule moves in a nearly circular arc. Its speed is constant at 7,769 m/sec.

Step 4. The rocket engines are fired again, increasing the capsule's speed so that it follows a much less curved path into space. (The minimum speed necessary to escape the earth completely is 11,027 m/sec.)

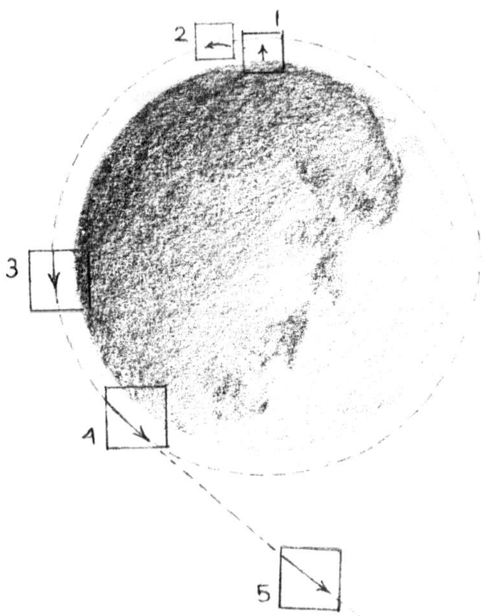

Step 5. In the flight between the earth and moon, occasional short bursts from the capsule's rockets keep it precisely on course. Between these correction thrusts, the capsule

moves under the influence of the gravitational forces of earth, moon, and sun. You know from Newton's first law that the capsule would move with constant velocity if it were not for these forces.

Step 6. On nearing the moon, the rocket engines are fired again to give the capsule the correct velocity to "inject" into a circular orbit around the moon.

Step 7. The capsule moves with a constant speed in a nearly circular path 80 km above the moon's surface.

Step 8. The rocket engines are fired into the direction of motion, to reduce the speed. The capsule then accelerates downward as it falls toward the surface of the moon. It follows an arcing path toward a landing at the chosen site. (Just before impact, the rocket engines fire a final time to reduce speed of fall and prevent a hard landing.)

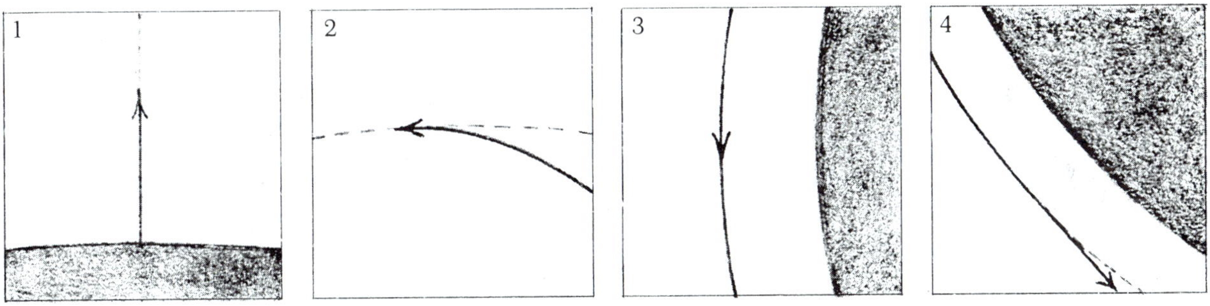

Motion along a straight line (as in Steps 1 and 5) is easy enough to describe. However, it is useful to analyze in greater detail other parts of this trip. Motion in a circular arc, as in Steps 3 and 7, and projectile motion, as in Step 8, are two important cases.

SG 2

How can you go about making this analysis? Following the example of Galileo and Newton, you can learn about motions beyond your reach, even on the moon or in the farthest parts of the universe, by studying motions near at hand. If physics is the same everywhere, the path of a lunar capsule moving as in Step 8 can be understood by studying a bullet fired from a horizontal rifle.

4.2 | Projectile motion

Imagine an experiment in which a rifle is mounted on a tower with its barrel parallel to the ground. The ground over which the bullet will travel is level for a great distance. At the instant a bullet leaves the rifle, an identical bullet is dropped from the height of the rifle's barrel. This second bullet has no horizontal motion relative to the ground; it goes only straight down. Which bullet will reach the ground first?

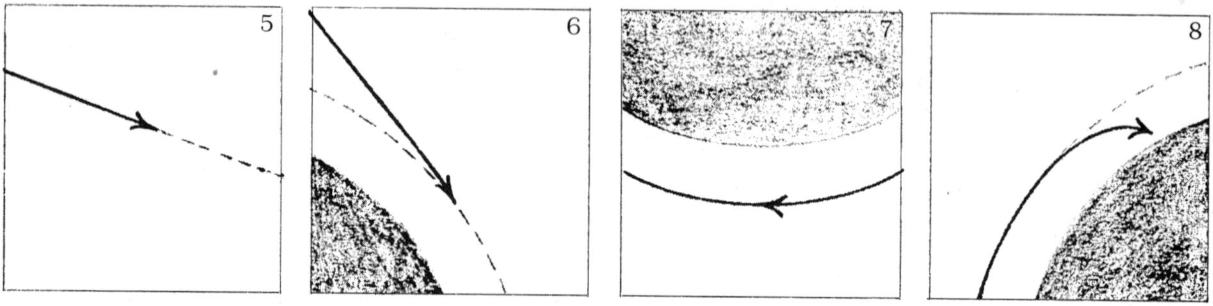

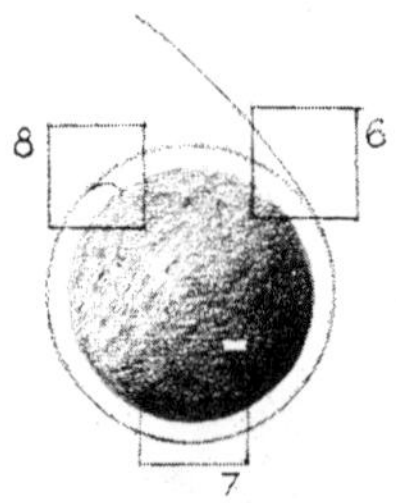

You do not need to know anything about the speed of the bullet or the height of the tower in order to answer this question. Consider first the motion of the second bullet, the one that is dropped. As a freely falling object, it accelerates toward the ground with constant acceleration. As it falls, the elapsed time interval Δt and the corresponding downward displacement Δy are related by the equation

$$\Delta y = \frac{1}{2} a_g \Delta t^2$$

where a_g is the acceleration due to gravity at that location. Following usual practice, you can now drop the Δ symbols, but keep in mind that y and t stand for displacement and time interval, respectively. So you can write the last equation as

$$y = \frac{1}{2} a_g t^2$$

Now consider the bullet that is fired horizontally from the rifle. When the gun fires, the bullet is driven by the force of expanding gases. It accelerates very rapidly until it reaches the muzzle of the rifle. On reaching the muzzle, the gases escape and no longer push the bullet. At that moment, however, the bullet has a great horizontal speed, v_x. The air slows the bullet slightly, but you can ignore that fact and imagine an ideal case with no air friction. As long as air friction is ignored, there is no horizontal force acting on the projectile. Therefore, the horizontal speed will remain constant. From the instant the bullet leaves the muzzle, its horizontal motion is described by the following equation involving the horizontal displacement Δx:

$$\Delta x = v_x \Delta t$$

or again dropping the Δ's,

$$x = v_x t$$

These equations describe the forward part of the motion of the bullet. There is, however, another part that becomes more and more important as t increases. From the moment the bullet leaves the gun, it falls toward the earth while it moves forward, like any other unsupported body. Can you use the same equation to describe its fall that you used to describe the fall of the dropped bullet? How will falling affect the bullet's horizontal motion? These questions raise a more fundamental one which goes beyond just the behavior of the bullets in this experiment. Is the vertical motion of an object affected by its horizontal motion or vice versa?

To answer these questions, you could carry out a real experiment similar to the thought experiment. A special laboratory device which fires a ball horizontally and at the same moment releases a second ball to fall freely can be used. Set up the apparatus so that both balls are the same height above a level floor and release them at exactly the same time. Although the motion of the balls may be too rapid to follow with the eye, your ears will tell you that they do in fact reach the floor at the same time. This result suggests that the vertical motion of the projected ball is unaffected by its horizontal motion.

In the margin is a stroboscopic photograph of this experiment. Equally spaced horizontal lines aid the examination of the two motions. Look first at the ball on the left, the one that was released without any horizontal motion. You can see that it accelerates because it moves a greater distance between successive flashes of the strobe's light. Careful measurement of the photograph shows that the acceleration is constant, within the uncertainty of the measurements.

Now, compare the vertical positions of the second ball, fired horizontally, with the vertical positions of the ball that is falling freely. The horizontal lines show that the distances of fall are the same for corresponding time intervals. The two balls obey the same law for motion in a vertical direction. That is, at every instant they both have the same constant acceleration a_g, the same downward velocity, and the same vertical displacement. Therefore, the experiment supports the idea that the vertical motion is the same whether or not the ball has a horizontal motion also. The horizontal motion does not affect the vertical motion.

You can also use the strobe photo to see if the vertical motion of the projectile affects its horizontal velocity. Do this by measuring the *horizontal* distance between images. You will find that the horizontal distances are practically equal. Since the time intervals between images are equal, you can conclude that the horizontal velocity v_x is constant. Therefore, the vertical motion does not affect the horizontal motion.

The experiment shows that the vertical and horizontal parts, or *components*, of the motion are independent of each other. This experiment can be repeated from different heights and with different horizontal velocities, but the results lead to the same conclusion.

In one of the most famous allegorical frontispieces of Renaissance science, from Nova Scientia *(1537) by Nicola Tartaglia, Euclid greets students at the outer gate of the circle of knowledge. The fired cannon and mortar show the trajectories defined by Tartaglia. Plato and Aristotle are shown in the inner circle.*

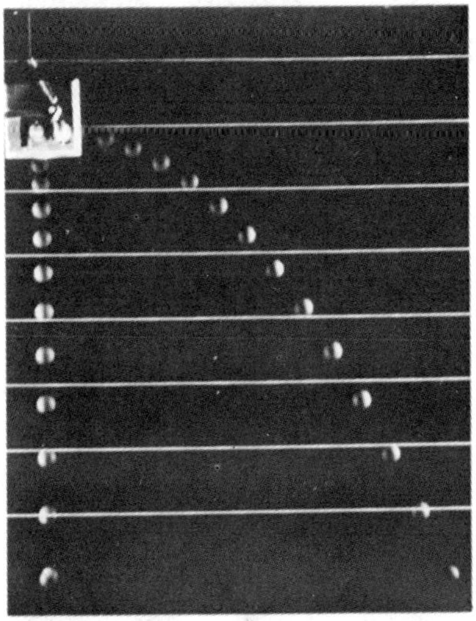

The two balls in this stroboscopic photograph were released simultaneously. The one on the left was simply dropped from a rest position; the one on the right was given an initial velocity in the horizontal direction.

SG 3

SG 4

The independence of horizontal and vertical motions has important consequences. For example, it allows you to predict the displacement and the velocity of a projectile at any time during its flight. You need merely to consider the horizontal and vertical aspects of the motion *separately*, and then add the results by the vector method. You can predict the magnitude of the components of displacement x and y and the components of velocity v_x and v_y at any instant by applying the appropriate equations. For the horizontal component of motion, the equations are

$$v_x = \text{constant}$$

and

$$x = v_x t$$

and for the vertical component of motion,

$$v_y = a_g t$$

$$y = \frac{1}{2} a_g t^2$$

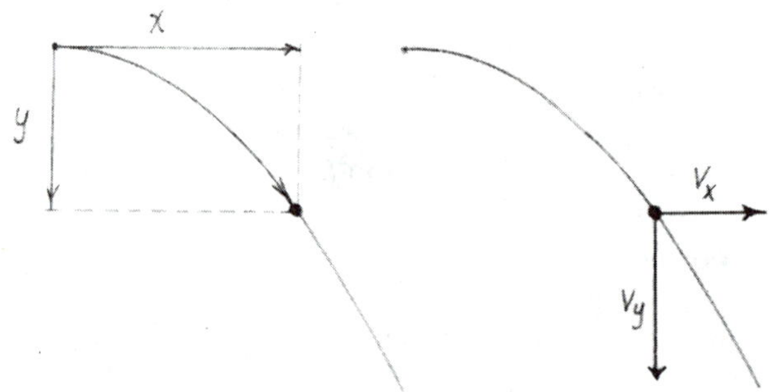

?

1. *How do you know that it is correct to simply break down complicated motion into separate vertical and horizontal components in the case described in Sec. 4.2?*
2. *If a body falls from rest with acceleration* $\mathbf{a}_g$, *with what acceleration will it fall if it also has an initial horizontal speed* v_x?

4.3 | What is the path of a projectile?

It is easy to see that a thrown object, such as a rock, follows a curved path. But it is not as easy to see just what kind of curve it traces. Arcs of circles, ellipses, parabolas, hyperbolas, and cycloids (to name only a few geometric figures) all provide likely looking curved paths.

Early scientists gained better knowledge about the path of a projectile when they applied mathematics to the problem. This was done by deriving the equation that expresses the shape of the path. Only a few steps are involved. First list equations you already know for a projectile launched horizontally:

$$x = v_x t$$

and

$$y = \tfrac{1}{2} a_g t^2$$

You could plot the shape of the path, or *trajectory* as it is often called in physics, if you had an equation that gave the value of y for each value of x. You could find the fall distance y for any horizontal distance x by combining these two equations in a way that eliminates the time variable. Solving the equation $x = v_x t$ for t gives

$$t = \frac{x}{v_x}$$

Because t means the same in both equations, you can substitute x/v_x for t in the equation for y:

$$y = \tfrac{1}{2} a_g t^2$$

and thus,

$$y = \tfrac{1}{2} a_g \left(\frac{x}{v_x}\right)^2$$

This last equation is a specialized equation of the kind that need not be memorized. It contains two variables of interest, x and y. It also contains three constant quantities: the number ½, the uniform acceleration of free fall a_g, and the horizontal speed v_x, which is constant for any one flight, from launching to the end. The vertical distance y that the projectile falls is thus a constant times the square of the horizontal displacement x. In other words, the two quantities y and x^2 are proportional: $y \propto x^2$.

Thus, there is a fairly simple relationship between x and y for the trajectory. For example, when the projectile goes twice as far horizontally from the launching point, it drops vertically four times as far.

The mathematical curve represented by this relationship between x and y is called a *parabola*. Galileo deduced the parabolic shape of trajectories by an argument similar to the one used here. (Even projectiles not launched horizontally, as in the photographs on page 106, have parabolic trajectories.) This discovery greatly simplified the study of projectile motion, because the geometry of the parabola had been established centuries earlier by Greek mathematicians.

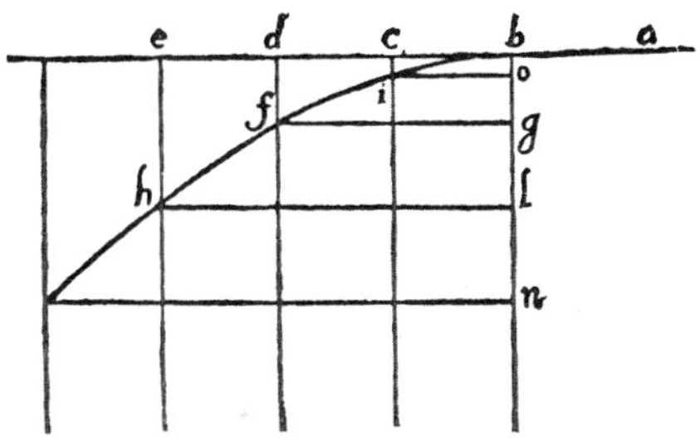

Drawing of a parabolic trajectory from Galileo's Two New Sciences.

Here is a clue to one of the important strategies in modern science. Whenever possible, scientists express the features of a phenomenon quantitatively and put the relations between them into equation form. Then the rules of mathematics can be used to shift and substitute terms, opening the way to unexpected insights.

SG 5–10

?

3. Rewrite the steps in Sec. 4.3 yourself, defining all the variables and explaining each step and each equation used.

4. Which of the conditions below must hold in order for the relationship $y \propto x^2$ to describe the path of a projectile? (a) a_g is a constant (b) a_g depends on t (c) a_g is straight down (d) v_x depends on t (e) air friction is negligible

5. How far, vertically and horizontally, will a projectile travel in 10 sec if it is launched with an initial horizontal speed of 4 m/sec? (For simplicity, use a_g as approximately 10 m/sec^2.)

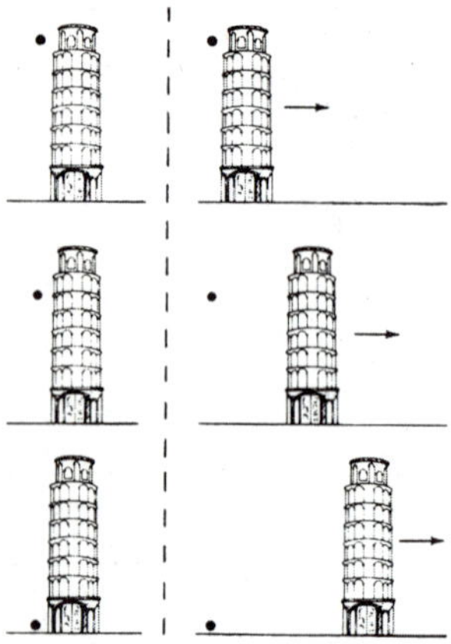

The critics of Galileo claimed that if the earth moved, a dropped stone would be left behind and land beyond the foot of the tower.

4.4 | Moving frames of reference: Galilean relativity

Galileo's work on projectiles leads to thinking about reference frames. As you will see in Unit 2, Galileo strongly supported the idea that the proper reference frame for discussing motions in our planetary system is one fixed to the sun, not the earth. From that point of view, the earth both revolves around the sun and rotates on its own axis. For many scientists of Galileo's time, this idea was impossible to accept, and they thought they could prove their case. If the earth rotated, they said, a stone dropped from a tower would not land directly at the tower's base. For if the earth rotates once a day, the tower would move onward hundreds of meters for every second the stone is falling. The stone would be left behind while falling through the air and so would land far away from the base of the tower. But this is *not* what happens. As nearly as one can tell, the stone lands directly below the point of release. Therefore, many of Galileo's critics believed that the tower and the earth could not possibly be in motion.

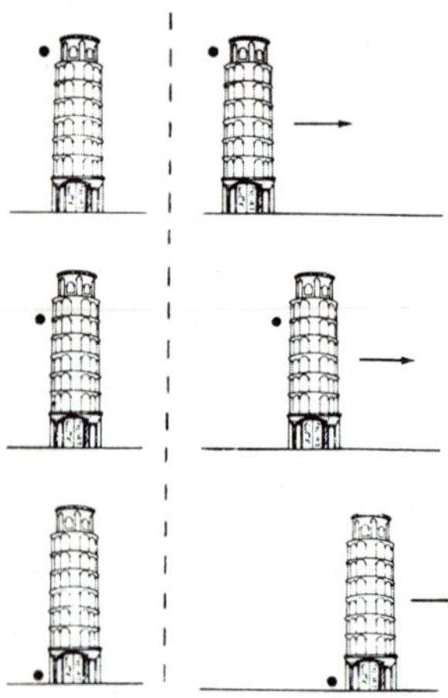

Galileo argued that the falling stone continued to share the motion of the earth, so that an observer on earth could not tell whether or not the earth moved by watching the stone.

To answer these arguments, Galileo used the same example to support his *own* view. During the time of fall, Galileo said, the tower and the ground supporting it move forward together with the same uniform velocity. While the stone is held at the top of the tower, it has the same horizontal velocity as the tower. Releasing the stone allows it to gain vertical speed. But by the principle of independence of v_x and v_y discussed in Sec. 4.2, this vertical component does not diminish any horizontal speed the stone had on being released. In other words, the falling stone behaves like any other projectile. The horizontal and vertical components of its motion are independent of each other. Since the stone and tower continue to have the same v_x throughout, the stone will not be left behind as it falls. Therefore, no matter what the speed of the earth, the stone will land at the foot of the tower. So the fact that falling stones are not left behind does *not* prove that the earth is standing still.

Similarly, Galileo said, an object released from a crow's nest at the top of a ship's mast lands at the foot of the mast, whether the ship is standing still or moving with constant velocity in calm water. This was actually tested by experiment in 1642 (and is also the subject of three *Project Physics* film loops). Many everyday observations support this view. For example, when you drop or throw a book in a bus or train or plane that is moving with constant velocity, you see the book move just as it would if the vehicle were standing still. Similarly, if the wind is small enough not to interfere, an object projected vertically *upward* from inside an open truck moving at constant velocity will fall back into the truck. A person in the truck sees the same thing happen whether the truck is moving at constant velocity or standing still.

At high speeds, air drag will affect the results considerably. The situation is still distinguishable from a car at rest, but in a high wind!

These and other observations lead to a valuable generalization: If Newton's laws hold in any one laboratory, then they will hold equally well in any other laboratory (or "reference frame") moving at constant velocity with respect to the first. This generalization is called the *Galilean relativity principle*. It holds true for all "classical" mechanical phenomena, that is, phenomena where the relative velocities are in the range from almost negligible up to millions of kilometers per hour.

Even if the laws of mechanics are the same for all reference frames moving with constant velocity with respect to each other, a problem still arises. Namely, there is no way to find the *speed* of one's *own* reference frame from any mechanical experiment done *within* that frame. Nor can one pick out any one reference frame as the "true" frame, the one that is, say, at absolute rest. Thus, there can be no such thing as the "absolute" velocity of a body. All measured velocities are only *relative*.

What about observations of phenomena outside of one's own frame of reference? Certainly some outside phenomena appear differently to observers in different reference frames. For example, the observed velocity of an airplane will have a value when measured from the earth different from that when measured from a moving ship. Other quantities, such as mass, acceleration, and time interval, have the *same* values when measured from different reference frames moving with constant velocity with respect to one another. Moreover, certain *relationships* among such measurements will be the same for these different reference frames. Newton's laws of motion are examples of such "invariant" relationships, and so are all the laws of mechanics that follow from them.

When relative speeds become a noticeable fraction of the speed of light (approximately 300,000 km/sec), some deviations from this simple relativity principle begin to appear. We will consider some of them in Unit 5.

Notice that the relativity principle, even in this restricted, classical form, does not say "everything is relative." On the contrary, it asks you to look for relationships that do *not* change when you transfer your attention from one frame to another.

SG 11–13

?

6. If the laws of mechanics are found to be the same in two reference frames, what must be true of the motions of those frames?

7. An outfielder running at constant speed under a falling ball sees the ball falling straight down to her. What is the path of the ball as seen by someone in the stands? Explain how this common experience supports (a) Galilean relativity and (b) the breakdown of projectile motion into independent horizontal and vertical parts.

4.5 | Circular motion

In discussing circular motion it is useful to keep clearly in mind a distinction between *revolution* and *rotation*. We define these terms differently: Revolution is the act of traveling along a circular or elliptical path; rotation is the act of spinning rather than traveling. A point on the rim of a phonograph turntable travels a long way; it is *revolving* about the axis of the turntable. But the turntable as a unit does not move from place to place; it merely *rotates*. In some situations both processes occur simultaneously; for example, the earth rotates about its own axis, while it also revolves (in a nearly circular path) around the sun.

A projectile launched horizontally from a tall tower strikes the earth at a point determined by three factors. These factors are the horizontal speed of the projectile, the height of the tower, and the acceleration due to the force of gravity. As the projectile's launch speed is increased, it strikes the earth at points farther and farther from the tower's base. It is then no longer true that the trajectory is a simple parabola, because the force of gravity does not pull on the projectile in the *same* direction throughout the path. Eventually, then, you would have to consider a fourth factor: The earth is not flat but curved. If the launch speed were increased even more, the projectile would strike the earth at points even farther from the tower, and at last it would rush around the earth in a nearly circular orbit. (See the quotation from Newton, page 112.) At the launching speed that puts the projectile into orbit, the fall of the projectile away from straight-line forward motion is just matched by the curvature of the earth's surface. Therefore, the projectile stays at a constant distance above the surface.

What horizontal launch speed is required to put an object into a circular orbit about the earth or the moon? You will be able to answer this question quite easily after you have learned about circular motion.

The simplest kind of circular motion is *uniform* circular motion, that is, motion in a circle at constant speed. If you are in a car that goes around a perfectly circular track so that at every instant the speedometer reading is 64 km/hr, you are in uniform circular motion. This is not the case if the track is of any shape other than circular or if your speed changes at any point.

How can you find out if an object in circular motion is moving at constant speed? You can apply the same test used in deciding whether or not an object traveling in a straight line does so with constant speed. That is, measure the instantaneous speed at many different moments and see whether the values are the same. If the speed is constant, you can describe the circular motion of any object by means of two numbers: the radius R of the circle and the speed v along the path. For regularly repeated circular motion, you can use a quantity more easily measured than speed: either the time required by an object to make one complete revolution, or the number of revolutions the object completes in a unit of time. The time required for an object to complete one revolution in a circular path is called the *period* (T) of the motion. The number of revolutions completed by the same object in a unit time interval is called the *frequency* (f) of the motion.

Note: always use t for time, T for period, and in general stick to symbols as given. Otherwise, you may confuse their meanings.

An an example, these terms are used to describe a car moving with uniform speed on a circular track. Suppose the car takes 20 sec to make one lap around the track. Thus, $T = 20$ sec. On the other hand, the car makes three laps in 1 min. Thus, $f = 3$ revolutions per minute, or $f = 1/20$ revolution per second. The relationship between frequency and period (*when the same time unit is used*) is $f = 1/T$. If the period of the car is 20 sec/rev, then the frequency is given by

1/20 revolution per second can be written 0.05 rev/sec, or more briefly, just 0.05 sec^{-1}. In this last expression the symbol sec^{-1} stands for 1/sec, or "per second." The unit sec^{-1} is called hertz, symbol Hz.

$$f = \frac{1}{20 \text{ sec/rev}} = \frac{1}{20} \text{ rev/sec} = \frac{1}{20} \text{ Hz}$$

SG 14

All units are a matter of convenience. Radius may be expressed in terms of centimeters, kilometers, or any other distance unit. Period may be expressed in seconds, minutes, years, or any other time unit. Frequency may be expressed as "per second," "per minute," "per year," and so on. The most widely used units of radius, period, and frequency in scientific work are, respectively, meter, second, and hertz (or per second).

The term "revolutions" is not assigned any units because it is a pure number, a count. There is no need for a standard as there is for distance, mass, and time. So, the unit for frequency is usually given without "rev." This looks strange, but one gets used to it. It is not very important, because it is merely a matter of terminology, not a fact of physics.

TABLE 4.1 COMPARISON OF THE FREQUENCY AND PERIOD FOR VARIOUS KINDS OF CIRCULAR MOTION.

PHENOMENON	*Period (T)*		*Frequency (f)*
Electron in circular accelerator	10^{-6}	sec	10^{6} Hz (per sec)
Ultra-centrifuge	0.00033	sec	3000 Hz
Hoover Dam turbine	0.33	sec	3 Hz
Rotation of earth	24	hr	0.0007 per min
Moon around the earth	27	days	0.0015 per hour
Earth around the sun	365	days	0.0027 per day

*Note the difference between units.

If you know the frequency of revolution f and the radius R of the path, you can compute the speed v of any object in uniform

"... the greater the velocity ... with which [a stone] is projected, the farther it goes before it falls to the earth. We may therefore suppose the velocity to be so increased, that it would describe an arc of 1, 2, 5, 10, 100, 1000 miles before it arrived at the earth, till at last, exceeding the limits of the earth, it should pass into space without touching it."—Newton, System of the World.

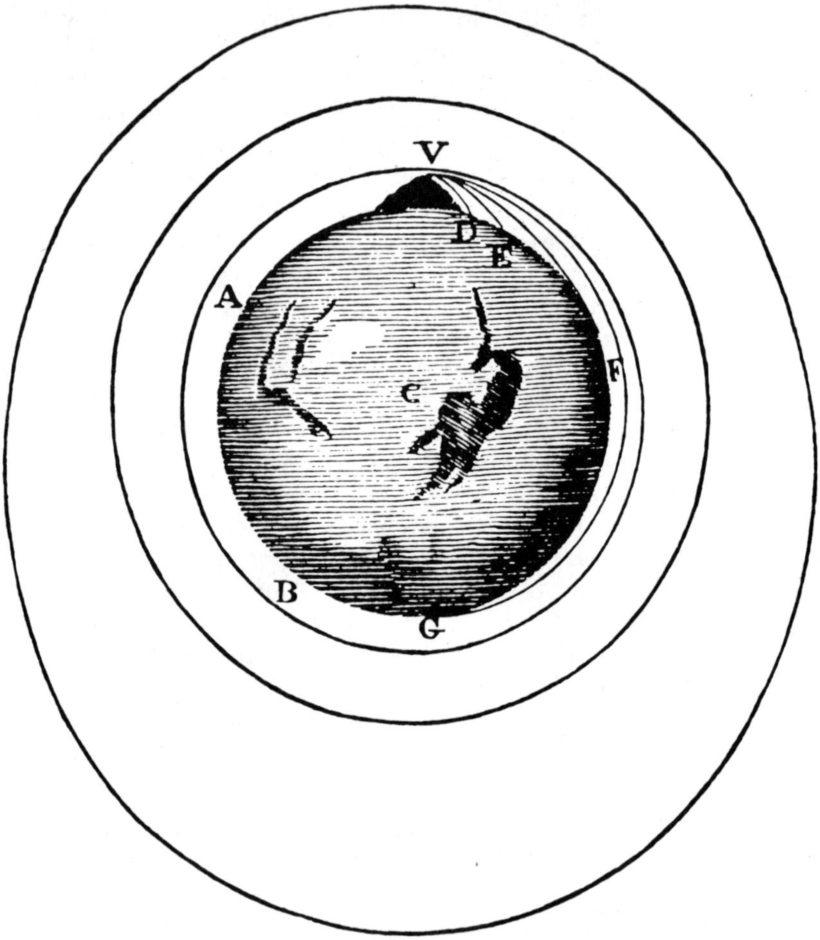

circular motion without difficulty. The distance traveled in one revolution is simply the perimeter of the circular path, that is, $2\pi R$. The time for one revolution is by definition the period T. For uniform motion along any path, it is always true that

$$\text{speed} = \frac{\text{distance traveled}}{\text{time elapsed}}$$

By substitution,

$$v = \frac{2\pi R}{T}$$

To express this equation for circular motion in terms of the frequency f, rewrite it as

$$v = 2\pi R \times \frac{1}{T}$$

Now, since by definition

$$f = \frac{1}{T}$$

you can write

$$v = 2\pi R \times f$$

If the body is in *uniform* circular motion, the speed computed by this equation is both the instantaneous speed and the average speed. If the motion is not uniform, the formula gives only the *average* speed. The *instantaneous* speed for any point on the circle can be determined if you find $\Delta d/\Delta t$ from measurements of very small portions of the path.

How can you best use the last equation above? You can, for example, calculate the speed of the tip of a helicopter rotor blade in its motion around the central shaft. On one model, the main rotor has a diameter of 7.50 m and a frequency of 480 revolutions/min under standard conditions. Thus, R = 3.75 m, and so

$$v = 2\pi Rf$$

$$v = 2\ (3.14)\ (3.75)\left(\frac{480}{60}\right)\ \text{m/sec}$$

$$v = 189\ \text{m/sec}$$

or about 680 km/hr.

?

8. If a phonograph turntable is running at 33.3 revolutions per minute,

(a) what is its period (in minutes)?

(b) what is its period (in seconds)?

(c) what is its frequency in hertz?

9. What is the period of the minute hand of an ordinary clock? If the hand is 3.0 cm long, what is the speed of the tip of the minute hand?

10. The terms frequency and period can also be used for any other regular, repetitive phenomenon. For example, if your heart beats 80 times per minute, what are the frequency and period for your pulse?

4.6 | Centripetal acceleration and centripetal force

The adjective "centripetal" means literally "moving, or directed, toward the center."

Assume that a stone on the end of a string is whirling about with uniform circular motion in a horizontal plane. The speed of the stone is constant. The velocity, however, is always changing. Velocity is a vector quantity which includes both speed and direction. Up to this point, you have dealt with accelerations in which only the speed was changing. In uniform circular motion the speed of the revolving object remains the same, while the

In uniform circular motion, the instantaneous velocity and the centripetal force at any instant of time are perpendicular, one being along the tangent, the other along the radius. So instantaneous velocity and the acceleration are also always at right angles.

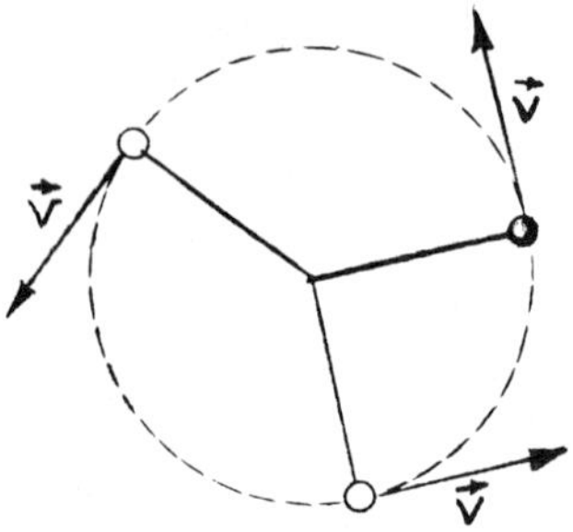

direction of motion changes continually. The top figure in the margin shows the whirling stone at three successive moments in its revolution. At any instant, the direction of the velocity vector is tangent to the curving path. Notice that the stone's *speed*, represented by the *length* of the velocity arrow, does not change. But its *direction* does change from moment to moment. Since acceleration is defined as a change in velocity per unit time, the stone is, in fact, accelerating.

To produce an acceleration, a net force is needed. In the case of the whirling stone, a force is exerted on the stone by the string. If you ignore the weight of the stone and air resistance, this force is the net force. If the string were suddenly cut, the stone would go flying off with the velocity it had at the instant that the string was cut. Its path would start off on a tangent to the circular path. But as long as the string holds, the stone is forced to move in the circular path.

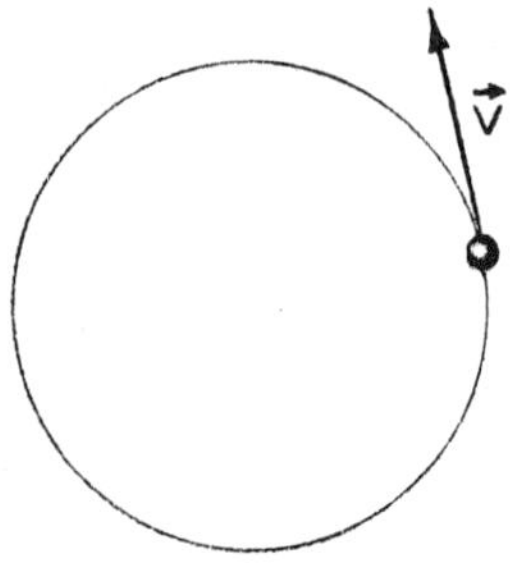

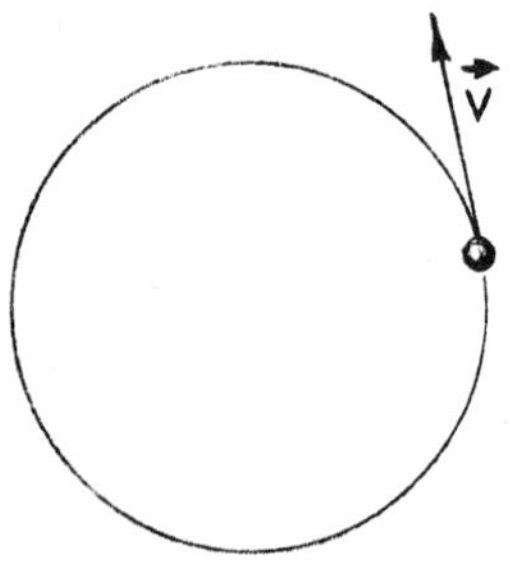

$\vec{a}_c$ and $\vec{F}_c$ are parallel, but $\vec{v}$ is perpendicular to $\vec{a}_c$ and $\vec{F}_c$. Note that usually one should not *draw different kinds of vector quantities on the same drawing.*

The direction of this force which is holding the stone in its circular path is along the string. Thus, the force vector always points toward the center of rotation. This kind of force, which is always directed toward the center of rotation, is called *centripetal* force.

From Newton's second law, you know that net force and corresponding acceleration are in the same direction. Therefore, the acceleration vector is also directed toward the center. This acceleration is called *centripetal* acceleration and has the symbol $\vec{a}_c$. Any object moving along a circular path has a centripetal acceleration.

You know now the direction of centripetal acceleration. What is its magnitude? An expression for a_c can be derived from the definition of acceleration $\vec{a}_c = \Delta v/\Delta t$. The details for such a derivation are given on the next page. The result shows that $\vec{a}_c$ depends on $\vec{v}$ and R. In fact, the magnitude of a_c is given by

$$a_c = \frac{v^2}{R}$$

You can verify this relationship with a numerical example. As sketched in the diagram on page 116, a car goes around a circular curve of radius $R = 100$ m at a uniform speed of $v = 20$ m/sec. What is the car's centripetal acceleration a_c toward the center of curvature? By the equation derived on the next page,

$$a_c = \frac{v^2}{R}$$

$$= \frac{(20 \text{ m/sec})^2}{100 \text{ m}}$$

Close Up

Derivation of the Equation $a_c = \frac{v^2}{R}$

Assume that a stone on the end of a string is moving uniformly in a circle of radius R. You can find the relationship between a_c, v, and R by treating a small part of the circular path as a combination of tangential motion and acceleration toward the center. To follow the circular path, the stone must accelerate toward the center through a distance h in the *same time* that it would move through a tangential distance d. The stone, with speed v, would travel a tangential distance d given by $d = v\Delta t$. In the same time Δt, the stone, with acceleration a_c, would travel toward the center through a distance h given by $h = \frac{1}{2}a_c\Delta t^2$. (You can use this last equation because at $t = 0$, the stone's velocity toward the center is zero.)

You can apply the Pythagorean theorem to the triangle in the figure below.

$$\begin{aligned} R^2 + d^2 &= (R + h)^2 \\ &= R^2 + 2Rh + h^2 \end{aligned}$$

When you subtract R^2 from each side of the equation, you are left with

$$d^2 = 2Rh + h^2$$

You can simplify this expression by making an approximation. Since h is very small compared to R, h^2 will be very small compared to Rh. And since Δt must be vanishingly small to get the instantaneous acceleration, h^2 will become vanishingly small compared to Rh. So you can neglect h^2 and write

$$d^2 = 2Rh$$

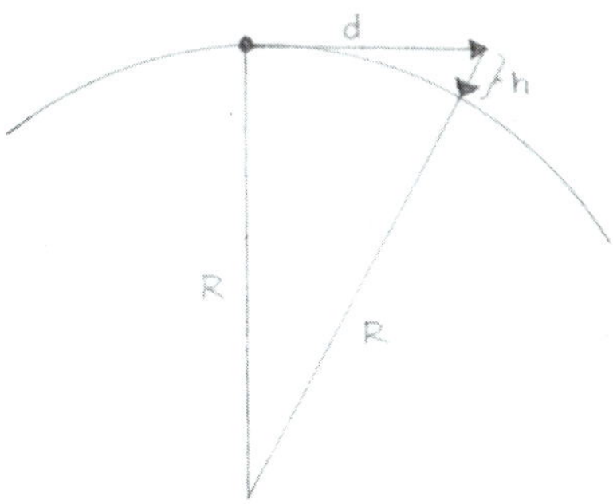

Also, $d = v\Delta t$ and $h = \frac{1}{2}a_c\Delta t^2$, so you can substitute for d^2 and for h accordingly. Thus,

$$(v\Delta t)^2 = 2R \cdot \frac{1}{2}a_c(\Delta t)^2$$

$$v^2(\Delta t)^2 = Ra_c(\Delta t)^2$$

$$v^2 = Ra_c$$

or

$$a_c = \frac{v^2}{R}$$

The approximation becomes better and better as Δt becomes smaller and smaller. In other words, v^2/R gives the magnitude of the *instantaneous* centripetal acceleration for a body moving on a circular arc of radius R. For uniform circular motion, v^2/R gives the magnitude of the centripetal acceleration at every point of the path. (Of course, it does not have to be a stone on a string. It can be a small particle on the rim of a rotating wheel, or a house on the rotating earth, or a coin sitting on a rotating phonograph disk, or a car in a curve on the road, or an electron in its path through a magnetic field.)

The relationship between a_c, v, and R was discovered by Christian Huygens and was published by him in 1673. Newton, however, must have known it in 1666, but he did not publish his proof until 1687, in the *Principia*.

We can substitute the relation $v = 2\pi R/f$ or $v = 2\pi R/T$ (derived in Chapter 3 for uniform circular motion) into the equation for a_c:

$$a_c = \frac{v^2}{R} \qquad a_c = \frac{v^2}{R}$$

$$= \frac{(2\pi Rf)^2}{R} \qquad = \frac{\left(\frac{2\pi R}{T}\right)^2}{R}$$

$$= 4\pi^2 Rf^2 \qquad = \frac{4\pi^2 R}{T^2}$$

These two resulting expressions for a_c are entirely equivalent.

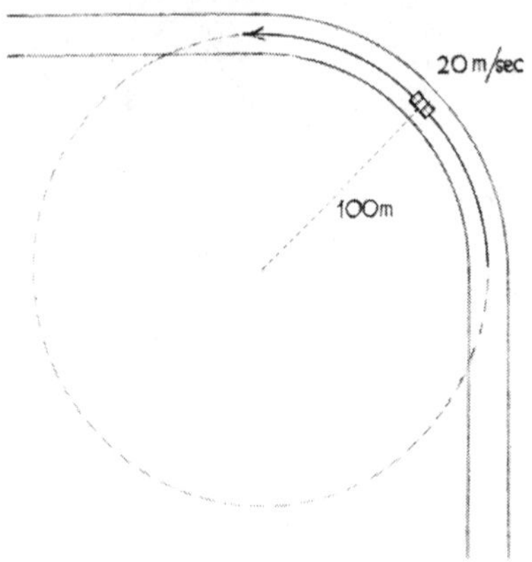

$$= \frac{400\ \text{m}^2/\text{sec}^2}{100\ \text{m}}$$

$$= 4.0\ \text{m/sec}^2\ \text{(about 0.4 g)}$$

Does this make sense? You can check the result by going back to the basic vector definition of acceleration: $\vec{a}_{av} = \Delta\vec{v}/\Delta t$. You will need a scale drawing of the car's velocity vector at two instants a short time Δt apart. Then, you can measure the change in velocity $\Delta\vec{v}$ between these points, and divide the magnitude of $\Delta\vec{v}$ by Δt to get $\vec{a}_{av}$ over the interval.

Consider a time interval of $\Delta t = 1$ sec. Since the car is moving at 20 m/sec, its position will change 20 m during Δt. Two positions **P** and **P′**, separated by 20 m, are marked in Diagram A.

SG 15
SG 16

Now, draw arrows representing velocity vectors. If you choose a scale of 1 cm = 10 m/sec, the velocity vector for the car will be represented by an arrow 2 cm long. These are drawn at **P** and **P′** in Diagram B.

A

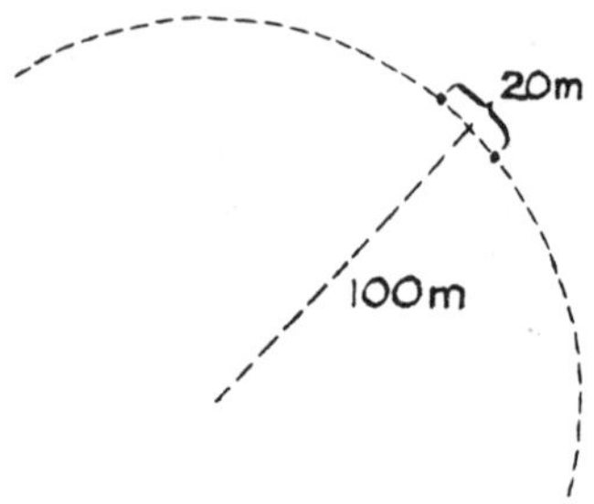

Now, put these two arrows together tail to tail as in Diagram C. It is easy to see what the change in the velocity vector has been during Δt. Notice that if you had drawn Δv halfway between **P** and **P′**, it would point directly toward the center of the curve. So the average acceleration between **P** and **P′** is indeed directed centripetally. The $\Delta\vec{v}$ arrow in the diagram is 0.40 cm long, so it represents a velocity change of 4.0 m/sec. This change occurred during $\Delta t = 1$ sec, so the rate of change is 4.0 m/sec². This is the same value found by using $a_c = v^2/R$!

The relation $a_c = v^2/R$ agrees completely with the mechanics developed in Unit 1. You can show this by doing some experiments to measure the centripetal force required to keep an object moving in a circle. Return to the example of the car. If its mass were 1,000 kg, you could find the centripetal force acting on it as follows:

B

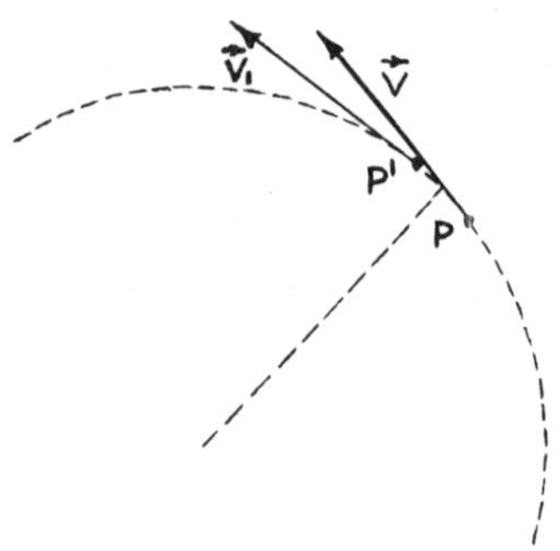

$$F_c = m \times a_c$$

$$= 1000\ \text{kg} \times 4.0\ \text{m/sec}^2$$

$$= 4000\ \text{kg m/sec}^2$$

$$= 4000\ \text{N}$$

C

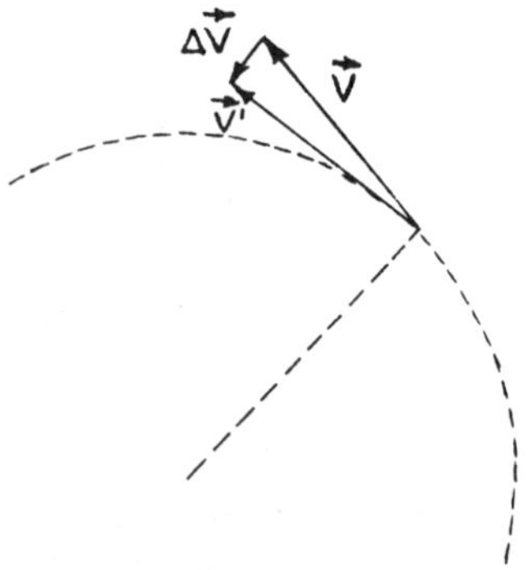

This force would be directed toward the center of curvature of the road. That is, it would always be sideways to the direction the car is moving. This force is exerted on the tires by the road. If the road is wet or icy and cannot exert the force of 4,000 N sideways on the tires, the centripetal acceleration will be less than 4.0 m/sec². Then the car will follow a less curved path as sketched in Diagram D. In situations where the car's path is less curved than the road, you would say the car "left the road." Of course, it might be just as appropriate to say the road left the car.

?

11. Is a body accelerating when it

(a) moves with constant speed?

(b) moves in a circle with constant radius?

(c) moves with constant velocity?

12. A car of mass m *going at speed* v *enters a curve of radius* R. *What is the force required to keep the car curving with the road?*

13. A rock of mass 1 kg is swung in a circle with a frequency of 1 revolution/sec on a string of length 0.5 m. What is the magnitude of the force that the string exerts on the rock? What is the magnitude and direction of the force on the string *that must be present according to Newton's third law? If the string were cut, what kind of path would the rock follow? How fast would it move?*

SG 17–20

4.7 | The motion of earth satellites

Nature and technology provide many examples of objects in uniform circular motion. The wheel has been a main characteristic of civilization, first appearing on crude carts and later forming essential parts of complex machines. The historical importance of rotary motion in the development of modern technology has been described by the historian V. Gordon Childe in *The History of Technology*:

> Rotating machines for performing repetitive operations, driven by water, by thermal power, or by electrical energy, were the most decisive factors of the industrial revolution, and, from the first steamship till the invention of the jet plane, it is the application of rotary motion to transport that has revolutionized communications. The use of rotary machines, as of any other human tools, has been cumulative and progressive. The inventors of the eighteenth and nineteenth centuries were merely extending the applications of rotary motion that had been devised in previous generations, reaching back thousands of years into the prehistoric past. . . .

SG 25

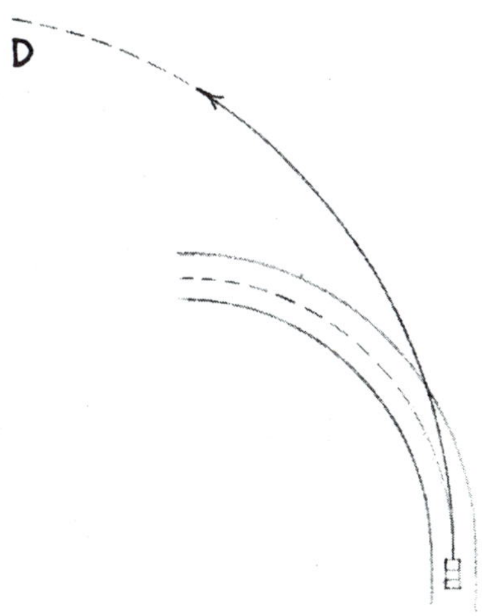

As you will see in Unit 2, there is another rotational motion that has concerned scientists throughout recorded history. This motion is the orbiting of planets around the sun and of the moon around the earth.

The kinematics and dynamics for *any* uniform circular motion are the same. Therefore, you can apply what you have learned so far to the motion of artificial earth satellites in circular (or nearly circular) paths. A typical illustration is Alouette I, Canada's first satellite, which was launched into a nearly circular orbit.

The same equation ($v + 2\pi R/T$) can be used to find the speed of any satellite in nearly or fully circular orbit, for example, that of our moon. The average distance from the center of the moon is approximately 3.82×10^5 km, and the moon takes an average of 27 days, 7 hr, 43 min to complete one revolution around the earth with respect to the fixed stars. Thus

$$v = \frac{2\pi(3.82 \times 10^5)\text{ km}}{3.93 \times 10^4\text{ min}}$$

$$= 61.1\text{ km/min}$$

or about 3,666 km/hr.

Tracking stations located in many places around the world maintain a record of any satellite's position in the sky. From the position data, the satellite's period of revolution and its distance above the earth at any time are found. By means of such tracking, scientists know that Alouette I moves at an average height of 1,010 km above sea level. It takes 105.4 min to complete one revolution.

You can now quickly calculate the orbital speed and the centripetal acceleration of Alouette I. The relation $v = 2\pi R/T$ gives the speed of any object moving uniformly in a circle if you know its period T and its distance R from the center of its path. In this case, the center of the path is the center of the earth. So, adding 1,010 km to the earth's radius of 6,380 km, you get R = 7,390 km. Thus,

$$v = \frac{2\pi R}{T}$$

$$= \frac{2\pi \times 7390\text{ km}}{105.4\text{ min}}$$

$$= \frac{46{,}432\text{ km}}{6324\text{ sec}}$$

$$= 7.34\text{ km/sec}$$

To calculate the centripetal acceleration of Alouette I, you can use this value of v along with the relationship $a_c = v^2/R$. Thus,

$$a_c = \frac{v^2}{R}$$

$$= \frac{(7.34\text{ km/sec})^2}{7390\text{ km}}$$

$$= 0.0073\text{ km/sec}^2$$

$$= 7.3\text{ m/sec}^2$$

An artist's impression of the Space Shuttle retrieving a satellite from orbit. When operational, the Shuttle Orbiter will function as the world's first reusable space cargo plane.

(You could just as well have used the values of R and T directly in the relationship $a_c = 4\pi^2R/T^2$.)

SG 29
SG 30

What is the origin of the force that gives rise to this acceleration? You will not find a good argument for the answer until Chapter 8, but you surely know already that it is due to the earth's attraction. Evidently the centripetal acceleration a_c of the satellite is just the gravitational acceleration a_g at that height. (Note that a_g at this height has a value about 25% less than a_g very near the earth's surface.)

Earlier we asked the question, "What speed is required for an object to stay in a circular orbit about the earth?" You can answer this question now for an orbit 1,010 km above the earth's surface. To get a general answer, you need to know how the acceleration due to gravity changes with distance. Chapter 8 will come back to the problem of injection speeds for orbits.

The same kind of analysis applies to an orbit around the moon. For example, consider the first manned orbit of the moon. The mission control group wanted to put the capsule into a circular orbit 110 km above the lunar surface. They knew (from other arguments) that the acceleration due to the moon's gravity at that height would be $a_g = 1.43\ \text{m/sec}^2$. What direction and speed would they have had to give the capsule to inject it into lunar orbit?

The direction problem is fairly easy to solve. To stay at a constant height above the surface, the capsule would have to be moving horizontally at the instant the orbit correction was completed. So injection would have to occur just when the capsule was moving on a tangent, at a height of 110 km, as shown in the sketch in the margin. What speed (relative to the moon, of course) would the capsule have to be given? The circular orbit has a radius 110 km greater than the radius of the moon, which is 1,740 km. So $R = 1{,}740\ \text{km} + 110\ \text{km} = 1{,}850\ \text{km} = 1.85 \times 10^6\ \text{m}$. The centripetal acceleration is just the acceleration caused by gravity, namely, $1.43\ \text{m/sec}^2$, so

$$a_c = a_g$$

$$\frac{v^2}{R} = a_g$$

$$v^2 = Ra_g$$

$$v = \sqrt{Ra_g}$$

$$= \sqrt{(1.85 \times 10^6\text{m}) \times 1.43\ \text{m/sec}^2}$$

$$= \sqrt{2.65 \times 10^6\ \text{m}^2/\text{sec}^2}$$

$$= 1.63 \times 10^3\ \text{m/sec}$$

The necessary speed for an orbit at 110 km above the surface is 1,630 m/sec. Knowing the capsule's speed, ground control could calculate the speed changes needed to reach 1,630 m/sec. Knowing the thrust force of the engines and the mass of the capsule, they could calculate the time of thrust required to make this speed change.

Chariot, *by the sculptor Alberto Giacometti, 1950.*

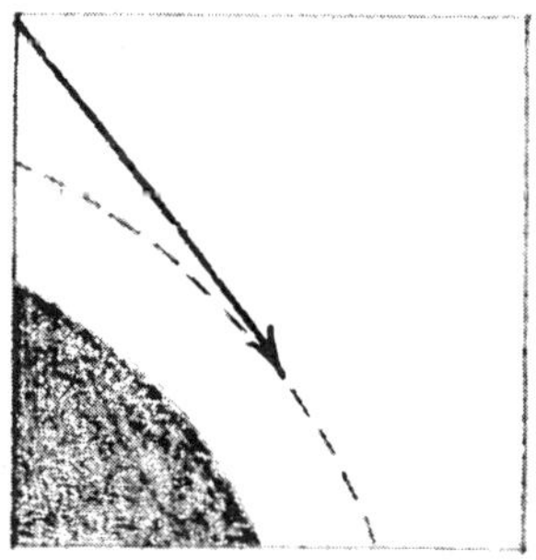

SG 31–33

Because the injection thrust is not applied instantaneously, the details are actually more difficult to calculate.

? *14. What information was necessary to calculate the speed for an orbit 110 km above the moon's surface?*

15. If you know the period and speed of a satellite, you also know the acceleration of gravity at the height of the satellite. Why?

16. What is the magnitude of the force that holds a 500-kg satellite in orbit if the satellite circles the earth every 3 hr at a height of 3,400 km above the surface?

TABLE 4.2 SOME INFORMATION ON A FEW EARLY ARTIFICIAL SATELLITES.

Name	*Launch Date*	*Weight (kg)*	*Period (min)*	*Height (km) Perigee–Apogee*	*Remarks (including purpose)*
Sputnik 1 1957 (USSR)	Oct. 4, 1957	83.4	96.2	229–947	First earth satellite. Internal temperature, pressure inside satellite.
Explorer 7 1958 (USA)	Jan. 31, 1958	14.0	114.8	301–2,532	Cosmic rays, micrometeorites, internal and shell temperatures, discovery of first Van Allen belts.
Lunik 3 1959 (USSR)	Oct. 4, 1959	433	22,300	48,300–468,510	Transmitted photographs of far side of moon.
Vostok 1 1961 (USSR)	Apr. 12, 1961	4,613	89.34	175–303	First manned orbital flight (Major Yuri Gagarin; one orbit).
Midas 3 1961 (USA)	July 12, 1961	1,590	161.5	3,428–3,466	Almost circular orbit.
Telestar 1 1962 (USA)	July 10, 1962	77	157.8	955–5,640	Successful transmission across the Atlantic: telephony, phototelegraphy, and television.
Alouette 1 1962 (USA–Canada)	Sept. 29, 1962	145.3	105.4	998–1,030	Joint project between NASA and Canadian Defense Research Board; measurement in ionosphere.
Luna 4 1963 (USSR)	Apr. 2, 1963	1,423	42,000	90,160–700,350	Passed 8,500 km from moon; very large orbit.
Vostok 6 1963 (USSR)	June 16, 1963	4,600	88.34	171–216	First orbital flight by a woman (Valentina Tereshkova; 48 orbits).
Syncom 2 1963 (USA)	July 26, 1963	39	1,460.4	35,721–35,729	Successfully placed in near-synchronous orbit (stays above same spot on earth).

4.8 What about other motions? Retrospect and prospect

So far we have described straight-line motion, projectile motion, and uniform circular motion. We have considered only examples where the acceleration was constant (in magnitude if not in direction) or very nearly constant. There is another basic kind of motion that is equally common and important in physics. In this kind of motion the acceleration is always changing. A common example is seen in playground swings or in vibrating guitar strings. Such back and forth motion, or *oscillation*, about a center position occurs when there is a force always directed toward the center position. When a guitar string is pulled aside, for example, a force arises which tends to restore the string to its undisturbed center position. If it is pulled to the other side, a similar restoring force arises in the opposite direction.

In very common types of such motion, the restoring force is proportional, or nearly proportional, to how far the object is displaced. This is true for the guitar string, if the displacements are not too large. Pulling the string aside 2 mm produces twice the restoring force that pulling it aside 1 mm does. Oscillation with a restoring force proportional to the displacement is called *simple harmonic motion*. The mathematics for describing simple harmonic motion is relatively simple, and many phenomena, from pendulum motion to the vibration of atoms, have aspects that are very close to simple harmonic motion. Consequently, the analysis of simple harmonic motion is used very widely in physics. The *Project Physics Handbook* describes a variety of activities you can do to become familiar with oscillations and their description.

The dynamics discussed in this chapter will cover most motions of interest. It provides a good start toward understanding apparently very complicated motions such as water ripples on a pond, a person running, the swaying of a tall building or bridge in the wind, a small particle zig-zagging through still air, an amoeba seen under a microscope, or a high-speed nuclear particle moving in the field of a magnet. The methods developed in this unit give you the means for dealing with any kind of motion whatsoever, on earth or anywhere in the universe.

SG 34
SG 35

When you considered the forces needed to produce motion, Newton's laws supplied the answers. Later, other motions, ranging from motion of the planets to the motion of an alpha particle passing near a nucleus inside an atom, will be discussed. You will continue to find in Newton's laws the tools for determining the magnitude and direction of the forces acting in each case.

If you know the magnitude and direction of the forces acting on an object, you can determine what its change in motion will be. If you also know the present position, velocity, and mass of an object, you can reconstruct how it moved in the past and predict how it will move in the future under these forces. Thus, Newton's laws provide a nearly unlimited view of forces and motion. It is not surprising that Newtonian mechanics became a model for many other sciences. They seemed to provide a method for understanding all motions, no matter how mysterious the motions appeared to be.

study guide

1. The *Project Physics* learning materials particularly appropriate for Chapter 4 include the following:

Experiments
Curves of Trajectories
Prediction of Trajectories
Centripetal Force
Centripetal Force on a Turntable

Activities
Projectile Motion Demonstration
Speed of a Stream of Water
Photographing a Waterdrop Parabola
Ballistic Cart Projectiles
Motion in a Rotating Reference Frame
Penny and Coat Hanger
Measuring Unknown Frequencies

Film Loops
A Matter of Relative Motion
Galilean Relativity: Ball Dropped from Mast of Ship
Galilean Relativity: Projectile Fired Vertically
Analysis of Hurdle Race. I
Analysis of Hurdle Race. II

2. The thrust developed by a Saturn Apollo rocket is 7,370,000 N and its mass is 540,000 kg. What is the acceleration of the vehicle relative to the earth's surface at lift-off? How long would it take for the vehicle to rise 50 m?

The acceleration of the vehicle increases greatly with time (it is 47 m/sec^2 at first stage burnout) even though the thrust force does not increase appreciably. Explain why the acceleration increases.

3. A person points a gun barrel directly at a bottle on a distant rock. Will the bullet follow the line of sight along the barrel? If the bottle falls off the rock at the very instant of firing, will it then be hit by the bullet? Explain.

4. It is helpful to consider the vertical and horizontal motions of a projectile separately. You can do the same for almost any motion, uniform or accelerated.

(a) What are the x and y displacements of a particle that travels uniformly with a velocity v_x = 4 m/sec and a velocity v_y = 3 m/sec after 2 sec of travel? Find the displacements after 5 sec and 10 sec.

(b) Add the two displacement vectors x and y that you found in part (a) to determine the distance of the particle from the origin after 2 sec, 5 sec, and 10 sec.

(c) Use your results from part (b) to find the velocity of the particle along its path. Can you relate this velocity to the velocity components v_x and v_y?

5. If you like algebra, try this general proof: If a body is launched with speed v at some angle other than 0°, it will initially have both a horizontal speed v_x and a vertical speed v_y. The equation for its horizontal displacement is $x = v_x t$, as before. But the equation for its vertical displacement has an additional term: $y = v_y t + \frac{1}{2} a_g t^2$. Show that the trajectory is still parabolic in shape.

6. A lunch pail is accidently kicked off a steel beam on a skyscraper under construction. Suppose the initial horizontal speed v_x is 1.0 m/sec. Where is the pail (displacement), and what is its speed and direction (velocity) 0.5 sec after launching?

7. **(a)** A ball is thrown from a roof with a horizontal speed of 10 m/sec toward a building 25 m away. How long will it take the ball to hit the building?

(b) How far will the ball have fallen when it hits the building?

(c) If the roof from which the ball is thrown is 20 m above the ground, what is the minimum speed with which the ball must be thrown to hit the building?

8. A shingle slides down a roof having a 30° pitch and falls off with a velocity of 2 m/sec. How long will it take to hit the ground 45 m below? Why can you ignore the horizontal velocity of the shingle in calculating the answer?

9. A projectile is launched with a horizontal speed of 8 m/sec.

(a) How much time will the projectile take to hit the ground 80 m below?

(b) How does this time change if the horizontal speed is doubled?

(c) What is the vertical speed of the projectile when it hits the ground?

(d) What is the horizontal speed of the projectile when it hits the ground?

10. In Galileo's drawing on page 107, the distances *bc, cd, de*, etc., are equal. What is the relationship among the distances *bo, og, gl*, and *ln*?

11. You are inside a van that is moving with a constant velocity. You drop a ball.

(a) What would be the ball's path relative to the van?

(b) Sketch its path relative to a person driving past the van at a high uniform speed.

(c) Sketch its path relative to a person standing on the road.

You are inside a moving van that is accelerating uniformly in a straight line. When the van is traveling at 10 km/hr (and still accelerating) you drop a ball from near the roof of the van onto the floor.

(d) What would be the ball's path relative to the van?

(e) Sketch its path relative to a person driving past the van at a high uniform speed.

(f) Sketch its path relative to a person standing on the road.

12. Two persons watch the same object move. One says it accelerates straight downward, but the other claims it falls along a curved path. Describe conditions under which each observation would be correct.

13. An airplane has a gun that fires bullets straight ahead at the speed of 1,000 km/hr while the plane is stationary on the ground. The plane takes off and flies due east at 1,000 km/hr. Which of the following describes what the pilot of the plane will see? In defending your answers, refer to the Galilean relativity principle.

(a) When fired directly ahead, the bullets moved eastward at a speed of 2,000 km/hr.

(b) When fired in the opposite direction, the bullets dropped vertically downward.

(c) When fired vertically downward, the bullets moved eastward at 1,000 km/hr while they fell.

Specify the frames of reference from which (a), (b), and (c) are the correct observations.

14. Many commercial record turntables are designed to rotate at frequencies of 16⅔ rpm (called transcription speed), 33⅓ rpm (long playing), 45 rpm (pop singles), and 78 rpm (old fashioned). What is the period corresponding to each of these frequencies?

15. Passengers on the right side of the car in a left turn have the sensation of being "thrown against the door." Explain what actually happens to the passengers in terms of force and acceleration.

16. The tires of the turning car in the example on page 116 were being pushed sideways by the road with a total force of 4 kN. Of course the tires would be pushing on the road with a force of 4 kN also.

(a) What happens if the road is covered with loose sand or gravel?

(b) How would softer (lower pressure) tires help?

(c) How would banking the road (that is, tilting the surface toward the center of the curve) help? (Hint: Consider the extreme case of banking in the bobsled photo on p. 114.)

17. Using a full sheet of paper, make and complete a table like the one below.

Name of Concept	*Symbol*	*Definition*	*Example*
		Length of a path between any two points, as measured along the path.	
			Straight-line distance and direction from Detroit to Chicago.
	v		
Instantaneous speed			

			An airplane flying west at 640 km/hr at constant altitude.
		Time rate of change of velocity.	
	a_g		
Centripetal acceleration			
			The drive shaft of some automobiles turns 600 rpm in low gear.
		The time it takes to make one complete revolution.	

18. Our sun is located at a point in our galaxy about 30,000 light years (1 light year = 9.46×10^{12} km) from the galactic center. It is thought to be revolving around the center at a linear speed of approximately 250 km/sec.

(a) What is the sun's centripetal acceleration with respect to the center of the galaxy?

(b) The sun's mass can be taken to be 1.98×10^{30} kg. What centripetal force is required to keep the sun moving in a circular orbit about the galactic center?

(c) Compare the centripetal force in **(b)** with that necessary to keep the earth in orbit about the sun. (The earth's mass is 5.98×10^{24} kg, and its average distance from the sun is 1.495×10^{8} km.)

19. The hammer thrower in the photograph on page 121 is exerting a large centripetal force to keep the hammer moving fast in a circle and applies it to the hammer through a connecting wire. The mass of the hammer is 7.27 kg.

(a) Estimate the radius of the circle and the period. Calculate a rough value for the amount of force required just to keep the hammer moving in a circle.

(b) What other components are there to the total force exerted on the hammer?

20. Contrast rectilinear motion, projectile motion, and uniform circular motion by

(a) defining each.

(b) giving examples.

(c) describing the relation between velocity and acceleration in each case.

21. An object of mass 2 kg is swung in a circle of radius 2 m, taking 2 sec to complete the circle. What is the speed of the object? What is the size and direction of the acceleration keeping the object moving in a circle? What is the size and direction of the force?

22. A force of 5 N is required to hold a stone of mass 2.5 kg in a circle of 5-m radius. What must be the speed of uniform circular motion?

23. An object 3 m away from the center of its uniform circular motion moves with an acceleration of 10 m/sec^2 toward the center. How long does the object take to complete one circle? Why is it not necessary to know the mass of the object in order to find the answer?

24. A fan blade takes 0.1 sec to go around. What is its frequency? What is the acceleration of a point 0.5 m from the center?

25. These questions refer to Table 4.2 on page 120.

(a) Which satellite has the most nearly circular orbit?

(b) Which satellite has the most eccentric orbit? How did you arrive at your answer?

(c) Which has the longest period?

(d) How does the position of Syncom 2 relative to a point on earth change over one day?

26. A satellite is put into an orbit 400 km above the surface of the earth (6,800 km from the center of the earth) where the acceleration owing to gravity is 8.7 m/sec^2. What is the satellite's speed? Based on the orbit, are there any restrictions on the mass of the satellite?

27. A satellite with a mass of 500 kg requires 380 min to circle the earth in an orbit 18,000 km from the center of the earth. What is the magnitude of the force holding the satellite in orbit?

28. A satellite with a mass of 500 kg orbits the moon (whose gravity is one sixth that of earth's) at a distance of 18,000 km from the center of the moon.

What is the magnitude of the force acting on the satellite? How long would it take to circle the moon?

29. If the earth had no atmosphere, what would be the period of a satellite skimming just above the earth's surface? What would its speed be?

30. Explain why it is impossible to have an earth satellite orbit the earth in 80 min. Does this mean that it is impossible for any object to go around the earth in less than 80 min?

31. What was the period of the "110-km" Apollo 8 lunar orbit?

32. Knowing a_g near the moon's surface and the orbital speed in an orbit near the moon's surface, we can now work an example of Part 8 of the earth–moon trip described in Sec. 4.1. The Apollo 8 capsule was orbiting about 100 km above the surface. The value of a_g near the moon's surface is about 1.5 m/sec^2.

If the rocket's retro-engines are fired, it will slow down. Consider the situation in which the rockets fire long enough to reduce the capsule's horizontal speed to 100 m/sec.

(**a**) About how long will the fall to the moon's surface take?

(**b**) About how far will it have moved horizontally during the fall?

(**c**) About how far in advance of the landing target might the "braking" maneuver be performed?

33. Assume that a capsule is approaching the moon along the right trajectory, so that it will be moving tangent to the desired orbit. Given the speed v_o necessary for orbit and the current speed v, how long should the engine with thrust F fire to give the capsule of mass m the right speed?

34. The intention of the first four chapters has been to describe "simple" motions and to progress to the description of more "complex" motions. Classify each of the following examples as "simplest motion," "more complex," or "very complex." Be prepared to defend your choices and state any assumptions you made.

(**a**) helicopter landing

(**b**) "human cannon ball" in flight

(**c**) car going from 50 km/hr to a complete stop

(**d**) tree growing

(**e**) child riding a Ferris wheel

(**f**) rock dropped 3 km

(**g**) person standing on a moving escalator

(**h**) climber ascending Mt. Everest

(**i**) person walking

(**j**) leaf falling from a tree

35. Write a short essay on the physics involved in the motions shown in the picture below, using the ideas on motion from Unit 1.

EPILOGUE

This unit dealt with the fundamental concepts of motion. We started by analyzing very simple kinds of motion. After learning the "ABC's" of physics, we expected to be able to turn our attention to some of the more complex features of the world. To what extent were our expectations fulfilled?

We did find that a relatively few basic concepts gave us a fairly solid understanding of motion. We could describe many motions of objects by using the concepts of distance, displacement, time, speed, velocity, and acceleration. To these concepts we added force and mass and the relationships expressed in Newton's three laws of motion. With this knowledge, we found we could describe most observed motion in an effective way. The surprising thing is that these concepts of motion, which were developed in very restricted circumstances, can be so widely applied. For example, our discussion of motion in the laboratory centered around the use of sliding dry-ice disks and steel balls rolling down inclined planes. These are not objects ordinarily found moving around in the everyday "natural" world. Yet we found that the ideas obtained from those specialized experiments led to an understanding of objects falling near the earth's surface, of projectiles, and of objects moving in circular paths. We started by analyzing the motion of a disk of dry ice moving across a smooth surface. We ended up analyzing the motion of a space capsule as it circles the moon and descends to its surface.

We have made quite a lot of progress in analyzing complex motions. On the other hand, we cannot be certain that we have here all the tools needed to understand all the phenomena that interest us. In Unit 3, we will add to our stock of fundamental concepts a few additional ones, particularly those of momentum, work, and energy. They will help us when we turn our attention from interactions involving a relatively few objects of easily measured size, to interactions involving countless numbers of submicroscopic objects such as molecules and atoms.

In this unit, we have dealt mainly with concepts that owe their greatest debts to Galileo, Newton, and their followers. If space had permitted, we should also have included the contributions of René Descartes and the Dutch scientist Christian Huygens. The mathematician and philosopher A. N. Whitehead, in *Science and the Modern World,* has summarized the role of these four men and the importance of the concepts we have been studying as follows:

> This subject of the formation of the three laws of motion and of the law of gravitation [which we shall take up in Unit 2] deserves critical attention. The whole development of thought occupied exactly two generations. It commenced with Galileo and ended with Newton's *Principia*; and Newton was born in

> the year that Galileo died. Also the lives of Descartes and Huygens fall within the period occupied by these great terminal figures. The issue of the combined labours of these four men has some right to be considered as the greatest single intellectual success which mankind has achieved.

The laws of motion that Whitehead mentions were the subject of this unit. They were important most of all because they suddenly allowed a new understanding of celestial motion. For at least 20 centuries people had been trying to reduce the complex motions of the stars, sun, moon, and planets to an orderly system. The genius of Galileo and Newton lay in their studying the nature of motion as it occurs on earth, and then assuming that the same laws would apply to objects in the heavens beyond human reach.

Unit 2 is an account of the centuries of preparation that paved the way for the great success of this idea. We will trace the line of thought, starting with the formulation of the problem of planetary motion by the ancient Greeks. We will continue through the work of Copernicus, Tycho Brahe, Kepler, and Galileo to provide a planetary model and the laws of planetary motion. Finally, we will discover Newton's magnificent synthesis of terrestrial and celestial physics through his law of universal gravitation.

UNIT

Motion in the Heavens

CHAPTER 5 **Where Is the Earth? The Greeks' Answers**
CHAPTER 6 **Does the Earth Move? The Work of Copernicus and Tycho**
CHAPTER 7 **A New Universe Appears: The Work of Kepler and Galileo**
CHAPTER 8 **The Unity of Earth and Sky: The Work of Newton**

PROLOGUE

Astronomy, the oldest science, deals with objects now known to lie vast distances from the earth. To early observers, the sun, moon, planets, and stars did not seem to be so far away. Yet always, even today, the majesty of celestial events has fired our imagination and curiosity. The ancients noted the great variety of objects visible in the sky, the regularity of their motions, the strangely slow changes in their position and brightness. This whole mysterious pattern of motions required some reason, some cause, some explanation.

To the eye, the stars and planets appear as very well-defined pinpoints of light. They are easy to observe and follow precisely, unlike most other naturally occurring phenomena. The sky therefore provided the natural "laboratory" for beginning a science based on the ability to abstract, measure, and simplify.

Astronomical events not only affected the imagination of the ancients, but also had a practical effect on everyday life. The working day began when the sun rose, and it ended when the sun set. Before electric lighting, human activity was dominated by the presence or absence of daylight and the sun's warmth, which changed season by season.

Of all time units commonly used, "one day" is probably the most basic and surely the most ancient. For counting longer intervals, a "moon" or month was an obvious unit. Over the centuries, clocks were devised to subdivide days into smaller units, and calendars were invented to record the passage of days into years.

The Aztec calendar, carved over 100 years before our calendar was adopted, divides the year into 18 months of 20 days each.

When the early nomadic tribes settled down to live in villages some 10,000 years ago, they became dependent upon agriculture for their food. They needed a calendar for planning their plowing and sowing. Throughout recorded history, most of the world's population has been involved in agriculture and so has depended on a calendar. If seeds were planted too early, they might rot in the ground, or the young shoots might be killed by a frost. If they were planted too late, the crops would not ripen before winter came. Therefore, a knowledge of the best times for planting and harvesting was important for survival. Because religious festivals were often related to the seasons, the job of making and improving the calendar often fell to priests. Such improvements required observation of the sun, planets, and stars. The first astronomers, therefore, were probably priests.

Even in modern times people who live much in the open use the sun by day and the stars by night as a clock. True south can be determined from the position of the sun, at local noon. The Pole Star gives a bearing on true north after dark.

Practical needs and imagination acted together to give astronomy an early importance. Many of the great buildings of ancient times were constructed and situated with a clear awareness of astronomy. The great pyramids of Egypt, tombs of the Pharaohs, have sides that run due north–south and east–west. The awesome circles of giant stones at Stonehenge in England appear to have been arranged about 2000 B.C. to permit accurate observations of the positions of the sun and moon. The Mayans and the Incas in America as well as the ancient civilizations of India and China put enormous effort into buildings from which they could measure changes in the positions of the sun, moon, and planets. At least as early as 1000 B.C. the Babylonians and Egyptians had developed considerable ability in timekeeping. Their recorded observations are still being unearthed.

The positions of Jupiter from 132 B.C. to 60 B.C. are recorded on this section of Babylonian clay tablet, now in the British Museum.

Thus, for thousands of years, the motions of the heavenly bodies were carefully observed and recorded. In all science, no other field has had such a long accumulation of data as astronomy.

But our debt is greatest to the Greeks, who began trying to deal in a new way with what they saw. The Greeks recognized the contrast between the apparently haphazard and short-lived

Stonehenge, England, was apparently a prehistoric observatory.

motions of objects on earth and the unending cycles of the heavens. About 600 B.C. they began to ask new questions: How can we explain these cyclic events in the sky in a simple way? What order and sense can we make of them? The Greeks' answers, discussed in Chapter 5, had an important effect on science. For example, the writings of Aristotle (about 330 B.C.) became widely studied and accepted in Western Europe after 1200 A.D., and they were important factors in the scientific revolution that followed.

After the conquests of Alexander the Great, the center of Greek thought and science shifted to Egypt. At the new city of Alexandria, founded in 332 B.C., a great museum similar to a modern research institute was created. It flourished for many centuries. But as Greek civilization gradually declined, the practical-minded Romans captured Egypt, and interest in science died out. In 640 A.D., Alexandria was captured by the Muslims as they swept along the southern shore of the Mediterranean Sea and moved northward through Spain to the Pyrenees. Along the way they seized and preserved many libraries of Greek documents, some of which were later translated into Arabic and carefully studied. During the following centuries, Muslim scientists made new and better observations of the heavens. However, they made no major changes in the explanations or theories of the Greeks.

In Western Europe during this period, the works of the Greeks were largely forgotten. Eventually Europeans rediscovered them through Arabic translations found in Spain after the Muslims were forced out. By 1130 A.D., complete manuscripts of at least one of Aristotle's books were known in Italy and France. After the founding of the University of Bologna in the late twelfth century, and of the University of Paris around 1200, many other writings of Aristotle were acquired. Scholars studied these writings both in Paris and at the new English universities, Oxford and Cambridge.

In the twelfth century, the Muslim scholar Ibn Rashd had attempted a similar union of Aristotelianism and Islam.

During the next century, the Dominican monk Thomas Aquinas blended major elements of Greek thought and Christian theology into a single philosophy. His work was widely studied and accepted in Western Europe for several centuries. In achieving this synthesis, Aquinas accepted the physics and astronomy of Aristotle. Because the science was blended with theology, any questioning of the science seemed also to question the theology. Thus, for a time there was little effective criticism of Aristotelian science.

The Renaissance movement, which spread out across Europe from Italy, brought new art and music. It also brought new ideas about the universe and humanity's place in it. Curiosity and a questioning attitude became acceptable, even prized. Scholars acquired a new confidence in their ability to learn about the world. Among those whose work introduced the new age were Columbus and Vasco da Gama, Gutenberg and da Vinci, Michelangelo and Raphael, Erasmus and Vesalius, Luther, Calvin, and Henry VIII. (The chart in Chapter 6 shows their life spans.) Within this emerging Renaissance culture lived Niklas Koppernigk, later called Copernicus, whose reexamination of astronomical theories is discussed in Chapter 6.

Further improvements in astronomical theory were made in the seventeenth century by Kepler, mainly through mathematical reasoning, and by Galileo, through his observations and writings. These contributions are discussed in Chapter 7. Chapter 8 deals with Newton's work in the second half of the seventeenth century. Newton's genius extended ideas about motion on earth to explain motion in the heavens, a magnificent synthesis of terrestrial and celestial dynamics. These men, and others like them in other sciences such as anatomy and physiology, literally changed the world. The results they obtained and the ways in which they went about their work had effects so far-reaching that we generally refer to their work as the *Scientific Revolution*.

Great scientific advances can, and often do, affect ideas outside science. For example, Newton's work helped to create a new feeling of self-confidence. It seemed possible to understand all things in the heavens and on the earth. This great change in attitude was a major characteristic of the eighteenth century,

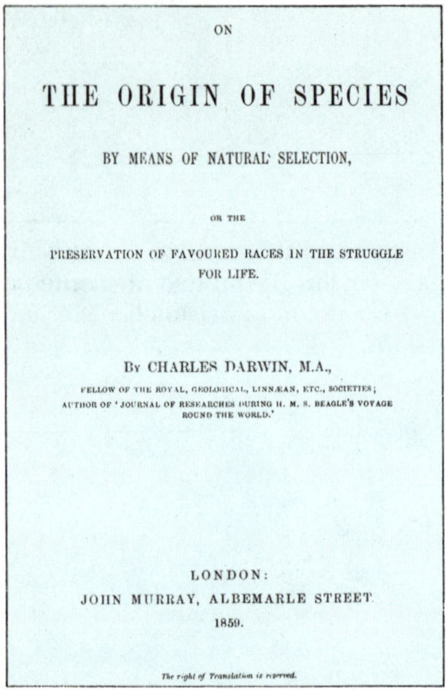

ON

THE ORIGIN OF SPECIES

BY MEANS OF NATURAL SELECTION,

OR THE

PRESERVATION OF FAVOURED RACES IN THE STRUGGLE FOR LIFE.

BY CHARLES DARWIN, M.A.,

FELLOW OF THE ROYAL, GEOLOGICAL, LINNÆAN, ETC., SOCIETIES;
AUTHOR OF 'JOURNAL OF RESEARCHES DURING H. M. S. BEAGLE'S VOYAGE ROUND THE WORLD.'

LONDON:
JOHN MURRAY, ALBEMARLE STREET.
1859.

The right of Translation is reserved.

which has been called the Age of Reason. To a degree, what we think today and how we run our affairs are still affected by scientific discoveries made centuries ago.

Broad changes in thought developed at the start of the Renaissance and grew for nearly a century, from the work of Copernicus to that of Newton. In a sense, this era of invention resembles the sweeping changes that have occurred during the past hundred years. This recent period might extend from the publication of Darwin's *Origin of Species* in 1859 to the first controlled release of nuclear energy in 1942. Within this interval lived such great scientists as Mendel and Pasteur, Planck and Einstein, Rutherford and Fermi. The ideas they and others introduced into science have become increasingly important. These scientific ideas are just as much a part of our time as the ideas and works of people such as Roosevelt and Ghandi, Martin Luther King and Pope John XXIII, Marx and Lenin, Freud and Dewey, Picasso and Stravinsky, Shaw and Joyce. If we understand how science influenced the people of past centuries, we can better understand how science influences our thought and lives today. This is clearly one of the basic aims of this course.

The material treated in this unit is historical as well as scientific. Even today, the history of science remains of great importance to anyone interested in understanding science. The reasons for presenting the science of astronomy in its historical framework include the following: The results that were finally obtained still hold true and rank among the oldest ideas used every day in scientific work. The characteristics of all scientific work are clearly visible. We can see the role of assumptions, of experiment and observations, of mathematical theory. We can note the social mechanisms for cooperation, teaching, and disputing. And we can appreciate the possibility of having one's scientific findings become part of the accepted knowledge of the time.

There is an interesting conflict between the rival theories used to explain the same set of astronomical observations. This conflict is typical of all such disputes down to our day. Thus, it helps us to see clearly what standards may be used to judge one theory against another.

This subject matter includes the main reasons for the rise of science as we understand it now. The story of the Scientific Revolution and its many effects outside science itself is as necessary to understanding our current age of science as is the story of the American Revolution to an understanding of America today.

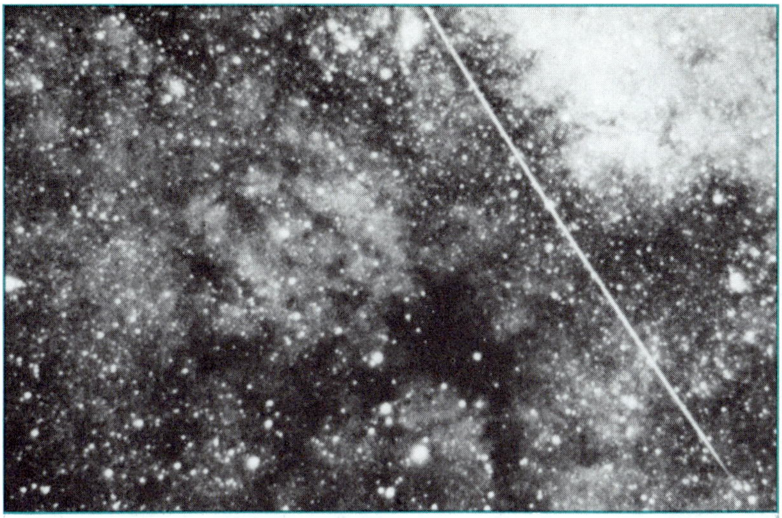

Where Is the Earth?
The Greeks' Answers

5.1 | Motions of the sun and stars

SG 1

The facts of everyday astronomy, the celestial events themselves, are the same now as in the times of the ancient Greeks. You can observe with your unaided eyes most of what these early scientists saw and recorded. You can discover some of the long-known cycles and rhythms: the seasonal changes of the sun's height at noon, the monthly phases of the moon, and the glorious spectacle of the slowly turning night sky. If you wish only to forecast eclipses, planetary positions, and the seasons, you could, like the Babylonians and Egyptians, focus your attention on recording the details of the cycles and rhythms. Suppose, however, like the Greeks, you wish to *explain* these cycles. Then you must also use your data and imagination to

The motions of these bodies, essentially the same as they were thousands of years ago, are not difficult to observe. You should make a point of doing so. The *Handbook* has many suggestions for observing the sky, both with the naked eye and with a small telescope.

The midnight sun, photographed at 5-min intervals over the Ross Sea in Antarctica.

construct some sort of simple model or theory with which you can predict the observed variations. Before you explore several theories proposed in the past, review the *major observations* which the theories tried to explain: the motions of the sun, moon, planets, and stars.

The most basic celestial cycle as seen from the earth is, of course, that of day and night. Each day the sun rises above the local horizon on the eastern side of the sky and sets on the western side. The sun follows an arc across the sky, as is sketched in part (a) of the diagram at the top of the next page. At noon, halfway between sunrise and sunset, the sun is highest above the horizon. Every day, it follows a similar path from sunrise to sunset. Indeed all the objects in the sky show this pattern of daily motion. They all rise in the east, reach a high point, and drop lower in the west. (However, some stars never actually sink below the horizon.)

This description is for observers in the northern hemisphere. For observers south of the equator, exchange "north" and "south."

As the seasons change, so do the details of the sun's path across the sky. In the northern hemisphere during winter, the sun rises and sets more to the south. Its altitude at noon is lower, and so its run across the sky lasts for a shorter period of time. In summer the sun rises and sets more toward the north. Its height at noon is greater, and its track across the sky lasts a longer time. The whole cycle takes a little less than 365¼ days. In the southern hemisphere the pattern is similar, but is displaced by half a year.

This year-long cycle north and south is the basis for the seasonal or "solar" year. Apparently, the ancient Egyptians once thought that the year had 360 days, but they later added five feast days to have a year of 365 days. This longer year agreed better with their observations of the seasons. Now we know that the solar year is 365.24220 days long. The decimal fraction 0.24220 raises a problem for the calendar maker, who works with whole days. If you used a calendar of just 365 days, after four years New Year's Day would come early by one day. In a century, you would be in error by almost a month. In a few centuries, the date called January 1 would come in the summertime! In ancient times, extra days or even whole months were inserted from time to time to keep a calendar of 365 days in fair agreement with the seasons.

Such a makeshift calendar is, however, hardly satisfactory. In 45 B.C., Julius Caesar decreed a new 365-day calendar (the Julian calendar) with one extra whole day (a "leap day") inserted each fourth year. Over many years, the average would therefore be 365¼ days per year. This calendar was used for centuries, during which the small difference between ¼ (0.25) and 0.24220 added up to several days. Finally, in 1582 A.D. Pope Gregory announced a new calendar (the Gregorian calendar). This calendar had only 97 leap days in 400 years, and the new approximation has lasted satisfactorily to this day without revision.

You may have noticed that a few stars are bright and many are faint. The brighter stars may seem to be larger, but if you look at them through binoculars, they still appear as points of light. Some bright stars show colors, but most appear whitish. People have grouped many of the brighter stars into patterns, called *constellations*. Examples of constellations include the familiar Big Dipper and Orion.

You may have also noticed a particular pattern of stars overhead and then several hours later noticed it low in the west. What was happening? More detailed observation, for example, by taking a time-exposure photograph, would show that the entire bowl of stars had moved from east to west. New stars had risen in the east, and others had set in the west. As seen from the northern hemisphere, during the night the stars appear to move counterclockwise around a point in the sky called the north celestial pole. This stationary point is near the fairly bright star Polaris (see the photograph at the top of page 136).

Some star patterns, such as Orion (the Hunter) and Cygnus (the Swan, also called the Northern Cross), were described and named thousands of years ago. Since the star patterns described by the ancients still fit, we can conclude that the star positions change very little, if at all, over the centuries. This constancy of relative positions has led to the term "fixed stars."

Thus, we observe in the heavens unchanging relationships over the centuries and smooth, orderly motions each day. But,

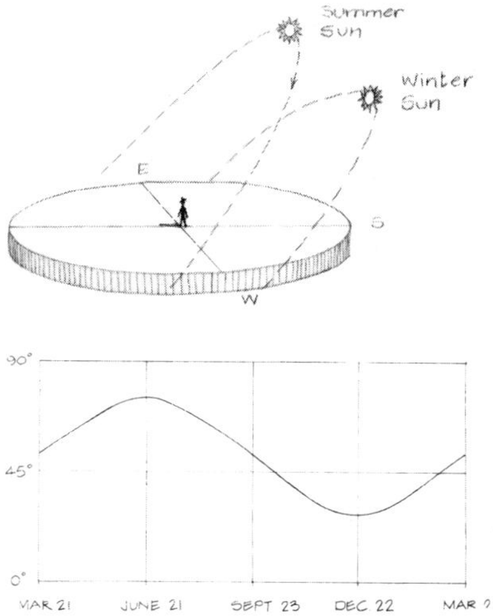

(a) Path of the sun through the sky for one day of summer and one day of winter.
(b) Noon altitude of the sun as seen from St. Louis, Missouri, throughout the year.

SG 2

SG 3

To reduce the number of the leap days from 100 to 97 in 400 years, century years not divisible by 400 were omitted as leap years. Thus, the year 1900 was not a leap year, but the year 2000 will be a leap year.

SG 4

SG 5

A very easy but precise way to time the motions of the stars is explained in the *Handbook*.

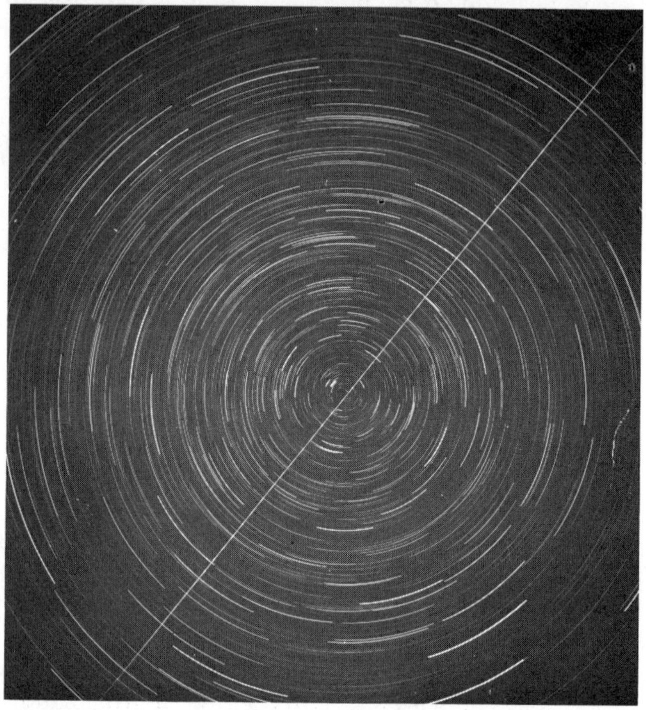

A combination trail and star photograph of the constellation Orion. The camera shutter was opened for several hours while the stars moved across the sky (leaving trails on the photographic plate). Then the camera was closed for a few minutes and reopened while the camera was moved to follow the stars.

Right:
Time-exposure showing star trails around the north celestial pole. The diagonal line was caused by the rapid passage of an artificial earth satellite.

You can use a protractor to determine the duration of the exposure; the stars appear to move about 15° per hour.

although the daily rising and setting cycles of the sun and stars are similar, they are not identical. Unlike the sun's path, the paths of the stars do not vary in altitude from season to season. Also, stars do not have quite the same rhythm of rising and setting as the sun, but go a little faster. Some constellations seen high in the sky soon after sunset appear low in the west at the same time several weeks later. As measured by sun-time, the stars set about 4 minutes earlier each day.

Thus far, the positions and motions of the sun and stars have been described in relation to the observer's horizon. But different observers have different horizons. Therefore, the horizon cannot be used as a frame of reference from which all observers will see the same positions and motions in the sky. However, the fixed stars provide a frame of reference which *is* the same for all observers. The positions of these stars relative to one another do not change as the observer moves over the earth. Also, their daily

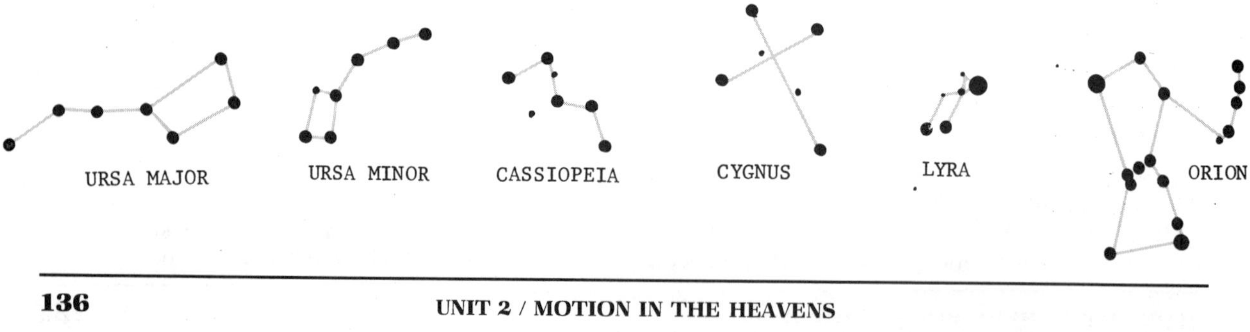

motions are simple circles with almost no changes during a year or through the years. For this reason, positions in the heavens are usually described in terms of a frame of reference defined by the stars.

A description of the sun's motion must include the daily crossing of the sky, the daily difference in rising and setting times, and the seasonal change in noon altitude. You have already seen that, as measured by sun-time, each star sets about 4 minutes earlier each day than it did the previous day; it goes ahead of the sun toward the west. You can just as well say that, measured by star-time, the sun sets about 4 minutes *later* each day; that is, the sun appears gradually to *slip behind* the daily east-to-west motion of the stars. In other words, the sun moves very slowly *eastward* against a background of "fixed" stars.

The difference in noon altitude of the sun during the year corresponds to a drift of the sun's path north and south on the background of stars. In the first diagram below, the middle portion of the sky is represented by a band around the earth. The sun's yearly path against this background of stars is represented by the dark line. If you cut and flatten out this band, as shown in the second and third diagrams, you get a chart of the sun's path during the year. (The 0° line is the *celestial equator*, an imaginary line in the sky directly above the earth's equator.) The sun's path against the background of the stars is called the *ecliptic*. Its drift north and south of the celestial equator is about 23.5°. You also need to define one point on the ecliptic so you can locate the sun or other celestial objects along it. For centuries this point has been the place where the sun crosses the equator from south to north on about March 21. This point is called the *vernal (spring) equinox*. It is the zero point from which positions among the stars usually are measured.

Thus, there are three apparent motions of the sun: (1) its daily westward motion across the sky, (2) its yearly drift eastward among the stars, (3) its yearly cycle of north–south drift in noon altitude. These cyclic events can be described by using a simple model to represent them.

The differences between the two frames of reference (the horizon and the fixed stars) are the basis for establishing a position on the earth, as in navigation.

In New Mexico, a construction of stone slabs directs beams of sunlight onto spiral carvings on a cliff face. The light forms changing patterns throughout the year and also marks the solstices and equinoxes. This unique astronomical marker makes use of the changing height of the midday sun throughout the year; it was apparently built by the Anasazi Indians. The above photo shows the light pattern at 11:13 on the morning of summer solstice.

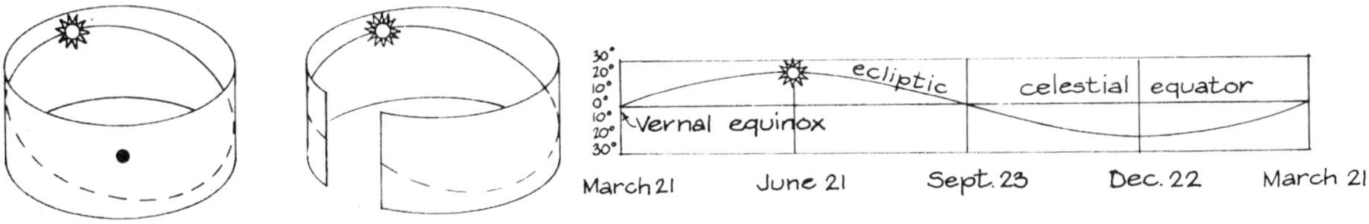

?

1. *If you told time by the stars, would the sun set earlier or later each day?*
2. *For what practical purposes were calendars needed?*
3. *What are the observed motions of the sun during one year?*

These end-of-section questions are intended to help you check your understanding before going on to the next section.

5.2 | Motions of the moon

The moon shares the general east-to-west daily motion of the sun and stars. But the moon slips eastward against the background of the stars faster than the sun does. Each night the moon rises nearly 1 hour later. When the moon rises in the east at sunset (opposite the sun in the sky), it is a bright, full disc (full moon). Each day after that, it rises later and appears less round. Finally, it wanes to a thin crescent low in the sky at dawn. After about 14 days, the moon is passing near the sun in the sky and rising with it. During this time (new moon), you cannot see the moon at all. After the new moon, you first see the moon as a thin crescent low in the western sky at sunset. As the moon moves rapidly eastward from the sun, its crescent fattens to a half disc at first quarter. Within another week, it reaches full moon again. After each full moon the cycle repeats itself.

The moon as it looks 26 days after new moon (left); 17 days after new moon (middle); and 3 days after new moon (right).

As early as 380 B.C., the Greek philosopher Plato recognized that the phases of the moon could be explained by thinking of the moon as a globe reflecting sunlight and moving around the earth in about 29 days. Because the moon appears so big and moves so rapidly compared to the stars, people in early times thought that it must be quite close to the earth.

The moon's path around the sky is close to the yearly path of the sun; that is, the moon is always near the ecliptic. But the moon's path tilts a bit with respect to the sun's path. If it did not, the moon would come exactly in front of the sun at every new moon, causing an eclipse of the sun. It would be exactly opposite the sun at every full moon, moving into the earth's shadow and causing an eclipse of the moon.

The motions of the moon have been studied with great care for centuries, partly because of interest in predicting eclipses. These motions are very complicated. The precise prediction of the moon's position is an exacting test for any theory of motion in the heavens.

SG 6

4. Draw a rough diagram to show the relative positions of the sun, earth, and moon during each of the moon's four phases.

5. Why do eclipses not occur each month?

6. (a) What are the observed motions of the moon during a month?

(b)) How do these motions change during the year?

(c) What do you think is the origin of the word "month"? Look it up in the dictionary.

5.3 | The "wandering" stars

Without a telescope you can see, in addition to the sun and moon, five rather bright objects that move among the stars. These are the "wanderers," or planets: Mercury, Venus, Mars, Jupiter, and Saturn. With the aid of telescopes, three more planets have been discovered: Uranus, Neptune, and Pluto. None of these three planets were known until nearly a century after the time of Isaac Newton. Like the sun and moon, all the planets rise daily in the east and set in the west. Also like the sun and moon, the planets generally move slowly eastward among the stars. But they have another remarkable and puzzling motion of their own. At certain times, each planet stops moving eastward among the stars and for some months loops back westward. This westward or "wrong-way" motion is called *retrograde motion*. The retrograde loops made by Mercury, Mars, and Saturn during 1 year are plotted on page 140.

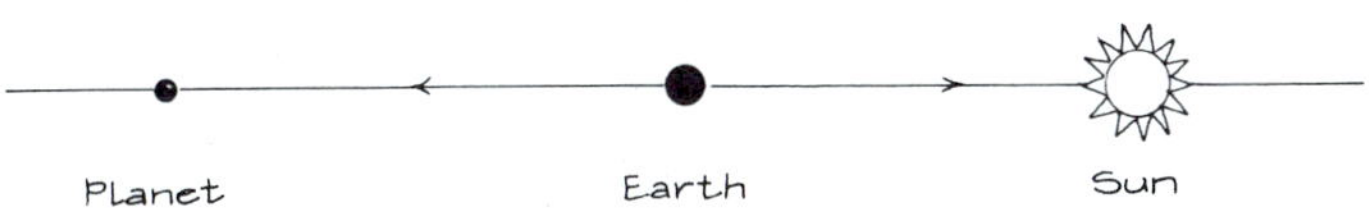

When a planet is observed directly opposite the sun, the planet is said to be in opposition. *Retrograde motions of Mars, Jupiter, and Saturn are observed about the time they are in opposition.*

The paths of all the planets are close to the sun's path among the stars—the ecliptic. Mercury and Venus are always fairly near the sun. The greatest angular distance east or west of the sun is 28° for Mercury and 48° for Venus. The westward, or retrograde, motions of Mercury and Venus begin after the planets are farthest east of the sun and visible in the evening sky. The other planets may have any position relative to the sun. Their westward retrograde motion occurs around the time when they are opposite the sun (highest in the sky at midnight).

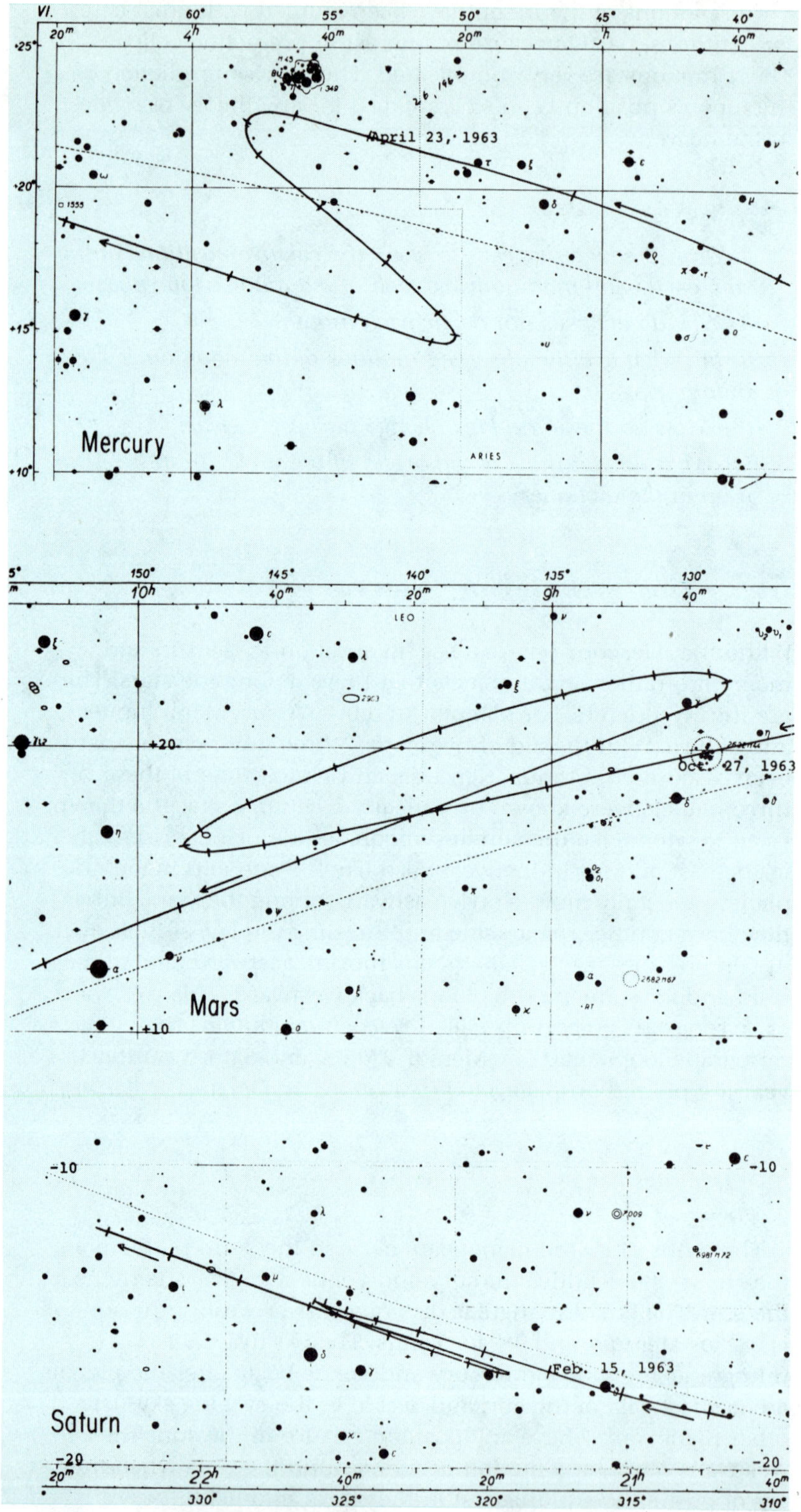

The retrograde motions of Mercury (marked at 5-day intervals), Mars (at 10-day intervals), and Saturn (at 20-day intervals) in 1963, plotted on a star chart. The dotted line is the annual path of the sun, called the ecliptic.

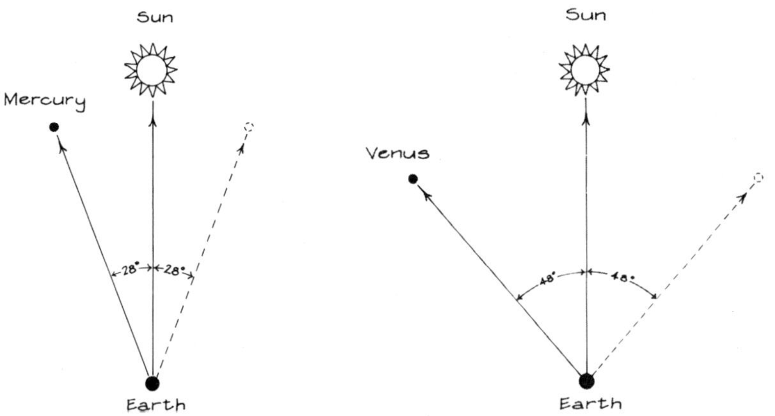

The maximum angles from the sun at which Mercury and Venus can be observed. Both planets can, at times, be observed at sunset or at sunrise. Mercury is never more than 28° from the sun, and Venus is never more than 48° from the sun.

SG 7

The planets change considerably in brightness. When Venus first appears in the evening sky as the "evening star," it is only fairly bright. During the following 4–5 months, it moves farther eastward from the sun. Gradually it becomes so bright that it often can be seen in daytime if the air is clear. A few weeks later, Venus moves westward toward the sun. It fades rapidly, passes the sun, and soon reappears in the morning sky before sunrise as the "morning star." Then it goes through the same pattern, but in the opposite order: bright at first, then gradually fading. Mercury follows much the same pattern. But because Mercury is seen only near the sun (that is, only during twilight), its changes are difficult to observe.

Mars, Jupiter, and Saturn are brightest about the time that they are in retrograde motion and opposite the sun. Yet over many years their maximum brightness differs. The change is most obvious for Mars; the planet is brightest when it is opposite the sun during August or September.

The sun, moon, and planets generally slip behind as the celestial sphere goes around the earth each day, and thus they appear to move eastward among the stars. Also, the moon and planets (except Pluto) are always found within a band, called the *Zodiac*, only 8° wide on either side of the sun's path.

These, then, are some of the main observations of celestial phenomena. All of them were known to the ancients. In their day as in ours, the puzzling regularities and variations seemed to cry out for some explanation.

?

7. In what part of the sky must you look to see the planets Mercury and Venus?

8. In what part of the sky would you look to see a planet that is in opposition to the sun?

9. When do Mercury and Venus show retrograde motion?

10. When do Mars, Jupiter, and Saturn show retrograde motion?

11. Can Mars, Jupiter, and Saturn appear any place in the sky?

12. What are the observed motions of the planets during the year?

5.4 | Plato's problem

In the fourth century B.C., Greek philosophers asked new questions: How can we explain the cycles of changes observed in the sky? What model can consistently and accurately account for the observed motions? Plato sought a theory to explain what was seen or, as he phrased it, "to save the appearances." The Greeks were among the first people to desire clear, rational explanations for natural events. Their attitude was an important step toward science as we know it today.

How did the Greeks begin their explanation of celestial motion? What were their assumptions?

Any answers to these questions must be partly guesswork. Many scholars over the centuries have devoted themselves to the study of Greek thought. But the documents on which our knowledge of the Greeks is based are mostly copies of copies and translations of translations. Many errors and omissions occur. In some cases, all we have are reports from later writers on what certain philosophers did or said. These accounts may be distorted or incomplete. The historians's task is difficult. Most of the original Greek writings were on papyrus or cloth scrolls which have decayed through the ages. Wars, plundering, and burning have also destroyed many important documents. Especially tragic was the burning of the famous library of Alexandria in Egypt, which contained several hundred thousand documents. (It was burned three times: in part by Caesar's troops in 47 B.C., in the fourth century A.D. by Christians, and about 640 A.D. by early Muslims when they overran the country.) The general picture of Greek culture is fairly clear, but many interesting details are missing.

Several centuries later, a more mature Islamic culture led to extensive study and scholarly commentary on the remains of Greek thought. Several centuries later still, a more mature Christian culture used the ideas preserved by the Muslims to evolve early parts of modern science.

The approach to celestial motion taken by the Greeks and their intellectual followers for many centuries was outlined by Plato in the fourth century B.C. He defined the problem to his students in terms of order and status. The stars, Plato said, represent eternal, divine, unchanging beings. They move at a uniform speed around the earth in the most regular and perfect of all paths, an endless circle. But the sun, moon, and planets wander across the sky by complex paths, including even retrograde motions. Yet, being heavenly bodies, surely they too are really moving in a way that suits their high status. Their motions, if not in a single perfect circle, must be in some combination of perfect circles. What combinations of circular motions at uniform speed could account for these strange variations?

Notice that the problem deals only with the changing *apparent* positions of the sun, moon, and planets. The planets appear to be only points of light moving against the background of stars. From two observations at different times, an observer can obtain a rate of motion—a value of so many degrees per day. The problem then is to invent a "mechanism," some combination of motions, that reproduces the observed angular motions and leads to accurate predictions. The ancient astronomers had no data for the distance of the planets from the earth. All they had were directions, dates, and rates of angular motion. They did know that the changes in brightness of the planets were related to their positions with respect to the sun. But these changes in brightness were not included in Plato's problem.

Plato and many other Greek philosophers assumed that there were only a few basic "elements." Mixed together, these few elements gave rise to the great variety of materials observed in the world. (See Unit 1, Chapter 2.) Perfection could only exist in the heavens, which were separate from the earth, and were the home of the gods. Just as motions in the heavens must be eternal and perfect, the unchanging heavenly objects could not contain elements normally found on or near the earth. Therefore, they were supposed to consist of a changeless fifth element (or *quintessence*).

Plato's problem in explaining the motion of planets remained the most important problem in astronomy for nearly 2,000 years. In later chapters, you will explore the different interpretations developed by Kepler, Galileo, and Newton. But in order to appreciate these efforts, you must first examine the solutions offered by the Greeks to Plato's problem. For their time, these solutions were useful, intelligent, and, indeed, beautiful.

?

13. What was Plato's problem of planetary motion?

14. Why is our knowledge of Greek science incomplete?

15. Why did the Greeks feel that they should use only uniform circular motions to explain celestial phenomena?

5.5 | The Greek idea of "explanation"

Plato's statement of this historic problem of planetary motion illustrates these major contributions of the Greek philosophers. With slight changes, these concepts are still basic to an understanding of the nature of physical theories:

1. A theory should be based on simple ideas. Plato regarded it not merely as simple, but also as self-evident, that heavenly

bodies must move uniformly along circular paths. Only in recent centuries have scientists learned that such common-sense beliefs may be misleading. While unproved assumptions are made at the outset, they must be examined closely and never accepted without reservation. As you will see often in this course, it has been very difficult to identify hidden assumptions in science. Yet in many cases, when the assumptions were identified and questioned, entirely new theories followed.

2. Physical theory must agree with the measured results of observation of *phenomena*, such as the motions of the planets. The purpose of a theory is to discover the uniformity of behavior, the hidden simplicity underlying apparent irregularities. For organizing observations, the language of numbers and geometry has become useful. Plato stressed the fundamental role of numerical data only in his astronomy, while Aristotle largely avoided detailed measurements. This was unfortunate because, as reported in the Prologue, Aristotle greatly influenced later scholars. His arguments, which gave little attention to the idea of measurement of change as a tool of knowledge, were adopted centuries afterward by such important philosophers as Thomas Aquinas.

3. To "explain" complex phenomena means to develop or invent a physical model, or a geometrical or other mathematical construction. This model must reproduce the same features as the phenomena to be explained. For example, with a model of interlocking spheres, a point on one of the spheres must have the same motions as the planet which the point represents.

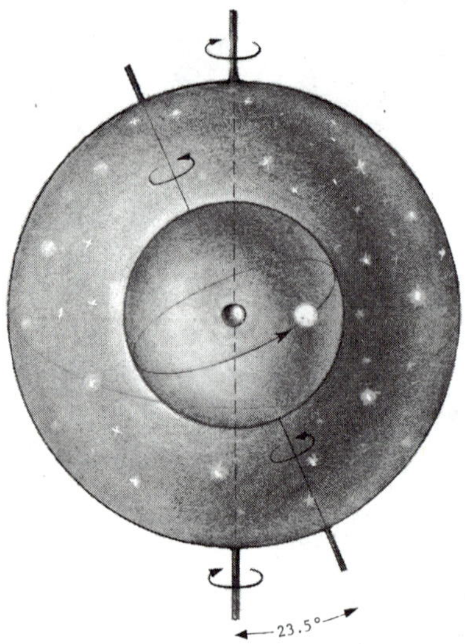

The annual north–south (seasonal) motion of the sun was explained by having the sun on a sphere whose axis was tilted 23.5° from the axis of the eternal sphere of the stars.

5.6 | The first earth-centered solution

The Greeks observed that the earth was large, solid, and permanent. Meanwhile the heavens seemed to be populated by small, remote objects that were continually in motion. What was more natural than to conclude that the big, heavy earth was the steady, unmoving center of the universe? Such an earth-centered viewpoint is called *geocentric*. From this viewpoint the daily motion of the stars could be explained easily: The stars were attached to, or were holes in, a large, dark, spherical shell surrounding the earth. They were all at the same distance from the earth. Daily, this celestial sphere turned once on an axis through the earth. As a result, all the stars fixed on it moved in circular paths around the pole of rotation. Thus, a simple model of a rotating celestial sphere and a stationary earth could explain the daily motions of the stars.

The three observed motions of the sun required a somewhat more complex model. To explain the sun's motion with respect to the stars, a separate, invisible shell was imagined. This shell was fixed to the celestial sphere and shared its daily motion. But it also had a slow, opposite motion of its own, amounting to one 360° cycle per year. The yearly north–south motion of the sun was accounted for by tilting the axis of its sphere. This adjustment matched the 23.5° tilt of the sun's path from the axis of the dome of stars.

The motions of the visible planets (Mercury, Venus, Mars, Jupiter, and Saturn) were more difficult to explain. These planets share generally the daily motion of the stars, but they also have peculiar motions of their own. Saturn moves most slowly among the stars, revolving once in 30 years. Therefore, its sphere was assumed to be largest and closest to the stars. Inside the sphere of Saturn were spheres carrying the faster-moving Jupiter (12 years) and Mars (687 days). Since they all require more than a year for a complete trip among the stars, these three planets were believed to lie beyond the sphere of the sun. Venus, Mercury, and the moon were placed between the sun and the earth. The fast-moving moon was assumed to reflect sunlight and to be closest to the earth.

This imaginary system of transparent shells or spheres provided a rough model for explaining the general motions of heavenly objects. By choosing the sizes of the spheres and their rates and direction of motions, one could roughly match the model with the observations. If additional observations revealed other cyclic variations, more spheres could be added to adjust the model. (An interesting description of this general system of cosmology appears in the *Divine Comedy*, written by the poet Dante about 1300 A.D. This was shortly after Aristotle's writings became known in Europe.)

You may feel that Greek science was bad science because it was different from our own or because it was less accurate. You should understand from your study of this chapter that such a conclusion is not justified. The Greeks were just beginning the development of scientific theories. Naturally, they made assumptions that appear odd or inaccurate to us today. Their science was not "bad science," but in many ways it was a different kind of science from ours. And our science is not the last word, either. You must realize that to scientists 2,000 years from now our efforts may seem clumsy and strange.

Even today's scientific theory does not and cannot claim to account for every detail of every specific situation. Scientific concepts are general ideas which treat only selected aspects of observations. They do not cover the whole mass of raw data and raw experience that exists in the universe. Also, each period in history puts its own limits on the range of human imagination. As you learned in Unit 1, important general concepts such as force and acceleration were invented specifically to help organize observations. Such concepts are human inventions.

The history of science contains many cases in which certain factors overlooked by one researcher later turn out to be very

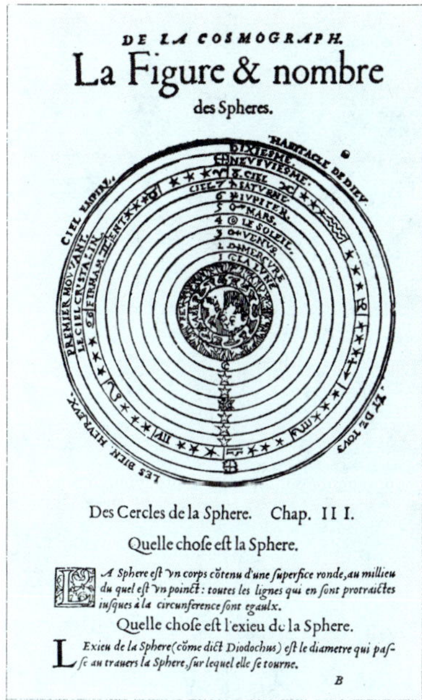
DE LA COSMOGRAPH.

La Figure & nombre des Spheres.

Des Cercles de la Sphere. Chap. III.

Quelle chofe eft la Sphere.

LA Sphere eft vn corps cõtenu d'une fuperfice ronde, au millieu du quel eft vn poinct: toutes les lignes qui en font protraictes iufques à la circunference font egaulx.

Quelle chofe eft l'exieu de la Sphere.

L'Exieu de la Sphere (cõme dict Diodochus) eft le diametre qui paffe au trauers la Sphere, fur lequel elle fe tourne.

B

A geocentric cosmological scheme. The earth is fixed at the center of concentric rotating spheres. The sphere of the moon (lune) *separates the terrestrial region (composed of concentric shells of the four elements Earth, Water, Air, and Fire) from the celestial region. In the latter are the concentric spheres carrying Mercury, Venus, Sun, Mars, Jupiter, Saturn, and the stars. To simplify the diagram, only one sphere is shown for each planet. (From the DeGolyer copy of Petrus Apianus'* Cosmographia, *1551.)*

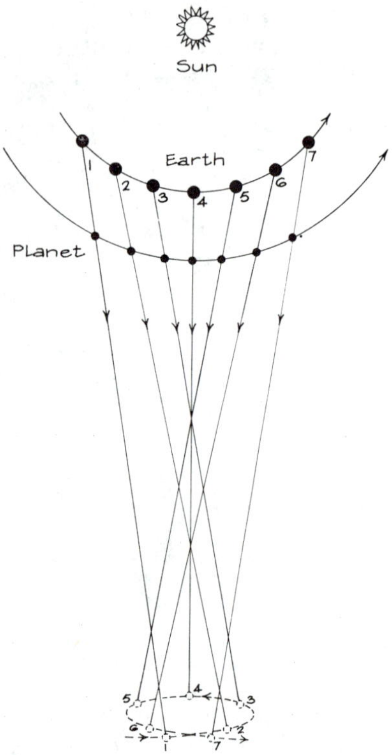

As the earth passes a planet in its orbit around the sun, the planet appears to move backwards in the sky. The arrows show the sight lines toward the planet for the different numbered positions of the earth. The lower-numbered circles symbolize the resulting apparent positions of the planet when seen against the background of the distant stars.

important. But how would better systems for making predictions be developed without first trials? Theories are improved through tests and revisions, and sometimes are completely replaced by better ones.

?

16. What is a geocentric system? How does it account for the motions of the sun?

17. Describe the first solution to Plato's problem.

5.7 | A sun-centered solution

For nearly 2,000 years after Plato and Aristotle, the basic geocentric model was generally accepted, though scholars debated certain details. But a very different model, based on different assumptions, had been proposed in the third century B.C. The astronomer Aristarchus, perhaps influenced by the earlier writings of Heracleides, offered this new model. Aristarchus suggested that a simpler explanation of heavenly motion would place the light-giving *sun* at the center, with the earth, planets, and stars all revolving around it. A sun-centered system is called *heliocentric.*

Aristarchus proposed that the celestial sphere is motionless and that the earth rotates once daily on an axis of its own. He believed that this assumption could explain all the daily motions observed in the sky. In this heliocentric system, the apparent tilt of the paths of the sun, moon, and all the planets results from the tilt of the earth's own axis. The yearly changes in the sky, including retrograde motions of planets, are explained by assuming that the earth and the planets revolve around the sun. In this model, the motion previously assigned to the sun around the earth is assigned to the earth moving around the sun. Also, the earth becomes just one among several planets. The planets are not the homes of gods, but are now considered to be bodies rather like the earth.

The diagram in the margin shows how such a system can explain the retrograde motions of Mars, Jupiter, and Saturn. An outer planet and the earth are assumed to be moving around the sun in circular orbits. The outer planet moves more slowly than the earth. As a result, when the earth is directly between the sun and the planet, the earth moves rapidly past the planet. To us the planet appears for a time to be moving backward in retrograde motion across the sky.

The interlocking spheres are no longer needed. The heliocentric hypothesis has one further advantage. It explains the observation that the planets are brighter during retrograde motion, since at that time the planets are nearer to the earth.

Even so, the proposal by Aristarchus was severely criticized for three basic reasons. The idea of a moving earth contradicted the philosophical doctrines that the earth is different from the celestial bodies and that its natural place is at the center of the universe. In fact, his contemporaries considered Aristarchus impious for even suggesting that the earth moved. Also, this new picture of the solar system contradicted common sense and everyday observations: The earth certainly *seemed* to be at rest rather than rushing through space.

Another criticism was that expected motions of the stars were not observed. If the earth moved in an orbit around the sun, it would also move back and forth under the fixed stars. As shown in the sketch in the margin, the angle from the vertical at which any star is seen would be different for various points in the earth's annual path. This shift is called *parallax*. This difference was not observed by the Greek astronomers. This awkward fact could be explained in two ways: either (1) the earth does *not* go around the sun and there is no shift, or (2) the earth does go around the sun but the stars are so far away that the shift is *too small to observe*. As the Greeks realized, for the shift to be too small to detect, the stars must be enormously far away.

Today, the annual shift of the stars can be observed with telescopes; thus, Aristarchus' model is in fact useful. The shift is so small that even with telescopes it was not measured until 1838. The largest annual shift is an angle of only 1/100 of the smallest angle observable by the human eye. The shift exists, but we can sympathize with the Greeks who rejected the heliocentric theory partly because they could not observe the required shift. Only Aristarchus imagined that the stars might be as immensely distant as we now know them to be.

Finally, Aristarchus was criticized because he did not develop his system in detail or use it to predict planetary positions. His work seems to have been purely qualitative, a general scheme of how things might be.

The geocentric and heliocentric systems offered two different ways of explaining the same observations. The heliocentric proposal required such a drastic change in people's image of the universe that Aristarchus' hypothesis had little influence on Greek thought. Fortunately, his arguments were recorded and handed down. Eighteen centuries later, they gained new life in the thoughts of Copernicus. Ideas are not bound by space or time.

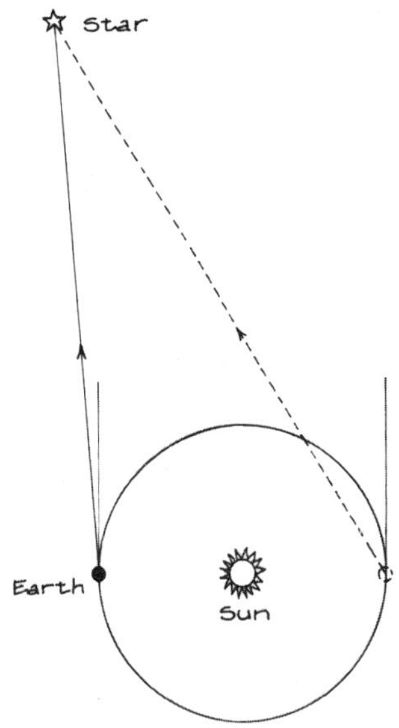

If the earth goes around the sun, then the direction in which we have to look for a star should change during the year. A shift in the relative observed positions of objects that is caused by a displacement of the observer is called a parallax. *The greatest observed parallax of a star caused by the earth's annual motion around the sun is about 1/2400°. This is explained by the fact that the distance to this nearest star is not just hundreds of millions of kilometers but 40 million million kilometers.*

?

18. What two new assumptions were made by Aristarchus? What simplification resulted?

19. How can the heliocentric model proposed by Aristarchus explain retrograde motion?

20. What change predicted by Aristarchus' theory was not observed by the Greeks?

21. Why was Aristarchus considered impious? Why was his system neglected?

5.8 | The geocentric system of Ptolemy

Disregarding the heliocentric model suggested by Aristarchus, the Greeks continued to develop their geocentric system. As noted, the first solutions in terms of interlocking spheres lacked accuracy. During the 500 years after Plato and Aristotle, astronomers began to seek more accurate predictions. To fit the observed data, a complex mathematical theory was required for each planet.

Several Greek astronomers made important contributions, which climaxed about 150 A.D. in the geocentric theory of Claudius Ptolemy of Alexandria. Ptolemy's book on the motions of heavenly objects is a masterpiece of analysis.

The Arabic title given to Ptolemy's book, the *Almagest*, means "the greatest."

Ptolemy wanted a system that would predict accurately the positions of each planet. The type of system and the motions he accepted were based on the assumptions of Aristotle. In the preface of his *Almagest*, Ptolemy defined the problem and stated his assumptions as follows:

> . . . we wish to find the evident and certain appearances from the observations of the ancients and our own, and applying the consequences of these conceptions by means of geometrical demonstrations.
>
> And so, in general, we have to state, that the heavens are spherical and move spherically; that the earth, in figure, is sensibly spherical . . .; in position, lies right in the middle of the heavens, like a geometrical center; in magnitude and distance, [the earth] has the ratio of a point with respect to the sphere of the fixed stars, having itself no local motion at all.

Ptolemy then argued that each of these assumptions was necessary and fit all the observations. The strength of his belief is illustrated by his statement ". . . it is once for all clear from the very appearances that the earth is in the middle of the world and all weights move towards it." Notice that he supported his interpretation of astronomical observations by citing the physics of falling bodies. Later, Ptolemy applied this mixture of astronomy and physics to the earth itself and to its place in the scheme. In doing so, Ptolemy tried to disprove Aristarchus' idea that the earth might rotate and revolve:

> Now some people, although they have nothing to oppose to these arguments, agree on something, as they think, more plausible, And it seems to them there is nothing against their

supposing, for instance, the heavens immobile and the earth as turning on the same axis [as the stars] from west to east very nearly one revolution a day....

But it has escaped their notice that, indeed, as far as the appearances of the stars are concerned, nothing would perhaps keep things from being in accordance with this simpler conjecture, but that in the light of what happens around us in the air such a notion would seem altogether absurd.

SG 8

Ptolemy believed that if the earth rotated, it would not pull its blanket of air around with it. As a result, all clouds would fly past toward the west. All birds and other things in the air also would be carried away to the west. Even if the earth did drag the air along with it, objects in the air would still tend to be left behind by the earth and air together.

The paragraphs quoted above contain a main theme of Unit 2. Ptolemy recognized that the two systems were equally successful in *describing* motion, that is, in terms of kinematics. He preferred the geocentric theory because it fit better the *causes* of motion, that is, the dynamics, as accepted at the time. Much later, when Newton developed a completely different dynamics, the choice fell the other way.

Ptolemy developed very clever and rather accurate procedures for predicting the positions of each planet on a geocentric model. He went far beyond the scheme of the earlier Greeks, constructing a model out of circles and three other geometrical devices. Each device provided for variations in the rate of angular motion as seen from the earth. In order to appreciate Ptolemy's solution, examine one of the very small variations he was attempting to explain.

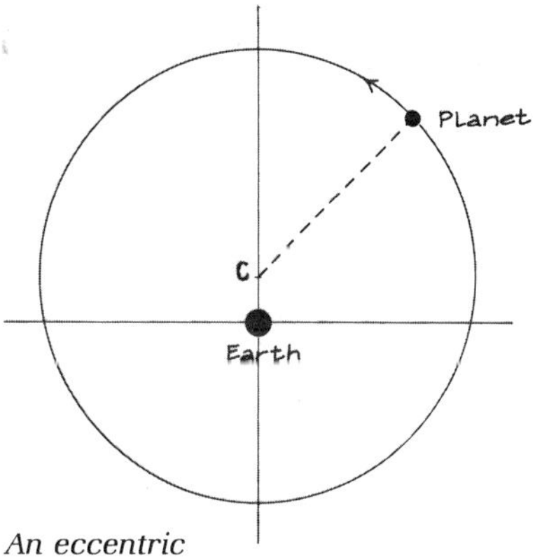

An eccentric

The sun's yearly 360° path across the background of stars can be divided into four 90° parts. If the sun is at the zero point on March 21, it will be 90° farther east on June 21, 90° farther still on September 23, another 90° farther on December 22, and back at the starting point on March 21, one whole year later. If the sun moves uniformly on a circle around the earth, the times between these dates ought to be equal. But, as you will find by consulting a calendar, they are not equal. The sun takes a few days longer to move 90° in spring or summer than it does in fall or winter. So any simple circular system based on motion with constant speed will not work for the sun.

The three devices that Ptolemy used to improve geocentric theory were the *eccentric*, the *epicycle*, and the *equant*.

Agreeing with Plato, astronomers had held previously that a celestial object must move at a uniform angular rate and at a constant distance from the center of the earth. Ptolemy, too, believed that the earth was at the center of the universe. But he did not insist that it stood at the geometrical centers of all the perfect circles. He proposed that the center C of a circle could be

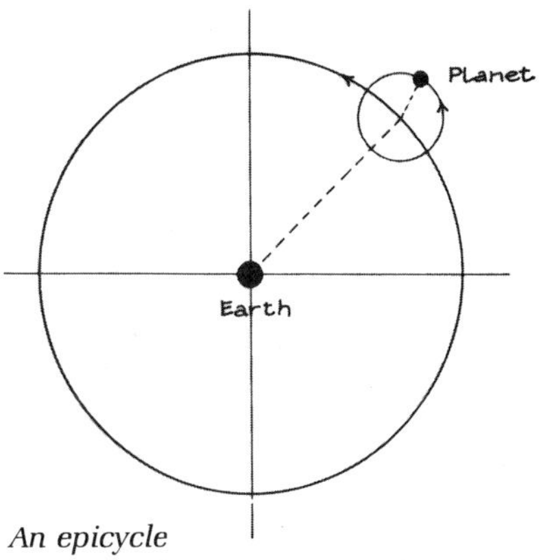

An epicycle

off-center from the earth, in an *eccentric* position. Thus, motion that was really uniform around the center C would not appear to be uniform when observed from the earth. An eccentric orbit of the sun would therefore account for the seasonal variation observed in the sun's rate of motion.

The eccentric can also account for small variations in the rate of motion of planets. However, it cannot describe such drastic changes as retrograde motion of the planets. To account for retrograde motion, Ptolemy used another device, the *epicycle* (see the figure on page 149). The planet is considered to be moving at a uniform rate on the small epicyle. The center of the epicycle moves at a uniform rate on a large circle, called the *deferent,* around the earth.

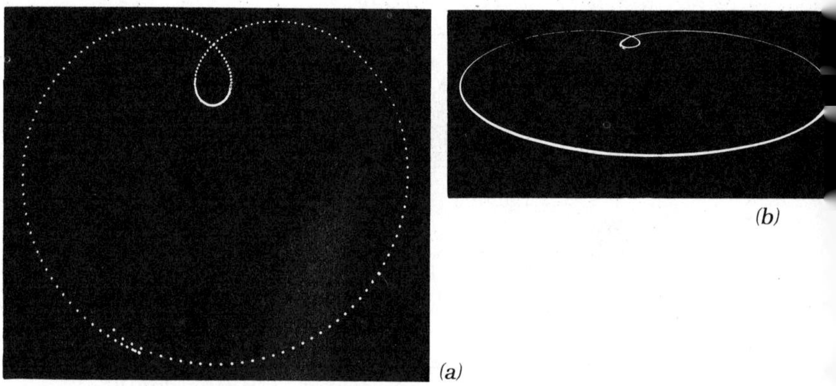

(a) (b)

Retrograde motion created by a simple epicycle machine: (a) Stroboscopic photograph of epicyclic motion. The flashes were made at equal time intervals. Note that the motion is slowest in the loop. (b) Loop seen from near its plane.

If a planet's speed on the epicycle is greater than the speed of the epicycle on the large circle, the planet as seen from *above* the system appears to move through loops. When observed from a location near the center of the system, these loops look like the retrograde motions actually observed for planets. The photographs above show two views of the motions produced by a simple mechanical model, an "epicycle machine." A small light takes the place of the planet. The photo on the left was taken from "above," like the diagram on page 149. The photo on the right was taken "on edge," almost in the plane of the motion. Thus, the loop looks much as it would if viewed from near the center.

With epicycles it was not too difficult to produce a system that had all the main features of observed planetary motion. Ptolemy's system included a unique pattern for the epicycles for the outer planets. All had the same period, exactly 1 year! Moreover, the positions of the outer planets on their epicycles always matched

SG 9 the position of the sun relative to the earth. See the sketches (p. 151) for this matching of epicyles to the relative motion of sun and earth. Fourteen centuries later, this peculiar feature became a key point of concern to Copernicus.

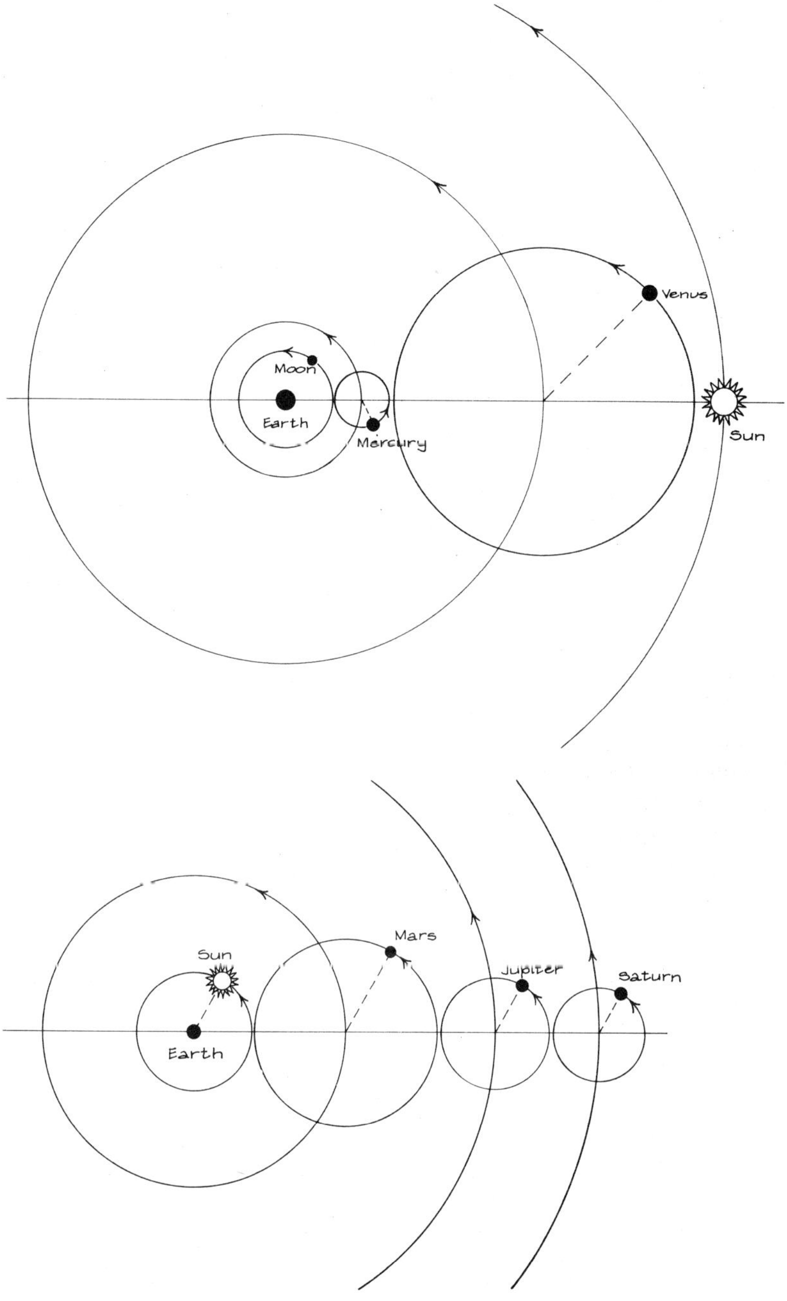

Simplified representation of the Ptolemaic system. The scale of the upper drawing, which shows the planets between the earth and the sun, is eight times that of the lower drawing, which shows the planets that are beyond the sun. The planets' epicycles are shown along one straight line to emphasize the relative sizes of the epicycles.

Mars plotted at 4-day intervals on three consecutive oppositions. Note the different sizes and shapes of the retrograde curves.

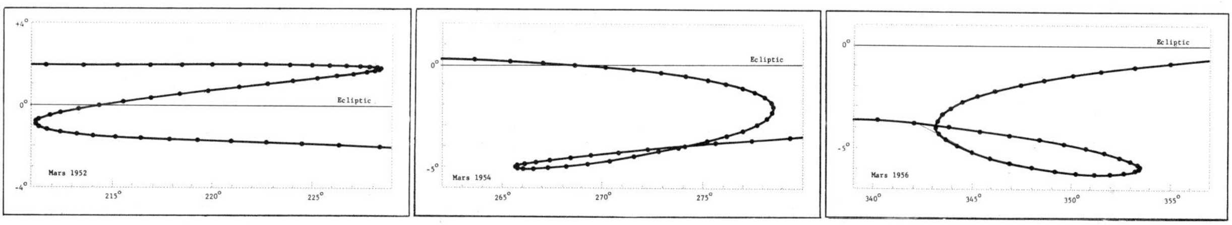

Ptolemy did not picture the planetary motions as those of an interlocking machine in which each planet determined the motion of the next. Because there was no information about the distances of the planets, Ptolemy adopted an old order of distances from the earth: stars being the most remote, then Saturn, Jupiter, Mars, the sun, Venus, Mercury, and the moon. The orbits were usually shown nested inside one another so that their epicycles did not overlap.

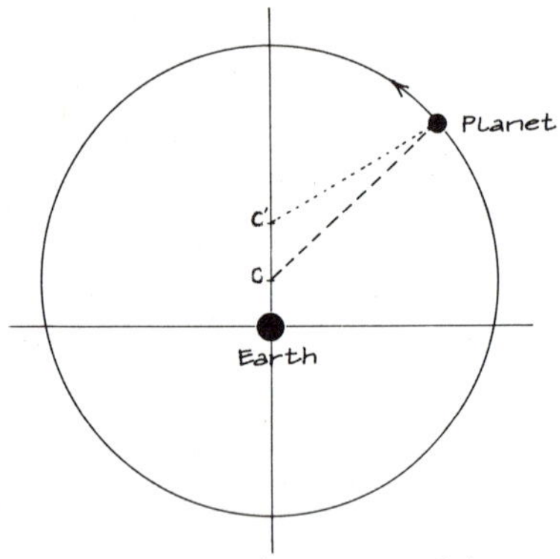

An equant. C is the center of the circle. The planet P moves at a uniform rate around the off-center point C′. The earth is also off center.

SG 10

Astronomical observations were all observations of *angles*. A small loop in the sky could be a small loop fairly near, or a larger loop much farther away.

SG 11

So far, the system of epicycles and deferents "works" well enough. It explains not only retrograde motion, but also the greater brightness of the planets when they are in retrograde motion. Then a planet is closest to the earth, and so appears brightest. This is an unexpected bonus, since the model was not designed to explain the brightness change.

Even with combinations of eccentrics and epicycles, Ptolemy could not fit the motions of the five planets exactly. For example, as you see in the three figures on page 151, the retrograde motion of Mars is not always of the same angular size or duration. To allow for such difficulties, Ptolemy used a third geometrical device, called the *equant.* The equant is a variation of the eccentric with the uniform motion about an off-center point C′.

5.9 | Successes and limitations of the Ptolemaic model

Ptolemy's model always used a uniform rate of angular motion around some center. To that extent, it stayed close to the assumptions of Plato. But, to fit the observations, Ptolemy was willing to displace the centers of motion from the center of the earth as much as necessary. By combining eccentrics, epicycles, and equants, he described the positions of each planet separately. For each planet, Ptolemy found a combination of motions that predicted its observed positions over long periods of time. These predictions were accurate to within about 2° (roughly four diameters of the moon). This accuracy was a great improvement over earlier systems.

Ptolemy's model was quite successful, especially in its unexpected explanation of varying brightness. Such success might be taken as proof that objects in the sky actually move on epicycles and equants around off-center points. Ptolemy did not believe he was providing an actual physical model of the universe. He created a mathematical model, like equations, for computing positions.

The Ptolemaic model was a series of mathematical devices meant to match and predict the motion of each planet separately. His geometrical analyses were like complicated equations of motion for each individual planet. But in the following centuries most scholars, including the poet Dante, accepted the model as real. They actually believed that the planets moved on transparent, invisible spheres. Also, they felt that somehow the motion of all these separate spheres should be related. In Ptolemy's original work, each planet was independent of the others.

Ptolemy proposed his model of the planetary system in 150 A.D. Although it is now discarded, it was used for about 1,500 years. There were good reasons for this long acceptance.

1. It predicted fairly accurately the positions of the sun, moon, and planets.

2. It explained why the fixed stars do not show an annual shift (parallax) when observed with the naked eye.

3. It agreed in most details with philosophies developed by the early Greeks, including the ideas of "natural motion" and "natural place."

4. It had common-sense appeal to all who saw the sun, moon, planets, and stars moving around them.

5. It agreed with the comforting assumption that we live on an unmoving earth at the center of the universe.

6. Later, it fitted into Thomas Aquinas' widely accepted synthesis of Christian belief and Aristotelian physics.

Yet Ptolemy's system eventually was displaced by a heliocentric one. Why did this occur? What advantages did the new theory have over the old? From this historic argument about competing theories, what can you learn about the relative value of rival theories in science today? These are some of the questions to consider in the next chapter.

SG 12

study guide

1. The *Project Physics* learning materials particularly appropriate for Chapter 5 include:

Experiments
Naked-Eye Astronomy
Size of the Earth
The Distance to the Moon
Height of Piton—A Mountain on the Moon
Retrograde Motion

Activities
Making Angular Measurements
Epicycles and Retrograde Motion
Celestial Sphere Model
How Long is a Sidereal Day?
Scale Model of the Solar System
Build a Sundial
Plot an Analemma
Stonehenge
Moon Crater Names
Literature

Film Strip
Retrograde Motion of Mars

Film Loops
Retrograde Motion—Geocentric Model

2. How could you use the shadow cast by a vertical stick on horizontal ground to find

(a) the local noon?

(b) which day was June 21st?

(c) the length of a solar year?

3. What is the difference between 365.24220 days and 365¼ days (a) in seconds (b) in percent?

4. (a) List the observations of the motions of heavenly bodies that you might make which would also have been possible in ancient Greek times.

(b) For each observation, list some reasons why the Greeks thought these motions were important.

sg

5. Which of the apparent motions of the stars could be explained by a flat earth and stars fixed to a bowl that rotated around it?

6. Describe the observed motion of the moon during one month, using drawings. (Use your own observations if possible.)

7. Mercury and Venus show retrograde motion after they have been farthest east of the sun and visible in the evening sky. Then they quickly move ahead westward toward the sun, pass it, and reappear in the morning sky. During this motion they are moving westward relative to the stars, as is shown by the plot of Mercury on page 140. Describe the rest of the cyclic motion of Mercury and Venus.

8. Center a protractor on point C in the top diagram on page 149 and measure the number of degrees in the four quadrants. Consider each 1° around C as one day. Make a table of the days needed for the planet to move through the four arcs as seen from the earth.

9. **(a)** How many degrees of terrestrial longitude does the sun move each hour?

(b) What rough value for the diameter of the earth can you obtain from the following information:

(1) Washington, D.C., and San Francisco have about the same latitude. How can one easily test this?

(2) A nonstop jet plane, going up wind at a ground speed of 800 km/hr from Washington, D.C., to San Francisco, takes 5 hr to get there.

(3) When it is just sunset in Washington, D.C., a person there turns on a TV set to watch a baseball game that is just beginning in San Francisco. The game goes into extra innings. After 3 hr the announcer notes that the last out occurred just as the sun set.

10. In Ptolemy's theory of the planetary motions there were, as in all theories, a number of assumptions. Which of the following did Ptolemy assume?

(a) The vault of stars is spherical in form.

(b) The earth has no motions.

(c) The earth is spherical.

(d) The earth is at the center of the sphere of stars.

(e) The size of the earth is extremely small compared to the distance to the stars.

(f) Uniform angular motion along circles (even if measured from an off-center point) is the only proper behavior for celestial objects.

11. As far as the Greeks were concerned, and indeed as far as we are concerned, a reasonable argument can be made for either the geocentric or the heliocentric theory of the universe.

(a) In what ways were both ideas successful?

(b) In terms of Greek science, what are some advantages and disadvantages of each system?

(c) What were the major contributions of Ptolemy?

12. Why was astronomy the first successful science, rather than, for example, meteorology or zoology?

Does the Earth Move?
The Work of Copernicus and Tycho

6.1 The Copernican system
6.2 New conclusions
6.3 Arguments for the Copernican system
6.4 Arguments against the Copernican system
6.5 Historical consequences
6.6 Tycho Brahe
6.7 Tycho's observations
6.8 Tycho's compromise system

Nicolas Copernicus (1473–1543). (In Polish his name was Koppernigk, but, in keeping with the scholarly tradition of the age, he gave it the Latin form Copernicus.)

SG 1

6.1 | The Copernican system

Nicolaus Copernicus (1473–1543) was a young student in Poland when America was discovered by Europeans. An outstanding astronomer and mathematician, Copernicus was also a talented and respected churchman, jurist, administrator, diplomat, physician, and economist. During his studies in Italy he read the writings of Greek and other early philosophers and astronomers. As Canon of the Cathedral of Frauenberg he was busy with civic and church affairs and also worked on calendar reform. It is said that on the day of his death in 1543, he saw the first copy of his great book, on which he had worked most of his life. It was this book which opened a whole new vision of the universe.

Copernicus titled his book *De Revolutionibus Orbium Coelestium*, or *On the Revolutions of the Heavenly Spheres*. This

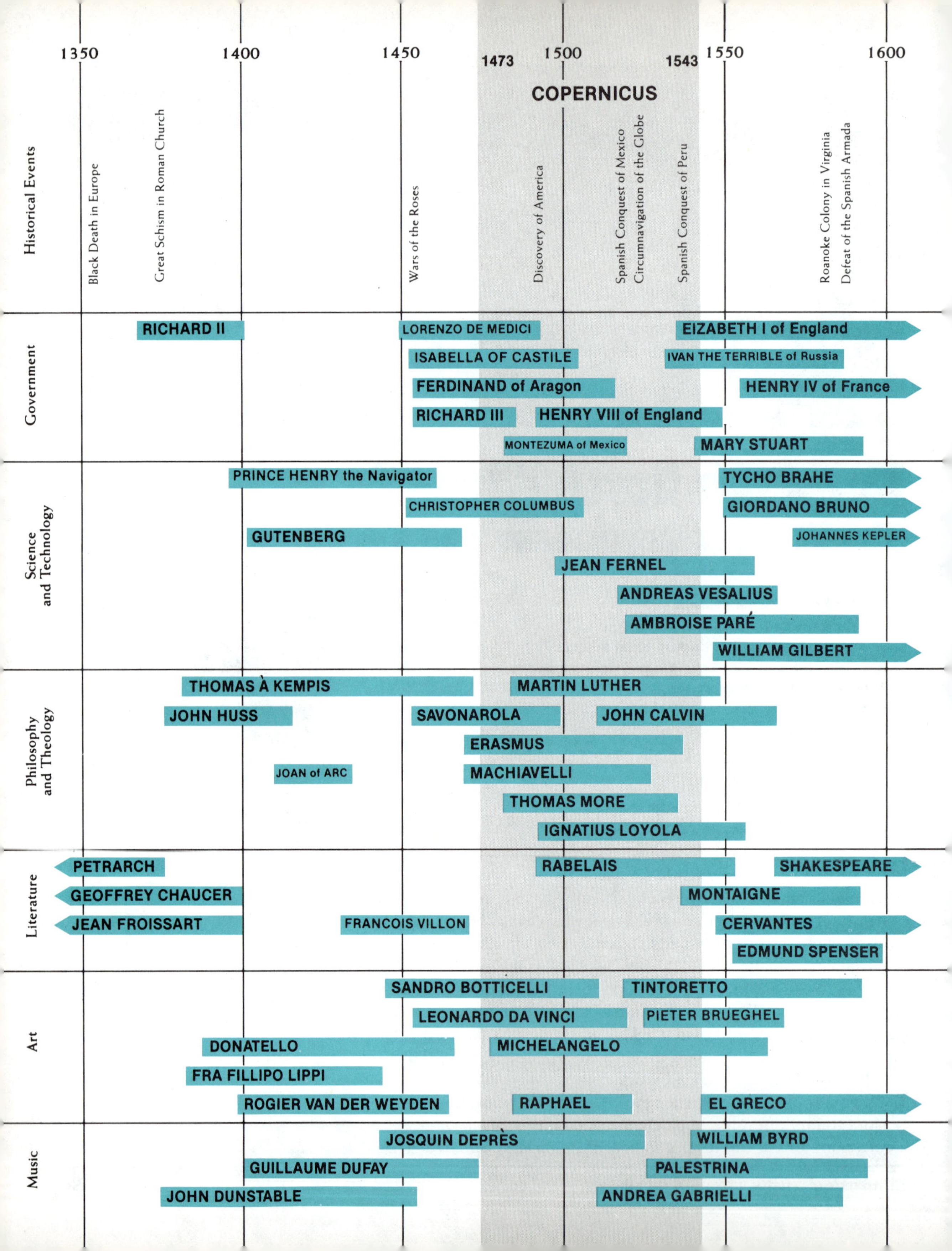
1350
1400
1450
1473
1500
1543
1550
1600
COPERNICUS
Historical Events
Black Death in Europe
Great Schism in Roman Church
Wars of the Roses
Discovery of America
Spanish Conquest of Mexico
Circumnavigation of the Globe
Spanish Conquest of Peru
Roanoke Colony in Virginia
Defeat of the Spanish Armada
Government
RICHARD II
LORENZO DE MEDICI
EIZABETH I of England
ISABELLA OF CASTILE
IVAN THE TERRIBLE of Russia
FERDINAND of Aragon
HENRY IV of France
RICHARD III
HENRY VIII of England
MONTEZUMA of Mexico
MARY STUART
Science and Technology
PRINCE HENRY the Navigator
TYCHO BRAHE
CHRISTOPHER COLUMBUS
GIORDANO BRUNO
GUTENBERG
JOHANNES KEPLER
JEAN FERNEL
ANDREAS VESALIUS
AMBROISE PARÉ
WILLIAM GILBERT
Philosophy and Theology
THOMAS À KEMPIS
MARTIN LUTHER
JOHN HUSS
SAVONAROLA
JOHN CALVIN
ERASMUS
JOAN of ARC
MACHIAVELLI
THOMAS MORE
IGNATIUS LOYOLA
Literature
PETRARCH
RABELAIS
SHAKESPEARE
GEOFFREY CHAUCER
MONTAIGNE
JEAN FROISSART
FRANCOIS VILLON
CERVANTES
EDMUND SPENSER
Art
SANDRO BOTTICELLI
TINTORETTO
LEONARDO DA VINCI
PIETER BRUEGHEL
DONATELLO
MICHELANGELO
FRA FILLIPO LIPPI
ROGIER VAN DER WEYDEN
RAPHAEL
EL GRECO
Music
JOSQUIN DEPRÈS
WILLIAM BYRD
GUILLAUME DUFAY
PALESTRINA
JOHN DUNSTABLE
ANDREA GABRIELLI

title suggests the early Greek notions of the spheres. Copernicus was indeed concerned with the old problem of Plato: how to construct a planetary system by combinations of the fewest possible uniform circular motions. He began his study to rid the Ptolemaic system of the equants, which seemed contrary to Plato's assumptions. In his words, taken from a short summary written about 1512,

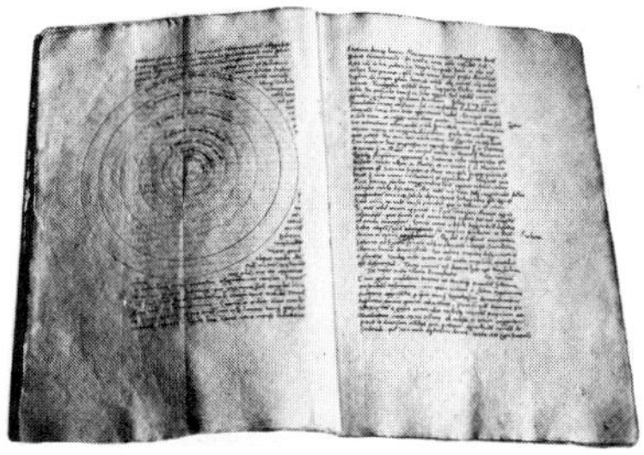

Copernicus' diagram of his heliocentric system (from his manuscript of De Revolutionibus, *1543). This simplified representation omits the many small epicycles actually used in the system.*

... the planetary theories of Ptolemy and most other astronomers, although consistent with the numerical data, seemed likewise to present no small difficulty. For these theories were not adequate unless certain equants were also conceived; it then appeared that a planet moved with uniform velocity neither on its deferent nor about the center of its epicycle. Hence a system of this sort seemed neither sufficiently absolute nor sufficiently pleasing to the mind.

Having become aware of these defects, I often considered whether there could perhaps be found a more reasonable arrangement of circles, from which every apparent inequality would be derived and in which everything would move uniformly about its proper center.

In *De Revolutionibus* he wrote:

We must however confess that these movements [of the sun, moon, and planets] are circular or are composed of many circular movements, in that they maintain these irregularities in accordance with a constant law and with fixed periodic returns, and that could not take place, if they were not circular. For it is only the circle which can bring back what is past and over with....

I found first in Cicero that Nicetas thought that the Earth moved. And afterwards I found in Plutarch that there were some others of the same opinion.... Therefore I also ... began to meditate upon the mobility of the Earth. And although the opinion seemed absurd, nevertheless, because I knew that others before me had been granted the liberty of constructing whatever circles they pleased in order to demonstrate astral phenomena, I thought that I too would be readily permitted to test whether or not, by the laying down that the Earth had some movements, demonstrations less shaky than those of my predecessors could be found for the revolutions of the celestial spheres.... I finally discovered by the help of long and numerous observations that if the movements of the other wandering stars are correlated with the circular movement of the Earth, and if the movements are computed in accordance with the revolution of each planet, not only do all their phenomena follow from that but also this correlation binds together so closely the order and magnitudes of all the planets and of their spheres or orbital circles and the heavens themselves that nothing can be shifted around in any part of them without disrupting the remaining parts and the universe as a whole.

The "wandering stars" are the planets.

In his final work, the result of nearly 40 years of study, Copernicus proposed a system of more than 30 eccentrics and epicycles. These would, he said, "suffice to explain the entire structure of the universe and the entire ballet of the planets." Like Ptolemy's *Almagest, De Revolutionibus* uses long geometrical analyses and is difficult to read. Comparison of the two books strongly suggests that Copernicus thought he was producing an improved version of the *Almagest*. He used many of Ptolemy's observations plus some more recent ones. Yet Copernicus' system differed from Ptolemy's in several fundamental ways. Above all, Copernicus adopted a sun-centered system which in general matched that of Aristarchus.

Like all scientists, Copernicus made a number of assumptions in his system. In his own words (using more modern terms in several places), his assumptions were:

1. There is no precise, geometrical center of all the celestial circles or spheres.

2. The center of the earth is not the center of the universe, but only of gravitation and of the lunar sphere.

3. All the spheres revolve about the sun ... and therefore the sun has a central location in the universe.

4. The distance from the earth to the sun is very small in comparison with the distance to the stars.

5. Whatever motion appears in the sky arises not from any motion of the sky, but from the earth's motion. The earth together with its water and air performs a complete rotation on its fixed poles in a daily motion, while the sky remains unchanged.

6. What appears to us as motions of the sun arise not from its motion but from the motion of the earth and ... we revolve about the sun like any other planet. The earth has, then, more than one motion.

SG 2

7. The apparent retrograde motion of the planets arises not from their motion but from the earth's. The motions of the earth alone, therefore, are enough to explain many apparent motions in the sky.

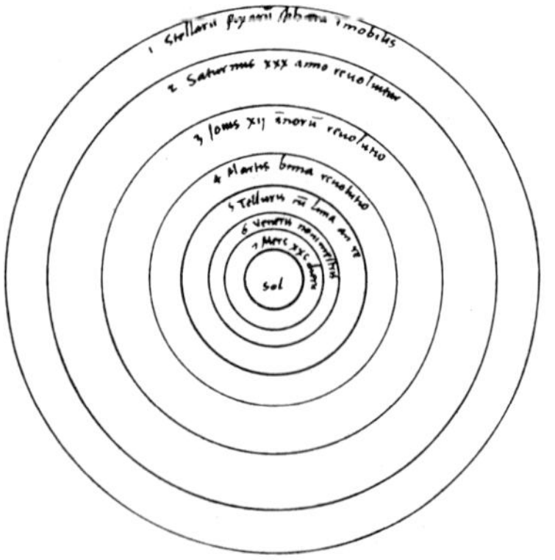

Compare this list with the assumptions of Ptolemy, given in Chapter 5. You will see close similarities and important differences. Notice that Copernicus proposed that the earth rotates daily. As Aristarchus and others had realized, this rotation would explain all the daily risings and settings seen in the sky. Copernicus also proposed, as had Aristarchus, that the sun was stationary and stood at the center of the universe. The earth and other planets each moved about a different central point near the sun.

The figure at the left shows the main spheres carrying the planets around the sun (sol). Copernicus' text explains the basic features of his system:

The ideas here stated are difficult, even almost impossible, to accept; they are quite contrary to popular notions. Yet with the help of God, we will make everything as clear as day in what follows, at least for those who are not ignorant of mathematics. . . .

The first and highest of all the spheres is the sphere of the fixed stars. It encloses all the other spheres and is itself self-contained; it is immobile; it is certainly the portion of the universe with reference to which the movement and positions of all the other heavenly bodies must be considered. If some people are yet of the opinion that this sphere moves, we are of contrary mind; and after deducing the motion of the earth, we shall show why we so conclude. Saturn, first of the planets, which accomplishes its revolution in thirty years, is nearest to the first sphere. Jupiter, making its revolution in twelve years, is next. Then comes Mars, revolving once in two years. The fourth place in the series is occupied by the sphere which contains the earth and the sphere of the moon, and which performs an annual revolution. The fifth place is that of Venus, revolving in nine months. Finally, the sixth place is occupied by Mercury, revolving in eighty days. . . . In the midst of all, the sun reposes, unmoving.

Already you can see an advantage in Copernicus' system that makes it "pleasing to the mind." The rates of rotation for the heavenly spheres increase progressively, from the motionless sphere of stars to speedy Mercury.

SG 3

?

1. What reasons did Copernicus give for rejecting the use of equants?

2. In the following list of propositions, mark with a P those made by Ptolemy and with a C those made by Copernicus.

(a) The earth is spherical.

(b) The earth can be thought of as a point in reference to the distance to the stars.

(c) The heavens rotate daily around the earth.

(d) The earth has one or more motions.

(e) Heavenly motions are circular.

(f) The observed retrograde motion of the planets results from the earth's motion around the sun.

6.2 | New conclusions

A new way of looking at old observations (a new theory) can suggest quite new kinds of observations to make, or new uses for old data. Copernicus used his moving-earth model to obtain two important results which were not possible with the Ptolemaic

Close Up

The Periods of Revolution of the Planets

The problem is to find the rate at which a planet moves around the sun by using observations made from the earth, which is itself moving around the sun. Say, for example, that a planet closer to the sun than the earth is goes around the sun at the frequency (rate) of 1¼ cycles per year. The earth moves around the sun also, in the same direction, at the rate of 1 cycle per year. Because the earth follows along behind the planet, the planet's motion around the sun, as seen from the earth, appears to be at a rate less than 1¼ cycles per year. In fact, as the diagrams below suggest, the planet's *apparent* rate of motion around the sun equals the difference between the planet's rate and the earth's rate: 1¼ cycle per year minus 1 cycle per year, or ¼ cycle per year. In general, if an inner planet moves around the sun at frequency f_p and the earth moves around the sun with frequency f_e, then the planet's apparent rate of motion, f_{pe}, as seen from the earth, is given by $f_{pe} = f_p - f_e$.

A similar argument holds for planets farther from the sun than the earth is. (See Diagram B.) Since these "outer planets" revolve about the sun more slowly than the earth does, the earth repeatedly leaves the planets behind. Consequently, for the outer planets, the sign in the equation for f_{pe} is reversed: $f_{pe} = f_p + f_e$.

The apparent frequency f_{pe} represents what is actually observed. Since f_e is by definition 1 cycle per year, either equation is easily solved for the unknown actual rate f_p:

For inner planets: $f_p = 1 \text{ cycle/yr} + f_{pe}$

For outer planets: $f_p = 1 \text{ cycle/yr} - f_{pe}$

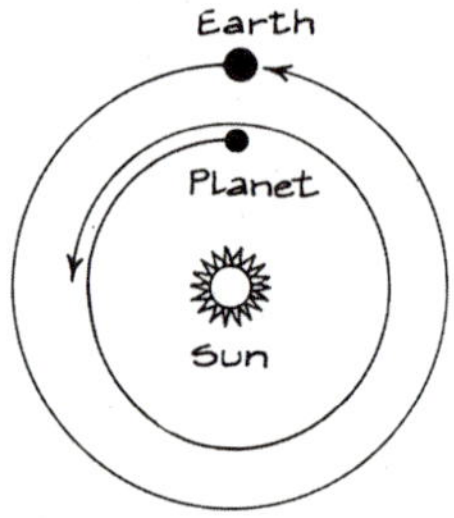

(A) A planet that is inside the Earth's orbit and moves 1¼ revolutions around the sun in a year would, as seen from the earth, appear to have made only a ¼ cycle relative to the sun.

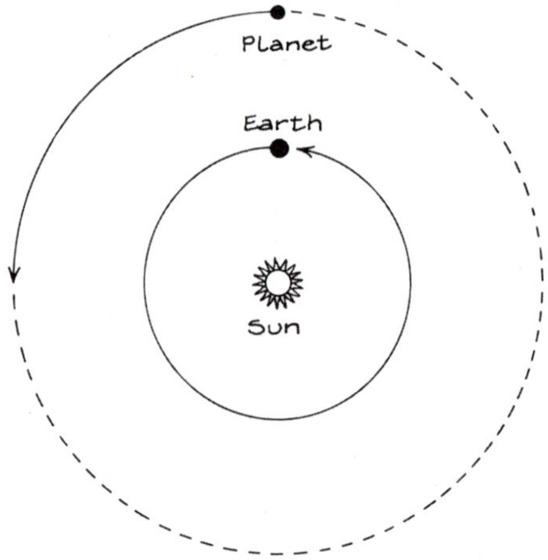

(B) A planet that is outside the Earth's orbit and moves only ¼ revolution around the sun in a year would, as seen from the earth, appear to make about 1¾ revolutions relative to the sun.

TABLE 6.1

	Number of Years of Observation (t)	*Apparent Number of Cycles with Respect to Sun During* t (n)	*Apparent Frequency f_{pe} in Cycles per Year* (n/t)	*Frequency f_p Around Sun in Cycles per Year*	*Period Around Sun ($1/f_p$) in Years*
Mercury	46	145	3.15	4.15	0.241
Venus	8	5	0.625	1.625	0.614
Mars	79	37	0.468	0.532	1.88
Jupiter	71	65	0.915	0.085	11.8
Saturn	59	57	0.966	0.034	29.4

theory. Copernicus was able to calculate: (a) the period of motion of each planet around the sun, and (b) the sizes of each planet's orbit compared to the size of the earth's orbit. These calculations, for the first time, gave a scale for the dimensions of the planetary system, based on observations.

SG 4, 7

To calculate the periods of the planets around the sun, Copernicus used observations recorded over many centuries. The method of calculation is similar to the "chase problem" of how often the hands of a clock pass one another. The details of this calculation are shown on page 160. In Table 6.2 below, Copernicus' results are compared with accepted values.

TABLE 6.2

Planet	Copernicus' Value	Modern Value
Mercury	0.241 y (88 d)	87.97 d
Venus	0.614 y (224 d)	224.70 d
Mars	1.88 y (687 d)	686.98 d
Jupiter	11.8 y	11.86 y
Saturn	29.5 y	29.46 y

Copernicus was also able, for the first time in history, to derive relative distances between the planets and the sun. Remember that the Ptolemaic system had no distance scale. It provided only a way of predicting the planets' angular motions and positions.

Ptolemy's system described the motions of the sun and five planets in terms of 1-yr epicycles on deferent circles. It gave only the *relative* sizes of epicycle and deferent circle, and gave them separately for each planet. Copernicus, on the other hand, described all these features of planetary motion in terms of the motion of the earth's yearly revolution *around the sun*. (The details of how this can be done are given on pages 162 and 163.) Thus, it became possible to compare the radii of the planets' orbits with the radius of the earth's orbit. Because all distances were compared to it, the average sun–earth distance is called one *astronomical unit*, abbreviated 1 AU.

Table 6.3 below compares Copernicus' values for the orbital radii (deferent circles only, the radii of the epicycles being relatively small) with the currently accepted values for the average distances to the sun.

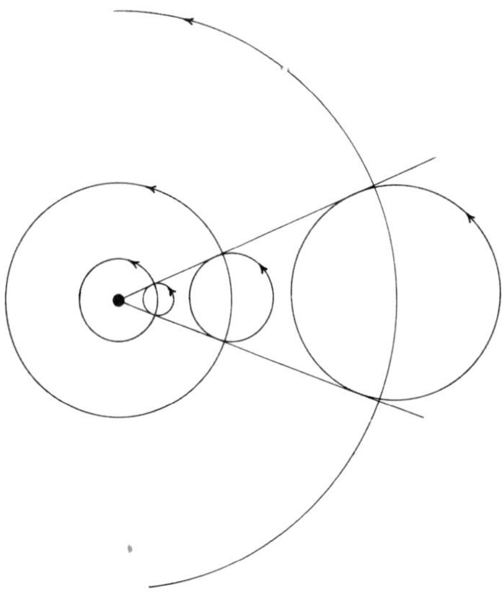

In the Ptolemaic system, only the relative sizes of epicycle and deferent were specified. Their sizes could be changed at will, as long as they kept the same proportions.

TABLE 6.3

Planet	Radii of Planetary Orbits Copernicus' Values	Modern Value
Mercury	0.38 AU	0.39 AU
Venus	0.72	0.72
Earth	1.00	1.00
Mars	1.52	1.52
Jupiter	5.2	5.20
Saturn	9.2	9.54

Close Up

Changing Frame of Reference from the Earth to the Sun

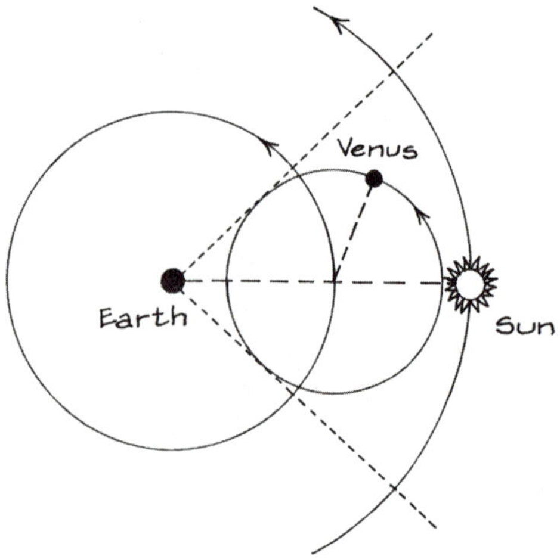

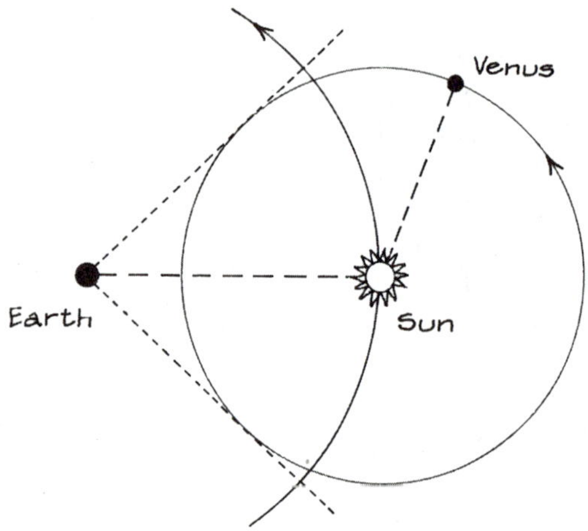

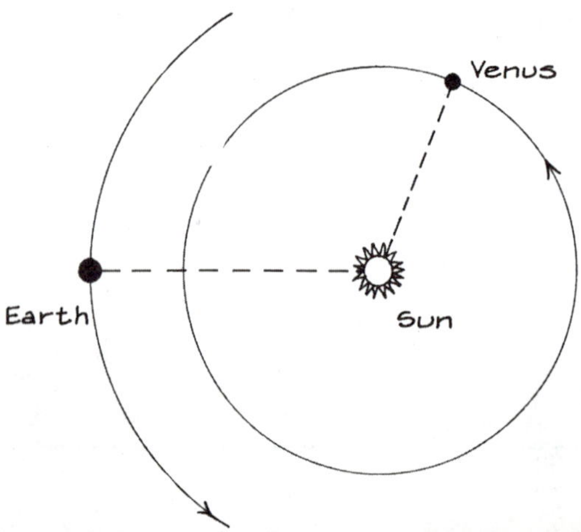

The change of viewpoint from Ptolemy's system to Copernicus' involved what today would be called a shift in frame of reference. The apparent motion previously attributed to the deferent circles and epicycles was attributed by Copernicus to the earth's orbit and the planets' orbits around the sun.

For example, consider the motion of Venus. In Ptolemy's earth-centered system, the center of Venus' epicycle was locked to the motion of the sun, as shown in the top diagram at the left. The size of Venus' deferent circle was thought to be smaller than the sun's. The epicycle was thought to be entirely between the earth and the sun. However, the observed motions to be explained by the system *required* only a certain *relative* size of epicycle and deferent. The deferent could be changed to any size, as long as the epicycle was changed proportionally.

The first step toward a sun-centered system was taken by moving the center of Venus' 1-yr deferent out to the sun. Venus' epicycle was enlarged proportionally, as shown in the middle diagram at the left. Now the planet moved about the sun, while the sun moved about the earth. Tycho Brahe later proposed such a system with all visible planets moving about the moving sun.

Copernicus went further. He accounted for the relative motion of the earth and sun by considering the earth to be moving around the sun, instead of the sun moving about the earth. In the Copernican system, Venus' enlarged epicycle became its orbit around the sun. Also, the sun's deferent was replaced by the earth's orbit around the sun. See the bottom diagram at the left. All three systems, Ptolemy's, Copernicus', and Tycho's, explain the same observations.

For the outer planets the argument was similar, but the roles of epicycle and deferent circle were reversed. For the outer planets in the Ptolemaic model, the epicycles instead of the deferent circles had 1-yr periods and moved in parallel with the sun in its orbit. The sizes of the deferents were chosen so that the epicycle of each planet would just miss the epicycles of the planets next nearest and next farthest from the sun. (This was a beautiful example of a simplifying assumption. It filled the space with no overlap and no gaps.) This system is represented in the top diagram at the right; the planets are shown in the unlikely condition of having their epicycle centers along a single line.

The first step in shifting to a sun-centered view for these planets involves adjusting the sizes of the deferent circles, keeping the epicycles in proportion. Eventually, the 1-yr epicycles are the same size as the sun's 1-yr orbit. See the middle diagram at the right. Next, the sun's apparent yearly motion around the earth is explained just as well by having the earth revolve around the sun. Also, the same earth orbit would explain the retrograde loops associated with the outer planets' matched 1-yr epicycles. So all the matched epicycles of the outer planets and the sun's orbit are replaced by the *single* device of the earth's orbit around the sun. This shift is shown in the bottom diagram at the right. The deferent circles of the outer planets become their orbits around the sun.

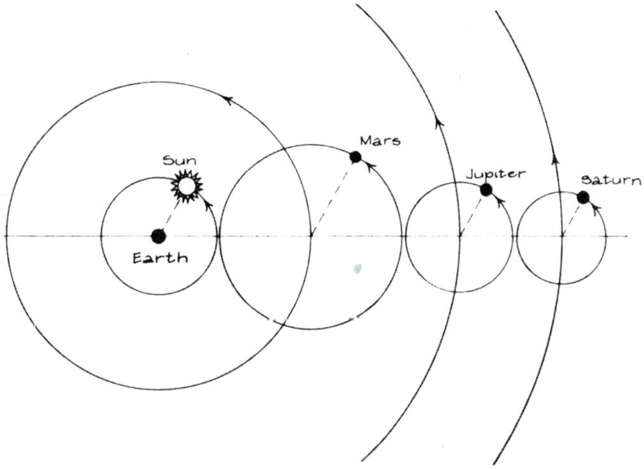

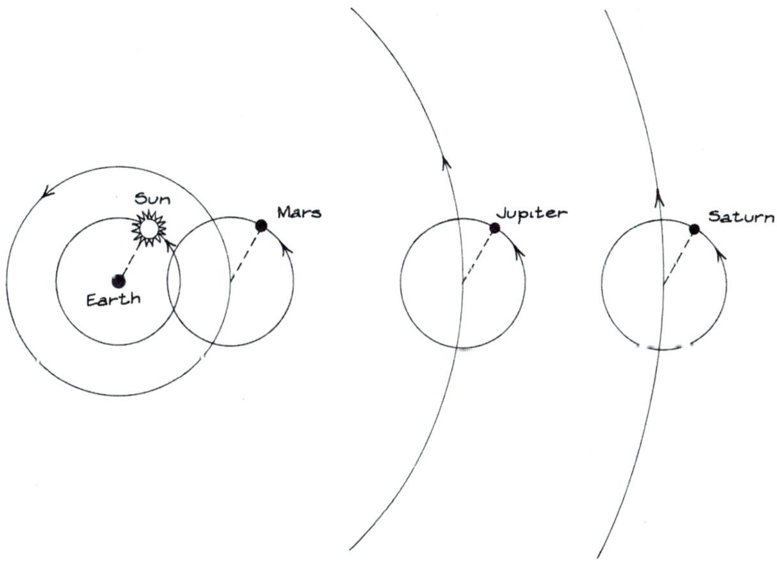

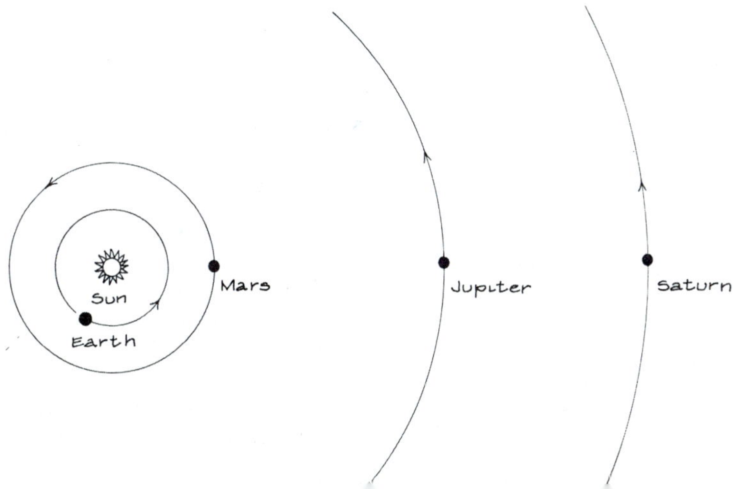

Notice that Copernicus now had one system which related the size of each planet's orbit to the sizes of all the other planets' orbits. Contrast this to Ptolemy's solutions, which were completely independent for each planet. No wonder Copernicus said that "nothing can be shifted around in any part of them without disrupting the remaining parts and the universe as a whole."

SG 8

? *3. What new kinds of results did Copernicus obtain with a moving-earth model which were not possible with a geocentric model for the planetary system?*

6.3 | Arguments for the Copernican system

Copernicus knew that to many his work would seem absurd, "nay, almost contrary to ordinary human understanding," so he tried in several ways to meet the old arguments against a moving earth.

1. Copernicus argued that his assumptions agreed with religious doctrine at least as well as Ptolemy's. Copernicus' book had many sections on the faults of the Ptolemaic system (most of which had been known for centuries). To Copernicus, as to many scholars, complex events were merely symbols of God's thinking. To find order and symmetry in them was an act of piety, for order and symmetry were proofs of God's existence. As a church official, Copernicus would have been stunned to think that, in Galileo's time, his theory would contribute to the conflict between religious doctrine and science.

2. Copernicus carefully calculated relative radii and speeds of the circular motions in his system. From these data, tables of planetary motion could be made. Actually, the theories of Ptolemy and Copernicus were about equally accurate in predicting planetary positions. Both theories often differed from the observed positions by as much as 2° (about four diameters of the moon).

3. Copernicus tried to answer several other objections. Most of them had been raised against Aristarchus' heliocentric system nearly 19 centuries earlier. One argument held that a rapidly rotating earth would surely fly apart. Copernicus replied, "Why does the defender of the geocentric theory not fear the same fate for his rotating celestial sphere—so much faster because so much larger?" It was argued that birds and clouds in the sky would be left behind by the earth's rotation and revolution. Copernicus answered this objection by indicating that the atmosphere is dragged along with the earth. To the lack of

observed annual shift for the fixed stars, he could only give the same kind of answer that Aristarchus had proposed:

> ...though the distance from the sun to the earth appears very large as compared with the size of the spheres of some planets, yet compared with the dimensions of the sphere of the fixed stars, it is as nothing.

4. Copernicus claimed that the greatest advantage of his scheme was its simple description of the general motions of the planets. There certainly is a basic overall simplicity to his system. Yet for precise calculations, because Copernicus would not use equants, he needed *more* small motions than did Ptolemy to explain the observations. A diagram from Copernicus' manuscript shows more detail (page 166).

5. Copernicus pointed out that the simplicity of his system was not merely convenient, but also beautiful and "pleasing to the mind." The pleasure which scientists find in the simplicity of their models is one of the most powerful experiences in science. Far from being a "cold," merely logical exercise, scientific work is full of such recognitions of harmony and beauty. Another sign of beauty that Copernicus saw in his system was the central place given to the sun, the biggest, brightest object in the heavens and the giver of light, warmth, and life. As Copernicus himself put it:

> In the midst of all, the sun reposes, unmoving. Who, indeed, in this most beautiful temple would place the light-giver in any other part than whence it can illumine all other parts? So we find underlying this ordination an admirable symmetry in the Universe and a clear bond of the harmony in the motion and magnitude of the spheres, such as can be discovered in no other wise.

Look again at SG 2.

?

4. *Which of these arguments did Copernicus use in favor of his system?*

(a) It was obvious to ordinary common sense.

(b) It was consistent with Christian beliefs.

(c) It was much more accurate in predicting planet positions.

(d) Its simplicity made it beautiful.

(e) The stars showed an annual shift in position due to the earth's motion around the sun.

5. *What were the largest differences between observed planetary positions and those predicted by Ptolemy? by Copernicus?*

6. *Did the Copernican system allow simple calculations of where the planets should be seen?*

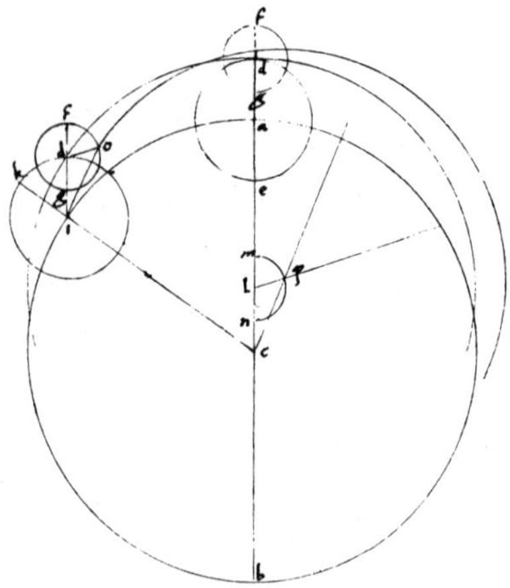

This drawing in Copernicus' manuscript of De Revolutionibus *shows details of some epicycles in his model.*

6.4 | Arguments against the Copernican system

Copernicus' hopes for acceptance of his theory were not quickly fulfilled. More than 100 years passed before the heliocentric system was generally accepted even by astronomers. Even then, the acceptance came on the basis of arguments quite different from those of Copernicus. In the meantime, the theory and its few defenders met powerful opposition. Most of the criticisms were the same as those used by Ptolemy against Aristarchus.

1. Apart from its apparent simplicity, the Copernican system had no clear *scientific* advantages over the geocentric theory. No known observation was explained by one system and not by the other. Copernicus had a different viewpoint. But he had no new types of observations, no experimental data that could not be explained by the old theory. Furthermore, the accuracy of his predictions of planetary positions was little better than that of Ptolemy's. As Francis Bacon wrote in the early seventeenth century: "Now it is easy to see that both they who think the earth revolves and they who hold the old construction are about equally and indifferently supported by the phenomena."

Basically, the rival systems differed in their choice of a reference frame for describing the observed motions. Copernicus himself stated the problem clearly:

Ptolemy, too, had recognized the possibility of alternative frames of reference. (Reread the quotation on page 148 in Chapter 5.) Most of Ptolemy's followers did not share this insight.

> Although there are so many authorities for saying that the Earth rests in the centre of the world that people think the contrary supposition . . . ridiculous; . . . if, however, we consider the thing attentively, we will see that the question has not yet been decided and accordingly is by no means to be scorned. For every apparent change in place occurs on account of the movement either of the thing seen or of the spectator, or on account of the necessarily unequal movement of both. For no movement is perceptible relatively to things moved equally in the same directions—I mean relatively to the thing seen and the spectator. Now it is from the Earth that the celestial circuit is beheld and presented to our sight. Therefore, if some movement should belong to the Earth . . . it will appear, in the parts of the universe which are outside, as the same movement but in the opposite direction, as though the things outside were passing over. And the daily revolution . . . is such a movement.

Here Copernicus invites the reader to shift the frame of reference from the earth to a remote position overlooking the whole system with the sun at its center. As you may know from personal experience, such a shift is not easy. We can sympathize with those who preferred to hold an earth-centered system for describing what they saw.

Physicists now generally agree that *any* system of reference may in principle be used for describing phenomena. Some systems are easier and others more complex to use or think about. Copernicus and those who followed him felt that the heliocentric system was right in some absolute sense: that the sun was really fixed in space. The same claim was made for the earth by his opponents. The modern attitude is that the best frame of reference is the one that allows the simplest discussion of the problem being studied. You should not speak of reference systems as being right or wrong, but rather as being convenient or inconvenient. (To this day, navigators use a geocentric model for their calculations. See the page of a navigation book in the margin.)

2. The lack of an observable annual shift for the fixed stars was contrary to Copernicus' model. His only possible reply was unacceptable because it meant that the stars were at an enormous distance from the earth. Naked-eye instruments allowed positions in the sky to be measured to a precision of about 0.10°. But for an annual shift to be less than 0.10°, the stars would have to be more than 1,000 times farther from the sun than the earth is! To us this is no shock, because we live in a society that accepts the idea of enormous extensions in space and in time. Even so, such distances strain the imagination. To the opponents of Copernicus, such distances were absurd. Indeed, even if an annual shift in star position had been observable, it might not have been accepted as unmistakable evidence against one and for the other theory. One can usually modify a theory more or less pleasingly to fit in a bothersome finding.

The Copernican system demanded other conclusions that puzzled or threatened its critics. Copernicus determined the distances between the sun and the planetary orbits. Perhaps, then, the Copernican system was not just a mathematical model for predicting the positions of the planets! Perhaps Copernicus was describing a real system of planetary orbits in space (as he thought he was). This would be difficult to accept, for the described orbits were far apart. Even the small epicycles which Copernicus still used to explain variations in planetary motions did not fill up the spaces between the planets. Then what did fill up these spaces? Because Aristotle had stated that "nature abhors a vacuum," it was agreed that something had to fill all that space. Even many of those who believed in Copernicus' system felt that space should contain something. Some of these scholars imagined various invisible fluids to fill up the emptiness. More recently, similar imaginary fluids were used in theories of chemistry and of heat, light, and electricity.

3. No definite decision between the Ptolemaic and the Copernican theories could be made based on the astronomical

CELESTIAL OBSERVATIONS

By PHILIP KISSAM, C. E.
Professor of Civil Engineering, Princeton University

I. The Principles upon which Celestial Observations are Based.

A. CONCEPTS.

1. **The Celestial Sphere.** To simplify the computations necessary for the determinations of the direction of the meridian, of latitude, and of longitude or time, certain concepts of the heavens have been generally adopted. They are the following:

a. The earth is stationary.

b. The heavenly bodies have been projected outward, along lines which extend from the center of the earth, to a sphere of infinite radius called the ***celestial sphere.***

The celestial sphere has the following characteristics:

a. Its center is at the center of the earth.

b. Its equator is on the projection of the earth's equator.

c. With respect to the earth, the celestial sphere rotates from east to west about a line which coincides with the earth's axis. Accordingly, the poles of the celestial sphere are at the prolongations of the earth's poles.

d. The speed of rotation of the celestial sphere is 360° 59.15′ per 24 hours.

e. With the important exception of bodies in the solar system, which change position slowly, all heavenly bodies remain practically fixed in their positions on the celestial sphere, never changing more than negligible amounts in 24 hours, and accordingly are often called *fixed stars.*

Celestial navigation involves comparing the apparent position of the sun (or star) with the "actual" position as given in a table called an "ephemeris." Above is an excerpt from the introduction to the tables in the Solar Ephemeris. *(Keuffel and Esser Co.)*

SG 9
SG 10

Galileo had this experience (see Chapter 7).

evidence. Therefore, attention was focused on the argument concerning the central, immovable position of the earth. Despite his efforts, Copernicus could not persuade most of his readers that his heliocentric system reflected the mind of God as closely as did the geocentric system.

4. The Copernican theory conflicted with the basic ideas of Aristotelian physics. This conflict is well described by H. Butterfield in *Origins of Modern Science:*

> . . . at least some of the economy of the Copernican system is rather an optical illusion of more recent centuries. We nowadays may say that it requires smaller effort to move the earth round upon its axis than to swing the whole universe in a twenty-four hour revolution about the earth; but in the Aristotelian physics it required something colossal to shift the heavy and sluggish earth, while all the skies were made of a subtle substance that was supposed to have no weight, and they were comparatively easy to turn, since turning was concordant with their nature. Above all, if you grant Copernicus a certain advantage in respect of geometrical simplicity, the sacrifice that had to be made for the sake of this was tremendous. You lost the whole cosmology associated with Aristotelianism—the whole intricately dovetailed system in which the nobility of the various elements and the hierarchical arrangement of these had been so beautifully interlocked. In fact, you had to throw overboard the very framework of existing science, and it was here that Copernicus clearly failed to discover a satisfactory alternative. He provided a neater geometry of the heavens, but it was one which made nonsense of the reasons and explanations that had previously been given to account for the movements in the sky.

All religious faiths in Europe, including the new Protestants, opposed Copernicus. They used biblical quotations (for example, *Joshua* 10:12–13) to assert that the Divine Architect must have worked from a Ptolemaic blueprint. Indeed, Martin Luther called Copernicus "the fool who would overturn the whole science of astronomy."

Eventually, in 1616, more storm clouds were raised by the case of Galileo. The Inquisition put *De Revolutionibus* on the *Index* of forbidden books as "false and altogether opposed to Holy Scriptures." Some Jewish communities also prohibited the teaching of Copernicus's theory. It seems that humanity, believing itself central to God's plan, had to insist that the earth stood at the center of the physical universe.

The assumption that the earth was not the center of the universe was offensive enough. Even worse, the Copernican system suggested that the other planets were similar to the earth. Thus, the concept of the distinctly different heavenly matter was threatened. What next? What if some rash person suggested that the sun and possibly even the stars were made of earthly materials? If other celestial bodies were similar to the earth, they might even be inhabited. And the inhabitants might be heathens, or beings as well-beloved by God as humans, possibly even more beloved! Thus, the whole Copernican scheme led to profound philosophical questions which the Ptolemaic scheme avoided.

In short, the sun-centered Copernican scheme was scientifically equivalent to the Ptolemaic scheme in explaining astronomical observations. But, philosophically, it seemed false, absurd, and dangerous. Most learned Europeans at that time recognized the Bible and the writings of Aristotle as their two supreme sources of authority. Both appeared to be challenged by the Copernican system. Although the freedom of thought that marked the Renaissance was just beginning, the old image of the universe provided security and stability to many. Belief in a sun-centered rather than an earth-centered universe allowed a gain in simplicity; but it also seemed to contradict all common sense and observation. It required a revolution in philosophy, religion, and the physical science of the time. No wonder Copernicus had so few believers!

Conflicts between accepted beliefs and the philosophical content of new scientific theories have occurred many times and are bound to occur again. During the last century there were at least two such conflicts. Neither is completely resolved today. In biology, the theory of evolution based on Darwin's work has caused major philosophical and religious reactions. In physics, developing theories of atoms, relativity, and quantum mechanics have challenged long-held assumptions about the nature of the world and our knowledge of reality. Units 4, 5, and 6 touch upon these new theories. As the dispute between the Copernicans and the Ptolemaists illustrates, the assumptions which "common sense" defends so fiercely are often only the remains of an earlier, less complete scientific theory.

SG 11

?

7. Why were many people, such as Francis Bacon, undecided about the correctness of the Ptolemaic and Copernican systems?

8. How did the astronomical argument become involved with religious beliefs?

9. From a modern viewpoint, was the Ptolemaic or the Copernican system of reference more valid?

6.5 | Historical consequences

Eventually Copernicus' moving-earth model was accepted. But acceptance came very slowly. John Adams, who later became the second president of the United States, wrote that he attended a lecture at Harvard College in which the correctness of the Copernican viewpoint was debated on June 19, 1753!

The Copernican model with moving earth and fixed sun opened a floodgate of new possibilities for analysis and description. According to this model the planets could be thought of as real bodies moving along actual orbits. Now Kepler and others could consider these planetary paths in quite new ways. In science, the sweep of possibilities usually cannot be foreseen by those who begin the revolution or by their critics.

SG 12 Today, Copernicus is honored not so much for the details of his theory, but for his successful challenge of the prevailing world-picture. His theory became a major force in the intellectual revolution which shook humanity out of its self-centered view of the universe. As people gradually accepted the Copernican system, they also had to accept the view that the earth was only one of several planets circling the sun. Thus, it became increasingly difficult to believe that all creation centered on human beings. At the same time, the new system stimulated a new self-reliance and curiosity about the world.

Acceptance of a revolutionary idea based on quite new assumptions, such as Copernicus' shift of the frame of reference, is always slow. Sometimes compromise theories are proposed as attempts to unite conflicting theories, to "split the difference." As you will see in later units, such compromises are rarely successful. But often they do stimulate new observations and concepts. In turn, these may lead to a very useful development or restatement of the original revolutionary theory.

Such a restatement of the heliocentric theory came during the 150 years after Copernicus. Many scientists provided observations and ideas. In Chapters 7 and 8, you will see the major contributions made by Kepler, Galileo, and Isaac Newton. First we will consider the work of Tycho Brahe, who devoted his life to improving the precision with which planetary positions were observed and to the working out of a compromise theory of planetary motion.

? *10. In terms of historical perspective, what were the greatest contributions of Copernicus to modern planetary theory?*

6.6 | Tycho Brahe

Tycho Brahe was born in 1546 of a noble, but not particularly rich, Danish family. By the time he was 13 or 14, he had become intensely interested in astronomy. Although he was studying law, Tycho secretly spent his allowance on astronomical tables and books. He read the *Almagest* and *De Revolutionibus*. Soon he discovered that both Ptolemy and Copernicus had relied upon tables of planetary positions that were inaccurate. He concluded that astronomy needed new observations of the highest possible precision gathered over many years. Only then could a satisfactory theory of planetary motion be created.

Although there were precision sighting instruments, all observations were with the naked eye. The telescope was not to be invented for another 50 years.

Tycho's interest in studying the heavens was increased by an exciting celestial event. Although the ancients had taught that the stars were unchanging, a "new star" appeared in the constellation Cassiopeia in 1572. It soon became as bright as Venus and could be seen even during the daytime. Then over several years it faded until it was no longer visible. To Tycho these changes in the starry sky were astonishing. Evidently at least one assumption of the ancients was wrong. Perhaps other assumptions were wrong, too.

After observing and writing about the new star, Tycho traveled through northern Europe. He met many other astronomers and collected books. Apparently he was considering moving to Germany or Switzerland where he could easily meet other astronomers. To keep the young scientist in Denmark, King Frederick II offered him an entire small island and also the income from various farms. This income would allow Tycho to build an observatory on the island and to staff and maintain it. He accepted the offer, and in a few years Uraniborg ("Castle of the Heavens") was built. It was an impressive structure with four large observatories, a library, a laboratory, shops, and living quarters for staff, students, and observers. There was even a complete printing plant. Tycho estimated that the observatory cost Frederick II more than a ton of gold. For its time, this magnificent laboratory was at least as important, complex, and expensive as some of today's great research centers. Uraniborg was a place where scientists, technicians, and students from many lands could gather to study astronomy. Here, a group effort under the leadership of an imaginative scientist was to advance the boundaries of knowledge in one science.

In 1577, Tycho observed a bright comet, a fuzzy object whose motion seemed irregular, unlike the orderly motions of the planets. To find the distance to the comet, Tycho compared its position as observed from Denmark with its positions as observed from elsewhere in Europe. Some of these observation points lay hundreds of kilometers apart. Yet, at any given time, all observers reported the comet as having the same position with respect to the stars. By contrast, the moon's position in the sky was measurably different when observed from places so far apart. Therefore, Tycho concluded, the comet must be at least several times farther away than the moon.

This was an important conclusion. Up to that time, people had believed that comets were some sort of local event, like clouds

Close Up

Tycho Brahe

At the top right is a plan of the observatory and gardens built for Tycho Brahe at Uraniborg, Denmark.

The cross section of the observatory, above center, shows where most of the important instruments, including large models of the celestial spheres, were housed.

The picture at the left shows the room containing Tycho's great quadrant. On the walls are pictures of some of his instruments. He is making an observation, aided by assistants.

Above is a portrait of Tycho, painted about 1597.

The bright comet of 1965.

or lightning. Now comets had to be considered distant astronomical objects from the realm of eternal things beyond the moon. Stranger still, they seemed to move right through the crystalline spheres that were still generally believed to carry the planets. Tycho's book on this comet was widely read and helped to weaken old beliefs about the nature of the heavens.

SG 13

?

11. What event stimulated Tycho's interest in astronomy?

12. In what ways was Tycho's observatory like a modern research institute?

13. Why were Tycho's conclusions about the comet of 1577 important?

6.7 | Tycho's observations

Tycho's fame results from his lifelong devotion to making unusually accurate observations of the positions of the stars, sun, moon, and planets. These observations were made before the telescope was invented. Over the centuries, many talented observers had recorded the positions of the celestial objects. But the accuracy of Tycho's work was much greater than that of the best astronomers before him. How was Tycho Brahe able to do what no others had done before?

For a more modern example of this same problem of instrumentation, you may wish to read about the development and construction of the 500-cm Hale telescope on Mt. Palomar.

Singleness of purpose certainly aided Tycho. He knew that highly precise observations must be made during many years. For this he needed improved instruments that would give consistent readings. Fortunately, he had the mechanical skill to devise such instruments. He also had the funds to pay for their construction and use.

Tycho's first improvement on the astronomical instruments of the day was to make them larger. Most of the earlier instruments had been rather small, of a size that could be moved by one person. In comparison, Tycho's instruments were gigantic. For instance, one of his early devices for measuring the angular altitude of planets had a radius of about 1.8 m. This wooden instrument, shown in the etching on page 174, was so large that it took several workers to set it into position. Tycho put his

Johannes Hevelius and his wife jointly using a quadrant in his observatory in Danzig (seventeenth century).

instruments on heavy, firm foundations or attached them to a wall that ran exactly north–south. By fixing the instruments so solidly, Tycho increased the reliability of the readings over long periods of time. Throughout his career Tycho created better sighting devices, more precise scales, and stronger support systems. He made dozens of other changes in design which increased the precision of the observations.

Tycho did more than just devise better instruments for making his observations. He also determined and specified the actual limits of precision of each instrument. He realized that merely making larger and larger instruments does not always result in greater precision. In fact, the very size of the instrument can cause errors, since the parts bend under their own weight. Tycho tried to make his instruments as large and strong as he could without introducing such errors. Furthermore, in modern style, he calibrated each instrument and determined its range of error. (Nowadays many commercial instrument makers supply a measurement report with scientific instruments designed for precision work. Such reports are usually in the form of a table of small corrections that have to be applied to the direct readings.)

Like Ptolemy and the Muslim astronomers, Tycho knew that the light coming from any celestial body was bent downward by the earth's atmosphere. He knew that this bending, or *refraction,* increased as the celestial object neared the horizon. To improve the precision of his observations, Tycho carefully determined the amount of refraction involved. Thus, each observation could be corrected for refraction effects. Such careful work was essential to the making of improved records.

Tycho worked at Uraniborg from 1576 to 1597. After the death of King Frederick II, the Danish government became less interested in helping to pay the cost of Tycho's observatory. Yet Tycho was unwilling to consider any reductions in the cost of his activities. Because he was promised support by Emperor Rudolph of Bohemia, Tycho moved his records and several instruments to Prague. There, fortunately, he hired as an assistant an able, imaginative young man named Johannes Kepler. After Tycho's death in 1601, Kepler obtained all his records of observations of the motion of Mars. As Chapter 7 reports, Kepler's analysis of Tycho's data solved many of the ancient problems of planetary motion.

One of Tycho's sighting devices. Unfortunately Tycho's instruments were destroyed in 1619, during the Thirty Years' War.

?

14. What improvements did Tycho make in astronomical instruments?

15. In what way did Tycho correct his observations to provide records of higher accuracy?

6.8 | Tycho's compromise system

Tycho hoped that his observations would provide a basis for a new theory of planetary motion, which he had outlined in an early book. He saw the simplicity of the Copernican system, in which the planets moved around the sun. Yet, because he observed no annual parallax of the stars, he could not accept an annual motion of the earth around the sun. In Tycho's system, all the planets except the earth moved around the sun. Meanwhile, the sun moved around the stationary earth, as shown in the sketch in the margin. Thus, Tycho devised a compromise model which, as he said, included the best features of both the Ptolemaic and the Copernican systems. However, he did not live to publish quantitative details of his theory.

The compromise Tychonic system was accepted by only a few people. Those who accepted the Ptolemaic model objected to having the planets moving around the sun. Those who accepted the Copernican model objected to having the earth held stationary. So the argument continued. Many scholars clung to the seemingly self-evident position that the earth was stationary. Others accepted, at least partially, the strange, exciting proposals of Copernicus that the earth might rotate and revolve around the sun. The choice depended mainly on one's philosophy.

All planetary theories up to that time had been developed only to provide some system for predicting the positions of the planets fairly precisely. In the terms used in Unit 1, these would be called kinematic descriptions. The causes of the motions, now called dynamics, had not been considered in any detail. Aristotle had described angular motions of objects in the heavens as "natural." Everyone, including Ptolemy, Copernicus, and Tycho, agreed. Celestial objects were still considered to be completely different from earthly materials and to behave in quite different ways. *That a single theory of dynamics could describe both earthly and heavenly motions was a revolutionary idea yet to be proposed.*

As long as there was no explanation of the causes of motion, a basic problem remained unsolved. Were the orbits proposed for the planets in the various systems actual paths of real objects in space? Or were they only convenient imaginary devices for making computations? The status of the problem in the early seventeenth century was later described well by the English poet John Milton in *Paradise Lost:*

... He his fabric of the Heavens
Hath left to their disputes, perhaps to move
His laughter at their quaint opinions wide
Hereafter, when they come to model Heaven
And calculate the stars, how they will wield
The mighty frame, how build, unbuild, contrive

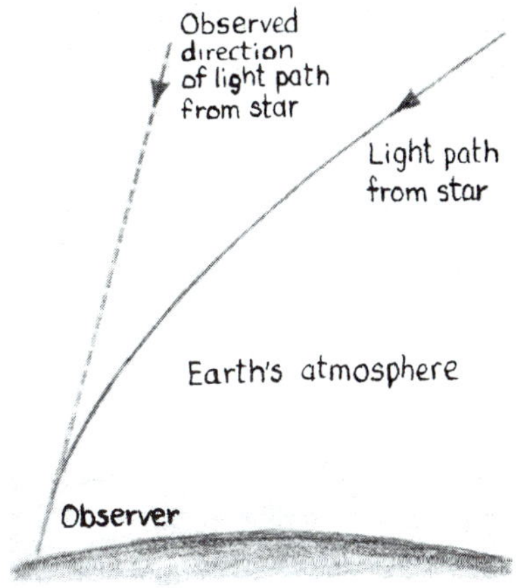

Refraction, or bending, of light from a star by the earth's atmosphere. The amount of refraction shown in the figure is a great exaggeration of what actually occurs.

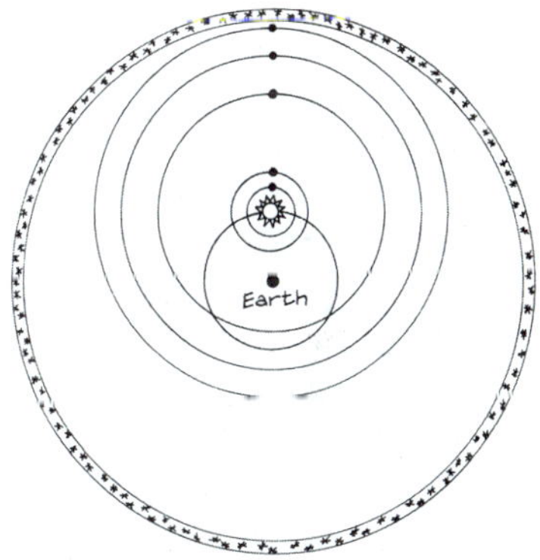

Main spheres in Tycho Brahe's system of the universe. The earth was fixed and was at the center of the universe. The planets revolved around the sun, while the sun, in turn, revolved around the fixed earth.

Close Up

Observatories

Observing instruments have changed dramatically in the years since Tycho's work at Uraniborg. Kitt Peak National Observatory is located in the mountains southwest of Tucson, Arizona. This center for ground-based optical astronomy in the northern hemisphere has the largest concentration of facilities for stellar, solar, and planetary research in the world. The largest of Kitt Peak's 14 instruments is the 4-m Mayall telescope, which is the second largest reflecting telescope in the United States (the largest is the 5-m Hale telescope on Palomar Mountain). An almost identical 4-m telescope is located at the Cerro Tololo Inter-American Observatory in Chile. The location of this observatory in the Andes Mountains provides a clear view of the southern sky not visible from Kitt Peak and other northern hemisphere observatories. Technical improvements over the 13 years necessary to design and build these two telescopes have led to a significant increase in the efficiency of the telescopes.

A different type of telescope is the Multiple Mirror Telescope (MMT), also located in the Arizona desert. The MMT represents the first major innovation in telescope design in a century. Light is collected by six 1.8-m mirrors and brought to a common focus (the alignment is corrected by lasers). Thus, the MMT has the light-gathering capacity of a conventional 4.5-m telescope making it the third largest telescope in the world. The telescope is housed in a four-story rotating structure, rather than the conventional dome. Both the building and the telescope move in unison when in operation.

The largest optical telescope in the world is the 6-m instrument located in the mountains of the North Caucasus in the Soviet Union.

Above left: The eye end of the 90-cm refractor at Lick Observatory showing the automatic camera for direct photography.

To save appearances, how gird the sphere
With centric and eccentric scribbled o'er
Cycle and epicycle, orb in orb.

The eventual success of Newton's universal dynamics led to the belief that scientists were describing the "real world." This belief was held confidently for about two centuries. Later chapters of this text deal with recent discoveries and theories which have lessened this confidence. Today, scientists and philosophers are much less certain that the common-sense notion of "reality" is very useful in science.

SG 14

This unit presents the first example of the highly successful trend of modern science toward synthesis, that is, not two or more kinds of science, but only one. For example, not a separate physics of energy in each branch, but one conservation law; not separate physics for optical, heat, electric, magnetic phenomena, but one (Maxwell's); not two kinds of beings (animal and human) but one (in Darwinian views); not space and time separately, but space–time; not mass and energy separately, but mass–energy, etc. To a point, at least, the great advances of science are the results of such daring extensions of one set of ideas into new fields. The danger is the false extrapolation that science by itself can solve all problems, including political, health, or educational, and "explain" all human emotions. The majority of scientists do not believe this extrapolation, but many nonscientists falsely believe that all scientists do.

? *16. In what ways did Tycho's system for planetary motions resemble the Ptolemaic and the Copernican systems?*

study guide

1. The *Project Physics* learning materials particularly appropriate for Chapter 6 include the following:

Experiments
The Shape of the Earth's Orbit
Using Lenses to Make a Telescope

Activity
Frames of Reference

Film Loop
Retrograde Motion: Heliocentric Model

sg

2. The first diagram on the next page shows numbered positions of the sun and Mars (on its epicycle) at equal time intervals in their motion around the earth, as described in the Ptolemaic system. You can easily redraw the relative positions to change from the earth's frame of reference to the sun's. Mark a sun-sized circle in the middle of a thin piece of paper; this will be a frame of reference centered on the sun. Place the circle over each successive position of the sun, and trace the corresponding numbered position of Mars and the position of the earth. (Be sure to keep the piece of paper straight.) When you have done this for all 15 positions, you will have a diagram of the motions of Mars and the earth as seen in the sun's frame of reference.

3. What reasons did Copernicus give for believing that the sun is fixed at or near the center of the planetary system?

4. Consider the short and long hands of a clock or watch. If, starting from 12:00 o'clock, you rode on the slow short hand, how many times in 12 hr would the long hand pass you? If you are not certain, slowly turn the hands of a clock or watch, and keep count. From this information, can you derive a relation by which you could conclude that the period of the long hand around the center is 1 hr?

5. Copernicus' theory is considered valuable because it allows new predictions and conclusions. What new conclusions resulted from Copernicus' theory? Why do these conclusions make this theory "better" than previous theories?

6. Section 6.4 states that Copernicus' heliocentric theory is scientifically equivalent in many ways to Ptolemy's geocentric theory and merely represents a change in the frame of reference from a fixed earth to a fixed sun. In what ways is the Copernican system more than just "a change in the frame of reference"?

7. The diagram at the upper right section of the next page shows the motions of Mercury and Venus east and west of the sun as seen from the earth. The time scale is indicated at 10-day intervals along the central line of the sun's position.

(a) Can you explain why Mercury and Venus appear to move from farthest east to farthest west more quickly than from farthest west to farthest east?

(b) From this diagram, can you find a period for Mercury's apparent position in the sky relative to the sun?

(c) Can you derive a period for Mercury's actual orbital motion around the sun?

(d) What are the major sources of uncertainty in the results you derived?

(e) Similarly, can you estimate the orbital period of Venus?

8. From the sequence of orbital radii from Mercury to Saturn, estimate what the orbital radius would be for a new planet if one were discovered. What is the basis for your estimate?

9. The largest observed annual shift in star position is about 1/2400 of a degree. What is the distance (in astronomical units) to this closest star?

10. How might a Ptolemaic astronomer have modified the geocentric system to account for observed stellar parallax?

11. What conflicts between scientific theories and common sense do you know of today?

12. How did the Copernican system encourage the suspicion that there might be life on objects other than the earth? Is such a possibility seriously considered today? What important questions would such a possibility raise?

13. How can you explain the observed motion of Halley's comet during 1909–1910, as shown on the star chart on the next page?

14. To what extent do you feel that the Copernican system, with its many motions in eccentrics and epicycles, reveals real paths in space, rather than provides only another way of computing planetary positions?

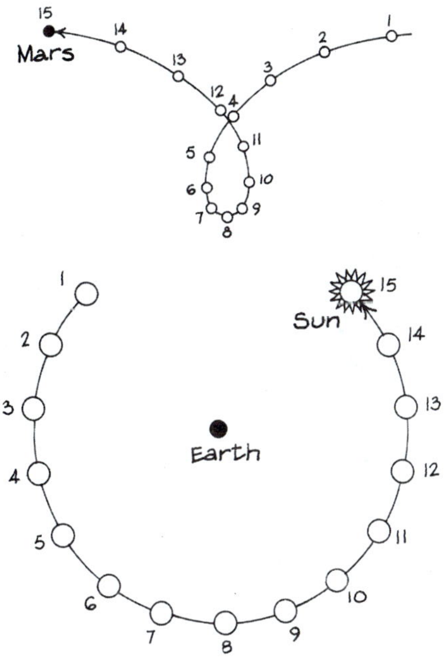

Apparent motion of Mars and the sun around the earth.

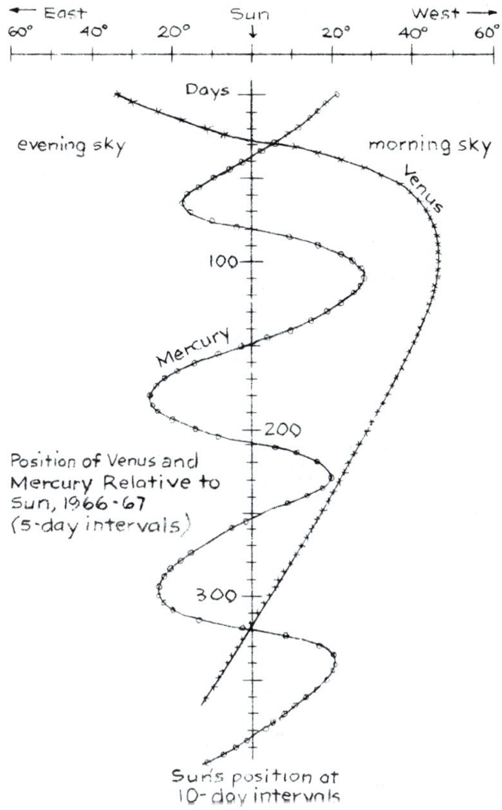

Positions of Venus and Mercury relative to the sun.

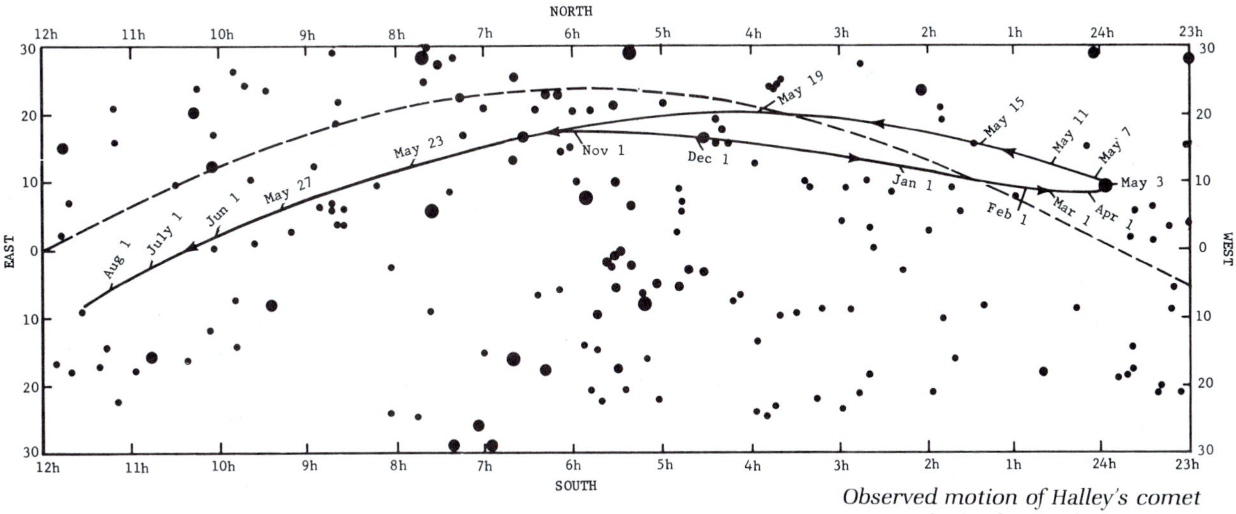

Observed motion of Halley's comet against the background of stars during 1909–1910.

A New Universe Appears

The Work of Kepler and Galileo

7.1 | The abandonment of uniform circular motion

SG 1 Kepler, who became Tycho's assistant, had the lifelong desire to perfect the heliocentric theory. He viewed the harmony and simplicity of that theory with "incredible and ravishing delight." To Kepler, such patterns of geometric order and numerical relation offered clues to God's mind. Kepler sought to unfold these patterns further through the heliocentric theory. In his first major work, he attempted to explain the spacing of the planetary orbits as calculated by Copernicus (page 161 of Chapter 6).

Kepler was searching for the reasons why there are just six

visible planets (including the earth) and why they are spaced as they are. These are excellent scientific questions, but even today they are too difficult to answer. Kepler thought that the key lay in geometry. He began to wonder whether there was any relation between the six known planets and the five *regular solids*. A regular solid is a polyhedron whose faces all have equal sides and angles. From the time of the Greeks, it was known that there are just five regular geometrical solids. Kepler imagined a model in which these five regular solids nested one inside the other, somewhat like a set of mixing bowls. Between the five solids would be spaces for four planetary spheres. A fifth sphere could rest inside the whole nest and a sixth sphere could lie around the outside. Kepler then sought some sequence of the five solids that, just touching the spheres, would space the spheres at the same relative distance from the center as were the planetary orbits. Kepler said:

> I took the dimensions of the planetary orbits according to the astronomy of Copernicus, who makes the sun immobile in the center, and the earth movable both around the sun and upon its own axis; and I showed that the differences of their orbits corresponded to the five regular Pythagorean figures. . . .

By trial and error Kepler found a way to arrange the solids so that the spheres fit within about 5% of the actual planetary distances. We now know that this arrangement was entirely accidental. But to Kepler it explained both the spacings of the planets and the fact that there were just six. Also, it had the unity he expected between geometry and scientific observations. Kepler's results, published in 1597, demonstrated his imagination and mathematical ability. Furthermore, his work came to the attention of major scientists such as Galileo and Tycho. In 1600, Kepler was invited to become one of Tycho's assistants at his new observatory in Prague.

There, Kepler was given the task of determining in precise detail the orbit of Mars. This unusually difficult problem had not been solved by Tycho and his other assistants. As it turned out, Kepler's study of the motion of Mars was only a starting point. From it, he went on to redirect the study of celestial motion. In the same way, Galileo used the motion of falling bodies to redirect the study of terrestrial motion.

Kepler began by trying to fit the observed motions of Mars with motions of an eccentric circle and an equant. Like Copernicus, Kepler eliminated the need for the large epicycle by putting the sun motionless at the center and having the earth move around it (see page 157). But Kepler made an assumption which differed from that of Copernicus. Recall that Copernicus had rejected the equant as an improper type of motion, using small epicycles instead. Kepler used an equant, but refused to use even a single small epicycle. To Kepler the epicycle seemed "unphysical." He

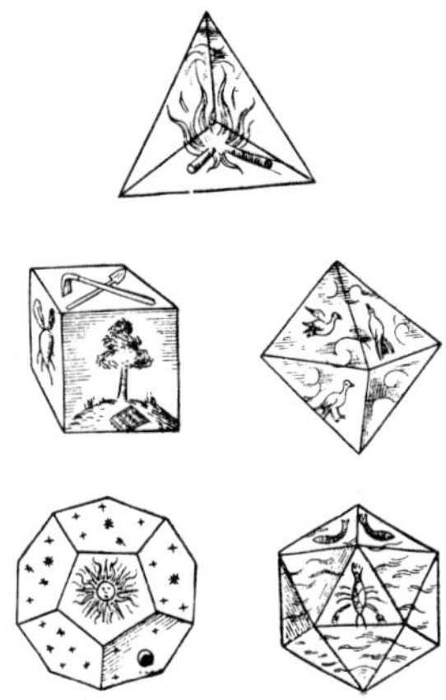

The five "regular solids" (also called Pythagorean figures or Platonic solids), taken from Kepler's Harmonices Mundi *(Harmony of the World). The* cube *is a regular solid with 6 square faces. The* dodecahedron *has 12 five-sided faces. The other three regular solids have faces that are equilateral triangles. The* tetrahedron *has 4 triangular faces, the* octahedron *has 8 triangular faces, and the* icosahedron *has 20 triangular faces.*

For Kepler, this geometric view was related to ideas of harmony.

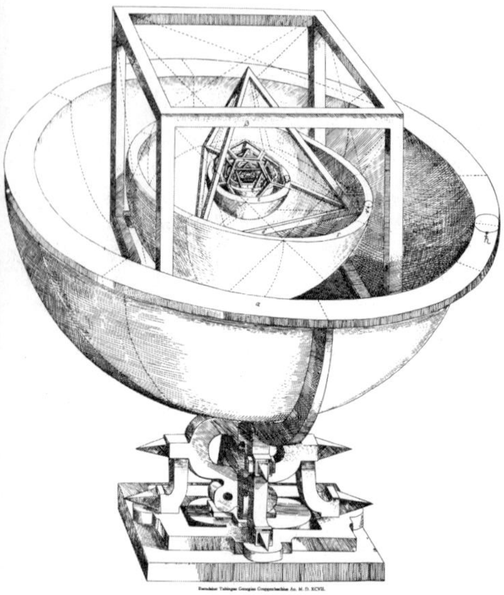

Kepler's model for explaining the spacing of the planetary orbits by means of the regular geometrical solids. Notice that the planetary spheres were thick enough to include the small epicycle used by Copernicus.

SG 2

In keeping with Aristotelian physics, Kepler believed that force was necessary to drive the planets along their circles, not to hold them in circles.

Fortunately, Kepler had made a major discovery earlier that was crucial to his later work. He found that the orbits of the earth and other planets were in planes that passed through the sun. Ptolemy and Copernicus required special explanations for the motion of planets north and south of the ecliptic, but Kepler found that these motions were simply the result of the orbits lying in planes tilted to the plane of the earth's orbit.

reasoned that the center of the epicycle was empty, and empty space could not exert any force on a planet. Thus, from the start, Kepler assumed that the orbits were real and that the motion resulted from physical causes, namely, the *action of forces* on the planets. Even his beloved teacher, Maestlin, advised the young man to stick to geometrical models and astronomical observation and to avoid physical assumptions. But Kepler stubbornly stuck to his idea that the motions must be produced and explained by forces. When he finally published his results on Mars in his book *Astronomia Nova (New Astronomy)*, it was subtitled *Celestial Physics.*

For a year and a half Kepler struggled to fit his findings with Tycho's observations of Mars by various arrangements of an eccentric and an equant. When, after 70 trials, success finally seemed near, he made a discouraging discovery. He could represent fairly well the motion of Mars in longitude (east and west along the ecliptic), but he failed markedly with the latitude (north and south of the ecliptic). Even in longitude his very best predictions still differed by 8 min of arc from Tycho's observed positions.

Eight minutes of arc, about one-fourth of the moon's diameter, may not seem like much of a difference. Others might have been tempted to explain it away as an observational error. But Kepler knew that Tycho's instruments and observations were rarely in error by even 2 min of arc. Those 8 min of arc meant to Kepler that his best system, using the old, accepted devices of eccentric and equant, would never work. In *New Astronomy,* Kepler wrote:

> Since divine kindness granted us Tycho Brahe, the most diligent observer, by whose observations an error of eight minutes in the case of Mars is brought to light in this Ptolemaic calculation, it is fitting that we recognize and honor this favor of God with gratitude of mind. Let us certainly work it out, so that we finally show the true form of the celestial motions (by supporting ourselves with these proofs of the fallacy of the suppositions assumed). I myself shall prepare this way for others in the following chapters according to my small abilities. For if I thought that the eight minutes of longitude were to be ignored, I would already have corrected the hypothesis which he had made earlier in the book and which worked moderately well. But as it is, because they could not be ignored, these eight minutes alone have prepared the way for reshaping the whole of astronomy, and they are the material which is made into a great part of this work.

Kepler concluded that the orbit was not a circle and that there was no point around which the motion was uniform. Plato's idea of fitting perfect circles to the heavens had guided astronomers for 20 centuries. Now, Kepler realized, this idea must be abandoned. Kepler had in his hands the finest observations ever made, but now he had no theory by which they could be explained. He would have to start over facing two altogether new questions. First, what *is* the shape of the orbit followed by Mars? Second, how does the speed of the planet change as it moves along the orbit?

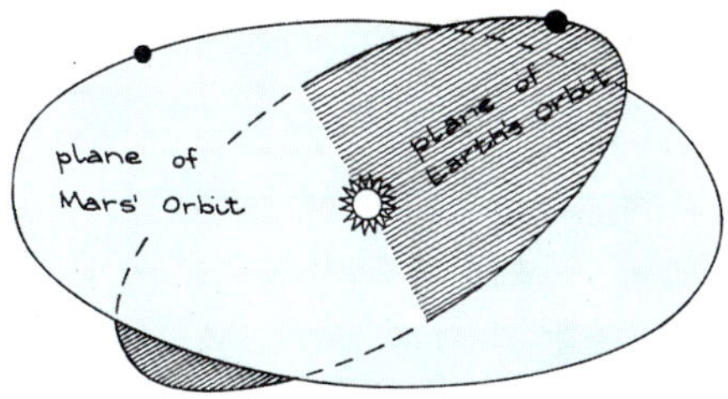

The diagram depicts a nearly edge-on view of the orbital planes of earth and Mars, both intersecting at the sun.

?

1. When Kepler joined Tycho Brahe what task was he assigned?

2. Why did Kepler reject the use of epicycles in his theory? Why is this reason important?

3. Why did Kepler conclude that Plato's problem, to describe the motions of the planets by combinations of circular motions, could not be solved?

7.2 | Kepler's law of areas

Kepler's problem was immense. To solve it would demand all of his imagination and skill. As the basis for his study, Kepler had Tycho's observed positions of Mars and the sun on certain dates. But these observations had been made from a moving earth whose orbit was not well known. Kepler realized that he must first determine more accurately the shape of the earth's orbit. This would allow him to calculate the earth's location on the dates of the various observations of Mars. Then he might be able to use the observations to determine the shape and size of the orbit of Mars. Finally, to predict positions for Mars, he would need to discover how fast Mars moved along different parts of its orbit.

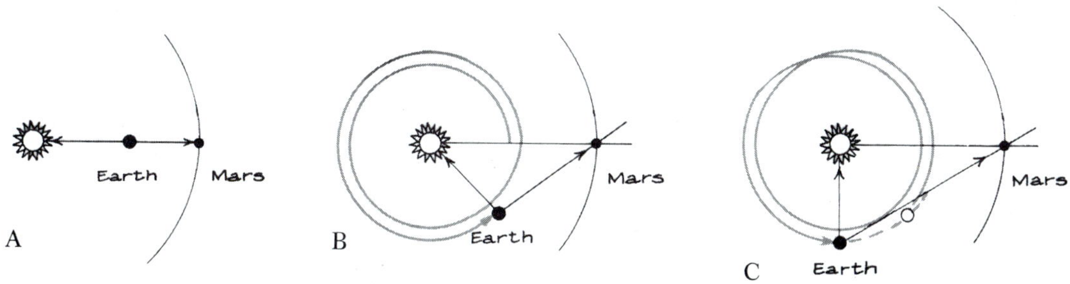

As you follow this brilliant analysis here, and particularly if you repeat some of his work in the laboratory, you will see the series of problems that Kepler solved.

To find the earth's orbit he began by considering the moments when the sun, earth, and Mars lie almost in a straight line (Figure A). After 687 days, as Copernicus had found, Mars would return to the same place in its orbit (Figure B). Of course, the earth at that time would *not* be at the same place in its own orbit as when the first observation was made. As Figures B and C indicate, the directions to the sun and Mars, as seen from the earth against the fixed stars, would be known. The crossing point of the sight-lines to the sun and to Mars must be a point on the earth's orbit. Kepler worked with several groups of observations made 687 days apart (one Mars "year"). In the end, he determined fairly accurately the shape of the earth's orbit.

SG 3

The orbit Kepler found for the earth appeared to be almost a circle, with the sun a bit off center. Knowing now the shape of the earth's path and knowing also the recorded apparent position of the sun as seen from the earth for each day of the year, he could locate the position of the earth on its orbit and its speed along the orbit. Now he had the orbit and the timetable for the earth's motion. You may have made a similar plot in the experiment "The Shape of the Earth's Orbit."

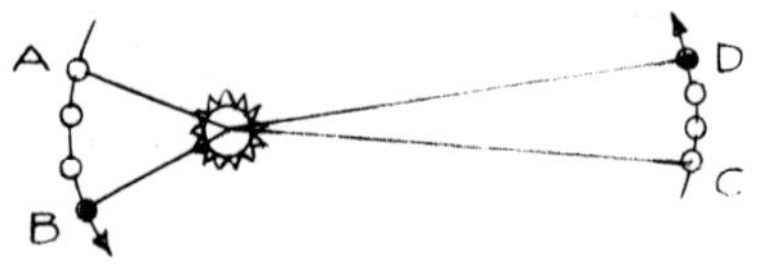

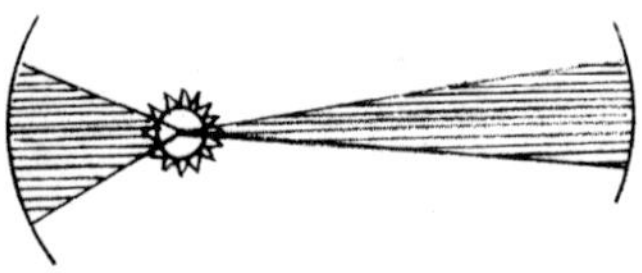

Kepler's plot of the earth's motion revealed that the earth moves fastest when nearest the sun. Kepler wondered why this occurred. He thought that the sun might exert some force that drove the planets along their orbits. This concern with the physical cause of planetary motion marked the change in attitude toward motion in the heavens.

The drawings at the left represent (with great exaggeration) the earth's motion for two parts of its orbit. The different positions on the orbit are separated by equal time intervals. Between points A and B there is a relatively large distance, so the planet is moving rapidly. Between points C and D it moves more slowly. Kepler noticed, however, that the two *areas* swept over by a line from the sun to the planet are equal. It is believed that he actually calculated such areas only for the nearest and farthest positions of two planets, earth and Mars. Yet the beautiful simplicity of the relation led him to conclude that it was generally true for all parts of orbits. In its general form, the law of areas states: *The line from the sun to the moving planet sweeps over areas that are proportional to the time intervals.* Later, when Kepler found the exact shape of orbits, his law of areas became a powerful tool for predicting positions.

SG 4

Another way to express this relationship for the nearest and farthest positions would be to say the speeds were inversely proportional to the distance; but this rule does *not* generalize to any other points on the orbit.

You may be surprised that the first rule Kepler found for the motions of the planets dealt with the areas swept over by the line from the sun to the planet. Scientists had been considering circles, eccentric circles, epicycles, and equants. This was a quite unexpected property: The area swept over per unit time is the first property of the orbital motion to remain constant. (As you will see in Chapter 8, this major law of nature applies to all orbits in the solar system and also to double stars.) Besides being new and different, the law of areas drew attention to the central role of the sun. Thus, it strengthened Kepler's faith in the still widely neglected Copernican idea of a heliocentric system.

As you will see, Kepler's other labors would have been of little use without this basic discovery. However, the rule does not give any hint why such regularity exists. It merely describes the relative rates at which the earth and Mars (and, Kepler thought, any other planet) move at any point of their orbits. Kepler could not fit the rule to Mars by assuming a circular orbit, so he set out to find the shape of Mars' orbit.

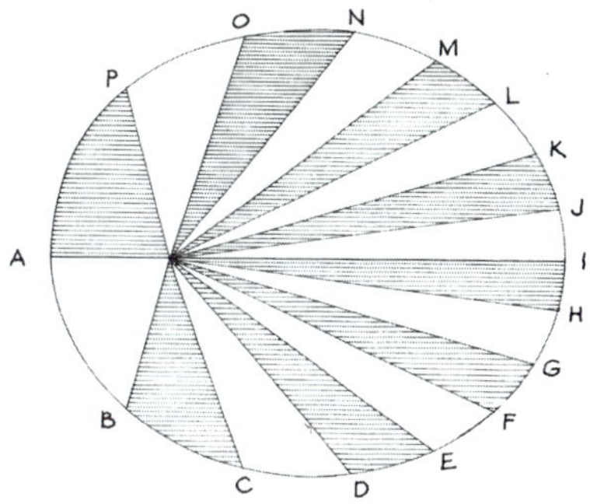

Kepler's law of areas. A planet moves along its orbit at a rate such that the line from the sun to the planet sweeps over areas which are proportional to the time intervals. The time taken to cover AB is the same as that for BC, CD, etc.

?

4. *What observations did Kepler use to plot the earth's orbit?*
5. *State Kepler's law of areas.*
6. *Where in its orbit does a planet move the fastest?*

7.3 | Kepler's law of elliptical orbits

Kepler knew the orbit and timetable of the earth. Now he could reverse his analysis and find the shape of Mars' orbit. Again he used observations separated by one Martian year. Because this interval is somewhat less than two earth years, the earth is at different positions in its orbit at the two times. Therefore, the two directions from the earth to Mars differ. Where they cross is a point on the orbit of Mars. From such pairs of observations Kepler fixed many points on the orbit of Mars. The diagrams below illustrate how two such points might be plotted. By drawing a curve through such points, Kepler obtained fairly

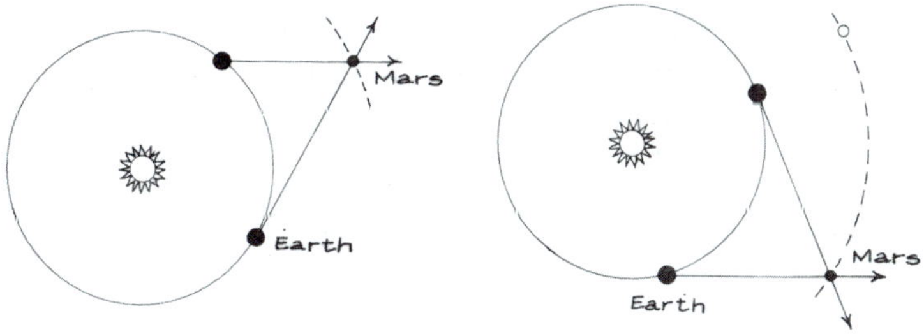

accurate values for the size and shape of Mars' orbit. Kepler saw at once that the orbit of Mars was not a circle around the sun. You will find the same result from the experiment, "The Orbit of Mars." What sort of path was this? How could it be described most simply? As Kepler said, "The conclusion is quite simply that the planet's path is not a circle—it curves inward on both sides and outward again at opposite ends. Such a curve is called an oval." But what kind of oval?

SG 5

Many different closed curves can be called ovals. Kepler thought for a time that the orbit was egg-shaped. However, this shape did not agree with his ideas of physical interaction between the sun and the planet. He concluded that there must be some better way to describe the orbit. For many months, Kepler struggled with the question. Finally, he realized that the orbit was a simple curve that had been studied in detail by the

In this experiment, the orbit of Mars was plotted from measurements made on pairs of sky photographs taken 1 Martian year apart.

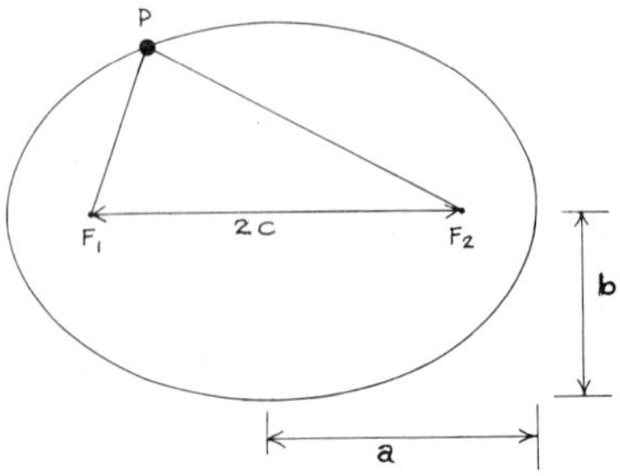

An ellipse showing the semimajor axis a, *the semiminor axis* b, *and the two foci* F_1 *and* F_2. *The shape of an ellipse is described by its eccentricity* e, *where* e = c/a.

SG 6

SG 10

In the "Orbit of Mercury" Experiment, you can plot the shape of Mercury's very eccentric orbit from observational data. See also SG 13.

SG 13

Greeks 2,000 years before. The curve is called an *ellipse.* It is the shape you see when you view a circle at a slant.

Ellipses can differ greatly in shape. They have many interesting properties. For example, you can draw an ellipse by looping a piece of string around two thumbtacks pinned to a drawing board at points F_1 and F_2 as shown at the left. Pull the loop taut with a pencil point (P) and run the pencil around the loop. You will have drawn an ellipse. (If the two thumb tacks had been together, on the same point, what curve would you have drawn? What results do you get as you separate the two tacks more?)

Each of the points F_1 and F_2 is called a *focus* of the ellipse. the greater the distance between F_1 and F_2, the flatter, or more *eccentric* the ellipse becomes. As the distance between F_1 and F_2 shrinks to zero, the ellipse becomes more nearly circular. A measure of the eccentricity of the ellipse is the ratio of the distance F_1F_2 to the long axis. Since the distance between F_1 and F_2 is $2c$ and the length of the long axis is $2a$, the eccentricity, e, is defined by the equation $e = c/a$.

The eccentricities are given for each of the ellipses shown in the margin of the next page. You can see that a circle is the special case of an ellipse with $e = 0$. Also, note that the greatest possible eccentricity for an ellipse is $e = 1.0$

TABLE 7.1 THE ECCENTRICITIES OF PLANETARY ORBITS

Planet	Orbital Eccentricity	Notes
Mercury	0.206	Too few observations for Kepler to study
Venus	0.007	Nearly circular orbit
Earth	0.017	Small eccentricity
Mars	0.093	Largest eccentricity among planets Kepler could study
Jupiter	0.048	Slow moving in the sky
Saturn	0.056	Slow moving in the sky
Uranus	0.047	Not discovered until 1781
Neptune	0.009	Not discovered until 1846
Pluto	0.249	Not discovered until 1930

Kepler's discovery that the orbit of Mars is an ellipse was remarkable enough in itself. But he also found that the sun is at one focus of the orbit. (The other focus is empty.) Kepler stated these results in his *law of elliptical orbits: The planets move in orbits that are ellipses and have the sun at one focus.*

As Table 7.1 shows, Mars has the largest orbital eccentricity of any planet that Kepler could have studied. (The other planets for which there were enough data at the time were Venus, Earth, Jupiter, and Saturn.) Had he studied any planet other than Mars, he might never have noticed that the orbit was an ellipse! Even for Mars, the difference between the elliptical orbit and an off-center circle is quite small. No wonder Kepler later wrote: "Mars alone enables us to penetrate the secrets of astronomy which otherwise would remain forever hidden from us."

Like Kepler, modern scientists believe that observations represent some aspects of reality more stable than the changing emotions of human beings. Like Plato and all scientists after him, scientists assume that nature is basically orderly and consistent. Therefore, it must be understandable in a simple way. This faith has led to great theoretical and technical gains. Kepler's work illustrates a basic scientific attitude: Wide varieties of phenomena are better understood when they can be summarized by a simple law, preferably expressed in mathematical form.

After Kepler's initial joy at discovering the law of elliptical paths, he may have asked himself a question. Why are the planetary orbits elliptical rather than in some other geometrical shape? While Plato's desire for uniform circular motions is understandable, nature's insistence on the ellipse is surprising.

In fact, there was no satisfactory answer to this question for almost 80 years until, at last, Newton showed that elliptical orbits were necessary results of a more general law of nature. You can accept Kepler's laws as rules that contain the observed facts about the motions of the planets. As *empirical laws,* they each summarize the data obtained by observing the motion of any planet. The law of orbits describes the paths of planets as ellipses around the sun. It gives all the possible positions each planet can have if the orbit's size and eccentricity are known. However, it does not tell *when* the planet will be at any particular position on its ellipse or how rapidly it will be moving then. The law of areas, on the other hand, does not specify the shape of the orbit. But it does describe how the angular speed changes as the distance from the sun changes. Clearly, these two laws complement each other. Using them together, you can determine both the position and angular speed of a planet at any time, past or future. To do so, you need only to know the values for the size and eccentricity of the orbit and to know the position of the planet at any one time on its orbit. You can also find the earth's position for the same instant. Thus, you can calculate the

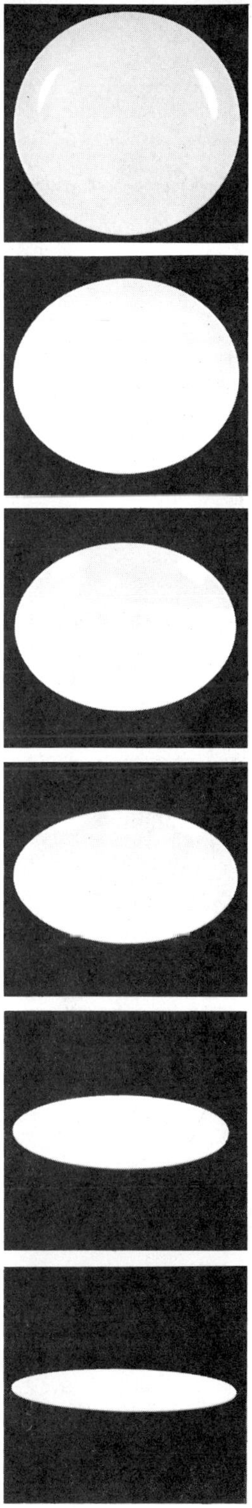

Ellipses of different eccentricities. (The pictures were made by photographing a saucer at different angles.)

Empirical means based on observation, not derived from theory.

position of the planet as it would have been or will be seen from the earth.

The elegance and simplicity of Kepler's two laws are impressive. Surely Ptolemy and Copernicus would have been amazed to see the problem of planetary motions solved in such short statements. You must remember, however, that these laws were distilled from Copernicus' idea of a moving earth and the great labors that went into Tycho's fine observations, as well as from the imagination and devotion of Kepler.

?

7. If you noticed a man walking to the store to buy a newspaper at 8:00 A.M. for three days in a row, what empirical law might you propose? What is an empirical law? If you later found that the man did not go to the store on Sunday, what conclusions could you make about empirical laws?

8. What special feature of Mars' orbit made Kepler's study of it so fortunate?

9. If the average distance and eccentricity of a planet's orbit are known, which of the following can be predicted from the law of areas alone? from the law of elliptical orbits alone? Which require both?

(a) all possible positions in the orbit

(b) speed at any point in an orbit

(c) position at any given time

Conic sections, *as shown in the diagram, are figures produced by cutting a cone with a plane. The eccentricity of a figure is related to the angle of the cut. In addition to circles and ellipses, parabolas and hyperbolas are conic sections, with eccentricities greater than ellipses. Newton eventually showed that all of these shapes are possible paths for a body moving under the gravitational attraction of the sun.*

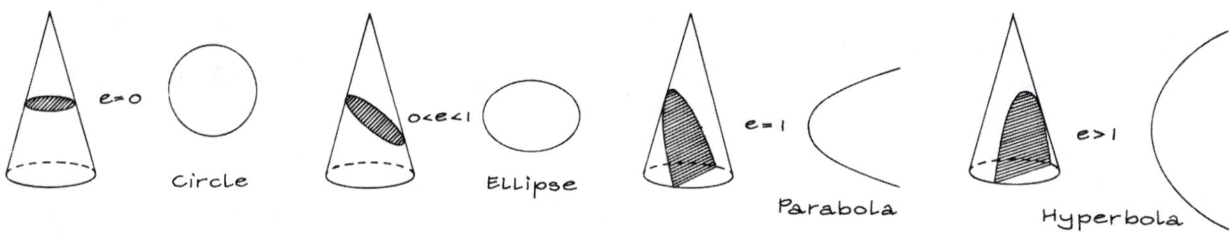

7.4 | Kepler's law of periods

Kepler published his first two laws in 1609 in his book *Astronomia Nova*. But he was still dissatisfied. He had not yet found any relation among the motions of the different planets. Each planet seemed to have its own elliptical orbit and speed. There appeared to be no overall pattern relating all planets to one another. Kepler had begun his career by trying to explain the number of planets and their spacing. He was convinced that the observed orbits and speeds could not be accidental. There *must* be some regularity linking all the motions in the solar system. His conviction was so strong that he spent years

As Einstein later put it: "The Lord is subtle, but He is not malicious."

examining possible combinations of factors by trial and error. Surely one combination would reveal a third law, relating all the planetary orbits. His long, stubborn search illustrates a belief that has run through the whole history of science: Despite apparent difficulties in getting a quick solution, nature's laws are rationally understandable. This belief is still a source of inspiration in science, keeping up one's spirit in periods of seemingly fruitless labor. For Kepler it made endurable a life of poverty, illness, and other personal misfortunes. Finally, in 1619, he wrote triumphantly in his *Harmony of the World:*

> . . . after I had by unceasing toil through a long period of time, using the observations of Brahe, discovered the true relation . . . [It] overcame by storm the shadows of my mind, with such fullness of agreement between my seventeen years' labor on the observations of Brahe and this present study of mine that I at first believed that I was dreaming. . . .

Kepler's law of periods, also called the *harmonic law,* relates the periods of planets to their average distances from the sun. The period is the time taken to go once completely around the orbit. The law states that *the squares of the periods of the planets are proportional to the cubes of their average distances from the sun.* Calling the period T and the average distance R_{av}, this law can be expressed as

$$T^2 \propto R_{av}^3 \quad \text{or} \quad T^2 = kR_{av}^3 \quad \text{or} \quad \frac{T^2}{R_{av}^3} = k$$

where k is a constant. This relation applies to all the planets as well as to comets and other bodies in orbit around the sun. You can use it to find the period of any planet if you know its average distance from the sun, and vice versa.

For the earth, T is 1 year. The average distance R_{av} of the earth from the sun is one astronomical unit (1 AU). So one way to express the value of the constant k is k = 1 year^2/AU^3.

SG 15–18

TABLE 7.2 VERIFICATION OF KEPLER'S LAW OF PERIODS

	Using Copernicus' Values			Using Modern Values		
Planet	*Period T (Years)*	*Average Distance* R_{av}*(AU)*	$\frac{T^2}{R_{av}^3}$	*Period T (Years)*	*Average Distance* R_{av}*(AU)*	$\frac{T^2}{R_{av}^3}$
Mercury	0.241	0.38	1.06	0.241	0.387	1.00
Venus	0.614	0.72	1.01	0.615	0.723	1.00
Mars	1.881	1.52	1.01	1.881	1.523	1.00
Jupiter	11.8	5.2	0.99	11.862	5.20	1.00
Saturn	29.5	9.2	1.12	29.458	9.54	1.00

Kepler's three laws are so simple that their great power may be overlooked. Combined with his discovery that each planet moves in a plane passing through the sun, their value is greater still. They let us derive the past and future history of each planet if we know six quantities about that planet. Two of these

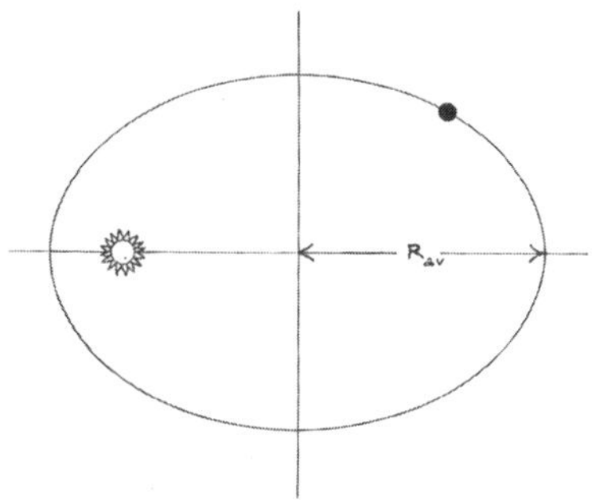

The value of R_{av} for an ellipse is just half the major axis.

quantities are the size (long axis, *a*) and eccentricity (*e*) of the orbit. Three others are angles that relate the plane of the orbit to that of the earth's orbit. The sixth quantity needed is the location of the planet in its orbit on any *one* date. These quantities are explained more fully in the Activities and Experiments listed in the *Handbook* for Chapters 7 and 8.

In this manner, the past and future positions of each planet and each comet can be found. Kepler's system was vastly simpler and more precise than the multitude of geometrical devices in the planetary theories of Ptolemy, Copernicus, and Tycho. With different assumptions and procedures Kepler had at last solved the problem which had occupied so many great scientists over the centuries. Although he abandoned the geometrical *devices* of Copernicus, Kepler did depend on the Copernican *viewpoint* of a sun-centered universe. None of the earth-centered models could have led to Kepler's three laws.

In 1627, after many troubles with his publishers and Tycho's heirs, Kepler published a set of astronomical tables. These tables combined Tycho's observations and the three laws in a way that permitted accurate calculations of planetary positions for any time, past or future. These tables remained useful for a century, until telescopic observations of greater precision replaced Tycho's data.

The tables, named for Tycho's and Kepler's patron, Emperor Rudolph II, were called the *Rudolphine tables*. They were also important for a quite different reason. In them, Kepler pioneered in the use of logarithms for making rapid calculations and included a long section, practically a textbook, on the nature and use of logarithms (first described in 1614 by Napier in Scotland). Kepler's tables spread the use of this computational aid, widely needed for nearly three centuries, until modern computing machines came into use.

Kepler's scientific interest went beyond the planetary problem. Like Tycho, who was fascinated by the new star of 1572, Kepler observed and wrote about new stars that appeared in 1600 and 1604. His observations and comments added to the impact of Tycho's earlier statement that changes did occur in the starry sky.

As soon as Kepler learned of the development of the telescope, he spent most of a year studying how the images were formed. He published his findings in a book titled *Dioptrice* (1611), which became the standard work on optics for many years. Kepler wrote other important books on mathematical and astronomical problems. One was a popular and widely read description of the Copernican system as modified by his own discoveries. This book added to the growing interest in and acceptance of the sun-centered system.

You know (see Question 11) that T^2/R^3_{av} equals 1 for any object in orbit around the sun. Therefore, if T = 5 years, then

$$\frac{T^2}{R^3_{av}} = 1$$

$$\frac{(5)^2}{R^3_{av}} = 1$$

$$R^3_{av} = 25$$

$$R^3_{av} = \text{approximately 3 AU}$$

Find R_{av} for an object whose period is 4 years.

?

10. State Kepler's law of periods.

11. Use the periods and average radii of Jupiter's and Saturn's orbits to show that T^2/R^3_{av} is the same for both. What do all objects in orbit around the sun have in common?

12. What is the average orbital radius of a planetoid that orbits the sun every 5 years?

13. Why is it that none of the earth-centered models of the solar system could have led to Kepler's law of periods?

7.5 | The new concept of physical law

One general feature of Kepler's lifelong work greatly affected the development of all the physical sciences. When Kepler began his studies, he still accepted Plato's assumptions about the importance of geometric models. He also agreed with Aristotle's emphasis on "natural place" to explain motion. But later he came to concentrate on algebraic laws describing how planets moved. His successful statement of empirical laws in mathematical form helped to establish the *equation* as a normal form of stating physical laws.

More than anyone before him, Kepler expected a theory to agree with precise and quantitative observation. From Tycho's observations he learned to respect the power of precise measurement. Models and theories can be modified by human inventiveness, but good data endure regardless of changes in assumptions or viewpoints.

Going beyond observation and mathematical description, Kepler attempted to explain motion in the heavens in terms of physical forces. In Kepler's system, the planets no longer moved by some divine nature or influence, or in "natural" circular motion caused by their spherical shapes. Rather, Kepler looked for physical laws, based on observed phenomena and describing the whole universe in a detailed quantitative manner. In an early letter to Herwart (1605), he expressed his guiding thought:

> I am much occupied with the investigation of the physical causes. My aim in this is to show that the celestial machine is to be likened not to a divine organism but rather to a clockwork ... insofar as nearly all the manifold movements are carried out by means of a single, quite simple magnetic force, as in the case of a clockwork, all motions are caused by a simple weight. Moreover, I show how this physical conception is to be presented through calculation and geometry.

Kepler's likening of the celestial machine to a clockwork driven by a single force was like a look into the future of scientific thought. Kepler had read William Gilbert's work on magnetism, published a few years earlier. Now he could imagine magnetic forces from the sun driving the planets along their orbits. This was a reasonable and promising hypothesis. As Newton later showed (Chapter 8), the basic idea that a single kind of force controls the motions of all the planets was correct. The force is not magnetism, however, and does not keep the planets moving forward, but rather bends their paths into closed orbits.

Even though Kepler did not understand correctly the nature of the forces responsible for celestial motion, his work illustrates an enormous change in outlook that had begun more than two centuries earlier. Although Kepler still shared the ancient idea that each planet had a "soul," he refused to base his explanation of planetary motion on this idea. Instead, he began to search for physical causes. Copernicus and Tycho were willing to settle for geometrical models by which planetary positions could be predicted. Kepler was one of the first to seek dynamic *causes* for the motions. This new desire for physical explanations marked the beginning of a chief characteristic of modern physical science.

Kepler's statement of empirical laws reminds us of Galileo's suggestion, made at about the same time. Galileo said that science should deal first with the *how* of motion in free fall and then with the *why*. A half-century later, Newton used the concept of *gravitational force* to tie together Kepler's three planetary laws with laws of terrestrial mechanics. This magnificent synthesis will be the subject of Chapter 8.

?

14. In what ways did Kepler's work exemplify a "new" concept of physical law?

7.6 | Galileo and Kepler

One of the scientists with whom Kepler corresponded about scientific developments was Galileo. Kepler's main contributions to planetary theory were his empirical laws based on Tycho's observations. Galileo contributed to both theory and observation. As reported in Chapters 2 and 3, Galileo based his theory of motion on observations of bodies moving on the earth's surface. His work in the new science of mechanics contradicted Aristotelian assumptions about physics and the nature of the heavens. Galileo's books and speeches triggered wide discussion about the differences or similarities of earth and heaven. Interest extended far outside of scientific circles. Some years after his visit to Galileo in 1638, the poet John Milton wrote:

(***Paradise Lost***, **Book V, line 574, published 1667.**)

. . . What if earth
Be but the shadow of Heaven, and things therein
Each to the other like, more than on earth is thought?

Galileo challenged the ancient interpretations of experience. As stated earlier, he focused attention on new concepts: time and distance, velocity and acceleration, forces and matter. In contrast, the Aristotelians spoke of essences, final causes, and fixed geometric models. In Galileo's study of falling bodies, he insisted on fitting the concepts to the observed facts. By seeking results that could be expressed in algebraic form, he paralleled the new style being used by Kepler.

The sharp break between Galileo and most other scientists of the time arose from the kind of questions he asked. To his opponents, many of Galileo's problems seemed trivial. His procedures for studying the world also seemed peculiar. What was important about watching pendulums swing or rolling balls down inclines when deep philosophical problems needed solving?

In recent times, similar receptions were initially given to such artists as the painter Picasso, the sculptor Giacometti, and the composers Stravinsky and Schönberg. The same has often been true in most fields, whether literature, mathematics, economics, or politics. While great creative novelty is often attacked at the start, it does not follow that, conversely, everything that is attacked must be creative.

Although Kepler and Galileo lived at the same time, their lives were quite different. Kepler lived in near poverty and was driven from city to city by the religious wars of the time. Few people, other than a handful of friends and correspondents, knew of or cared about his studies and results. He wrote long, complex books that demanded expert knowledge from his readers.

Galileo, on the other hand, wrote his essays and books in Italian. His language and style appealed to many readers who did not know scholarly Latin. Galileo was a master at publicizing his work. He wanted as many people as possible to know of his studies and to accept the Copernican theory. He wrote not only to small groups of scholars, but to the nobles and to civic and religious leaders. His arguments included humorous attacks on individuals or ideas. In return, Galileo's efforts to inform and persuade on such a "dangerous" topic as cosmological theory stirred up ridicule and even violence. Those who have a truly new point of view often must face such a reaction.

?

15. Which of the following would you associate more with Galileo's work than with that of his predecessors: qualities and essences, popular language, concise mathematical expression, final causes?

7.7 The telescopic evidence

Like Kepler, Galileo was surrounded by scholars who believed the heavens were eternal and could not change. Galileo therefore took special interest in the sudden appearance in 1604 of a new star, one of those observed by Kepler. Where there had been nothing visible in the sky, there was now a brilliant star. Like Tycho and Kepler, Galileo realized that such events conflicted with the old idea that the stars could not change. This new star awakened in Galileo an interest in astronomy that lasted all his life.

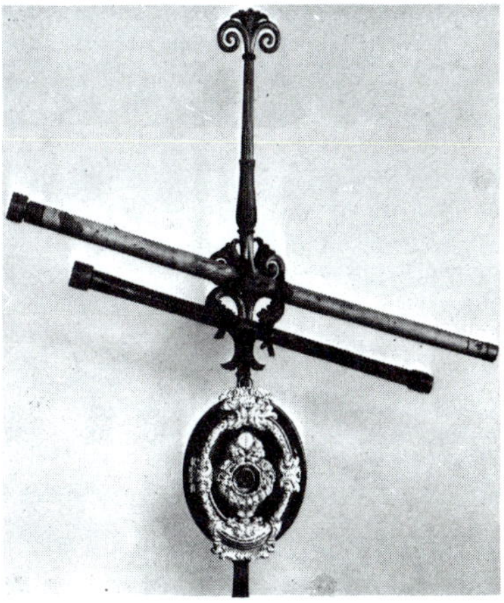

Two of Galileo's telescopes, displayed at the Museum of Science in Florence.

Four or 5 years later, Galileo learned that a Dutchman "had constructed a spy glass by means of which visible objects, though very distant from the eye of the observer, were distinctly seen as if nearby." Galileo (as he tells it) quickly worked out some of the optical principles involved. He then set to work to grind the lenses and build such an instrument himself. Galileo's first telescope made objects appear three times closer than when seen with the naked eye. He reported on his third telescope in his book *The Starry Messenger:*

> Finally, sparing neither labor nor expense, I succeeded in constructing for myself so excellent an instrument that objects seen by means of it appeared nearly one thousand times larger and over thirty times closer than when regarded with our natural vision.

Galileo meant that the *area* of the object was nearly 1,000 times greater. The area is proportional to the *square* of the magnification (or "power") as we define it now.

What would you do if you were handed "so excellent an instrument"? Like the scientists of Galileo's time, you probably would put it to practical uses. "It would be superfluous," Galileo agreed,

> . . . to enumerate the number and importance of the advantages of such an instrument at sea as well as on land. But forsaking terrestrial observations, I turned to celestial ones, and first I saw the moon from as near at hand as if it were scarcely two terrestrial radii away. After that I observed often with wondering delight both the planets and the fixed stars. . . .

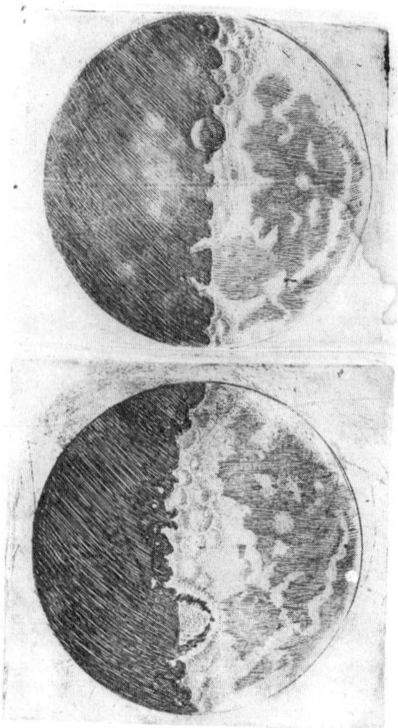

Two of Galileo's early drawings of the moon from Siderius Nuncius (The Starry Messenger).

In a few short weeks in 1609 and 1610, Galileo used his telescope to make several major discoveries. First, he pointed his telescope at the moon. What he saw convinced him that

> . . . the surface of the moon is not smooth, uniform, and precisely spherical as a great number of philosophers believe it (and other heavenly bodies) to be, but is uneven, rough, and full of cavities and prominences, being not unlike the face of the earth, relieved by chains of mountains and deep valleys.

Galileo did not stop with that simple observation, so contrary to the Aristotelian idea of heavenly perfection. He supported his conclusions with several kinds of evidence, including careful measurement. For instance, he worked out a method for determining the height of mountains on the moon from the length of their shadows. (His value of about 6.4 km for the height of some lunar mountains is not far from modern results. For example, try the experiment, "The Height of Piton—A Mountain on the Moon," in the *Handbook*.)

Next Galileo looked at the stars. To the naked eye, the Milky Way had seemed to be a continuous blotchy band of light. But through the telescope it was seen to consist of thousands of faint stars. Wherever Galileo pointed his telescope in the sky, he saw many more stars than appeared to the unaided eye. This observation clashed with the old argument that the stars were created to help humans to see at night. By this argument, there should not be stars invisible to the naked eye. But Galileo found thousands.

Galileo soon made another discovery which, in his opinion, "... deserves to be considered the most important of all—the disclosure of four *Planets* never seen from the creation of the world up to our time." He was referring to his discovery of four of the satellites which orbit Jupiter. Here before his eyes was a miniature solar system with its own center of revolution. Today, as to Galileo so long ago, it is a sharp thrill to see the moons of Jupiter through a telescope for the first time. Galileo's discovery strikingly contradicted the Aristotelian notion that the earth is the center of the universe and the chief center of revolution.

The manner in which Galileo discovered Jupiter's "planets" is a tribute to his ability as an observer. On each clear night during this period he was discovering dozens if not hundreds of new stars never before seen. On the evening of January 7, 1610, he was looking in the vicinity of Jupiter. He noticed "... that beside the planet there were three starlets, small indeed, but very bright. Though I believe them to be among the hosts of fixed stars, they aroused my curiosity somewhat by appearing to lie in an exact straight line...." (The sketches in which he recorded his observations are reproduced in the margin.) When Galileo looked again on the following night, the "starlets" had changed position with reference to Jupiter. Each clear evening for weeks he observed and recorded their positions in drawings. Within days he had concluded that there were four "starlets" and that they were indeed satellites of Jupiter. Galileo continued observing until he could estimate their periods of revolution around Jupiter.

Of all of Galileo's discoveries, that of the satellites of Jupiter caused the most stir. His book *The Starry Messenger* was an immediate success. Copies sold as fast as they could be printed. For Galileo, the result was a great demand for telescopes and great fame.

Galileo continued to use his telescope with remarkable results. By projecting an image of the sun on a screen, he observed sunspots. This seemed to indicate that the sun, like the moon, was not perfect in the Aristotelian sense. It was disfigured rather than even and smooth. Galileo also noticed that the sunspots moved across the face of the sun in a regular pattern. He concluded that the sun rotated with a period of about 27 days.

Galileo also found that Venus showed phases, just as the moon does (see photos on page 196). Therefore, Venus could not stay always between the earth and the sun, as Ptolemaic astronomers assumed. Rather, it must move completely around the sun as Copernicus and Tycho had believed. Saturn seemed to carry bulges around its equator, as indicated in the drawings on the next page. Galileo's telescopes were not strong enough to show that these were rings. (He called them "ears.") With his telescopes, Galileo collected an impressive array of new

Telescopic photograph of Jupiter and its four bright satellites. This is approximately what Galileo saw and what you see through the simple telescope described in the Handbook.

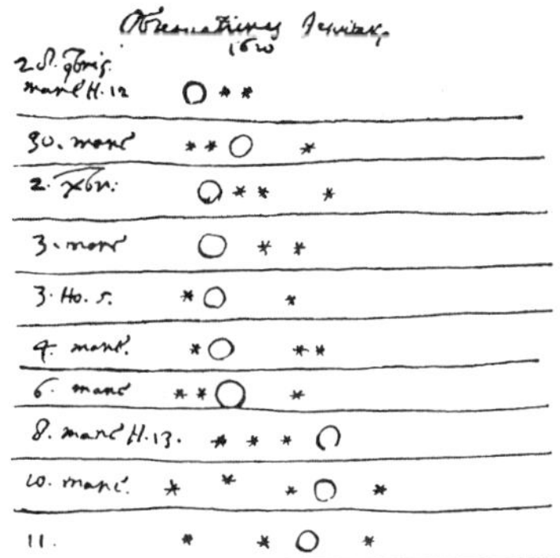

These sketches of Galileo's are from the first edition of The Starry Messenger.

Thirteen satellites of Jupiter have now been observed.

SG 23

information about the heavens, all of which seemed to contradict the basic assumptions of the Ptolemaic world scheme.

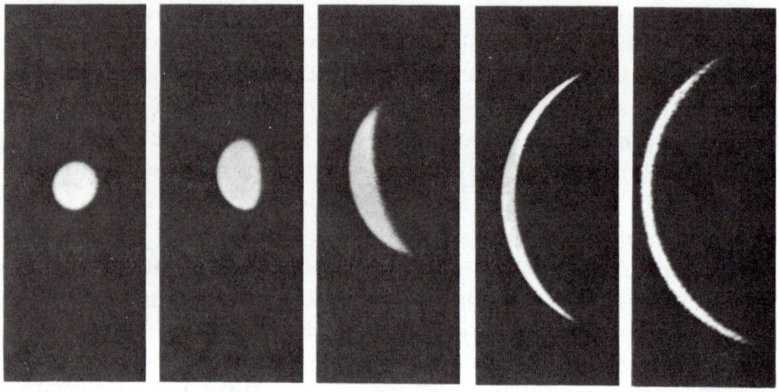

Photographs of Venus at various phases with a constant magnification.

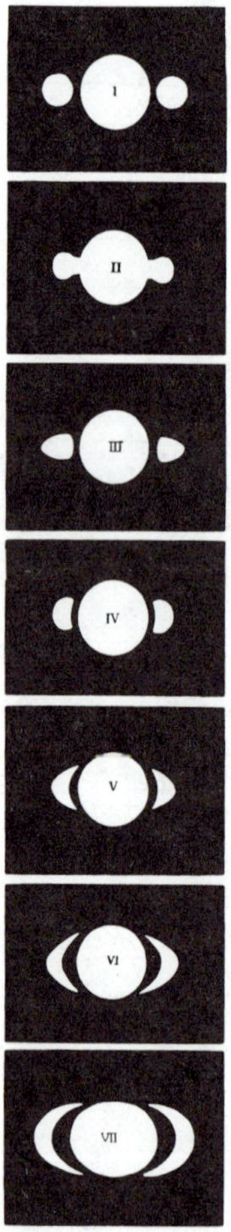

Drawings of Saturn, made in the seventeenth century.

16. Which commonly held beliefs of Galileo's contemporaries were contradicted by these observations of Galileo?

(a) A "new" star appeared in the heavens.

(b) The moon has mountains.

(c) There are many stars in dark areas of the sky.

(d) Jupiter has several satellites.

(e) The sun has spots and rotates.

(f) Venus has phases like the moon.

17. Could Galileo's observations of all phases of Venus support the heliocentric theory, the Tychonic system, or Ptolemaic system?

18. In what way did telescopic observation of the moon and sun weaken the earth-centered view of the universe?

7.8 | Galileo focuses the controversy

Galileo's observations supported his belief in the heliocentric Copernican system, but they were not the *cause* of his belief. His great work, *Dialogue Concerning the Two Chief World Systems* (1632), was based more on assumptions that seemed self-evident to him than on observations. Galileo recognized, as Ptolemy and Copernicus had, that the observed motions of planets do not prove either the heliocentric or the geocentric hypothesis right and the other one wrong. With proper adjustments of the systems, said Galileo, "The same phenomena would result from either hypothesis." He accepted the earth's motion as real because the heliocentric system seemed simpler and more pleasing. The support for the heliocentric view provided by the "observed facts" was of course necessary, but the "facts" by themselves were not sufficient. Elsewhere in this course you will find other cases like this. Scientists quite often accept or reject an idea because of some strong belief or feeling that, at the time, cannot be proved decisively by experiment.

In the *Dialogue Concerning the Two Chief World Systems*, Galileo presents his arguments in a systematic and lively way. Like his later book, *Discourses Concerning Two New Sciences*, mentioned in Chapter 2, it is in the form of a discussion between three learned men. Salviati, the voice of Galileo, wins most of the arguments. His opponent is Simplicio, an Aristotelian who defends the Ptolemaic system. The third member, Sagredo, represents the objective and intelligent citizen not yet committed to either system. However, Sagredo usually accepts Galileo's arguments in the end.

In *Two Chief World Systems*, Galileo's arguments for the Copernican system are mostly those given by Copernicus. Oddly enough, Galileo made no use of Kepler's laws, although Galileo's observations did provide new evidence for Kepler's laws. In studying Jupiter's four moons, Galileo found that the larger the orbit of the satellite, the longer was its period of revolution. Copernicus had noted that the periods of the planets increased with their average distances from the sun. Kepler's law of periods had stated this relation in detailed, quantitative form. Now Jupiter's satellite system showed a similar pattern, reinforcing the challenge to the old assumptions of Plato, Aristotle, and Ptolemy.

Two Chief World Systems relies upon Copernican arguments, Galilean observations, and attacks on basic assumptions of the geocentric model. In response, Simplicio desperately tries to dismiss all of Galileo's arguments with a typical counter argument:

> . . . with respect to the power of the Mover, which is infinite, it is just as easy to move the universe as the earth, or for that matter a straw.

To this argument, Galileo makes a very interesting reply. Notice how he quotes Aristotle against the Aristotelians:

> . . . what I have been saying was with regard not to the Mover, but only the movables . . . Giving our attention, then, to the movable bodies, and not questioning that it is a shorter and readier operation to move the earth than the universe, and paying attention to the many other simplifications and conveniences that follow from merely this one, it is much more probable that the diurnal motion belongs to the earth alone than to the rest of the universe excepting the earth. This is supported by a very true maxim of Aristotle's which teaches that . . . "it is pointless to use many to accomplish what may be done with fewer."

Galileo thought his telescopic discoveries would soon demolish the assumptions that prevented wide acceptance of the Copernican theory. But people cannot believe what they are not ready to believe. The Aristotelians firmly believed that the heliocentric theory was obviously false and contrary to observation and common sense. The evidences provided by the telescope could be distorted; after all, glass lenses change the path of light rays. Even if telescopes seemed to work on earth, nobody could be sure they worked when pointed at the vastly distant stars.

Most Aristotelians really could not even consider the Copernican system as a possible theory. To do so would involve questioning too many of their own basic assumptions, as you saw in Chapter 6. It is nearly humanly impossible to give up all of one's common-sense ideas and find new bases for one's religious and moral doctrines. The Aristotelians would have to admit that the earth is not at the center of creation. Then perhaps the universe was not created especially for humanity. No wonder Galileo's arguments stirred up a storm of opposition!

Galileo's observations intrigued many, but were unacceptable to Aristotelian scholars. Most of these critics had reasons one can respect. But a few were driven to positions that must have seemed silly even then. For example, the Florentine astronomer Francesco Sizzi argued in 1611 that there could not possibly be any satellites around Jupiter:

> There are seven windows in the head, two nostrils, two ears, two eyes and a mouth; so in the heavens there are two favorable stars, two unpropitious, two luminaries, and Mercury alone undecided and indifferent. From which and many other similar phenomena of nature such as the seven metals, etc., which it were tedious to enumerate, we gather that the number of planets is necessarily seven [including the sun and moon but excluding the earth].... Besides, the Jews and other ancient nations, as well as modern Europeans, have adopted the division of the week into seven days, and have named them from the seven planets; now if we increase the number of planets, this whole system falls to the ground.... Moreover, the satellites are invisible to the naked eye and therefore can have no influence on the earth, and therefore would be useless, and therefore do not exist.

A year after his discoveries, Galileo wrote to Kepler:

> You are the first and almost the only person who, even after a but cursory investigation, has ... given entire credit to my statements.... What do you say of the leading philosophers here to whom I have offered a thousand times of my own accord to show my studies, but who with the lazy obstinacy of a serpent who has eaten his fill have never consented to look at the planets, or moon, or telescope?

Some of the arguments brought forward against the new discoveries sound silly to the modern mind. "One of his [Galileo's] opponents, who admitted that the surface of the moon looked rugged, maintained that it was actually quite smooth and spherical as Aristotle had said, reconciling the two ideas by saying that the moon was covered with a smooth transparent material through which mountains and craters inside it could be discerned. Galileo, sarcastically applauding the ingenuity of this contribution, offered to accept it gladly—provided that his opponent would do him the equal courtesy of allowing him then to assert that the moon was even more rugged than he had thought before, its surface being covered with mountains and craters of this invisible substance ten times as high as any he had seen." [Quoted from *Discoveries and Opinions of Galileo*, by Stillman Drake.]

?

19. Did Galileo's telescopic observations cause him to believe in the Copernican viewpoint?

20. What reasons did Galileo's opponents give for ignoring telescopic observations?

7.9 | Science and freedom

The political and personal tragedy that struck Galileo is described at length in many books. Only some of the major events are mentioned here. Galileo was warned in 1616 by the highest officials of the Roman Catholic Church to cease teaching the Copernican theory as true. It could be taught only as just one of several possible methods for computing the planetary motions. The Inquisitors held that the theory was contrary to Holy Scripture. At the same time, Copernicus' book was placed on the *Index of Forbidden Books* "until corrected." As you saw before, Copernicus had used Aristotelian doctrine whenever possible to support his theory. Galileo had reached a new point of view. He urged that the heliocentric system be accepted on its merits alone. Although Galileo himself was a devoutly religious man, he deliberately ruled out questions of religious faith from scientific discussions. This was a fundamental break with the past.

Cardinal Barberini, once a close friend of Galileo, was elected in 1623 to be Pope Urban VIII. Galileo talked with him about the decree against Copernican ideas. As a result of the discussion, Galileo considered it safe to write again on the topic. In 1632, having made some required changes, Galileo obtained consent to publish *Two Chief World Systems*. This book presented very persuasively the Ptolemaic and Copernican viewpoints and their relative merits. After its publication, his opponents argued that Galileo had tried to get around the warning of 1616. Furthermore, Galileo sometimes spoke and acted without tact. This fact, and the Inquisition's need to demonstrate its power over suspected heretics, combined to mark Galileo for punishment.

Among the many factors in this complex story, it is important to remember that Galileo considered himself religiously faithful. In letters of 1613 and 1615, Galileo wrote that God's mind contains all the natural laws. Consequently, the occasional glimpses of these laws that scientists might gain are direct revelations of God, just as true as those in the Bible: "From the Divine Word, the Sacred Scripture and Nature did both alike proceed. . . . Nor does God less admirably discover himself to us in Nature's action than in the Scripture's sacred dictions." These opinions are held by many people today, whether scientists or not. Few people think of them as conflicting with religion. In

Pantheism refers to the idea that God is no more (and no less) than the forces and laws of nature.

According to a well-known, but probably apocryphal story, at the end of these proceedings Galileo muttered, *"E pur si muove"* (but it does move).

Galileo's time, such ideas were widely regarded as symptoms of pantheism. Pantheism was one of the religious "crimes" or heresies for which Galileo's contemporary, Giordano Bruno, was burned at the stake. The Inquisition was alarmed by Galileo's seeming denial of the Bible as a certain source of knowledge about natural science. In reply, arrogant as Galileo often was, he quoted Cardinal Baronius: "The Holy Spirit intended to teach us how to go to heaven, not how the heavens go."

Though he was old and sick, Galileo was called to Rome and confined for a few months. The records of his trial are still partly secret. It is known that he was tried, threatened with torture, and forced to make a formal confession for holding and teaching forbidden ideas. He was also forced to deny the Copernican theory. In return for his confessions and denial, Galileo was sentenced only to perpetual house arrest. Galileo's friends in Italy did not dare to defend him publicly. His book was placed on the *Index*. It remained there, along with that of Copernicus and one of Kepler's, until 1835. Thus, Galileo was used as a warning to all people that demands for spiritual conformity also required intellectual conformity.

Without intellectual freedom, science cannot flourish for long. Italy had given the world many outstanding scholars. But for 2 centuries after Galileo, Italy produced hardly a single great scientist, while elsewhere in Europe many appeared. Today, scientists are acutely aware of this famous part of the story of the development of planetary theories. Teachers and scientists in our time have had to face strong enemies of open-minded inquiry and of unrestricted teaching. Today, as in Galileo's time, men and women who create or publicize new thoughts must be ready to stand up for them. There still are people who fear and wish to stamp out the open discussion of new ideas and new evidence.

Over 200 years after his confinement in Rome, opinions had changed so that Galileo was honored as in the fresco "Galileo presenting his telescope to the Venetian Senate," by Luigi Sabatelli (1772–1850).

Plato knew that a government that wishes to control its people totally is threatened by new ideas. To prevent the spread of such ideas, Plato recommended the now well-known treatment: reeducation, prison, or death. Not long ago, Soviet geneticists were required to discard well-established theories. They did so, not on the basis of new scientific evidence, but because party "philosophers" accused them of conflicts with political doctrines. Similarly, the theory of relativity was banned from textbooks in Nazi Germany because Einstein's Jewish background was said to make his work worthless. Another example of intolerance was the prejudice that led to the "Monkey Trial" held in 1925 in Tennessee. At that trial, the teaching of Darwin's theory of biological evolution was attacked because it conflicted with certain types of biblical interpretation.

On two points, one must be cautious not to romanticize the lessons of this episode. First, while a Galileo sometimes still may be neglected or ridiculed, not everyone who feels neglected or ridiculed is for that reason a Galileo. The person may in fact be just wrong. Second, it has turned out that, at least for a time, science in some form can continue to live in the most hostile surroundings. When politicians decide what may be thought and what may not, science will suffer like everything else. But it will not necessarily be extinguished. Scientists can take comfort from the judgment of history. Less than 50 years after Galileo's trial, Newton's great book, the *Principia*, appeared. Newton brilliantly united the work of Copernicus, Kepler, and Galileo with his own new statement of the principles of mechanics. Without Kepler and Galileo, there probably could have been no Newton. As it was, the work of these three, and of many others working in the same spirit, marked the triumphant beginning of modern science. Thus, the hard-won new laws of science and new views of humanity's place in the universe were established. What followed has been termed by historians The Age of Enlightenment.

SG 24

Palomar Observatory houses the 500-cm Hale reflecting telescope. It is located on Palomar Mountain in southern California.

?

21. Which of the following appears to have contributed to Galileo's being tried by the Inquisition?

(a) He did not believe in God.

(b) He was arrogant.

(c) He separated religious and scientific questions.

(d) He wrote in Italian.

study guide

1. The *Project Physics* learning materials particularly appropriate for Chapter 7 include the following:

Experiments
The Orbit of Mars
Inclination of Mars' Orbit
The Orbit of Mercury

Activities
Three-Dimensional Models of Two Orbits
Demonstrating Satellite Orbits
Galileo
Conic-Section Models
Challenging Problems: Finding Earth–Sun Distance
Measuring Irregular Areas

Film Loop
Jupiter Satellite Orbit

2. How large is an error of 8 min of arc in degrees? What fraction of the moon's observed diameter does 8 min of arc represent?

3. Summarize the steps Kepler used to determine the orbit of the earth.

4. For the orbit positions nearest and farthest from the sun, a planet's speeds are inversely proportional to the distances from the sun. What is the percentage change between the earth's slowest speed in July, when it is 1.02 AU from the sun, and its greatest speed in January, when it is 0.98 AU from the sun?

5. Summarize the steps Kepler used to determine the orbit of Mars.

6. In any ellipse, the sum of the distances from the two foci to a point on the curve equals the length of the major axis, or $(F_1P + F_2P) = 2a$. This property of ellipses allows us to draw them by using a loop of string around two tacks at the foci. What should the length of the looped string be?

7. Draw an ellipse by looping a string around two thumbtacks in a piece of paper and pulling the loop taut with a pencil. If the two tacks were on the same point, what kind of geometrical figure would you draw? As the two points are separated more and more, what shapes do you draw?

8. What is the eccentricity of an ellipse if the two points (foci) are 5 cm apart and the ellipse is 9 cm long at its widest part?

9. **(a)** Draw an ellipse and label the following distances: a, c, perihelion, aphelion.

(b) State the algebraic expression for c and a in terms of the perihelion and aphelion; state the expression for perihelion and aphelion in terms of a and c.

(c) What is R_{av} in terms of c and a? in terms of perihelion and aphelion?

10. Kepler found that the sun is at one focus of the ellipse that describes the orbit of a planet. What is at the other focus?

11. In describing orbits around the sun, the point nearest the sun is called the *perihelion point*, and the point farthest from the sun is called the *aphelion point*. The distances of these two points from the sun are called the *perihelion distance* and the *aphelion distance*, respectively. The terms perihelion and aphelion come from the Greek, in which *helios* is the sun, *peri* means near, and *apo* means away from.

(a) List some other words in which the prefixes *peri* and *apo* or *ap* have similar meanings.

(b) In describing earth satellite orbits, the terms *apogee* and *perigee* are often used. What do they mean?

(c) What would such points for satellites orbiting the moon be called?

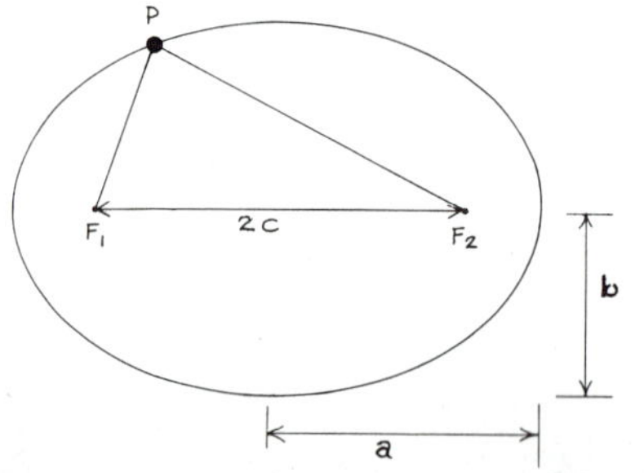

12. (a) An ellipse is 5 cm from one focus at its farthest point and 2 cm from the same focus at its nearest point. Find c, a, and the eccentricity of the ellipse.

(b) An ellipse has an eccentricity of 0.5 and $a = 10$ cm. What are the distances of nearest and farthest approach from one of the foci?

(c) An ellipse is 5 cm from one of its foci at its nearest approach and has an eccentricity of 0.8. What is its greatest distance from that focus? Find c and a for the ellipse.

13. For the planet Mercury, the perihelion distance (closest approach to the sun) has been found to be about 45.8×10^6 km, and the aphelion distance (greatest distance from the sun) about 70.0×10^6 km. What is the eccentricity of the orbit of Mercury?

14. The eccentricity of Pluto's orbit is 0.254. What is the ratio of the minimum orbital speed to the maximum orbital speed of Pluto?

15. Halley's comet has a period of 76 years, and its orbit has an eccentricity of 0.97.

(a) What is its average distance from the sun?

(b) What is its greatest distance from the sun?

(c) What is its least distance from the sun?

(d) How does its greatest speed compare with its least speed?

16. The mean distance of the planet Pluto from the sun is 39.6 AU. What is the orbital period of Pluto?

17. Three major planets have been discovered since Kepler's time. Their orbital periods and mean distances from the sun are given in the table below. Determine whether Kepler's law of periods holds for these planets also.

	Discovery Date	*Orbital Period*	*Average Distance From Sun*	*Eccentricity of Orbit*
Uranus	1781	84.013 yr	19.19 AU	0.047
Neptune	1846	164.783	30.07	0.009
Pluto	1930	248.420	39.52	0.249

18. Considering the data available to him, do you think Kepler was justified in concluding that the ratio T^2/R_{av}^3 is a constant?

19. What is T^2/R_{av}^3 for a satellite orbiting the earth if the average orbital radius is 18,000 km and the satellite orbits the earth every 380 min?

20. A satellite already in orbit around a planet is put into a new orbit whose radius is 4 times as large as the old radius. How many times longer is the new period than the old?

21. Using the value of T^2/R_{av}^3 that you found in question 19, what is the average distance of a satellite from the center of the earth if its period is 28 days?

22. Using the table of periods and orbital radii of earth satellites on p. 120 of Chapter 4, verify that Kepler's law of periods holds for these satellites.

23. The chart on p. 204 is reproduced from the January, 1979, issue of *Sky and Telescope.*

(a) Make a sketch of how Jupiter and its satellites appeared at one-week intervals, beginning with day "0."

(b) Make measurements of the chart to find R_{av} and T for each satellite. (For this problem, R_{av} can be to any convenient scale, such as centimeters on the diagram.)

(c) Does Kepler's law of periods, T^2/R_{av}^3 = constant, hold for Jupiter's satellites?

24. What are the current procedures by which the public is informed of new scientific theories? Do you think they are adequate? To what extent do news media emphasize clashes of points of view? Bring in some examples from news magazines.

25. Kepler discovered his three laws because he believed that they should exist; that is, he believed that nature exhibits simple uniformity in motion.

(a) Explain the hard work and courage that were necessary for Kepler to be successful.

(b) Is it fair to say that Kepler was "lucky" to have studied Mars, whose orbit is the most elliptical?

(c) Discuss Kepler's reliance on the work of those who preceded him, particularly Copernicus and Brahe.

(d) In what ways was Kepler independent enough not to rely completely on Copernicus and Brahe, but to go beyond the limits of their work?

26. List at least three ways in which Kepler's approach to science differed from that of his predecessors.

27. Kepler's laws are empirical laws. What is an empirical law? What are its limitations? Why are empirical laws important?

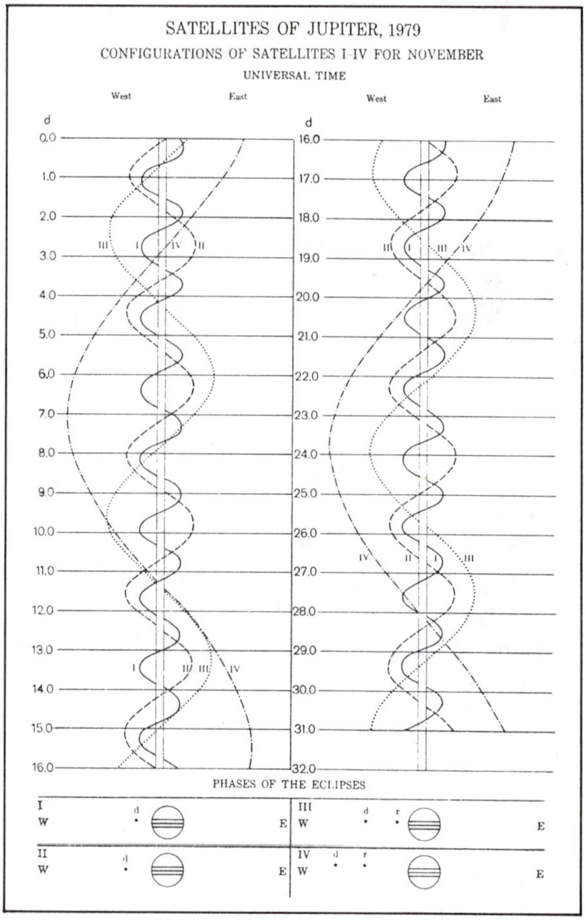

Jupiter's satellites. The four curving lines represent Jupiter's four bright (Galilean) satellites: (I) Io, (II) Europa, (III) Ganymede, (IV) Callisto. The location of the planet's disk is indicated by the pairs of vertical lines. If a moon is invisible because it is behind the disk (that is, occulted by Jupiter), the curve is broken. For successive dates, the horizontal lines mark 0^{h} Universal time, or 7 P.M. Eastern standard time (or 4 P.M. Pacific standard time) on the preceding date. Along the vertical scale, 0.16 cm is almost 7 hours. In this chart, west is to the left, as in an inverting telescope for a northern hemisphere observer. At the bottom, "d" is the point of disappearance of a satellite in the shadow of Jupiter; "r" is the point of reappearance. From the American Ephemeris and Nautical Almanac.

The Unity of Earth and Sky
The Work of Newton

8.1 | Newton and seventeenth-century science

SG 1

Forty-five years passed between the death of Galileo in 1642 and the publication of Newton's *Principia* in 1687. In those years, major changes occurred in the social organization of scientific studies. The new philosophy of experimental science, applied with enthusiasm and imagination, produced a wealth of new results. Scholars began to work together and organize scientific societies in Italy, France, and England. One of the most famous, the Royal Society of London for Improving Natural Knowledge, was founded in 1662. Through these societies, scientific experimenters exchanged information, debated new ideas, argued against opponents of the new experimental activities, and

Isaac Newton (1642–1727)

In referring to the time period between 1500 and 1600, the forms "1500's" and "sixteenth century" are often used interchangeably, although the latter is preferable.

published technical papers. Each society sought public support for its work and published studies in widely read scientific journals. Through the societies, scientific activities became well-defined, strong, and international.

This development was part of the general cultural, political, and economic change occurring in the sixteenth and seventeenth centuries. (See the time chart on page 207.) Artisans and people of wealth and leisure became involved in scientific studies. Some sought to improve technological methods and products. Others found the study of nature through experiment a new and exciting hobby. However, the availability of money and time, the growing interest in science, and the creation of

1600
1650
1700
1750
1642
NEWTON
1727
Settlement of Jamestown
Settlement of Plymouth
Puritan Revolution
Glorious Revolution
Peace of Utrecht
CARDINAL RICHELIEU
WILLIAM III of England
MARIA THERESA
OLIVER CROMWELL
PETER THE GREAT of Russia
JEAN COLBERT
CATHERINE THE GREAT
LOUIS XIV of France
Science and Technology
RENÉ DESCARTES
EDMUND HALLEY
WILLIAM HARVEY
JEAN BERNOULLI
JOHANNES KEPLER
ROBERT HOOKE
BLAISE PASCAL
ROBERT BOYLE
CHRISTIAN HUYGENS
GOTTFRIED LEIBNITZ
Philosophy and Theology
THOMAS HOBBES
GEORGE BERKELEY
SPINOZA
MONTESQUIEU
JOHN LOCKE
DAVID HUME
VOLTAIRE
Literature
BEN JONSON
JONATHAN SWIFT
JOHN MILTON
ALEXANDER POPE
MOLIÈRE
HENRY FIELDING
JOHN DRYDEN
DANIEL DEFOE
Art
PETER PAUL RUBENS
JEAN WATTEAU
VELAZQUEZ
WILLIAM HOGARTH
GIOVANNI BERNINI
FRANCOIS BOUCHER
REMBRANDT VAN RIJN
CHRISTOPHER WREN
Music
JEAN LULLY
HENRY PURCELL
ANTONIO VIVALDI
JOHANN SEBASTIAN BACH
GEORGE FREDERICK HANDEL

Newton entered Trinity College, Cambridge University, in 1661 at the age of 18. He was doing experiments and teaching while still a student. This early engraving shows the quiet student wearing a wig and heavy academic robes, as was customary.

organizations are not enough to explain the growing success of scientific studies. This rapid growth also depended upon able scientists, well-formulated problems, and good experimental and mathematical tools. Some of the important scientists who lived between 1600 and 1750 are shown in the time chart for the Age of Newton. The list includes amateurs as well as university professors.

Many well-formulated problems appear in the writings of Galileo and Kepler. Their studies showed how useful mathematics could be when combined with experimental observation. Furthermore, their works raised exciting new questions. For example, what forces act on the planets and cause the paths actually observed? Why do objects fall as they do near the earth's surface?

Good experimental and mathematical tools were being created. With mathematics being applied to physics, studies in each field stimulated development in the others. Similarly, the instrument maker and the scientist aided each other.

Another factor of great importance was the rapid build-up of scientific knowledge itself. From the time of Galileo, scientists had reported repeatable experiments in books and journals. Theories could now be tested, modified, and applied. Each study built on those done previously.

Newton, who lived in this new scientific age, is the central person in this chapter. However, in science as in any other field,

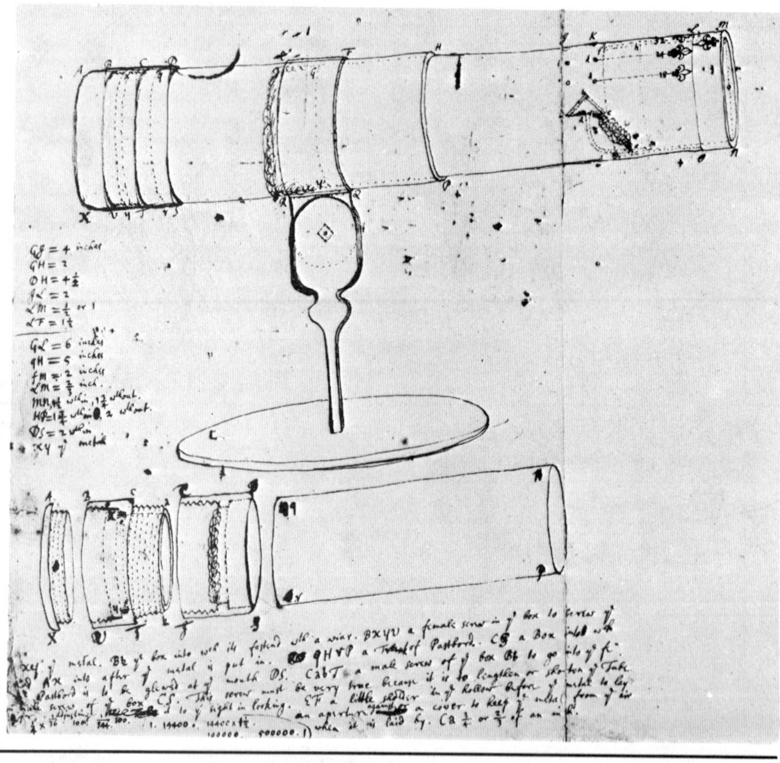

This drawing of the reflecting telescope he invented was done by Newton while he was still a student.

many workers made useful contributions. The structure of science depends not only upon recognized geniuses, but also upon many lesser-known scientists. As Lord Rutherford, one of the founders of modern atomic theory, said:

> It is not in the nature of things for any one man to make a sudden violent discovery; science goes step by step, and every man depends upon the work of his predecessors. . . . Scientists are not dependent on the ideas of a single man, but on the combined wisdom of thousands. . . .

To tell the story properly, each scientist's debt to others who worked previously and in the same age, and each scientist's influence upon future scientists should be traced. Within the space available, we can only briefly hint at these relationships.

Isaac Newton was born on Christmas Day, 1642, in the small English village of Woolsthorpe in Lincolnshire. He was a quiet farm boy. Like young Galileo, Newton loved to build mechanical gadgets and seemed to have a liking for mathematics. With financial help from an uncle, he went to Trinity College of Cambridge University in 1661. There he enrolled in the study of mathematics and was a successful student. In 1665, the Black Plague swept through England. The college was closed, and Newton went home to Woolsthorpe. There, by the time he was 24, he had made spectacular discoveries. In mathematics, he developed the binomial theorem and differential calculus. In optics, he worked out a theory of colors. In mechanics, he already had formulated a clear concept of the first two laws of motion and the law of gravitational attraction. He also had discovered the equation for centripetal acceleration. However, Newton did not announce this equation until many years after Huygens' equivalent statement.

This period at Woolsthorpe must have been the time of the famous and disputed fall of the apple. One version of the apple story appears in a biography of Newton written by his friend William Stukeley. In it we read that on a particular occasion Stukeley was having tea with Newton. They were sitting under some apple trees in a garden, and Newton said that

> . . . he was just in the same situation, as when formerly, the notion of gravitation came into his mind. It was occasion'd by the fall of an apple, as he sat in a contemplative mood. Why should that apple always descend perpendicularly to the ground, thought he to himself. Why should it not go sideways or upwards, but constantly to the earth's centre?

The main emphasis in this story probably should be placed on the "contemplative mood" and not on the apple. You have seen this pattern before: A great puzzle (here, that of the forces acting on planets) begins to be solved when a clear-thinking person contemplates a familiar event (the fall of an object on earth). Where others had seen no relationship, Newton did. Referring to the plague years, Newton once wrote;

> I began to think of gravity extending to the orb of the moon, and . . . from Kepler's rule [third law, law of periods] . . . I deduced that the forces which kept the planets in their orbs must be reciprocally as the squares of their distances from the centers about which they revolve: and thereby compared the force requisite to keep the moon in her orb with the force of gravity at the surface of the earth, and found them to answer pretty nearly. All this was in the two plague years of 1665 and 1666, for in those days [at age 21 or 22] I was in the prime of my age for invention, and minded mathematics and philosophy more than at any time since.

Soon after Newton's return to Cambridge, he was chosen to follow his former teacher as professor of mathematics. Newton taught at the university and contributed papers to the Royal Society. At first, his contributions were mainly on optics. His *Theory of Light and Colors*, finally published in 1672, fired a long and bitter controversy with certain other scientists. Newton, a private and complex man, resolved never to publish anything more.

In 1684, Newton's devoted friend Halley, a noted astronomer, came to ask his advice. Halley was involved in a controversy with Christopher Wren and Robert Hooke about the force needed to cause a body to move along an ellipse in accord with Kepler's laws. This was one of the most debated and interesting scientific problems of the time. Halley was pleasantly surprised to learn that Newton had already solved this problem ("and much other matter"). Halley then persuaded his friend to publish these important studies. To encourage Newton, Halley became responsible for all the costs of publication. Less than 2 years later, Newton had the *Principia* ready for the printer. Publication of the *Principia* in 1687 quickly established Newton as one of the greatest thinkers in history.

Several years afterward, Newton appears to have had a nervous breakdown. He recovered, but from then until his death, 35 years later, Newton made no major scientific discoveries. He rounded out earlier studies on heat and optics and turned more and more to writing on theology. During those years, he received many honors. In 1699, Newton was appointed Master of the Mint, partly because of his great knowledge of the chemistry of metals. In this position, he helped to reestablish the value of British coins, in which lead and copper had been introduced in place of silver and gold. In 1689 and 1701, Newton represented Cambridge University in Parliament, and he was knighted in 1705 by Queen Anne. He was president of the Royal Society from 1703 until his death in 1727. Newton is buried in Westminster Abbey.

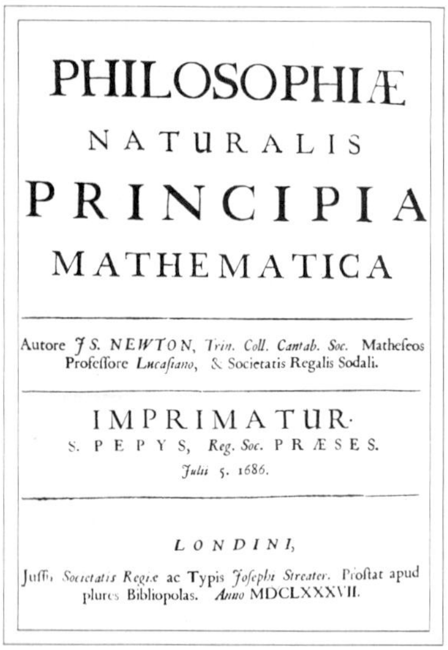
PHILOSOPHIÆ
NATURALIS
PRINCIPIA
MATHEMATICA

Autore J S. NEWTON, *Trin. Coll. Cantab. Soc.* Matheſeos Profeſſore *Lucaſiano*, & Societatis Regalis Sodali.

IMPRIMATUR.
S. PEPYS, *Reg. Soc.* PRÆSES.
Julii 5. 1686.

LONDINI,
Juſſu *Societatis Regiæ* ac Typis *Joſephi Streater*. Proſtat apud plures Bibliopolas. *Anno* MDCLXXXVII.

Title page of Newton's Principia Mathematica. *Because the Royal Society sponsored the book, the title page includes the name of the Society's president, Samuel Pepys, famous for his diary, which describes life during the seventeenth century.*

? 1. *List five characteristics of the society during Newton's lifetime that fostered scientific progress.*

8.2 | Newton's *Principia*

The original preface to Newton's *Principia*, parts of which you have already studied, gives an outline of the book:

> Since the ancients (as we are told by Pappus) esteemed the science of mechanics of greatest importance in the investigation of natural things, and the moderns, rejecting substantial forms and occult qualities, have endeavored to subject the phenomena of nature to the laws of mathematics, I have in this treatise cultivated mathematics as far as it relates to philosophy [we would say 'physical science'] . . . for the whole burden of philosophy seems to consist in this—from the phenomena of motions to investigate [induce] the forces of nature, and then from these forces to demonstrate [deduce] the other phenomena, and to this end the general propositions in the first and second Books are directed. In the third Book I give an example of this in the explication of the system of the World; for by the propositions mathematically demonstrated in the former Books, in the third I derive from the celestial phenomena the forces of gravity with which bodies tend to the sun and the several planets. Then from these forces, by other propositions which are also mathematical, I deduce the motions of the planets, the comets, the moon, and the sea [tides]. . . .

The work begins with the definitions of mass, momentum, inertia, and force. Next come the three laws of motion and the principles of addition for forces and velocities (discussed in Unit 1). Newton also included an equally important and remarkable passage on "Rules of Reasoning in Philosophy." The four rules, or assumptions, reflect Newton's profound faith in the uniformity of all nature. Newton intended the rules to guide scientists in making hypotheses. He also wanted to make clear to the reader his own philosophical assumptions. These rules had their roots in ancient Greece and are still useful. The first has been called a principle of parsimony, the second and third, principles of unity. The fourth rule expresses a faith needed to use the process of logic.

In a brief form, and using some modern language, Newton's four rules of reasoning are:

These rules are stated by Newton at the beginning of Book III of the *Principia*.

1. "Nature does nothing . . . in vain, and more is in vain when less will serve." Nature is essentially simple. Therefore, scientists ought not to introduce more hypotheses than are needed to explain observed facts. This fundamental faith of all scientists

had been also expressed in Galileo's "Nature . . . does not that by many things, which may be done by few." Galileo in turn was reflecting an opinion of Aristotle. Thus, the belief in simplicity has a long history.

2. "Therefore to the same natural effects we must, as far as possible, assign the same causes. As to respiration in a man and in a beast; the descent of stones in Europe and in America; . . . the reflection of light in the earth, and in the planets."

3. Properties common to all bodies within reach of experiments are assumed (until proved otherwise) to apply to all bodies in general. For example, all physical objects known to experimenters had always been found to have mass. So, by this rule, Newton proposed that *every* object has mass, even those beyond our reach in the celestial region.

Notice that Newton's assumption denies the distinction between terrestrial and celestial matter.

4. In "experimental philosophy," hypotheses or generalizations based on experience should be accepted as "accurately or very nearly true, notwithstanding any contrary hypotheses that may be imagined." Scientists must accept such hypotheses until they have additional evidence by which the hypotheses may be made more accurate or revised.

You should restate these rules in your own words before going on to the next section. (A good topic for an essay would be whether Newton's rules of reasoning are applicable outside of science.)

The *Principia* is an extraordinary document. Its three main sections contain a wealth of mathematical and physical discoveries. Overshadowing everything else is the theory of universal gravitation, with the proofs and arguments leading to it. Newton used a form of argument patterned after that of Euclid. You may have encountered this type of proof in studying geometry. But the style of detailed geometrical steps used in the *Principia* is unfamiliar today. Therefore, many of the steps Newton used in his proofs will be more understandable when restated in modern terms.

The central idea of universal gravitation can be simply stated: *Every object in the universe attracts every other object.* Moreover, the amount of attraction depends in a simple way on the masses of the objects and the distance between them.

This was Newton's great synthesis, boldly combining terrestrial laws of force and motion with astronomical laws of motion. Gravitation is a *universal* force. It applies to the earth and apples, to the sun and planets, and to all other bodies (such as comets) moving in the solar system. Heaven and earth were united in one grand system dominated by the law of universal gravitation. The general astonishment and awe were reflected in the words of the English poet Alexander Pope:

Nature and Nature's laws lay hid in night:
God said, Let Newton be! and all was light.

The *Principia*, written in Latin, was filled with long geometrical arguments and was difficult to read. Happily, several gifted writers wrote summaries that allowed a wide circle of readers to learn of Newton's arguments and conclusions. One of the most popular of these books was published in 1736 by the French philosopher and reformer Voltaire.

Readers of these books must have been excited and perhaps puzzled by the new approach and assumptions. For 2,000 years, from the time of the ancient Greeks until well after Copernicus, the ideas of natural place and natural motion had been used to explain the general position and movements of the planets. From the time of the Greeks, scholars had widely believed that the planets' orbits were their "natural motion." However, to Newton the natural motion of a body was at a uniform rate along a straight line. Motion in a curve showed that a net force was continuously accelerating the planets away from their natural straight-line motion. Yet the force acting on the planets was entirely natural and acted between all bodies in heaven and on earth. Furthermore, it was the same force that caused bodies on the earth to fall. What a reversal of the old assumptions about what was "natural"!

?

2. *Explain Newton's concept of the "whole burden of philosophy," that is, the job of the scientist.*

3. *In your own words, state Newton's four rules of reasoning and give an example of each.*

4. *State, in your own words, the central idea of universal gravitation.*

5. *How did Newton differ from Aristotle, who believed that the rules of motion on earth are different from the rules of motion in the heavens?*

8.3 | The inverse-square law of planetary force

Newton believed that the natural straight-line path of a planet was forced into a curve by the influence of the sun. He demonstrated that Kepler's law of areas could be true if, and only if, forces exerted on the planets were always directed toward a single point. (Details of his argument for this "central" force are given on the special pages entitled "Motion under a central force.") Newton also showed that the single point was the location of the sun. The law of areas is obeyed no matter what *magnitude* the force has, as long as the force is always directed to the same point. Newton still had to show that a central gravitational force would cause the exact relationship observed between orbital radius and period. How great was the gravitational force and how did it differ for different planets?

The combination of Kepler's laws with Newton's laws provides a fine example of the power of logical reasoning. Compare these laws:

Newton's Laws

1. A body continues in a state of rest, or of uniform motion in a straight line, unless acted upon by a net force (law of inertia).
2. The net force acting on an object is directly proportional to and in the same direction as the acceleration.
3. To every action there is an equal and opposite reaction.

Kepler's Laws

1. The planets move in orbits that are ellipses and have the sun at one focus.
2. The line from the sun to a planet sweeps over areas that are proportional to the time intervals.
3. The squares of the periods of the planets are proportional to the cubes of their mean distances from the sun ($T^2 = kR^3_{av}$).

According to Newton's first law, a change in motion, either in direction or in magnitude (speed), requires the action of a net force. According to Kepler, the planets move in orbits that are ellipses, that is, curved orbits. Therefore, a net force must be acting to change their motion. Notice that this conclusion does not specify the type or direction of the net force.

Combining Newton's second law with the first two laws of Kepler clarifies the direction of the force. According to Newton's second law, the net force is exerted in the direction of the observed acceleration. What is the direction of the force acting on the planets? Newton employed the geometrical analysis described on pages 218–219, "Motion under a central force."

Newton's analysis indicated that a body moving under a central force will, when viewed from the center of the force, move according to Kepler's law of areas. Kepler's law of areas relates to the distance of the planets from the sun. Therefore, Newton could conclude that the sun at one focus of each ellipse was the source of the central force acting on the planets.

Newton then found that motion in an elliptical path would occur only when the central force was an inverse-square force, $F \propto 1/R^2$. Thus, only an inverse-square force exerted by the sun would result in the observed ellipitcal orbits described by Kepler. Newton then proved the argument by showing that such a force law would also result in Kepler's third law, the law of periods, $T^2 = kR^3_{av}$.

From this analysis, Newton concluded that one general law of universal gravitation applied to all bodies moving in the solar system. This is the central argument of Newton's great synthesis.

Consider the motions of the six then-known planets in terms of their centripetal acceleration toward the sun. By Newton's proof, mentioned above, this acceleration decreases inversely as the square of the planets' average distances from the sun. The proof for circular orbits is very short. The expression for centripetal acceleration a_c of a body moving uniformly in a circular path, in terms of the radius R and the period T, is

$$a_c = \frac{4\pi^2 R}{T^2}$$

(This expression was derived in Chapter 4.) Kepler's law of periods stated a definite relation between the orbital periods of the planets and their average distances from the sun:

$$\frac{T^2}{R_{av}^3} = \text{constant}$$

Using the symbol k for constant,

$$T^2 = kR_{av}^3$$

For circular orbits, R_{av} is just R. Substituting kR^3 for T^2 in the centripetal force equation gives

$$a_c = \frac{4\pi^2 R}{kR^3} = \frac{4\pi^2}{kR^2}$$

Since $4\pi^2/k$ is a constant,

$$a_c \propto \frac{1}{R^2}$$

This conclusion follows necessarily from Kepler's law of periods and the definition of acceleration. If Newton's second law, $F \propto a$, holds for planets as well as for bodies on earth, then there must be a centripetal force F_c acting on a planet. Furthermore, this force must decrease in proportion to the square of the distance of the planet from the sun:

$$F_c \propto \frac{1}{R^2}$$

Newton showed that the same result holds for all ellipses. Indeed, it holds for any object moving in an orbit around a center of force. [The possible orbital shapes are circle, ellipse, parabola, or hyperbola. These shapes are all conic sections (see page 188)]. Any such object is being acted upon by a centripetal force that varies inversely with the square of the distance from the center of force.

In Newton's time, four of Jupiter's satellites and four of Saturn's satellites had been observed.

SG 6

Newton had still more evidence from telescopic observations of Jupiter's satellites and Saturn's satellites. The satellites of each planet obeyed Kepler's law of areas around the planet as a center. For Jupiter's satellites, Kepler's law of periods, T^2/R^3 = *constant*, held. But the *value* of the constant was different from that for the planets around the sun. The law held also for Saturn's satellites, but with still a different constant. Therefore, Jupiter's satellites were acted on by a central force directed toward Jupiter and decreasing with the square of the distance from Jupiter. The same held true for Saturn's satellites and Saturn. These observed interactions of astronomical bodies supported Newton's proposed $1/R^2$ central attractive force.

?

6. What can be proved from the fact that the planets sweep out equal areas with respect to the sun in equal times?

7. With what relationship can T^2/R_{av}^3 = constant *be combined to prove that the gravitational attraction varies as* $1/R^2$*?*

8. What simplifying assumption was made in the derivation given in this section?

9. Did Newton limit his own derivation by the same assumption?

10. If two objects are moved twice as far away from one another, by how much is the gravitational force between them decreased? if they are moved three times as far? five times as far? By how much is the gravitational force increased if the objects are moved together to one-fourth their original separation?

11. The moon is 60 times farther from the center of the earth than objects at the earth's surface. How much less is the gravitational attraction of the earth acting on the moon than on objects at its surface? Express this value as a fraction of 9.8 m/sec^2.

8.4 Law of universal gravitation

Subject to further evidence, you can now accept the idea that a central force is holding the planets in their orbits. Furthermore, the strength of this central force changes inversely with the square of the distance from the sun. This strongly suggests that the sun is the source of the force but it does not necessarily require this conclusion. Newton's results so far describe the force in mathematical terms but do not provide any mechanism for its transmission.

SG 7

The French philosopher Descartes (1596–1650) had proposed that all space was filled with a thin, invisible fluid. This fluid carried the planets around the sun in a huge whirlpool-like motion. This interesting idea was widely accepted at the time. However, Newton proved by a precise argument that this mechanism could not explain the details of planetary motion summarized in Kepler's laws.

Kepler had made a different suggestion some years earlier. He proposed that some magnetic force reached out from the sun to keep the planets moving. Kepler was the first to regard the sun as the controlling mechanical agent behind planetary motion. But Kepler's magnetic model was inadequate. The problem remained: Was the sun actually the source of the force? If so, on what properties of the sun or planets did the amount of the force depend?

As you read in Sec. 8.1, Newton had begun to think about planetary force while living at home during the Black Plague. There an idea came to him, perhaps when he saw an apple fall, and perhaps not. Newton's idea was that the planetary force was the same kind of force that caused objects near the earth's surface to fall. He first tested this idea on the earth's attraction for the moon. The data available to him fixed the distance between the center of the earth and the center of the moon at nearly 60 times the radius of the earth. Newton believed that the attractive force varies as $1/R^2$. Therefore, the gravitational acceleration the earth exerts on the moon should be only $1/(60)^2 = 1/3{,}600$ of that exerted upon objects at the earth's surface. Observations of falling bodies had long established gravitational acceleration at the earth's surface as about 9.80 m/sec/sec. Therefore, the moon *should* fall at 1/3,600 of that acceleration value: $9.80 \text{ m/sec}^2 \times (1/3{,}600) = 2.72 \times 10^{-3} \text{ m/sec}^2$.

Newton started from the knowledge that the orbital period of the moon was very nearly 27.33 days. The centripetal acceleration a_c of a body moving uniformly with period T in a circle of radius R is $a_c = 4\pi^2R/T^2$. (This equation was developed in Sec. 4.6, Unit 1.) When you insert values for the known quantities R and T (in meters and seconds) and do the arithmetic, you find that the *observed* acceleration is

$$a_c = 2.74 \times 10^{-3} \text{ m/sec}^2$$

This is in very good agreement with the value of 2.72×10^{-3} m/sec^2 predicted above. From the values available to Newton, which were close to these, he concluded that he had

> ... compared the force requisite to keep the moon in her orbit with the force of gravity at the surface of the earth, and found them to answer pretty nearly.
>
> Therefore, the force by which the moon is retained in its orbit becomes, at the very surface of the earth, equal to the force of gravity which we observe in heavy bodies there. And, therefore, (by rules of reasoning 1 and 2) the force by which the moon is retained in its orbit is that very same force which we commonly call gravity. . . .

This was really a triumph. The same gravity that brings apples down from trees also holds the moon in its orbit. This assertion

To you, who have heard about gravity from your earliest years, this may not seem to have been a particularly clever idea. But in Newton's time, after centuries of believing celestial events to be completely different from earthly events, it was the mental leap of a genius. Newton had already assumed the planets to be subject to the earth's laws of motion when he derived a $1/R^2$ force law using the formula for a_c. It was a still greater step to guess that the force on planets was not some special celestial force, but nothing other than the earth's gravitational pull, which gave apples and everything else on earth their weight.

Close Up

Motion under a Central Force

How will a moving body respond to a central force? In order to follow Newton's analysis, remember that the area of a triangle equals ½ *base* × *altitude*. Any of the three sides can be chosen as the base, and the altitude is the perpendicular distance to the opposite corner.

Suppose that a body passing some point P, was moving at uniform speed *v* along the straight line through PQ. (See Figure A below.) If it goes on with no force acting, then in equal intervals of time Δt it will continue to move equal distances, PQ, QR, RS, etc.

(A)

How will its motion appear to an observer at some point O? Consider the triangles OPQ and OQR in Figure B below.

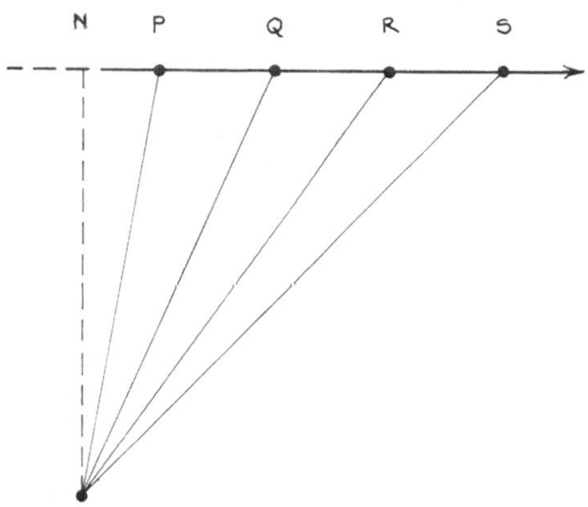

(B)

The triangles have equal bases, PQ = QR = RS, and also equal altitudes, ON, for all three. Therefore, the triangles OPQ and OQR have equal areas. And therefore the line drawn from an observer at point O to the body moving at a uniform speed in a straight line PQR will sweep over equal areas in equal times.

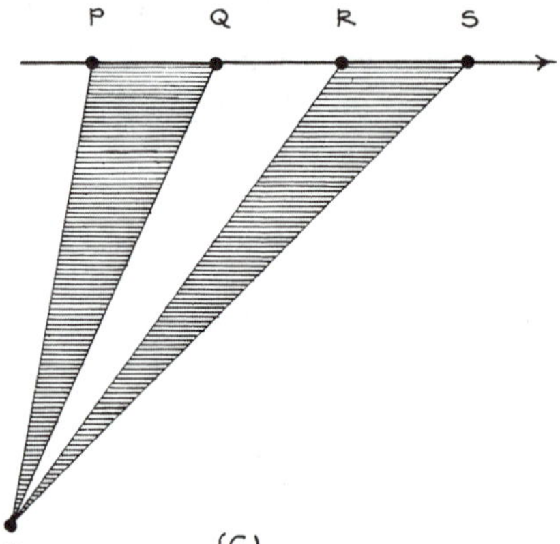

(C)

So, strange as it may seem at first, Kepler's law of areas applies even to a body on which the net force has the value *zero* and which therefore is moving uniformly along a straight line.

Suppose that the object discussed in Figure A, while passing through point Q, is briefly exposed to a force, such as a blow. If this force is directed toward point O, how will the object's motion change? (Refer to Figure D below.)

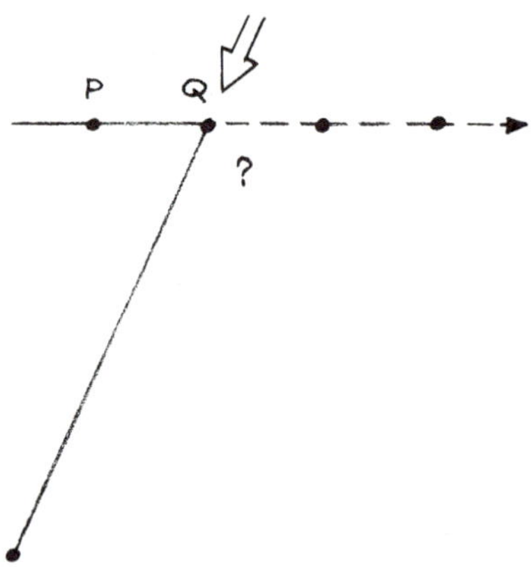

(D)

First, consider what happens if a body initially at rest at point Q were exposed to the same blow. The body would be accelerated during the blow toward O. It would then continue to move toward O at constant speed. After some definite time interval Δt, it would have moved a definite distance to a new point Q′. (See Figure E on the next page.)

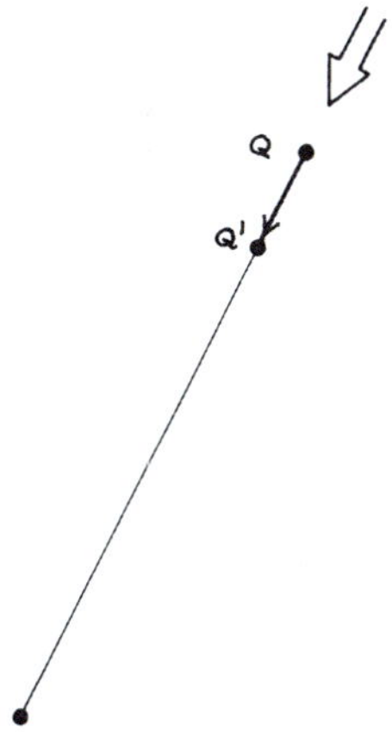

(E)

Now, consider the effect of the blow on the object that was initially moving toward point R. The resultant motion is the combination of these two components, and the object moves to point R′. (See Figure F below.)

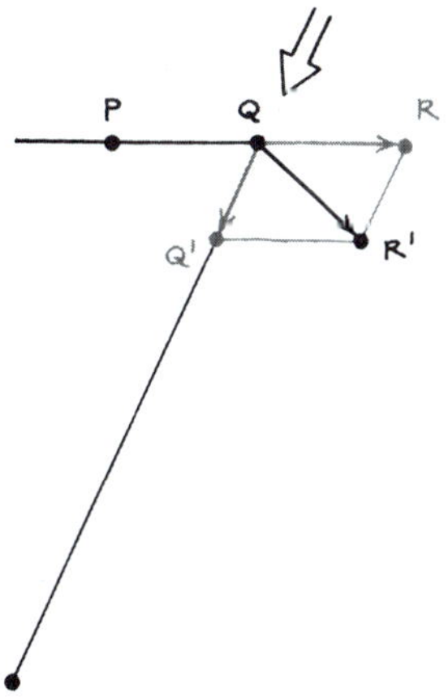

(F)

Earlier we found that the areas of the triangles OPQ and OQR were equal. Is the area of the triangle OQR′ the same? Both triangles OQR and OQR′ have a common base, OQ. Also, the altitudes of both triangles are the perpendicular distance from line OQ to line RR′. (See Figure G.) Therefore, the areas of triangles OQR and OQR′ are equal.

If another blow directed toward O were applied at point R′, the body would move to some point S″,

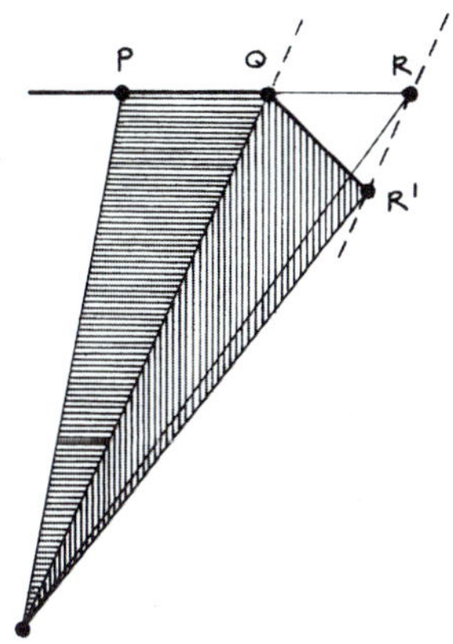

(G)

as indicated in Figure H below. By a similar analysis you can find that the areas of triangles OR′S″ and OR′S′ are equal. Their areas also equal the area of triangle OPQ.

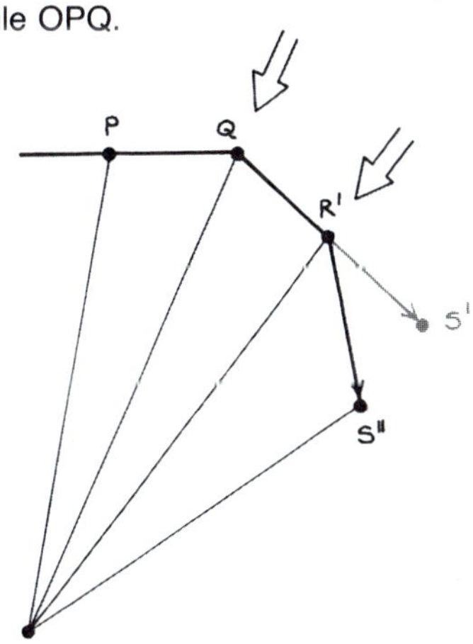

(H)

In this geometrical argument we have always applied the force toward the same point, O. A force always directed toward a single point is called a *central* force. (Notice that the proof has nothing to do with the *magnitude* of the force or with how it changes with distance from O.) Also, we have applied the force at equal intervals Δt. If each time interval Δt were made vanishingly small, the force would appear to be applied continuously. The argument would still hold. We then have an important conclusion: *If a body is acted upon by any central force, it will move in accordance with Kepler's law of areas.*

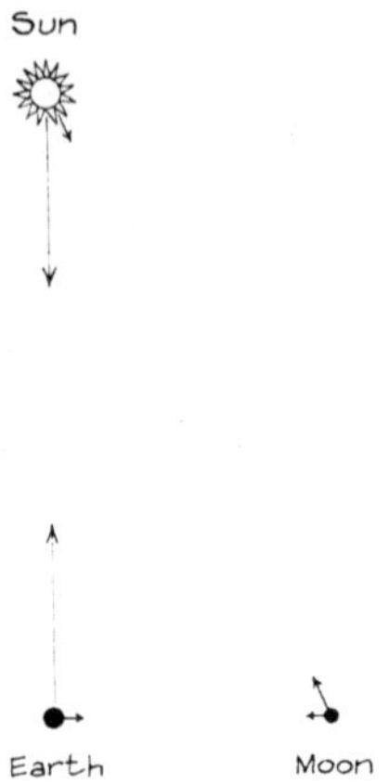

The sun, moon, and earth each pull on the other. The forces are in matched pairs, in agreement with Newton's third law of motion. As the moon moves through space, the gravitational attraction of the earth causes the moon to "fall" toward the earth. The continuous combination of its straight-line inertial motion and its "fall" produces the curved orbit.

is the first portion of what is known as the law of universal gravitation: *Every object in the universe attracts every other object with a gravitational force.* If this is so, there must be gravitational forces not only between a rock and the earth, but also between the earth and the moon, between Jupiter and its satellites, and between the sun and each of the planets.

Newton did not stop at saying that a gravitational force exists between the planets and the sun. He further claimed that the force is exactly the right size to explain *completely* the motion of every planet. No other mechanism (whirlpools of invisible fluids or magnetic forces) is needed. Gravitation, and gravitation alone, underlies the dynamics of the heavens.

This concept is so commonplace that you might be in danger of passing it by without really understanding what Newton was claiming. First, he proposed a truly universal physical law. Following his rules of reasoning, Newton extended to the whole universe what he found true for its observable parts. He excluded no object in the universe from the effect of gravity.

The idea that terrestrial laws and forces were the same as those that regulated the whole universe had stunning impact. Less than a century before, it would have been dangerous even to suggest such a thing. Kepler and Galileo had laid the foundation for combining the physics of the heavens and earth. Newton carried this work to its conclusion. Today, Newton's extension of the mechanics of terrestrial objects to the motion of celestial bodies is called the *Newtonian synthesis.*

Newton's claim that a planet's orbit is determined by the gravitational attraction between it and the sun had another effect. It moved science away from geometrical explanations and towards physical ones. Most philosophers and scientists before Newton had been occupied mainly with the question "What are the motions?" Newton asked instead "What force explains the motions?" In both the Ptolemaic and Copernican systems, the planets moved about *points* in space rather than about *objects*. The planets moved as they did because of their "nature" or geometrical shape, not because forces acted on them. Newton, on the other hand, spoke not of points, but of things, of objects, of physical bodies. Unless the gravitational attraction to the sun deflected them continuously from straight-line paths, the planets would fly out into the darkness of deep space. Thus, it was the physical sun that was important, not the point at which the sun happened to be located.

Newton's synthesis centered on the idea of gravitational force. By calling it a force of gravity, Newton knew that he was not explaining *why* it existed. When you hold a stone above the surface of the earth and release it, it accelerates to the ground. The laws of motion tell you that there must be a force acting on the stone to accelerate it. You know the *direction* of the force.

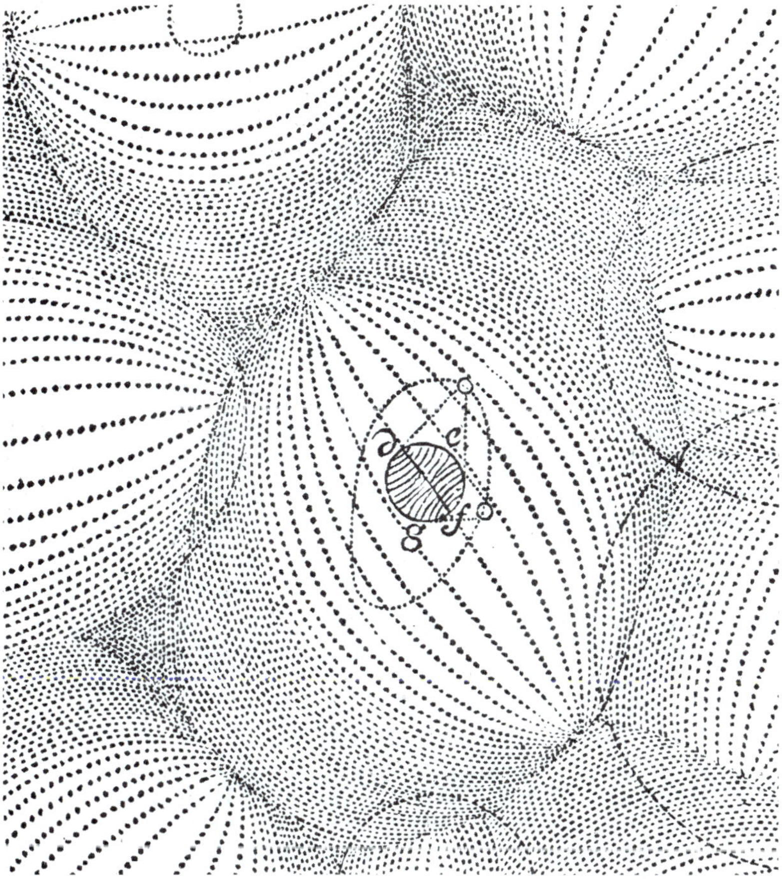

A drawing by which Descartes (1596–1650) illustrated his theory of space being filled with whirlpools of matter that drive the planets along their orbits.

You can find the *magnitude* of the force by multiplying the mass of the stone by the acceleration. You know that this force is weight, or gravitational attraction to the earth. But why such an interaction between bodies exists remains a puzzle. It is still an important problem in physics today.

?

12. What idea came to Newton while he was thinking about falling objects and the moon's acceleration?

13. Kepler, too, believed that the sun exerted forces on the planets. How did his view differ from Newton's?

14. The central idea of Chapter 8 is the "Newtonian synthesis." What did Newton synthesize (bring together)?

8.5 | Newton and hypotheses

Newton's claim that there is a mutual force (gravitational interaction) between a planet and the sun raised a new question:

How can a planet and the sun act upon each other at enormous distances without any visible connections between them? On earth you can exert a force on an object by pushing it or pulling it. You are not surprised to see a cloud or a balloon drifting across the sky, even though nothing seems to be touching it. Air is invisible, but you know that it is actually a material substance that you can feel when it moves. Falling objects and iron objects being attracted to a magnet are harder to explain, but the distances are small. However, the earth is over 144 million kilometers, and Saturn more than 1 billion kilometers, from the sun. How could there possibly be any physical contact between such distant objects? How can we account for such "action at a distance"?

In Newton's time and for a long time afterward, scholars advanced suggestions for solving this problem. Most solutions involved imagining space to be filled with some invisible substance (an "ether") that transmitted force. Newton himself privately guessed that such an ether was involved. But he could find no way to test this belief. Therefore, at least in public, he refused to speculate on possible mechanisms. As Newton said in a famous passage which he added in the second edition of the *Principia* (1713):

> ... Hitherto I have not been able to discover the cause of those properties of gravity from phenomena, and I frame no hypotheses; for whatever is not deduced from the phenomena is to be called an hypothesis; and hypotheses, whether metaphysical or physical, whether of occult qualities or mechanical, have no place in experimental philosophy.... And to us it is enough that gravity does really exist, and acts according to the laws which we have explained, and abundantly serves to account for all the motions of the celestial bodies, and of our sea.

Newton is quoted at length here because one particular phrase is often taken out of context and misinterpreted. The original Latin reads: *hypotheses non fingo*. This means "I frame no hypotheses" or "I do not feign hypotheses." The sense is, "I do not make *false* hypotheses." Newton in fact made many hypotheses in his publications. Also, his letters to friends contain many speculations which he did not publish. So his stern denial of "framing" hypotheses must be properly interpreted.

The fact is that there are two main kinds of hypotheses or assumptions. The most common hypothesis is a proposal of some hidden mechanism to explain observations. For example, you observe the moving hands of a watch. You might propose or imagine some arrangement of gears and springs that causes the motion. This would be a *hypothesis that is directly or indirectly testable, at least in principle, by reference to phenomena*. The hypothesis about the watch, for example, can be tested by opening the watch or by making an X-ray photograph of it. In this context, consider an invisible fluid that transmitted gravitational force, the so-called "ether." Newton and others thought that certain direct tests might establish the presence of this substance. Many experimenters tried to "catch" the ether. A common approach involved pumping the air from a bottle. Then tests were made to see if any wind, pressure, or friction due to the ether remained to affect objects in the bottle. Nothing of this sort worked (nor has it since). So Newton wisely avoided making public any hypothesis for which he could not also propose a test.

A quite different type of assumption is often made in published scientific work. It involves a hypothesis which everyone knows is not directly testable, but which still is necessary *just to get started on one's work*. An example is such a statement as "nature is simple" or any other of Newton's four rules of reasoning. Acceptance of either the heliocentric system or the geocentric system is another example. In choosing the heliocentric system, Copernicus, Kepler, and Galileo made the hypothesis that the sun is at the center of the universe. They knew that this hypothesis was not directly testable and that either system seemed to explain "the phenomena" equally well. Yet they adopted the point of view that seemed to them simpler, more convincing, and more "pleasing to the mind." It was this kind of hypothesis that Newton used without apology in his published work. Every scientist's work involves both kinds of hypothesis. The popular image of the scientist is of a person who uses only deliberate, logical, objective thoughts, and immediately tests them by definitive experiments. But, in fact, the working scientist feels quite free to entertain any guess, imaginative speculation, or hunch, provable or not, that might be helpful. (Sometimes these hunches are dignified by the phrase "working hypotheses." Without them there would be little progress!) Like Newton, however, most scientists today do not like to *publish* something that is still only an unproven hunch.

Stephen W. Hawking, an astrophysicist at Cambridge University, is considered by many scientists to be the equal of Newton and Einstein. Professor Hawking is searching for a quantum theory of gravitation by studying the phenomena associated with black holes. The achievement of such a theory would be the last step in the theoretical unification of all the forces in the universe.

15. *Did Newton explain the gravitational attraction of all bodies?*

16. *What was the popular type of explanation for "action at a distance"? Why did Newton not use this type of explanation?*

17. *What are two main types of hypotheses used in science?*

18. *Newton's claim to "frame no hypotheses" seems to refer to hypotheses that cannot be tested. Which of the following claims are not testable?*

(a) Plants need sunlight to grow, even on other planets.

(b) *This bandage is guaranteed to be free from germs unless the package is opened.*

(c) *Virtual particles exist for a time that is too short for them to affect anything.*

(d) *Life exists in the distant galaxies.*

(e) *The earth really does not move, since you would feel the motion if it did.*

(f) *Universal gravitation holds between every pair of objects in the universe.*

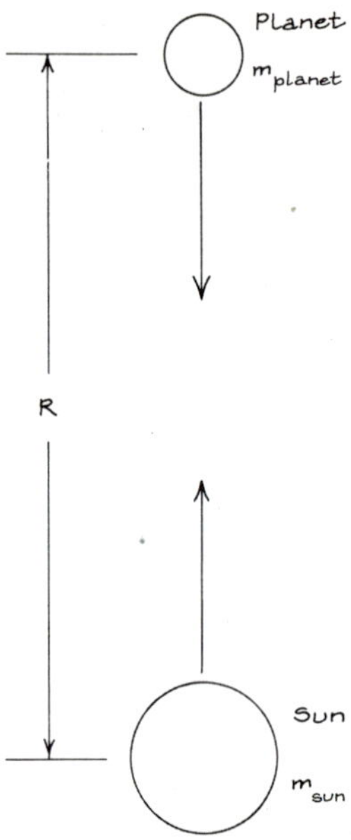

The gravitational force on a planet owing to the sun's pull is equal and opposite to the gravitational force on the sun owing to the planet.

8.6 | The magnitude of planetary force

The general statement that gravitational forces exist universally must now be turned into a quantitative law. An expression is needed for both the *magnitude* and *direction* of the forces any two objects exert on each other. It was not enough for Newton to assert that a mutual gravitational attraction exists between the sun and Jupiter. To be convincing, he had to specify what quantitative factors determine the magnitudes of those mutual forces. He had to show how they could be measured, either directly or indirectly.

The first problem was defining precisely the distance R. Should it, for example, be taken as the distance between the surface of the earth and the surface of the moon? For many astronomical problems, the sizes of the interacting bodies are extremely small compared to the distances between them. In such cases, the distance between the surfaces is practically the same as the distance between the centers. (For the earth and the moon, the distance between centers is only about 2% greater than the distance between surfaces.) Yet, some historians believe Newton's uncertainty about a proper answer to this problem led him to drop the study for many years.

Eventually, Newton solved the problem. The gravitational force exerted *by* a spherical body is the same as if all its mass were concentrated at its center. The gravitational force exerted *on* a spherical body by another body is the same as would be exerted on it if all its mass were concentrated at its center. Therefore, the distance R in the law of gravitation is the distance between centers.

This was a very important discovery. The gravitational attraction between spherical bodies can be considered as though their masses were concentrated at single points. Thus, in thought, the objects can be replaced by *mass points*.

Newton's third law states that action equals reaction. If this is universally true, the amount of force the sun exerts on a planet must exactly equal the amount of force the planet exerts on the sun. For such a very large mass and such a relatively small mass, this may seem contrary to common sense. But the equality is easy to prove. First, assume only that Newton's third law holds between small pieces of matter. For example, a 1-kg piece of Jupiter pulls *on* a 1-kg piece of the sun as much as it is pulled *by* it. Now consider the total attraction between Jupiter and the sun, whose mass is about 1,000 times greater than Jupiter's. As the figure in the right margin indicates, you can consider the sun as a globe containing 1,000 Jupiters. Define one unit of force as the force that two Jupiter-sized masses exert on each other when separated by the distance of Jupiter from the sun. Then Jupiter pulls on the *sun* (a globe of 1,000 Jupiters) with a total force of 1,000 units. Each of the 1,000 parts of the sun also pulls on the planet Jupiter with 1 unit. Therefore, the total pull of the sun on *Jupiter* is also 1,000 units. Each part of the massive sun not only pulls *on* the planet, but is also pulled upon *by* the planet. The more mass there is to *attract*, the more there is to be *attracted*. (Although the mutual attractive forces are equal in magnitude, the resulting *accelerations* are not. Jupiter pulls on the sun as hard as the sun pulls on Jupiter, but the sun *responds* to the pull with only 1/1,000 of the acceleration, because its *inertia* is 1,000 times Jupiter's.)

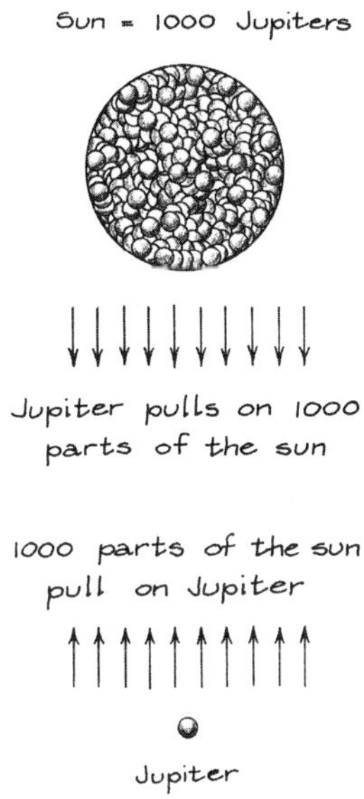

Sec. 3.8 of Unit 1 explained why bodies of different mass fall with the same acceleration near the earth's surface. The greater the inertia of a body, the more strongly it is acted upon by gravity; that is, near the earth's surface, the gravitational force on a body is directly proportional to its mass. Like Newton, extend this earthly effect to all gravitation. You then can assume that the gravitational force exerted on a planet by the *sun* is proportional to the mass of the planet. Similarly, the gravitational force exerted on the sun by the *planet* is proportional to the mass of the sun. You have just seen that the forces the sun and planet exert on each other are equal in magnitude. It follows that the magnitude of the gravitational force is proportional to the mass of the sun *and* to the mass of the planet; that is, the gravitational attraction between two bodies is proportional to the *product* of their masses. If the mass of either body is tripled, the force is tripled. If the masses of both bodies are tripled, the force is increased by a factor of 9. Using the symbol F_{grav} for the magnitude of the forces, $F_{grav} \propto m_{planet} m_{sun}$.

SG 8

The conclusion is that the amount of attraction between the sun and a planet is proportional to the product of their masses. Earlier you saw that the attraction also depends on the square of the distance between the centers of the bodies. Combining these two proportionalities gives *one* force law, which now includes mass and distance:

SG 9

$$F_{grav} \propto \frac{m_{planet} m_{sun}}{R^2}$$

Such a proportionality can be written as an equation by introducing a constant. (The constant allows for the units of measurement used.) Using G for the proportionality constant, the law of planetary forces can be written as

$$F_{grav} = G\frac{m_{planet} m_{sun}}{R^2}$$

This equation asserts that the force between the sun and any planet depends *only* upon three factors. These factors are the masses of the sun and planet and the distance between them. The equation seems unbelievably simple when you remember how complex the observed planetary motions seemed. Yet every one of Kepler's empirical laws of planetary motion agrees with this relation. In fact, you can even *derive* Kepler's empirical laws from this force law and Newton's second law of motion. More important still, details of planetary motion not obtainable with Kepler's laws alone can be calculated using this force law.

Newton's proposal that this simple equation describes completely the forces between the sun and planets was not the final step. He saw nothing to limit this mutual force to the sun and planets, or to the earth and apples. Rather, Newton insisted that an identical relation should apply *universally*. This relation would hold true for *any two bodies* separated by a distance that is large compared to their dimensions. It would apply equally to two atoms or two stars. In short, Newton proposed a *general law of universal gravitation:*

SG 10

$$F_{grav} = G\frac{m_1 m_2}{R^2}$$

where m_1 and m_2 are the masses of the bodies and R is the distance between their centers. The numerical constant G is called the *constant of universal gravitation*. Newton assumed it to be the same for all gravitational interactions, whether between two grains of sand, two members of a solar system, or two stars in different parts of the sky. As you will see, the successes made possible by this simple relationship have been very great. In fact, scientists have come to assume that this equation applies everywhere and at all times, past, present, and future.

Even before you gather more supporting evidence, the sweeping majesty of Newton's theory should command your wonder and admiration. It also leads to the question of how such a bold universal theory can be proved. There is no complete proof, of course, for that would mean examining every interaction between all bodies in the universe! But the greater the variety of single tests made, the greater will be the belief in the correctness of the theory.

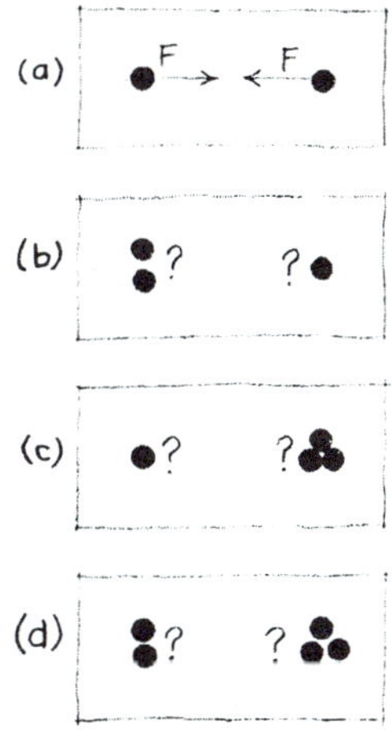

19. *According to Newton's law of action and reaction, the earth should experience a force and accelerate toward a falling stone.*

(a) How does the force on the earth compare with the force on the stone?

(b) How does the earth's acceleration compare with the stone's acceleration?

20. *Diagram A at the right represents two bodies of equal mass that exert gravitational forces of magnitude* F *on one another. What is the magnitude of the gravitational attractions in each of the other cases?*

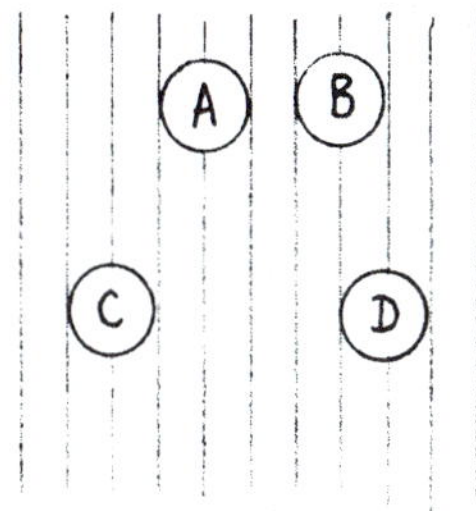

21. *A, B, C, and D are bodies with equal masses. How do the forces of attraction that A and B exert on each other compare with the forces that C and D exert on each other?*

(a) $F_{AB} = 3 \times F_{CD}$

(b) $F_{AB} = 4 \times F_{CD}$

(c) $F_{AB} = 9 \times F_{CD}$

(d) $F_{AB} = 16 \times F_{CD}$

22. *Why is it a great simplification to use the distance between the centers of spherical objects in the formula for gravitational force? How is the use of the distance between centers justified? What does it mean to say that the law of gravitation is important because it is simple?*

8.7 | Planetary motion and the gravitational constant

Suppose that a planet of mass m_p is moving along an orbit of radius R and period T. According to Newton's mechanics, there is a continual centripetal acceleration $a_c = 4\pi^2R/T^2$. Therefore, there must be a continual force $F_c = m_pa_c = 4\pi^2Rm_p/T^2$. If gravity is the central force, then

$$F_{grav} = F_c$$

or

$$G\frac{m_pm_{sun}}{R^2} = \frac{4\pi^2Rm_p}{T^2}$$

Simplifying this equation and rearranging some terms gives an expression for G:

$$G = \frac{4\pi^2}{m_{sun}}\left(\frac{R^3}{T^2}\right)$$

This photograph, of the surface of the moon, shows some latter-day evidence that the laws of mechanics for heavenly bodies are at least similar to those applying on earth. The trails of two huge boulders that rolled about 300 m down a lunar slope are shown.

You know from Kepler that for the planets' motion around the sun, the ratio R^3/T^2 is a constant; $4\pi^2$ is a constant also. If the mass of the sun is assumed to be constant, then all factors on the right of the equation for G are constant. So G must be a constant for the gravitational effect of the sun on the planets. By similar reasoning, the value of G must be a constant for the effect of Jupiter on its moons. It must also be a constant for Saturn and its moons, for earth and its moon, and for an apple falling to the earth. But is it the same value of G for all these cases?

It is impossible to *prove* that G is the same for the gravitational interaction of *all* bodies. But by *assuming* that G is a universal constant, the relative masses of the sun and the planets can be obtained.

Begin by again equating the centripetal force on the planets with the gravitational attraction to the sun. This time solve the equation for m_{sun}:

$$F_{grav} = F_c$$

$$\frac{Gm_p m_{sun}}{R^2} = \frac{4\pi^2 R m_p}{T^2}$$

$$m_{sun} = \frac{4\pi^2 R^3}{GT^2}$$

Writing k_{sun} for the constant ratio T^2/R^3 gives

$$m_{sun} = \frac{4\pi^2}{Gk_{sun}}$$

By similar derivation,

$$m_{Jupiter} = \frac{4\pi^2}{Gk_{Jupiter}}, \quad m_{Saturn} = \frac{4\pi^2}{Gk_{Saturn}}, \quad m_{earth} = \frac{4\pi^2}{Gk_{earth}}$$

Here $k_{Jupiter}$, k_{Saturn}, and k_{earth} are the known values of the constant ratios T^2/R^3 for the satellites of Jupiter, Saturn, and the earth.

To compare Jupiter's mass to the mass of the sun, simply divide the formula for $m_{Jupiter}$ by the formula for m_{sun}:

$$\frac{m_{Jupiter}}{m_{sun}} = \frac{\dfrac{4\pi^2}{Gk_{Jupiter}}}{\dfrac{4\pi^2}{Gk_{sun}}} \quad \text{or} \quad \frac{m_{Jupiter}}{m_{sun}} = \frac{k_{sun}}{k_{Jupiter}}$$

Similarly, you can compare the masses of any two planets if you know the values of T^2/R^3 for them both; that is, both must have satellites whose motion has been carefully observed.

These comparisons are based on the *assumption* that G is a universal constant. Calculations based on this assumption have led to *consistent* results for a wide variety of astronomical data. One example is the successful orbiting and landing of a space vehicle on the moon. Results consistent with this assumption also appeared in difficult calculations of the small disturbing effects that the planets have on each other. There is still no way of proving G is the same everywhere and always. But it is a reasonable working assumption until evidence to the contrary appears.

Masses Compared to Earth	
Earth	1
Saturn	95
Jupiter	318
Sun	333,000

If the numerical value of G were known, the *actual* masses of the earth, Jupiter, Saturn, and the sun could be calculated. G is defined by the equation $F_{grav} = Gm_1m_2/R^2$. To find the value of G, you must know values for all the other variables; that is, you must measure the force F_{grav} between two measured masses m_1 and m_2, separated by a measured distance R. Newton knew this. In his time there were no instruments sensitive enough to measure the very tiny force expected between masses small enough for experimental use.

?

23. What information can be used to compare the masses of two planets?

24. What additional information is necessary for calculation of the actual masses?

Calculation of G from approximate experimental values:

$$F_{grav} = G\frac{Mm}{R^2}$$

$$G = \frac{F_{grav}R^2}{Mm}$$

$$= \frac{(10^{-6}\ \text{N})\,(0.1\ \text{m})^2}{(100\ \text{kg})\,(1\ \text{kg})}$$

$$= \frac{10^{-6} \times 10^{-2}}{10^2}\ \text{N} = \text{m}^2/\text{kg}^2$$

$$= 10^{-10}\ \text{N}\cdot\text{m}^2/\text{kg}^2$$

8.8 | The value of *G* and the actual masses of the planets

The masses of small solid objects can be found easily enough from their weights. Measuring the distance between solid objects of spherical shape presents no problem. But how can one measure the tiny mutual gravitational force between relatively small objects in a laboratory? (Remember that each object is also experiencing separately a huge gravitational force toward the tremendously massive earth.)

This serious technical problem was eventually solved by the English scientist, Henry Cavendish (1731–1810). For measuring gravitational forces, he employed a torsion balance. In this device, the gravitational attraction between two pairs of lead spheres twisted a wire holding up one of the pairs. The twist of the wire could be calibrated from the twist produced by small known forces. A typical experiment might involve a 100-kg sphere and a 1-kg sphere at a center-to-center distance of 0.1 m. The resulting force would be about one-millionth of a newton (0.000001 N)! As the calculations in the margin show, these data lead to a value for G of about 10^{-10} ($N \cdot m^2/kg^2$). This experiment has been steadily improved, and the accepted value of G is now

$N \cdot m^2/kg^2$ can be expressed as $m^3/kg \cdot sec^2$.

$$G = 6.67 \times 10^{-11}\ N \cdot m^2/kg^2$$

SG 14–19

Evidently, gravitation is a weak force that becomes important only when at least one of the masses is very great. The gravitational force on a 1-kg mass at the surface of the earth is 9.8 N; if released, a 1-kg mass falls with an acceleration of 9.8 m/sec^2. Substituting 9.8 N for F_{grav} and the radius of the earth for R, you can calculate the mass of the earth! (See SG 11.)

Actual Masses (in units of 10^{24} kg)

Sun	1,980,000
Mercury	0.328
Venus	4.83
Earth	5.98
Mars	0.637
Jupiter	1,900
Saturn	567
Uranus	88.0
Neptune	103
Pluto	0.014

Assume that the same value for G applies to all gravitational interaction. Now you can calculate the masses of the planets from the known values of T^2/R^3 for their satellites. Since Newton's time, satellites have been discovered around all of the outer planets. The values of these planets' masses, calculated from $m = 4\pi^2/G \times R^3/T^2$, are given in the table in the margin. Venus and Mercury have no satellites. Their masses are found by analyzing the slight disturbing effects each has on other planets. Modern values for the actual masses of the planets are listed in the margin. Notice that the planets taken together make up not much more than 1/1000 of the mass of the solar system. By far, most of the mass is in the sun. For this reason, the sun dominates the motion of the planets, acting almost like an infinitely massive, fixed object.

In light of Newton's third law, this picture must be modified a little. For every pull the sun exerts on a planet, the sun itself experiences an equally strong pull in the opposite direction. Of course, the very much greater mass of the sun keeps its acceleration to a correspondingly smaller value. But some slight acceleration does exist. Therefore, the sun cannot really be fixed in space even within the solar system, if we accept Newtonian dynamics. Rather, it moves a little about the point that forms the common center of mass of the sun and each moving planet. This is true for every one of the nine planets. Since the planets rarely move all in one line, the sun's motion is actually a combination

SG 20–22

of nine small ellipses. Such motion might be important in a solar system in which the planets were very heavy compared to their sun. In our solar system, it is not large enough to be of interest for most purposes.

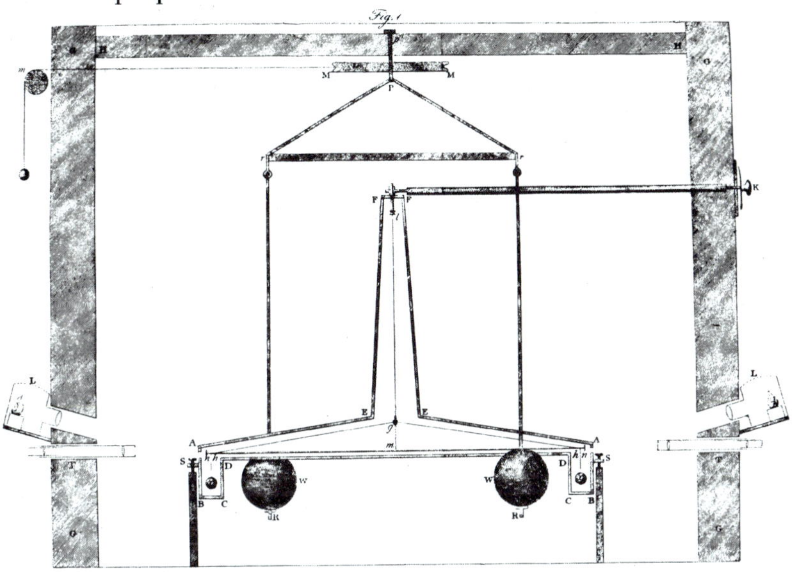

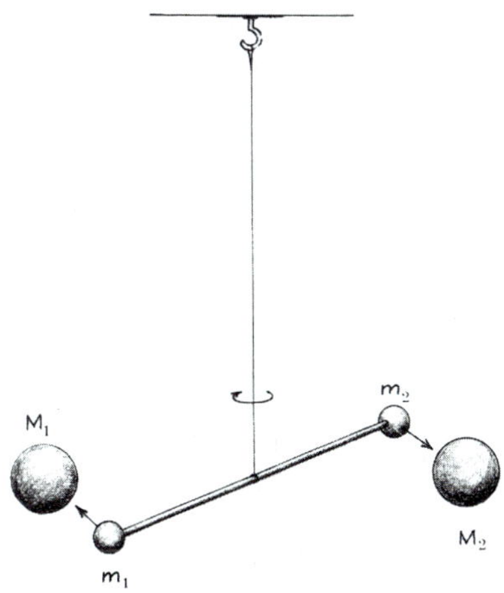

Schematic diagram of the device used by Cavendish for determining the value of the gravitational constant G. Large lead balls of masses M_1 and M_2 were brought close to small lead balls of masses m_1 and m_2. The mutual gravitational attraction between M_1 and m_1 and between M_2 and m_2 caused the vertical wire to be twisted by a measurable amount.

Right:
Cavendish's original drawing of his apparatus for determining the value of G. To prevent disturbance from air currents, he enclosed it in a sealed case. Cavendish observed the deflection of the balance rod from outside with telescopes.

?

25. *Which of the quantities in the equation* $F_{grav} = Gm_1 m_2/R^2$ *did Cavendish* measure?
26. *Knowing a value for* G, *what other information can be used to find the mass of the earth?*
27. *Knowing a value for* G, *what other information can be used to find the mass of Saturn?*
28. *The mass of the sun is about 1,000 times the mass of Jupiter. How does the sun's acceleration due to Jupiter's attraction compare with Jupiter's acceleration due to the sun's attraction?*

8.9 | Further successes

SG 23

Newton did not stop with the fairly direct demonstrations described so far. In the *Principia*, he showed that his law of universal gravitation could explain other complicated gravitational interactions. Among these were the tides of the sea and the peculiar drift of comets across the sky.

The tides: Knowledge of the tides had been vital to navigators, traders, and explorers through the ages. The *cause* of the tides had remained a mystery despite the studies of such scientists as Galileo. However, by applying the law of gravitation, Newton was

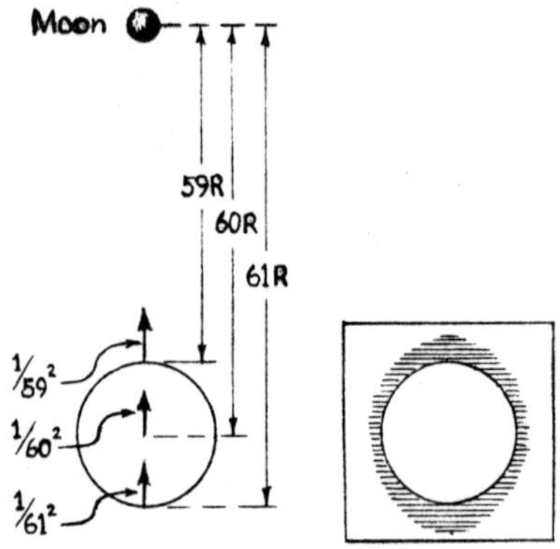

Tidal forces. The earth–moon distance indicated in the figure is greatly reduced because of the space limitations.

able to explain the main features of the ocean tides. He found them to result from the attraction of the moon and the sun upon the waters of the earth. As you can calculate in SG 16, the moon's tide-raising force is greater than the sun's. Each day, two high tides normally occur. Also, twice each month, the moon, sun, and earth are in line with each other. At these times the tidal changes are greater than average.

Two questions about tidal phenomena demand special attention. First, why do high tides occur on both sides of the earth, including the side away from the moon? Second, why does high tide occur at a given location some hours after the moon is highest in the sky?

Newton knew that the gravitational attractions of the moon and sun accelerate the whole solid earth. These forces also accelerate the fluid water at the earth's surface. Newton realized that the tides result from the *difference* in acceleration of the earth and its waters. The moon's distance from the earth's center is 60 earth radii. On the side of the earth nearer the moon, the distance of the water from the moon is only 59 radii. On the side of the earth away from the moon, the water is 61 earth radii from the moon. The accelerations are shown in the figure at the left. On the side nearer the moon, the acceleration of the water toward the moon is greater than the acceleration of the earth as a whole. The net effect is that the water is accelerated away from the earth. On the side of the earth away from the moon, the acceleration of the water toward the moon is less than that of the earth as a whole. The net result is that the earth is accelerated away from the water there.

Perhaps you have watched the tides change at the seashore or examined tide tables. If so, you know that high tide occurs some hours *after* the moon is highest in the sky. To understand this, even qualitatively, you must remember that on the whole the oceans are not very deep. The ocean waters moving in from more distant parts of the oceans in response to the moon's attraction are slowed by friction with the ocean floors, especially in shallow water. Thus, the time of high tide is delayed. In any particular place, the amount of delay and the height of the tides depends greatly upon how easily the waters can flow. No general theory can account for all the particular details of the tides. Most local predictions in the tide tables are based on empirical rules using the tidal patterns recorded in the past.

Since there are tides in the seas, you may wonder if the atmosphere and the earth itself undergo tides. They do. The earth is not completely rigid, but bends somewhat, like steel. The tide in the earth is about 30 cm high. The atmospheric tides are generally masked by other weather changes. However, at altitudes of about 160 km, satellites have recorded considerable rises and falls in the thin atmosphere.

Comets: From earliest history through the Middle Ages, comets have been interpreted as omens of disaster. Halley and Newton showed them to be only shiny, cloudy masses moving around the sun according to Kepler's laws, just as planets do. They found that most comets are visible only when closer to the sun than the distance of Jupiter. Several very bright comets have orbits that take them well inside the orbit of Mercury. Such comets pass within a few million kilometers of the sun, as the figure at the right indicates. Many orbits have eccentricities near 1.0 and are almost parabolas; these comets have periods of thousands or even millions of years. Some other faint comets have periods of only 5-10 years.

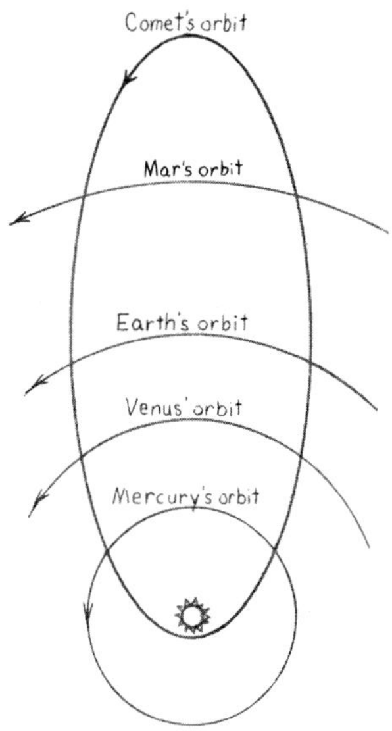

Schematic diagram of the orbit of a comet projected onto the ecliptic plane; comet orbits are tilted at all angles.

Unlike the planets, all of whose orbits lie nearly in a single plane, the planes of comet orbits tilt at all angles. Yet, like all members of the solar system, they obey all the laws of dynamics, including the law of universal gravitation.

Edmund Halley (1656–1742) applied Newton's concepts of celestial motion to the motion of bright comets. Among the comets he studied where those seen in 1531, 1607, and 1682. Halley found the orbits for these comets to be very nearly the same. He suspected that they might be *one* comet, moving in a closed orbit with a period of about 75 years. He predicted that the comet would return in about 1757, which it did, although Halley did not live to see it. Halley's comet appeared in 1833 and 1909 and is due to be near the sun and visible again in 1985–86.

SG 24

With the period of this bright comet known, its dates of appearance could be tracked back in history. Ancient Indian, Chinese, and Japanese documents record all expected appearances except one since 240 B.C. Almost no European records of this great comet exist. This is a sad comment upon the level of culture in Europe during the so-called Dark Ages. One of the few European records is the famous Bayeux tapestry, embroidered with 72 scenes of the Norman Conquest of England in 1066. One scene shows the comet overhead while King Harold of England and his court cower below. A major triumph of Newtonian science was its explanation of comets. Now they were seen to be regular members of the solar system, instead of unpredictable, fearful events.

SG 25

The scope of the principle of universal gravitation: Newton applied the law of universal gravitation to many other problems which cannot be considered in detail here. For example, he investigated the causes of the somewhat irregular motion of the moon. He showed that these motions are explained by the gravitational forces acting on the moon. As the moon moves around the earth, the moon's distance from the sun changes continually. This changes the resultant force of the earth and the sun on the orbiting moon. Newton also showed that other changes in the moon's motion occur because the earth is not a

A scene from the Bayeux tapestry, which was embroidered about 1070. The bright comet of 1066 can be seen at the top of the figure. This comet was later identified as Halley's comet. At the right, Harold, pretender to the throne of England, is warned that the comet is an ill omen. Later that year, at the Battle of Hastings, Harold was defeated by William the Conqueror.

perfect sphere. (The earth's diameter at the equator is 43.2 km greater than the diameter through the poles.) On the problem of the moon's motion Newton commented that "the calculation of this motion is difficult." Even so, he obtained predicted values reasonably close to the observed values available at that time. He even predicted some details of the motion which had not been noticed before.

Newton investigated the variations of gravity at different latitudes on the spinning and bulging earth. He noted differences in the rates at which pendulums swing at different latitudes. From these data, he derived an approximate shape for the earth.

In short, Newton created a whole new quantitative approach to the study of astronomical motion. Because some of his predicted variations had not been observed, improved instruments were built. These instruments improved the old observations that had been fitted together under the grand theory. Many new theoretical problems also clamored for attention. For example, what were the predicted and observed influences among the planets themselves upon their motions? Although the planets are small compared to the sun and are very far apart, their interactions are observable. As precise data accumulated, the Newtonian theory permitted calculations about the past and future of the planetary system. For past or future intervals beyond some hundreds of billions of years, such extrapolations become too uncertain. But for shorter intervals, Newtonian theory says that the planetary system has been and will continue to be about as it is now.

Newton's greatness went beyond the scope and genius of his work in mechanics. It went beyond the originality and elegance of his proofs. It had another dimension: the astonishing detail in which he developed the full meaning of each of his ideas. Sure of his law of universal gravitation, Newton applied it successfully to a vast range of terrestrial and celestial problems. As a result, the theory became more and more widely accepted. Newton's theory has been the chief tool for solving all of the new problems concerning motion in the solar system. For example, the motion of every artificial satellite and space probe is calculated according to Newton's law of universal gravitation. You can well agree with the reply given to ground control as Apollo 8 returned from the first trip to the moon. Ground control : "Who's driving up there?" Captain of Apollo 8: "I think Isaac Newton is doing most of the driving right now."

Tiny variations from a $1/R^2$ centripetal acceleration of satellites in orbit around the moon have led to a mapping of "mascons" on the moon. Mascons are unusually dense concentrations of mass under the surface.

Beyond the solar system: You have seen how Newton's laws explain motions and other physical events on the earth and in the solar system. Now consider a new and even broader question: Do Newton's laws also apply at greater distances, for example, among the stars?

Over the years following publication of the *Principia*, several sets of observations provided an answer to this important question. One observer was William Herschel, a British musician turned amateur astronomer. In the late 1700's, with the help of his sister Caroline, Herschel made a remarkable series of observations. Using homemade, high-quality telescopes, Herschel hoped to measure the parallax of stars due to the earth's motion around the sun. Occasionally he noticed that one star seemed quite close to another. Of course, this might mean only that two stars happened to lie in the same line of sight. But Herschel suspected that some of these pairs were actually double stars held together by their mutual gravitational attractions. He continued to observe the directions and distances from one star to the other in such pairs. In some cases, one star moved during a few years through a small arc of a curved path around the other. (The figure shows the motion of one of the two stars in a system.) Other astronomers gathered more information about these double stars, far removed from the sun and planets. Eventually, it was clear that they move around each other according to Kepler's laws. Therefore, their motions also agree with Newton's law of universal gravitation. Using the same equation as that used for planets (see page 228), astronomers have calculated the masses of these stars. They range from about 0.1 to 50 times the sun's mass.

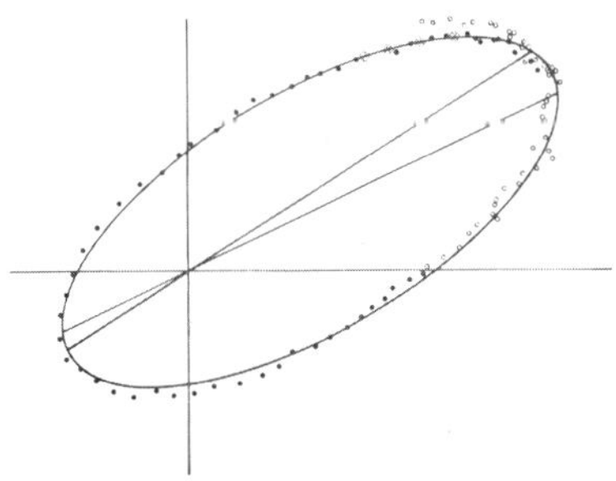

The motion over many years for one of the two components of a binary star system. Each circle indicates the average of observations made over an entire year.

A theory can never be completely proven. But theories become increasingly acceptable as they are found useful over a wider and wider range of problems. No theory has stood this test better than Newton's theory of universal gravitation as applied to the planetary system. It took nearly a century for physicists and astronomers to comprehend, verify, and extend Newton's work on planetary motion. As late as the nineteenth century, most of what had been accomplished in mechanics since Newton's day was but a development or application of his work.

?

29. How does the moon cause the water level to rise on both *sides of the earth?*

30. In which of the following does the moon produce tides? (a) *the seas* (b) *the atmosphere* (c) *the solid earth*

31. Why is the calculation of the moon's motion so difficult?

32. How are the orbits of comets different from the orbits of the planets?

33. Do these differences affect the validity of Newton's law of universal gravitation for comets?

8.10 | Some effects and limitations of Newton's work

Today Newton and his system of mechanics are honored for many reasons. The *Principia* formed the basis for the development of much of our physics and technology. Also, the success of Newton's approach made it the model for all the physical sciences for the next 2 centuries.

Throughout Newton's work, you will find his basic belief that celestial phenomena can be explained by applying quantitative earthly laws. Newton felt that his laws had real physical meaning, that they were not just mathematical conveniences behind which unknowable laws lay hidden. The natural physical laws governing the universe *could* be known. The simple mathematical forms of the laws were evidence of their reality.

Newton combined the skills and approaches of both the experimental and the theoretical scientist. He invented pieces of equipment, such as the first reflecting telescope. He performed skillful experiments, especially in optics. Yet he also applied his great mathematical and logical powers to the creation of specific, testable predictions.

Many of the concepts that Newton used came from earlier scientists and those of his own time. Galileo and Descartes had contributed the first steps leading to a proper idea of inertia, which became Newton's first law of motion. Kepler's planetary laws were central in Newton's consideration of planetary motions. Huygens, Hooke, and others clarified the concepts of force and acceleration, ideas that had been evolving for centuries.

In addition to his own experiments, Newton selected and used data from a large number of sources. Tycho Brahe was only one of several astronomers whose observations of the motion of the moon he used. When Newton could not complete his own measurements, he knew whom he could ask.

Last, recall how completely and how fruitfully he used and expanded his own specific contributions. A good example is his theory of universal gravitation. In developing it, Newton used his laws of motion and his various mathematical inventions again and again. Yet Newton was modest about his achievements. He once said that if he had seen further than others "it was by standing upon the shoulders of Giants."

Scientists recognize today that Newton's mechanics hold true only within a well-defined region of science. For example, the forces within each galaxy appear to be Newtonian. But this may not be true for forces acting between one galaxy and another. At the other end of the scale are atoms and subatomic particles. Entirely non-Newtonian concepts had to be developed to explain the observed motions of these particles.

Even within the solar system, there are several small differences between the predictions and the observations. The most famous involves the angular motion of the axis of Mercury's orbit. This motion is greater than the value predicted from Newton's laws by about 1/80° per century. What causes this difference? For a while, it was thought that gravitational force might not vary inversely *exactly* with the square of the distance. Perhaps, for example, the law is $F_{grav} = 1/R^{2.000001}$.

Such difficulties should not be hastily assigned to some minor mathematical imperfection. The law of gravitation applies with unquestionable accuracy to all other planetary motions. It may be that the basic assumptions in the theory make it too limited, as with the Ptolemaic system of epicycles. Many studies have shown that there is no way to modify the details of Newtonian mechanics to explain certain observations. Instead, these observations can be accounted for only by constructing *new* theories based on some very different assumptions. The predictions from these theories are almost identical to those from Newton's laws for familiar phenomena. But they are also accurate in some extreme cases where the Newtonian predictions begin to show inaccuracies. Thus, Newtonian science is linked at one end with *relativity theory*, which is important for bodies with very great mass or moving at very high speeds. At the other end Newtonian science approaches *quantum mechanics*, which is important for particles of extremely small mass and size, for example, atoms, molecules, and nuclear particles. For a vast range of problems between these extremes, Newtonian theory gives accurate results and is far simpler to use. Moreover, it is in Newtonian mechanics that relativity theory and quantum mechanics have their roots.

Newtonian mechanics refers to the science of the motion of bodies, based on Newton's work. It includes his laws of motion and of gravitation as applied to a range of bodies from microscopic size to stars and incorporates developments of mechanics for over two centuries after Newton's own work.

SG 26

SG 29, 30

study guide

1. The *Project Physics* learning materials particularly appropriate for Chapter 8 include:

Experiments
Stepwise Approximation to an Orbit
Model of the Orbit of Halley's comet

Activities
Other Comet Orbits
Forces on a Pendulum
Trial of Copernicus
Discovery of Neptune and Pluto

Film Loops
Jupiter Satellite Orbit
Program Orbit I
Program Orbit II
Central Forces: Iterated Blows
Kepler's Laws
Unusual Orbits

2. Complete the following statements:

(a) Newton believed that the natural path of any moving object in the absence of forces is

____________________ .

(b) Since Kepler's first law claims that planets travel along elliptical paths, Newton hypothesized a force that ____________________ .

(c) Newton then discovered that if the motion of the planets follows Kepler's law of areas, this force must be ____________________

(d) Newton also discovered that planets that obey Kepler's harmonic law of periods require a force that

____________________.

3. **(a)** People have always known that apples fall to the ground. What particular thing about the fall of an apple led Newton to compare the apple's fall to the motion of the moon?

(b) If the moon is 60 times farther from the center of the earth than an apple, what is the moon's acceleration toward the earth's center?

(c) Use the formula for centripetal acceleration to find the acceleration of the moon in its orbit. Compare this value to your answer in part **(b)**.

4. How would you answer the following question: What keeps the moon up?

5. State the law of universal gravitation in words and symbols, defining each symbol. What is the direction of the force of gravity? Give the value of G. Does this value ever change?

6. In the table below are the periods and distances from Jupiter of the four large satellites, as measured by telescopic observations. Does Kepler's law of periods apply to the Jupiter system?

Satellite	*Period*	*Distance from Jupiter's Center (in terms of Jupiter's radius, r)*
I	1.77 days	6.04r
II	3.55	9.62
III	7.15	15.3
IV	16.7	27.0

7. Give some reasons why Descartes' theory of planetary motion might have been "a useful idea."

8. On p. 225 it was claimed that the dependence of the gravitational force on the masses of both interacting bodies could be expressed as $m_{sun}m_{planet}$.

(a) Using a diagram similar to that for Question 19 on p. 227, show that this is correct.

(b) To test alternatives to using the product, consider the possibilities that the force could depend upon the masses in either of two ways:

(1) total force depends on $(m_{sun} + m_{planet})$, or

(2) total force depends on (m_{sun}/m_{planet}).

With these relationships, what would happen to the force if either mass were reduced to zero? Would there still be a force even though there were only one mass left? Could you speak of a gravitational force when there was no body to be accelerated?

9. Use the values for the mass and size of the moon to show that the "surface gravity" (acceleration due to gravity near the moon's surface) is only about 0.16 of what it is near the earth's. (Mass of moon = 7.34×10^{22} kg; radius of moon = 1.74×10^{6} m.)

10. Use the equation for centripetal force and the equation for gravitational force to derive an expression for the period of a satellite orbiting a planet in terms of the radius of the orbit and mass of the planet.

11. Using the formula for the gravitational force between two objects, calculate the force between a 100-kg sphere and a 1,000-kg sphere placed 10 m apart. What are the accelerations of the two spheres? Could these accelerations ever be measured?

12. Why were the discoveries of Neptune and Pluto triumphs for Newton's theory?

13. By Newton's time, telescopic observations of Jupiter led to values for the orbital periods and radii of Jupiter's four large satellites. For example, the one named Callisto was found to have a period of 16.7 days, and the radius of its orbit was calculated as 1/80 AU.

(a) From these data calculate the value of k_{Jupiter}. (First convert days to years.)

(b) Show that Jupiter's mass is about 1/1000 the mass of the sun.

(c) How was it possible to have a value for the orbital radius of a satellite of Jupiter?

14. What orbital radius must an earth satellite be given to keep it always above the same place on the earth, that is, in order to have a 24-hr period? (Hint: See Question 10.)

15. Calculate the mass of the earth from the fact that a 1-kg object at the earth's surface is attracted to the earth with a force of 9.8 N. The distance from the earth's center to its surface is 6.4×10^6 m. How many times greater is this than the greatest masses that you have had some experience in accelerating (for example, cars)?

16. The mass of the earth can be calculated also from the distance and period of the moon. Show that the value obtained in this way agrees with the value calculated from measurements at the earth's surface.

17. If tides are caused by the pull of the moon, why are there also tides on the side of the earth opposite the moon?

18. The moon's orbit around the earth is a combination of two separate motions: a straight line and a fall toward the earth's center. Using diagrams, discuss each part of the moon's motion.

19. Cavendish's value for *G* made it possible to calculate the mass of the earth and therefore its average density. The "density" of water is 1,000 kg/m^3; that is, for any sample of water, dividing the mass of the sample by its volume gives 1,000 kg/m^3.

(a) What is the earth's average density?

(b) The densest kind of rock known has a density of about 5,000 kg/m^3. Most rock has a density of about 3,000 kg/m^3. What do you conclude from this about the internal structure of the earth?

20. The manned Apollo 8 capsule was put into a nearly circular orbit 112 km above the moon's surface. The period of the orbit was 120.5 min. From these data, calculate the mass of the moon. (The radius of the moon is 1,740 km. Use a consistent set of units.)

21. Mars has two satellites, Phobos and Deimos (Fear and Panic). A science-fiction story was once written in which the natives of Mars showed great respect for a groove in the ground. The groove turned out to be the orbit of Mars' closest moon "Bottomos."

(a) If such an orbit were possible, what would the period be?

(b) What speed would it need to have in order to go into such an orbit?

(c) What would you expect to happen to an object in such an orbit?

22. Using the values given in the table on p. 230, make a table of relative masses compared to the mass of the earth.

23. The sun's mass is about 27,000,000 times greater than the moon's mass; the sun is about 400 times farther from the earth than the moon is. How does the gravitational force exerted on the earth by the sun compare with that exerted by the moon?

24. The period of Halley's comet is about 75 yr. What is its average distance from the sun? The eccentricity of its orbit is 0.967. How far from the sun does it go? How close does it come to the sun?

25. Accepting the validity of $F_{grav} = Gm_1m_2/R^2$ and recognizing that G is a universal constant, we are able to derive, and therefore to understand better, many particulars that previously seemed separate. For example, we can conclude:

(a) that a_g for a body of any mass m_o should be constant at a particular place on earth.

(b) that a_g might be different at places on earth at different distances from the earth's center.

(c) that at the earth's surface the weight of a body is related to its mass.

(d) that the ratio R^3/T^2 is a constant for all the satellites of a body.

(e) that tides occur about 6 hr apart.

Describe briefly how each of these conclusions can be derived from the equation.

26. The making of theories to account for observations is a major purpose of scientific study. Therefore, some reflection upon the theories encountered thus far in this course will be useful. Comment in a paragraph or more, with examples from Units 1 and 2, on some of the statements below. Look at all the statements and select at least six, in any order you wish.

(a) A good theory should summarize and not conflict with a body of tested observations. (Take, for example, Kepler's unwillingness to explain away the difference of 8 min of arc between his predictions and Tycho's observations.)

(b) There is nothing more practical than a good theory.

(c) A good theory should permit predictions of new observations which sooner or later can be made.

(d) A good new theory should give almost the same predictions as older theories for the range of phenomena where the older theories worked well.

(e) Every theory involves assumptions. Some involve also esthetic preferences of the scientist.

(f) A new theory relates some previously unrelated observations.

(g) Theories often involve abstract concepts derived from observation.

(h) Empirical laws or "rules" organize many observations and reveal how changes in one quantity vary with changes in another, but such laws provide no explanation of the causes or mechanisms.

(i) A theory never fits all data exactly.

(j) Predictions from theories may lead to the observation of new effects.

(k) Theories that later had to be discarded may have been useful because they encouraged new observations.

(l) Theories that permit quantitative predictions are preferred to qualitative theories.

(m) An "unwritten text" lies behind the statement of every law of nature.

(n) Communication between scientists is an essential part of the way science grows.

(o) Some theories seem initially so strange that they are rejected completely or accepted only very slowly.

(p) Models are often used in the making of a theory or in describing a theory.

(q) The power of theories comes from their generality.

27. What happened to Plato's problem? Was it solved?

28. Why do we believe today in a heliocentric system? Is it the same as either Copernicus' or Kepler's? What is the experimental evidence? Has the geocentric system been disproved?

29. Is Newton's work only of historical interest, or is it useful today? Explain.

30. What were some of the major consequences of Newton's work on scientists' views of the world?

EPILOGUE

This unit started at the beginning of recorded history and followed human attempts to explain the cyclic motions observed in the heavens. We saw the long, gradual change from an earth-centered view to the modern one in which the earth is just another planet moving around the sun. We examined some of the difficulties encountered in making this change of viewpoint. We also tried to put into perspective Newton's synthesis of earthly and heavenly motions. From time to time, we suggested that there was an interaction of these new world views with the general culture. We stressed that all scientists are products of their times. They are limited in the degree to which they can abandon the teachings on which they were raised. Gradually, through the work of many scientists over the centuries, a new way of looking at heavenly motions arose. This in turn opened new possibilities for even more new ideas, and the end is not in sight.

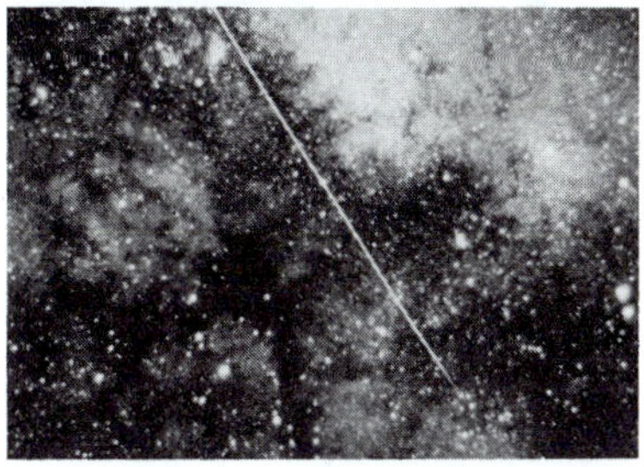

In addition, we looked at how theories are made and tested. We discussed the place of assumption and experiment, of mechanical models and mathematical description. In later parts of the course, you will come back to this discussion in more recent contexts. You will find that attitudes developed toward theory-making during the seventeenth-century scientific revolution remain immensely helpful today.

In our study, we have referred to scientists in Greece, Egypt, Poland, Denmark, Austria, Italy, England, and other countries. Each, as Newton said of himself, stood on the shoulders of those who came earlier. For each major success there are many lesser advances or, indeed, failures. Science is a cumulative intellectual activity not restricted by national boundaries or by time. It is not constantly and unfailingly successful, but grows as a forest grows. New growth replaces and draws nourishment from the old, sometimes with unexpected changes in the different parts. Science is not a cold, calculated pursuit. It may involve passionate controversy, religious convictions, judgments of what beauty is, and sometimes wild private speculation.

It is also clear that the Newtonian synthesis opened whole new lines of investigation, both theoretical and observational. In fact, much of our present science and also our technology had their effective beginnings with the work of Newton. New models, new mathematical tools, and a new self-confidence encouraged those who followed to attack new problems. A never-ending series of questions, answers, and more questions was well launched. The modern view of science is that it is a continuing exploration of ever more interesting fields.

One problem remaining after Newton's work was the study of objects interacting not by gravitational forces, but by friction and collision. This study led, as the next unit will show, to the concepts of momentum and energy. It brought about a much broader view of the connection between different parts of science, such as physics, chemistry, and biology. Eventually, this line of study produced other statements as grand as Newton's law of universal gravitation. Among them were the conservation laws on which much of modern science and technology is based. An important part of these laws describes how systems consisting of many interacting bodies work. That account will be the main subject of Unit 3.

The engraving of the French Academy by Sébastian LeClerc (1698) reflects the activity of learned societies at that time. The picture does not depict an actual scene, of course, but in allegory shows the excitement of communication that grew in an informal atmosphere. The dress is symbolic of the Greek heritage of the sciences. Although all the sciences are represented, the artist has put anatomy, botany, and zoology, symbolized by skeletons and dried leaves, toward the edges, along with alchemy and theology. Mathematics and the physical sciences, including astronomy, occupy the center stage.

Newton's influence was not limited to science alone. The century following his death in 1727 was a period of further understanding and application of his discoveries and methods. His influence was felt especially in philosophy and literature, but also in many other fields outside science. Let us round out our view of Newton by considering some of these effects.

The eighteenth century is often called the Age of Reason or Century of Enlightenment. "Reason" was the motto of the eighteenth-century philosophers. However, their theories about improving religion and society were not convincingly connected. Newtonian physics, religious toleration, and republican government were all advanced by the same movement. This does not mean there was really a logical link among these concepts. Nor were many eighteenth-century thinkers in any field or nation much bothered by other gaps in logic and feeling. For example, they believed that "all men are created equal." Yet they did little

to remove the chains of black slaves, the ghetto walls imprisoning Jews, or the laws that denied rights to women.

Still, compared with the previous century, the dominant theme of the eighteenth century was *moderation*, the "happy medium." The emphasis was on toleration of different opinions, restraint of excess, and balance of opposing forces. Even reason was not allowed to question religious faith too strongly. Atheism, which some philosophers thought would logically result from unlimited rationality, was still regarded with horror by most Europeans.

The Constitution of the United States of America is one of the most enduring achievements of this period. Its system of "checks and balances" was designed specifically to prevent any one group from getting too much power. It attempted to establish in politics a state of equilibrium of opposing trends. This equilibrium, some thought, resembled the balance between the sun's gravitational pull and the tendency of a planet to fly off in a straight line. If the gravitational attraction upon the planet increased without a corresponding increase in planetary speed, the planet would fall into the sun. If the planet's speed increased without a corresponding increase in gravitational attraction, it would escape from the solar system.

Political philosophers, some of whom used Newtonian physics as a model, hoped to create a similar balance in government. They tried to devise a system that would avoid the extremes of dictatorship and anarchy. According to James Wilson (1742–1798), who played a major role in writing the American Constitution,

> In government, the perfection of the whole depends on the balance of the parts, and the balance of the parts consists in the independent exercise of their separate powers, and, when their powers are separately exercised, then in their mutual influence and operation on one another. Each part acts and is acted upon, supports and is supported, regulates and is regulated by the rest. It might be supposed, that these powers, thus mutually checked and controlled, would remain in a state of inaction. But there is a necessity for movement in human affairs; and these powers are forced to move, though still to move in concert. They move, indeed, in a line of direction somewhat different from that, which each acting by itself would have taken; but, at the same time, in a line partaking of the natural directions of the whole—the true line of public liberty and happiness.

Both Newton's life and his writings seemed to support the idea of political democracy. A former farm boy had penetrated to the outermost reaches of the human imagination. What he had found there meant, first of all, that only one set of laws governed heaven and earth. This smashed the old beliefs about "natural place" and extended a new democracy throughout the universe. Newton had shown that all matter, whether the sun or an ordinary stone, was created equal; that is to say, all matter had the same standing before "the Laws of Nature and of Nature's God." (This phrase was used at the beginning of the Declaration of Independence to justify the desire of the people in the colonies to throw off their oppressive political system and to become an independent people.) All political thought at this time was heavily influenced by Newtonian ideas. The *Principia* seemed to offer a parallel to theories about democracy. It seemed logical that all people, like all natural objects, are created equal before nature's creator.

In literature, too, many welcomed the new scientific viewpoint. It supplied many new ideas, convenient figures of speech, parallels, and concepts which writers used in poems and essays. Newton's discovery that white light is composed of colors was referred to in many poems of the 1700's. (See Unit 4.) Samuel Johnson advocated that words drawn from the vocabulary of the natural sciences be used in literary works. He defined many such words in his *Dictionary* and illustrated their application in his "Rambler" essays.

The first really powerful reaction against Newtonian cosmology was the Romantic movement. Romanticism was started in Germany about 1780 by young writers inspired by Johann Wolfgang von Goethe. The most familiar examples of Romanticism in English literature are the poems and novels of Blake, Coleridge, Wordsworth, Shelley, Byron, and Scott. The Romantics scorned the mathematical view of nature. They believed that any whole thing, whether a single human being or the entire universe, is filled with a unique spirit. This spirit cannot be explained by reason; it can only be *felt*. The Romantics insisted that phenomena cannot be meaningfully analyzed and reduced to their separate parts by mechanical explanations.

The Romantic philosophers in Germany regarded Goethe as their greatest scientist as well as their greatest poet. They pointed in particular to his theory of color, which flatly contradicted Newton's theory of light. Goethe held that white light does not consist of a mixture of colors and that it is useless to "reduce" a beam of white light by passing it through a prism to study its separate spectral colors. Rather, he charged, the colors of the spectrum are artificially produced by the prism, acting on and changing the light which is itself pure.

In the judgment of all modern scientists, Newton was right and Goethe wrong. This does not mean that Nature Philosophy, introduced by Friedrich Schelling in the early 1800's, was without any value. It encouraged speculation about ideas so general that they could not be easily tested by experiment. At the time, it was condemned by most scientists for just this reason. Today, most historians of science agree that Nature Philosophy eventually played an important role in making possible certain scientific discoveries. Among these was the general principle of conservation of energy, which is described in Chapter 10. This principle asserted that all the "forces of nature," that is, the phenomena of heat, gravity, electricity, magnetism, and so forth, are forms of one underlying "force" (which we now call energy). This idea agreed well with the viewpoint of Nature Philosophy. It also could eventually be put in a scientifically acceptable form.

Some modern artists, some intellectuals, and most members of the "counterculture" movements express a dislike for science. Their reasoning is similar to that of the Romantics. It is based on the mistaken notion that scientists claim to be able to find a mechanical explanation for *everything*.

Even the Roman philosopher Lucretius (100–55 B.C.), who supported the atomic theory in his poem *On the Nature of Things*, did not go this far. To preserve some trace of "free will" in the universe, Lucretius suggested that atoms might swerve randomly in their paths. This was not enough for Romantics and also for some scientists. For example, Erasmus Darwin, a scientist and grandfather of evolutionist Charles Darwin, asked

> Dull atheist, could a giddy dance
>   Of atoms lawless hurl'd
> Construct so wonderful, so wise,
>   So harmonised a world?

The Nature Philosophers thought they could discredit the Newtonian scientists by forcing them to answer this question. To say "yes," they argued, would be absurd, and to say "no" would be disloyal to Newtonian beliefs. But the Newtonians succeeded quite well without committing themselves to any definite answer to Erasmus Darwin's question. They went on to discover immensely powerful and valuable laws of nature, which are discussed in the next units.

UNIT

The Triumph of Mechanics

PROLOGUE

Isaac Newton's development of mathematical principles of natural philosophy marks a turning point in the growth of human knowledge. The simple, elegant laws that rule not only terrestrial but also astronomical phenomena were soon recognized as the premier example of the power of human reason. The fact that Newton had done so much to show that there is a rational order in natural events made it seem possible that any problem could be solved by reason. It is not surprising that after his death in 1727 Newton was looked upon almost as a god, especially in England. Many poems like this one appeared:

Newton the unparalleld'd, whose Name
No Time will wear out of the Book of Fame,
Celestial Science has promoted more,
Than all the Sages that have shone before.
Nature compell'd his piercing Mind obeys,
And gladly shows him all her secret Ways;
'Gainst Mathematics she has no defence,
And yields t' experimental Consequence;

His tow'ring Genius, from its certain Cause
Ev'ry Appearance *a priori* draws
And shews th' Almighty Architect's unalter'd Laws.

(From J. T. Desagulier, *The Newtonian System of the World, the Best Model of Government, an Allegorical Poem.*)

Newton's success in mechanics altered profoundly the way in which scientists viewed the universe. The motions of the sun and planets could now be considered as purely mechanical. As for any machine, whether a clock or the solar system, the motions of the parts were completely determined once the system had been put together.

This model of the solar system is called the *Newtonian world machine.* As is true of any model, certain things are left out. The mathematical equations that govern the motions of the model cover only the main properties of the real solar system. The masses, positions, and velocities of the parts of the system, and the gravitational forces among them, are well described. But the Newtonian model neglects the internal structure and chemical composition of the planets, heat, light, and electric and magnetic forces. Nevertheless, it serves splendidly to deal with observed motions. Moreover, Newton's approach to science and many of his concepts became useful later in the study of those aspects he had to leave aside.

The idea of a world machine does not trace back only to Newton's work. In *Principles of Philosophy* (1644), René Descartes, the most influential French philosopher of the seventeenth century, had written:

> I do not recognize any difference between the machines that artisans make and the different bodies that nature alone composes, unless it be that the effects of the machines depend only upon the adjustment of certain tubes or springs, or other instruments, that, having necessarily some proportion with the hands of those who make them, are always so large that their shapes and motions can be seen, while the tubes and springs that cause the effects of natural bodies are ordinarily too small to be perceived by our senses. And it is certain that all the laws of Mechanics belong to Physics, so that all the things that are artificial, are at the same time natural.

Robert Boyle (1627–1691), a British scientist, is known particularly for his studies of the properties of air. (See Chapter 11.) Boyle, a pious man, expressed the "mechanistic" viewpoint even in his religious writings. He argued that a God who could design a universe that ran by itself like a machine was more wonderful than a God who simply created several different kinds of matter and gave each a natural tendency to behave as it does. Boyle also thought it was insulting to God to believe that the world machine would be so badly designed as to require any further divine adjustment once it had been created. He suggested that an engineer's skill in designing "an elaborate engine" is

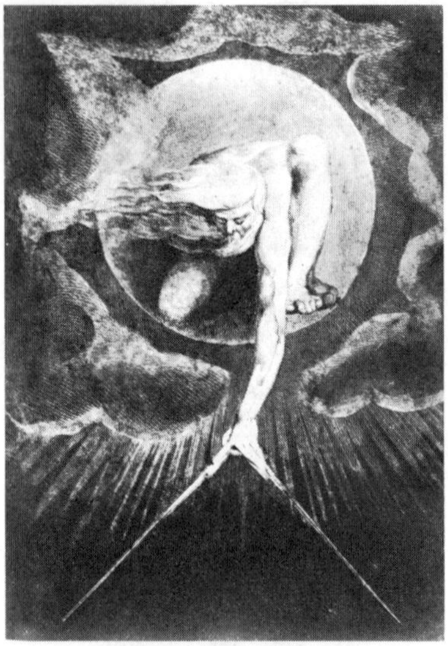

"The Ancient of Days" by William Blake, an English poet who had little sympathy with the Newtonian style of "Natural Philosophy."

Ironically, Newton himself explicitly rejected the deterministic aspects of the "world machine" which his followers had popularized.

A small area from the center of the picture has been enlarged to show what the picture is "really" like. Is the picture only a collection of dots? Knowing the underlying structure does not spoil your other reactions to the picture, but rather gives you another dimension of understanding it.

more deserving of praise if the engine never needs supervision or repair. "Just so," he continued,

> . . . it more sets off the wisdom of God in the fabric of the universe, that He can make so vast a machine perform all those many things, which He designed it should, by the meer contrivance of brute matter managed by certain laws of local motion, and upheld by His ordinary and general concourse, than if He employed from time to time an intelligent overseer, such as nature is fancied to be, to regulate, assist, and controul the motions of the parts. . . .

Boyle and many other scientists in the seventeenth and eighteenth centuries tended to think of God as a supreme engineer and physicist. God had set down the laws of matter and motion. Human scientists could best glorify the Creator by discovering and proclaiming these laws.

This unit is mainly concerned with physics as it developed after Newton. In mechanics, Newton's theory was extended to cover a wide range of phenomena, and new concepts were introduced. The conservation laws discussed in Chapters 9 and 10 became increasingly important. These powerful principles offered a new way of thinking about mechanics. They opened up new areas to the study of physics, for example, heat and wave motion.

Newtonian mechanics treated directly only a small range of experiences. It dealt with the motion of simple bodies or those largely isolated from others, as are planets, projectiles, or sliding discs. Do the same laws work when applied to complex phenomena? Do real solids, liquids, and gases behave like machines or mechanical systems? Can their behavior be explained by using the same ideas about matter and motion that Newton used to explain the solar system?

At first, it might seem unlikely that everything can be reduced to matter and motion, the principles of mechanics. What about temperature, colors, sounds, odors, hardness, and so forth? Newton himself believed that the mechanical view would essentially show how to investigate these and all other properties. In the preface to the *Principia* he wrote:

> I wish we could derive the rest of the phenomena of Nature by the same kind of reasoning from mechanical principles, for I am induced by many reasons to suspect that they may all depend upon certain forces by which the particles of bodies, by some causes hitherto unknown, are mutually impelled towards one another, and cohere according to regular figures, or are repelled and recede from one another. These forces being unknown, philosophers have hitherto attempted the search of Nature in vain; but I hope the principles here laid down will afford some light either to this or some truer method of Philosophy.

Conservation of Mass and Momentum

9.1 Conservation of mass
9.2 Collisions
9.3 Conservation of momentum
9.4 Momentum and Newton's laws of motion
9.5 Isolated systems
9.6 Elastic collisions
9.7 Leibniz and the conservation law

9.1 | Conservation of mass

The idea that despite ever-present, obvious change all around us the total amount of material in the universe does not change is really very old. The Roman poet Lucretius restated (in the first century B.C.) a belief held in Greece as early as the fifth century B.C.:

SG 1

> . . . and no force can change the sum of things; for there is no thing outside, either into which any kind of matter can emerge out of the universe or out of which a new supply can arise and burst into the universe and change all the nature of things and alter their motions. [*On the Nature of Things*]

Just 24 years before Newton's birth, the English philosopher Francis Bacon included the following among his basic principles of modern science in *Novum Organum* (1620):

> There is nothing more true in nature than the twin propositions that "nothing is produced from nothing" and

"nothing is reduced to nothing" . . . the sum total of matter remains unchanged, without increase or diminution.

This view agrees with everyday observation to some extent. While the form in which matter exists may change, in much of our ordinary experience matter appears somehow indestructible. For example, you may see a large boulder crushed to pebbles and not feel that the amount of matter in the universe has diminished or increased. But what if an object is burned to ashes or dissolved in acid? Does the amount of matter remain unchanged even in such chemical reactions? What of large-scale changes such as the forming of rain clouds or seasonal variations?

To test whether the total quantity of matter actually remains constant, you must know how to measure that quantity. Clearly, it cannot simply be measured by its volume. For example, you might put water in a container, mark the water level, and then freeze the water. If so, you will find that the volume of the ice is greater than the volume of the water you started with. This is true even if you carefully seal the container so that no water can possibly come in from the outside. Similarly, suppose you compress some gas in a closed container. The volume of the gas decreases even though no gas escapes from the container.

Following Newton, we regard the *mass* of an object as the proper measure of the amount of matter it contains. In all the examples in Units 1 and 2, we assumed that the mass of a given object does not change. However, a burnt match has a smaller mass than an unburnt one; an iron nail increases in mass as it rusts. Scientists had long assumed that something escapes from the match into the atmosphere and that something is added from the surroundings to the iron of the nail. Therefore, nothing is really "lost" or "created" in these changes. Not until the end of the eighteenth century was sound experimental evidence for this assumption provided. The French chemist Antoine Lavoisier produced this evidence.

Lavoisier caused chemical reactions to occur in *closed* flasks (a "closed system"). He carefully weighed the flasks and their contents before and after the reaction. For example, he burned iron in a closed flask. The mass of the iron oxide produced equalled the sum of the masses of the iron and oxygen used in the reaction. With experimental evidence like this at hand, he could announce with confidence in *Traité Élémentaire de Chimie* (1789):

We may lay it down as an incontestable axiom that in all the operations of art and nature, nothing is created; an equal quantity of matter exists both before and after the experiment, . . . and nothing takes place beyond changes and modifications in the combinations of these elements. Upon this principle, the whole art of performing chemical experiments depends.

In some open-air chemical reactions, the mass of objects seems to decrease, while in others it seems to increase.

Note the closed flask shown in his portrait on page 252.

SG 2

Lavoisier knew that if he put some material in a well-sealed bottle and measured its mass, he could return at any later time and find the same mass. It would not matter what had happened to the material inside the bottle. It might change from solid to liquid or liquid to gas, change color or consistency, or even undergo violent chemical reactions. At least one thing would remain unchanged: the *total* mass of all the different materials in the bottle.

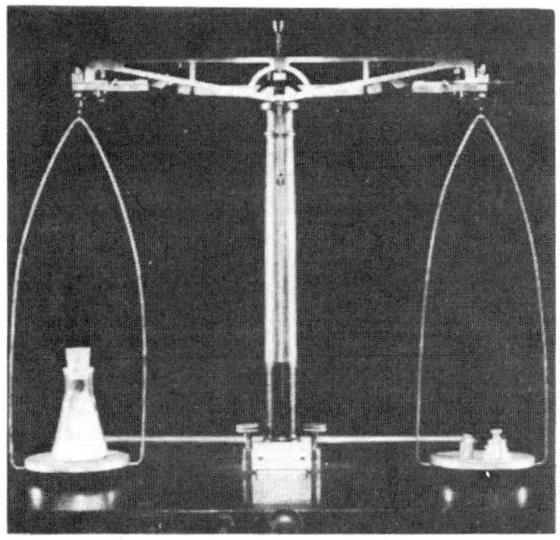

Conservation of mass was demonstrated in experiments on chemical reactions in closed flasks.

In the years after Lavoisier's pioneering work, a vast number of similar experiments were performed with ever-increasing accuracy. The result was always the same. As far as scientists now can measure with sensitive balances (having a precision of better than 0.000001%), mass is *conserved,* that is, remains constant, in chemical reactions.

To sum up, despite changes in location, shape, chemical composition, and so forth, *the mass of any closed system remains constant.* This is the statement of the *law of conservation of mass.* This law is basic to both physics and chemistry.

The meaning of the phrase "closed system" will be discussed in more detail in Sec. 9.5.

"The change in the total mass is zero" can be expressed symbolically as $\Delta\Sigma_i m_i = 0$ where Σ_i represents the sum of the masses of m_i in all parts of the system.

Obviously, one must know whether a given system is closed or not before applying this law to it. For example, it is perhaps surprising that the earth itself is not exactly a closed system within which all mass would be conserved. Rather, the earth, including its atmosphere, gains and loses matter constantly. The most important addition occurs in the form of dust particles. These particles are detected by their impacts on satellites that are outside most of the atmosphere, and by the light and ionization they create when they pass through the atmosphere and are slowed down by it. The great majority of these particles are very small (less than 10^{-6} m in diameter) and cannot be detected individually when they enter the atmosphere. Particles larger than several millimeters in diameter appear as luminous meteorite trails when they vaporize in the upper atmosphere; these particles are only a small fraction of the total, both in terms of numbers and in terms of mass. The total estimated inflow of mass of all these particles, large and small, is about 10^5 g/sec over the whole surface of the earth. (The mass of the earth is about 6×10^{27}g.) This gain is not balanced by any loss of dust or larger particles, not counting an occasional spacecraft and its debris. The earth also collects some of the hot gas evaporating from the sun, but the amount is comparatively small.

Meteorites have been found in all parts of the world. This meteorite fragment was one of several found in the Atacama desert of Chile in 1822.

The earth does lose mass by evaporation of molecules from the top of the atmosphere. The rate of this evaporation depends on how many molecules are near enough to the top of the atmosphere to escape without colliding with other molecules. Also, such molecules must have velocities high enough to escape the earth's gravitational pull. The velocities of the molecules are determined by the temperature of the upper atmosphere. Therefore, the rate of evaporation depends greatly on this

Close Up

The Father of Modern Chemistry

Antoine Laurent Lavoisier (1743–1794) is known as the "father of modern chemistry" because he showed the decisive importance of quantitative measurements, confirmed the principle of conservation of mass in chemical reactions, and helped develop the present system of nomenclature for the chemical elements. He also showed that organic processes such as digestion and respiration are similar to burning.

To earn money for his scientific research, Lavoisier invested in a private company which collected taxes for the French government. Because the tax collectors were allowed to keep any extra tax which they could collect from the public, they became one of the most hated groups in France. Lavoisier was not directly engaged in tax collecting, but he had married the daughter of an important executive of the company, and his association with the company was one of the reasons why Lavoisier was guillotined during the French Revolution.

Also shown in the elegant portrait by J. L. David is Madame Lavoisier. She assisted her husband by taking data, translating scientific works from English into French, and making illustrations. About 10 years after her husband's execution, she married another scientist, Count Rumford, who is remembered for his experiments which cast doubt on the caloric theory of heat.

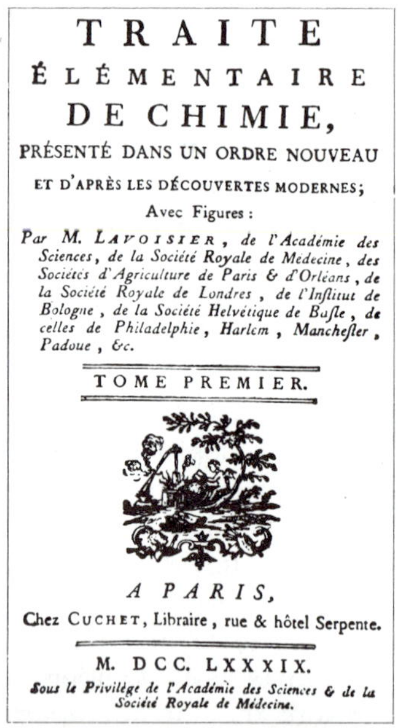

TRAITE
ÉLÉMENTAIRE
DE CHIMIE,
PRÉSENTÉ DANS UN ORDRE NOUVEAU
ET D'APRÈS LES DÉCOUVERTES MODERNES;
Avec Figures :

Par M. LAVOISIER, de l'Académie des Sciences, de la Société Royale de Medecine, des Sociétés d'Agriculture de Paris & d'Orléans, de la Société Royale de Londres, de l'Inſtitut de Bologne, de la Société Helvétique de Baſle, de celles de Philadelphie, Harlem, Mancheſter, Padoue, &c.

TOME PREMIER.

A PARIS,
Chez CUCHET, Libraire, rue & hôtel Serpente.

M. DCC. LXXXIX.
Sous le Privilège de l'Académie des Sciences & de la Société Royale de Médecine.

temperature. At present, the rate is probably less than 5×10^3 g/sec over the whole earth. This loss is very small compared to the addition of dust. (No water molecules are likely to be lost directly by atmospheric "evaporation"; they would first have to be dissociated into hydrogen and oxygen molecules.)

Try these end-of-section questions before going on.

?

1. True or false: *Mass is conserved in a closed system only if there is no chemical reaction in the system.*

2. *If 50 cm^3 of alcohol is mixed with 50 cm^3 of water, the mixture amounts to only 98 cm^3. An instrument pack on the moon weighs much less than on earth. Are these examples of contradictions to the law of conservation of mass?*

SG 3–7

3. *Which one of the following statements is true?*

(a) Lavoisier was the first person to believe that the amount of material in the universe does not change.

(b) Mass is measurably increased when heat enters a system.

(c) A closed system was used to establish the law of conservation of mass experimentally.

4. *Five grams (5 g) of a red fluid at 12° C with a volume of 4 mL are mixed with 10 g of a blue fluid at 5° C with a volume of 8 mL. On the basis of this information* only, *and assuming that the fluids are mixed in a closed system, what can you be sure of about the resulting mixture?*

9.2 | Collisions

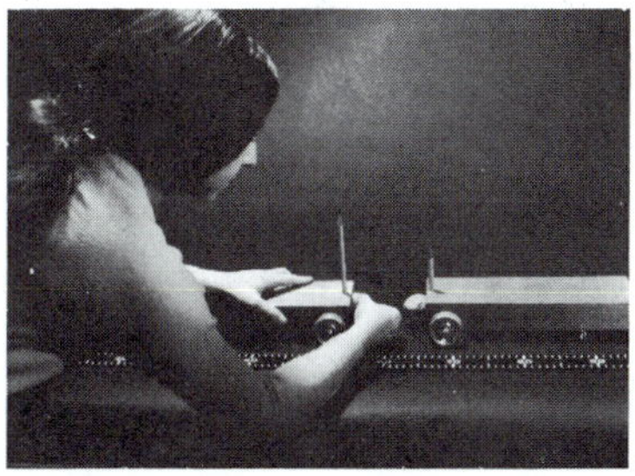

Looking at moving things in the world around us easily leads to the conclusion that everything set in motion eventually stops. Every machine eventually runs down. It appears that the amount of motion in the universe must be decreasing. The universe, like any machine, must be running down.

Many philosophers of the 1600's could not accept the idea of a universe that was running down. The concept clashed with their idea of the perfection of God, who surely would not construct such an imperfect mechanism. Some definition of "motion" was needed that would permit one to make the statement that "the quantity of motion in the universe is constant."

Is there a constant "quantity of motion" that keeps the world machine going? To suggest an answer to this question, you can do some simple laboratory experiments. Use a pair of identical carts with nearly frictionless wheels; even better are two dry-ice discs or two air-track gliders. In the first experiment, a lump of putty is attached so that the carts will stick together when they collide. The carts are each given a push so that they approach each other with equal speeds and collide head-on. As you will

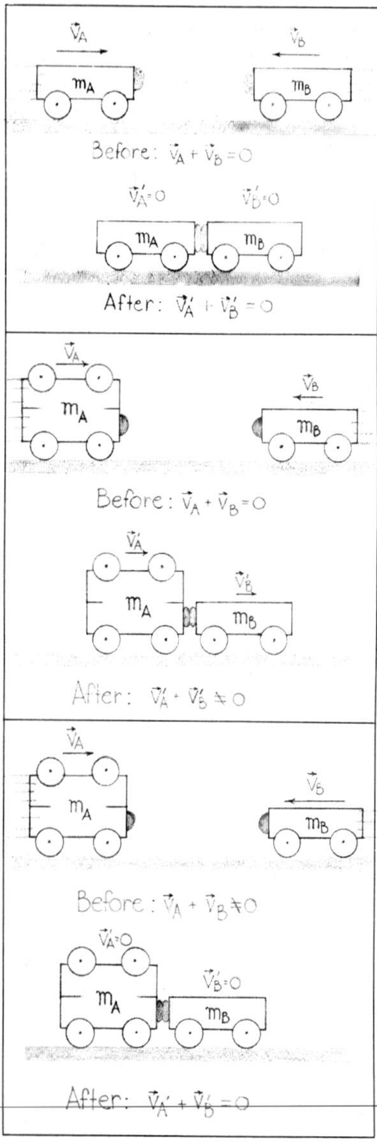

In symbols, $\Delta \Sigma_i v_i = \Delta \Sigma_f v_f = 0$ in this particular case.

In general symbols, $\Delta\Sigma\, m_i\vec{v}_i = 0$.

see when you do the experiment, both carts stop in the collision; their motion ceases. But is there anything related to their motions that does not change?

Yes, there is. If you add the velocity $\vec{v}_A$ of one cart to the velocity $\vec{v}_B$ of the other cart, you find that the *vector sum* does not change. The vector sum of the velocities of these oppositely moving carts is zero *before* the collision. It is also zero for the carts at rest *after* the collision.

Does this finding hold for all collisions? In other words, is there a "law of conservation of velocity"? The example above was a very special circumstance. Carts with equal masses approach each other with equal speeds. Suppose the mass of one of the carts is twice the mass of the other cart. (You can conveniently double the mass of one cart by putting another cart on top of it.) Now let the carts approach each other with equal speeds and collide, as before. This time the carts do *not* come to rest. There is some motion remaining. Both objects move together in the direction of the initial velocity of the more massive object. The vector sum of the velocities is not conserved in all collisions.

Another example of a collision will confirm this conclusion. This time let the first cart have twice the mass of the second, but only half the speed. When the carts collide head-on and stick together, they stop. The vector sum of the velocities is equal to zero *after* the collision. But it was not equal to zero *before* the collision. Again, there is no conservation of velocity.

These examples show that the "quantity of motion" is always the same before and after the collision. The results indicate that the proper definition of "quantity of motion" may involve the *mass* of a body as well as its speed. Descartes had suggested that the proper measure of a body's quantity of motion was the product of its mass and its speed. Speed does not involve direction and is considered always to have a positive value. The examples above, however, show that this product (a scalar and always positive) is not a conserved quantity. In the second and third collisions, for example, the products of mass and speed are zero for the stopped carts *after* the collision. They obviously are *not* equal to zero *before* the collision.

If we make one very important change in Descartes' definition, we do obtain a conserved quantity. Instead of defining "quantity of motion" as the product of mass and *speed, mv*, we can define it (as Newton did) as the product of the mass and *velocity, $m\vec{v}$*. In this way we include the idea of the *direction* of motion as well as the speed. In all three collisions, the motion of both carts before and after collision is described by the equation

$$\underbrace{m_A\vec{v}_A + m_B\vec{v}_B}_{\text{before collision}} = \underbrace{m_A\vec{v}_A' + m_B\vec{v}_B'}_{\text{after collision}}$$

if m_A and m_B represent the masses of the carts, $\vec{v}_A$ and $\vec{v}_B$ represent their velocities before the collision, and $\vec{v}_A'$ and $\vec{v}_B'$ represent their velocities after the collision.

In other words: The *vector sum of the quantities mass × velocity is constant, or conserved, in all these collisions.* This is a very important and useful equation, leading directly to a powerful law.

In Unit 1, initial and final velocities were represented as $\vec{v}_i$ and $\vec{v}_f$. Here they are represented by $\vec{v}$ and $\vec{v}'$ because we now need to add subscripts such as *A* and *B*.

SG 8, 9

5. *Descartes defined the quantity of motion of an object as the product of its mass and its speed. Is his quantity of motion conserved as he believed it was? If not, how would you modify his definition so the quantity of motion would be conserved?*

6. *Two carts collide head-on and stick together. In which of the following cases will the carts be at rest immediately after the collision?*

	Cart A		*Cart B*	
	mass (kg)	*speed before (m/sec)*	*mass (kg)*	*speed before (m/sec)*
(a)	2	3	2	3
(b)	2	2	3	3
(c)	2	3	3	2
(d)	2	3	1	6

9.3 | Conservation of momentum

The product of mass and velocity often plays an interesting role in mechanics. It therefore has been given a special name. Instead of being called "quantity of motion," as in Newton's time, it is now called *momentum*. The total momentum of a system of objects (for example, the two carts) is the vector sum of the momenta of all objects in the system. Consider each of the collisions examined. The momentum of the system as a whole, that is, the vector sum of the individual parts, is the same before and after collision. Thus, the results of the experiments can be summarized briefly: *The momentum of the system is conserved.*

Since the momentum of a system is the vector sum of the momentum of its parts, it is sometimes called the "total momentum" of the system. We will assume that "total" is understood.

SG 10, 11

This rule (or law, or principle) follows from observations of the special case of collisions between two carts that stuck together after colliding. In fact, this *law of conservation of momentum* is a completely general, universal law. The momentum of *any* system is conserved *if one condition is met:* that no net force is acting on the system.

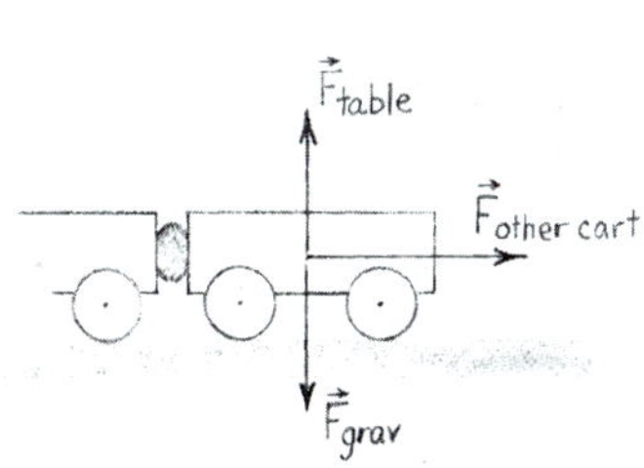

Forces on cart B during collision.

To see just what this condition means, examine the forces acting on one of the carts. Each cart experiences three main forces. There is, of course, a downward pull $\vec{F}_{grav}$ exerted by the earth and an upward push $\vec{F}_{table}$ exerted by the table. During the collision, there is also a push $\vec{F}_{from\ other\ cart}$ exerted by the other cart. The first two forces evidently cancel, since the cart is not accelerating up or down. Thus, the net force on each cart is just the force exerted on it by the other cart as they collide. (Assume that frictional forces exerted by the table and the air are small enough to neglect. That was the reason for using dry-ice disks, air-track gliders, or carts with "frictionless" wheels. This assumption makes it easier to discuss the law of conservation of momentum. Later, you will see that the law holds whether friction exists or not.)

The two carts form a *system* of bodies, each cart being a part of the system. The force exerted by one cart on the other cart is a force exerted by one part of the system on another part. It is *not* a force on the system as a whole. The outside forces acting on the carts (by the earth and by the table) exactly cancel. Thus, there is no *net* outside force. The system is "isolated." This condition must be met in order for the momentum of a system to stay constant or be conserved.

In general, for n objects the law can be written

$$\sum_{j=1}^{n} (m_j\vec{v})_{\text{before}} = \sum_{j=1}^{n} (m_j\vec{v}_j)_{\text{after}}$$

If the net force on a system of bodies is zero, the momentum of the system will not change. This is the *law of conservation of momentum* for systems of bodies that are moving with linear velocity $\vec{v}$.

So far, you have considered only cases in which two bodies collide directly and stick together. The remarkable thing about the law of conservation of momentum is how universally it applies. For example:

1. It holds true no matter what *kind* of forces the bodies exert on each other. They may be gravitational forces, electric or magnetic forces, tension in strings, compression in springs, attraction or repulsion. The sum of the $m\vec{v}$'s before is equal to the sum of $m\vec{v}$'s after any interaction.

2. It does not matter whether the bodies stick together or scrape against each other or bounce apart. They do not even have to touch. When two strong magnets repel or when an alpha particle is repelled by a nucleus, conservation of momentum still holds.

3. The law is not restricted to systems of only two objects; there can be any number of objects in the system. In those cases, the basic conservation equation is made more general simply by adding a term for each object to both sides of the equation.

SG 12–15

4. The size of the system is not important. The law applies to a galaxy as well as to an atom.

5. The angle of the collision does not matter. All of the examples so far have involved collisions between two bodies moving along the same straight line. They were "one-dimensional collisions." If two bodies make a glancing collision rather than a head-on collision, each will move off at an angle to the line of approach. The law of conservation of momentum applies to such "two-dimensional collisions" also. (Remember that momentum

Close Up

Conservation of Momentum

(1) A space capsule at rest in space, far from the sun or planets, has a mass of 1,000 kg. A meteorite with a mass of 0.1 kg moves toward it with a speed of 1,000 m/sec. How fast does the capsule (with the meteorite stuck in it) move after being hit?

m_A (mass of the meteorite) = 0.1 kg
m_B (mass of the capsule) = 1,000 kg
v_A (initial speed of meteorite) = 1,000 m/sec
v_B (initial speed of capsule) = 0

$$\left.\begin{array}{l} v_A' \text{ (final speed of meteorite)} \\ v_B' \text{ (final speed of capsule)} \end{array}\right\} = ?$$

The law of conservation of momentum states that

$$m_A\vec{v}_A + m_B\vec{v}_B = m_A\vec{v}_A{}' + m_B\vec{v}_B{}'$$

Inserting the values given,

$$(0.1 \text{ kg})\,(1{,}000 \text{ m/sec}) + (1{,}000 \text{ kg})\,(0) = (0.1 \text{ kg})\vec{v}_A{}' + (1{,}000 \text{ kg})\vec{v}_B{}'$$

$$100 \text{ kg} \cdot \text{m/sec} = (0.1 \text{ kg})\vec{v}_A{}' + (1{,}000 \text{ kg})\vec{v}_B{}'$$

Since the meteorite sticks to the capsule, $\vec{v}_B{}' = \vec{v}_A{}'$; so we can write

$$100 \text{ kg} \cdot \text{m/sec} = (0.1 \text{ kg})v_A{}' + (1{,}000 \text{ kg})v_A{}'$$

$$100 \text{ kg} \cdot \text{m/sec} = (1{,}000.1 \text{ kg})v_A{}'$$

Therefore,

$$v_A{}' = \frac{100 \text{ kg} \cdot \text{m/sec}}{1{,}000.1 \text{ kg}} = 0.1 \text{ m/sec}$$

(in the original direction of the motion of the meteorite). Thus, the capsule (with the stuck meteorite) moves on with a speed of 0.1 m/sec.

Another approach to the solution is to handle the symbols first, and substitute the values as a final step. Substituting $\vec{v}_A{}'$ for $\vec{v}_B{}'$ and letting $\vec{v}_B{}' = 0$ would leave the equation $m_A\vec{v}_A = m_A\vec{v}_A{}' + m_B\vec{v}_A{}' = (m_A + m_B)\vec{v}_A{}'$. Solving for $\vec{v}_A{}'$

$$\vec{v}_A{}' = \frac{m_A\vec{v}_A}{(m_A + m_B)}$$

This equation holds true for any projectile hitting (and staying with) a body initially at rest that moves on in a straight line after collision.

(2) An identical capsule at rest nearby is hit by a meteorite of the same mass as the other. However, this meteorite, hitting another part of the capsule, does not penetrate. Instead, it bounces straight

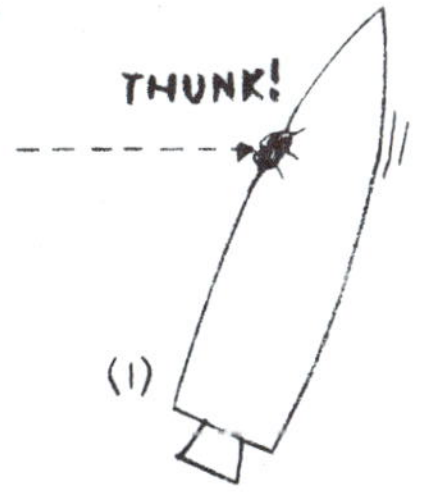

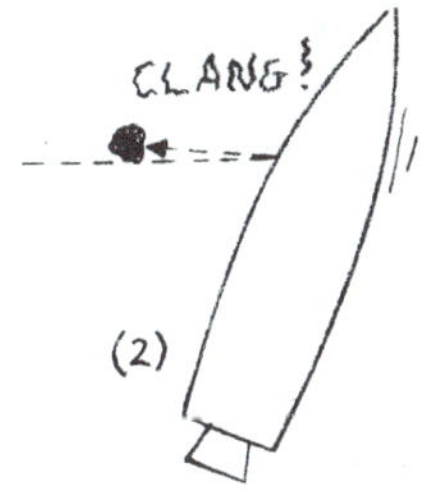

back with almost no change of speed. (Some support for the reasonableness of this claim is given in SG 24.) How fast does the capsule move after being hit? Since all these motions are along a straight line, we can drop the vector notation from the symbols and indicate the reversal in direction of the meteorite with a minus sign.

The same symbols are appropriate as in (1):

$m_A = 0.1$ kg	$v_B = 0$
$m_B = 1{,}000$ kg	$v_A{}' = -1{,}000$ m/sec
$v_A = 1{,}000$ m/sec	$v_B{}' = ?$

The law of conservation of momentum started that $m_A\vec{v}_A + m_B\vec{v}_B = m_A\vec{v}_A{}' + m_B\vec{v}_B{}'$. Here,

$$(0.1 \text{ kg})\,(1{,}000 \text{ m/sec}) + (1{,}000 \text{ kg})\,(0) = (0.1 \text{ kg})\,(-1{,}000 \text{ m/sec}) + (1{,}000 \text{ kg})v_B{}'$$

$$100 \text{ kg} \cdot \text{m/sec} = -\,100 \text{ kg} \cdot \text{m/sec} + (1{,}000 \text{ kg})v_B{}'$$

$$v_B{}' = \frac{200 \text{ kg} \cdot \text{m/sec}}{1{,}000 \text{ kg}} = 0.2 \text{ m/sec}$$

Thus, the struck capsule moves on with about twice the speed of the capsule in (1). (A general symbolic approach to this solution can be taken, too. The result is valid only for the special case of a projectile rebounding perfectly elastically from a body of much greater mass.)

There is a general lesson here. It follows from the law of conservation of momentum that a struck object is given less momentum if it *absorbs* the projectile than if it *reflects* it. (A goalie who catches the soccer ball is pushed back less than one who lets the ball bounce off.) Some thought will help you to understand this idea: An interaction that merely stops the projectile is not as great as an interaction that first stops it and then propels it back again.

is a vector quantity.) The law of conservation of momentum also applies in *three* dimensions. The vector sum of the momenta is still the same before and after the collision.

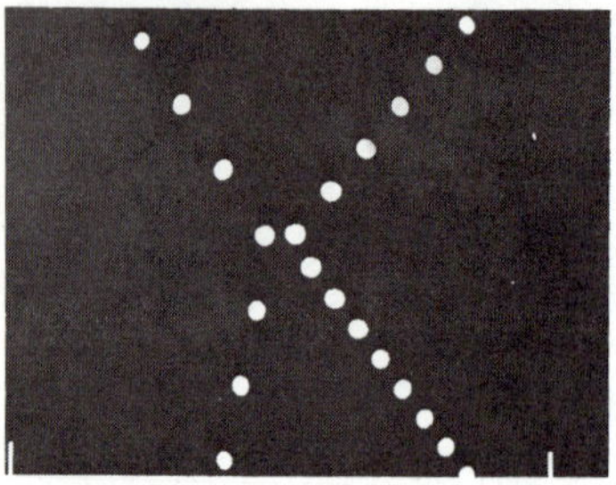

One of the stroboscopic photographs of two colliding objects that appears in the Handbook.

On page 257 is a worked-out example that will help you become familiar with the law of conservation of momentum. On page 259 is an analysis of a two-dimensional collision. There are also stroboscopic photographs in the *Project Physics Handbook* and film loops of colliding bodies and exploding objects. These include collisions and explosions in two dimensions. The more of them you analyze, the more convinced you will be that the law of conservation of momentum applies to *any* isolated system.

The worked-out example on page 257 displays a characteristic feature of physics. Again and again, physics problems are solved by applying the expression of a *general* law to a specific situation. Both the beginning student and the veteran research physicist find it helpful, but also somewhat mysterious, that one can *do* this. It seems strange that a few general laws enable one to solve an almost infinite number of specific individual problems. Everyday life seems different. There you usually cannot calculate answers from general laws. Rather, you have to make quick decisions, some based on rational analysis, others based on "intuition." The use of general laws to solve scientific problems will become, with practice, quite natural also.

?

7. State the law of conservation of momentum in terms of

(a) change in the total momentum of a system;

(b) the total initial momentum and final momentum;

(c) the individual parts of a system.

8. Which of the following has the least momentum? Which has the greatest momentum?

(a) a pitched baseball

(b) a jet plane in flight

(c) a jet plane taxiing toward the terminal

9. A girl on ice skates is at rest on a horizontal sheet of smooth ice. As a result of catching a rubber ball moving horizontally toward her, she moves at 2 cm/sec. Give a rough estimate of what her speed would have been

(a) if the rubber ball were thrown twice as fast.

(b) if the rubber ball had twice the mass.

(c) if the girl had twice the mass.

(d) if the rubber ball were not caught by the girl, but bounced off and went straight back with no change of speed.

Close Up

A Collision in Two Dimensions

The stroboscopic photograph shows a collision between two wooden discs on a "frictionless horizontal table" photographed from straight above the table. The discs are riding on tiny plastic spheres which make their motion nearly frictionless. Body B (marked x) is at rest before the collision. After the collision it moves to the left, and Body A (marked –) moves to the right. The mass of Body B is known to be twice the mass of Body A: $m_B = 2m_A$. We will analyze the photograph to see whether momentum was conserved. (*Note*: The size reduction factor of the photograph and the [constant] stroboscopic flash rate are not given here. But as long as all velocities for this test are measured in the same units, it does not matter what those units are.)

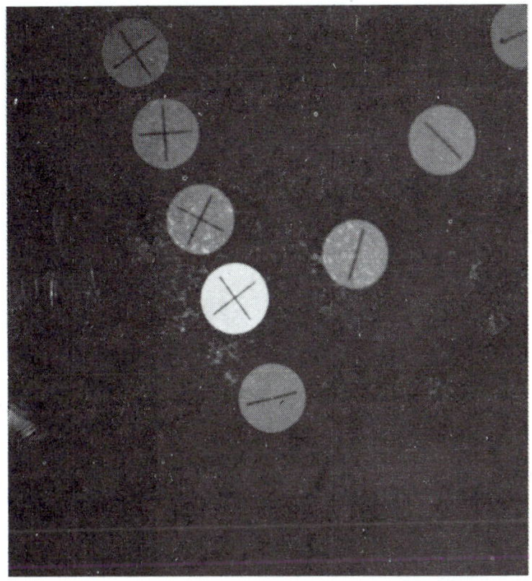

In this analysis, we will measure in centimeters the distance the discs moved on the photograph. We will use the time between flashes as the unit of time. Before the collision, Body A (coming from the lower part of the photograph) traveled 36.7 mm in the time between flashes: $\vec{v}_A$ = 36.7 speed-units. Similarly, we find that $\vec{v}_A{}'$ = 17.2 speed-units, and $\vec{v}_B{}'$ = 11.0 speed units.

The total momentum before the collision is just $m_A\vec{v}_A{}'$. It is represented by an arrow 36.7 momentum-units long, drawn at right.

The vector diagram shows the momenta $m_A\vec{v}_A{}'$ and $m_B\vec{v}_B{}'$ after the collision; $m_A\vec{v}_A{}'$ is represented by an arrow 17.2 momentum-units long. Since $m_B = 2m_A$, the $m_B\vec{v}_B{}'$ arrow is 22.0 momentum-units long.

The dotted line represents the vector sum of $m_A\vec{v}_A{}'$ and $m_B\vec{v}_B{}'$, that is, the total momentum after the collision. Measurement shows it to be 34.0 momentum-units long. Thus, our measured values of the total momentum before and after the collision differ by 2.7 momentum-units. This is a difference of about 7%. We can also verify that the *direction* of the total is the same before and after the collision to within a small uncertainty.

Have we now demonstrated that momentum was conserved in the collision? Is the 7% difference likely to be due entirely to measurement inaccuracies? Or is there reason to expect that the total momentum of the two discs after the collision is really a bit less than before the collision?

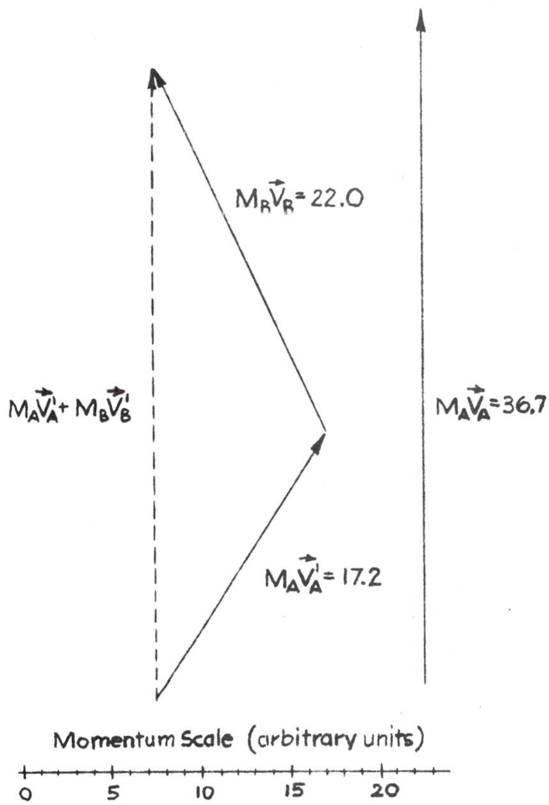

9.4 | Momentum and Newton's laws of motion

Section 9.2 developed the concept of momentum and the law of conservation of momentum by considering experiments with colliding carts. The law was an "empirical" law; that is, it was discovered (perhaps "invented" or "induced" are better terms) as a generalization from experiment.

SG 16

We can show, however, that the law of conservation of momentum also follows directly from Newton's laws of motion. It takes only a little algebra; that is, we can *deduce* the law from an established theory. It would also be possible to derive Newton's laws from the conservation law. Which is the fundamental law and which the conclusion is therefore a bit arbitrary. Newton's laws used to be considered the fundamental ones, but since about 1900 the conservation law has been assumed to be the fundamental one.

Newton's second law expresses a relation between the net force $\vec{F}_{net}$ acting on a body, the mass m of the body, and its acceleration a. We wrote this as $\vec{F}_{net} = m\vec{a}$. We can also write this law in terms of *change of momentum* of the body. Recalling that acceleration is the rate-of-change of velocity, $\vec{a} = \Delta\vec{v}\Delta t$, we can write

See Chapter 20.

$$\vec{F}_{net} = m\frac{\Delta\vec{v}}{\Delta t}$$

or

$$\vec{F}_{net}\Delta t = m\Delta\vec{v}$$

If m is a constant,

$$\begin{aligned}\Delta(m\vec{v}) &= m\vec{v}' - m\vec{v}\\ &= m(\vec{v}' - \vec{v})\\ &= m\,\Delta\,\vec{v}\end{aligned}$$

$\vec{F}\Delta t$ is called the "impulse."

If the mass of the body is constant, the change in its momentum, $\Delta(m\vec{v})$, is the same as its mass times its change in velocity, $m(\Delta\vec{v})$. Then we can write

$$\vec{F}_{net}\,\Delta t = \Delta(m\vec{v})$$

SG 17–20

that is, *the product of the net force on a body and the time interval during which this force acts equals the change in momentum of the body.*

In Newton's second law, "change of motion" meant change of momentum. See Definition II at the beginning of the *Principia*.

This statement of Newton's second law is more nearly what Newton used in the *Principia*. Together with Newton's third law, it enables us to derive the law of conservation of momentum for the cases we have studied. The details of the derivation are given on page 262. Thus, Newton's laws and the law of conservation of momentum are not separate, independent laws of nature.

In all the examples considered so far and in the derivation above, we have considered each piece of the system to have a constant mass. But the definition permits a change of momentum to arise from a change of mass as well as from a change of velocity. In many cases, it is convenient to consider objects whose mass is changing. For example, as a rocket spews out exhaust gases its mass is decreasing; the mass of a train of coal cars increases as it rolls past a hopper that drops coal into the cars. These cases can be analyzed using objects that do have constant masses, but this requires, for example, analyzing separately the case of each lump of coal that falls into the train. However, the law of conservation of momentum also is valid when the masses of the objects involved are not constant, as long as no net forces act on the system as a whole. Problems such as the one of the coal train or the rocket are much easier to analyze in this way than in the other.

SG 21–24

In one form or another, the law of conservation of momentum can be derived from Newton's second and third laws. Nevertheless, the law of conservation of momentum is often the preferred tool because it enables us to solve many problems that would be difficult to solve using Newton's laws directly. For example, suppose a cannon that is free to move fires a shell horizontally. Although it was initially at rest, the cannon is forced to move while firing the shell; it *recoils*. The expanding gases in the cannon barrel push the cannon backward just as hard as they push the shell forward. Suppose you had a continuous record of the magnitude of the force. You could then apply Newton's second law separately to the cannon and to the shell to find their respective accelerations. After a few more steps (involving calculus), you could find the speed of the shell and the recoil speed of the cannon. In practice, it is very difficult to get a continuous record of the magnitude of the force. For one thing, the force almost certainly decreases as the shell moves toward the end of the barrel. So it would be very difficult to use Newton's laws to find the final speeds.

However, you can use the law of conservation of momentum even if you know nothing about the force. The law of conservation of momentum is a law of the kind that says "before = after." Thus, it works in cases where you do not have enough information to apply Newton's laws during the whole interval between "before" and "after." In the case of the cannon and shell, the momentum of the system (cannon plus shell) is zero initially. Therefore, by the law of conservation of momentum, the momentum will also be zero after the shell is fired. If you know the masses and the speed of one, after firing you can calculate the speed of the other. (The film loop titled "Recoil" provides just such an event for you to analyze.) On the other hand, if both speeds can be measured afterwards, then the ratio of the masses can be calculated. In Unit 6, "The Nucleus," you will see how just such an approach was used to find the mass of the neutron when it was originally discovered.

SG 25

SG 26
SG 27

Close Up

Deriving Conservation of Momentum from Newton's Laws

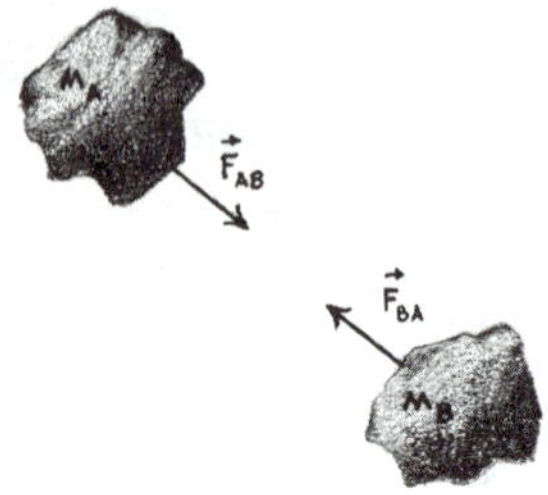

Suppose two bodies with masses m_A and m_B exert forces on each other (by gravitation or by mutual friction, etc.). $\vec{F}_{AB}$ is the force exerted on body A by body B, and $\vec{F}_{BA}$ is the force exerted on body B by body A. No other unbalanced force acts on either body; they form an isolated system. By Newton's third law, the forces $\vec{F}_{AB}$ and $\vec{F}_{BA}$ are at every instant equal in magnitude and opposite in direction. Each body acts on the other for exactly the same time Δt. Newton's second law, applied to each of the bodies, says

$$\vec{F}_{AB}\Delta t = \Delta(m_A\vec{v}_A)$$

and

$$\vec{F}_{BA}\Delta t = \Delta(m_B\vec{v}_B)$$

By Newton's third law,

$$\vec{F}_{AB} = -\vec{F}_{BA}$$

so that

$$\vec{F}_{AB}\Delta t = -\vec{F}_{BA}\Delta t$$

Therefore,

$$\Delta(m_A\vec{v}_A) = -\Delta(m_B\vec{v}_B)$$

Suppose that each of the masses m_A and m_B are constant. Let $\vec{v}_A$ and $\vec{v}_B$ stand for the velocities of the two bodies at some instant, and let $\vec{v}_A{}'$ and $\vec{v}_B{}'$ stand for their velocities at some later instant. Then we can write the last equation as

$$m_A\vec{v}_A{}' - m_A\vec{v}_A = -\ (m_B\vec{v}_B{}' - m_B\vec{v}_B)$$

A little rearrangement of terms leads to

$$m_A\vec{v}_A{}' - m_A\vec{v}_A = -\ m_B\vec{v}_B{}' + m_B\vec{v}_B$$

and

$$m_A\vec{v}_A{}' + m_B\vec{v}_B{}' = m_A\vec{v}_A + m_B\vec{v}_B$$

You will recognize this as our original expression of the law of conservation of momentum.

Here we are dealing with a system consisting of two bodies. This method works equally well for a system consisting of any number of bodies. For example, SG 21 shows you how to derive the law of conservation of momentum for a system of three bodies.

Globular clusters of stars like this one contain tens of thousands of suns held together by gravitation attraction.

? *10. Since the law of conservation of momentum can be derived from Newton's laws, what good is it?*

11. What force is required to change the momentum of an object by 50 kg · m/sec in 15 sec?

9.5 | Isolated systems

There are important similarities between the conservation law of mass and that of momentum. Both laws are tested by observing systems that are in some sense isolated from the rest of the universe. When testing or using the law of conservation of *mass*, an isolated system such as a sealed flask is used. Matter can neither enter nor leave this system. When testing or using the law of conservation of *momentum*, another kind of isolated system, one which experiences no net force from outside the system, is used.

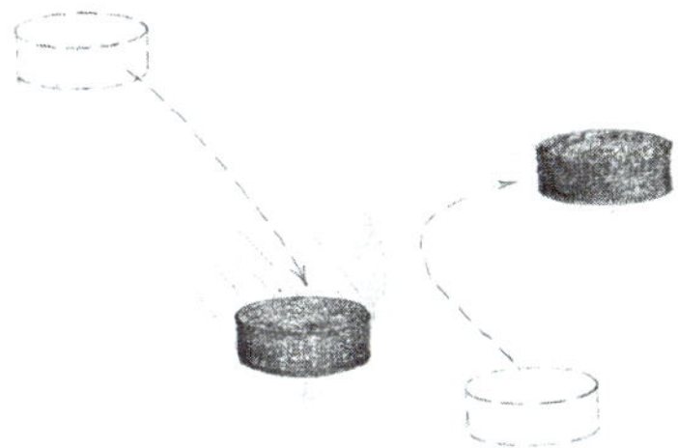

Consider, for example, two dry-ice pucks colliding on a smooth horizontal table. The very low-friction pucks form a very nearly closed or isolated system. The table and the earth do not have to be included since their individual effects on each puck cancel. Each puck experiences a downward gravitational force exerted by the earth. The table exerts an equally strong upward push.

Even in this artificial example, the system is not entirely isolated. There is always a little friction with the outside world. The layer of gas under the puck and air currents, for example, exert friction. All outside forces are not *completely* balanced, and so the two pucks do not form a truly isolated system. Whenever this is unacceptable, one can expand or extend the system so that it *includes* the bodies that are responsible for the external forces. The result is a new system on which the unbalanced forces are small enough to ignore.

For example, picture two cars skidding toward a collision on an icy road. The frictional forces exerted by the road on each car may be several hundred newtons. These forces are very small compared to the immense force (thousands of newtons) exerted by each car on the other when they collide. Thus, for many purposes, the action of the road can be ignored. For such purposes, the two skidding cars *during the collision* are nearly enough an isolated system. However, if friction with the road (or the table on which the pucks move) is too great to ignore, the law of conservation of momentum still holds, but in a larger system—one which includes the road or table. In the case of the skidding cars or the pucks, the road or table is attached to the earth. So the entire earth would have to be included in a "closed system."

SG 28–33

? *12. Define what is meant by "closed" or "isolated" system for the purpose of the law of conservation of mass; for the purpose of the law of conservation of momentum.*

13. Explain whether or not each of the following can be considered an isolated system.

(a) a baseball thrown horizontally

(b) an artificial earth satellite

(c) the earth and the moon

14. Three balls in a closed system have the following masses and velocities:

ball A: 4 kg, 8 m/sec left
ball B: 10 kg, 3 m/sec up
ball C: 8 kg, 4 m/sec right

Using the principles of mass and momentum conservation, what can you discover about the final condition of the system? What cannot be discovered?

9.6 | Elastic collisions

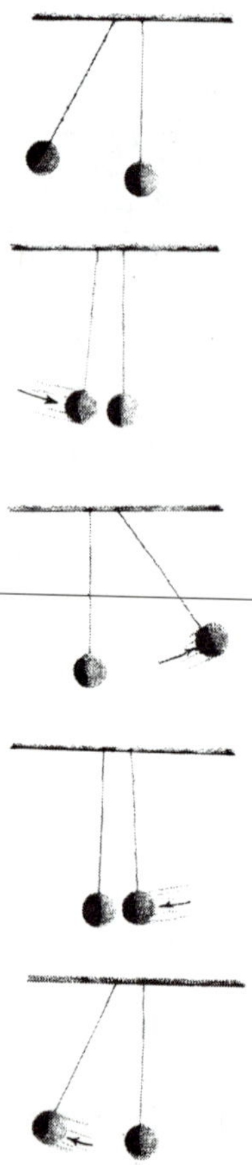

In 1666, members of the recently formed Royal Society of London witnessed a demonstration. Two hardwood balls of equal size were suspended at the ends of two strings to form two pendula. One ball was released from rest at a certain height. It swung down and struck the other, which had been hanging at rest.

After impact, the first ball stopped at the point of impact while the second ball swung from this point to the same height as that from which the first ball had been released. When the second ball returned and struck the first, it was now the second ball which stopped at the point of impact as the first swung up to almost the same height from which it had started. This motion repeated itself for several swings.

This demonstration aroused great interest among members of the Society. In the next few years, it also caused heated and often confusing arguments. Why did the balls rise each time to nearly the same height after each collision? Why was the motion "transferred" from one ball to the other when they collided? Why did the first ball not bounce back from the point of collision, or continue moving forward after the second ball moved away from the collision point?

The law of conservation of momentum explains what is observed, but it would also allow quite different results. The law says only that the momentum of ball A just before it strikes ball B is equal to the total momentum of A and B just after collision. It does not say how A and B share the momentum. The actual result is just one of infinitely many different outcomes that would all agree with conservation of momentum. For example, suppose (though it has never been observed to happen) that ball A bounced back with 10 times its initial speed. Momentum would still be conserved *if* ball B went ahead at 11 times A's initial speed.

In 1668, three men reported to the Royal Society on the whole matter of impact. The three men were the mathematician John Wallis, the architect and scientist Christopher Wren, and the physicist Christian Huygens. Wallis and Wren offered partial answers for some of the features of collisions; Huygens analyzed the problem in complete detail.

Huygens explained that in such collisions *another conservation law* in addition to the law of conservation of momentum, also holds. Not only is the vector sum of $m\vec{v}$'s conserved, but so is the ordinary arithmetic sum of mv^2s! In modern algebraic form, the relationship he discovered can be expressed as

In general symbols, $\Delta\Sigma_j \frac{1}{2}m_jv_j^2 = 0$.

Compare this equation with the conservation of momentum equation on page 254.

$$\frac{1}{2}m_Av_A^2 + \frac{1}{2}m_Bv_B^2 = \frac{1}{2}m_Av_A'^2 + \frac{1}{2}m_Bv_B'^2$$

The scalar quantity $\frac{1}{2}mv^2$ has come to be called *kinetic energy*. (The reason for the $\frac{1}{2}$, which does not really affect the rule here, will become clear in the next chapter.) The equation stated above, then, is the mathematical expression of the conservation of kinetic energy. This relationship holds for the collision of two "perfectly hard" objects similar to those observed at the Royal Society meeting. There, ball A stopped and ball B went on at A's initial speed. A little algebra will show that this is the *only* result that agrees with *both* conservation of momentum and conservation of kinetic energy. (See SG 33.)

SG 34–37

But is the conservation of kinetic energy as general as the law of conservation of momentum? Is the total kinetic energy present conserved in *any* interaction occurring in *any* isolated system?

It is easy to see that it is not. Consider the first example of Sec. 9.2. Two carts of equal mass (and with putty between the bumping surfaces) approach each other with equal speeds. They meet, stick together, and stop. The kinetic energy of the system after the collision is 0, since the speeds of both carts are zero. Before the collision the kinetic energy of the system was $\frac{1}{2}m_Av_A^2 + \frac{1}{2}m_Bv_B^2$. Both $\frac{1}{2}m_Av_A^2$ and $\frac{1}{2}m_Bv_B^2$ are always positive numbers. Their sum cannot possibly equal zero (unless both v_A and v_B are zero, in which case there would be no collision and not much of a problem). Kinetic energy is *not* conserved in this collision in which the bodies stick together. In fact, *no* collision in which the bodies stick together will show conservation of kinetic energy. It applies only to the collision of "perfectly hard" bodies that bounce back from each other.

Christian Huygens (1629–1695) was a Dutch physicist. He devised an improved telescope with which he discovered a satellite of Saturn and saw Saturn's rings clearly. Huygens was the first to obtain the expression for centripetal acceleration (v^2/R); he worked out a wave theory of light; and he invented a pendulum-controlled clock. His scientific contributions were major, and his reputation would undoubtedly have been greater had he not been overshadowed by his contemporary, Newton.

The law of conservation of kinetic energy, then, is *not* as general as the law of conservation of momentum. If two bodies collide, the kinetic energy may or may not be conserved, depending on the type of collision. It *is* conserved if the colliding bodies do not crumple or smash or dent or stick together or heat up or change physically in some other way. Bodies that rebound without any such change are *perfectly elastic*. Collisions between them are *perfectly elastic collisions*. In perfectly elastic collisions, *both* momentum and kinetic energy are conserved.

Huygens, and others after him for about a century, did not use the factor ½. The quantity mv^2 was called *vis viva*, Latin for "living force." Seventeenth- and eighteenth-century scientists were greatly interested in distinguishing and naming various "forces." They used the word loosely; it meant sometimes a push or a pull (as in the colloquial modern use of the word force), sometimes what is now called "momentum," and sometimes what is now called "energy." The term *vis viva* is no longer used.

Most collisions are not perfectly elastic, and kinetic energy is not conserved. Thus, the sum of the $\frac{1}{2}mv^2$'s after the collision is *less* than that before the collision. Depending on how much kinetic energy is "lost," such collisions might be called "partially elastic" or "perfectly inelastic." The loss of kinetic energy is greatest in perfectly inelastic collisions, when the colliding bodies remain together.

Collisions between steel ball bearings, glass marbles, hardwood balls, billiard balls, or some rubber balls (silicone rubber) are almost perfectly elastic, if the colliding bodies are not damaged in the collision. The total kinetic energy after the collision might be as much as, say, 96% of this value before the collision. Examples of perfectly elastic collisions are found only in collisions between atoms or subatomic particles.

?

15. Which phrases correctly complete the statement? Kinetic energy is conserved

(a) in all collisions.

(b) whenever momentum is conserved.

(c) in some collisions.

(d) when the colliding objects are not too hard.

16. Kinetic energy is never negative because

(a) scalar quantities are always positive.

(b) it is impossible to draw vectors with negative length.

(c) speed is always greater than zero.

(d) it is proportional to the square of the speed.

Descartes (1596–1650) was the most important French scientist of the seventeenth century. In addition to his early contribution to the idea of momentum conservation, he is remembered by scientists as the inventor of coordinate systems and the graphical representation of algebraic equations. Descartes' system of philosophy, which used the deductive structure of geometry as its model, is still influential.

9.7 | Leibniz and the conservation law

René Descartes believed that the total quantity of motion in the universe did not change. He wrote in *Principles of Philosophy:*

> It is wholly rational to assume that God, since in the creation of matter He imparted different motions to its parts, and preserves all matter in the same way and conditions in which He created it, so He similarly preserves in it the same quantity of motion.

Descartes proposed to define the quantity of motion of an object as the product of its mass and its speed. As you saw in Sec. 1.1, this product is a conserved quantity only in very special cases.

Gottfried Wilhelm Leibniz was aware of the error in Descartes' ideas on motion. In a letter in 1680 he wrote:

> M. Descartes' physics has a great defect; it is that his rules of motion or laws of nature, which are to serve as the basis, are for the most part false. This is demonstrated. And his great principle, that the quantity of motion is conserved in the world, is an error.

Leibniz, however, was as sure as Descartes had been that *something* involving motion was conserved. Leibniz called this something, which he identified as "force," the quantity mv^2 (which he called *vis viva*). This is just twice the quantity now called kinetic energy. (Of course, whatever applies to mv^2 applies equally to $\frac{1}{2}mv^2$.)

Leibniz (1646–1716), a contemporary of Newton, was a German philosopher and diplomat, an advisor to Louis XIV of France and Peter the Great of Russia. Independently of Newton, Leibniz invented the method of mathematical analysis called calculus. A long public dispute resulted between the two great men concerning charges of plagiarism of ideas.

As Huygens had pointed out, the quantity $\frac{1}{2}mv^2$ is conserved only in perfectly elastic collisions. In other collisions, the total quantity of $\frac{1}{2}mv^2$ after collision is always *less* than before the collision. Still, Leibniz was convinced that $\frac{1}{2}mv^2$ is *always* conserved. In order to save his conservation law, Leibniz invented an explanation for the apparent loss of *vis viva*. He maintained that the *vis viva* is *not* lost or destroyed. Rather, it is merely "dissipated among the small parts" of which the colliding bodies are made. This was pure speculation, and Leibniz offered no supporting evidence. Nonetheless, his explanation anticipated modern ideas about the connection between energy and the motion of molecules.

Leibniz extended conservation ideas to phenomena other than collisions. For example, when a stone is thrown straight upward, its quantity of $\frac{1}{2}mv^2$ decreases as it rises, even without any collision. At the top of the trajectory, $\frac{1}{2}mv^2$ is zero for an instant. Then it reappears as the stone falls. Leibniz wondered whether something applied or given to a stone at the start is somehow *stored* as the stone rises, instead of being lost. His idea would mean that $\frac{1}{2}mv^2$ is just one part of a more general and really conserved quantity.

?

17. According to Leibniz, Descartes' principle of conservation of mv *was*

(a) correct, but trivial.

(b) another way of expressing the conservation of vis viva.

(c) incorrect.

(d) correct only in elastic collisions.

18. How did Leibniz explain the apparent disappearance of the quantity $\frac{1}{2}mv^2$

(a) during the upward motion of a thrown object?

(b) when the object strikes the ground?

study guide

1. The *Project Physics* learning materials particularly appropriate for Chapter 9 include:

Experiments
Collisions in One Dimension
Collisions in Two Dimensions

Film Loops
One-Dimensional Collisions. I
One-Dimensional Collisions. II
Inelastic One-Dimensional Collisions
Two-Dimensional Collisions. I
Two-Dimensional Collisions. II
Inelastic Two-Dimensional Collisions
Scattering of a Cluster of Objects
Explosion of a Cluster of Objects

2. Certainly Lavoisier did not investigate every possible interaction. What justification did he have for claiming mass was conserved "in all the operations of art and nature"?

3. It is estimated that every year at least 1,800 metric tons of meteoric dust fall to the earth. The dust is mostly debris that was moving in orbits around the sun.

(a) Is the earth (whose mass is about 5.4×10^{21} tons) reasonably considered to be a closed system with respect to the law of conservation of mass?

(b) How large would the system, including the earth, have to be in order to be completely closed?

4. Would you expect that in your lifetime, when more accurate balances are built, you will see experiments which show that the law of conservation of mass does not entirely hold for chemical reactions in closed systems?

5. Dayton C. Miller, a renowned experimenter at Case Institute of Technology, was able to show that two objects placed side by side on an equal-arm pan balance did not exactly balance two otherwise identical objects placed one on top of the other. (The reason is that the pull of gravity decreases with distance from the center of the earth.) Does this experiment contradict the law of conservation of mass?

6. A children's toy known as a Snake consists of a tiny pill of mercuric thiocyanate. When the pill is ignited, a large, serpent-like foam curls out almost from nothingness. Devise and describe an experiment by which you would test the law of conservation of mass for this demonstration.

7. Consider the following chemical reaction, which was studied by Landolt in his tests of the law of conservation of mass. In a closed container, a solution of 19.4 g of potassium chromate in 100.0 g of water is mixed with a solution of 33.1 g of lead nitrate in 100.0 g of water. A bright yellow solid precipitate forms and settles to the bottom of the container. When removed from the liquid, this solid is found to have a mass of 32.3 g and is found to have properties different from either of the reactants.

(a) What is the mass of the remaining liquid? (Assume the combined mass of all substances in the system is conserved.)

(b) After removal of the yellow precipitate, the remaining liquid is heated to 95° C. The water evaporates, leaving a white solid with a mass of 20.2 kg. Why does this result imply that the water did not react with anything in either (a) or (b)?

8. (a) Ten grams (10 g) of a solid are added to 50 g of a liquid on earth. What is their total mass? What would be their total mass on the moon?

(b) A mixture that weighs 50 N on earth weighs about 8 N on the moon. Why is this fact not a violation of conservation of mass?

(c) Ten cubic centimeters (10 cm^3) of a solid are added to 50 mL (cm^3) of a liquid. The resulting mixture has a volume of 54 cm^3. Why does this result not violate conservation of mass?

9. (a) For an isolated system, state the principle of conservation of momentum in terms of the change in total momentum $\Delta\vec{P}$ and of the total initial and final momenta.

(b) For the following disks, find the momentum of each disk and the total momentum of the system:

disk A: 5 kg, 8 m/sec west
disk B: 6 kg, 25 m/sec north
disk C: 10 kg, 2 m/sec east
disk D: 4 kg, 5 m/sec east

(c) If the disks in (b) all collide at the same instant and stick together, what is the final momentum of the system? What is the final speed of the group of disks?

(d) Explain how a system could have zero total momentum and yet still have many massive objects in motion.

10. A freight car of mass 10^5 kg travels at 2.0 m/sec and collides with a motionless freight car of mass 1.5×10^5 kg on a horizontal track. The two cars lock and roll together after impact. Find the velocity of the two cars after collision. Hints: The general equation for conservation of momentum for a two-body system is:

$$m_A\vec{v}_A + m_B\vec{v}_B = m_A\vec{v}_A{}' + m_B\vec{v}_B{}'$$

(a) What quantities in the problem can be substituted in the equation?

(b) Rearrange terms to get an expression for $\vec{v}_A{}'$.

(c) Find the value of $\vec{v}_A{}'$. (Note: $\vec{v}_A{}' = \vec{v}_B{}'$.)

11. You have been given a precise technical definition for the word *momentum*. Look it up in a large dictionary and record its various uses. Can you find anything similar to our definition in these more general meanings? How many of the uses seem to be consistent with the technical definition given here?

12. Benjamin Franklin, in correspondence with his friend James Bowdoin (founder and first president of the American Academy of Arts and Sciences), objected to the corpuscular theory of light by saying that a particle traveling with such immense speed (3×10^8 m/sec) would have the impact of a 10-kg ball fired from a cannon at 100 m/sec. What mass did Franklin assign to the "light particle"?

13. If powerful magnets are placed on top of each of two carts, and the magnets are so arranged that like poles face each other when one cart is pushed toward the other, the carts bounce away from each other without actually making contact.

(a) In what sense can this be called a collision?

(b) Will the law of conservation of momentum apply?

(c) Describe an arrangement for testing your answer to **(b)**.

14. A person throws a fast ball vertically. Clearly, the momentum of the ball is not conserved; it first loses momentum as it rises, then gains it as it falls. How large is the "closed system" within which the ball's momentum, *together* with that of other bodies (tell which), *is* conserved. What happens to the rest of the system as the ball rises? as it falls?

15. Did Newton arrive at the law of conservation of momentum in the *Principia*? If a copy of the *Principia* is available, read Corollary III and Definition II (just before and just after the three laws).

16. (a) For how long would a force of 20 N have to be applied to cause a system to gain a momentum of 80 kg · m/sec?

(b) A 5-kg object initially travels with a momentum of 50 kg · m/sec. If a 10-N force acts on the object for 5 sec, what is the final speed? (Solve this problem first using Newton's second law directly and then using momentum formulas.)

(c) What are the advantages of momentum formulas over Newton's laws? Is one more basic than the other?

17. (a) Why can ocean liners or planes not turn corners sharply?

(b) In the light of your knowledge of the relationship between momentum and force, comment on reports about unidentified flying objects (UFO's) turning sharp corners in full flight.

18. A girl on skis (mass of 60 kg including skis) reaches the bottom of a hill going 20 m/sec. What is her momentum? She strikes a snowdrift and stops within 3 sec. What force does the snow exert on the girl? How far does she penetrate the drift? What happens to her momentum?

19. During sports, the forces exerted on parts of the body and on the ball, etc., can be astonishingly large. To illustrate this, consider the forces in hitting a golf ball. Assume the ball's mass is 0.046 kg. From the strobe photo on p. 27 of Unit 1, in which the time interval between strobe flashes was 0.01 sec, estimate:

(**a**) the speed of the ball after impact;

(**b**) the magnitude of the ball's momentum after impact;

(**c**) how long the impact lasted;

(**d**) the average force exerted on the ball during impact.

20. We derived the law of conservation of momentum for two bodies from Newton's third and second laws. Is the principle of the conservation of mass essential to this derivation? If so, where does it it enter?

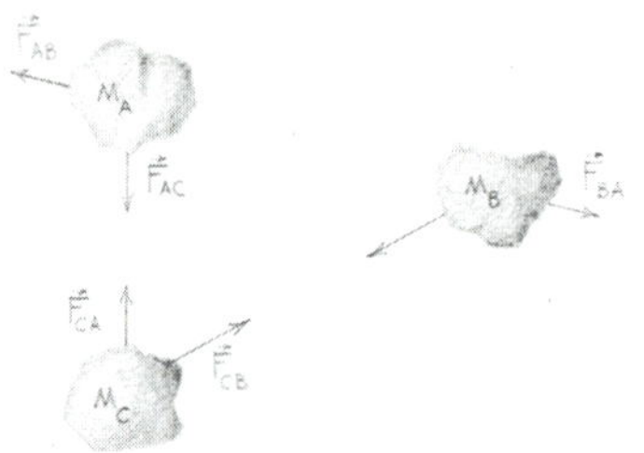

21. Consider an isolated system of three bodies, A, B, and C. The forces acting among the bodies can be indicated by subscripts; for example, the force exerted on body A by body B can be given the symbol $\vec{F}_{AB}$. By Newton's third law of motion, $\vec{F}_{BA} = -\vec{F}_{AB}$. Since the system is isolated, the only force on each body is the sum of the forces exerted on it by the other two; for example, $\vec{F}_A = \vec{F}_{AB} + \vec{F}_{AC}$. Using these principles, show that the total momentum change of the system will be zero.

22. In Chapter 4, SG 33 was about putting an Apollo capsule into an orbit around the moon. The question was: "Given the speed v_0 necessary for orbit and the current speed v, how long should the engine with thrust F fire to give the capsule of mass m the right speed?" There you solved the problem by considering the acceleration.

(**a**) Answer the question more directly by considering change in momentum.

(**b**) What would be the total momentum of all the exhaust from the rocket?

(**c**) If the "exhaust velocity" were v_e, about what mass of fuel would be required?

23. (a) Show that when two bodies collide their changes in velocity are inversely proportional to their masses; that is, if m_A and m_B are the masses and $\Delta\vec{v}_A$ and $\Delta\vec{v}_B$ the velocity changes, show that numerically,

$$\frac{\Delta v_A}{\Delta v_B} = \frac{m_B}{m_A}$$

(**b**) Show how it follows from conservation of momentum that if a light particle (like a B.B. pellet) bounces off a massive object (like a bowling ball), the velocity of the light particle is changed much more than the velocity of the massive object.

(**c**) For a head-on elastic collision between a body of mass m_A moving with velocity v_A and a body of mass m_B at rest, combining the equations for conservation of momentum and conservation of kinetic energy leads to the relationship $v_A' = v_A(m_A - m_B)/(m_A + m_B)$. Show that if body B has a much greater mass than body A, then v_A' is almost exactly the same as v_A; that is, body A bounces back with virtually no loss in speed.

24. The equation $m_A\vec{v}_A + m_B\vec{v}_B = m_A\vec{v}_A' + m_B\vec{v}_B'$ is a general equation applicable to countless separate situations. For example, consider a 10-kg shell fired from a 1,000-kg cannon. If the shell is given a speed of 1,000 m/sec, what would be the recoil speed of the cannon? (Assume the cannon is on an almost frictionless mount.) Hint: Your answer could include the following steps:

(**a**) If A refers to the cannon and B to the shell, what are $\vec{v}_A$ and $\vec{v}_B$ (before firing)?

(**b**) What is the total momentum before firing?

(**c**) What is the total momentum after firing?

(**d**) Compare the magnitudes of the momenta of the cannon and of the shell after firing.

(**e**) Compare the ratios of the speeds and of the masses of the shell and cannon after firing.

25. The engines of the first stage of the Apollo/Saturn rocket develop an average thrust of 35 million

N for 150 sec. (The entire rocket weighs 28 million N near the earth's surface.)

(a) How much momentum will be given to the rocket during that interval?

(b) The final speed of the vehicle is 9,760 km/hr. What would one have to know to compute its mass?

26. Newton's second law can be written $\vec{F}\Delta t = \Delta(m\vec{v})$. Use the second law to explain the following:

(a) It is safer to jump into a fire net or a load of hay than onto the hard ground.

(b) When jumping down from some height, you should bend your knees as you come to rest, instead of keeping your legs stiff.

(c) Hammer heads are generally made of steel rather than rubber.

(d) Some cars have plastic bumpers which, temporarily defomed under impact, slowly return to their original shape. Others are designed to have a somewhat pointed front-end bumper.

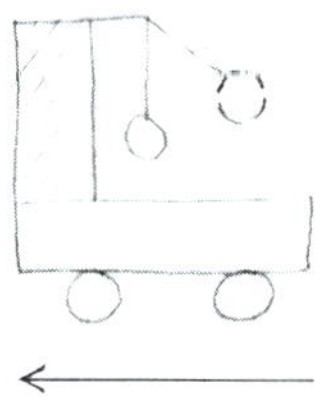

27. A student in a physics class, having learned about elastic collisions and conservation laws, decides that he can make a self-propelled car. He proposes to fix a pendulum on a cart, using a "superball" as a pendulum bob. He fixes a block to the cart so that when the ball reaches the bottom of the arc, it strikes the block and rebounds elastically. It is supposed to give the cart a series of bumps that propel it along.

(a) Will his scheme work? (Assume the "superball" is perfectly elastic.) Give reasons for your answer.

(b) What would happen if the cart had an initial velocity in the forward direction?

(c) What would happen if the cart had an initial velocity in the backward direction?

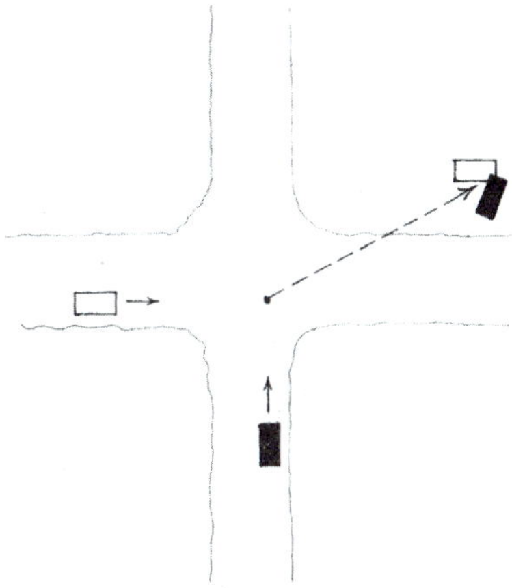

28. A police report of an accident describes two vehicles colliding (inelastically) at an icy intersection of country roads. The cars slid to a stop in a field as shown in the diagram. Suppose the masses of the cars are approximately the same.

(a) How did the speeds of the two cars compare just before collision?

(b) What information would you need in order to calculate the actual speeds of the automobiles?

(c) What simplifying assumptions have you made in answering (b)?

29. A person fires a gun horizontally at a target fixed to a hillside. Describe the changes of momentum to the person, the bullet, the target, and the earth. Is momentum conserved

(a) when the gun is fired?

(b) when the bullet hits?

(c) during the bullet's flight?

30. A billiard ball moving 0.8 m/sec collides with the cushion along the side of the table. The collision is head-on and can here be regarded as perfectly elastic. What is the momentum of the ball

(a) before impact?

(b) after impact?

(c) What is the change in momentum of the ball?

(d) Is momentum conserved?

31. Discuss conservation of momentum for the system shown in this sketch from *Le Petit Prince*. What happens

(**a**) if he leaps in the air?

(**b**) if he runs around?

32. (a) State the principle of conservation of kinetic energy. For what systems is this law applicable?

(**b**) What does this principle tell you about the change in the total kinetic energy, and about the total initial and final kinetic energies, of an isolated system of elastic interactions?

(**c**) What is the total kinetic energy of a system of two carts with masses of 5 kg and 10 kg, traveling toward one another at 4 m/sec and 3 m/sec, respectively?

(**d**) If the two carts in (**c**) rebound elastically, what is the total final kinetic energy?

33. Fill in the blanks for the following motions:

Object	*m* (*kg*)	*v* *m/s*	*mv* *kg · m/s*	*½mv²* *kg · m²/s²*
baseball	0.14	30.0	—	—
hockey puck	—	50.0	8.55	—
superball	0.050	1.5	—	—
light car	1460	—	—	1.79×10^6
mosquito	—	—	2.0×10^5	4.0×10^{-6}
football player	—	—	—	—

34. A system consists of three particles with masses of 4 g, 6 g, and 8 g traveling toward a single point with velocities of 20 cm/sec north, 3 cm/sec east, and 10 cm/sec south.

(**a**) Calculate the total initial mass, momentum, and kinetic energy of the system.

(**b**) If the particles reach the point at the same instant, calculate where possible the final total mass, momentum, and kinetic energy of the system in each of the following situations: (1) the system is not closed and the collision is elastic; (2) the system is not closed and the collision is inelastic; (3) the system is closed and the collision is elastic; and (4) the system is closed and the collision is inelastic.

35. Two balls, one of which has three times the mass of the other, collide head-on, each moving with the same speed. The more massive ball stops; the other rebounds with twice its original speed. Show that both momentum and kinetic energy are conserved.

36. A ball of mass *m* moving at speed *v* strikes, elastically, head-on, a second ball of mass 3*m* which is at rest. Using the principle of conservation of momentum and kinetic energy, find the speeds of the two balls after collision.

37. When one ball collides with a stationary ball of the same mass, the first ball stops and the second goes on with the speed the first ball had. The claim is made on p. 265 that this result is the only possible result that will be consistent with conservation of both momentum and kinetic energy. (That is, if $m_A = m_B$ and $v_B = 0$, then the result must be $v_A' = 0$ and $v_B' = v_A$.) Combine the equations that express the two conservation laws and show that this is actually the case. (Hint: Rewrite the equations with m for m_A and m_B' and $v_B = 0$; solve the simplified momentum equation for v_A'; substitute in the simplified kinetic energy equation; solve for v_B'.)

Energy

10.1 | Work and kinetic energy

In everyday language, pitching, catching, and running on the baseball field are "playing," while sitting at a desk, reading, writing, and thinking are "working." However, in the language of physics, "work" has been given a rather special definition, one that involves physical concepts of force and displacement instead of the subjective ones of reward or accomplishment. It is more closely related to the simple sense of effort or labor. The work done on an object is defined as the *product of the force exerted on the object times the displacement of the object along the direction of the force.*

When you throw a baseball, you exert a large force on it while it moves forward for about 1 m. In doing so, you do a large amount of work. By contrast, in writing or in turning the pages of a book you exert only a small force over a short distance. This does not require much work, as the term "work" is understood in physics.

SG 1

Suppose you are employed in a factory to lift boxes from the floor straight upward to a conveyor belt at waist height. Here the language of common usage and that of physics both agree that you are doing work. If you lift two boxes at once, you do twice as much work as you do if you lift one box. If the conveyor belt were twice as high above the floor, you would do twice as much work to lift a box to it. The work you do depends on both the *magnitude* of the force you must exert on the box and the *distance* through which the box moves in the direction of the force.

Note that work you do on a box does not depend on how *fast* you do your job.

We can now define the work W done on an object by a force $\vec{F}$ as the product of the magnitude F of the force and the distance d that the object moves *in the direction of* $\vec{F}$ while the force is being exerted; in symbols,

$$W = Fd$$

The way d is defined here, the $W = Fd$ is correct. It does not, however, explicitly tell how to compute W if the motion is not in exactly the same direction as the force. The definition of d implies that it would be the component of the displacement along the direction of F; and this is entirely correct.

To lift a box weighing 100 N upward through 0.8 m requires you to apply a force of 100 N. The work you do on the box is 100 N × 0.8 m = 80 N-m.

From the definition of work, it follows that no work is done if there is no displacement. No matter how hard you push on a wall, no work is done if the wall does not move. Also, no work is done if the only motion is perpendicular to the direction of the force. For example, suppose you are carrying a book bag. You must pull up against the downward pull of gravity to keep the bag at a constant height. But as long as you are standing still you do no work on the bag. Even if you walk along with it steadily in a horizontal line, the only work you do is in moving it forward against the small resisting force of the air.

Note that work is a scalar quantity.

Work is a useful concept in itself. The concept is most useful in understanding the concept of *energy*. There are a great many forms of energy. A few of them will be discussed in this chapter. We will define them, in the sense of describing how they can be measured and how they can be expressed algebraically. We will also discuss how energy changes from one form to another. The *general* concept of energy is very difficult to define. But to define some *particular* forms of energy is easy enough. The concept of work helps greatly in making such definitions.

The equation $W = Fd$ implies that work is always a positive quantity. However, by convention, when the force on a body and its displacement are in opposite directions, the work is negative. This implies that the body's energy would be decreased. The sign convention follows naturally from the more rigorous definition of mechanical work as $W = Fl \cos \theta$, where θ is the angle between $\vec{F}$ and $\vec{l}$.

The chief importance of the concept of work is that work represents an amount of energy transformed from one form to another. For example, when you throw a ball you do work on it. You also transform chemical energy, which your body obtains from food and oxygen, into energy of motion of the ball. When you lift a stone (doing work on it), you transform chemical energy into gravitational potential energy. If you release the stone, the earth pulls it downward (does work on it); gravitational potential energy is transformed into energy of motion. When the stone strikes the ground, it compresses the ground below it (does work on it); energy of motion is transformed into heat. In each case, the work is a measure of how much energy is transferred.

SG 2

The form of energy we have been calling "energy of motion" is perhaps the simplest to deal with. We can use the definition of work, $W = Fd$, together with Newton's laws of motion to get an expression of this form of energy. Imagine that you exert a constant net force F on an object of mass m. This force accelerates the object over a distance d from rest to a speed v. Using Newton's second law of motion, we can show in a few steps of algebra that

$$Fd = \frac{1}{2}mv^2$$

The details of this derivation are given on the first half of page 276.

Fd is the expression for the work done on the object by whatever exerted the force F. The work done on the object equals the amount of energy transformed from some form into energy of motion of the object. So $\frac{1}{2}mv^2$ is the expression for the energy of motion of the object. The energy of motion of an object at any instant is given by the quantity $\frac{1}{2}mv^2$ at that instant and is called *kinetic energy*. The symbol KE is used to represent kinetic energy. By definition then,

The Greek word *kinetos* means "moving."

$$KE = \frac{1}{2}mv^2$$

Now it is clearer why $\frac{1}{2}mv^2$ instead of just mv^2 was used in Chapter 9: $\frac{1}{2}mv^2$ relates directly to the concept of work and so provides a useful expression for energy of motion.

The speed of an object must be measured relative to some reference frame, so kinetic energy is a relative quantity also. See SG 3.

The equation $Fd = \frac{1}{2}mv^2$ was obtained by considering the case of an object initially at rest. In other words, the object had an initial kinetic energy of zero. The relation also holds for an object already in motion when the net force is applied. In that case, the work done on the object still equals the change in its kinetic energy:

SG 3–8

$$Fd = \Delta(KE)$$

The quantity $\Delta(KE)$ is by definition equal to $(\frac{1}{2}mv^2)_{final} - (\frac{1}{2}mv^2)_{initial}$. The proof of this general equation appears on the second half of page 276.

Work is defined as the product of a force and a distance. Therefore, its units in the mks system are *newtons* × *meters* or newton-meters. A newton-meter is also called a *joule* (symbol *J*). The joule is the unit of work or of energy.

The name of the unit of energy and work commemorates J. P. Joule, nineteenth-century English physicist famous for his experiments showing that heat is a form of energy (see Sec. 10.7). There is no general agreement today whether the name should be pronounced like "jool" or like "jowl." The majority of physicists favor the former.

Close Up

Doing Work on a Sled

Suppose a loaded sled of mass m is initially at rest on low-friction ice. You, wearing spiked shoes, exert a constant horizontal force F on the sled. The weight of the sled is balanced by the upward push exerted by the ice, so F is effectively the net force on the sled. You keep pushing, running faster and faster as the sled accelerates, until the sled has moved a total distance d.

Since the net force F is constant, the acceleration of the sled is constant. Two equations that apply to motion starting from rest with constant acceleration are

$$v = at$$

and

$$d = \tfrac{1}{2}at^2$$

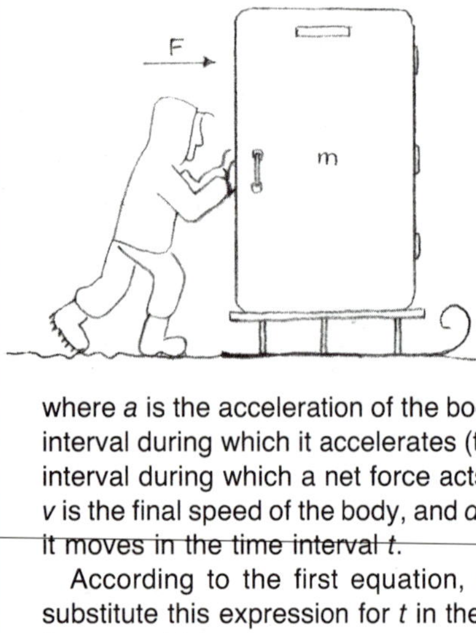

where a is the acceleration of the body, t is the time interval during which it accelerates (that is, the time interval during which a net force acts on the body), v is the final speed of the body, and d is the distance it moves in the time interval t.

According to the first equation, $t = v/a$. If we substitute this expression for t in the second equation, we obtain

$$d = \tfrac{1}{2}at^2 = \tfrac{1}{2}a\frac{v^2}{a^2} = \tfrac{1}{2}\frac{v^2}{a}$$

The work done on the sled is $W = Fd$. From Newton's second law, $F = ma$, so

$$W = Fd$$
$$= ma \times \tfrac{1}{2}\frac{v^2}{a}$$

The acceleration cancels out, giving

$$W = \tfrac{1}{2}mv^2$$

Therefore, the work done in this case can be found from just the mass of the body and its final speed. With more advanced mathematics, it can be shown that the result is the same whether the force is constant or not.

More generally, we can show that the change in kinetic energy of a body already moving is equal to the work done on the body. By the definition of average speed,

$$d = v_{av}t$$

If we consider a uniformly accelerated body whose speed changes from v_o to v, the average speed (v_{av}) during t is $\tfrac{1}{2}(v + v_o)$. Thus,

$$d = \frac{v + v_0}{2} \times t$$

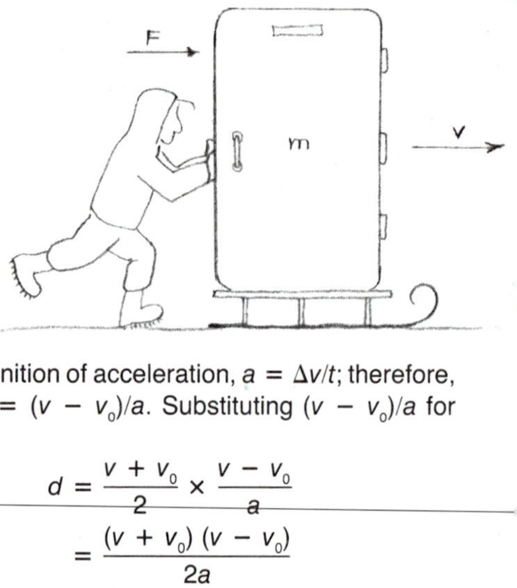

By the definition of acceleration, $a = \Delta v/t$; therefore, $t = \Delta v/a = (v - v_0)/a$. Substituting $(v - v_0)/a$ for t gives

$$d = \frac{v + v_0}{2} \times \frac{v - v_0}{a}$$
$$= \frac{(v + v_0)(v - v_0)}{2a}$$
$$= \frac{v^2 - v_0^2}{2a}$$

The work (W) done is $W = Fd$, or, since $F = ma$,

$$W = ma \times d$$
$$= ma \times \frac{v^2 - v_0^2}{2a}$$
$$= \frac{m}{2}(v^2 - v_0^2)$$
$$= \tfrac{1}{2}mv^2 - \tfrac{1}{2}mv_0^2$$

? 1. *If a force* F *is exerted on an object while the object moves a distance* d *in the direction of the force, the work done on the object is:*
(a) F *(b)* Fd *(c)* F/d *(d)* $\frac{1}{2}Fd^2$

2. *The kinetic energy of a body of mass* m *moving at a speed* v *is:*
(a) $\frac{1}{2}mv$ *(b)* $\frac{1}{2}mv^2$ *(c)* mv^2 *(d)* $2mv^2$ *(e)* m^2v^2

10.2 | Potential energy

As you saw in the previous section, doing work on an object can increase its kinetic energy. Work also can be done on an object *without* increasing its kinetic energy. For example, you might lift a book straight up at a small, constant speed, so that its kinetic energy stays the same. But you are still doing work on the book. By doing work you are using your body's store of chemical energy. Into what form of energy is it being transformed?

The answer, as Leibniz suggested, is that there is "energy" associated with height above the earth. This energy is called *gravitational potential energy*. Lifting the book higher and higher increases the gravitational potential energy. You can see clear evidence of this effect when you drop the book. The gravitational potential energy is transformed rapidly into kinetic energy of fall. In general terms, suppose a force $\vec{F}$ is used to displace an object upwards a distance d, without changing its KE. Then, the increase in gravitational potential energy, $\Delta(PE)_{grav}$, is

$$\Delta(PE)_{grav} = F_{applied} \cdot d$$
$$= -F_{grav} \cdot d$$

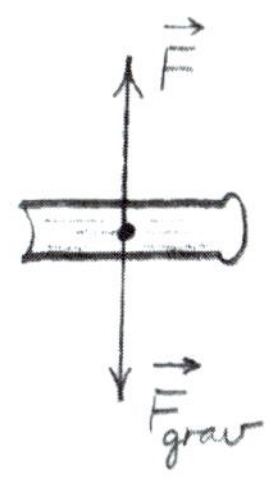

To lift the book at constant speed, you must exert an upward force $\vec{F}$ equal in magnitude to the weight F_{grav} of the book. The work you do in lifting the book through distance d is Fd, which is numerically equal to $F_{grav}d$. See SG 9 and 10.

Potential energy can be thought of as *stored* energy. As the book falls, its gravitational potential energy decreases while its kinetic energy increases correspondingly. When the book reaches its original height, all of the gravitational potential energy stored during the lift will have been transformed into kinetic energy.

Many useful applications follow from this idea of potential or stored energy. For example, the steam hammer used by construction crews is driven by high-pressure steam ("pumping in" energy). When the hammer drops, the gravitational potential energy is converted to kinetic energy. Another example is the proposal to use extra available energy from electric power plants during low-demand periods to pump water into a high reservoir. When there is a large demand for electricity later, the water is allowed to run down and drive the electric generators.

A stretched bow contains elastic potential energy. When released, the resulting kinetic energy propels the arrow to the target.

There are forms of potential energy other than gravitational. For example, if you stretch a rubber band or a spring, you increase its *elastic potential energy*. When you release the rubber band, it can deliver the stored energy to a projectile in the form of kinetic energy. Some of the work done in blowing up an elastic balloon is also stored as potential energy.

Other forms of potential energy are associated with other kinds of forces. In an atom, the negatively charged electrons are attracted by the positively charged nucleus. If an externally applied force pulls an electron *away* from the nucleus, the *electric potential energy* increases. If the electron is pulled back and moves *toward* the nucleus, the potential energy decreases as the electron's kinetic energy increases.

If two magnets are pushed together with north poles facing, the *magnetic potential energy* increases. When released, the magnets will move apart, gaining kinetic energy as they lose potential energy.

Where is the potential energy located in all these cases? It might seem at first that it "belongs" to the body that has been
SG 11 moved. But without the other object (the earth, the nucleus, the
SG 12 other magnet) the work would not increase any potential form
SG 13 of energy. Rather, it would increase only the kinetic energy of the object on which work was done. The potential energy belongs not to *one* body, but to the whole system of interacting bodies! This is evident in the fact that the potential energy is available to any one or to all of these interacting bodies. For example, you could give either magnet all the kinetic energy just by releasing one magnet and holding the other in place. Or suppose you could fix the book somehow to a hook that would hold it at one point in space. The earth would then "fall" up toward the book. Eventually the earth would gain just as much kinetic energy at the expense of stored potential energy as the book would if it were free to fall.

The work you have done on the earth–book system is equal to the energy you have given up from your store of chemical energy.

The increase in gravitational potential energy "belongs" to the earth–book *system*, not the book alone. The work is done by an "outside" agent (you), increasing the total energy of the earth–book system. When the book falls, it is responding to forces exerted by one part of the system on another. The *total energy* of the system does not change; it is converted from *PE* to *KE*. This is discussed in more detail in the next section.

?

3. *If a stone of mass* m *falls a vertical distance* d, *pulled by its weight* $F_{grav} = ma_g$, *the decrease in gravitational potential energy is:*

(a) md *(b)* ma_g *(c)* ma_gd *(d)* $\frac{1}{2}md^2$ *(e)* d

4. *When you compress a coil spring, you do work on it. The elastic potential energy:*

(a) disappears (b) breaks the spring (c) increases (d) decreases

5. *Two electrically charged objects repel one another. To increase the electric potential energy, you must*

(a) *make the objects move faster.*
(b) *move one object in a circle around the other object.*
(c) *attach a rubber band to the objects.*
(d) *pull the objects farther apart.*
(e) *push the objects closer together.*

10.3 | Conservation of mechanical energy

In Sec. 10.1, you learned that the amount of work done on an object *equals* the amount of energy transformed from one form to another. For example, the chemical energy of a muscle is transformed into the kinetic energy of a thrown ball. This statement implies that the *amount* of energy involved does not change, only its *form* changes. This is particularly obvious in motions where no "outside" force is applied to a mechanical system.

While a stone falls freely, for example, the gravitational potential energy of the stone-earth system is continually transformed into kinetic energy. Neglecting air friction, the *decrease* in gravitational potential energy is, for any portion of the path, equal to the *increase* in kinetic energy. Consider a stone thrown upward. Between any two points in its path, the *increase* in gravitational potential energy equals the *decrease* in kinetic energy. For a stone falling or rising (without external forces such as friction), the only force applied is F_{grav}. The work done by this force is (with d positive for upward displacements)

The equations in this section are true only if friction is negligible. We shall extend the range later to include friction, which can cause the conversion of mechanical energy into heat energy.

SG 14

$$-F_{grav}d = (\Delta PE)_{grav}$$
$$= -\Delta KE$$

This relationship can be rewritten as

$$\Delta(KE) + \Delta(PE)_{grav} = 0$$

or still more concisely as

$$\Delta(KE + PE_{grav}) = 0$$

If $(KE + PE_{grav})$ represents the *total mechanical energy* of the system, then the *change* in the system's total mechanical energy is *zero*. In other words, the total mechanical energy, $\Delta(KE + PE_{grav})$, remains constant; it is *conserved*.

A similar statement can be made for a vibrating guitar string. While the string is being pulled away from its unstretched position, the string–guitar system gains elastic potential energy. When the string is released, the elastic potential energy decreases while the kinetic energy of the string increases. The string coasts through its unstretched position and becomes stretched in the other direction. Its kinetic energy then decreases as the elastic potential energy increases. As it vibrates, there is a repeated transformation of elastic potential energy into kinetic energy and back again. The string loses some mechanical energy; for example, sound waves radiate away. Otherwise, the decrease in elastic potential energy over any part of the string's motion would be accompanied by an equal increase in kinetic energy, and *vice versa*:

$$\Delta(PE)_{\text{elastic}} = -\Delta(KE)$$

In such an ideal case, the total mechanical energy ($KE + PE_{\text{elastic}}$) remains constant; it is conserved.

Up to this point, we have always considered only *changes* in PE. There is some subtlety in defining an actual value of PE. See SG 15.

Galileo's experiment with the pendulum (Sec. 3.5) can also be described in these terms. The gravitational potential energy is determined by the height to which the pendulum was originally pulled. That potential energy is converted to kinetic energy at the bottom of the swing and back to potential energy at the other side. Since the pendulum retains its initial energy, it will stop ($KE = 0$, $PE = \text{max}$) only when it returns to its initial height.

SG 16

You have seen that the potential energy of a system can be transformed into the kinetic energy of some part of the system, and *vice versa*. Suppose that an amount of work W is done on part of the system by some external force. Then the energy of the system is increased by an amount equal to W. Consider, for example, a suitcase–earth system. You must do work on the suitcase to pull it away from the earth up to the second floor. This work increases the total mechanical energy of the earth–suitcase system. If you yourself are included in the system, then your internal chemical energy decreases in proportion to the work you do. Therefore, the *total* energy of the lifter + suitcase + earth system does not change.

The law of conservation of energy can be derived from Newton's laws of motion. Therefore, it tells nothing that could not, in principle, be computed directly from Newton's laws of motion. However, there are situations where there is simply not enough information about the forces involved to apply Newton's laws. It is in these cases that the law of conservation of mechanical energy demonstrates its usefulness.

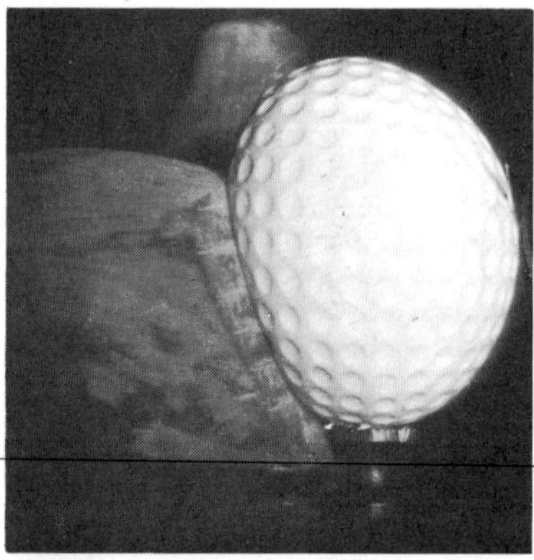

During its contact with a golf club, a golf ball is distorted, as is shown in the high-speed photograph. As the ball moves away from the club, the ball recovers its normal spherical shape, and elastic potential energy is transformed into kinetic energy.

A perfectly elastic collision is a good example of a situation where we often cannot apply Newton's laws of motion. In such collisions, we do not know and cannot easily measure the force that one object exerts on the other. We do know that during the actual collision, the objects distort one another. (See the photograph of the golf ball in the margin.) The distortions are produced against elastic forces. Thus, some of the combined kinetic energy of the objects is transformed into elastic potential energy as they distort one another. Then elastic potential energy is transformed back into kinetic energy as the objects separate. In an ideal case, both the objects and their surroundings are exactly the same after colliding as they were before.

This is useful but incomplete knowledge. The law of conservation of mechanical energy gives only the *total* kinetic energy of the objects after the collision. It does not give the kinetic energy of each object separately. (If enough information were available, we could apply Newton's laws to get more detailed results, namely, the speed of *each* object.) You may recall that the law of conservation of momentum also supplies useful but incomplete knowledge. It can be used to find the *total* momentum, but not the *individual* momentum vectors, of elastic objects in collision. In Chapter 9, you saw how conservation of momentum and conservation of mechanical energy together limit the possible outcomes of perfectly elastic collisions. For two colliding objects, these two restrictions are enough to give an exact solution for the two velocities after collision. For more complicated systems, conservation of energy remains important. Scientists usually are not interested in the detailed motion of every part of a complex system. They are not likely to care, for example, about the motion of every molecule in a rocket exhaust. Rather, they probably want to know only about the overall thrust and temperature. These can be found from the overall conservation laws.

?

6. As a stone falls frictionlessly,

(a) its kinetic energy is conserved.

(b) its gravitational potential energy is conserved.

(c) its kinetic energy changes into gravitational potential energy.

(d) no work is done on the stone.

(e) there is no change in the total energy.

7. In which position is the elastic potential energy of the vibrating guitar string greatest? In which position is its kinetic energy greatest?

8. If a guitarist gives the same amount of elastic potential energy to a bass string and to a treble string, which one will gain more speed when released? (The mass of 1 m of bass string is greater than that of 1 m of treble string.)

9. How would you compute the potential energy stored in the system shown in the margin made up of the top boulder and the earth?

10.4 | Forces that do no work

In Sec. 10.1, the *work* done on an object was defined as the product of the magnitude of the force $\vec{F}$ applied to the object

The reason that your arm gets tired even though you do no work on the book is that muscles are not rigid, but are constantly relaxing and tightening up again. That requires chemical energy. When you carry a heavy load on your back or shoulders, the supporting force is mostly provided by bones, not muscles. As a result, you can carry much bigger loads for greater distances.

and the magnitude of the distance $\vec{d}$ in the direction of $\vec{F}$ through which the object moves while the force is being applied. In all the examples so far, the object moved in the same direction as that of the force vector.

Usually, the direction of motion and the direction of the force are *not* the same. For example, suppose you carry a book at constant speed horizontally, so that its kinetic energy does not change. Since there is no change in the book's energy, you are doing no work on the book (by the definition of work). You do apply a force on the book, and the book does move through a distance. But here the applied force and the distance are at right angles. You exert a vertical force on the book upward to balance its weight. But the book moves horizontally. Here, an applied force $\vec{F}$ is exerted on an object while the object moves at right angles to the direction of the force. Therefore, $\vec{F}$ has no component in the direction of $\vec{d}$, and so the force *does no work*. This statement agrees entirely with the idea of work as *energy being transformed from one form to another*. Since the book's speed is constant, its kinetic energy is constant. Since its distance from the earth is constant, its gravitational potential energy is constant. Therefore, there is no transfer of mechanical energy.

A similar reasoning, but not so obvious, applies to a satellite in a circular orbit. The speed and the distance from the earth are both constant. Therefore, the kinetic energy and the gravitational potential energy are both constant, and there is no energy transformation. For a circular orbit, the centripetal force vector is perpendicular to the tangential direction of motion at any instant. No work is being done. To put an artificial satellite into a circular orbit requires work. Once it is in orbit, however, the *KE* and *PE* stay constant, and no further work is done on the satellite.

When the orbit is eccentric, the force vector is usually not perpendicular to the direction of motion. In such cases, energy is continually transformed between kinetic and gravitational potential forms. The total energy of the system remains constant, of course.

Situations where the net force is exactly perpendicular to the motion are as rare as situations where the force and motion are in exactly the same direction. What about the more usual case, involving some angle between the force and the motion?

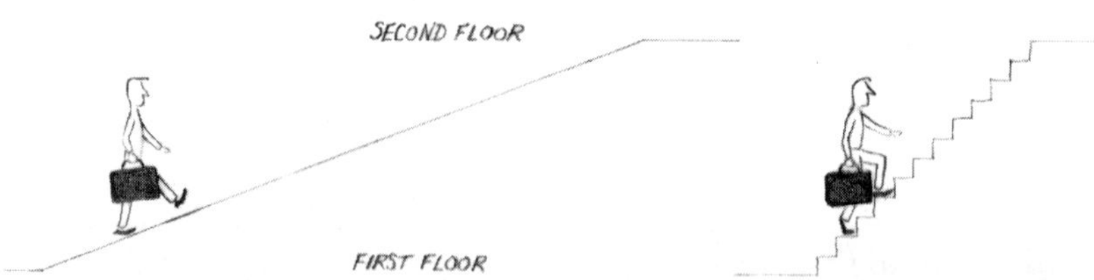

In general, the work done on an object depends on how far the body moves *in the direction of the force*. As stated before, the equation $W = Fd$ properly defines work only if d is the distance moved in the direction of the force. The gravitational force $\vec{F}_{grav}$ is directed *down*. So only the distance *down* determines the amount of work done by $\vec{F}_{grav}$. Change in gravitational potential energy depends *only* on change in height, near the earth's surface, at least. For example, consider raising a suitcase from the first floor to the second floor. The same increase in PE_{grav} of the suitcase–earth system occurs regardless of the path by which the suitcase is raised. Also, each path requires the same amount of work.

More generally, change in PE_{grav} depends only on change of position. The details of the path followed in making the change make no difference at all. The same is true for changes in elastic potential energy and electric potential energy. The changes depend only on the initial and final positions, and not on the path taken between these positions.

If frictional forces also have to be overcome, additional work will be needed, and that additional work may depend on the path chosen, for example, whether it is long or short.

SG 17

An interesting conclusion follows from the fact that change in PE_{grav} depends only on change in height. For example, consider a child on a slide. The gravitational potential energy decreases as her altitude decreases. If frictional forces are vanishingly small, all the work goes into transforming PE_{grav} into KE. Therefore, the increases in KE depend only on the decreases in altitude. In other words, the child's speed when she reaches the ground will be the same whether she slides down or jumps off the top. A similar principle holds for satellites in orbit and for electrons in TV tubes. In the absence of friction, the change in kinetic energy depends only on the initial and final positions, and not on the path taken between them. This principle gives great simplicity to some physical laws, as you will see when you study gravitational and electric fields in Chapter 14.

?

10. *How much work is done on a satellite during each revolution if its mass is* m, *its period is* T, *its speed is* v, *and its orbit is a circle of radius* R?

11. *Two skiers were together at the top of a hill. While one skier skied down the slope and went off the jump, the other rode the ski-lift down. Compare their changes in gravitational potential energy.*

12. *A third skier went directly down a straight slope. How would this skier's speed at the bottom compare with that of the skier who went off the jump?*

13. *No work is done when*

(a) *a heavy box is pushed at constant speed along a rough horizontal floor.*

(b) a nail is hammered into a board.

(c) there is no component of force parallel to the direction of motion.

(d) there is no component of force perpendicular to the direction of motion.

See, for example, Experiment 3.11 on mixing hot and cold liquids.

10.5 Heat as energy

Suppose that a book on a table has been given a push and is sliding across the tabletop. If the surface is rough, it will exert a fairly large frictional force, and the book will stop quickly. The book's kinetic energy will rapidly disappear. No corresponding increase in potential energy will occur, since there is no change in height. It appears that, in this example, mechanical energy is not conserved.

However, close examination of the book and the tabletop shows that they are warmer than before. The disappearance of kinetic energy of the book is accompanied by the appearance of *heat*. This suggests, but by no means proves, that the kinetic energy of the book was transformed into heat. If so, heat must be one form of energy. This section deals with how the idea of heat as a form of energy gained acceptance during the nineteenth century. You will see how theory was aided by practical knowledge of the relation of heat and work.

Until the middle of the nineteenth century, heat was generally thought to be some kind of fluid, called *caloric fluid*. No heat is lost or gained overall when hot and cold bodies are mixed. (Mixing equal parts of boiling and nearly freezing water produces water at just about 50°C.) One could therefore conclude that the caloric fluid is conserved in that kind of experiment. Some substances, like wood or coal, seem to lock up that "fluid" and can release it during combustion.

Although the idea that the heat content of a substance is represented by a quantity of conserved fluid was an apparently useful one, it does not adequately explain some phenomena involving heat. Friction, for example, was known to produce heat. But it was difficult to understand how the conservation of caloric fluid applied to friction.

Benjamin Thompson was born in Woburn, Massachusetts, in 1753. After several years as a shopkeeper's apprentice, he married a rich widow and moved to Concord (then called Rumford). During the Revolution, Thompson was a Tory; he left with the British army for England when Boston was taken by the rebels. In 1783, Thompson left England and ultimately settled in Bavaria. There he designed fortifications and built munitions, and served as an administrator. The King of Bavaria was sufficiently impressed to make him a Count in 1790, and Thompson took the name Rumford. In 1799 he returned to England and continued to work on scientific experiments. Rumford was one of the founders of the Royal Institution. In 1804 he married Lavoisier's widow; the marriage was an unhappy one, and they soon separated. Rumford died in France in 1814, leaving his estate to institutions in the United States.

In the late eighteenth century, while boring cannon for the Elector of Bavaria, Benjamin Thompson, Count Rumford, observed that a great deal of heat was generated. Some of the cannon shavings were hot enough to glow. Rumford made some careful measurements by immersing the cannon in water and measuring the rate at which the temperature rose. His results showed that so much heat evolved that the cannon would have melted had it not been cooled. From many such experiments, Rumford concluded that heat is not a conserved fluid but is generated when work is done and continues to appear without limit as long as work is done. His estimate of the ratio of heat to work was within an order of magnitude of the presently accepted value.

Rumford's experiments and similar work by Davy at the Royal Institution did not convince many scientists at the time. The reason may have been that Rumford could give no clear suggestion of just what heat is, at least not in terms that were compatible with the accepted models for matter at that time.

Nearly a half-century later, James Prescott Joule repeated on a smaller scale some of Rumford's experiments. Starting in the 1840's and continuing for many years, Joule refined and elaborated his apparatus and his techniques. In all cases, the more careful he was, the more exact was the proportionality of the quantity of heat (as measured by a change in temperature and the amount of work done).

James Prescott Joule (1818–1889). Joule was the son of a wealthy Manchester brewer. He is said to have become first interested in his arduous experiments by the desire to develop more efficient engines for the family brewery.

For one of his early experiments, Joule constructed a simple electric generator, which was driven by a falling weight. The electric current that was generated heated a wire. The wire was immersed in a container of water, which it heated. From the distance that the weight descended Joule calculated the work done (the decrease in gravitational potential energy). The product of the mass of the water and its temperature rise gave him a measure of the amount of heat produced. In another experiment, he compressed gas in a bottle immersed in water, measuring the amount of work done to compress the gas. He then measured the amount of heat given to the water as the gas grew hotter on compression.

Joule's most famous experiments involved an apparatus in which slowly descending weights turned a paddle wheel in a container of water. Owing to the friction between the wheel and the liquid, work was done on the liquid, raising its temperature.

Joule repeated this experiment many times, constantly improving the apparatus and refining his analysis of the data. He learned to take very great care to insulate the container so that heat was not lost to the room. He measured the temperature rise with a precision of a small fraction of a degree. He even allowed for the small amount of kinetic energy the descending weights had when they reached the floor.

Joule published his results in 1849. He reported:

> 1st. That the quantity of heat produced by the friction of bodies, whether solid or liquid, is always proportional to the quantity of [energy] expended. And 2nd. That the quantity of heat capable of increasing the temperature of a pound of water . . . by 1° Fahr. requires for its evolution the expenditure of a mechanical energy represented by the fall of 772 lb through the distance of one foot.

Joule used the word "force" instead of "energy." The currently used scientific vocabulary was still being formed.

This unit is called a *British Thermal Unit* (BTU).

The first statement is the evidence that heat is a form of energy, contrary to the caloric theory. The second statement gives the numerical magnitude of the ratio he had found. This ratio related a unit of mechanical energy (the foot-pound) and a unit of heat (the heat required to raise the temperature of 1 lb of water by 1° on the Fahrenheit scale).

By the time Joule did his experiments, the idea of a caloric fluid seemed to have outlived its usefulness. The idea that heat is a form of energy was slowly being accepted. Joule's experiments were a strong argument in favor of that idea. You will look at its development more closely in Sec. 10.9.

Until recently, it was traditional to measure heat in units based on temperature changes, and mechanical energy in units based on work. This made comparison of results in one kind of experiment easy, but it obscured the fundamental equivalence of heat and other forms of energy. All types of energy are expressed in joules (J):

$$1\ \text{J} = 1\ \text{kg}^2/\text{sec}^2 = 1\ \text{N-m (newton-meter)}$$

However, you will often see energies expressed in terms of other units. A few of them are listed here.

Unit Name	Symbol	Definition	Conversion
kilowatt hour	kWh	A watt (W) is 1 J per second, so 1 J = 1 W · sec. A kWh is the amount of energy delivered in an hour if 1 kJ is delivered per second.	1 kWh = 3.60 MJ
Calorie (or kilocalorie)	Cal (or kcal)	The energy required to heat 1 kg of water by 1°C.	4.18 kJ
British Thermal Unit	BTU	The energy required to heat 0.454 kg by 0.556°C.	1.06 kJ

The metric system uses prefixes to specify some multiplication factors for the fundamental units. The following table shows the common prefixes.

Factor	Prefix	Symbol
10	deca-	da
10^2	hecto-	h
10^3	kilo-	k
10^6	mega-	M
10^9	giga-	G
10^{12}	tera-	T
10^{15}	peta-	P
10^{-1}	deci-	d
10^{-2}	centi-	c
10^{-3}	milli-	m
10^{-6}	micro-	μ
10^{-9}	nano-	n
10^{-12}	pico-	p
10^{-15}	femto-	f

?

14. When a book slides to a stop on the horizontal rough surface of a table

(a) the kinetic energy of the book is transformed into potential energy.

(b) heat is transformed into mechanical energy.

(c) the kinetic energy of the book is transformed into heat energy.

(d) the momentum of the book itself is conserved.

15. The kilocalorie is

(a) a unit of temperature.

(b) a unit of energy.

(c) 1 kg of water at 1°C.

16. In Joule's paddle-wheel experiment, was all the change of gravitational potential energy used to heat the water?

10.6 | The steam engine and the Industrial Revolution

Until about 200 years ago, most work was done by people or animals. Work was obtained from wind and water also, but these were generally unreliable as sources of energy. For one thing, they were not always available when and where they were needed. In the eighteenth century, miners began to dig deeper and deeper in search of a greater coal supply. Water tended to seep in and flood these deeper mines. The need arose for an economical method of pumping water out of mines. The steam engine was developed initially to meet this very practical need.

The steam engine is a device for converting the energy of some kind of fuel into heat energy. For example, the chemical energy of coal or oil, or the nuclear energy of uranium, is converted to heat. The heat energy in turn is converted into mechanical energy. This mechanical energy can be used directly to do work, as in a steam locomotive, or transformed into electrical energy. In typical twentieth-century industrial societies, most of the energy used in factories and homes comes from electrical energy. Falling water is used to generate electricity in some parts of the country. But steam engines still generate most of the electrical energy used in the United States today. There are other heat engines, such as internal combustion engines, for example. The steam engine remains a good model for the basic operation of this whole family of engines.

Since ancient times, it has been known that heat can be used to produce steam, which can then do mechanical work. The *aeolipile*, invented by Heron of Alexandria about 100 A.D., worked on the principle of Newton's third law. (See margin.) The rotating lawn sprinkler works the same way except that the driving force is water pressure instead of steam pressure.

Heron's aeolipile was a toy, meant to entertain rather than to do any useful work. Perhaps the most "useful" application of steam to do work in the ancient world was another of Heron's inventions. This steam-driven device astonished worshippers in a temple by causing a door to open when a fire was built on the altar. Not until late in the eighteenth century, however, were commercially successful steam engines invented.

The first commercially successful steam engine was invented by Thomas Savery (1650–1715), an English military engineer. In Savery's engine, water is lifted out of a mine by alternately filling a tank with high-pressure steam, driving the water up and out of the tank, and then condensing the steam, drawing more water into the tank.

Unfortunately, inherent in the Savery engine's use of high-pressure steam was a serious risk of boiler or cylinder explosions. This defect was remedied by Thomas Newcomen (1663–1729), another Englishman. Newcomen invented an engine

A model of Heron's aeolipile. Steam produced in the boiler escapes through the nozzles on the sphere, causing the sphere to rotate.

that used steam at lower pressure. His engine was superior in other ways also. For example, it could raise loads other than water. Instead of using the steam to force water into and out of a cylinder, Newcomen used the steam to force a piston back and forth. The motion of the piston could then be used to drive a pump or other engine.

The Newcomen engine was widely used in Britain and other European countries throughout the eighteenth century. By modern standards, it was not a very good engine. It burned a large amount of coal but did only a small amount of work at a slow, jerky rate. But the great demand for machines to pump water from mines produced a good market, even for Newcomen's uneconomical engine.

The work of Scotsman James Watt led to a greatly improved steam engine. Watt's father was a carpenter who had a successful business selling equipment to shipowners. Watt was in poor health much of his life and gained most of his early education at home. In his father's attic workshop, Watt developed considerable skill in using tools. He wanted to become an instrument maker and went to London to learn the trade. Upon his return to Scotland in 1757, he obtained a position as instrument maker at the University of Glasgow.

In the winter of 1763–1764, Watt was asked to repair a model of Newcomen's engine that was used for demonstration lectures at the university. This assignment had immense worldwide consequences. In acquainting himself with the model, Watt was impressed by how much steam was required to run the engine. He undertook a series of experiments on the behavior of steam and found that a major problem was the temperature of the cylinder walls. Newcomen's engine wasted most of its heat in warming up the walls of its cylinders. The walls were then cooled again every time cold water was injected to condense the steam.

Early in 1765, Watt remedied this wasteful defect by devising a modified type of steam engine. In retrospect, it sounds like a simple idea. After pushing the piston up, the steam was admitted to a *separate* container to be condensed. With this system, the cylinder could be kept hot all the time, and the condenser could be kept cool all the time.

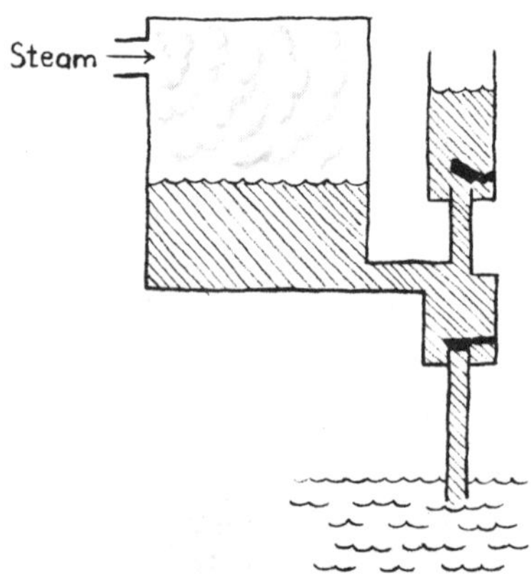

The separate condenser allowed huge fuel savings. Watt's engine could do twice as much work as Newcomen's with the same amount of fuel. Watt also added many other refinements, such as automatically controlled valves that were opened and closed by the reciprocating action of the piston itself, as well as a governor that controlled the amount of steam reaching the engine, to maintain a constant speed for the engine.

This idea, of using part of the output of the process to regulate the process itself, is called *feedback*. It is an essential part of the

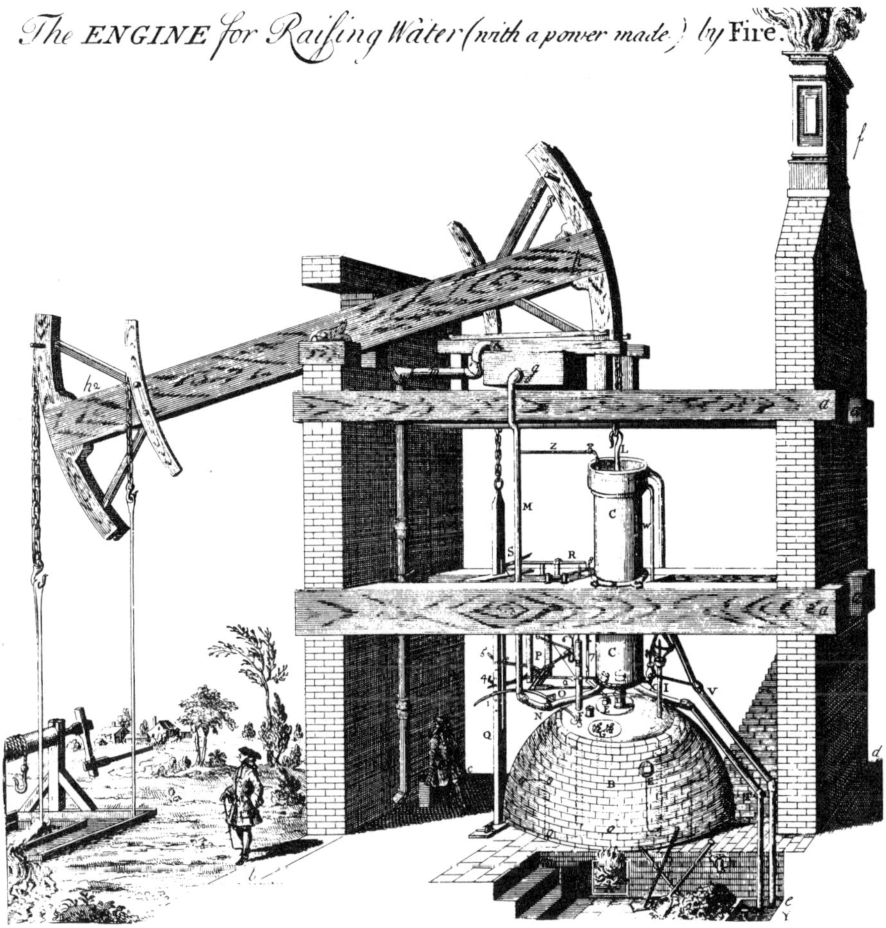

Above, a contemporary engraving of a working Newcomen steam engine. In July 1698, Savery was granted a patent for "A new invention for raising of water and occasioning motion to all sorts of mill work by the impellant force of fire, which will be of great use and advantage for drayning mines, serving townes with water, and for the working of all sorts of mills where they have not the benefitt of water nor constant windes." The patent was good for 35 years and prevented Newcomen from making much money from his superior engine during this period.

design of many modern mechanical and electronic systems.

Watt and his associates were good businessmen as well as good engineers. They made a fortune manufacturing and selling the improved steam engines.

Watt's inventions stimulated the development of engines that could do many other jobs. Steam drove machines in factories, railway locomotives, steamboats, and so forth. Watt's engine gave an enormous stimulus to the growth of industry in Europe and America. It thereby helped transform the economic and social structure of Western civilization.

The widespread development of engines and machines revolutionized mass production of consumer goods, construction, and transportation. The average standard of living in Western Europe and the United States rose sharply. It is difficult for most people in the industrially "developed" countries to imagine what life was like before the Industrial Revolution. But not all the effects of industrialization have been beneficial. The nineteenth-century factory system provided an opportunity for

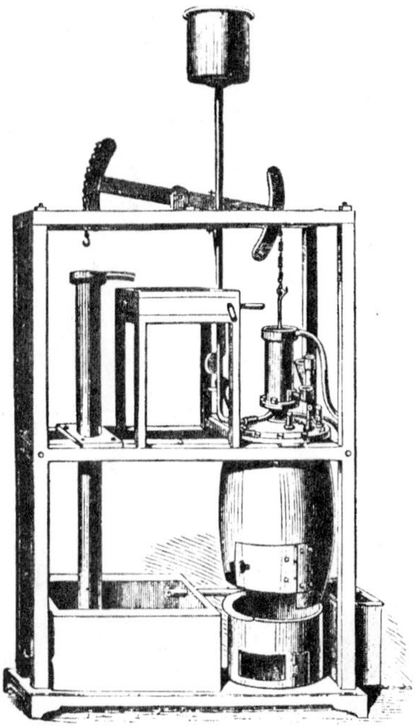

The actual model of the Newcomen engine that inspired Watt to conceive of the separation of condenser and piston.

some greedy and cruel employers to treat workers almost like slaves. These employers made huge profits, while keeping employees and their families on the edge of starvation. This situation, which was especially serious in England early in the nineteenth century, led to demands for reform. New laws eventually eliminated the worst excesses.

More and more people left the farms, either voluntarily or forced by poverty and new land laws, to work in factories. Conflict grew intense between the working class, made up of employees, and the middle class, made up of employers and professionals. At the same time, some artists and intellectuals began to attack the new tendencies of their society. They saw this society becoming increasingly dominated by commerce, machinery, and an emphasis on material goods. In some cases, they confused research science itself with its technical applications (as is still done today). Scientists were sometimes accused of explaining away all the awesome mysteries of nature. These artists denounced both science and technology, while often refusing to learn anything about them. A poem by William Blake contains the questions:

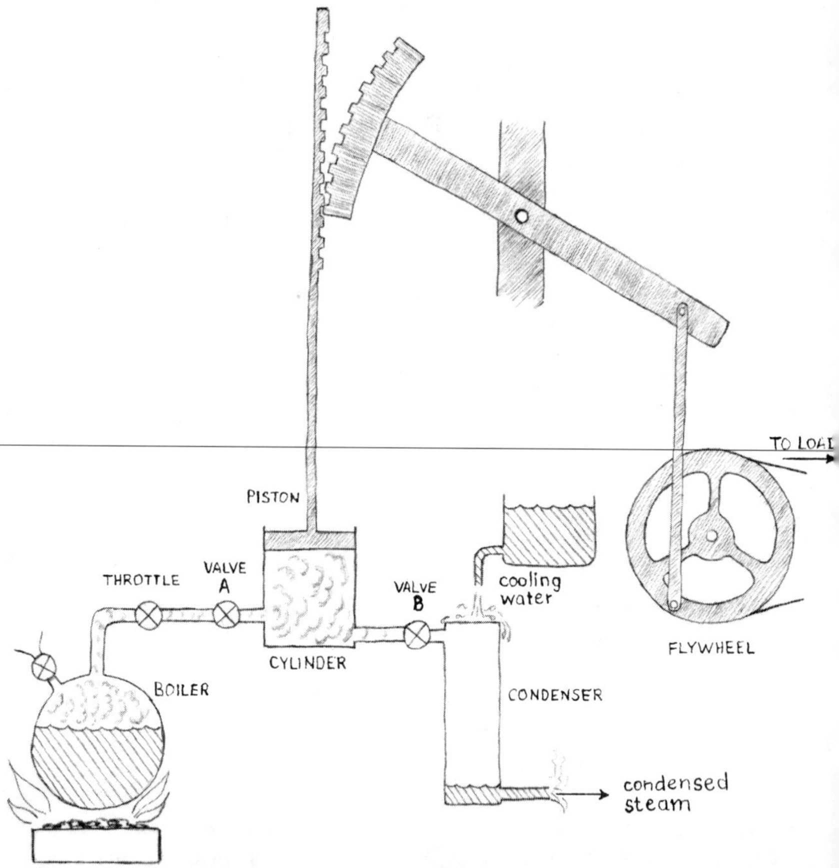

With valve A open and valve B closed, steam under pressure enters the cylinder and pushes the piston upward. When the piston nears the top of the cylinder, valve A is closed to shut off the steam supply. Then valve B is opened, so that steam leaves the cylinder and enters the condenser. The condenser is kept cool by water flowing over it, so the steam condenses. As steam leaves the cylinder, the pressure there decreases. Atmospheric pressure (helped by the inertia of the flywheel) pushes the piston down. When the piston reaches the bottom of the cylinder, valve B is closed, and valve A is opened, starting the cycle again. Valves A and B are connected to the piston directly, so that the motion of the piston itself operates them.

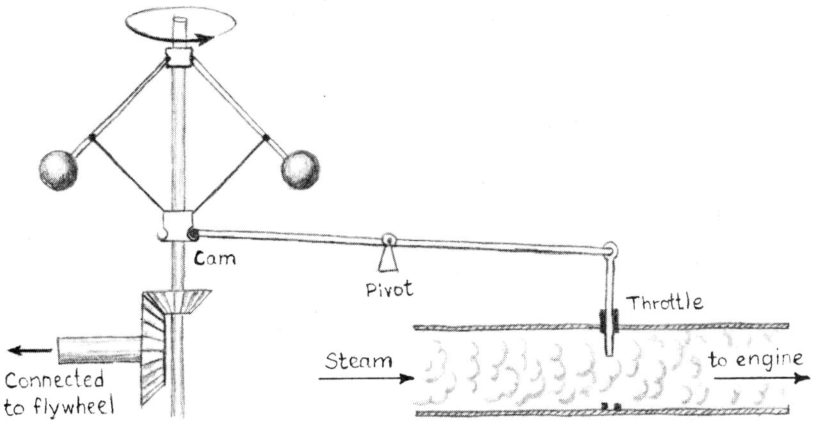

Watt's "governor." If the engine speeds up for some reason, the heavy balls swing out to rotate in larger circles. They are pivoted at the top, so the sleeve below is pulled up. The cam that fits against the sleeve is therefore also pulled up; this forces the throttle to move down and close a bit. The reduction in steam reaching the engine thus slows it down again. The opposite happens when the engine starts to slow down. The net result is that the engine tends to operate at nearly a stable level.

And did the Countenance Divine
   Shine forth upon our clouded hills?
And was Jerusalem builded here
   Among these dark Satanic mills?

Elsewhere, Blake advised his readers "To cast off Bacon, Locke, and Newton." John Keats was complaining about science when he included in a poem the line: "Do not all charms fly / At the mere touch of cold philosophy?" These attitudes are part of an old tradition, going back to the ancient Greek opponents of Democritus' atomism. Galilean and Newtonian physics also were attacked for distorting values. The same type of accusation can still be heard today.

Steam engines are no longer widely used as direct sources of power in industry and transportation. Indirectly, however, steam is still the major source of power. The steam turbine, invented by the English engineer Charles Parsons in 1884, has now largely replaced older kinds of steam engines. At present, steam turbines drive the electric generators in most electric-power stations. These steam-run generators supply most of the power for the machinery of modern civilization. Even in nuclear power stations, the nuclear energy is generally used to produce steam, which then drives turbines and electric generators.

A steam locomotive from the early part of the twentieth century.

The basic principle of the Parsons turbine is simpler than that of the Newcomen and Watt engines. A jet of high-pressure steam strikes the blades of a rotor, driving the rotor around at high speed. The steam expands after passing through the rotor, so the next rotor must be larger. This accounts for the characteristic shape of turbines. Large electric-power station turbines, such as "Big Allis" in New York City, use more than 500,000 kg of steam an hour and generate electrical energy at a rate of 150 million joules per second.

The usefulness of an engine for many tasks is given by the rate at which it can deliver energy, that is, by its power. The unit of power is the watt, symbol W. It is defined as 1 W = 1 J/sec

Watt, of course, used pounds and feet to express these results.

In some contexts, the horsepower is defined as 735.36 W. More often, it is defined as 745.56 W. This ambiguity of traditional units is one of the reasons for replacing them with metric ones.

Matthew Boulton (Watt's business partner) proclaimed to Boswell (the biographer of Samuel Johnson): "I sell here, Sir, what all the world desires to have: POWER!"

As with energy, there are many common units of power with traditional definitions. Before the steam engine, the standard source of power was the workhorse. Watt, in order to rate his engines in a unit people could understand, measured the power output of a horse. He found that a strong horse, working steadily, could lift an object of 75 kg mass, which weighed about 10 N, at a rate of about 1 m/sec. The horse in this case thus did work at a rate of about 750 W. The "horsepower" unit is still used, but its value is now given by definition (see margin), not by experiment.

SOME POWER RATINGS

Person turning a crank	0.06 h.p.	50 W
Overshot waterwheel	3	2 kW
Turret windmill	10	7 kW
Savery steam engine (1702)	1	0.7 kW
Newcomen engine (1732)	12	9 kW
Smeaton's Long Benton engine (1772)	40	30 kW
Watt engine (of 1778)	14	10 kW
Cornish engine for London water works (1837)	135	100 kW
Electric power station engines (1900)	1,000	0.7 MW
Nuclear power station turbine (1970)	300,000	200 MW

(Adapted from R. J. Forbes, in C. Singer *et al.*, *History of Technology.*)

Richard Trevithick's railroad at Euston Square, London, 1809.

A nineteenth-century French steam cultivator.

The "Charlotte Dundas," the first practical steamboat, built by William Symington, an engineer who had patented his own improved steam engine. It was tried out on the Forth and Clyde Canal in 1801.

17. *The purpose of the separate condenser in Watt's steam engine is to*

(a) *save the water so it can be used again.*

(b) *save fuel by not having to cool and reheat the cylinder.*

(c) *keep the steam pressure as low as possible.*

(d) *make the engine more compact.*

18. *The history of the steam engine suggests that the social and economic effects of technology are*

(a) *always beneficial to everyone.*

(b) *mostly undesirable.*

(c) *unimportant one way or another.*

Above, a 200,000-kilowatt turbine being assembled. Notice the thousands of blades on the rotor.

10.7 | The efficiency of engines

Joule's finding a value for the "mechanical equivalent of heat" made it possible to describe engines in a new way. The concept of *efficiency* applies to an engine or to any device that transforms energy from one form to another. Efficiency is defined as the percentage of the input energy that appears as useful output. Since energy is conserved, the greatest possible efficiency is 100% when *all* of the input energy appears as useful output. Obviously, efficiency must be considered as seriously as power output in

A schematic model of an integrated energy supply.

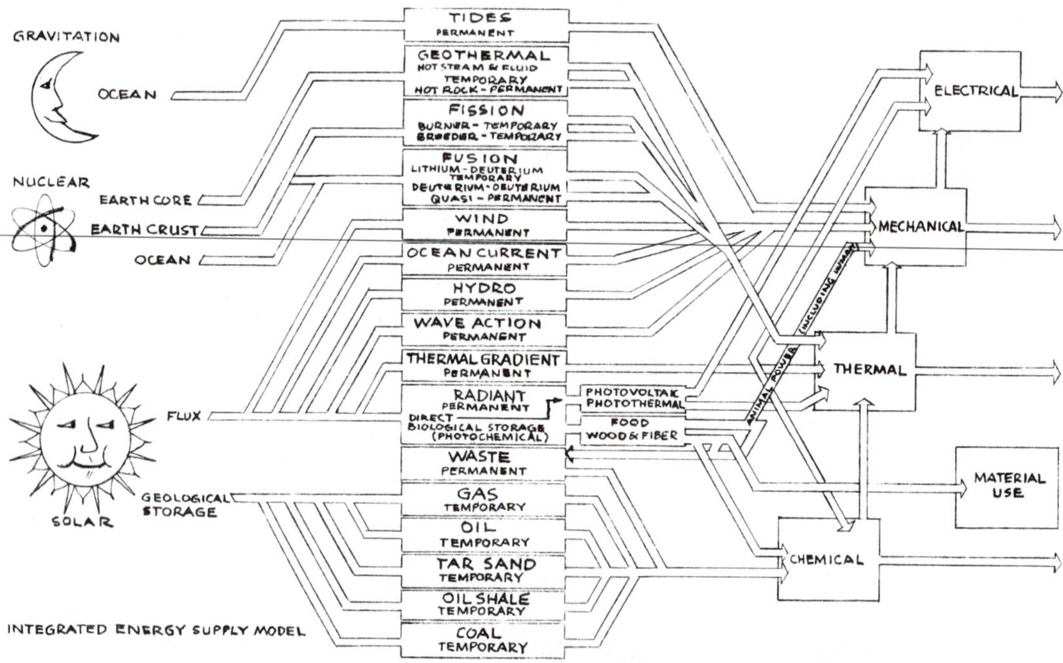

designing engines. Fuel is, after all, a part of the cost of running an engine, and the more efficient an engine is, the cheaper it is to run.

Watt's engine was more efficient than Newcomen's, which in turn was more efficient than Savery's. Is there any limit to improvements in efficiency?

The law of energy conservation clearly imposes a limit of 100%. No engine can put out more work than is put into it. Even before that law had been formulated, a young French engineer, Sadi Carnot, established that there is in practice an even lower limit. The reasons for this limit are just as fundamental as the law of energy conservation.

Carnot started with the postulate that heat does not by itself flow from a cold body to a hot one. It then follows that if heat does flow from cold to hot, some other change must take place elsewhere. Some work must be done. Using an elegant argument, which is summarized on page 300, Carnot showed that no engine can be more efficient than an ideal, reversible engine and that all such engines have the same efficiency.

Since all reversible engines have the same efficiency, one has only to choose a simple engine and calculate its efficiency to find an upper limit to the efficiency of any engine. Carnot did the calculation and found that the ratios of heat and work in a reversible engine depend only on the temperature of the hot substance from which the engine obtains its heat and on the temperature of the cold substance that extracts the waste heat from the engine. The temperatures used in this case are called *absolute*, or *Kelvin*, temperatures:

$$T \text{ (absolute, in °K)} = T \text{ (Celsius, in °C)} + 273$$

On the Kelvin scale, water freezes at 273°K. Absolute zero (0°K) is −273°C.

The expression found by Carnot for the efficiency of reversible engines is

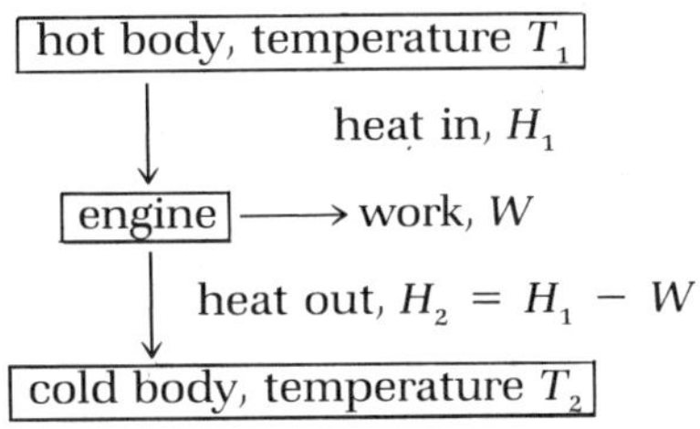

$$\text{efficiency} = \frac{\text{work out}}{\text{heat in}} = \frac{W}{H_1} = \left(1 - \frac{T_2}{T_1}\right)$$

Sadi Carnot (1796–1832). Son of one of Napoleon's most trusted generals, Sadi Carnot was one of the new generation of expert administrators who hoped to produce a new enlightened order in Europe. He died of cholera in Paris at the age of 36.

See also Sec. 11.5.

Although Carnot did not write the formula this way, we are making use of the fact that heat and energy are equivalent.

SG 24–29

The relation involved can also be written as follows:

$$\frac{H_1}{T_1} = \frac{H_2}{T_2} = \frac{W}{T_1 - T_2}$$

You can feel some of the waste heat by feeling the exhaust from a car.

This result is called Carnot's theorem. Notice that unless T_2 is zero, an unattainably low temperature, no engine can have an efficiency of 1 (or 100%); that is, every engine *must* return some "waste" heat to the outside before returning to get more energy from the hot body.

In steam engines, the "hot body" is the steam fresh from the boiler, and the waste heat is extracted from the condenser. In an internal combustion engine (a car engine, for example), the hot body is the gas inside the cylinder just as it explodes, and the cold substance is the exhaust. Any engine that derives its mechanical energy from heat must also be cooled to remove the "waste" heat. If there is any friction, or other inefficiency, in the engine, it will add further heat to the waste and reduce the efficiency to below the theoretical limit.

In the MKS system of units, the coefficient of performance yields a "pure" number (without units). Many American engineers measure heat in BTU and electric energy in kilowatt-hours (kWh), so that coefficients of performance for air conditioners or refrigerators are often rated in BTU/kWh.

A refrigerator or air conditioner is also called a "heat engine." It uses work (in the form of electrical or mechanical power) to move heat from a cold body (from inside the freezing compartment) to a hot one (the outside room). Carnot's relations also provide an upper limit to how much heat can be extracted by such an engine for a given amount of work, or in practical terms, for how big your electric bill will be.

hot body, T_1
↑ heat out, $H_1 = W + H_2$
engine ← work, W
↑ heat in, H_2
cold body, T_2

$$\text{coefficient of performance} = \frac{\text{heat in}}{\text{work in}} = \frac{H_2}{W} = \frac{T_2}{T_1 - T_2}$$

See Sec. 10.10.

The generalization of Carnot's theorem is now known as the *second law of thermodynamics*. This law is recognized as one of the most powerful laws of physics. Even in simple situations it can help explain natural phenomena and the fundamental limits of technology.

SG 25

The coefficient of performance of an air conditioner depends on the reciprocal of the temperature difference between the inside of a house and the outside. The bigger that difference, the more work it will take to move the same amount of heat from inside to outside. An air conditioner mounted in the sun therefore needs more electricity than one mounted in the shade on the same house.

If you burn oil at home, the furnace requires some inefficiency to burn cleanly, so some heat is lost out the chimney. (Also, a very efficient furnace would be unaffordably expensive.) Usually about half the energy content of the oil is converted to useful heat in a house.

If you install "flameless electric heat," which works just like an electric blanket or toaster put in under the floor, the power company has to burn oil in a boiler, use the steam to generate electricity, and deliver the electricity to your home. Because metals melt above a certain temperature and because the cooling water never gets below freezing, Carnot's theorem makes it impossible to make the efficiency greater than about 0.6. Since the power company's boiler also loses some of its energy out the chimney, and since the electricity loses some of its energy on the way from the power plant, only about one-quarter to one-third of the energy originally in the oil actually makes it to your home. Obviously, electric heating wastes a lot of oil or coal.

The coldest temperature feasible for T_2 is about 280°K. (Why?) The hottest possible temperature for T_1 is about 780°K. So the maximum efficiency is 0.64.

Because of the limits placed by Carnot's theorem on heat engines, it is sometimes important not only to give the actual efficiency of a heat engine but also to specify how close it comes to the maximum possible. One freezer might have a much larger coefficient of performance than another, but only because it does not operate at as low a temperature as the other.

The more carefully you look at a process, the more information is seen to be important. At first, you could probably be satisfied with any sort of engine. A closer examination will lead you to understand the equivalence of heat and energy, so that energy use and efficiency become important criteria in choosing an engine. Carnot's investigation showed the importance not only of energy but also of temperature.

?

19. The efficiency of a heat engine is the ratio of

(a) the work output to the heat input.

(b) the work output to the heat output.

(c) the heat output to the heat input.

20. A heat engine is most efficient when it works between objects that have

(a) a large temperature difference.

(b) a small temperature difference.

(c) a large size.

10.8 | Energy in biological systems

All living things need a supply of energy to maintain life and to carry on their normal activities. Human beings are no exception;

like all animals, we depend on food to supply us with energy.

Most human beings are omnivores; that is, they eat both animal and plant materials. Some animals are herbivores, eating only plants, while others are carnivores, eating only animal flesh. But all animals, even carnivores, ultimately obtain their food energy from plant material. The animal eaten by the lion has previously dined on plant material or on another animal that has eaten plants.

Carbohydrates are molecules made of carbon, hydrogen, and oxygen. A simple example is the sugar glucose, the chemical formula for which is $C_6H_{12}O_6$.

Green plants obtain energy from sunlight. Some of that energy is used by the plant to perform the functions of life. Much of the energy is used to make carbohydrates out of water (H_2O) and carbon dioxide (CO_2). The energy used to synthesize carbohydrates is not lost; it is stored in the carbohydrate molecules as chemical energy.

The process by which plants synthesize carbohydrates is called *photosynthesis*. This process is still not completely understood, and research in this field is lively. The synthesis takes place in many small steps, and many of the steps are well understood. It is conceivable that scientists may learn how to photosynthesize carbohydrates without plants, thus producing food economically for the rapidly increasing world population. The overall process of producing carbohydrates (the sugar glucose, for example) by photosynthesis can be represented as follows:

carbon dioxide + water + energy → glucose + oxygen

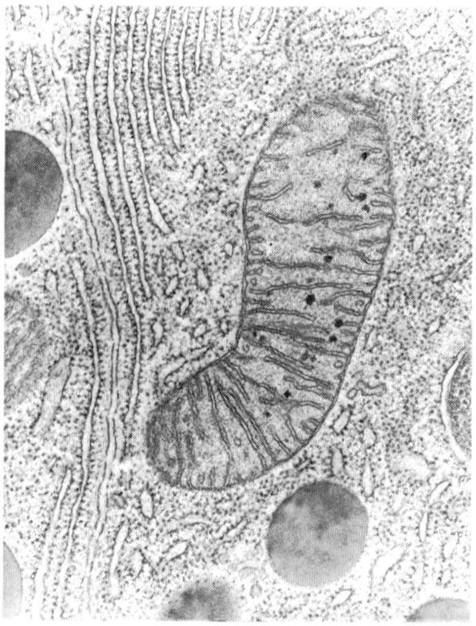

Electron micrograph of an energy-converting mitochondrion in a bat cell (200,000 times actual size). A chloroplast, which is the part of a plant cell that converts CO_2 and H_2O to carbohydrates, looks very much the same, except it is green, not colorless.

The energy stored in the glucose molecules is used by the animal that eats the plant. This energy maintains the body temperature, keeps the heart, lungs, and other organs operating, and enables various chemical reactions to occur in the body. The animal also uses the energy to do work on external objects. The process by which the energy stored in sugar molecules is made available to the cell is very complex. It takes place mostly in tiny bodies called *mitochondria*, which are found in all cells. Each mitochondrion contains enzymes which, in a series of about 10 steps, split glucose molecules into simpler molecules. In another sequence of reactions, these molecules are oxidized (combined with oxygen), thereby releasing most of the stored energy and forming carbon dioxide and water:

glucose + oxygen → carbon dioxide + water + energy

Proteins and fats are used to build and restore tissue and enzymes, and to pad delicate organs. They also can be used to provide energy. Both proteins and fats can enter into chemical reactions that produce the same molecules as the split carbohydrates. From that point, the energy-releasing process is the same as in the case of carbohydrates.

The released energy is used to change a molecule called *adenosine diphosphate* (ADP) into *adenosine triphosphate* (ATP).

In short, chemical energy originally stored in glucose molecules in plants is eventually stored as chemical energy in ATP molecules in animals. The ATP molecules pass out of the mitochondrion into the body of the cell. Wherever energy is needed in the cell, it can be supplied by an ATP molecule. As it releases its stored energy, the ATP changes back to ADP. Later, back in a mitochondrion, the ADP is reconverted to energy-rich ATP.

The overall process in the mitochondrion involves breaking glucose, in the presence of oxygen, into carbon dioxide and water. The energy released is transferred to ATP and stored there until needed by the animal's body.

The chemical and physical operations of the living body are in some ways like those of an engine. Just as a steam engine uses chemical energy stored in coal or oil, the body uses chemical energy stored in food. In both cases, the fuel is oxidized to release its stored energy. The oxidation is vigorous in the steam engine, and gentle, in small steps, in the body. In both the steam engine and the body, some of the input energy is used to do work; the rest is used up internally and eventually "lost" as heat to the surroundings.

Some foods supply more energy per unit mass than others. The energy stored in food is usually measured in Calories. However, it could just as well be measured in joules. The table on page 302 gives the energy content of some foods. (The "Calorie" or "large calorie" used by dieticians is identical to the kilocalorie of chemists.)

The chemical energy stored in food can be determined by burning the food in a closed container immersed in water and measuring the temperature rise of water.

Much of the energy you obtain from food keeps your body's internal "machinery" running and keeps your body warm. Even when asleep, your body uses about 1 Cal every minute. This amount of energy is needed just to keep alive.

To do work, you need more energy. Yet only a fraction of this energy can be used to do work; the rest is wasted as heat. Like any engine, the body of humans or other animals is not 100% efficient. Its efficiency when it does work varies with the job and the physical condition and skill of the worker. Efficiency probably never exceeds 25% and usually is less. Studies of this sort are carried out in *bioenergetics*, one of many fascinating and useful fields where physics and biology overlap.

The table on page 302 gives the results of experiments done in the United States to determine the rate at which a healthy young person of average build and metabolism uses energy in various activities. The estimates were made by measuring the amount of carbon dioxide exhaled. Thus, they show the total amount of food energy used, including the amount necessary just to keep the body functioning.

SG 29–31

According to this table, if you did nothing but sleep for eight hours a day and lie quietly the rest of the time, you would still

Close Up

Carnot's Proof

Carnot's proof of maximum efficiency of ideal, reversible engines starts with the premise that when a cold object is in touch with a warmer one, the cold object does not spontaneously cool itself further and so give more heat to the warm object. However, an engine placed between the two bodies *can* move heat from a cold object to a hot one. Thus, a refrigerator can cool a cold bottle further, ejecting heat into the hot room. You will see that this is not simple. Carnot proposed that during any such experiment, the net result cannot be *only* the transfer of a given quantity of heat from a cold body to a hot one.

The engines considered in this case all work in cycles. At the end of each cycle, the engine itself is back to where it started. During each cycle, it has taken up and given off heat, and it has exerted forces and done work.

Consider an engine, labeled R in the figure, which suffers no internal friction, loses no heat because of poor insulation, and runs so perfectly that it can work backwards in exactly the same way as forwards (Fig. A).

A

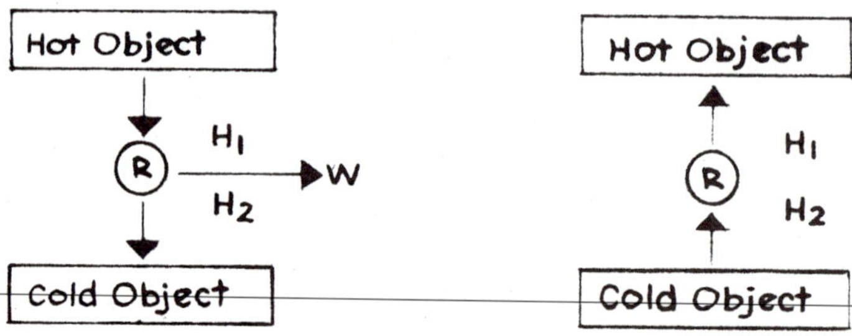

Now suppose someone claims to have invented an engine, labeled Z in the next figure, which is even more efficient than the ideal engine R. That is, in one cycle it makes available the same amount of work, W, as the R engine does, but takes less heat energy, H', from the hot object to do it ($H'_1 < H_1$). Since heat and energy are equivalent and since $H_2 = H_1 - W$ and $H'_2 = H'_1 - W$, it will also be true that $H'_2 < H_2$ (Fig. B).

B

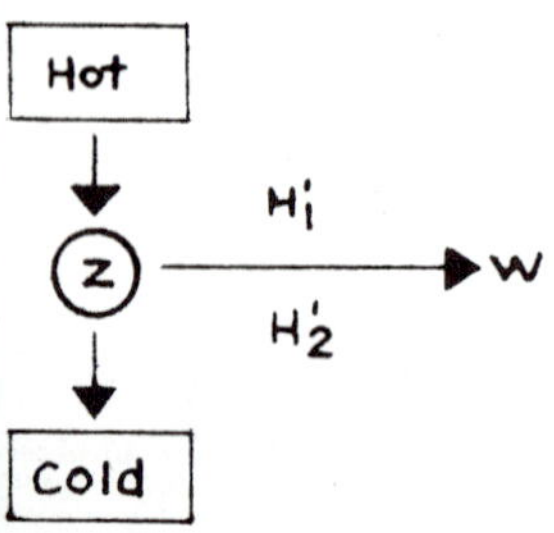

Suppose the two engines are connected so that the work from one can be used to drive the other. For example, the Z engine can be used to make the R engine work like a refrigerator (Fig. C).

At the end of one cycle, both Z and R are back where they started. No work has been done; the *Z* engine has transferred some heat to the cold object; and the R engine has transferred some heat to the hot object. The *net* heat transferred is $H_1 - H'_1$, and the net heat taken from the cold object is $H_2 - H'_2$. These are, in fact, the same:

$$\begin{aligned} H_2 - H'_2 &= (H_1 - W) - (H'_1 - W) \\ &= H_1 - H'_1 \end{aligned}$$

Because Z is supposed to be more efficient than R, this quantity should be positive; that is, heat has been transferred from the cold object to the hot object. Nothing else has happened. But, according to the fundamental premise, this is impossible.

The only conclusion is that the Z engine was improperly "advertised" and that it is either impossible to build or in fact it is *less* efficient than R.

As for two different reversible engines, they must have the same efficiency. Suppose the efficiencies were different; then one would have to be more efficient than the other. What happens when the more efficient engine is used to drive the other reversible engine as a refrigerator? The same argument just used shows that heat would be transferred from a cold body to a hot one. This is impossible. Therefore, the two reversible engines must have the same efficiency.

To actually compute that efficiency, you must know the properties of one reversible engine; all reversible engines working between the same temperatures must have that same efficiency. (Carnot computed the efficiency of an engine that used an ideal gas instead of steam.)

C

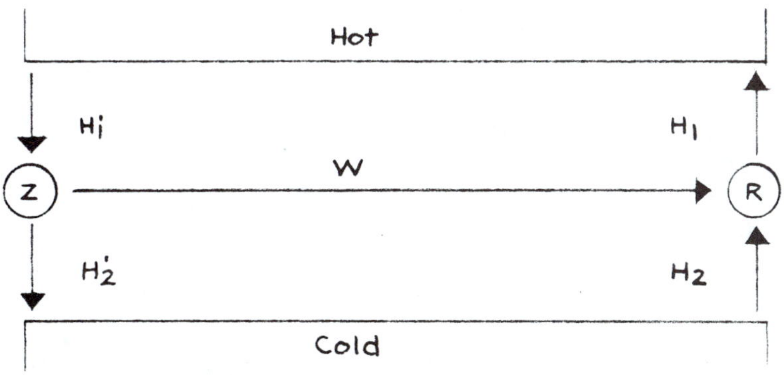

need at least 1,700 Calories of energy each day. There are countries where large numbers of working people exist on less than 1,700 Calories a day. The U.N. Yearbook of National Accounts Statistics shows that in India the average food intake was about 1,600 Calories per day. The United States average was 3,100 Calories per day. About half the population of Southeast Asia is at or below the starvation line. Vast numbers of people elsewhere in the world, including some parts of the United

SG 32 States, are also close to that line. It is estimated that if the available food were equally distributed among all the earth's inhabitants, each would have about 2,400 Calories a day on the average. This is only a little more than the minimum required by a working person.

APPROXIMATE ENERGY CONTENT OF VARIOUS FOODS

	Cal/kg	*MJ/kg*
Butter	7,000	29
Chocolate (sweetened)	5,000	21
Beef (hamburger)	4,000	17
Bread	2,600	11
Milk (whole)	700	3
Apples (raw)	500	2
Lettuce	150	0.6

Adapted from a U.S. Department of Agriculture handbook.

APPROXIMATE RATES OF USING ENERGY DURING VARIOUS ACTIVITIES

	Cal/hr	*W*
Sleeping	70	80
Lying down (awake)	80	90
Sitting still	100	120
Standing	120	140
Typewriting rapidly	140	160
Walking (5 km/hr)	220	250
Digging a ditch	400	460
Running fast	600	700
Rowing in a race	1,200	1,400

Adapted from a U.S. Department of Agriculture handbook.

It is now estimated that at the current rate of increase, the population of the world may double in 30 years. Thus, by the year 2000 it would be 7 billion or more. Furthermore, the *rate* at which the population is increasing is itself increasing. Meanwhile, the production of food supply per person has not increased markedly on a global scale. For example, in the last 10 years the increase in crop yield per acre in the poorer countries has averaged less than 1% per year—far less than the increase in population. The problem of supplying food energy for the world's hungry is one of the most difficult problems facing humanity today.

In this problem of life-and-death importance, what are the roles science and technology can play? Obviously, better agricultural practices should help, both by opening up new land for farming and by increasing production per acre on existing land. The application of fertilizers can increase crop yields, and factories that make fertilizers are not too difficult to build. However, it is important to study *all* the consequences before applying science through technology; otherwise you may create two new problems for every old one that you wish to "fix."

In any particular country, the questions to ask include these: How will fertilizers interact with the plant being grown and with the soil? Will some of the fertilizer run off and spoil rivers and lakes and the fishing industry in that locality? How much water will be required? What variety of the desired plant is the best to use within the local ecological framework? How will the ordinary farmer be able to learn the new techniques? How will the farmer be able to pay for using them?

Upon study of this sort it may turn out that in addition to fertilizer, a country may need just as urgently a better system of bank loans to small farmers and better agricultural education to help the farmer. Such training has played key roles in the rapid rise of productivity in the richer countries. Japan, for example, produces 7,000 college graduate agriculturalists each year. All of Latin America produces only 1,100 per year. In Japan there is one farm advisor for each 600 farms. Compare this with perhaps one advisor for 10,000 farms in Colombia, and one advisor per 100,000 farms in Indonesia.

For long-term solutions, the problem of increasing food production in the poorer countries goes far beyond changing agricultural practices. Virtually all facets of the economies and cultures of the affected countries are involved. Important factors range from internal economic aid and internal food pricing policies to urbanization, industrial growth, public health, and family-planning practice.

Where, in all this, can the research scientist's contribution help? It is usually true that one of the causes of some of the worst social problems is ignorance, including the absence of specific scientific knowledge. For example, knowledge of how food plants can grow efficiently in the tropics is lamentably sparse. Better ways of removing salt from seawater or brackish groundwater are needed to allow irrigating fields with water from these plentiful sources. Before this will be economically possible, more basic knowledge will be needed on just how molecules move through membranes of the sort usable in desalting equipment. Answers to such questions, and many like them, can only come through research in "pure" science by trained research workers having access to adequate research facilities.

The physics of energy transformations in biological processes is one example of a lively interdisciplinary field, namely biophysics (where physics, biology, chemistry, and nutrition all enter). Another connection to physics is provided by the problem of inadequate world food supply; here, too, many physicists, with others, are presently trying to provide solutions through work using their special competence.

21. *Animals obtain the energy they need from food, but plants*

(a) obtain energy from sunlight.

(b) obtain energy from water and carbon dioxide.

(c) obtain energy from seeds.

(d) do not need a supply of energy.

22. *The human body has an efficiency of about 20%. This means that*

(a) only one-fifth of the food you eat is digested.

(b) four-fifths of the energy you obtain from food is destroyed.

(c) one-fifth of the energy you obtain from food is used to run the "machinery" of the body.

(d) you should spend 80% of each day lying quietly without working.

(e) only one-fifth of the energy you obtain from food can be used to enable your body to do work on external objects.

23. *Explain this statement: "The repast of the lion is sunlight."*

"The Repast of the Lion" by Henri Rousseau. The Metropolitan Museum of Art.

10.9 | Arriving at a general conservation law

In Sec. 10.3, the law of conservation of *mechanical* energy was introduced. This law applies only in situations where no mechanical energy is transformed into heat energy or vice versa. Early in the nineteenth century, developments in science, engineering, and philosophy suggested new ideas about energy. It appeared that all forms of energy (including heat) could be transformed into one another with no loss. Therefore, the total amount of energy in the universe must be constant.

Volta's invention of the electric battery in 1800 showed that chemical reactions could produce electricity. It was soon found that electric currents could produce heat and light. In 1820, Hans Christian Oersted, a Danish physicist, discovered that an electric current produces magnetic effects. In 1831, Michael Faraday, the great English scientist, discovered electromagnetic induction: When a magnet moves near a coil or a wire, an electric current is produced in the coil or wire. To some thinkers, these discoveries suggested that all the phenomena of nature were somehow united. Perhaps all natural events result from the same basic "force." This idea, though vague and imprecise, later bore fruit in the form of the law of conservation of energy. All natural events involve a transformation of energy from one form to another. But the total *quantity* of energy does not change during the transformation.

The invention and use of steam engines helped to establish the law of conservation of energy by showing how to measure energy changes. Almost from the beginning, steam engines were rated according to a quantity termed their "duty." This term referred to how heavy a load an engine could lift using a given supply of fuel. In other words, the test was how much *work* an engine could do for the price of a ton of coal. This very practical approach is typical of the engineering tradition in which the steam engine was developed.

Joule began his long series of experiments by investigating the "duty" of electric motors. In this case, duty was measured by the work the motor could do when a certain amount of zinc was used up in the battery that ran the motor. Joule's interest was to see whether motors could be made economically competitive with steam engines.

The concept of work began to develop about this time as a measure of the amount of energy transformed from one form to another. (The actual words "work" and "energy" were not used until later.) This concept made possible quantitative statements about the transformation of energy. For example, Joule used the work done by descending weights as a measure of the amount of gravitational potential energy transformed into heat energy.

In 1843, Joule stated that whenever a certain amount of mechanical energy seemed to disappear, a definite amount of heat always appeared. To him, this was an indication of the conservation of what we now call energy. Joule said that he was

> . . . satisfied that the grand agents of nature are by the Creator's fiat *indestructible*; and that, wherever mechanical [energy] is expended, an exact equivalent of heat is *always* obtained.

Julius Robert von Mayer (1814–1878) was one of the first to realize that heat is a form of energy. He worked out the mechanical equivalent of heat.

Friedrich von Schelling (1775–1854)

One of the great successes of the *Naturphilosophie* was Oersted's discovery of the connection between electricity and magnetism (see Unit 4, Sec. 14.11). See also Unit 2 Epilogue.

Having said this, Joule got back to his work in the laboratory. He was basically a practical man who had little time to speculate about a deeper philosophical meaning of his findings. But others, though using speculative arguments, were also concluding that the total amount of energy in the universe is constant.

A year before Joule's remark, for example, Julius Robert Mayer, a German physician, had proposed a general law of conservation of energy. Mayer had done no quantitative experiments, but he had observed body processes involving heat and respiration. He had also used other scientists' published data on the thermal properties of air to calculate the mechanical equivalent of heat. (Mayer obtained about the same value that Joule did.)

Mayer had been influenced by the German philosophical school now known as *Naturphilosophie* or "Nature Philosophy." This school flourished during the late eighteenth and early nineteenth centuries. According to Nature Philosophy, the various phenomena and forces of nature—such as gravity, electricity, and magnetism—are not really separate from one another but are all manifestations of some unifying "basic" natural force. This philosophy therefore encouraged experiments searching for that underlying force and for connections between different kinds of forces observed in nature.

The most influential thinkers of the school of Nature Philosophers were Johann Wolfgang von Goethe and Friedrich von Schelling. Neither of these men is known today as a scientist. Goethe is generally considered Germany's greatest poet and dramatist, while Schelling is remembered as a minor philosopher. Both men had great influence on the generation of German scientists educated at the beginning of the nineteenth century. The Nature Philosophers were closely associated with the Romantic movement in literature, art, and music. The Romantics protested against the idea of the universe as a great machine. This idea seemed morally empty and artistically worthless to them. The Nature Philosophers also detested the mechanical world view. They refused to believe that the richness of natural phenomena, including human intellect, emotions, and hopes, could be understood as the result of the motions of particles.

The Nature Philosophers claimed that nature could be understood as it "really" is only by direct observation. But no complicated "artificial" apparatus must be used, only feelings and intuitions. For Goethe the goal of his philosophy was: "That I may detect the inmost force which binds the world, and guides its course."

Although its emphasis on the unity of nature led the followers of *Naturphilosophie* to some very useful insights, such as the general concept of the conservation of energy, its romantic and antiscientific bias made it less and less influential. Scientists who had previously been influenced by it, including Mayer, now strongly opposed it. In fact, some hard-headed scientists at first doubted the law of conservation of energy simply because of their distrust of Nature Philosophy. For example, William Barton Rogers, founder of the Massachusetts Institute of Technology, wrote in 1858:

> To me it seems as if many of those who are discussing this question of the conservation of force are plunging into the fog of mysticism.

However, the law was so quickly and successfully put to use in physics that its philosophical origins were soon forgotten.

This episode is a reminder of a familiar lesson: In the ordinary day-to-day work of scientists, experiment and mathematical theory are the usual guides. But in making a truly major advance in science, philosophical speculation often also plays an important role.

Hermann von Helmholtz (1821–1894)

Mayer and Joule were only two of at least a dozen people who, between 1832 and 1854, proposed in some form the idea that energy is conserved. Some expressed the idea vaguely; others expressed it quite clearly. Some arrived at the belief mainly through philosophy; others from a practical concern with engines and machines or from laboratory investigations; still others from a combination of factors. Many, including Mayer and Joule, worked quite independently of one another. The idea of energy conservation was somehow "in the air," leading to essentially simultaneous, separate discoveries.

The initial wide acceptance of the law of conservation of energy owed much to the influence of a paper published in 1847. This was 2 years before Joule published the results of his most precise experiments. The author, a young German physician and physicist named Hermann von Helmholtz, entitled his work "On the Conservation of Force." Helmholtz boldly asserted the idea that others were only vaguely expressing, namely, "that it is impossible to create a lasting motive force out of nothing." He restated this theme even more clearly many years later in one of his popular lectures:

> We arrive at the conclusion that Nature as a whole possesses a store of force which cannot in any way be either increased or diminished, and that, therefore, the quantity of force in Nature is just as eternal and unalterable as the quantity of matter. Expressed in this form, I have named the general law 'The Principle of the Conservation of Force.'

Helmholtz's paper, "Zur Erhaltung der Kraft," was tightly reasoned and mathematically sophisticated. It related the law of conservation of energy to the established principles of Newtonian mechanics and thereby helped make the law scientifically respectable.

Any machine or engine that does work (provides energy) can do so only by drawing from some source of energy. The machine cannot supply more energy than it obtains from the source. When the source runs out, the machine will stop working. Machines and engines can only *transform* energy; they cannot create it or destroy it.

Energy Conservation on Earth

Nuclear reactions inside the earth produce energy at a rate of 3×10^{13}W

The nuclear reactions in the sun produce energy at a rate of 3.5×10^{27}W

The earth receives about 17×10^{16}W from the sun, of which about 33% is immediately reflected, mostly by clouds and the oceans; the rest is absorbed, converted to heat, and ultimately radiated into outer space as infrared radiation. Of that part of the solar energy that is not reflected,...

...5×10^{16}W heats dry land

...3×10^{16}W heats the air, producing winds, waves, etc.

...4×10^{16}W evaporates water

...1.5×10^{13}W is used by marine plants for photosynthesis

...3×10^{13}W is used by land plants for photosynthesis

Most of the energy given to water is given up again when the water condenses to clouds and rain; but every second about 10^{15} J remains as gravitational potential energy of the fallen rain.

Ancient green plants have decayed and left a store of about 2.2×10^{23} J in the form of oil, gas, and coal. This store is being used at a rate of 5×10^{12}W.

Present-day green plants are being used as food for people and animals, at a rate of 2×10^{13}W. Agriculture uses about 10% of this, and people ultimately consume 3×10^{11}W as food.

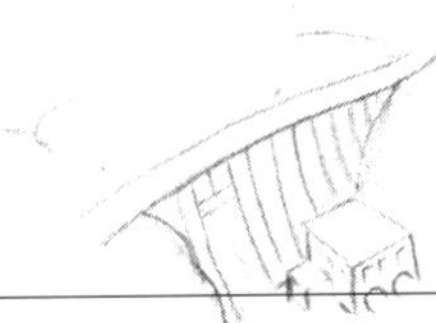

Some of this energy is used to produce 10^{11}W of hydroelectric power

12×10^{11}W is used in generating 4×10^{11}W of electrical power

9×10^{11}W is used in combustion engines. About 75% of this is wasted as heat; less than 3×10^{11}W appears as mechanical power

3×10^{12}W is used for heating; this is equally divided between industrial and domestic uses.

Direct use as raw materials for plastics and chemicals accounts for 2×10^{11}W

Controlled nuclear reactions produce 2×10^{10}W in electrical power

5×10^{11}W

electrochemistry light communication mechanical power

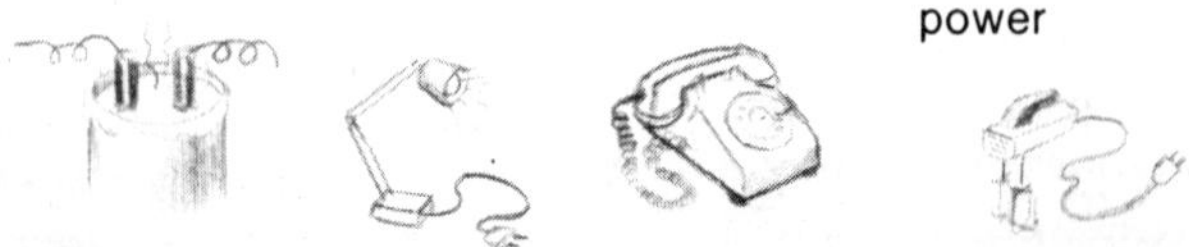

SG 33

24. *The significance of German Nature Philosophy in the history of science is that it*

(a) was the most extreme form of the mechanistic viewpoint.

(b) was a reaction against excessive speculation.

(c) stimulated speculation about the unity of natural phenomena.

(d) delayed progress in science by opposing Newtonian mechanics.

25. *Discoveries in electricity and magnetism early in the nineteenth century contributed to the discovery of the law of conservation of energy because*

(a) they attracted attention to the transformation of energy from one form to another.

(b) they made it possible to produce more energy at less cost.

(c) they revealed what happened to the energy that was apparently lost in steam engines.

(d) they made it possible to transmit energy over long distances.

26. *The development of steam engines helped the discovery of the law of conservation of energy because*

(a) steam engines produced a large amount of energy.

(b) the caloric theory could not explain how steam engines worked.

(c) the precise idea of work was developed to rate steam engines.

(d) the internal energy of a steam engine was always found to be conserved.

10.10 | The laws of thermodynamics

Two laws, one precise and one general, summarize many of the ideas in this chapter. Both of these laws are called laws of thermodynamics.

The first law of thermodynamics is a general statement of the conservation of energy and is based on Joule's finding that heat and energy are equivalent. It would be pleasingly simple to call heat "internal" energy associated with temperature. We could then add heat to the potential and kinetic energy of a system, and call this sum the total energy that is conserved. In fact, this solution works well for a great variety of phenomena, including the experiments of Joule. Difficulties arise with the idea of the heat "content" of a system. For example, when a solid is heated to its melting point, further heat input causes melting *without*

If you do not want to know what the detailed difficulties are, you can skip to the conclusion in the last paragraph on the next page.

Special case of an isolated system:

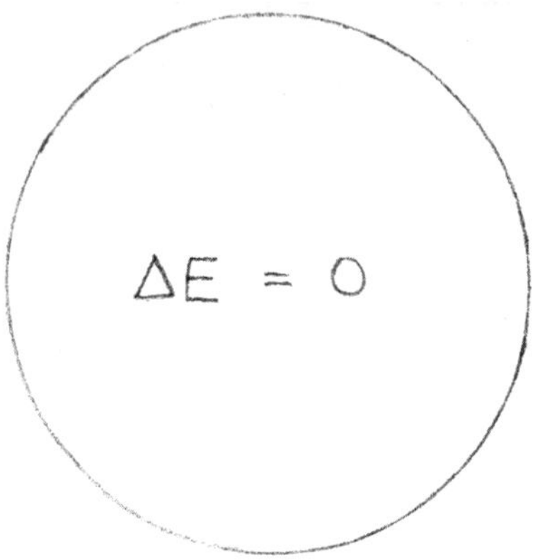

increasing the temperature. (You may have seen this in the experiment "Calorimetry.") Simply adding the idea of heat as one form of a system's energy will not give a complete general law.

Instead of "heat," we can use the idea of an *internal energy*—energy in the system that may take forms not directly related to temperature. We can then use the word "heat" to refer only to a *transfer* of energy between a system and its surroundings. (In a similar way, the term *work* is not used to describe something contained in the system. Rather, it describes the transfer of energy from one system to another.)

Even these definitions do not permit a simple statement such as "Heat input to a system increases its internal energy, and work done on a system increases its mechanical energy." Heat input to a system can have effects other than increasing internal energy. In a steam engine, for example, heat input increases the mechanical energy of the piston. Similarly, *work* done on a system can have effects other than increasing mechanical energy. In rubbing your hands together, for example, the work you do increases the internal energy of the skin of your hands.

In general:

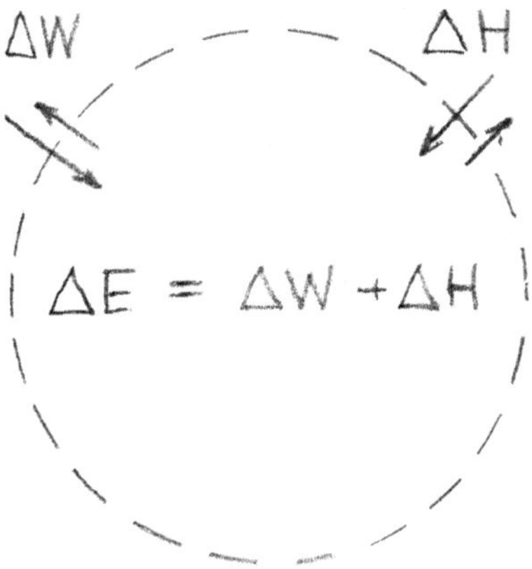

Therefore, a general conservation law of energy must include *both* work and heat transfer. Further, it must deal with change in the *total energy* of a system, not with a "mechanical" part and an "internal" part.

In an isolated system, that is, a system that does not exchange energy with its surroundings, the total energy must remain constant. If the system exchanges energy with its surroundings, it can do so in only one of two ways: Work can be done on or by the system, or heat can be passed to or from the system. In the latter case, the change in energy of the system must equal the net energy gained or lost by the surroundings. More precisely, ΔW stands for the net work on the system, which is all the work done *on* the system minus all the work done *by* the system. Similarly, ΔH represents the net heat transfer to the system, or the heat added to the system minus the heat lost by the system. Then the change in the *total* energy of the system, ΔE, is given by

The word "heat" is used rather loosely, even by physicists. This restriction on its meaning is not necessary in most contexts, but it is important for the discussion in this section.

$$\Delta E = \Delta W + \Delta H$$

This is the mathematical formulation of the first law of thermodynamics.

This general expression includes as special cases the preliminary versions of the conservation law given earlier in the chapter. If there is no heat transfer at all, then $\Delta H = 0$, and so $\Delta E = \Delta W$. In this case, the change in energy of a system equals the net work done on it. On the other hand, if work is done neither on nor by a system, then $\Delta W = 0$, and $\Delta E = \Delta H$. Here the change in energy of a system is equal to the net heat transfer.

We still need a description of that part of the total energy of a system called "heat" (or better, "internal" energy). So far, we have seen only that an increase in internal energy is sometimes associated with an increase in temperature. We also mentioned the long-held suspicion that internal energy involves the motion of the "small parts" of bodies. We will take up this problem in detail in Chapter 11.

SG 34–38

The second law of thermodynamics is a general statement of the limits of the heat engine and is based on Carnot's theorem. You saw that a reversible engine is the most efficient engine and the most effective refrigerator. Any other engine is not as efficient or effective. In order to formulate that idea generally and precisely, a new function, the *entropy*, must be introduced.

The change in entropy of a system, ΔS, is defined as the heat gained by the system, ΔH, divided by the temperature of the system, T:

$$\Delta S = \Delta H/T$$

Although this equation defines only changes of entropy, once a standard state for the system for which $S = 0$ is chosen, the total entropy for any state of the system can be determined.

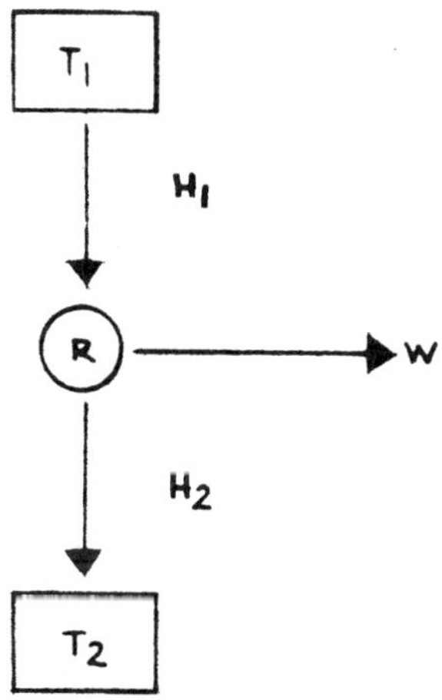

Since the entropy is defined for any state of the system, an engine that works in a cycle (as any heat engine does) must have the same entropy at the end of a cycle as it does at the start. This is not necessarily true of the boiler (hot object) and the condenser (cold object), since these are not returned to their initial states.

If the engine is a reversible one, the change in entropy of the cold object is

$$\Delta S_2 = H_2/T_2$$

Since the hot object loses heat, its change in entropy is

$$\Delta S_1 = -H_1/T_1$$

The total entropy change of the whole universe in this operation is

$$\Delta S_{universe} = \Delta S_1 + \Delta S_2 + \Delta S_{engine}$$
$$= \frac{H_2}{T_2} - \frac{H_1}{T_1} + 0$$

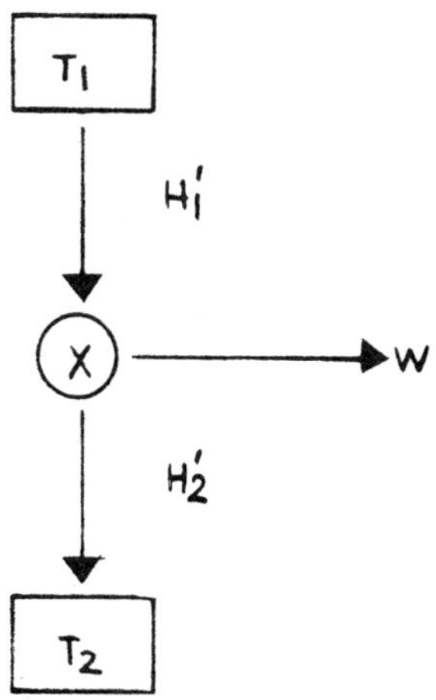

Carnot's theorem (page 296) says that the difference must be zero:

$$\Delta S_{universe} = 0$$

What about a less ideal engine? You know it must be less efficient than the reversible one; so for the engine X,

$$H'_1 > H_1 \quad \text{and} \quad H'_2 > H_2$$

The total entropy change of the universe is now

$$\Delta S_{universe} = \frac{H'_2}{T_2} - \frac{H'_1}{T_1} + 0$$

Rudolf Clausius (1822–1888)

It is a fairly straightforward matter to show that this time

$$\Delta S_{\text{universe}} > 0$$

the entropy *increases*.

Although proven only for these simple heat engines, the results that

$$\Delta S = 0 \quad \text{(reversible processes)}$$
$$\Delta S > 0 \quad \text{(any other process)}$$

are general ones. They are, in fact, mathematical formulations of the second law of thermodynamics.

Rudolf Clausius, who first formulated the second law in the form given here, paraphrased the two laws in 1850, as follows: "The energy of the universe remains constant, but its entropy seeks to reach a maximum."

?

27. The first law of thermodynamics is

(a) true only for steam engines.

(b) true only when there is no friction.

(c) a completely general statement of conservation of energy.

(d) the only way to express conservation of energy.

28. Define ΔE, ΔW, ΔH, and ΔS for a system.

29. What two ways are there for changing the total energy of a system?

30. The second law of thermodynamics says that the entropy of the universe

(a) cannot increase.

(b) cannot decrease.

(c) must increase.

(d) must decrease.

10.11 | Faith in the laws of thermodynamics

For over a century, the law of conservation of energy has stood as one of the most fundamental laws of science. You will encounter it again and again in this course, in studying electricity and magnetism, the structure of the atom, and nuclear physics. Throughout the other sciences, from chemistry to biology, and throughout engineering studies, the same law applies. Indeed, no other law so clearly brings together the various scientific fields, giving all scientists a common set of concepts.

The principle of conservation of energy has been immensely successful. It is so firmly believed that it seems almost impossible that any new discovery could disprove it. Sometimes energy seems to appear or disappear in a system, without being accounted for by changes in known forms of energy. In such cases, physicists prefer to assume that some hitherto unknown kind of energy is involved, rather than to consider seriously the possibility that energy is not conserved. You have already read Leibniz's proposal that energy could be dissipated among "the small parts" of bodies. He advanced this idea specifically in order to maintain the principle of conservation of energy in inelastic collisions and frictional processes. Leibniz's faith in energy conservation was justified. Other evidence showed that "internal energy" changed by just the right amount to explain observed changes in external energy.

SG 39, 40

Another similar example is the "invention" of the neutrino by the physicist Wolfgang Pauli in 1933. Experiments had suggested that energy disappeared in certain nuclear reactions. Pauli proposed that a tiny particle, named the "neutrino" by Enrico Fermi, was produced in these reactions. Unnoticed, the neutrino carried off some of the energy. Physicists accepted the neutrino theory for more than 20 years even though neutrinos could not be detected by any method. Finally, in 1956, neutrinos were detected in experiments using the radiation from a nuclear reactor. (The experiment could not have been done in 1933, since no nuclear reactor existed until nearly a decade later.) Again, faith in the law of conservation of energy turned out to be justified.

The theme of "conservation" is so powerful in science that scientists believe it will always be justified. Any apparent exceptions to the law will sooner or later be understood in a way which does not require us to give up the law. At most, these exceptions may lead us to discover new forms of energy, making the law even more general and powerful.

The French mathematician and philosopher Henri Poincaré expressed this idea in 1903 in his book *Science and Hypothesis:*

> . . . the principle of conservation of energy signifies simply that there is *something* which remains constant. Indeed, no matter what new notions future experiences will give us of the world, we are sure in advance that there will be something which will remain constant, and which we shall be able to call *energy.*

Today, it is agreed that the discovery of conservation laws was one of the most important achievements of science. These laws are powerful and valuable tools of analysis. All of them basically affirm that, whatever happens within a system of interacting bodies, certain measurable quantities will remain constant as long as the system remains isolated.

The list of known conservation laws has grown in recent years. The area of fundamental (or "elementary") particles has yielded much of this new knowledge. Some of the newer laws are imperfectly and incompletely understood. Others are on uncertain ground and are still being argued.

Below is a list of conservation laws to date. This list is not complete or eternal, but it does include the conservation laws that make up the working tool-kit of physicists today. (Those laws that are starred are discussed in the basic text portions of this course. The others are treated in optional supplemental units, for example, the supplemental unit entitled "Elementary Particles.")

Conservation Laws

1. Linear momentum*
2. Energy (including mass)*
3. Angular momentum (including spin)
4. Charge*
5. Electron-family number
6. Muon-family number
7. Baryon-family number
8. Strangeness number
9. Isotopic spin

The first law of thermodynamics, or the general law of conservation of energy, does not forbid the full conversion of heat into mechanical energy. The second law is an additional constraint on what can happen in nature.

Numbers 5–9 result from work in nuclear physics, high-energy physics, or elementary or fundamental particle physics.

The second law of thermodynamics has a status rather different from the conservation laws. It, too, is an extremely successful and powerful law. It, too, has continued to stand as one of the fundamental laws of science. Unlike the conservation laws or the laws of motion, the second law of thermodynamics gives no precise results; it only says certain things are impossible. For example, it is impossible to make the entropy of the universe (or of an isolated system) decrease; it is impossible to make heat flow from a cold body to a hot one without doing work on something.

See the film loop on the irreversibility of time (Film Loop 36).

In other words, the processes involving heat happen in one direction only: The entropy increases; heat flows from hot objects to cold ones. Thus, the second law is connected in some fundamental way with the notion that time proceeds in one direction only. To word it differently, when a movie taken of real events is run backward, what you see cannot, in detail, be found to occur in the real world. These ideas will be examined in more detail in the next chapter.

study guide

1. The *Project Physics* materials particularly appropriate for Chapter 10 include:

Experiments
Conservation of Energy
Measuring the Speed of a Bullet
Temperature and Thermometers
Calorimetry
Ice Calorimetry

Activities
Student Horsepower
Steam Powered Boat
Predicting the Range of an Arrow

Film Loops
Finding the Speed of a Rifle Bullet. I
Finding the Speed of a Rifle Bullet. II
Recoil
Colliding Freight Cars
Dynamics of a Billiard Ball
A Method of Measuring Energy—Nail Driven into Wood
Gravitational Potential Energy
Kinetic Energy
Conservation of Energy: Pole Vault
Conservation of Energy: Aircraft Takeoff

2. A person carries a heavy load across the level floor of a building. Draw an arrow to represent the force applied to the load, and one to represent the direction of motion. By the definition of work given, how much work is done on the load? Do you feel uncomfortable about this result? Why?

3. Kinetic energy, like speed, is a *relative* quantity; that is, kinetic energy is different when measured in different frames of reference. An object of mass m is accelerated uniformly in a straight line by a force F through a distance d. Its speed changes from v_1 to v_2. The work done is equal to the change in kinetic energy: $Fd = \frac{1}{2} mv_2^2 - \frac{1}{2} mv_1^2$. Describe this event from a frame of reference that is itself moving with speed u along the same direction.

(a) What are the speeds as observed in the new reference frame?

(b) Are the kinetic energies observed to have the same value in both reference frames?

(c) Does the *change* in kinetic energy have the same value?

(d) Is the calculated amount of work the same? Hint: By the Galilean relativity principle, the magnitude of the acceleration, and therefore force, will be the same when viewed from frames of reference moving uniformly relative to each other.

(e) Is the change in kinetic energy still equal to the work done?

(f) Which of the following are "invariant" for changes in reference frame (moving uniformly relative to one another)?

(1) the quantity $\frac{1}{2}mv^2$

(2) the quantity Fd

(3) the relationship $Fd = \Delta(\frac{1}{2}mv^2)$

(g) Explain why it is misleading to consider kinetic energy as something a body *has,* instead of only a quantity calculated from measurements.

4. An electron of mass about 9.1×10^{-31} kg is traveling at a speed of about 2×10^8 m/sec toward the screen of a television set. What is its kinetic energy? How many electrons like this one would be needed for a total kinetic energy of 1 J?

5. A 5-kg object travels uniformly at 4 m/sec. Over what distance must a 4-N force be applied to give the object a total kinetic energy of 80 J?

6. Estimate the kinetic energy of each of the following:

(a) a pitched baseball

(b) a jet plane

(c) a sprinter in a 90-m dash

(d) the earth in its motion around the sun.

7. A 200-kg iceboat is supported by the smooth surface of a frozen lake. The wind exerts on the boat a constant force of 400 N while the boat moves 900 m. Assume that frictional forces are negligible and that the boat starts from rest. Find the speed attained at the end of a 900-m run by each of the following methods:

(a) Use Newton's second law to find the acceleration of the boat. How long does it take to move 900 m? How fast will it be moving then?

(b) Find the final speed of the boat by equating the work done on it by the wind and the increase in its kinetic energy. Compare your result with your answer in (a).

8. A 2-g bullet is shot into a tree stump. It enters at a speed of 300 m/sec and comes to rest after having penetrated 5 cm in a straight line.

(a) What was the change in the bullet's kinetic energy?

(b) How much work did the tree do on the bullet?

(c) What was the average force during impact?

9. Refer to SG 20 in Chapter 9. How much work does the golf club do on the golf ball? How much work does the golf ball do on the golf club?

10. A penny has a mass of about 3.0 g and is about 1.5 mm thick. You have 50 pennies which you pile one on top of the other.

(a) How much more gravitational potential energy has the top penny than the bottom one?

(b) How much more gravitational potential energy have all 50 pennies together than the bottom one alone?

11. (a) How high can you raise a book weighing 5 N if you have available 1 J of energy?

(b) How many joules of energy are needed just to lift a jet airliner weighing 7×10^5 N (fully loaded) to its cruising altitude of 10,000 m?

12. There are standards for length, time, and mass (for example, a standard meter). But energy is a "derived quantity" for which no standards need be kept. Nevertheless, assume someone asks you to supply 1 J of energy. Describe in as much detail as you can how you would do it.

13. As a home experiment, hang weights on a rubber band and measure its elongation. Plot the force versus stretch on graph paper.

(a) How can you measure the stored energy?

(b) Show that over a straight section of the graph the stored energy is equal to $\frac{1}{2} k(\Delta x)^2$, where Δx is the change in length of the rubber band over that section of the graph and k is the slope of the line (the change in force divided by the change in length).

(c) If the weights on the rubber band bob up and down, discuss the "flow of energy" from kinetic, gravitational potential energy, and the rubber band's potential energy.

14. (a) Estimate how long it would take for the earth to fall up 1 m to a 1-kg stone if this stone were somehow rigidly fixed in space.

(b) Estimate how far the earth will actually move up while a 1-kg stone falls 1 m from rest.

(c) Why is the gravitational potential energy assigned to the system rather than to the rock alone?

15. The photograph below shows a massive lead wrecking ball being used to demolish a wall. Discuss the transformations of energy involved.

16. This discussion will show that the *PE* of an object is relative to the frame of reference in which it is measured. The boulder in the photograph on page 281 was not lifted to its perch. Rather, the rest of the land has eroded away, leaving the boulder where it may have been almost since the formation of the earth. Consider the question "What is the gravitational potential energy of the boulder–earth system?" You can easily calculate what the change in potential energy would be if the rock fell. It would be the product of the rock's weight and the distance it fell. But would that be the actual value of the gravitational energy that had been stored in the boulder–earth system? Imagine that there happened to be a deep mine shaft nearby and the boulder fell into the shaft. It would then fall much farther, reducing the gravitational potential energy much more. Apparently, the amount of energy stored depends on how far you imagine the boulder can fall.

(a) What is the greatest possible decrease in gravitational potential energy the isolated boulder–earth system could have?

(b) Is the boulder–earth system really isolated?

(c) Is there a true absolute bottom of gravitational potential energy for any system that includes the boulder and the earth?

The value of *PE* depends on the location of the (resting) frame of reference from which it is measured. This is not a serious problem, because we are concerned only with *changes* in energy. In any given problem, physicists will choose some convenient reference for the "zero-level" of potential energy, usually one that simplifies calculations. What would be a convenient zero-level for the gravitational potential energy of

(a) a pendulum?

(b) a roller coaster?

(c) a weight oscillating up and down a spring?

(d) a planet in orbit around the sun?

17. The figure below (not drawn to scale) shows a model of a carnival "loop-the-loop." A car starting from a platform above the top of the loop coasts down and around the loop without falling off the track. Show that to traverse the loop successfully, the car must start from a height at least one-half a radius above the top of the loop. Hint: The car's weight must not be greater than the centripetal force required to keep it on the circular path at the top of the loop.

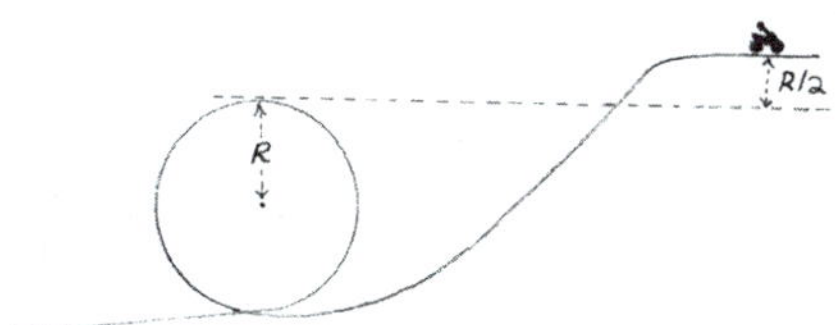

18. Discuss the conversion between kinetic and potential forms of energy in the system of a comet orbiting the sun.

19. Sketch an addition to one of the steam-engine diagrams of a mechanical linkage that would open and close the valves automatically (page 290).

20. Show that if a constant propelling force *F* keeps a vehicle moving at a constant speed *v* (against the friction of the surroundings) the power required is equal to *Fv*.

21. The Queen Mary, one of Britain's largest steamships, has been retired after completing 1,000 crossings of the Atlantic. Its mass is 75 million kilograms. A maximum engine power of 174 million watts is allowed the Queen Mary to reach a maximum speed of 30.63 knots (16 m/sec).

(a) What is the kinetic energy at full speed?

(b) Assume that at maximum speed all the power output of the engines goes into overcoming water drag. If the engines are suddenly stopped, how far will the ship coast before stopping? (Assume water drag is constant.)

(c) What constant force would be required to bring the ship to a stop from full speed within 1 nautical mile (2,000 m)?

(d) The assumptions made in (b) are not valid for the following reasons:

(1) Only about 60% of the power delivered to the propeller shafts results in a forward thrust to the ship; the rest results in friction and turbulence, eventually warming the water.

(2) Water drag is less for lower speed than for high speed.

(3) If the propellers are not free-wheeling, they add an increased drag.

Which of the above factors tends to increase, which to decrease the coasting distance?

(e) Explain why tugboats are important for docking big ships.

22. Devise an experiment to measure the power output of

(a) a person riding a bicycle.

(b) a motorcycle.

(c) an electric motor.

23. (a) A skier of 70 kg mass experiences a pull on a ski lift from an engine transmitting 140 W to the cable. Neglecting friction, how high can the engine pull the skier in 500 sec?

(b) What size light bulb (in watts) produces as much heat as the human body at rest?

24. One hundred joules (100 J) of heat is put into two engines. Engine A can lift 5 N a distance of 10 m in 10 sec. Engine B pulls with a force of 2 N for 5 sec a distance of 20 m. Calculate the efficiency and power of each engine.

25. Refer to the table of "Typical Power Ratings" on page 292.

(a) What advantages would Newcomen's engine have over a "turret windmill"?

(b) What advantage would you expect Watt's engine (1778) to have over Smeaton's engine (1772)?

26. Besides "horsepower," another term used in Watt's day to describe the performance of steam engines was "duty." The duty of a steam engine was defined as the distance in feet that an engine could lift a load of 1 million pounds, using 1 bushel of coal as fuel. For example, Newcomen's engine had a duty of 4.3; it could perform 4.3 million foot-pounds of work by burning a bushel of coal.

A bushel of coal contains about 900 MJ of energy. A bushel is 36 liters (L). What was the efficiency of Newcomen's engine?

27. The introduction of the steam engine had both positive and negative effects, although all of these effects were not predicted at the time.

(a) List several *actual* effects, both beneficial and undesirable, of the steam engine and of the gasoline internal combustion engine.

(b) List several *predicted* effects, both beneficial and undesirable, of nuclear power and of solar power.

28. (a) Find the maximum efficiency of an engine that makes use of the temperature differences in the ocean. In the tropics, the surface waters are about 15°C, and the bottom waters are about 5°C.

(b) Cooling 1 metric ton of water by 1°C produces about 4 MJ of energy. At what rate must warm surface water be pumped through an engine that is cooled by bottom water in order that the engine produce 1 MW of mechanical power?

29. (a) Find the maximum coefficient of performance of an air conditioner operating on a day when it is 40°C outside and 21°C inside.

(b) The coils on the outside of an air conditioner have to be considerably warmer than 40°C and those inside must be cooler than 21°C. Otherwise, heat would be exchanged too slowly. Suppose the coils are, respectively, 10°C warmer and cooler than the ideal. How much is the coefficient of performance decreased?

(c) What happens to the coefficient of performance of this air conditioner when the outside coils are put into the sun and heat up another 10°C?

30. A table of rates for truck transportation is given below. How does the charge depend on the amount of work done?

TRUCK TRANSPORTATION

Weight (kg)	*Moving rates (including pickup and delivery) from Boston to:*		
	Chicago (1,550 km)	Denver (3,150 km)	Los Angeles (4,800 km)
50	$ 18.40	$ 24.00	$ 27.25
225	92.00	120.00	136.25
450	128.50	185.50	220.50
900	225.00	336.00	406.00
1,800	383.00	606.00	748.00
2,700	576.00	909.00	1,122.00

31. Beginning with the postulate that heat does not flow by itself from a cold body to a hot body, arrange the following steps in the order used by Clausius in his formula for the maximum efficiency of an engine:

(a) No engine can be more efficient than an ideal reversible engine.

(b) Choose one ideal reversible engine and calculate its efficiency.

(c) Work must be done to cause heat to flow from a cold to a hot body.

(d) The ratio of heat to work depends only on the temperature differences of the reservoirs for ideal reversible engines.

(e) All reversible engines have the same efficiency.

(f) The efficiency of an ideal reversible engine is the maximum possible for any real engine; in reality, the efficiency is much less.

32. Explain why all ideal reversible engines have the same efficiency and why this efficiency is the maximum possible for an engine. What is an ideal engine? What is a reversible engine?

33. Assuming that no real engine can be perfectly reversible, why does the formula for the maximum efficiency of an engine imply that absolute zero can never be reached?

34. Consider the following hypothetical values for a paddle-wheel experiment like Joule's: A 1-kg weight descends through a distance of 1 m, turning a paddle wheel immersed in 5 kg of water.

(a) About how many times must the weight be allowed to fall in order that the temperature of the water increase by 0.5°C?

(b) How could you modify the experiment so that the same temperature rise would be produced with fewer falls of the weight? (Hint: There are at least three possible ways.)

35. While traveling in Switzerland, Joule attempted to measure the difference in temperature of the water at the top and at the bottom of a waterfall. Assuming that the amount of heat produced at the bottom is equal to the decrease in gravitational potential energy, calculate roughly the temperature difference you would expect to observe between the top and bottom of a waterfall about 50 m high, such as Niagara Falls. Does it matter how much water goes down the fall?

36. Because a nuclear power plant's interior must be kept to much closer tolerances than fossil-fueled plants, its operating temperature is kept lower. Compare the efficiencies of a nuclear power plant that produces steam at 600°K with a fossil-fueled plant that produces steam at 750°K. Both are cooled by water at 300°K.

(a) If both plants produce 1 MW of electrical power, at what rate does each plant dump heat into the environment?

(b) It takes 4.2 MJ to raise the temperature of 1 metric ton of water by 1°C (or 1°K). How many tons of water must flow through each of the two plants if the emerging water is to be no warmer than 303°K?

37. About how many kilograms of hamburgers would you have to eat to supply the energy for 30 min of digging? Assume that your body is 20% efficient.

38. If your food intake supplies less energy than you use, you start "burning" your own stored fat for energy. The oxidation of 0.45 kg of animal fat provides about 4,300 Calories of energy. Suppose that on your present diet of 4,000 Calories a day, you neither gain nor lose weight. If you cut your diet to 3,000 Calories and maintain your present physical activity, how long would it take to reduce your mass by 22.5 kg?

39. In order to engage in normal light work, a person in India has been found to need on the average about 1,950 Calories of food energy a day, whereas an average West European needs about 3,000 Calories a day. Explain how each of the following statements makes the difference in energy need understandable.

(a) The average adult Indian weighs about 49.5 kg; the average adult West European weighs about 67.5 kg.

(b) India has a warm climate.

(c) The age distribution of the population for which these averages have been obtained is different in the two areas.

40. No other concept in physics has the economic significance that "energy" does. Discuss the statement: "We could express energy in dollars just as well as in joules or calories."

41. Show how the conservation laws for energy and for momentum can be applied to a rocket during the period of its lift-off.

42. Discuss the following statement: "During a typical trip, all the chemical energy of the gasoline used in an automobile is used to heat up the car, the road, and the air."

43. If you place a hot body and a cold one in thermal contact, heat will flow spontaneously. Suppose an amount of heat H flows from a body at temperature T_1 to a body at T_2. What is the entropy change of the universe?

44. (a) Describe the procedure by which a space capsule can be changed from a high circular orbit to a lower circular orbit.

(b) How does the kinetic energy in the lower orbit compare with that in the higher orbit?

(c) How does the gravitational potential energy for the lower orbit compare with that of the higher orbit?

(d) It can be shown (by using calculus) that the change in gravitational potential energy in going from one circular orbit to another will be twice the change in kinetic energy. How, then, will the total energy for the lower circular orbit compare with that for the higher orbit?

(e) How do you account for the change in total energy?

45. Any of the terms in the equation $\Delta E = \Delta H + \Delta W$ can have negative values.

(a) What would be true of a system for which

(1) ΔE is negative?

(2) ΔH is negative?

(3) ΔW is negative?

(b) Which terms would be negative for the following systems?

(1) a person digging a ditch

(2) a car battery while starting a car

(3) an electric light bulb just after it is turned on

(4) an electric light bulb an hour after it is turned on

(5) a running refrigerator

(6) an exploding firecracker

46. In each of the following, trace the chain of energy transformations from the sun to the energy in its final form.

(a) A pot of water is boiled on an electric stove.

(b) An automobile accelerates from rest on a level road, climbs a hill at constant speed, and comes to a stop at a traffic light.

(c) A windmill pumps water out of a flooded field.

47. Review what you read about the second law of thermodynamics in Sec. 10.10. For the engine X considered in the discussion of that law,

$$H_1' > H_1 \quad \text{and} \quad H_2' - H_2$$

Refer to the difference between H'_1 and H_1 as h:

$$H_1' - H_1 = h$$

thus, h is a positive number.

(a) Use conservation of energy to show that $H'_2 - H_2 = h$ also.

(b) Use the following equation (Carnot's theorem)

$$\frac{H_1}{T_1} = \frac{H_2}{T_2}$$

to show that for one cycle of engine X,

$$\Delta S_{\text{universe}} = \frac{h}{T_2} - \frac{h}{T_1}$$

(c) Prove that $\Delta S_{\text{universe}}$ must be positive.

48. An ice cube (10 g) melts in a glass of water (100 g). Both are nearly at 0°C. Neglect temperature changes. Melting the ice requires 3.4 MJ of energy (which comes from cooling the water). What is the entropy change of the ice? of the water? of the universe?

The Kinetic Theory of Gases

11.1 | An overview of the chapter

SG 1

During the 1840's, many scientists recognized that heat is not a substance, but a form of energy that can be converted into other forms. Two of these scientists, James Prescott Joule and Rudolf Clausius, went a step further. Heat can produce mechanical energy, and mechanical energy can produce heat; therefore, they reasoned, the "heat energy" of a substance is simply the kinetic energy of its atoms and molecules. In this chapter, you will see that this idea is largely correct. This idea forms the basis of the *kinetic–molecular theory of heat.*

Molecules are the smallest pieces of a substance; they may be combinations of atoms of simpler substances.

However, even the idea of atoms and molecules was not completely accepted in the nineteenth century. If such small bits of matter really existed, they would be too small to observe even under the most powerful microscopes. Since scientists could not observe molecules, they could not check directly the hypothesis that heat is molecular kinetic energy. Instead, they had to derive from this hypothesis predictions about the behavior of measurably large samples of matter. Then they could test these predictions by experiment. For reasons that will be explained, it is easiest to test such hypotheses by observing the properties of gases. Therefore, this chapter deals mainly with the kinetic theory as applied to gases.

The development of the kinetic theory of gases in the nineteenth century led to one of the last major triumphs of Newtonian mechanics. The method involved using a simple theoretical model of a gas. In this model Newton's laws of motion were applied to the gas molecules which were pictured as tiny balls. This method produced equations that related the easily observable properties of gases, such as pressure, density, and temperature, to properties not directly observable, such as the sizes and speeds of molecules. For example, the kinetic theory:

1. explained rules that had been found previously by trial-and-error methods. (An example is "Boyle's law," which relates the pressure and the volume of a gas.)
2. predicted new relations. (One surprising result was that the friction between layers of gas moving at different speeds increases with temperature, but is independent of the density of the gas.)
3. led to values for the sizes and speeds of gas molecules.

Thus, the successes of kinetic theory showed that Newtonian mechanics provided a way for understanding the effects and behavior of invisible molecules.

Applying Newtonian mechanics to a mechanical model of gases resulted in some predictions that did *not* agree with the facts; that is, the model is not valid for all phenomena. According to kinetic theory, for example, the energy of a group of molecules should be shared equally among all the different motions of the molecules and their atoms. The properties of gases predicted from this "equal sharing" principle clearly disagreed with experimental evidence. Newtonian mechanics could be applied successfully to a wide range of motions and collisions of molecules in a gas. But it did not work for the motions of atoms inside molecules. It was not until the twentieth century that an adequate theory of the behavior of atoms, "quantum mechanics," was developed. (Some ideas from quantum mechanics are discussed in Unit 5.)

Kinetic theory based on Newtonian mechanics also had trouble dealing with the fact that most phenomena are not reversible. An inelastic collision is an irreversible process. Other examples are the mixing of two gases or scrambling an egg. In Newtonian theory, however, the reverse of any event is just as reasonable as the event itself. Can irreversible processes be described by a theory based on Newtonian theory? Or do they involve some new fundamental law of nature? In studying this problem from the viewpoint of kinetic theory, you will see how the concept of "randomness" entered physics.

Balloon for carrying weather-forecasting apparatus.

?

1. *Early forms of the kinetic molecular theory were based on the assumption that heat energy is*

(a) *a liquid.*

(b) *a gas.*

(c) *the kinetic energy of molecules.*

(d) *made of molecules.*

2. *True or false: In the kinetic theory of gases, as developed in the nineteenth century, it was assumed that Newton's laws of motion apply to the motion and collisions of molecules.*

3. *True or false: In the twentieth century, Newtonian mechanics was found to be applicable not only to molecules but also to the atoms inside molecules.*

11.2 | A model for the gaseous state

What are the differences between a gas and a liquid or solid? You know by observation that liquids and solids have definite volume. Even if their shapes change, they still take up the same amount of space. A gas, on the other hand, will expand to fill any container (such as a room). If not confined, it will leak out and spread in all directions. Gases have low densities compared to those of liquids and solids, typically about 1,000 times smaller. Gas molecules are usually relatively far apart from one another, and they only occasionally collide. In the kinetic theory model, forces between molecules act only over very short distances. Therefore, gas molecules are considered to be moving freely most of the time. In liquids, the molecules are closer together; forces act among them continually and keep them from flying apart. In solids, the molecules are usually even closer together, and the forces between them keep them in a definite orderly arrangement.

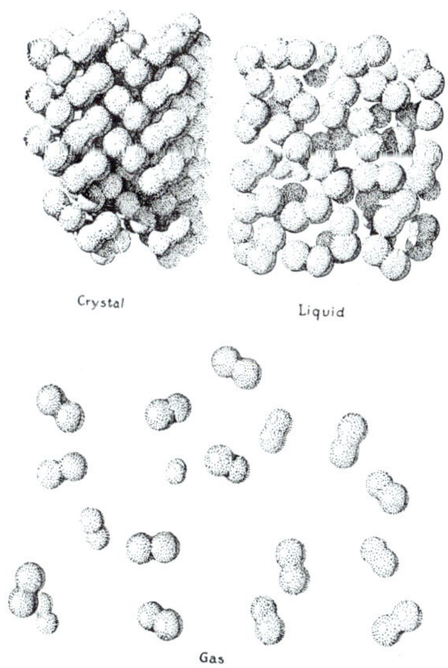

A very simplified "model" of the three states of matter: solid, liquid, gas. (From **General Chemistry**, *second edition, by Linus Pauling, W. H. Freeman and Company, © 1953.)*

The initial model of a gas is very simple. The molecules are considered to behave like miniature balls, that is, tiny spheres or clumps of spheres that exert no force at all on each other except when they make contact. Moreover, all the collisions of these spheres are assumed to be perfectly elastic. Thus, the total kinetic energy of two spheres is the same before and after they collide.

Note that the word "model" is used in two different senses in science. In Chapter 10, we mentioned the model of Newcomen's

Gases can be confined without a container. A star, for example, is a mass of gas confined by gravitational force. Another example is the earth's atmosphere.

The word "gas" was originally derived from the Greek word *chaos*; it was first used by the Belgian chemist Jan Baptista van Helmont (1580–1644).

engine which James Watt was given to repair. That was a *working model.* It actually did function, although it was much smaller than the original engine, and contained some parts made of different materials. Now we are discussing a *theoretical model* of a gas. This model exists only in the imagination. Like the points, lines, triangles, and spheres studied in geometry, this theoretical model can be discussed mathematically. The results of such a discussion may help you to understand the real world.

This theoretical model represents the gas as consisting of *a large number of very small particles in rapid, disordered motion.* "A large number" means something like a billion billion (10^{18}) or more particles in a sample as small as a bubble in a soft drink. "Very small" means a diameter about a hundred-millionth of a centimeter (10^{-10} m). "Rapid motion" means an average speed of a few hundred kilometers per hour. What is meant by "disordered" motion? Nineteenth-century kinetic theorists assumed that each individual molecule moved in a definite way, determined by Newton's laws of motion. Of course, in practice it is impossible to follow a billion billion particles at the same time. They move in all directions, and each particle changes its direction and speed during collisions with other particles. Therefore, we cannot make a definite prediction of the motion of any one *individual* particle. Instead, we must be content with describing the *average* behavior of large collections of particles. From moment to moment, each individual molecule behaves according to the laws of motion. But it is easier to describe the *average* behavior if we assume complete ignorance about any *individual* motions.

To see why this is so, consider the results of flipping a large number of coins all at once. It would be very hard to predict how a single coin would behave. But if you assume the coins behave randomly, you can confidently predict that flipping a million coins will give approximately 50% heads and 50% tails. The same principle applies to molecules bouncing around in a container. You can safely assume that about as many are moving in one direction as in another. Furthermore, the molecules are equally likely to be found in any cubic centimeter of space inside the container. This is true no matter where such a region is located, and even though we do not know where a given molecule is at any given time. "Disordered," then, means that velocities and positions are distributed randomly. Each molecule is just as likely to be moving to the right as to the left (or in any other direction). It is just as likely to be near the center as near the edge (or any other position).

? 4. *What kind of a model is a test model of a bridge made of balsa wood? a computer program that simulates the forces*

Averages and Fluctuations

Molecules are too small, too numerous, and too fast for us to measure the speed of any one molecule, its kinetic energy, or how far it moves before colliding with another molecule. For this reason, the kinetic theory of gases concerns itself with making predictions about *average* values. The theory enables us to predict quite precisely the *average* speed of the molecules in a sample of gas, the *average* kinetic energy, or the *average* distance the molecules move between collisions.

Any measurement made on a sample of gas reflects the combined effect of billions of molecules, averaged over some interval of time. Such average values measured at different times, or in different parts of the sample, will be slightly different. We assume that the molecules are moving randomly. Thus, we can use the mathematical rules of statistics to estimate just how different the averages are likely to be. We will call on two basic rules of statistics for random samples:

1. Large variations away from the average are less likely to occur than are small variations. (For example, if you toss 10 coins, you are less likely to get 9 heads and 1 tail than to get 6 heads and 4 tails.)
2. Percentage variations are likely to be smaller for large samples. (For example, you are likely to get nearer to 50% heads by flipping 1,000 coins than by flipping just 10 coins.)

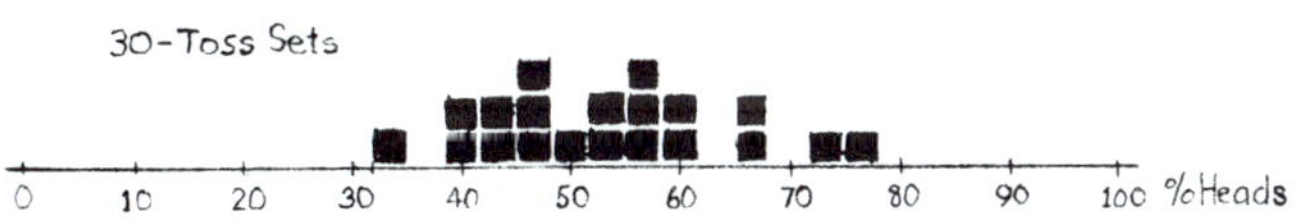

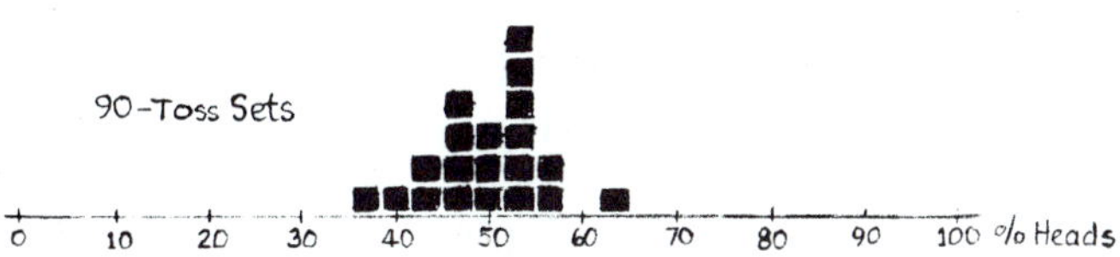

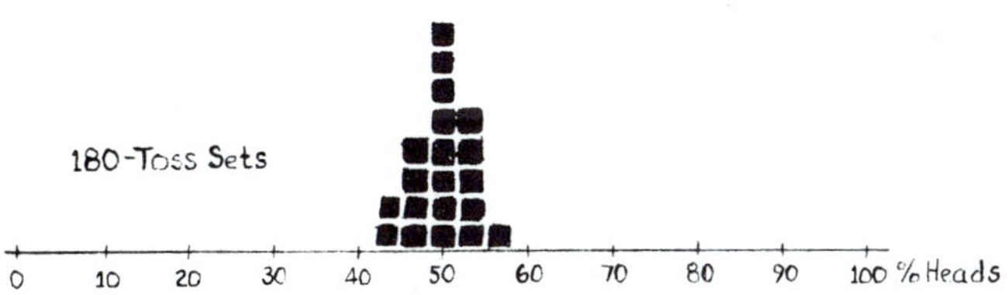

A simple statistical prediction is the statement that if a coin is tossed many times, it will land "heads" 50% of the time and "tails" 50% of the time. For small sets of tosses there will be many "fluctuations" (variations) to either side of the predicted average of 50% heads. Both statistical rules are evident in the charts at the right. The top chart shows the percentage of heads in sets of 30 tosses each. Each of the 10 black squares represents a set of 30 tosses. Its position along the horizontal scale indicates the percentage of heads. As we would expect from rule 1, there are more values near the theoretical 50% than far from it. The second chart is similar to the first, but here each square represents a set of 90 tosses. As before, there are more values near 50% than far from it. And, as we would expect from rule 2, there are fewer values far from 50% than in the first chart.

The third chart is similar to the first two, but now each square represents a set of 180 tosses. Large fluctuations from 50% are less common still than for the smaller sets.

Statistical theory shows that the *average* fluctuation from 50% shrinks in proportion to the square root of the number of tosses. We can use this rule to compare the average fluctuation for sets of, say, 30,000,000 tosses with the average fluctuation for sets of 30 tosses. The 30,000,000-toss sets have 1,000,000 times as many tosses as the 30-toss sets. Thus, their average fluctuation in percent of "heads" should be 1,000 times smaller!

These same principles hold for fluctuations from average values of any randomly distributed quantities, such as molecular speed or distance between collisions. Since even a small bubble of air contains about a quintillion (10^{18}) molecules, fluctuations in the average value for any isolated sample of gas are not likely to be large enough to be measurable. A measurably large fluctuation is not *impossible*, but extremely unlikely.

acting on a bridge? What are the differences between "theoretical" and "working" models?

5. Read the following description of a model of a gas and give a suitable numerical estimation for each underlined phrase: "a large number of small particles in rapid, disordered motion."

6. In the kinetic theory, particles are thought to exert significant forces on one another

(a) only when they are far apart.

(b) only when they are close together.

(c) all the time.

(d) never.

7. Why was the kinetic theory first applied to gases rather than to liquids or solids?

11.3 | The speeds of molecules

Daniel Bernoulli (1700–1782)

The basic idea of the kinetic theory is that heat is related to the kinetic energy of molecular motion. This idea had been frequently suggested in the past. However, many difficulties stood in the way of its general acceptance. Some of these difficulties are well worth mentioning. They show that not all good ideas in science (any more than outside of science) are immediately successful.

SG 3

In 1738, the Swiss mathematician Daniel Bernoulli showed how a kinetic model could explain a well-known property of gases. This property is described by Boyle's law: As long as the temperature does not change, the pressure of a gas is proportional to its density. Bernoulli assumed that the pressure of a gas is simply a result of the impacts of individual molecules striking the wall of the container. If the density of the gas were twice as great, there would be twice as many molecules per cubic centimeter. Thus, Bernoulli said, there would be twice as many molecules striking the wall per second and hence twice the pressure. Bernoulli's proposal seems to have been the first step toward the modern kinetic theory of gases. Yet it was generally ignored by other scientists in the eighteenth century. One reason for this was that Newton had proposed a different theory in his *Principia* (1687). Newton showed that Boyle's law *could* be explained by a model in which particles at rest exert forces that repel neighboring particles. Newton did not claim that he had proved that gases really *are* composed of such repelling particles. But most scientists, impressed by Newton's discoveries, simply assumed that his treatment of gas pressure was also right. (It was not.)

Pressure is defined as the perpendicular force on a surface divided by the area of the surface. The unit of pressure, N/m^2, has been given the name pascal (symbol Pa) after an eighteenth-century physicist. Atmospheric pressure is about 100 kPa.

The kinetic theory of gases was proposed again in 1820 by English physicist John Herapath. Herapath rediscovered

Bernoulli's findings on the relations between pressure and density of a gas and the speeds of the particles. Herapath's work also was ignored by most other scientists.

James Prescott Joule, however, did see the value of Herapath's work. In 1848, he read a paper to the Manchester Literary and Philosophical Society in which he tried to revive the kinetic theory. Joule showed how the speed of a hydrogen molecule could be computed (as Herapath had done). He reported a value of 2,000 m/sec at 0°C, the freezing temperature of water. This paper, too, was ignored by other scientists. For one thing, physicists do not generally look in the publications of a "literary and philosophical society" for scientifically important papers. However, evidence for the equivalence of heat and mechanical energy continued to mount. Several other physicists independently worked out the consequences of the hypothesis that heat energy in a gas is the kinetic energy of molecules. Rudolf Clausius in Germany published a paper in 1856 on "The Nature of the Motion we call Heat." This paper established the basic principles of kinetic theory essentially in the form accepted today. Soon afterward, James Clerk Maxwell in Britain and Ludwig Boltzmann in Austria set forth the full mathematical details of the theory.

Ludwig Boltzmann (1844–1906)

The Maxwell velocity distribution. It did not seem likely that all molecules in a gas would have the same speed. In 1859, Maxwell applied the mathematics of probability to this problem. He suggested that the speeds of molecules in a gas are distributed over all possible values. Most molecules have speeds not very far from the average speed. Some have much lower speeds and some much higher speeds.

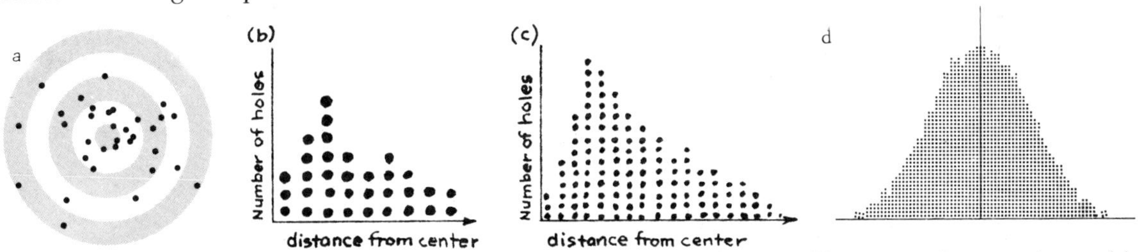

Target practice experiment: (a) scatter of holes in target; (b) graph showing number of holes in each half-ring of the bull's eye; (c) graph showing that the distribution becomes smooth for a very large number of shots and for very narrow rings.

A simple example will help you to understand Maxwell's distribution of molecular speeds. Suppose a person shoots a gun at a practice target many times. Some bullets will probably hit the bull's-eye. Others will miss by smaller or larger amounts, as shown in (a) in the sketch above. The number of bullets scattered at various distances from the center are counted. A graph of the results is shown in (b). This graph shows the distribution of hits for one set of shots. Another target may give a different distribution, but it is likely to have the same general shape. If you plot the distribution of hits for a very large number of shots, you will get a distribution like the one in (c).

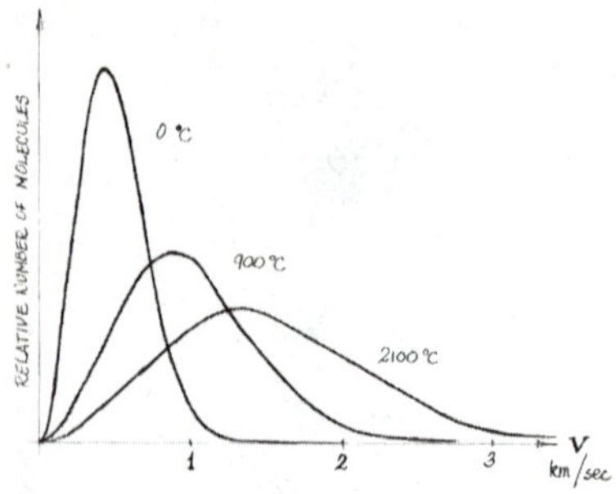

Maxwell's distribution of speeds in gases at different temperatures.

For still larger numbers of shots, the distribution spread you see in (c) will be too small to be noticed. Since the number of molecules in a gas is very large indeed, the graph showing the distribution of molecular speeds is smooth at any scale that can be drawn.

The actual shape of the curve in (c) is determined by many things about the gun, the person, and so on. Other processes give rise to other shapes of curves. The speeds of the molecules in a gas are determined by the collisions they have with each other. Maxwell used a very clever mathematical argument to deduce a distribution of molecular speeds. He was then able to argue, not rigorously, that the average number of molecules at each speed is not changed by the molecular collisions. Somewhat later, Ludwig Boltzmann published a rigorous proof that Maxwell's distribution is the only one that remains unchanged by collisions.

Maxwell's distribution law for molecular speeds in a gas is shown in the margin in graphical form for three different temperatures. For a gas at any given temperature, the "tail" of each curve is much longer on the right (high speeds) than on the left (low speeds). As the temperature increases, the peak of the curve shifts to higher speeds, and the speed distribution becomes more broadly spread out.

What evidence do we have that Maxwell's distribution law really applies to molecular speeds? Several successful predictions based on this law gave indirect support to it. Not until the 1920's was a direct experimental check possible. Otto Stern in Germany, and later Zartmann in the United States, devised a method for measuring the speeds in a beam of molecules. (See the illustration of Zartmann's method on page 330.) Stern, Zartmann,
SG 4 and others found that molecular speeds are indeed distributed
SG 5 according to Maxwell's law.

?

8. In the kinetic theory of gases, it is assumed that the pressure of a gas on the walls of the container is due to

(a) gas molecules colliding with one another.

(b) gas molecules colliding against the walls of the container.

(c) repelling forces exerted by molecules on one another.

9. The idea of speed distribution for gas molecules means that

(a) each molecule always has the same speed.

(b) there is a wide range of speeds of gas molecules.

(c) molecules are moving fastest near the center of the gas.

11.4 | The sizes of molecules

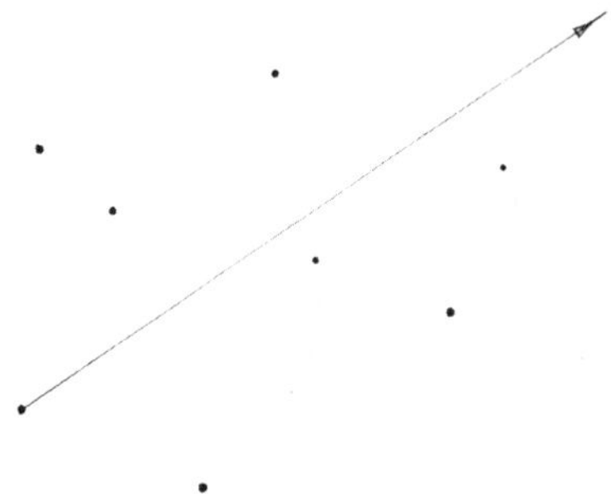

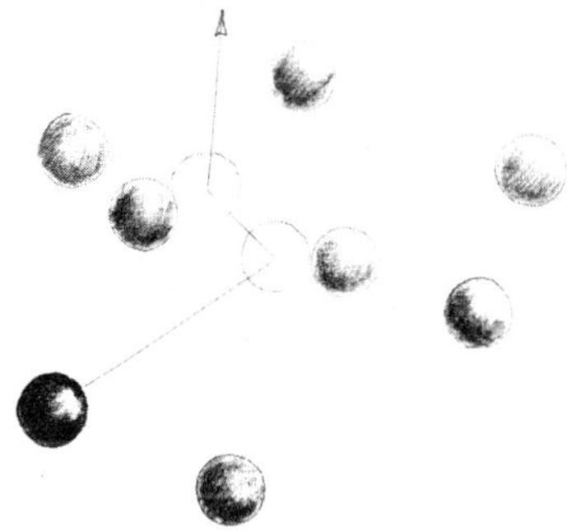

The larger the molecules are, the more likely they are to collide with each other.

Is it reasonable to suppose that gases consist of molecules moving at speeds up to several hundred meters per second? If this model were correct, gases should mix with each other very rapidly. But anyone who has studied chemistry knows that they do not. Suppose hydrogen sulfide or chlorine is generated at the end of a large room. Several minutes may pass before the odor is noticed at the other end. According to kinetic-theory calculations, each of the gas molecules should have crossed the room hundreds of times by then. Therefore, something must be wrong with the kinetic-theory model.

Rudolf Clausius recognized this as a valid objection to his own version of the kinetic theory. His 1856 paper had assumed that the particles are so small that they can be treated like mathematical points. If this were true, particles would almost never collide with one another. However, the observed *slowness* of diffusion and mixing convinced Clausius to change his model. He thought it likely that the molecules of a gas are not vanishingly small, but of a finite size. Particles of finite size moving very rapidly would often collide with one another. An individual molecule might have an instantaneous speed of several hundred meters per second, but it changes its direction of motion every time it collides with another molecule. The more often it collides with other molecules, the less likely it is to move very far in any one direction. How often collisions occur depends on how crowded the molecules are and on their size. For most purposes, you can think of molecules as being relatively far apart and of very small size. But they are just large enough and crowded enough to get in one another's way. Realizing this, Clausius was able to modify his model to explain why gases mix so slowly. In addition, he derived a precise quantitative relationship between the molecules' size and the average distance they moved between collisions.

Clausius now was faced with a problem that plagues every theoretical physicist. If a simple model is modified to explain better the observed properties, it becomes more complicated. Some plausible adjustment or approximation may be necessary in order to make any predictions from the model. If the predictions disagree with experimental data, is this because of a flaw in the model or a calculation error introduced by the approximations? The development of a theory often involves a compromise between adequate explanation of the data and mathematical convenience.

Nonetheless, it soon became clear that the new model was a great improvement over the old one. It turned out that certain other properties of gases also depend on the size of the molecules. By combining data on several such properties, it was

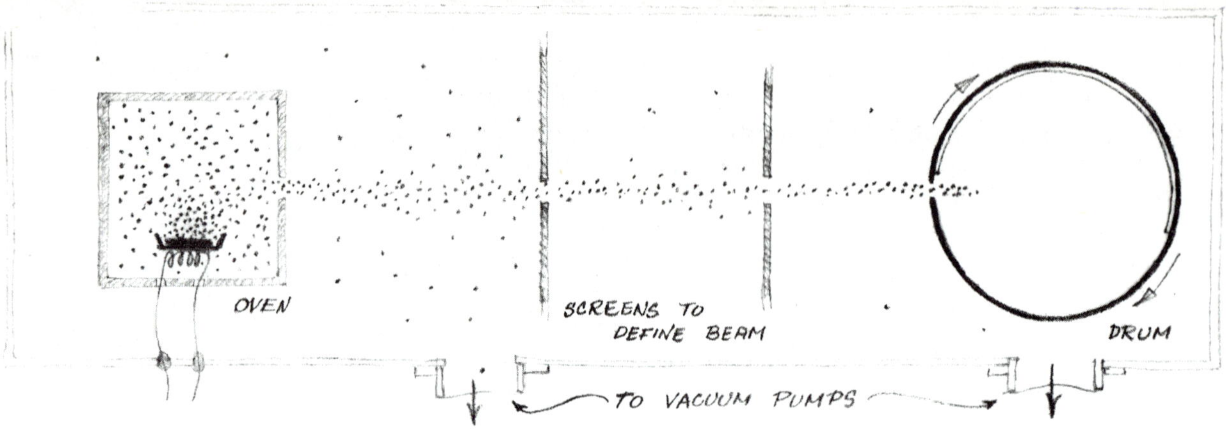

Direct Measurement of Molecular Speeds

A narrow beam of molecules is formed by letting molecules of a hot gas pass through a series of slits. In order to keep the beam from spreading out, collisions with randomly moving molecules must be avoided. Therefore, the source of gas and the slits are housed in a highly evacuated chamber. The molecules are then allowed to pass through a slit in the side of a cylindrical drum that can be spun very rapidly. The general scheme is shown in the drawing above.

As the drum rotates, the slit moves out of the beam of molecules. No more molecules can enter until the drum has rotated through a whole revolution. Meanwhile, the molecules in the drum continue moving to the right, some moving quickly and some moving slowly.

Fastened to the inside of the drum is a sensitive film that acts as a detector. Any molecule striking the film leaves a mark. The faster molecules strike the film first, before the drum has rotated very far.

The slower molecules hit the film later, after the drum has rotated farther. In general, molecules of different speeds strike different parts of the film. The darkness of the film at any point is proportional to the number of molecules that hit it there. Measurement of the darkening of the film shows the relative distribution of molecular speeds. The speckled strip at the right represents the unrolled film, showing the impact position of molecules over many revolutions of the drum. The heavy band indicates where the beam struck the film before the drum started rotating. (It also marks the place to which infinitely fast molecules would get once the drum was rotating.)

A comparison of some experimental results with those predicted from theory is shown in the graph. The dots show the experimental results, and the solid line represents the predictions from the kinetic theory.

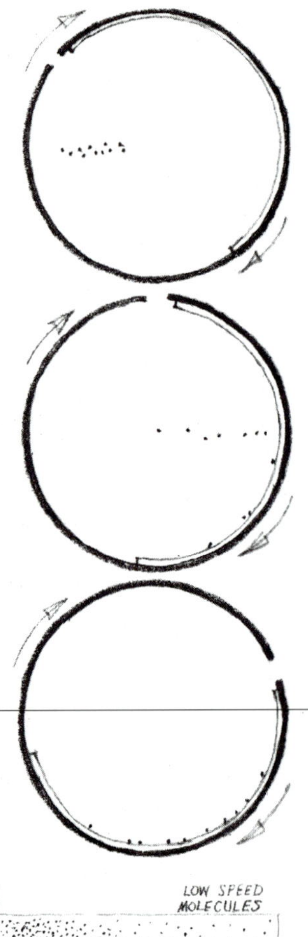

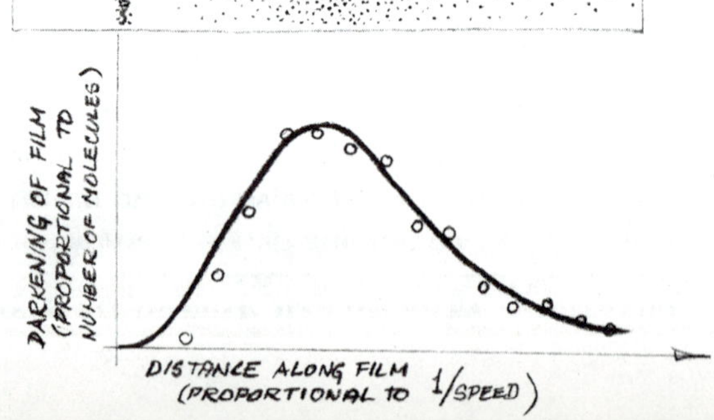

possible to work backwards and find fairly reliable values for molecular sizes. Here, only the result of these calculations is reported. Typically, the diameter of gas molecules came out to be of the order of 10^{-10} to 10^{-9} m. This is not far from the modern values, an amazingly good result. After all, no one previously had known whether a molecule was 1,000 times smaller or bigger than that. In fact, as Lord Kelvin remarked:

SG 6

> The idea of an atom has been so constantly associated with incredible assumptions of infinite strength, absolute rigidity, mystical actions at a distance and indivisibility, that chemists and many other reasonable naturalists of modern times, losing all patience with it, have dismissed it to the realms of metaphysics, and made it smaller than 'anything we can conceive.'

SG 7
SG 8

Kelvin showed that other methods could also be used to estimate the size of atoms. None of these methods gave results as reliable as did the kinetic theory. But it was encouraging that they all led to the same order of magnitude (within about 50%).

?

10. In his revised kinetic-theory model Clausius assumed that the particles have a finite size, instead of being mathematical points, because

(a) obviously everything must have some size.

(b) it was necessary to assume a finite size in order to calculate the speed of molecules.

(c) the size of a molecule was already well known before Clausius' time.

(d) a finite size of molecules could account for the slowness of diffusion.

11. Why were many people skeptical of the existence of atoms at the time of Clausius? How did Clausius' estimate of the size of atoms reinforce atomic theory?

11.5 | Predicting the behavior of gases from the kinetic theory

One of the most easily measured characteristics of a confined gas is pressure. Experience with balloons and tires makes the idea of air pressure seem obvious, but it was not always so.

Galileo, in his book on mechanics, *Two New Sciences* (1638), noted that a lift-type pump cannot raise water more than 10 m. This fact was well known. Such pumps were widely used to obtain drinking water from wells and to remove water from mines. You already have seen one important consequence of this

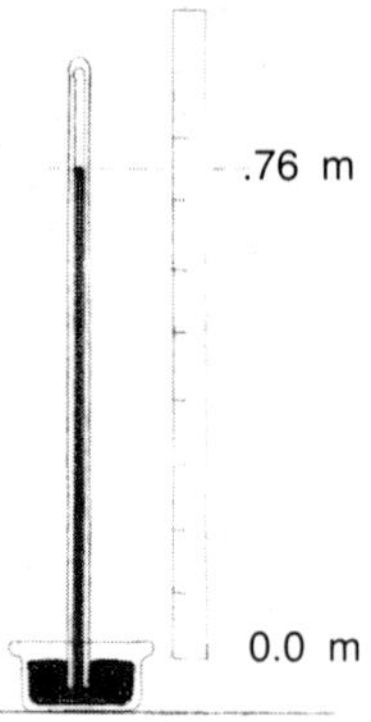

Torricelli's barometer is a glass tube standing in a pool of mercury. The topmost part of the tube is empty of air. The air pressure on the pool supports the weight of the column of mercury in the tube up to a height of about 0.76 m. The MKS unit of pressure is the N/m^2, which has been given the name pascal (symbol Pa).

limited ability of pumps to lift water out of deep mines. This need provided the initial stimulus for the development of steam engines. Another consequence was that physicists became curious about why the lift pump worked at all. Also, why should there be a limit to its ability to raise water?

Air pressure. The puzzle was solved as a result of experiments by Torricelli (a student of Galileo), Guericke, Pascal, and Boyle. By 1660, it was fairly clear that the operation of a "lift" pump depends on the pressure of the air. The pump merely reduces
SG 9 the pressure at the top of the pipe. It is the pressure exerted by
SG 10 the atmosphere on the pool of water below which forces water up the pipe. A good pump can reduce the pressure at the top of the pipe to nearly zero. Then the atmospheric pressure can force water up to about 10 m above the pool, but no higher. Atmospheric pressure at sea level is not great enough to support a column of water any higher. Mercury is almost 14 times as dense as water. Thus, ordinary pressure on a pool of mercury can support a column only 1/14 as high, about 0.76 m. This is a more convenient height for laboratory experiments. Therefore, much of the seventeenth-century research on air pressure was done with a column of mercury, or mercury "barometer." The first of these barometers was designed by Torricelli.

The height of the mercury column that can be supported by air pressure does not depend on the diameter of the tube; that is, it depends not on the total amount of mercury, but only on its height. This may seem strange at first. To understand it, you must understand the difference between *pressure* and *force.* Pressure is defined as the magnitude of the force acting perpendicularly on a surface divided by the area of that surface: $P = F_{\perp}/A$. Thus, a large force may produce only a small pressure if it is spread over a large area. For example, you can walk on snow without sinking in it if you wear snowshoes. On the other hand, a small force can produce a very large pressure if it is concentrated on a small area. Women's spike heel shoes have ruined many a wooden floor or carpet. The pressure at the place where the heel touched the floor was greater than that under an elephant's foot.

Because the force acts on a very small surface, the pressure under a thin, high heel is greater than that under an elephant's foot.

In 1661, two English scientists, Richard Towneley and Henry Power, discovered an important basic relation. They found that *the pressure exerted by a gas is directly proportional to the density of that gas.* Using P for pressure and D for density, this relationship is $P \propto D$ or $P = kD$ where k is some constant. For
SG 11 example, if the density of a given quantity of air is doubled (say,
SG 36, 37 by compressing it), its pressure also doubles. Robert Boyle confirmed this relation by extensive experiments. It is an
SG 12 empirical rule, now generally known as *Boyle's law.* The law holds true only under special conditions.

The effect of temperature on gas pressure. Boyle recognized that if the temperature of a gas changes during an experiment, the relation $P = kD$ no longer applies. For example, the pressure exerted by a gas in a closed container increases if the gas is heated, even though its density stays constant.

Many scientists throughout the eighteenth century investigated the expansion of gases by heat. The experimental results were not consistent enough to establish a quantitative relation between density (or volume) and temperature. Eventually, evidence for a surprisingly simple general law appeared. The French chemist Joseph-Louis Gay-Lussac (1778–1850) found that all the gases he studied (air, oxygen, hydrogen, nitrogen, nitrous oxide, ammonia, hydrogen chloride, sulfur dioxide, and carbon dioxide) changed their volume in the same way. If the pressure remained constant, then the change in volume was proportional to the change in temperature. On the other hand, if the volume remained constant, the change in pressure was proportional to the change in temperature.

A single equation summarizes all the experimental data obtained by Boyle, Gay-Lussac, and many other scientists. It is known as the *ideal gas law:*

$$P = kD(t + 273°)$$

Here, t is the temperature on the Celsius scale. The proportionality constant k depends only on the kind of gas (and on the units used for P, D, and t).

This equation is called the *ideal* gas law because it is not completely accurate for real gases except at very low pressures. Thus, it is not a law of physics in the same sense as the law of conservation of momentum. Rather, it simply gives an experimental and approximate summary of the observed properties of real gases. It does not apply when pressure is so high, or temperature so low, that the gas nearly changes to a liquid.

On the Celsius scale, water freezes at 0° and boils at 100°, when the pressure is equal to normal atmospheric pressure. On the Fahrenheit scale, water freezes at 32° and boils at 212°. Some of the details involved in defining temperature scales are part of the experiment "Temperature and Thermometers" in the *Handbook.*

The number 273 appears in the ideal gas law simply because temperature is measured on the Celsius scale. The fact that the number is 273 has no great importance. It just depends on the choice of a particular scale for measuring temperature. However, it *is* important to note what would happen if t were decreased to $-273°C$. Then the entire factor involving temperature would be zero. And, according to the ideal gas law, the pressure of any gas would also fall to zero at this temperature. Real gases become liquid long before a temperature of $-273°C$ is reached. Both experiment and thermodynamic theory indicate that it is impossible actually to cool anything, gas, liquid, or solid, down to precisely this temperature. However, a series of cooling operations has produced temperatures less than 0.0001° above this limit.

If the pressure were kept constant, then according to the ideal gas law, the *volume* of a sample of gas would shrink to zero at $-273°C$.

This "absolute zero" point on the temperature scale has been found to be $-273.16°$ Celsius.

In view of the unique meaning of this lowest temperature, Lord Kelvin proposed a new temperature scale. He called it the

For our purposes, it is sufficiently accurate to say the absolute temperature of any sample (symbolized by the letter T and measured in degrees Kelvin, or °K) is equal to the Celsius temperature t plus 273°:

$$T = t + 273°$$

The boiling point of water, for example, is 373°K on the absolute scale.

SG 13, 14

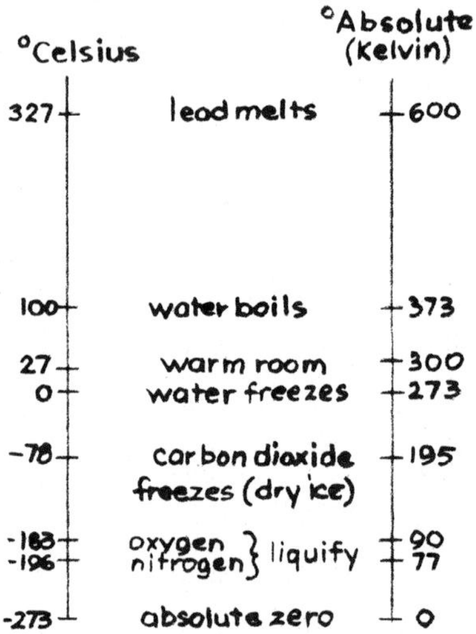

Comparison of the Celsius and absolute temperature scales.

absolute temperature scale and put its zero at −273°C. The absolute scale is sometimes called the Kelvin scale. The temperature of −273°C is now referred to as 0°K on the absolute scale and is called the *absolute zero* of temperature.

The ideal gas law may now be written in simpler form:

$$P = kDT$$

T is the temperature in degrees Kelvin, and k is the proportionality constant.

The equation $P = kDT$ summarizes *experimental facts* about gases. Now you can see whether the kinetic-theory model offers a *theoretical* explanation for these facts.

Kinetic explanation of gas pressure. According to the kinetic theory, the pressure of a gas results from the continual impacts of gas particles against the container wall. This explains why pressure is proportional to density: The greater the density, the greater the number of particles colliding with the wall. But pressure also depends on the *speed* of the individual particles. This speed determines the force exerted on the wall during each impact and the frequency of the impacts. If the collisions with the wall are perfectly elastic, the law of conservation of momentum will describe the results of the impact. The detailed reasoning for this procedure is worked out on pages 336 and 337. This is a beautifully simple application of Newtonian mechanics. The result is clear: Applying Newtonian mechanics to the kinetic molecular model of gases leads to the conclusion that $P = \frac{1}{3}D(v^2)_{av}$ where $(v^2)_{av}$ is the average of the squared speed of the molecules.

So there are two expressions for the pressure of a gas. One summarizes the experimental facts, $P = kDT$. The other is derived by Newton's laws from a theoretical model, $P = \frac{1}{3}D(v^2)_{av}$. The *theoretical* expression will agree with the *experimental* expression only if $kT = \frac{1}{3}(v^2)_{av}$. This would mean that *the temperature of a gas is proportional to* $(v^2)_{av}$. The mass m of each molecule is a constant, so the temperature is also proportional to $\frac{1}{2}m(v^2)_{av}$. Thus, the kinetic theory leads to the conclusion that the temperature of a gas is proportional to the average kinetic energy of its molecules! We already had some idea that raising the temperature of a material somehow affected the motion of its "small parts." We were aware that the higher the temperature of a gas, the more rapidly its molecules are moving. But the conclusion $T \propto \frac{1}{2}m(v^2)_{av}$ is a precise quantitative relationship derived from the kinetic model and empirical laws.

This relationship makes possible other quantitative predictions from kinetic theory. We know by experience that when a gas is compressed or condensed rapidly, the temperature changes, and the general gas law ($P = kDT$) applies. Can the model explain this result?

In the model used on pages 336-37, particles were bouncing back and forth between the walls of a box. Every collision with the wall was perfectly elastic, so the particles rebounded with no loss in speed. Suppose the outside force that holds one wall in place is suddenly reduced. What will happen to the wall? The force exerted on the wall by the collisions of the particles will now be greater than the outside force. Therefore, the wall will move outward.

As long as the wall was stationary, the particles did no work on it, and the wall did no work on the particles. Now the wall moves in the same direction as the force exerted on it by the particles. Thus, the particles must be doing work on the wall. The energy needed to do this work must come from somewhere. The only available source of energy here is the kinetic energy ($\frac{1}{2}mv^2$) of the particles. In fact, it can be shown that molecules colliding perfectly elastically with a receding wall rebound with slightly less speed. Therefore, the kinetic energy of the particles must decrease. The relationship $T \propto \frac{1}{2}m(v^2)_{av}$ implies that the temperature of the gas will drop. This is exactly what happens!

If the outside force on the wall is increased instead of decreased, just the opposite happens. The gas is suddenly compressed as the wall moves inward, doing work on the particles and increasing their kinetic energy. As $\frac{1}{2}mv^2$ goes up, the temperature of the gas should rise, which is just what happens when a gas is compressed quickly.

Many different kinds of experimental evidence support this conclusion and therefore also support the kinetic-theory model. Perhaps the best evidence is the motion of microscopic particles suspended in a gas or liquid, called *Brownian motion*. The gas or liquid molecules themselves are too small to be seen directly, but their effects on a larger particle (for example, a particle of smoke) can be observed through the microscope. At any instant, molecules moving at very different speeds are striking the larger particle from all sides. Nevertheless, so many molecules are taking part that their total effect *nearly* cancels. Any remaining effect changes in magnitude and direction from moment to moment. Therefore, the impact of the invisible molecules makes the visible particle "dance" in the viewfield of the microscope. The hotter the gas, the more lively the motion, as the equation $T \propto \frac{1}{2}m(v^2)_{av}$ predicts.

This experiment is simple to set up and fascinating to watch. You should do it as soon as you can in the laboratory. It gives visible evidence that the smallest parts of all matter in the universe are in a perpetual state of lively, random motion.

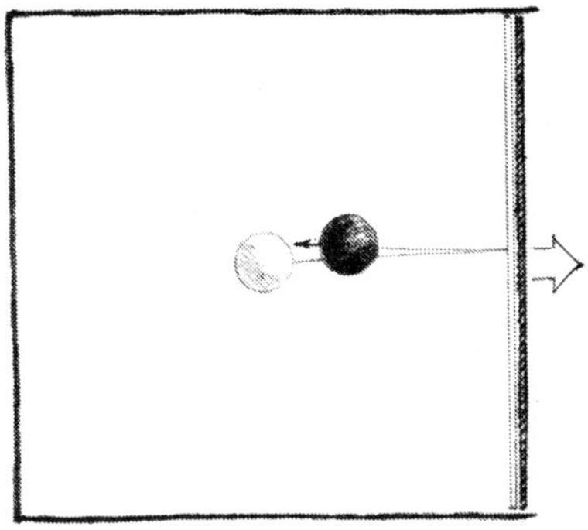

SG 16

This phenomenon can be demonstrated by means of the expansion cloud chamber, cooling of CO_2 fire extinguisher, etc. Here the "wall" is the air mass being pushed away.

Diesel engines have no spark plugs; ignition is produced by temperature rise during the high compression of the air–fuel vapor mixture.

SG 15

Brownian motion was named after the English botanist, Robert Brown, who in 1827 observed the phenomenon while looking at a suspension of the microscopic grains of plant pollen. The same kind of motion of particles ("thermal motion") exists also in liquids and solids, but there the particles are far more constrained than in gases.

? **12.** *The relationship between the density and pressure of a gas expressed by Boyle's law, P = kD, holds true*

Close Up

Deriving an Expression for Pressure from the Kinetic Theory

We begin with the model of a gas described in Sec. 11.2: "a large number of very small particles in rapid, disordered motion." We can assume here that the particles are points with vanishingly small size, so that collisions between them can be ignored. If the particles did have finite size, the results of the calculation would be slightly different. But the approximation used here is accurate enough for most purposes.

The motions of particles moving in all directions with many different velocities are too complex as a starting point for a model. So we fix our attention first on one particle that is simply bouncing back and forth between two opposite walls of a box. Hardly any molecules in a real gas would actually move like this. But we will begin here in this simple way and later in this chapter extend the argument to include other motions. This later part of the argument will require that one of the walls be movable. Therefore, we will arrange for that wall to be movable, but to fit snugly into the box.

In SG 24 of Chapter 9, you saw how the laws of conservation of momentum and energy apply to cases like this. When a very light particle hits a more massive object, like the wall, very little kinetic energy is transferred, If the collision is elastic, the particle will reverse its direction with very little change in speed. In fact, if a force on the outside of the wall keeps it stationary against the impact from inside, the wall will not move during the collisions. Thus *no work* is done on it, and the particles rebound without any change in speed.

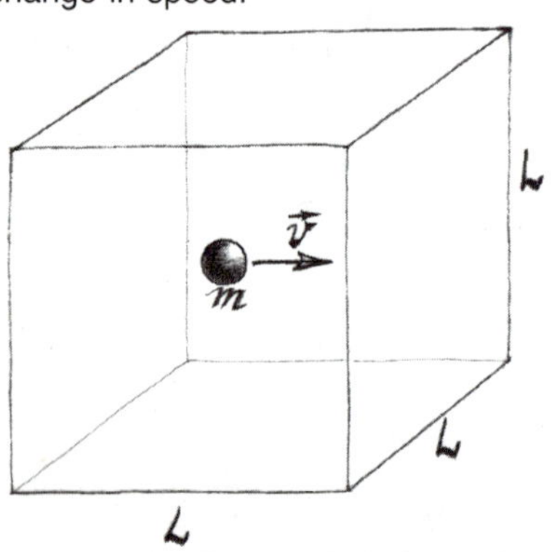

How large a force will these particles exert on the wall when they hit it? By Newton's third law the average force acting on the wall is equal and opposite to the average force with which the wall acts on the particles. The force on each particle is equal to the product of its mass times its acceleration ($\vec{F} = m\vec{a}$), by Newton's second law. As shown in Sec. 9.4, the force can also be written as

$$\vec{F} = \frac{\Delta(m\vec{v})}{\Delta t}$$

where $\Delta(m\vec{v})$ is the change in momentum. Thus, to find the average force acting on the wall we need to find the change in momentum per second due to molecule-wall collisions.

Imagine that a particle, moving with speed v_x (the component of $\vec{v}$ in the x direction) is about to collide with the wall at the right. The component of the particle's momentum in the x direction is $m\vec{v}_x$. Since the particle collides elastically with the wall, it rebounds with the same speed. Therefore, the momentum in the x direction after the collision is

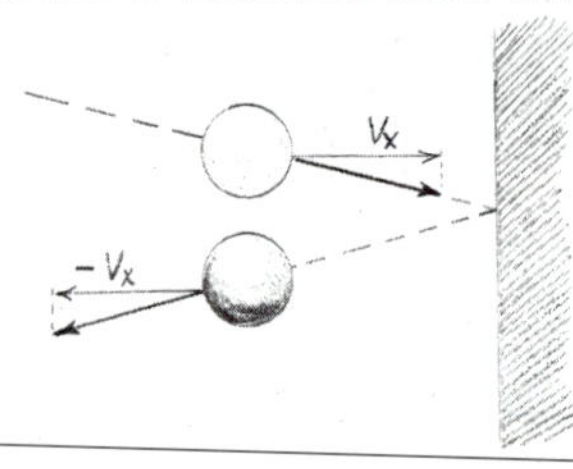

$m(-\vec{v}_x) = -m\vec{v}_x$. The change in the momentum of the particle as a result of this collision is

final momentum	−	initial momentum	=	change in momentum
$(-mv_x)$	−	(mv_x)	=	$(-2mv_x)$

Note that all the vector quantities considered in this derivation have only two possible directions: to

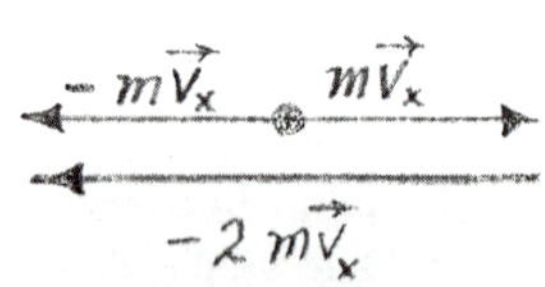

the right or to the left. We can therefore indicate direction by using a + or a − sign, respectively.

Now think of a single particle of mass m moving in a cubical container of volume L^3 as shown in the figure.

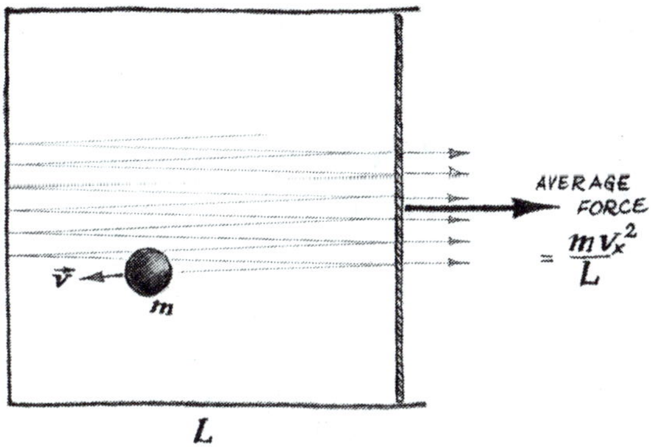

The time between collisions of one particle with the right-hand wall is the time required to cover a distance $2L$ at a speed of v_x; that is, $2L/v_x$. If $2L/v_x$ equals the time between collisions, then $v_x/2L$ equals the number of collisions per second. Thus, the change in momentum per second is given by

(change in momentum in one collision)	×	(number of collisions per second)	=	(change in momentum per second)
$(-2mv_x)$	×	$(v_x/2L)$	=	$\frac{-mv_x^2}{L}$

The net force equals the rate of change of momentum. Thus, the average force acting on the molecule (due to the wall) is equal to $-mv_x^2/L$, and by Newton's third law, the average force acting on the wall (due to the molecule) is equal to $+mv_x^2/L$. So the average pressure on the wall due to the collisions made by one molecule moving with speed v_x is

$$P = \frac{F}{A} = \frac{F}{L^2} = \frac{mv_x^2}{L^3} = \frac{mv_x^2}{V}$$

where v (here L^3) is the volume of the cubical container.

Actually, there are not one but N molecules in the container. They do not all have the same speed, but we need only the average speed in order to find the pressure they exert. More precisely, we need the average of the square of their speeds in the x direction. We call this quantity $(v_x^2)_{av}$. The pressure on the wall due to N molecules will be N times the pressure due to one molecule, or

$$P = \frac{Nm(v_x^2)_{av}}{V}$$

In a real gas, the molecules will be moving in all directions, not just in the x direction; that is, a mol-

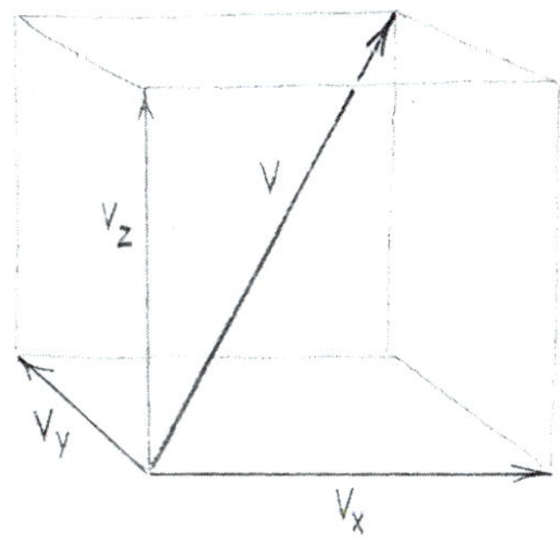

ecule moving with speed v will have three components: v_x, v_y, and v_z. If the motion is random, then there is no preferred direction of motion for a large collection of molecules, and $(v_x^2)_{av} = (v_y^2)_{av} = (v_z^2)_{av}$. It can be shown from Pythagoras' theorem that $v^2 = v_x^2 + v_y^2 + v_z^2$. These last two expressions can be combined to give

$$(v^2)_{av} = 3(v_x^2)_{av}$$

or

$$(v_x^2)_{av} = \tfrac{1}{3}(v^2)_{av}$$

By substituting this expression for $(v_x^2)_{av}$ in the pressure formula, we get

$$P = \frac{Nm \times \frac{1}{3}(v^2)_{av}}{V}$$

$$= \frac{1}{3}\frac{Nm}{V}(v^2)_{av}$$

Notice now that Nm is the total mass of the gas, and therefore Nm/V is just the density D. So

$$P = \tfrac{1}{3}D(v^2)_{av}$$

This is our theoretical expression for the pressure P exerted on a wall by a gas in terms of its density D and the molecular speed v.

(a) *for any gas under any conditions.*

(b) *for some gases under any conditions.*

(c) *only if the temperature is kept constant.*

(d) *only if the density is constant.*

13. *Using the concept of work and the kinetic theory of gases, explain why the temperature of a gas and the kinetic energy of its molecules both increase if a piston is suddenly pushed into the container.*

14. *What are the limits under which the ideal gas law describes the behavior of real gases?*

11.6 The second law of thermodynamics and the dissipation of energy

You have seen that the kinetic-theory model can explain the way a gas behaves when it is compressed or expanded, warmed or cooled. In the late nineteenth century, the model was refined to take into account many effects we have not discussed. There proved to be limits beyond which the model breaks down. For example, radiated heat comes from the sun through the vacuum of space. This is not explainable in terms of the thermal motion of particles. But in most cases the model worked splendidly, explaining the phenomenon of heat in terms of the ordinary motions of particles. This was indeed a triumph of Newtonian mechanics. It fulfilled much of the hope Newton had expressed in the *Principia* that all phenomena of nature could be explained in terms of the motion of the small parts of matter.

SG 23

"Our life runs down in sending up the clock.
The brook runs down in sending up our life.
The sun runs down in sending up the brook.
And there is something sending up the sun.
It is this backward motion toward the source,
Against the stream, that most we see ourselves in.
It is from this in nature we are from.
It is most us."
[Robert Frost, *West-Running Brook*]

A basic philosophical theme of the Newtonian cosmology is the idea that the world is like a machine whose parts never wear out and which never runs down. This idea inspired the search for conservation laws applying to matter and motion. So far in this text, you have seen that this search has been successful. We can measure "matter" by mass, and "motion" by momentum or by kinetic energy. By 1850, the law of conservation of mass had been firmly established in chemistry. In physics, the laws of conservation of momentum and of energy had been equally well established.

Yet these successful conservation laws could not banish the suspicion that somehow the world *is* running down, the parts of the machine *are* wearing out. Energy may be conserved in burning fuel, but it loses its *usefulness* as the heat goes off into the atmosphere. Mass may be conserved in scrambling an egg, but the organized *structure* is lost. In these transformations, something is conserved, but something is also lost. Some processes are irreversible; that is, they will not run backward. There is no way to *un*scramble an egg, although such a change would not violate mass conservation. There is no way to draw smoke and hot fumes back into a blackened stick, forming a new, unburned match.

Section 10.10 discussed one type of irreversible motion, that involving heat engines, which was governed by the second law of thermodynamics. That law can be stated in several equivalent ways: Heat will not by itself flow from a cold body to a hot one. It is impossible to fully convert a given amount of heat into work. The entropy of an isolated system, and therefore of the universe, tends to increase.

SG 24–26

The processes of scrambling an egg, of mixing smoke and air, or of wearing down a piece of machinery do not, at first sight, seem to obey the same laws as do heat engines. However, these processes also are governed by the second law. Heat, as you have seen, is represented by the *disordered* motions of atoms and molecules. Converting (ordered) mechanical work into heat thus leads to an increase in disordered motion.

The second law says that it is never possible to reverse this increase entirely. Heat can never be turned entirely into work. When the entropy of a system increases, the disordered motion in the system increases.

For example, think of a falling ball. If its temperature is very low, the random motion of its parts is very low, too. Thus, the motion of all particles during the falling is mainly downward ("ordered"). The ball strikes the floor and bounces several times. During each bounce, the mechanical energy of the ball decreases, and the ball warms up. Now the random thermal motion of the parts of the heated ball is far more vigorous. Finally, the ball as a whole lies still (no "ordered" motion). The disordered motion of its molecules (and of the molecules of the floor where it bounced) is the only motion left. As with the bouncing ball, all motions tend from ordered to disordered. In fact, entropy can be defined mathematically as a measure of the disorder of a system (though it is not necessary to go into the mathematics here).

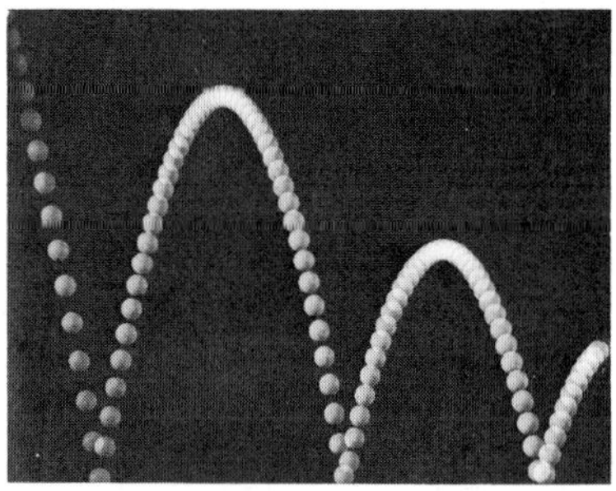

Irreversible processes are processes for which entropy increases. For example, heat will not flow by itself from cold bodies to hot bodies. A ball lying on the floor will not somehow gather the kinetic energy of its randomly moving parts and suddenly leap up. An egg will not unscramble itself. An ocean liner cannot be powered by an engine that takes heat from the ocean water and ejects ice cubes. All these and many other events could occur without violating any principles of Newtonian mechanics, including the law of conservation of energy. But they do not happen; they are "forbidden" by the second law of thermodynamics. (They are "forbidden" in the sense that such things do not happen in nature.)

All familiar processes are to some degree irreversible. Thus, Lord Kelvin predicted that all bodies in the universe would

Two illustrations from Flammarion's novel, **La Fin du Monde:** *(top) "La misérable race humaine périra par le froid." (bottom) "Ce sera la fin."*

SG 27

eventually reach the same temperature by exchanging heat with one another. When this happened, it would be impossible to produce any useful work from heat. After all, work can only be done by means of heat engines when heat flows from a hot body to a cold body. Finally, the sun and other stars would cool, all life on earth would cease, and the universe would be dead.

This general "heat-death" idea, based on predictions from thermodynamics, aroused some popular interest at the end of the nineteenth century. The idea appeared in several books of that time, such as H. G. Wells' *The Time Machine.* The French astronomer Camille Flammarion wrote a book describing ways in which the world would end.

?

15. The presumed "heat death of the universe" refers to a state in which

(a) all mechanical energy has been transformed into heat energy.

(b) all heat energy has been transformed into other forms of energy.

(c) the temperature of the universe decreases to absolute zero.

(d) the supply of coal and oil has been used up.

16. What is a reversible process?

17. Which of the following statements agrees with the second law of thermodynamics?

(a) Heat does not naturally flow from cold bodies to hot bodies.

(b) Energy tends to transform itself into less useful forms.

(c) No engine can transform all its heat input into mechanical energy.

(d) Most processes in nature are reversible.

18. If a pot of water placed on a hot stove froze, Newton's laws would not have been violated. Why would this event violate the second law of thermodynamics? If an extremely small group of water molecules in the pot cooled for a moment, would this violate the second law?

11.7 | Maxwell's demon and the statistical view of the second law of thermodynamics

Is there any way of avoiding the "heat death"? Is irreversibility a basic law of physics, or is it only an approximation based on limited experience of natural processes?

The Austrian physicist Ludwig Boltzmann investigated the theory of irreversibility. He concluded that the tendency toward dissipation of energy is not an *absolute* law of physics that always holds. Rather, it is only a *statistical* law. Think of a container of air containing about 10^{22} molecules. Boltzmann argued that, of all conceivable arrangements of the gas molecules at a given instant, nearly all would be almost completely "disordered." Only a relatively few arrangements would have most of the molecules moving in the same direction. Even if a momentarily ordered arrangement of molecules occurred by chance, it would soon become less ordered by collisions.

Consider also a pool table. The ordered motion of a cue ball moving into a stack of resting ones soon gets "randomized."

To illustrate Boltzmann's argument, consider a pack of cards when it is shuffled. Most possible arrangements of the cards after shuffling are fairly disordered. If you start with an ordered arrangement, for example, the cards sorted by suit and rank, then shuffling would almost certainly lead to a more disordered arrangement. (Nevertheless, it does occasionally happen that a player is dealt 13 spades, even if no one has stacked the deck.)

Fluctuations from complete disorder will, of course, occur. But the greater the fluctuations, the less likely they are to occur. For collections of particles as large as 10^{22}, the chance of a fluctuation large enough to be measurable is vanishingly small. It is *conceivable* that a cold kettle of water will heat up on its own after being struck by only the most energetic molecules in the surrounding air. It is also *conceivable* that air molecules will "gang up" and strike only one side of a rock, pushing it uphill. But such events, while conceivable, are *utterly improbable.*

For *small* collections of particles, however, it is a different story. For example, it is quite probable that the average height of people on a bus will be considerably greater or less than the national average. In the same way, it is probable that more molecules will hit one side of a microscopic particle than the other side. Thus, we can observe the "Brownian" motion of microscopic particles. Fluctuations are an important aspect of the world of very small particles. However, they are virtually undetectable for any large collection of molecules familiar in the everyday world.

The second law is different in character from all the other fundamental laws of physics you have studied so far. The difference is that it deals with probabilities, not certainties.

Maxwell proposed an interesting "thought experiment" to show a possible violation of the second law. Suppose a container

A living system like this tree appears to contradict the second law of thermodynamics by bringing order out of disorder. (See page 342).

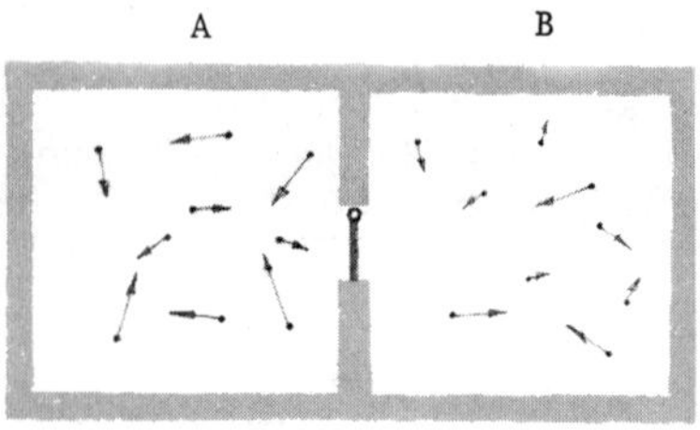

Initially the average *KE* of molecules is greater in *A*.

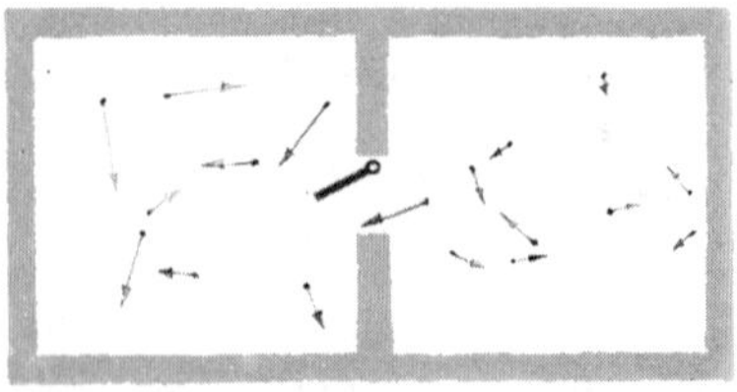

Only fast molecules are allowed to go from *B* to *A*.

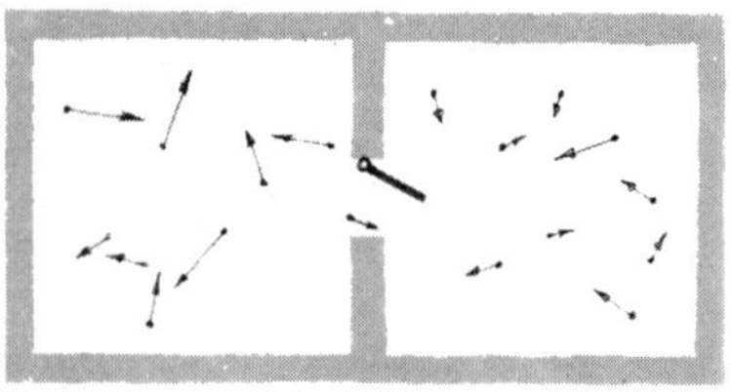

Only slow molecules are allowed to go from *A* to *B*.

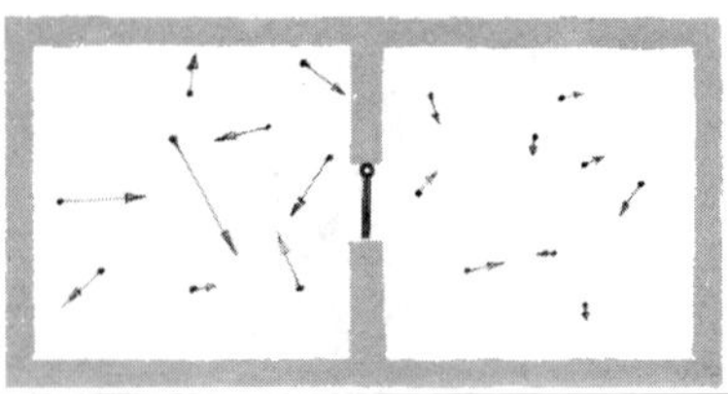

As this continues, the average *KE* in *A* increases and the average *KE* in *B* decreases.

How Maxwell's "demon" could use a small, massless door to increase *the order of a system and make heat flow from a cold gas to a hot gas.*

of gas is divided by a diaphragm into two parts, the gas in one part being hotter than in the other. "Now conceive of a finite being," Maxwell suggested, "who knows the paths and velocities of all the molecules but who can do no work except open and close a hole in the diaphragm." This "finite being," now known as "Maxwell's demon," can make the hot gas hotter and the cold gas cooler just by letting fast molecules move in only one direction through the hole (and slow molecules in the other), as is shown in the diagram.

There is, of course, no such fanciful demon (even in machine form) that can observe and keep track of all the molecules in a gas. (If somehow it could be made to exist, one might find that the demon's entropy is affected by its actions. For example, its entropy might increase enough to compensate for the decrease in entropy of the gas. This is what happens in other systems where local order is created; the entropy elsewhere must increase.)

Some biologists have suggested that certain large molecules, such as enzymes, may function as "Maxwell's demons." Large molecules may influence the motions of smaller molecules to build up the ordered structures of living systems. This result is different from that of lifeless objects and is in apparent violation of the second law of thermodynamics. This suggestion, however, shows a misunderstanding of the law. The second law does not say that the order can *never* increase in any system. It makes that claim only for closed or isolated systems. Any system that can exchange energy with its surroundings can increase its order.

There is some evidence, in fact, that the flow of energy through a system that is not closed tends to produce order in the system. The law governing these processes seems to be, again, the second law of thermodynamics. The existence of life therefore may be a result of the energy flow from the sun to the earth. Far from being a violation of the second law, life would be a manifestation of it. Life does have its cost in terms of the effect on the rest of the total system. This point is expressed vividly in the following passage from a UNESCO document on environmental pollution.

> Some scientists used to feel that the occurrence, reproduction, and growth of order in living systems presented an exception to the second law. This is no longer believed to be so. True, the living system may increase in order, but only by diffusing energy to the surroundings and by converting complicated molecules (carbohydrates, fats) called food into simple molecules (CO_2, H_2O). For example, to maintain a healthy human being at constant weight for one year requires the degradation of about 500 kilograms (one half ton) of food, and the diffusion into the surroundings (from the human and the food) of about 500,000 kilocalories (two million kilojoules) of

SG 28

energy. The "order" in the human may stay constant or even increase, but the order in the surroundings decreases much, much more. Maintenance of life is an expensive process in terms of generation of disorder, and no one can understand the full implications of human ecology and environmental pollution without understanding that first.

?

19. In each of the following pairs, which situation is more ordered?

(a) an unbroken egg; a scrambled egg.

(b) a glass of ice and warm water; a glass of water at uniform temperature.

20. True or false?

(a) Maxwell's demon was able to get around the second law of thermodynamics.

(b) Scientists have made a Maxwell's demon.

(c) Maxwell believed that his demon actually existed.

11.8 | Time's arrow and the recurrence paradox

Late in the nineteenth century, a small but influential group of scientists began to question the basic philosophical assumptions of Newtonian mechanics. They even questioned the very idea of atoms. The Austrian physicist Ernst Mach argued that scientific theories should not depend on assuming the existence of things (such as atoms) which could not be directly observed. Typical of the attacks on atomic theory was the argument used by the mathematician Ernst Zermelo and others against kinetic theory. Zermelo believed that (1) the second law of thermodynamics is an absolutely valid law of physics because it agrees with all the experimental data. However, (2) kinetic theory allows the possibility of exceptions to this law (due to large fluctuations). Therefore, (3) kinetic theory must be wrong. It is an interesting historical episode on a point that is still not quite settled.

The critics of kinetic theory pointed to two apparent contradictions between kinetic theory and the principle of dissipation of energy. These contradictions were the *reversibility paradox* and the *recurrence paradox*. Both paradoxes are based on possible exceptions to the second law; both could be thought to cast doubt on the kinetic theory.

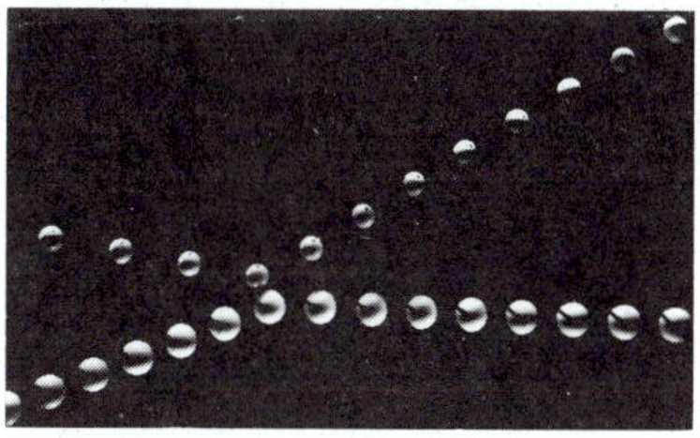

The reversibility paradox: Can a model based on reversible events explain a world in which so many events are irreversible? (Also see photographs on next page.)

The *reversibility* paradox was discovered in the 1870's by Lord Kelvin and Josef Loschmidt, both of whom supported atomic theory. It was not regarded as a serious objection to the kinetic theory until the 1890's. The paradox is based on the simple fact that Newton's laws of motion are reversible in time. For example,

if you watch a motion picture of a bouncing ball, it is easy to tell whether the film is being run forward or backward. You know that the collisions of the ball with the floor are inelastic and that the ball rises less high after each bounce. If, however, the ball made perfectly elastic bounces, it would rise to the same height after each bounce. Then you could not tell whether the film was being run forward or backward. In the kinetic theory, molecules *are* assumed to make perfectly elastic collisions. Imagine that you could take a motion picture of gas molecules colliding elastically according to this assumption. When showing this motion picture, there would be no way to tell whether it was being run forward or backward. Either way would show valid sequences of collisions. But here is the paradox: Consider motion pictures of interactions involving large objects, containing many molecules. You can immediately tell the difference between forward (true) and backward (impossible) time direction. For example, a smashed light bulb does not reassemble itself in real life, though a movie run backward can make it appear to do so.

The kinetic theory is based on laws of motion that are reversible for each individual molecular interaction. How, then, can it explain the existence of *irreversible* processes on a large scale? The existence of such processes seems to indicate that time flows in a definite direction, that is, from past to future. This contradicts the possibility, implied in Newton's laws of motion, that it does not matter whether we think of time as flowing forward or backward. As Lord Kelvin expressed the paradox,

> If . . . the motion of every particle of matter in the universe were precisely reversed at any instant, the course of nature would be simply reversed for ever after. The bursting bubble of foam at the foot of a waterfall would reunite and descend into the water; the thermal motions would reconcentrate their energy, and throw the mass up the fall in drops reforming into a close column of ascending water. Heat which had been generated by the friction of solids and dissipated by conduction, and radiation with absorption, would come again to the place of contact, and throw the moving body back against the force to which it had previously yielded. Boulders would recover from the mud the materials required to rebuild them into their previous jagged forms, and would become reunited to the mountain peak from which they had formerly broken away. And if also the materialistic hypothesis of life were true, living creatures would grow backwards, with conscious knowledge of the future, but no memory of the past, and would become again unborn. But the real phenomena of life infinitely transcend human science; and speculation regarding consequences of their imagined reversal is utterly unprofitable.

Kelvin himself, and later Boltzmann, used statistical probability to explain why we do not observe such large-scale reversals.

There are almost infinitely many possible disordered arrangements of water molecules at the bottom of a waterfall. Only an extremely small number of these arrangements would lead to the process described above. Reversals of this kind are possible *in principle,* but for all practical purposes they are out of the question.

The answer to Zermelo's argument is that his first claim is incorrect. The second law of thermodynamics is not an *absolute* law, but a *statistical* law. It assigns a very low probability to ever detecting any overall increase in order, but does not declare it impossible.

SG 29

However, another small possibility allowed in kinetic theory leads to a situation that seems unavoidably to contradict the dissipation of energy. The *recurrence paradox* revived an idea that appeared frequently in ancient philosophies and is present also in Hindu philosophy to this day: the myth of the "eternal return." According to this myth, the long-range history of the world is cyclic. All historical events eventually repeat themselves, perhaps many times. Given enough time, even the matter that people were made of will eventually reassemble by chance. Then people who have died may be born again and go through the same life. The German philosopher Friedrich Nietzsche was convinced of the truth of this idea. He even tried to prove it by appealing to the principle of conservation of energy. Nietzsche wrote:

The World's great age begins anew,
The golden years return.
The earth doth like a snake renew
His winter weeds outworn . . .
Another Athens shall arise
And to remoter time
Bequeath, like sunset to the skies,
The splendour of its prime . . .
[Percy Bysshe Shelley, "Hellas" (1822)]

> If the universe may be conceived as a definite quantity of energy, as a definite number of centres of energy—and every other concept remains indefinite and therefore useless—it follows therefrom that the universe must go through a calculable number of combinations in the great game of chance which constitutes its existence. In infinity [of time], at some moment or other, every possible combination must once have been realized; not only this, but it must have been realized an infinite number of times.

Lord Kelvin (1824–1907)

If the number of molecules is finite, there is only a finite number of possible arrangements of molecules. Therefore, somewhere in infinite time the same combination of molecules is bound to come up again. At the same point, all the molecules in the universe would reach exactly the same arrangement they had at some previous time. All events following this point would have to be exactly the same as the events that followed it before. That is, if any single instant in the history of the universe is ever *exactly* repeated, then the entire history of the universe will be repeated. As a little thought shows, it would then be repeated over and over again to infinity. Thus, energy would *not* endlessly become dissipated. Nietzsche claimed that this view of the eternal return disproved the "heat-death" theory.

At about the same time, in 1889, the French mathematician Henri Poincaré published a theorem on the possibility of recurrence in mechanical systems. According to Poincaré, even though the universe might undergo a heat death, it would ultimately come alive again:

SG 30–32

> A bounded world, governed only by the laws of mechanics, will always pass through a state very close to its initial state. On the other hand, according to accepted experimental laws (if one attributes absolute validity to them, and if one is willing to press their consequences to the extreme), the universe tends toward a certain final state, from which it will never depart. In this final state, from which will be a kind of death, all bodies will be at rest at the same temperature.
>
> ... the kinetic theories can extricate themselves from this contradiction. The world, according to them, tends at first toward a state where it remains for a long time without apparent change; and this is consistent with experience; but it does not remain that way forever; ... it merely stays there for an enormously long time, a time which is longer the more numerous are the molecules. This state will not be the final death of the universe, but a sort of slumber, from which it will awake after millions of centuries.
>
> According to this theory, to see heat pass from a cold body to a warm one, it will not be necessary to have the acute vision, the intelligence, and the dexterity of Maxwell's demon; it will suffice to have a little patience.

SG 33

Poincaré was willing to accept the possibility of a violation of the second law after a very long time. Others refused to admit even this possibility. In 1896, Zermelo published a paper attacking not only the kinetic theory but the mechanistic world view in general. This view, he asserted, contradicted the second law. Boltzmann replied, repeating his earlier explanations of the statistical nature of irreversibility.

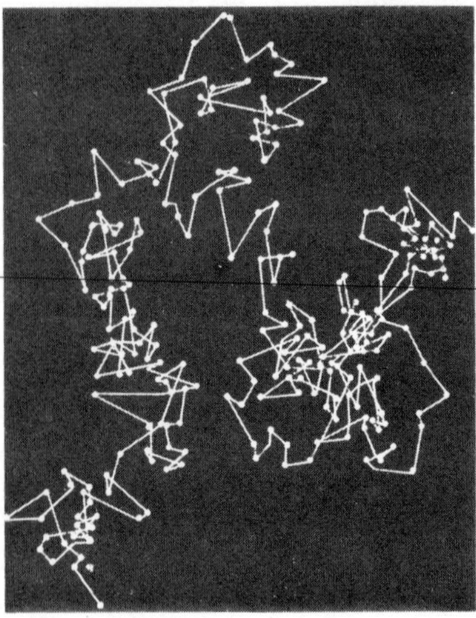

Record of a particle in Brownian motion. Successive positions, recorded every 20 sec, are connected by straight lines. The actual paths between recorded positions would be as erratic as the overall path.

The final outcome of the dispute between Boltzmann and his critics was that both sides were partly right and partly wrong. Mach and Zermelo were correct in believing that Newton's laws of mechanics cannot fully describe molecular and atomic processes. (We will come back to this subject in Unit 5.) For example, it is only approximately valid to describe gases in terms of collections of frantic little balls. But Boltzmann was right in defending the usefulness of the molecular model. The kinetic theory is very nearly correct except for those properties of matter that involve the structure of molecules themselves.

In 1905, Albert Einstein pointed out that the fluctuations predicted by kinetic theory could be used to calculate the rate of displacement for particles in "Brownian" motion. Precise quantitative studies of Brownian motion confirmed Einstein's theoretical calculations. This new success of kinetic theory, along with discoveries in radioactivity and atomic physics, persuaded

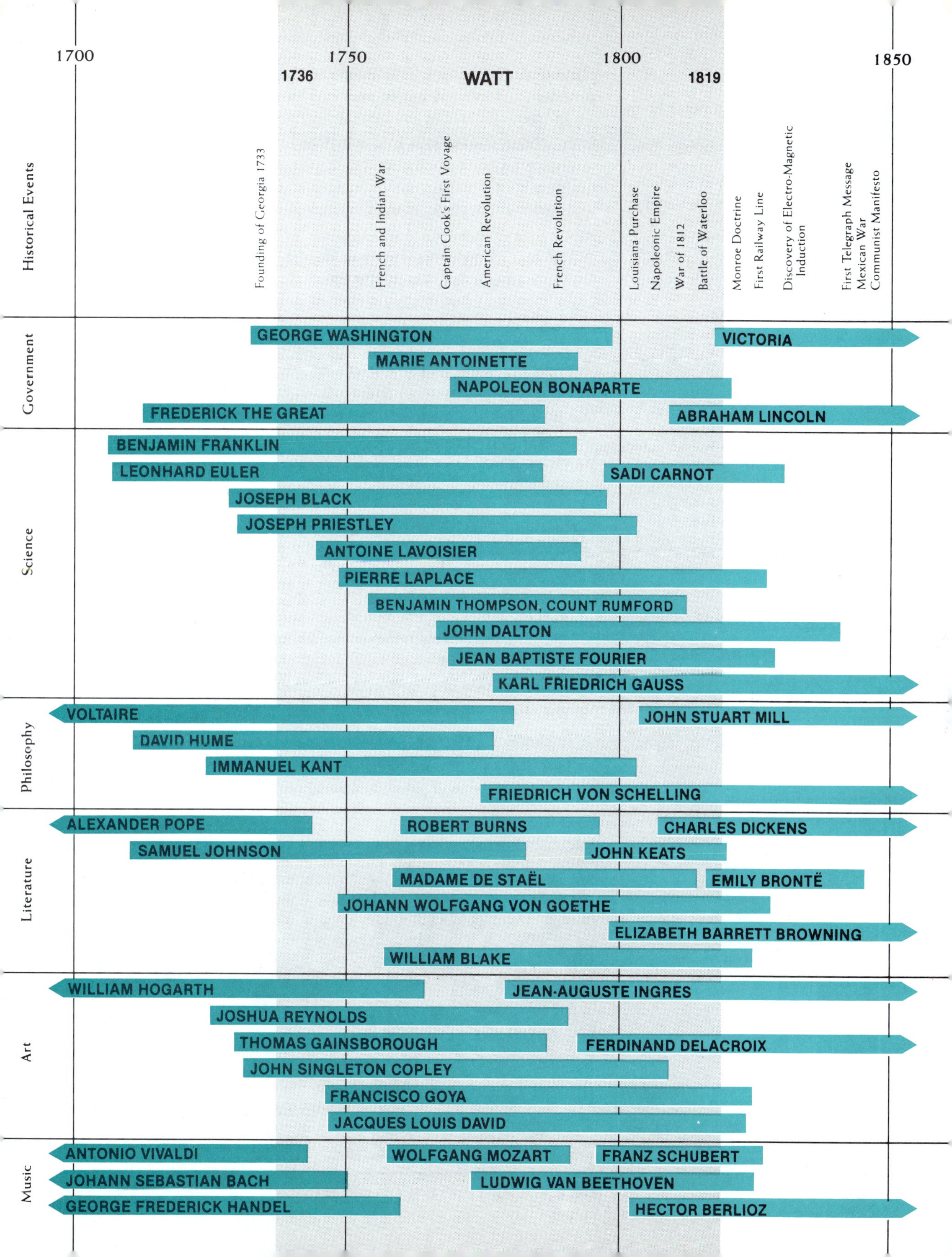
1700
1750
1800
1850
1736
WATT
1819
Historical Events
Founding of Georgia 1733
French and Indian War
Captain Cook's First Voyage
American Revolution
French Revolution
Louisiana Purchase
Napoleonic Empire
War of 1812
Battle of Waterloo
Monroe Doctrine
First Railway Line
Discovery of Electro-Magnetic Induction
First Telegraph Message
Mexican War
Communist Manifesto
Government
GEORGE WASHINGTON
VICTORIA
MARIE ANTOINETTE
NAPOLEON BONAPARTE
FREDERICK THE GREAT
ABRAHAM LINCOLN
Science
BENJAMIN FRANKLIN
LEONHARD EULER
SADI CARNOT
JOSEPH BLACK
JOSEPH PRIESTLEY
ANTOINE LAVOISIER
PIERRE LAPLACE
BENJAMIN THOMPSON, COUNT RUMFORD
JOHN DALTON
JEAN BAPTISTE FOURIER
KARL FRIEDRICH GAUSS
Philosophy
VOLTAIRE
JOHN STUART MILL
DAVID HUME
IMMANUEL KANT
FRIEDRICH VON SCHELLING
Literature
ALEXANDER POPE
ROBERT BURNS
CHARLES DICKENS
SAMUEL JOHNSON
JOHN KEATS
MADAME DE STAËL
EMILY BRONTË
JOHANN WOLFGANG VON GOETHE
ELIZABETH BARRETT BROWNING
WILLIAM BLAKE
Art
WILLIAM HOGARTH
JEAN-AUGUSTE INGRES
JOSHUA REYNOLDS
THOMAS GAINSBOROUGH
FERDINAND DELACROIX
JOHN SINGLETON COPLEY
FRANCISCO GOYA
JACQUES LOUIS DAVID
Music
ANTONIO VIVALDI
WOLFGANG MOZART
FRANZ SCHUBERT
JOHANN SEBASTIAN BACH
LUDWIG VAN BEETHOVEN
GEORGE FREDERICK HANDEL
HECTOR BERLIOZ

almost all the critics that atoms and molecules do exist. But the problems of irreversibility and of whether the laws of physics must distinguish between past and future survived. In a new form, these issues still interest physicists today.

This chapter concludes the application of Newtonian mechanics to individual particles. The story was mainly one of triumphant success. However, like all theories, Newtonian mechanics has serious limitations. These will be explored later.

The last chapter in this unit covers the successful use of Newtonian mechanics in the case of mechanical wave motion. Wave motion completes the list of possibilities of particle motion. In Unit 1, you studied the motion of single particles or isolated objects. The motion of a system of objects bound by a force of interaction, such as the earth and sun, was treated in Unit 2 and in Chapters 9 and 10 of this unit. In this chapter, you observed the motions of a system of a very large number of separate objects. Finally, in Chapter 12 you will study the action of many particles going back and forth together as a wave passes.

?

21. *The kinetic energy of a falling stone is transformed into heat when the stone strikes the ground. Obviously, this is an irreversible process; you never see the heat transform into kinetic energy of the stone, so that the stone rises off the ground. Scientists believe that the process is irreversible because*

SG 34
SG 35

(a) Newton's laws of motion prohibit the reversed process.

(b) the probability of such a sudden ordering of molecular motion is extremely small.

(c) the reversed process would not conserve energy.

(d) the reversed process would violate the second law of thermodynamics.

The ruins of a Greek temple at Delphi are an elegant testimony to the continual encroachment of disorder.

study guide

1. The *Project Physics* materials particularly appropriate for Chapter 11 include:

Experiments
Monte Carlo Experiment on Molecular Collisions
Behavior of Gases

Activities
Drinking Duck
Mechanical Equivalent of Heat
A Diver in a Bottle
Rockets
How to Weigh a Car with a Tire Pressure Gauge
Perpetual-Motion Machines

Film Loop
Reversibility of Time

2. The idea of randomness can be used in predicting the results of flipping a large number of coins. Give some other examples where randomness is useful.

3. The examples of early kinetic theories given in Sec. 11.3 include only *quantitative* models. Some of the underlying ideas are thousands of years old. Compare the kinetic molecular theory of gases to these Greek ideas expressed by the Roman poet Lucretius in about 60 B.C.:

> If you think that the atoms can stop and by their stopping generate new motions in things, you are wandering far from the path of truth. Since the atoms are moving freely through the void, they must all be kept in motion either by their own weight or on occasion by the impact of another atom. For it must often happen that two of them in their course knock together and immediately bounce apart in opposite directions, a natural consequence of their hardness and solidity and the absence of anything behind to stop them. . . .
>
> It clearly follows that no rest is given to the atoms in their course through the depths of space. Driven along in an incessant but variable movement, some of them bounce far apart after a collision while others recoil only a short distance from the impact. From those that do not recoil far, being driven into a closer union and held there by the entanglement of their own interlocking shapes, are composed firmly rooted rock, the stubborn strength of steel and the like. Those others that move freely through larger tracts of space, springing far apart and carried far by the rebound—these provide for us thin air and blazing sunlight. Besides these, there are many other atoms at large in empty space which have been thrown out of compound bodies and have nowhere even been granted admittance so as to bring their motions into harmony.

4. What is a distribution? Under what conditions would you expect a measured distribution to be similar to an ideal or predicted distribution?

5. Consider these aspects of the curves showing Maxwell's distribution of molecular speeds:

(a) All show a peak.

(b) The peaks move toward higher speed at higher temperatures.
Explain these characteristics on the basis of the kinetic model.

6. The measured speed of sound in a gas turns out to be nearly the same as the average speed of the gas molecules. Is this a coincidence? Discuss.

7. How did Clausius modify the simple kinetic model for a gas? What was he able to explain with this new model?

8. Benjamin Franklin observed in 1765 that a teaspoonful of oil would spread out to cover half an acre of a pond. This helps to give an estimate of the upper limit of the size of a molecule. Suppose that 1 cm^3 of oil forms a continuous layer one molecule thick that just covers an area on water of 1,000 m^2.

(a) How thick is the layer?

(b) What is the size of a single molecule of the oil (considered to be a cube for simplicity)?

9. Knowing the size of molecules allows us to compute the number of molecules in a sample of material. If we assume that molecules in a solid or liquid are packed close together, something like apples in a bin, then the total volume of a material is approximately equal to the volume of one molecule times the number of molecules in the material.

(a) Roughly how many molecules are there in 1 cm^3 of water? (For this approximation, you can take the volume of a molecule to be d^3 if its diameter is d.)

(b) The density of a gas (at atmospheric pressure and 0°C) is about 1/1000 the density of a liquid. Roughly how many molecules are there in 1 cm^3 of gas? Does this estimate support the kinetic model of a gas as described on p 336?

10. How high could water be raised with a lift pump on the moon?

11. At sea level, the atmospheric pressure of air ordinarily can balance a barometer column of mercury of height 0.76 m or 10.5 m of water. Air is approximately a thousand times less dense than liquid water. What can you say about the minimum height to which the atmosphere goes above the earth?

12. (a) The pressure of a gas is 100 N/m^2. If the temperature is doubled while the density is cut to one-third, what is the new pressure?

(b) The temperature of a gas is 100°C. If the pressure is doubled and the density is also doubled by cutting the volume in half, what is the new temperature?

13. What pressure do you exert on the ground when you stand on flat-heeled shoes? skis? skates?

14. From the definition of density, $D = M/V$ (where M is the mass of a sample and V is its volume), write an expression relating pressure P and volume V of a gas.

15. State the ideal gas law. What three proportionalities are contained in this law? What are the limitations of this law?

16. Show how all the proportionalities describing gas behavior on p. 332 are included in the ideal gas law: $P = kD\,(t + 273°)$.

17. The following information appeared in a pamphlet published by an oil company:

HOW'S YOUR TIRE PRESSURE?

If you last checked the pressure in your tires on a warm day, one cold morning you may find your tires seriously underinflated.

The Rubber Manufacturers Association warns that tire pressures drop approximately 12 kPa for every 10-deg dip in outside air. If your tires register 165 kPa pressure on a 30°C day, for example, they'll have only 130 kPa pressure when the outside air plunges to 0°C.

If you keep your car in a heated garage at 15°C, and drive out into a −30°C morning, your tire pressure drops from 165 kPa to 125 kPa.

Are these statements consistent with the ideal gas law? (*Note*: The pressure registered on a tire gauge is the pressure *above* normal atmospheric pressure of about 100 kPa.)

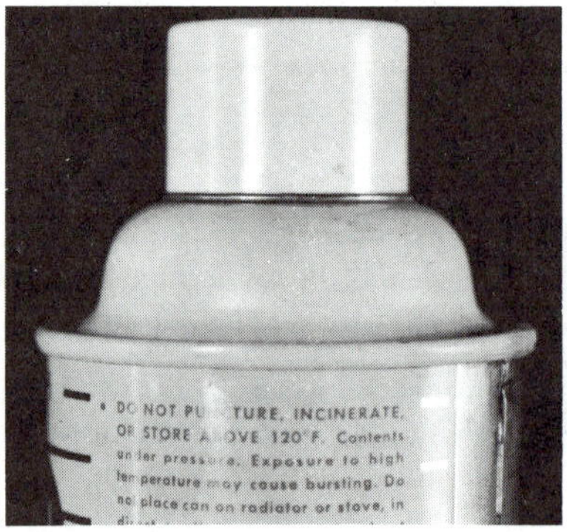

18. Distinguish between two uses of the word "model" in science.

19. If a light particle rebounds from a massive, stationary wall with almost no loss of speed, then, according to the principle of Galilean relativity, it would still rebound from a *moving* wall without changing speed as seen in *the frame of reference of the moving wall.* Show that the rebound speed as measured *in the laboratory* would be less from a retreating wall (as is claimed at the bottom of p. 335). (Hint: First write an expression relating the particle's speed relative-to-the-wall to its speed relative-to-the-laboratory.)

20. What would you expect to happen to the temperature of a gas that was released from a container in empty space (that is, with nothing to push back)?

21. List some of the directly observable properties of gases.

22. What aspects of the behavior of gases can the kinetic molecular theory be used to explain successfully?

23. Many products are now sold in spray cans. Explain in terms of the kinetic theory of gases why it is dangerous to expose the cans to high temperatures.

24. When a gas in an enclosure is compressed by pushing in a piston, its temperature increases. Explain this fact in two ways:

(**a**) by using the first law of thermodynamics.

(**b**) by using the kinetic theory of gases.

The compressed air eventually cools down to the same temperature as the surroundings. Describe this heat transfer in terms of molecular collisions.

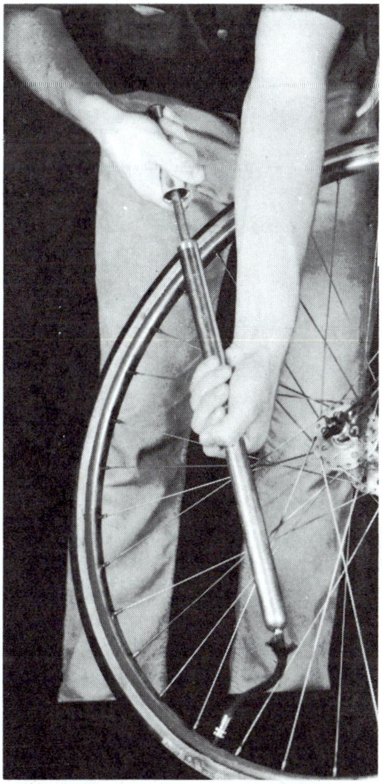

25. From the point of view of the kinetic theory, how can you explain: (a) that a hot gas would not cool itself down while in a perfectly insulated container? (b) how a kettle of cold water, when put on the stove, reaches a boiling temperature. (Hint: At a given temperature the molecules in and on the walls of the solid container are also in motion, although, being part of a solid, they do not often get far away.)

26. In the *Principia,* Newton expressed the hope that all phenomena could be explained in terms of the motion of atoms. How does Newton's view compare with this Greek view expressed by Lucretius in about 60 B.C.?

> I will now set out in order *the stages by which the initial concentration of matter laid the foundations of earth and sky,* of the ocean depths and the orbits of sun and moon. Certainly the atoms did not post themselves purposefully in due order by an act of intelligence, nor did they stipulate what movements each should perform. But multitudinous atoms, swept along in multitudinous courses through infinite time by mutual clashes and their own weight, have come together in every possible way and realized everything that could be formed by their combinations. So it comes about that a voyage of immense duration, in which they have experienced every variety of movement and conjunction, has at length brought together those whose sudden encounter normally forms the starting-point of substantial fabrics—earth and sea and sky and the races of living creatures.

27. In Sec. 11.6, three statements of the second law of thermodynamics are given. In Chapter 10, we showed that the first and third laws are equivalent. Show that the first and second laws are also equivalent, by using an argument analogous to Carnot's; that is, show that if an engine violates either statement, then this engine (perhaps together with a reversible engine) also violates the other.

28. There is a tremendous amount of internal energy in the oceans and in the atmosphere. What would you think of an invention that purported to draw on this source of energy to do mechanical work? (For example, a ship that sucked in seawater and exhausted blocks of ice, using the heat from the water to run the ship.)

29. Compute the entropy change associated with the bouncing ball of page 339. Assume that the ball has a mass of 0.1 kg and falls from a height of 1 m. How much energy is converted to heat? If the process happens nearly at room temperature (a bit less than 300°K), what is the entropy change of the universe? Does it matter how much of the heat goes to the ball and how much goes elsewhere?

30. Since there is a tendency for heat to flow from hot to cold, will the universe eventually reach absolute zero?

31. Does Maxwell's demon get around the second law of thermodynamics? List the assumptions in Maxwell's argument. Which of them do you believe are likely to be true?

32. Since all the evidence is that molecular motions are random, you might expect that any given arrangement of molecules will recur if you just wait long enough. Explain how a paradox arises when this prediction is compared with the second law of thermodynamics.

33. (a) Explain what is meant by the statement that Newton's laws of motion are time-reversible.

(b) Describe how a paradox arises when the time-reversibility of Newton's laws of motion is compared with the second law of thermodynamics.

34. Why is the melting of an ice cube an irreversible process even though it could easily be refrozen?

35. If there is a finite probability of an exact repetition of a state of the universe, there is also a finite probability of its exact opposite, that is, a state where molecules are in the same position but with reversed velocities. What would this imply about the subsequent history of the universe?

36. List the assumptions in the "recurrence" theory. Which of them do you believe to be true?

37. Some philosophical and religious systems of the Far East and the Middle East include the idea of the eternal return. If you have read about some of these philosophies, discuss what analogies exist to some of the ideas in the last part of this chapter. Is it appropriate to take the existence of such analogies to mean that there is some direct connection between these philosophical and physical ideas?

38. Where did Newtonian mechanics run into difficulties in explaining the behavior of molecules?

39. What are some advantages and disadvantages of theoretical models?

40. At any point in a fluid, the upward force on a column of fluid at rest must be sufficient to support the weight of the fluid above it. The pressure of a fluid must therefore increase with the depth of the fluid. Consider a fluid whose density is D. By how much does the pressure increase if the depth below the surface is increased by x? (Consider the force on a column 1 m^2 in area.)

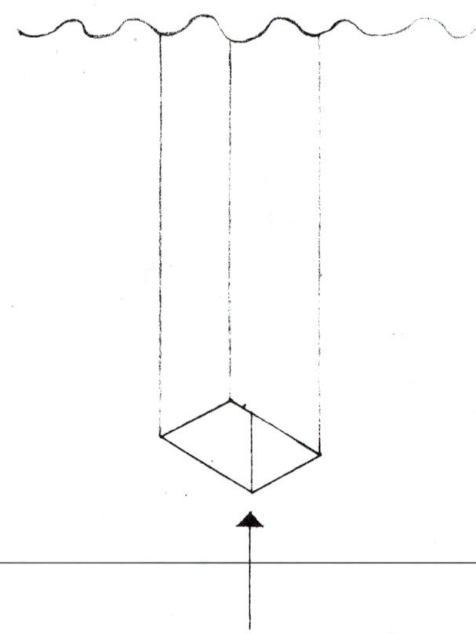

41. *Archimedes' principle*: The fact that pressure increases with depth implies that an object (say, a balloon) forcibly immersed in a fluid will experience a strong upward force from the fluid. The reason is that the fluid touching the bottom of the balloon has a greater pressure than has the fluid touching the top, so the upward force on the object exerted by the fluid below is greater than the downward force exerted by the fluid above it. This difference is called the "buoyant force."

You can find the buoyant force as follows. Consider a cube, length L on a side, immersed in a fluid of density D. In SG 36 you showed that the pressure increases by Dgx as the depth increases by x.

(**a**) Suppose the pressure at the top surface of the cube is p. What is the pressure at the bottom?

(**b**) What is the force exerted on the top of the cube (magnitude and direction)?

(**c**) What is the force exerted on the bottom of the cube (magnitude and direction)?

(**d**) What is the net force exerted by the fluid on the cube?

(**e**) Show that this force is equal to the weight of the fluid displaced by the object.

CHAPTER 12

Waves

12.1 What is a wave?

SG 1

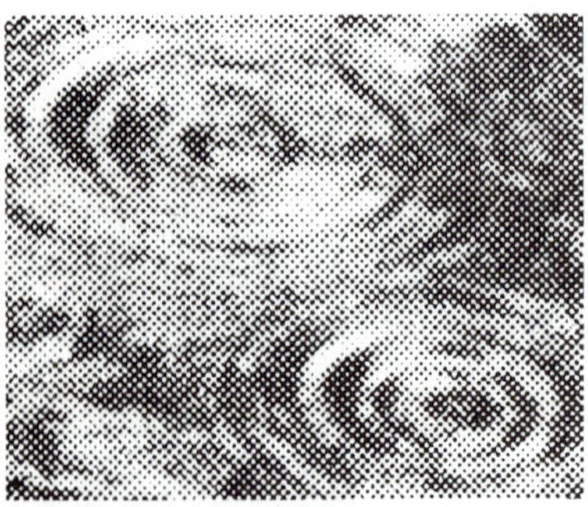

A small section from the lower right of the photograph on page 364.

The world is continually criss-crossed by waves of all sorts. Water waves, whether giant rollers in the middle of the ocean or gently formed rain ripples on a still pond, are sources of wonder or pleasure. If the earth's crust shifts, violent waves in the solid earth cause tremors thousands of kilometers away. A musician plucks a guitar string, and sound waves pulse against the ears. Wave disturbances may come in a concentrated bundle, like the shock front from an airplane flying at supersonic speeds. Or the disturbances may come in succession like the train of waves sent out from a steadily vibrating source, such as a bell or a string.

All of these examples are *mechanical* waves, in which bodies or particles physically move back and forth. There are also wave disturbances in electric and magnetic fields. In Unit 4, you will learn that such waves are responsible for what your senses experience as light. In all cases involving waves, however, the effects produced depend on the flow of energy as the wave moves forward.

So far in ths text, you have considered motion in terms of individual particles. In this chapter, you will study the cooperative motion of collections of particles in "continuous media" moving in the form of mechanical waves. You will see how closely related are the ideas of particles and waves used to describe events in nature.

A comparison will help here. Look at a black and white photograph in a newspaper or magazine with a magnifying glass. You will see that the picture is made up of many little black dots printed on a white page (up to 3,000 dots per square centimeter). Without the magnifier, you do not see the individual dots. Rather, you see a pattern with all possible shadings between completely black and completely white. These two views emphasize different aspects of the same thing. In much the same way, the physicist can sometimes choose between two (or more) ways of viewing events. For the most part, a particle view has been emphasized in the first three units of *Project Physics*. In Unit 2, for example, each planet was treated as a particle undergoing the sun's gravitational attraction. The behavior of the solar system was described in terms of the positions, velocities, and accelerations of point-like objects. For someone interested only in planetary motions, this is fine. But for someone interested in, say, the chemistry of materials on Mars, it is not very helpful.

Waves should be studied in the laboratory. Most of this chapter is only a summary of some of what you will learn there. Film loops on waves are listed in SG 1.

In the last chapter, you saw two different descriptions of a gas. One was in terms of the behavior of the individual particles making up the gas. Newton's laws of motion described what each *individual* particle does. Then average values of speed or energy described the behavior of the gas. You also studied concepts such as pressure, temperature, heat, and entropy. These concepts refer directly to a sample of gas *as a whole*. This is the viewpoint of thermodynamics, which does not depend on assuming Newton's laws or even the existence of particles. Each of these viewpoints served a useful purpose and helped you to understand what you cannot directly see.

To study waves, you once again see the possibility of using different points of view. Most of the waves discussed in this chapter can be described in terms of the behavior of particles. But you also can understand waves as disturbances traveling in a continuous medium. You can, in other words, see both the forest and the trees; the picture as a whole, not only individual dots.

12.2 | Properties of waves

Suppose that two people are holding opposite ends of a rope. Suddenly one person snaps the rope up and down quickly once.

That "disturbs" the rope and puts a hump in it which travels along the rope toward the other person. The traveling hump is one kind of a wave, called a *pulse*.

Originally, the rope was motionless. The height of each point on the rope depended only upon its position along the rope and did not change in time. When one person snaps the rope, a rapid change is created in the height of one end. This disturbance then moves away from its source. The height of each point on the rope depends upon time as well as position along the rope.

The disturbance is a pattern of *displacement* along the rope. The motion of the displacement pattern from one end of the rope toward the other is an example of a *wave*. The hand snapping one end is the *source* of the wave. The rope is the *medium* in which the wave moves.

Consider another example. When a pebble falls into a pool of still liquid, a series of circular crests and troughs spreads over the surface. This moving displacement pattern of the liquid surface is a wave. The pebble is the source; the moving pattern of crests and troughs is the wave; and the liquid surface is the medium. Leaves, sticks, or other objects floating on the surface of the liquid bob up and down as each wave passes. But they do not experience any net displacement on the average. No *material* has moved from the wave source, either on the surface or among the particles of the liquid. The same holds for rope waves, sound waves in air, etc.

As any one of these waves moves through a medium, the wave produces a changing displacement of the successive parts of the medium. Thus, we can refer to these waves as *waves of displacement*. If you can see the medium and recognize the displacements, then you can see waves. But waves also may exist in media you cannot see (such as air) or may form as disturbances of a state you cannot detect with your eyes (such as pressure or an electric field).

You can use a loose spring coil to demonstrate three different kinds of motion in the medium through which a wave passes. First, move the end of the spring from side to side, or up and down as in sketch (a) in the margin. A wave of side-to-side or up-and-down displacement will travel along the spring. Now push the end of the spring back and forth, along the direction of the spring itself, as in sketch (b). A wave of back-and-forth displacement will travel along the spring. Finally, twist the end of the spring clockwise and counterclockwise, as in sketch (c). A wave of angular displacement will travel along the spring. Waves like those in (a), in which the displacements are perpendicular to the direction the wave travels, are called *transverse* waves. Waves like those in (b), in which the displacements are in the direction the wave travels, are called *longitudinal* waves. Waves

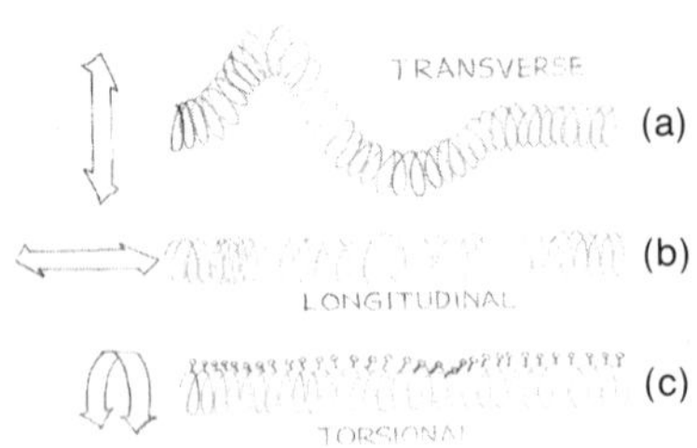

"Snapshots" of three types of waves on a spring. In (c), small markers have been put on the top of each coil in the spring.

like those in (c), in which the displacements are twisting in a plane perpendicular to the direction the wave travels, are called *torsional* waves.

All three types of wave motion can be set up in solids. In fluids, however, transverse and torsional waves die out very quickly and usually cannot be produced at all. Therefore, sound waves in air and water are longitudinal. The molecules of the medium are displaced back and forth along the direction in which the sound travels.

It is often useful to make a graph of wave patterns in a medium. However, a graph on paper always has a transverse appearance, even if it represents a longitudinal or torsional wave. For example, the graph at the right represents the pattern of compressions in a sound wave in air. The sound waves are longitudinal, but the graph line goes up and down. This is because the graph represents the increasing and decreasing density of the air. It does *not* represent an up-and-down motion of the air.

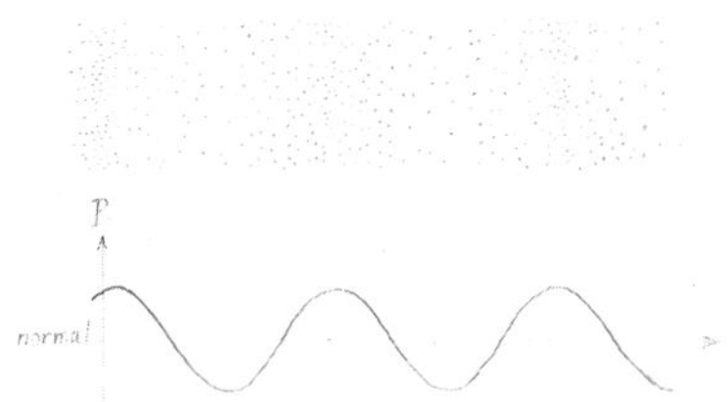

(a)
(b)

(a) "Snapshot" representation of a sound wave progressing to the right. The dots represent the density of air molecules. (b) Graph of air pressure P *versus position* x *at the instant of the snapshot.*

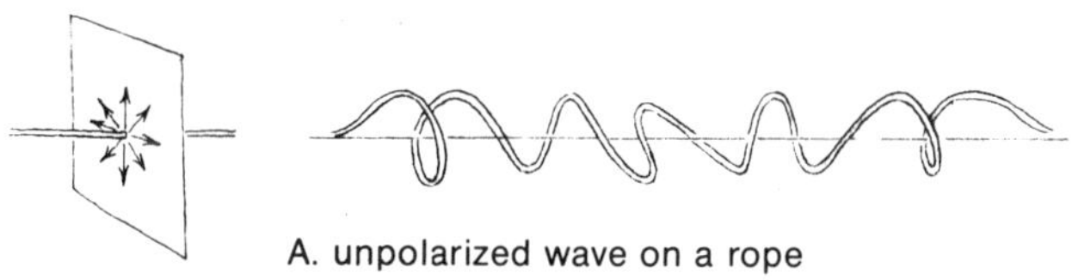
A. unpolarized wave on a rope

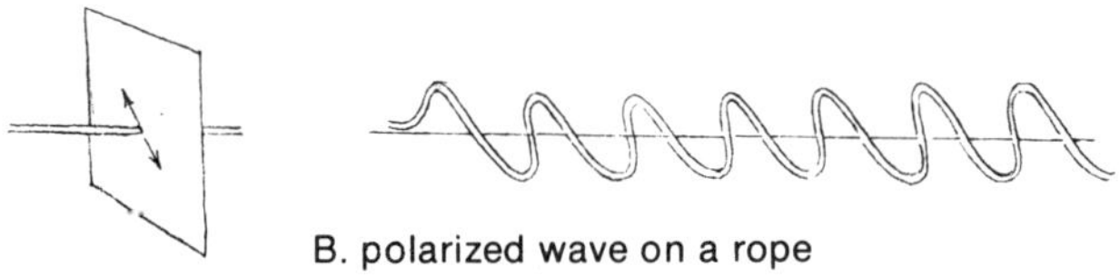
B. polarized wave on a rope

To describe completely transverse waves, such as those in ropes, you must specify the *direction* of displacement. When the displacement pattern of a transverse wave lies in a single plane, the wave is *polarized*. For waves on ropes and springs, you can observe the polarization directly. Thus, in the photograph on the previous page, the waves the person makes are in the horizontal plane. Although there are few special properties associated with polarized waves on ropes, you will see (in Sec. 13.7) that for light waves, for example, polarization can have important effects.

All three kinds of wave (longitudinal, transverse, and torsional) have an important characteristic in common. The disturbances move away from their sources through the media and *continue on their own*. We stress this particular characteristic by saying that these waves *propagate*. This means more than just that they "travel" or "move." An example will clarify the difference between waves that propagate and those that do not. You probably have read some description of the great wheat plains of the Middle West, Canada, or Central Europe. Such descriptions usually mention the "beautiful, wind-formed waves that roll for miles across the fields." The medium for such a wave is the wheat, and

An engine starting abruptly can start a displacement wave along a line of cars.

A very important point: Energy transfer can occur without matter transfer.

the disturbance is the swaying motion of the wheat. This disturbance does indeed travel, but it does *not* propagate; that is, the disturbance does not originate at a source and then go on by *itself*. Rather, it must be continually fanned by the wind. When the wind stops, the disturbance does not roll on, but stops, too. The traveling "waves" of swaying wheat are not at all the same as rope and water waves. This chapter will concentrate on waves that do originate at sources and propagate themselves. For the purposes of this chapter, *waves are disturbances which propagate in a medium*.

?

1. *What kinds of mechanical waves can propagate in a solid?*
2. *What kinds of mechanical waves can propagate in a fluid?*
3. *What kinds of mechanical waves can be polarized?*
4. *Suppose that a mouse runs along under a rug, causing a bump in the rug that travels with the mouse across the room. Is this moving disturbance a propagating wave?*

12.3 | Wave propagation

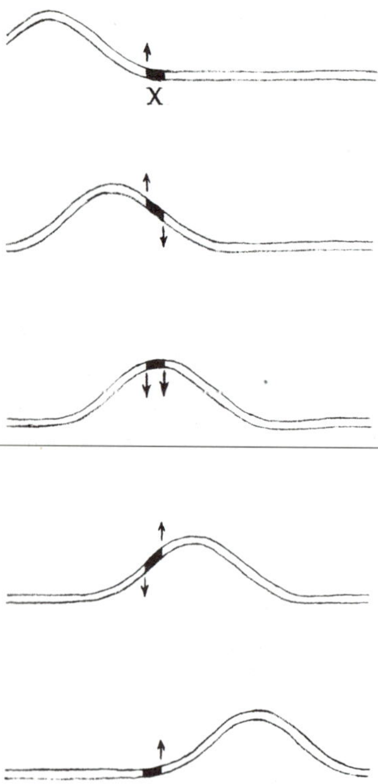

A rough representation of the forces at the ends of a small section of rope as a transverse pulse moves past.

Waves and their behavior are perhaps best studied by beginning with large mechanical models and focusing your attention on pulses. Consider, for example, a freight train, with many cars attached to a powerful locomotive, but standing still. If the locomotive starts abruptly, it sends a displacement wave running down the line of cars. The shock of the starting displacement proceeds from locomotive to caboose, clacking through the couplings one by one. In this example, the locomotive is the source of the disturbance, while the freight cars and their couplings are the medium. The "bump" traveling along the line of cars is the wave. The disturbance proceeds all the way from end to end, and with it goes *energy* of displacement and motion. Yet no particles of matter are transferred that far; each car only jerks ahead a bit.

How long does it take for the effect of a disturbance created at one point to reach a distant point? The time interval depends upon the speed with which the disturbance or wave propagates. This speed, in turn, depends upon the type of wave and the characteristics of the medium. In any case, the effect of a disturbance is never transmitted instantly over any distance. Each part of the medium has inertia, and each portion of the medium is compressible. So time is needed to transfer energy from one part to the next.

The same comments apply also to transverse waves. The series of sketches in the margin represents a wave on a rope. Think of the sketches as frames of a motion picture film, taken at equal time intervals. The material of the rope does *not* travel along with the wave. But each bit of the rope goes through an up-and-down motion as the wave passes. Each bit goes through exactly the same motion as the bit to its left, except a little later.

Consider the small section of rope labeled X in the diagrams. When the pulse traveling on the rope first reaches X, the section of rope just to the left of X exerts an upward force on X. As X is moved upward, a restoring downward force is exerted by the next section. The further upward X moves, the greater the restoring forces become. Eventually, X stops moving upward and starts down again. The section of rope to the left of X now exerts a restoring (downward) force, while the section to the right exerts an upward force. Thus, the trip down is similar, but opposite, to the trip upward. Finally, X returns to the equilibrium position, and both forces vanish.

The time required for X to go up and down, that is, the time required for the pulse to pass by that portion of the rope, depends on two factors. These factors are the *magnitude of the forces* on X, and the *mass* of X. To put it more generally: The speed with which a wave propagates depends on the *stiffness* and on the *density* of the medium. The stiffer the medium, the greater will be the force each section exerts on neighboring sections. Thus, the greater will be the propagation speed. On the other hand, the greater the density of the medium, the less it will respond to forces. Thus, the slower will be the propagation. In fact, the speed of propagation depends on the *ratio* of the stiffness factor and the density factor.

SG 2

?

5. What is transferred along the direction of wave motion?

6. On what two properties of a medium does wave speed depend?

7. If a spring is heated to make it less stiff, does it carry waves faster or slower? If the boxcars in a train are unloaded, does the longitudinal start-up wave travel faster or slower?

The exact meaning of stiffness and density factors is different for different kinds of waves and different media. For tight strings, for example, the stiffness factor is the tension T in the string, and the density factor is the mass per unit length, m/l. The propagation speed v is given by

$$v = \sqrt{\frac{T}{m/l}}$$

12.4 | Periodic waves

Many of the disturbances considered up to now have been sudden and short-lived. They were set up by a single disturbance like snapping one end of a rope or suddenly bumping one end of a train. In each case, you see a single wave running along the medium with a certain speed. This kind of wave is a *pulse*.

Now consider *periodic waves*, continuous regular rhythmic disturbances in a medium, resulting from *periodic vibrations* of a source. A good example of a periodic vibration is a swinging pendulum. Each swing is virtually identical to every other swing,

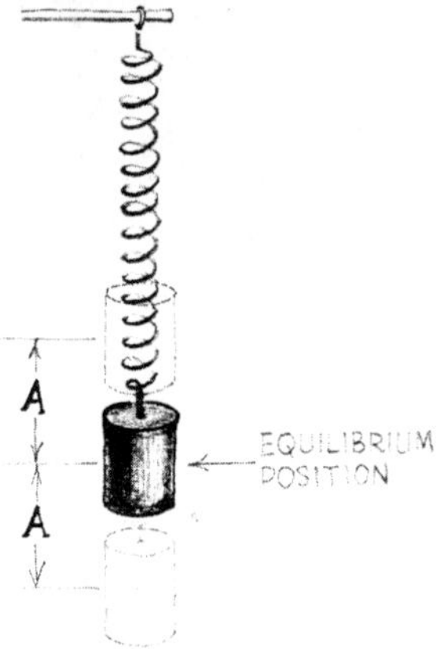

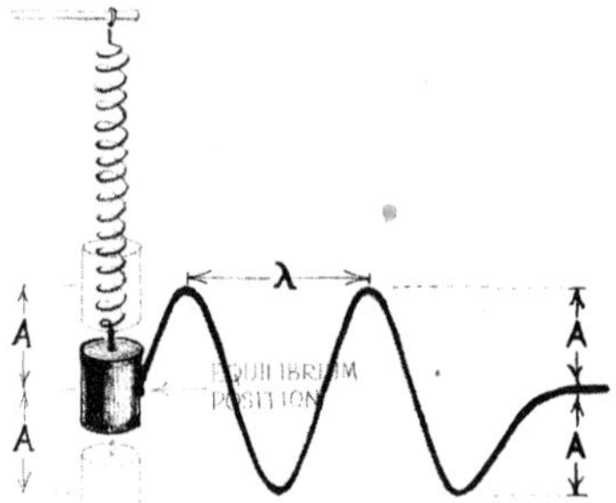

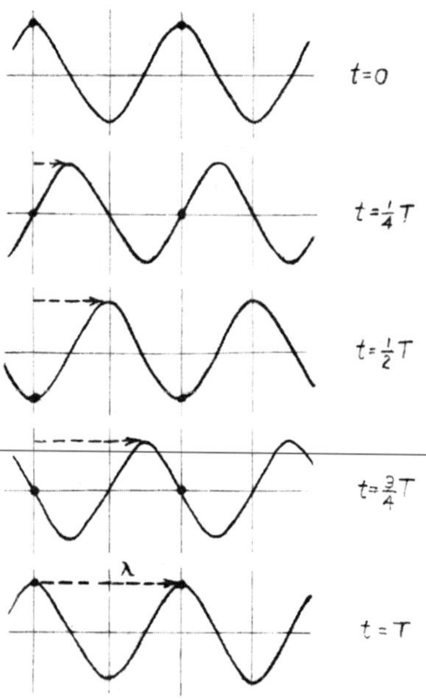

The wave generated by a simple harmonic vibration is a sine wave. A "snapshot" of the displacement of the medium would show it has the same shape as a graph of the sine function familiar in trigonometry. This shape is frequently referred to as "sinusoidal."

and the swing repeats over and over again in time. Another example is the up-and-down motion of a weight at the end of a spring. The maximum displacement from the position of equilibrium is called the *amplitude, A,* as shown in the margin. The time taken to complete one vibration is called the *period, T.* The number of vibrations per second is called the *frequency, f.*

What happens when such a vibration is applied to the end of a rope? Suppose that one end of a rope is fastened to the oscillating (vibrating) weight. As the weight vibrates up and down, you observe a wave propagating along the rope. The wave takes the form of a series of moving crests and troughs along the length of the rope. The source executes "simple harmonic motion" up and down. Ideally, every point along the length of the rope executes simple harmonic motion in turn. The wave travels to the right as crests and troughs follow one another. Each point along the rope simply oscillates up and down at the same frequency as the source. The amplitude of the wave is represented by A. The distance between any two consecutive crests or any two consecutive troughs is the same all along the length of the rope. This distance, called the *wavelength* of the periodic wave, is represented by the Greek letter λ (lambda).

If a single pulse or a wave crest moves fairly slowly through the medium, you can easily find its *speed.* In principle all you need is a clock and a meter stick. By timing the pulse or crest over a measured distance, you can get the speed. But it is not always simple to observe the motion of a pulse or a wave crest. As is shown below, however, the speed of a periodic wave can be found indirectly from its frequency and wavelength.

As a wave progresses, each point in the medium oscillates with the frequency and period of the source. The diagram in the margin illustrates a periodic wave moving to the right, as it might look in snapshots taken every ¼ period. Follow the progress of the crest that started out from the extreme left at $t = 0$. The time it takes this crest to move a distance of one wavelength is *equal* to the time required for one complete oscillation; that is, the crest moves one wavelength λ in one period of oscillation T. The speed v of the crest is therefore

$$v = \frac{\text{distance moved}}{\text{corresponding time interval}} = \frac{\lambda}{T}$$

All parts of the wave pattern propagate with the same speed. Thus, the speed of any one crest is just the speed of the wave. Therefore, the speed v of the wave is

$$v = \frac{\text{wavelength}}{\text{period of oscillation}} = \frac{\lambda}{T}$$

But $T = 1/f$, where f = frequency (see *Project Physics,* Chapter 4, page 112). Therefore, $v = f\lambda$, or wave speed = frequency × wavelength.

We can also write this relationship as $\lambda = v/f$ or $f = v/\lambda$. These expressions imply that, for waves of the same speed, the frequency and wavelength are inversely proportional; that is, a wave of twice the frequency would have only half the wavelength, and so on. This inverse relation of frequency and wavelength will be useful in other parts of this course.

The diagram below represents a periodic wave passing through a medium. Sets of points are marked that are moving "in step" as the periodic wave passes. The crest points C and C′ have reached maximum displacement positions in the upward direction. The trough points D and D′ have reached maximum displacement positions in the downward direction. The points C and C′ have identical displacements and velocities at any instant of time. Their vibrations are identical and in unison. The same is true for the points D and D′. Indeed there are infinitely many such points along the medium that are vibrating identically when this wave passes. Note that C and C′ are a distance λ apart, and so are D and D′.

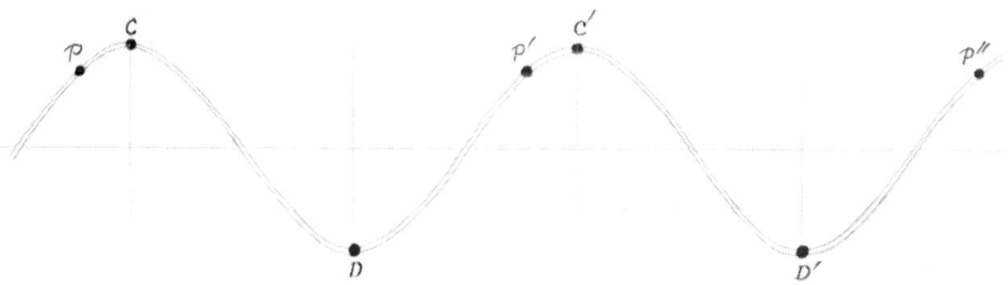

A "snapshot" of a periodic wave moving to the right. Letters indicate sets of points with the same phase.

SG 3

Points that move "in step," such as C and C′, are said to be *in phase* with one another. Points D and D′ also move in phase. Points separated from one another by distances of λ, 2λ, 3λ, . . ., and $n\lambda$ (where n is any whole number) are all in phase with one another. These points can be anywhere along the length of the wave. They need not correspond with only the highest or lowest points. For example, points such as P, P′, P″, are all in phase with one another. Each point is separated from the next one by a distance λ.

Some of the points are exactly *out* of step. For example, point C reaches its maximum upward displacement at the same time that D reaches its maximum downward displacement. At the instant that C begins to go down, D begins to go up. Points such as these are one-half period *out of phase* with respect to one another; C and D′ also are one-half period out of phase. Any two points separated from one another by distances of $\lambda/2$, $3\lambda/2$, $5\lambda/2$, . . . are one-half period out of phase.

? *8. Of the wave variables frequency, wavelength, period, amplitude, and polarization, which ones describe*

(a) space properties of waves?

(b) time properties of waves?

9. A wave with the displacement as smoothly and simply varying from point to point as that shown in the illustration on page 361 is called a "sine" wave. How might the "wavelength" be defined for a periodic wave that is not a sine wave?

10. A vibration of 100 Hz (cycles per second) produces a wave.

(a) What is the wave frequency?

(b) What is the period of the wave?

(c) If the wave speed is 10 m/sec, what is the wavelength? (If necessary, look back to find the relationship you need to answer this.)

11. If points X and Y on a periodic wave are one-half period "out of phase" with each other, which of the following must be true?

(a) X oscillates at half the frequency at which Y oscillates.

(b) X and Y always move in opposite directions.

(c) X is a distance of one-half wavelength from Y.

12.5 When waves meet: the superposition principle

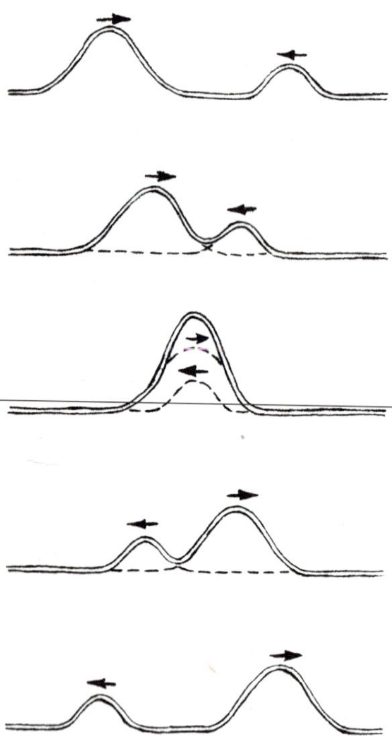

The superposition of two rope waves at a point. The dashed curves are the contributions of the individual waves.

So far, you have considered single waves. What happens when two waves encounter each other in the same medium? Suppose two waves approach each other on a rope, one traveling to the right and one traveling to the left. The series of sketches in the margin shows what would happen if you made this experiment. The waves pass through each other without being modified. After the encounter, each wave looks just as it did before and is traveling just as it was before. This phenomenon of passing through each other unchanged can be observed with all types of waves. You can easily see that it is true for surface ripples on water. (Look, for example, at the photograph on page 364.) You could reason that it must be true for sound waves also, since two conversations can take place across a table without distorting each other. (Note that when *particles* encounter each other, they collide. Waves can pass through each other.)

What happens during the time when the two waves overlap? The displacements they provide add up. At each instant, the displacement of each point in the overlap region is just the *sum* of the displacements that would be caused by each of the two waves separately. An example is shown in the margin. Two waves travel toward each other on a rope. One has a maximum displacement of 0.4 cm upward and the other a maximum displacement of 0.8 cm upward. The total maximum upward displacement of the rope at a point where these two waves pass each other is 1.2 cm.

What a wonderfully simple behavior, and how easy it makes everything! Each wave proceeds along the rope making its own contribution to the rope's displacement no matter what any other wave is doing. You can easily determine what the rope looks like at any given instant. All you need to do is add up the displacements caused by each wave at each point along the rope at that instant. This property of waves is called the *superposition principle*. Another illustration of wave superposition is shown in the margin. Notice that when the displacements are in opposite directions, they tend to cancel each other. One of the two directions of displacement may always be considered negative.

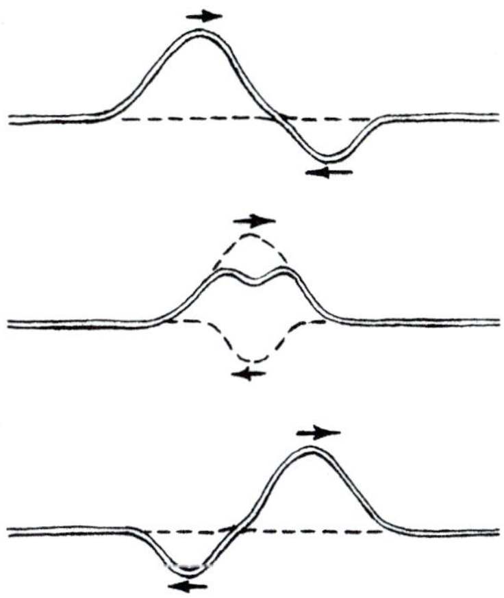

The superposition principle applies no matter how many separate waves or disturbances are present in the medium. In the examples just given, only two waves are present. But you would find by experiment that the superposition principle works equally well for three, ten, or any number of waves. Each wave makes its own contribution, and the net result is simply the sum of all the individual contributions.

If waves add as just described, then you can think of a complex wave as the sum of a set of simpler waves. In 1807, the French mathematician Jean-Baptiste Fourier advanced a very useful theorem. Fourier stated that any continuing periodic oscillation, however complex, could be analyzed as the sum of simpler, regular wave motions. This, too, can be demonstrated by experiment. The sounds of musical instruments have been analyzed in this way also. Such analysis makes it possible to "imitate" instruments electronically by combining just the right proportions of simple vibrations.

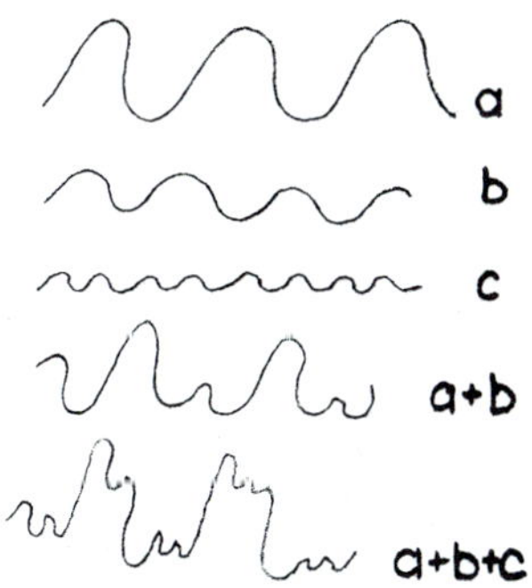

SG 4–8

12. *Two periodic waves of amplitudes A_1 and A_2 pass through a point P. What is the greatest possible displacement of P?*

13. *What is the displacement of a point produced by two waves together if the displacements produced by the waves separately at that instant are +5 cm and −6 cm? What is the special property of waves that makes this simple result possible?*

12.6 | A two-source interference pattern

The photograph on page 364 (center) shows ripples spreading from a vibrating source touching the water surface in a "ripple tank." The drawing to the left of it shows a "cut-away" view of

Close Up

Waves in a Ripple Tank

When something drops in the water, produces periodic wave trains of cre and troughs, somewhat as shown in t "cut-away" drawing at the left below.

The center figure below is an insta taneous photograph of the shadows ripples produced by a vibrating po source. The crests and troughs on t water surface show up in the image bright and dark circular bands. In t photo below right, there were two po sources vibrating in phase. The overla ping waves create an interference pa tern.

the water level pattern at a given instant. The third photograph (far right) introduces a phenomenon that will play an important role in later parts of the course. It shows the pattern of ripples on a water surface disturbed by *two* vibrating sources. The two small sources go through their up and down motions together; that is, they are in phase. Each source creates its own set of circular, spreading ripples. The photograph catches the pattern made by the overlapping sets of waves at one instant. This pattern is called an *interference pattern*.

You can interpret what you see in this photograph in terms of what you already know about waves. You can predict how the pattern will change with time. First, tilt the page so that you are viewing the interference pattern from a glancing direction. You will see more clearly some nearly straight gray bands. This feature can be explained by the superposition principle.

Suppose that two sources produce identical pulses at the same instant. Each pulse contains one crest and one trough. (See margin.) In each pulse the height of the crest above the undisturbed or average level is equal to the depth of the trough below. The sketches show the patterns of the water surface after equal time intervals. As the pulses spread out, the points at which they overlap move too. In the figure, a completely darkened circle indicates where a crest overlaps another crest. A half-darkened circle marks each point where a crest overlaps a trough. A blank circle indicates the meeting of two troughs. According to the superposition principle, the water level should be highest at the completely darkened circles (where the crests overlap). It should be lowest at the blank circles, and at average height at the half-darkened circles. Each of the sketches in the margin represents the spatial pattern of the water level at a given instant.

At the points marked with darkened circles in the figure, the two pulses arrive in phase. At the points indicated by blank circles, the pulses also arrive in phase. In either case, the waves reinforce each other, causing a *greater* amplitude of either the crest or the trough. Thus, the waves are said to *interfere constructively*. In this case, all such points are at the same distance from each source. As the ripples spread, the region of maximum disturbance moves along the central dotted line in (a).

At the points marked with half-darkened circles, the two pulses arrive completely out of phase. Here the waves cancel and so are said to interfere *destructively*, leaving the water surface undisturbed. The lines *N* show the path along which the overlapping pulses meet when they are just out of phase. All along these lines there is no change or displacement of the water level.

When periodic waves of equal amplitude are sent out instead of single pulses, overlap occurs all over the surface, as is shown

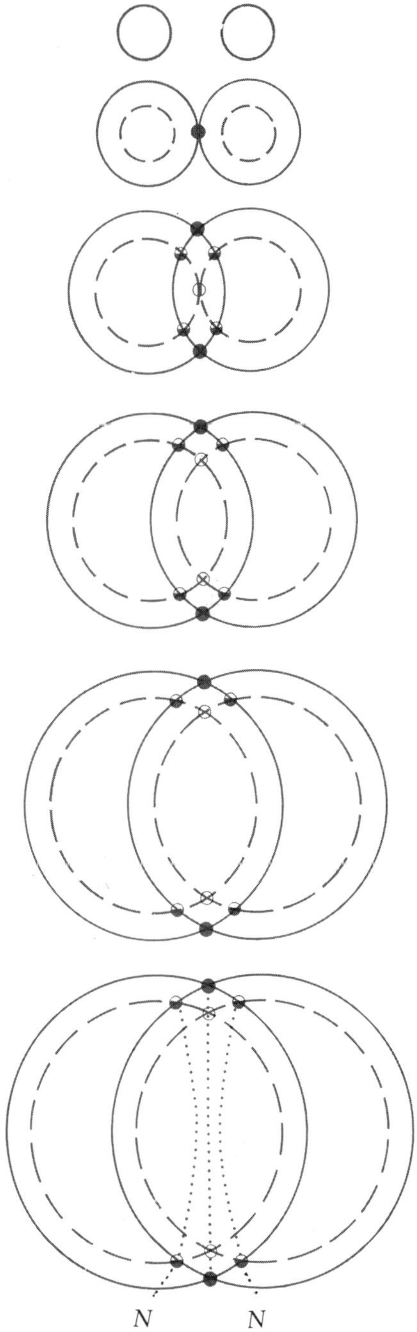

Pattern produced when two circular pulses, each of a crest and a trough, spread through each other. The small circles indicate the net displacement:
● = *double height peak*
◒ = *average level*
○ = *double depth trough*

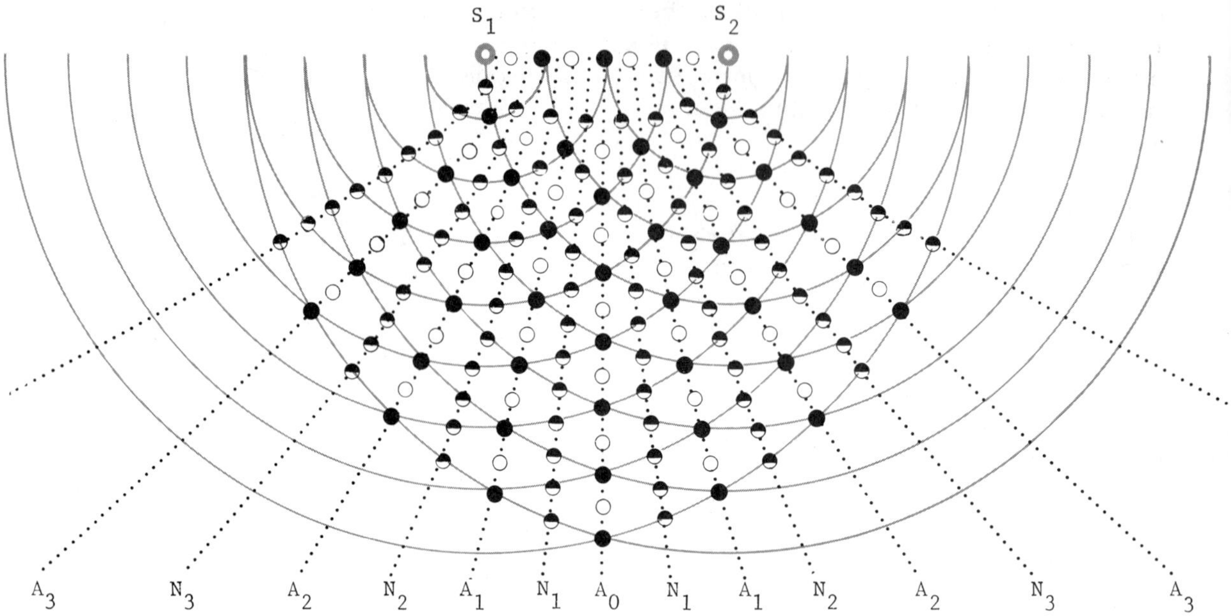

Analysis of interference pattern similar to that of the lower right photograph on page 364 set up by two in-phase periodic sources. (Here S_1 and S_2 are separated by four wavelengths.) The dark circles indicate where crest is meeting crest, the blank circles where trough is meeting trough, and the half-dark circles where crest is meeting trough. The other lines of maximum constructive interference are labeled A_0, A_1, A_2, etc. Points on these lines move up and down much more *than they would because of waves from either source alone. The lines labeled N_1, N_2, etc., represent bands along which there is maximum destructive interference. Points on these lines move up and down* much less *than they would because of waves from either source alone. Compare the diagram with the photograph and identify antinodal lines and nodal lines.*

on this page. All along the central dotted line, there is a doubled disturbance amplitude. All along the lines labeled *N*, the water height remains undisturbed. Depending on the wavelength and the distance between the sources, there can be many such lines of constructive and destructive interference.

Now you can interpret the ripple tank interference pattern shown at the lower right on page 364. The gray bands are areas where waves cancel each other, called *nodal* lines. These bands correspond to lines *N* in the simple case of pulses instead of periodic waves. Between these bands are other bands where crest and trough follow one another, where the waves reinforce. These are called *antinodal* lines.

Such an interference pattern is set up by overlapping waves from two sources. For water waves, the interference pattern can be seen directly. But whether visible or not, all waves, including earthquake waves, sound waves, or X rays, can set up interference patterns. For example, suppose two loudspeakers are working at the same frequency. By changing your position in front of the loudspeakers, you can find the nodal regions where destructive interference causes only little sound to be heard. You also can find the antinodal regions where a strong signal comes through.

The beautiful symmetry of these interference patterns is not accidental. Rather, the whole pattern is determined by the wavelength λ and the source separation S_1S_2. From these, you could calculate the angles at which the nodal and antinodal lines radiate out to either side of A_0. Conversely, you might know S_1S_2 and might have found these angles by probing around in the two-source interference pattern. If so, you can calculate the wavelength even if you cannot see the crests and troughs of the waves directly. This is very useful, for most waves in nature cannot be directly seen. Their wavelength has to be found by letting waves set up an interference pattern, probing for the nodal and antinodal lines, and calculating λ from the geometry.

SG 9

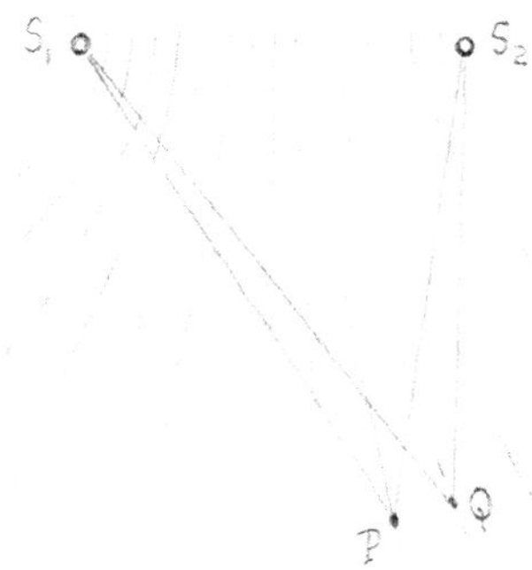

The figure at the right shows part of the pattern of the diagram on the opposite page. At any point P on an *antinodal* line, the waves from the two sources arrive *in phase*. This can happen only if P is equally far from S_1 and S_2, or if P is some whole number of wavelengths farther from one source than from the other. In other words, the difference in distances $(S_1P - S_2P)$ must equal $n\lambda$, λ being the wavelength and n being zero or any whole number. At any point Q on a *nodal* line, the waves from the two sources arrive exactly *out of phase*. This occurs because Q is an odd number of half-wavelengths ($\frac{1}{2}\lambda$, $\frac{3}{2}\lambda$, $\frac{5}{2}\lambda$, etc.) farther from one source than from the other. This condition can be written $S_1Q - S_2Q = (n + \frac{1}{2})\lambda$.

Since the sound wave patterns in space are three-dimensional, the nodal or antinodal regions in this case are two-dimensional surfaces; that is, they are *planes*, not lines.

The distance from the sources to a detection point may be much larger than the source separation d. In that case, there is a simple relationship between the node position, the wavelength λ, and the separation d. The wavelength can be calculated from measurements of the positions of nodal lines. The details of the relationship and the calculation of wavelength are described on page 369.

This analysis allows you to calculate from simple measurements made on an interference pattern the wavelength of any wave. It applies to water ripples, sound, light, etc. You will find this method very useful in later units. One important thing you can do now is find λ for a real case of interference of waves in the laboratory. This practice will help you later in finding the wavelengths of other kinds of waves.

?

14. Are nodal lines in interference patterns regions of cancellation or reinforcement?

15. What are antinodal lines? antinodal points?

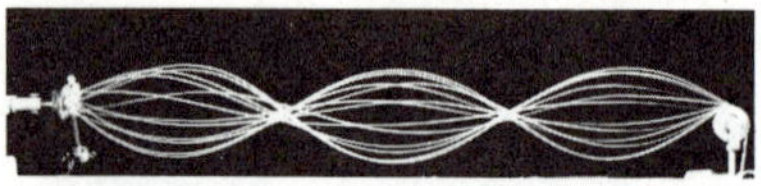

A vibrator at the left produces a wave train that runs along the rope and reflects from the fixed end at the right. The sum of the oncoming and the reflected waves is a standing wave pattern.

16. Nodal points in an interference pattern are places where

(a) the waves arrive "out of phase."

(b) the waves arrive "in phase."

(c) the point is equidistant from the wave sources.

(d) the point is one-half wavelength from both sources.

17. Under what circumstances do waves from two in-phase sources arrive at a point out of phase?

12.7 | Standing waves

If both ends of a rope are shaken with the same frequency and same amplitude, an interesting thing happens. The interference of the identical waves coming from opposite ends results in certain points on the rope not moving at all! In between these nodal points, the rope oscillates up and down. But there is no apparent propagation of wave patterns in either direction along the rope. This phenomenon is called a *standing wave* or *stationary wave*. The important thing to remember is that the standing oscillation you observe is really the effect of two *traveling* waves.

To make standing waves on a rope, there do not have to be two people shaking the opposite ends. One end can be tied to a hook on a wall. The train of waves sent down the rope by shaking one end will reflect back from the fixed hook. These reflected waves interfere with the new, oncoming waves and can produce a standing pattern of nodes and oscillation. In fact, you can go further and tie both ends of a string to hooks and pluck (or bow) the string. From the plucked point a pair of waves go out in opposite directions and then reflect back from the ends. The interference of these reflected waves traveling in opposite directions can produce a standing pattern just as before. The strings of guitars, violins, pianos, and all other stringed instruments act in just this fashion. The energy given to the strings sets up standing waves. Some of the energy is then transmitted from the vibrating string to the body of the instrument. The sound waves sent forth from the body are at essentially the same frequency as the standing waves on the string.

Lyre player painted on a Greek vase in the 5th century B.C.

The vibration frequencies at which standing waves can exist depend on two factors. One is the speed of wave propagation along the string. The other is the length of the string. The connection between length of string and musical tone was recognized over 2,000 years ago. This relationship contributed greatly to the idea that nature is built on mathematical principles. Early in the development of musical instruments, people learned how to produce certain pleasing harmonies by plucking strings. These harmonies result if the strings are of

Close Up

Calculating λ from an Interference Pattern

d = (S_1S_2) = separation between S_1 and S_2. (S_1 and S_2 may be actual sources that are in phase, or two slits through which a previously prepared wave front passes.)

ℓ = OQ = distance from sources to a far-off line or screen placed parallel to the two sources.

x = distance from center axis to point P along the detection line.

L = OP = distance to point P on detection line measured from sources.

Waves reaching P from S_1 have traveled farther than waves reaching P from S_2. If the extra distance is λ (or 2λ, 3λ, etc.), the waves will arrive at P in phase. Then P will be a point of strong wave disturbance. If the extra distance is ½λ (or 3/2λ, 5/2λ, etc.), the waves will arrive out of phase. Then P will be a point of weak or no wave disturbance.

With P as center, draw an arc of a circle of radius PS_2; it is indicated on the figure by the dotted line S_2M. Then line segment PS_2 equals line segment PM. Therefore, the extra distance that the wave from S travels to reach P is the length of the segment SM.

Now if d is very small compared to ℓ, as you can easily arrange in practice, the circular arc S_2M will be a very small piece of a large-diameter circle, or nearly a straight line. Also, the angle S_1MS_2 is very nearly 90°. Thus, the triangle S_1S_2M can be regarded as a right triangle. Furthermore, angle S_1S_2M is equal to angle POQ. Then the right triangle S_1S_2M is similar to triangle POQ.

$$\frac{S_1M}{S_1S_2} = \frac{x}{OP} \quad \text{or} \quad \frac{S_1M}{d} = \frac{x}{L}$$

If the distance ℓ is large compared to x, the distances ℓ and L are nearly equal. Therefore,

$$\frac{S_1M}{d} = \frac{x}{\ell}$$

But S_1M is the extra distance traveled by the wave from source S_1. For P to be a point of maximum wave disturbance, S_1M must be equal to $n\lambda$ (where n = 0 if P is at Q, and n = 1 if P is at the first maximum of wave disturbance found to one side of Q, etc.). So the equation becomes

$$\frac{n\lambda}{d} = \frac{x}{\ell}$$

and

$$\lambda = \frac{dx}{n\ell}$$

This important result says that if you measure the source separation d, the distance ℓ, and the distance x from the central line to a wave disturbance maximum, you can calculate the wavelength λ.

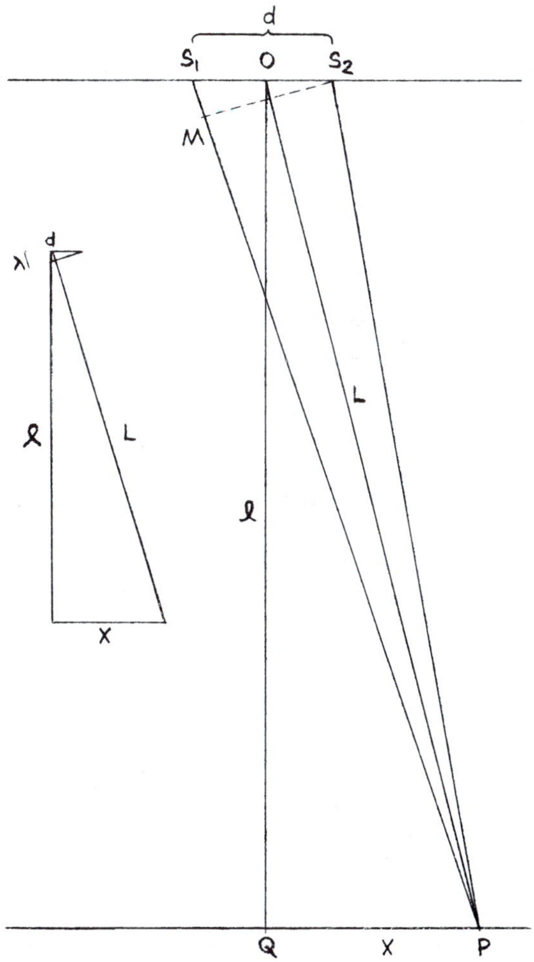

equal tautness and diameter and if their lengths are in the ratios of small whole numbers. Thus, the length ratio 2:1 gives the octave, 3:2 the musical fifth, and 4:3 the musical fourth. This striking connection between music and numbers encouraged the Pythagoreans to search for other numerical ratios or harmonies in the universe. The Pythagorean ideal strongly affected Greek science and many centuries later inspired much of Kepler's work. In a general form, the ideal flourishes to this day in many beautiful applications of mathematics to physical experience.

SG 13

Using the superposition principle, we can now define the harmonic relationship much more precisely. First, we must stress an important fact about standing wave patterns produced by reflecting waves from the boundaries of a medium. You can imagine an unlimited variety of waves traveling back and forth. But, in fact, *only certain wavelengths (or frequencies) can produce standing waves* in a given medium. In the example of a stringed instrument, the two ends are fixed and so must be nodal points. This fact puts an upper limit on the length of standing waves possible on a fixed rope of length L. Such waves must be those for which one-half wavelength just fits on the rope ($L = \lambda/2$). Shorter waves also can produce standing patterns having more nodes. But *always*, some whole number of one-half wavelengths must just fit on the rope ($L = \lambda/2, 2\lambda/2, 3\lambda/2$, etc.) so that $L = n\lambda/2$.

This relationship can be used to give an expression for all possible wavelengths of standing waves on a fixed rope:

$$\lambda_n = \frac{2L}{n}$$

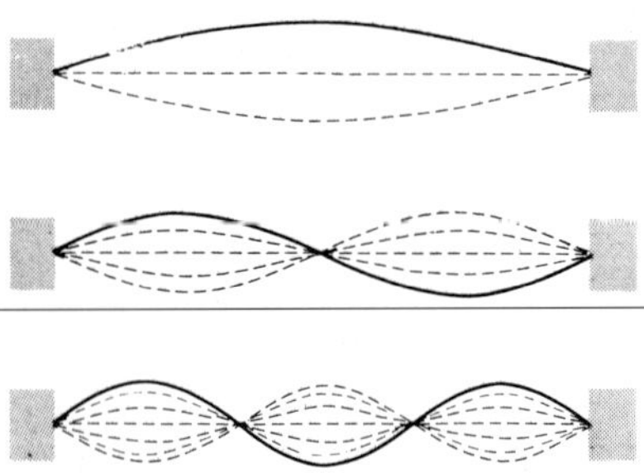

or simply $\lambda_n \propto 1/n$. That is, if λ_1 is the longest wavelength possible, the other possible wavelengths will be $\frac{1}{2}\lambda_1, \frac{1}{3}\lambda_1, \ldots, 1/n\lambda_1$. Shorter wavelengths correspond to higher frequencies. Thus, *on any bounded medium, only certain frequencies of standing waves can be set up.* Since frequency f is inversely proportional to wavelength, $f \propto 1/\lambda$, we can rewrite the expression for all possible standing waves on a plucked string as

$$f_n \propto n$$

In other circumstances, f_n may depend on n in some other way. The lowest possible frequency of a standing wave is usually the one most strongly present when the string vibrates after being plucked or bowed. If f_1 represents this lowest possible frequency, then the other possible standing waves would have frequencies $2f_1, 3f_1, \ldots, nf_1$. These higher frequencies are called "overtones" of the "fundamental" frequency f_1. On an "ideal" string, there are in principle an unlimited number of such frequencies, all simple multiples of the lowest frequency.

In real media, there are practical upper limits to the possible frequencies. Also, the overtones are not exactly simple multiples of the fundamental frequency; that is, the overtones are not strictly "harmonic." This effect is still greater in systems more complicated than stretched strings. In a saxophone or other wind instrument, an *air column* is put into standing wave motion. The overtones produced may not be even approximately harmonic.

Film Loops 38–43 show a variety of standing waves, including waves on a string, a drum, and in a tube of air.

As you might guess from the superposition principle, standing waves of different frequencies can exist in the same medium at the same time. A plucked guitar string, for example, oscillates in a pattern which is the superposition of the standing waves of many overtones. The relative oscillation energies of the different instruments determine the "quality" of the sound they produce. Each type of instrument has its own balance of overtones. This is why a violin sounds different from a trumpet, and both sound different from a soprano voice, even if all are sounding at the same fundamental frequency.

Mathematically inclined students are encouraged to pursue the topic of waves and standing waves, for example, Science Study Series paperbacks *Waves and the Ear* and *Horns, Strings, and Harmony.*

SG 16

?

18. When two identical waves of the same frequency travel in opposite directions and interfere to produce a standing wave, what is the motion of the medium at

(a) the nodes of the standing wave?

(b) the places between nodes, called antinodes or loops, of the standing wave?

19. If the two interfering waves have wavelength λ, what is the distance between the nodal points of the standing wave?

20. What is the wavelength of the longest traveling waves that can produce a standing wave on a string of length L?

21. Can standing waves of any frequency, as long as it is higher than the fundamental, be set up in a bounded medium?

12.8 | Wave fronts and diffraction

Waves can go around corners. For example, you can hear a voice coming from the other side of a hill, even though there is nothing to reflect the sound to you. You are so used to the fact that sound waves do this that you scarcely notice it. This spreading of the energy of waves into what you might expect to be "shadow" regions is called *diffraction*.

Once again, water waves will illustrate this behavior most clearly. From among all the arrangements that can result in diffraction, we will concentrate on two. The first is shown in the second photograph in the margin on page 373. Straight water waves (coming from the top of the picture) are diffracted as they pass through a narrow slit in a straight barrier. Notice that the slit is less than one wavelength wide. The wave emerges and

Close Up

Vibration of a Drum

In the Film Loop "Vibration of a Drum," *a marked rubber "drumhead" is seen vibrating in several of its possible modes. Below are pairs of still photographs from three of the symmetrical modes and from an antisymmetrical mode.*

spreads in all directions. Also notice the *pattern* of the diffracted wave. It is basically the same pattern a vibrating point source would set up if it were placed where the slit is.

The bottom photograph shows a second barrier arrangement. Now there are two narrow slits in the barrier. The pattern resulting from superposition of the diffracted waves from both slits is the same as that produced by two point sources vibrating in phase. The same kind of result is obtained when many narrow slits are put in the barrier; that is, the final pattern just matches that which would appear if a point source were put at the center of each slit, with all sources in phase.

You can describe these and all other effects of diffraction if you understand a basic characteristic of waves. This characteristic was first stated by Christian Huygens in 1678 and is now known as *Huygens' principle*. In order to understand the principle, you first need the definition of a *wave front*.

For a water wave, a wave front is an imaginary line along the water's surface and every point along this line is in exactly the same stage of vibration; that is, all points on the line are *in phase*. Crest lines are wave fronts, since all points on the water's surface along a crest line are in phase. Each has just reached its maximum displacement upward, is momentarily at rest, and will start downward an instant later.

Since a sound wave spreads not over a surface but in three dimensions, its wave fronts are not lines but surfaces. The wave fronts for sound waves from a very small source are very nearly spherical surfaces, just as the wave fronts for a very small source of water waves are circles.

Huygens' principle, as it is generally stated today, is that *every point on a wave front may be considered to behave as a point source for waves generated in the direction of the wave's propagation*. As Huygens said:

> There is the further consideration in the emanation of these waves, that each particle of matter in which a wave spreads, ought not to communicate its motion only to the next particle which is in the straight line drawn from the (source), but that it also imparts some of it necessarily to all others which touch it and which oppose themselves to its movement. So it arises that around each particle there is made a wave of which that particle is the center.

The diffraction patterns seen at slits in a barrier are certainly consistent with Huygens' principle. The wave arriving at the barrier causes the water in the slit to oscillate. The oscillation of the water in the slit acts as a source for waves traveling out from it in all directions. When there are two slits and the wave reaches both slits in phase, the oscillating water in each slit acts like a point source. The resulting interference pattern is similar

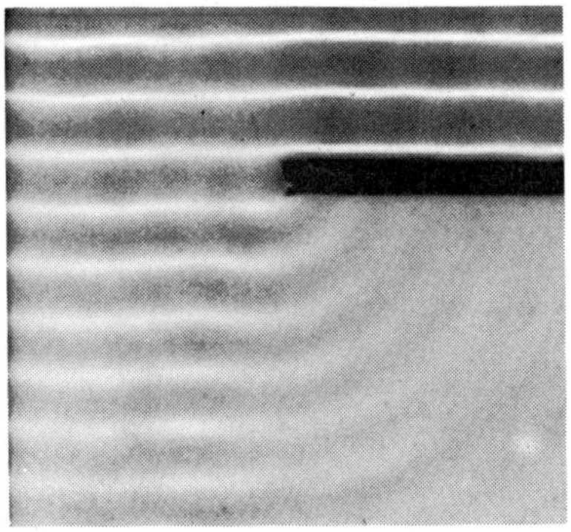

Diffraction of ripples around the edge of a barrier.

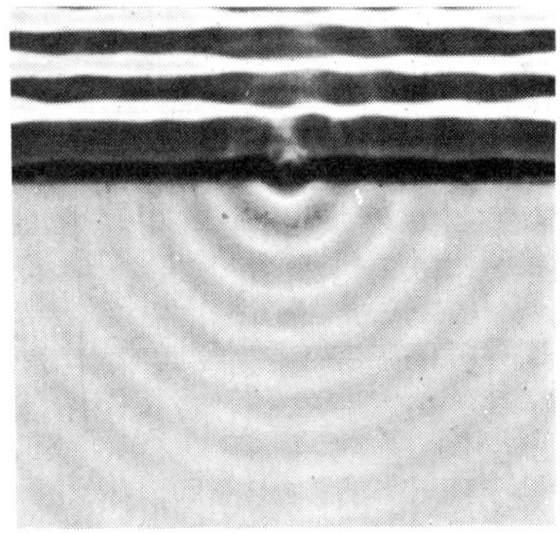

Diffraction of ripples through a narrow opening.

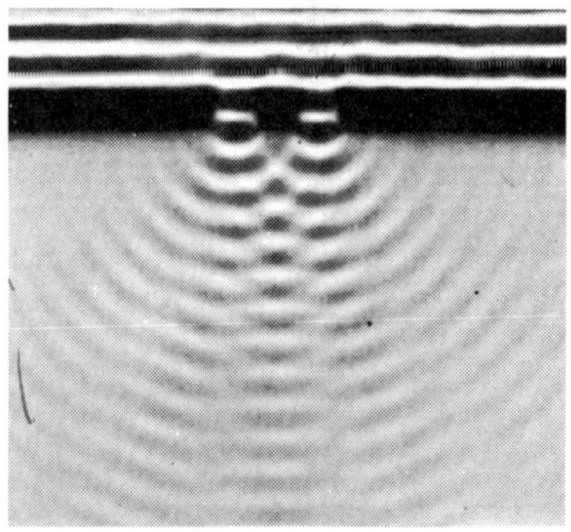

Diffraction of ripples through two narrow openings.

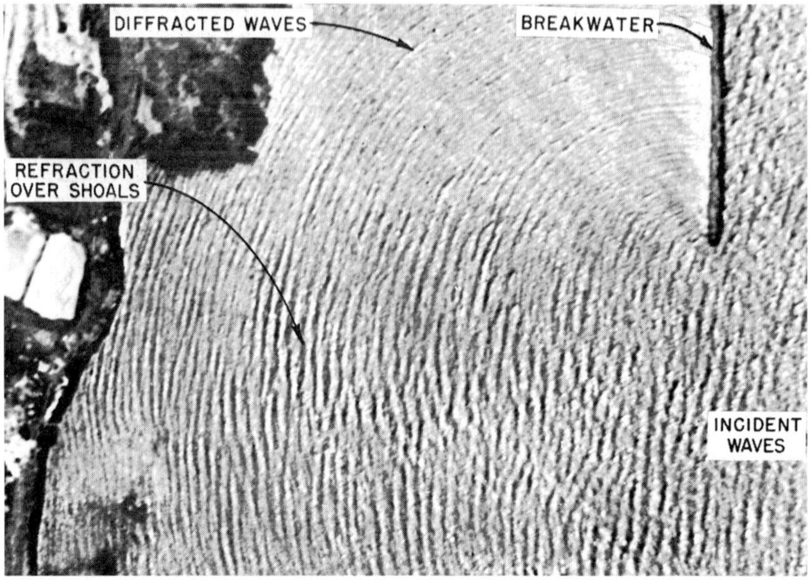

to the pattern produced by waves from two point sources oscillating in phase.

Consider what happens behind a breakwater wall as in the aerial photograph of the harbor above. By Huygens' principle, water oscillation near the end of the breakwater sends circular waves propagating into the "shadow" region.

You can understand all diffraction patterns if you keep both Huygens' principle and the superposition principle in mind. For example, consider a slit wider than one wavelength. In this case the pattern of diffracted waves contains nodal lines (see the series of four photographs in the margin on page 375).

The figure on page 375 helps to explain why nodal lines appear. There must be points like P that are just λ farther from side A of the slit than from side B; that is, there must be points P for which AP differs from BP by exactly λ. For such a point, AP and OP differ by one-half wavelength, λ/2. By Huygens' principle, you may think of points A and O as in-phase point sources of circular waves. But since AP and OP differ by λ/2, the two waves will arrive at P completely out of phase. So, according to the superposition principle, the waves from A and O will cancel at point P.

This argument also holds true for the pair of points consisting of the first point to the right of A and the first to the right of O. In fact, it holds true for *each* such matched pair of points, all the way across the slit. The waves originating at each such pair of points all cancel at point P. Thus, P is a nodal point, located on a nodal line. On the other hand, if the slit width is less than λ, then there can be *no* nodal point. This is obvious, since no point can be a distance λ farther from one side of the slit than from

Each point on a wave front can be thought of as a point source of waves. The waves from all the point sources interfere constructively only along their envelope, which becomes the new wave front.

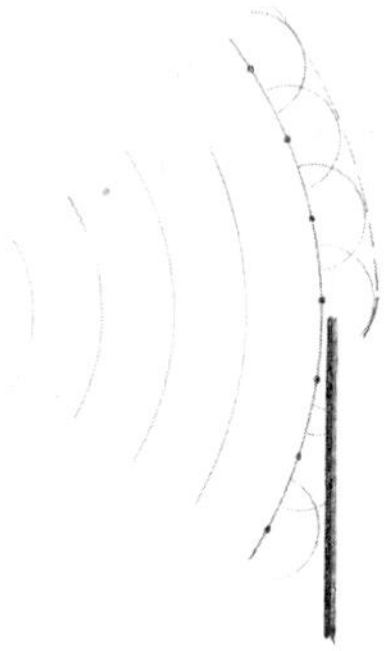

When part of the wave front is blocked, the constructive interference of waves from points on the wave front extends into the "shadow" region.

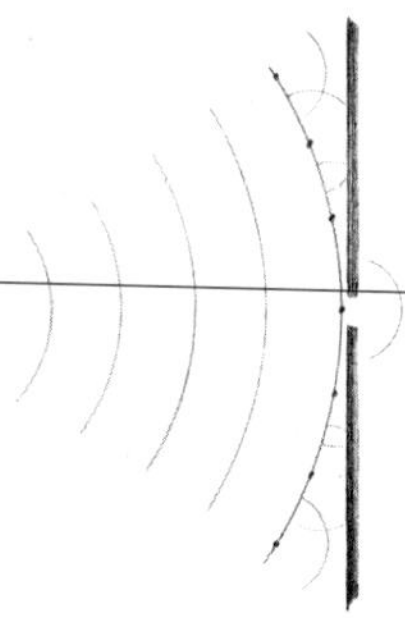

When all but a very small portion of a wave front is blocked, the wave propagating away from that small portion is nearly the same as that from a point source.

the other. Slits of widths less than λ behave nearly as point sources. The narrower they are, the more nearly their behavior resembles that of point sources.

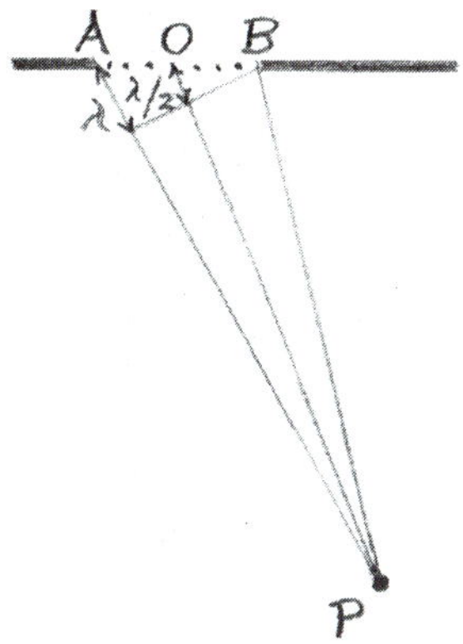

You can easily compute the wavelength of a wave from the interference pattern set up where diffracted waves overlap. For example, you can analyze the two-slit pattern on page 373 exactly as you analyzed the two-source pattern in Sec. 12.6. This is one of the main reasons for interest in the interference of diffracted waves. By locating nodal lines formed beyond a set of slits, you can calculate λ even for waves that you cannot see.

For two-slit interference, the larger the wavelength compared to the distance between slits, the more the interference pattern spreads out. That is, as λ increases or d decreases, the nodal and antinodal lines make increasingly large angles with the straight-ahead direction. Similarly, for single-slit diffraction, the pattern spreads when the ratio of wavelength to the slit width increases. In general, diffraction of longer wavelengths is more easily detected. Thus, when you hear a band playing around a corner, you hear the bass drums and tubas better than the piccolos and cornets, even though they actually are playing equally loudly.

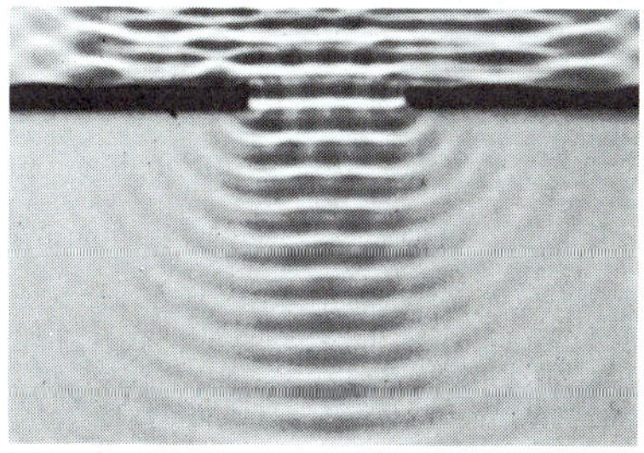

?

22. *What characteristic do all points on a wave front have in common?*

23. *State Huygens' principle.*

24. *Can there be nodal lines in a diffraction pattern from an opening less than one wavelength wide? Explain.*

25. *What happens to the diffraction pattern from an opening as the wavelength of the wave increases?*

26. *Can there be diffraction without interference? interference without diffraction?*

SG 21

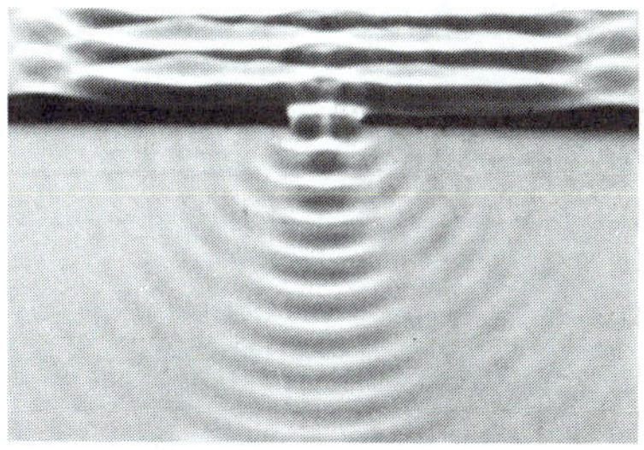

12.9 | Reflection

You have seen that waves can pass through one another and spread around obstacles in their paths. Waves also are reflected, at least to some degree, whenever they reach any boundary of the medium in which they travel. Echoes are familiar examples of the reflection of sound waves. All waves share the property of reflection. Again, the superposition principle will help you understand what happens when reflection occurs.

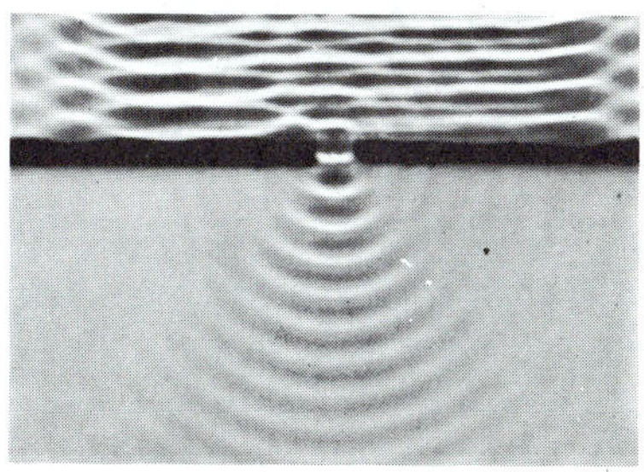

Suppose that one end of a rope is tied tightly to a hook securely fastened to a massive wall. From the other end, a pulse wave is sent down the rope toward the hook. Since the hook cannot move, the force exerted by the rope wave can do no work on the hook. Therefore, the energy carried in the wave cannot leave the rope at this fixed end. Instead, the wave bounces back, is *reflected*, ideally with the same energy.

What does the wave look like after it is reflected? The striking result is that the wave seems to flip upside down on reflection. As the wave comes in from left to right and encounters the fixed hook, it pulls up on it. By Newton's third law, the hook must exert a force on the rope in the opposite direction while reflection is taking place. The details of how this force varies in time are complicated. The net effect is that an inverted wave of the same form is sent back down the rope.

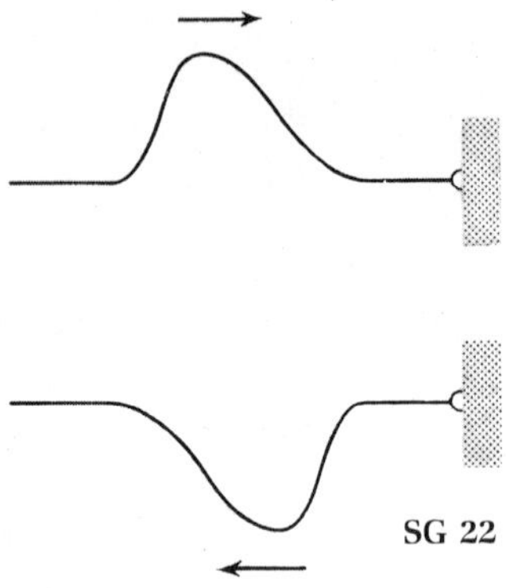

SG 22

The sketches in the margin on the left show the results of reflection of water waves from a straight wall. You can see if the sketches are accurate by trying to reproduce the effect in your sink or bathtub. Wait until the water is still, then dip your fingertip briefly into the water or let a drop fall into the water. In the upper sketch, the outer crest is approaching the barrier at the right. The next two sketches show the positions of the crests after first one and then two of them have been reflected. Notice the dashed curves in the last sketch. They show that the reflected wave appears to originate from a point S′ that is as far behind the barrier as S is in front of it. The imaginary source at point S′ is called the *image* of the source S.

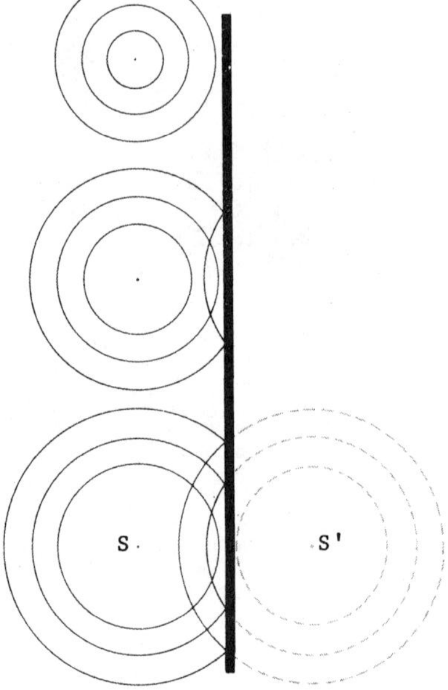

Reflection of circular waves is studied first, because that is what you usually notice first when studying water waves. But it is easier to see a general principle for explaining reflection by observing a straight wave front, reflected from a straight barrier. The ripple-tank photograph on page 377 shows one instant during such a reflection. (The wave came in from the upper left at an angle of about 45°.) The sketches below show in more detail what happens as the wave crests reflect from the straight barrier.

The description of wave behavior is often made easier by drawing lines perpendicular to the wave fronts. Such lines, called *rays*, indicate the direction of propagation of the wave. Notice the last drawing in the margin, for example. Rays have been drawn for a set of wave crests just before reflection and just after reflection from a barrier. The straight-on direction, perpendicular to the reflecting surface, is shown by a dotted line. The ray for the *incident* crests makes an angle θ_i with the straight-on direction. The ray for the *reflected* crests makes an angle θ_r with it. The *angle of reflection* θ_r is equal to the *angle of incidence* θ_i; that is, $\theta_r = \theta_i$. This is an experimental fact, which you can verify for yourself.

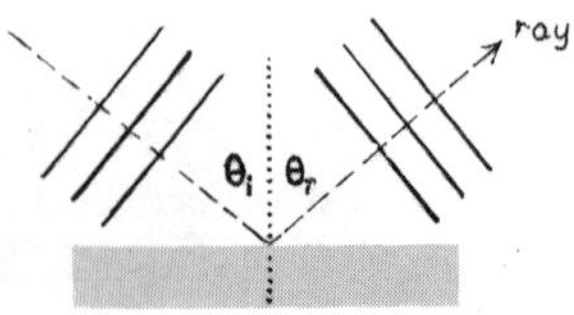

Many kinds of wave reflectors are in use today, from radar antennae to infrared heaters. Figures (a) and (b) below show how straight-line waves reflect from two circular reflectors. A few incident and reflected rays are shown. (The dotted lines are perpendicular to the barrier surface.) Rays reflected from the half-circle (a) head off in all directions. However, rays reflected from a small segment of the circle (b) come close to meeting at a single point. A barrier with the shape of a parabola (c) focuses straight-line waves precisely at a point. Similarly, a parabolic surface reflects plane waves to a sharp focus. An impressive example is a radio telescope. Its huge parabolic surface reflects faint radio waves from space to focus on a detector.

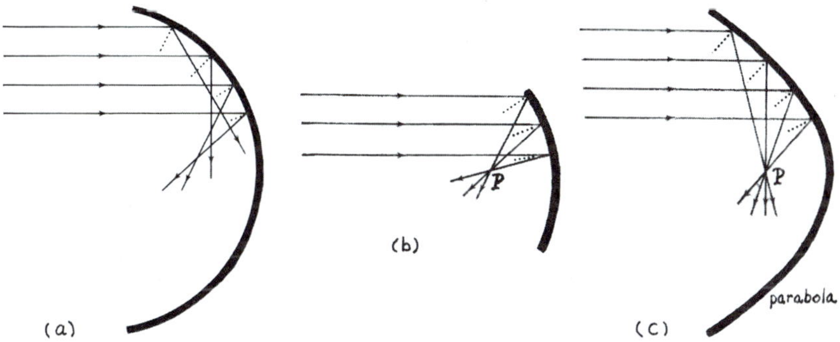

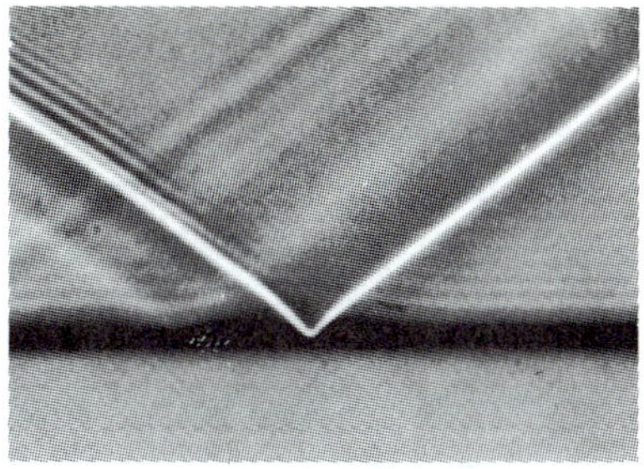

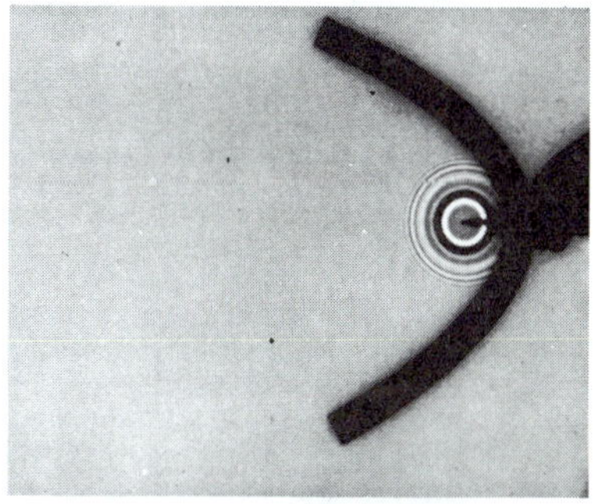

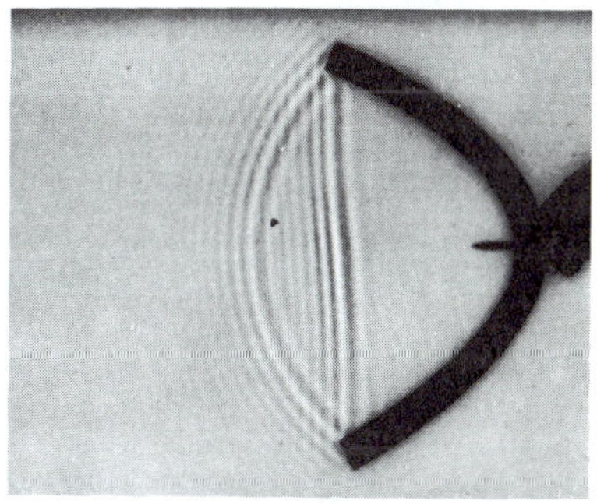

Above: A ripple tank shadow showing how circular waves produced at the focus of a parabolic wall are reflected from the wall into straight waves.

The wave paths indicated in the sketches could just as well be reversed. For example, spherical waves produced at the focus become plane waves when reflected from a parabolic surface. The flashlight and automobile headlamp are familiar applications of this principle. In them, white-hot wires placed at the focus of parabolic reflectors produce almost parallel beams of light.

?

27. What is a "ray"?

28. What is the relationship between the angle at which a wave front strikes a barrier and the angle at which it leaves?

29. What shape of reflector can reflect parallel wave fronts to a sharp focus?

30. What happens to wave fronts originating at the focus of such a reflecting surface?

12.10 | Refraction

What happens when a wave propagates from one medium to another medium in which its speed of propagation is different? Look at the simple situation pictured on page 378. Two one-

Above: The filament of a flashlight bulb is at the focus of a parabolic mirror, so the reflected light forms a nearly parallel beam.

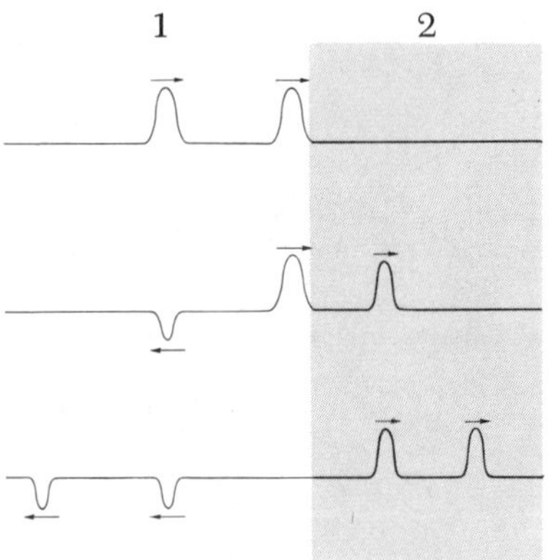

Pulses encountering a boundary between two different media. The speed of propagation is less in medium 2.

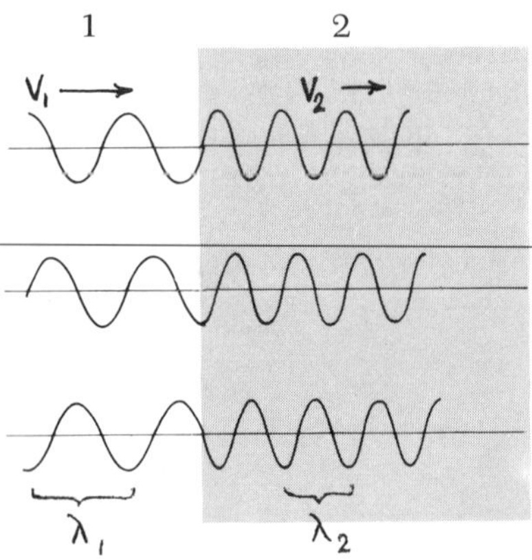

Continuous wave train crossing the boundary between two different media. The speed of propagation is less in medium 2.

SG 26

dimensional pulses approach a boundary separating two media. The speed of the propagation in medium 1 is greater than it is in medium 2. Imagine the pulses to be in a light rope (medium 1) tied to a relatively heavy rope (medium 2). Part of each pulse is reflected at the boundary. This reflected component is flipped upside down relative to the original pulse. Recall the inverted reflection at a hook in a wall discussed earlier. The heavier rope here tends to hold the boundary point fixed in just the same way. But what happens to that part of the wave that continues into the second medium?

As shown in the figure, the transmitted pulses are closer together in medium 2 than they are in medium 1. The speed of the pulses is less in the heavier rope. So the second pulse is catching up with the first while the second pulse is still in the light rope and the first is already in the heavy rope. In the same way, each separate pulse is itself squeezed into a narrower form; that is, while the front of the pulse is entering the region of less speed, the back part is still moving with greater speed.

Something of the same sort happens to a periodic wave at such a boundary. This situation is pictured in the figure below. For the sake of simplicity, assume that all of the wave is transmitted and none of it is reflected. Just as the two pulses were brought closer and each pulse was squeezed narrower, the periodic wave pattern is squeezed together, too. Thus, the wavelength λ_2 of the transmitted wave is shorter than the wavelength λ_1 of the incoming, or incident, wave.

Although the wavelength changes when the wave passes across the boundary, the frequency of the wave cannot change. If the rope is unbroken, the pieces immediately on either side of the boundary must go up and down together. The frequencies of the incident and transmitted waves must, then, be equal. We can simply label both of them f.

The wavelength, frequency, and speed relationship for both the incident and transmitted waves can be written separately:

$$\lambda_1 f = v_1 \quad \text{and} \quad \lambda_2 f = v_2$$

Dividing one of these equations by the other and eliminating the f's,

$$\frac{\lambda_1}{\lambda_2} = \frac{v_1}{v_2}$$

This equation tells that the ratio of the wavelengths in the two media equals the ratio of the speeds.

The same sort of thing happens when water ripples cross a boundary. Experiments show that the ripples move more slowly in shallower water. A piece of plate glass is placed on the bottom of a ripple tank to make the water shallower there. This creates a boundary between the deeper and shallower part (medium 1

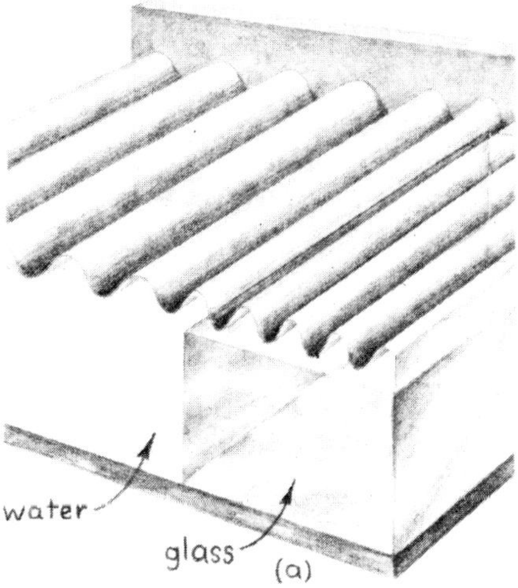

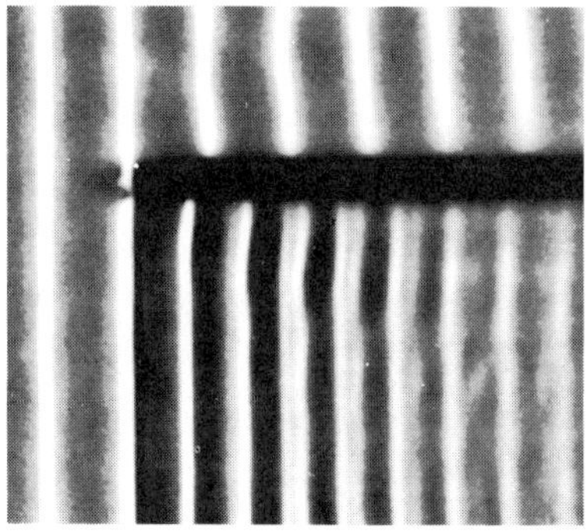
(b)

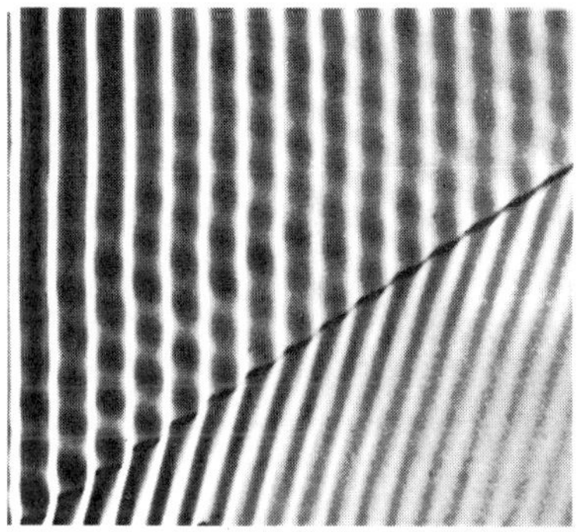
(c)

and medium 2). Figure (a) above shows the case where this boundary is parallel to the crest lines of the incident wave. As with rope waves, the wavelength of water waves in a medium is proportional to the speed in that medium.

Water waves offer a possibility not present for rope waves. The crest lines can approach the boundary at any angle, not only head-on. Photograph (c) shows such an event. A ripple-tank wave approaches the boundary at an angle. The wavelength and speed, of course, change as the wave passes across the boundary. The *direction* of the wave propagation also changes. As each part of a crest line in medium 1 enters medium 2, its speed decreases, and it starts to lag behind. In time, the

Top: Ripples on water (coming from the left) encounter the shallow region over the corner of a submerged glass plate. Bottom: Ripples on water (coming from the left) encounter a shallow region over a glass plate placed at an angle to the wave fronts.

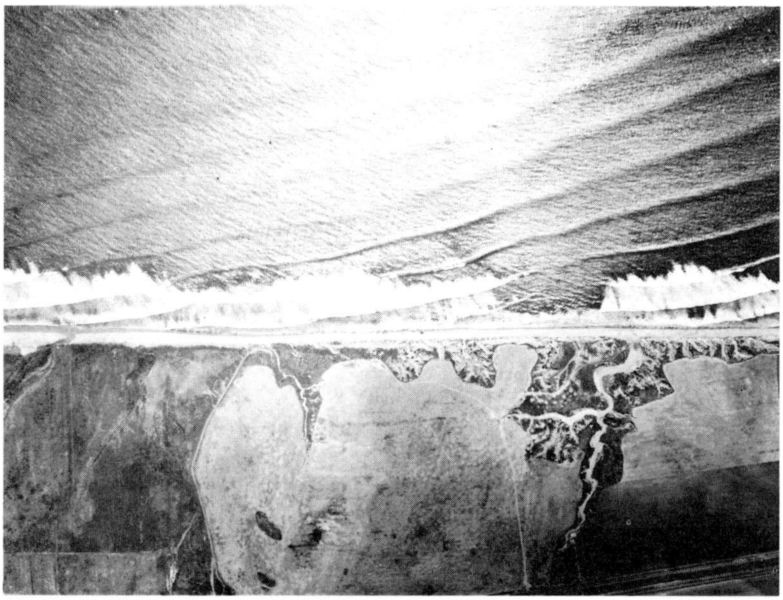
Aerial photograph of the refraction of ocean waves approaching shore.

SG 27–31

directions of the whole set of crest lines in medium 2 are changed from their directions in medium 1.

This phenomenon is called *refraction*. Refraction occurs whenever a wave passes into a medium in which the wave velocity is reduced. The wave fronts are turned (refracted) so that they are more nearly parallel to the boundary. See photographs (b) and (c). This accounts for something that you may have noticed if you have been at an ocean beach. No matter in what direction the waves are moving far from the shore, when they come near the beach their crest lines are nearly parallel to the shoreline. A wave's speed is steadily reduced as it moves into water that gets gradually more shallow. So the wave is refracted continuously as if it were always crossing a boundary between different media, as indeed it is. The refraction of sea waves is so great that wave crests can curl around a small island with an all-beach shoreline and provide surf on all sides.

The slowing of star light by increasingly dense layers of the atmosphere produces refraction that changes the apparent position of the star.

?

31. If a periodic wave slows down on entering a new medium, what happens to (a) its frequency? (b) its wavelength? (c) its direction?

32. Complete the sketch in the margin to show roughly what happens to a wave train that enters a new medium in which its speed is greater.

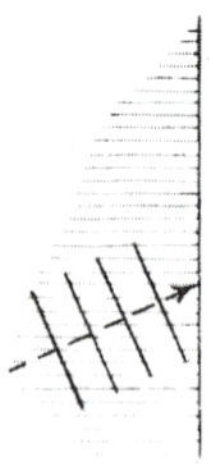

12.11 Sound waves

Look again at the bottom figure in the margin of page 357.

Sound waves are mechanical disturbances that propagate through a medium, such as the air. Typically, sound waves are longitudinal waves, producing changes of density and pressure in the medium through which they travel. The medium can be a solid, liquid, or gas. If the waves strike the ear, they can produce the sensation of hearing. The biology and psychology of hearing, as well as the physics of sound, are important to the science of acoustics. Here, of course, we will concentrate on sound as an example of wave motion. Sound has all the properties of wave motion considered so far. It exhibits refraction, diffraction, and the same relations among frequency, wavelength, and propagation speed and interference. Only the property of polarization is missing, because sound waves are longitudinal, not transverse.

Vibrating sources for sound waves may be as simple as a tuning fork or as complex as the human larynx with its vocal cords. Tuning forks and some special electronic devices produce a steady "pure tone." Most of the energy in such a tone is in simple harmonic motion at a single frequency. The "pitch" of a

1 Hz = 1/sec, or one cycle (or oscillation) per second.

sound goes up as the frequency of the wave increases.

People can hear sound waves with frequencies between about 20 and 20,000 Hz. Dogs can hear over a much wider range (15–50,000 Hz). Bats, porpoises, and whales generate and respond to frequencies up to about 120,000 Hz.

Loudness (or "volume") of sound is, like pitch, a psychological variable. Loudness is strongly related to the *intensity* of the sound. Sound intensity is a physical quantity. It is defined in terms of the energy carried by the wave and is measured in the number of watts per square centimeter transmitted through a surface perpendicular to the direction of motion of a wave front. The human ear can perceive a vast range of intensities of sound. The table below illustrates this range. It begins at a level of 10^{-16} watts per square centimeter (relative intensity = 1). Below this "threshold" level, the normal ear does not perceive sound.

Relative Intensity	*Sound*
1	Threshold of hearing
10^1	Normal breathing
10^2	Leaves in a breeze
10^3	
10^4	Library
10^5	Quiet restaurant
10^6	Two-person conversation
10^7	Busy traffic
10^8	Vacuum cleaner
10^9	Roar of Niagara Falls
10^{10}	Subway train
10^{11}	
10^{12}	Propeller plane at takeoff
10^{13}	Machine-gun fire
10^{14}	Small jet plane at takeoff
10^{15}	
10^{16}	Wind tunnel
10^{17}	Space rocket at lift-off

It is customary to measure loudness in decibels (db). The number of decibels is 10 times the exponent in the relative intensity of the sound. Thus, a jet plane at takeoff makes noise at the 140-db level.

Levels of noise intensity about 10^{12} times threshold intensity can be felt as a tickling sensation in the ear. Beyond 10^{13} times threshold intensity, the sensation changes to pain and may damage the unprotected ear.

Since many popular music concerts produce steady sound levels near this intensity (and above it for the performers), there are many cases of impaired hearing among young people.

Often the simplest way of reducing noise is by *absorbing* it after it is produced but before it reaches your ears. Like all sound, noise is the energy of back and forth motion of the medium through which the noise travels. Noisy machinery can be muffled by padded enclosures in which the energy of noise is changed to heat energy, which then dissipates. In a house, a thick rug on the floor can absorb 90% of room noise. (Thirty centimeters of fresh snow is an almost perfect absorber of noise outdoors. Cities and countrysides are remarkably hushed after a snowfall.)

It has always been fairly obvious that sound takes time to travel from source to receiver. By timing echoes over a known distance, the French mathematician Marin Mersenne in 1640 first

Close Up

The Sonic Boom

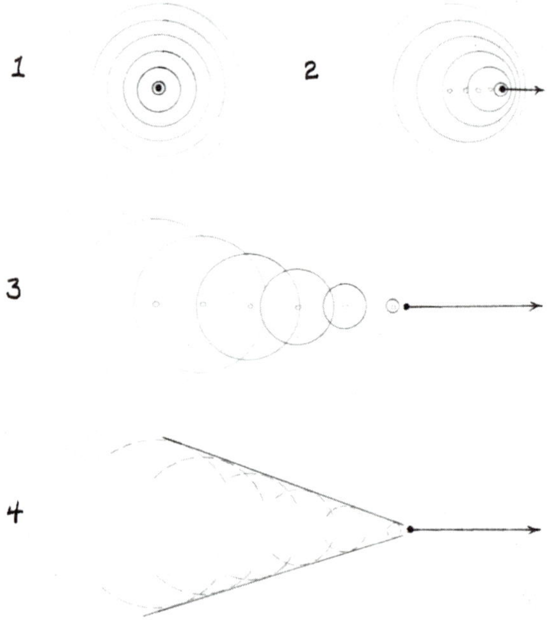

In the last decades a new kind of noise has appeared: the sonic boom. An explosion-like sonic boom is produced whenever an object travels through air at a speed greater than the speed of sound (supersonic speed). Sound travels in air at about 340 m/sec. Many types of military airplanes can travel at two or three times this speed. Flying at such speeds, the planes unavoidably and continually produce sonic booms. SST (Supersonic Transport) planes such as the *Concorde* are now in civilian use in some countries. The unavoidable boom raises important questions. What are the consequences of technological "progress"? Who gains, and what fraction of the population do they represent? Who and how many pay the price? *Must* we pay it; must SST's be used? How much say has the citizen in decisions that affect the environment so violently?

The formation of a sonic boom is similar to the formation of a wake by a boat. Consider a simple point source of waves. If the source remains in the same position in a medium, the wave it produces spreads out symmetrically around it, as in Diagram 1. If the source of the disturbance is *moving* through the medium, each new crest starts from a different point, as in Diagram 2.

Notice that the wavelength has become shorter in front of the object and longer behind it. This is called the *Doppler effect*. The Doppler effect is the reason that the sound an object makes seems to have a higher pitch when it is moving toward you and a lower pitch when it is moving away from you. In Diagram 3, the source is moving through the medium *faster than the wave speed*. Thus, the crests and the corresponding troughs overlap and interfere with one another. The interference is mostly destructive everywhere except on the line tangent to the wave fronts, indicated in Diagram 4. The result is a wake that spreads like a wedge away from the moving source, as in the photograph.

All these concepts apply not only to water waves but also to sound waves, including those disturbances set up in air by a moving plane as the wind and body push the air out of the way. If the source of sound is moving faster than the speed of sound wave, then there is a cone-shaped wake (in three dimensions) that spreads away from the source.

Actually, two cones of sharp pressure change are formed. One cone originates at the front of the airplane and one at the rear, as indicated in the graph at the right.

Because the double shock wave follows along behind the airplane, the region on the ground where people and houses may be struck by the boom (the "sonic-boom carpet") is as long as the supersonic flight path itself. In such an area, typically thousands of kilometers long and 80 km wide, there may be millions of people. Tests made with airplanes flying at supersonic speed have shown that a single such cross-country flight by a 315-ton supersonic transport plane would break many thousands of dollars worth of windows, plaster walls, etc., and cause fright and annoyance to millions of people. Thus, the supersonic flight of such planes has been confined to over-ocean use. It may even turn out that the annoyance to people on shipboard, on islands, and on coastal areas near the flight paths is so great that over-ocean flights, too, will have to be restricted.

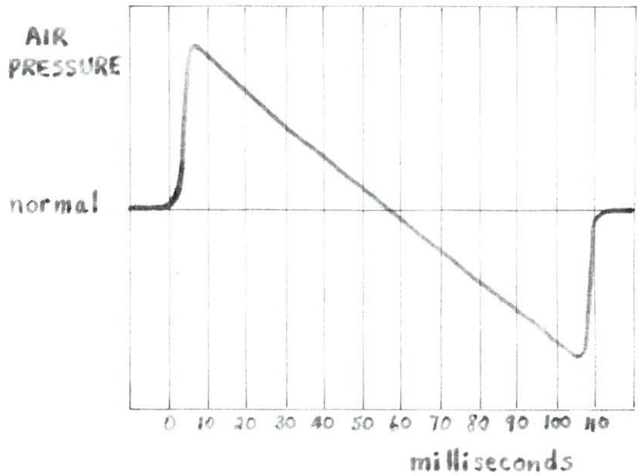

This curve represents the typical sonic boom from an airplane flying at supersonic speed (speed greater than about 340 m/sec). The pressure rises almost instantly, then falls relatively slowly to below-normal pressure, then rises again almost instantaneously. The second pressure rise occurs about 0.1 sec after the first one, making the boom sound "double."

Double-cone shock wave, or sonic boom, produced by an airplane that is traveling (at 21-km altitude) at three times the speed of sound. Building B is just being hit by the shock wave, building A was struck a few seconds ago, and building C will be hit a few seconds later. (Drawing Is not to scale.)

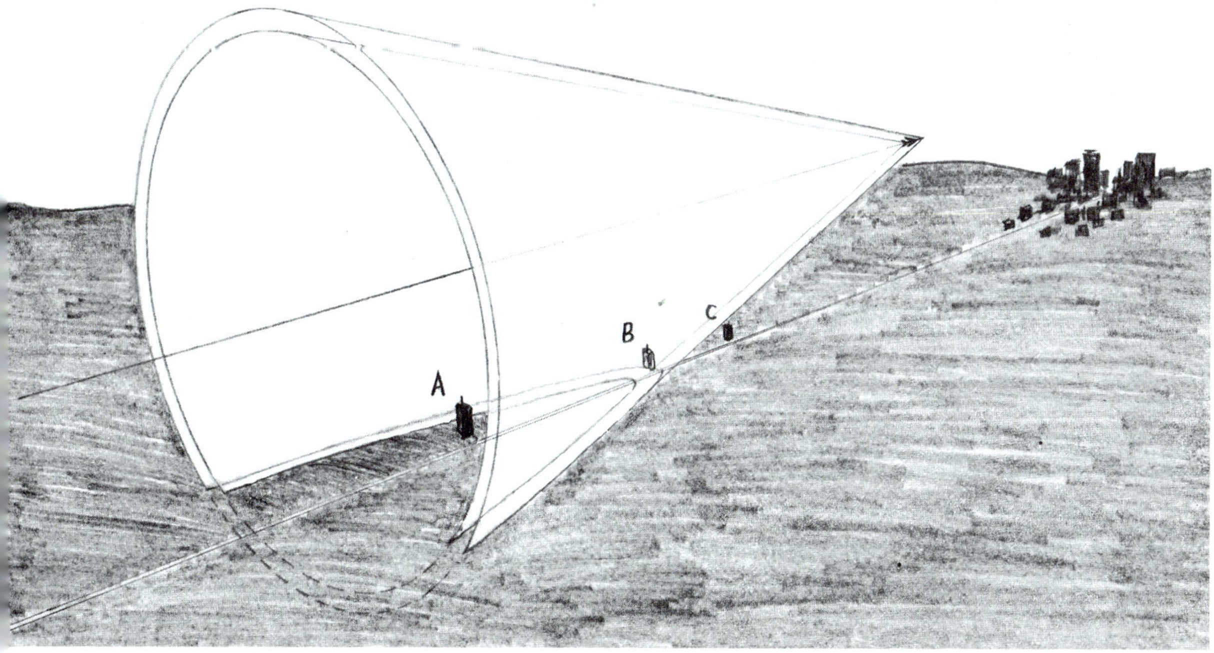

SG 32

computed the speed of sound in air. It took another 70 years before William Derham in England, comparing the flash and noise from cannons across 20 km, came close to the modern measurements.

SG 33–35

Sound in air at 20°C moves at about 344 m/sec. As for all waves, the speed of sound waves depends on the properties of the medium: the temperature, density, and elasticity. Sound waves generally travel faster in liquids than in gases, and faster still in solids. In seawater, their speed is about 1,500 m/sec; in steel, about 5,000 m/sec; in quartz, about 5,500 m/sec.

Since the two frequencies are different, the waves cannot stay in phase or out of phase. They therefore alternate between destructive and constructive interference. See SG 8.

Interference of sound waves can be shown in a variety of ways. In a large hall with hard, sound-reflecting surfaces, there will be "dead" spots. At these spots, sound waves coming together after reflection cancel each other. Acoustic engineers must consider this in designing the shape, position, and materials of an auditorium. Another interesting and rather different example of sound interference is the phenomenon known as *beats*. When two notes of slightly different frequency are heard together, they interfere. This interference produces beats, a rhythmic pulsing of the sound. Piano tuners and string players use this fact to tune two strings to the same pitch. They simply adjust one string or the other until the beats disappear.

Refraction of sound by different layers of air explains why you sometimes see lightning without hearing thunder. Similar refraction of sound occurs in layers of water of different temperatures. Geologists use the refraction of sound waves to study the earth's deep structure and to locate fossil fuels and minerals. Very intense sound waves are produced in the ground (as by dynamite blasts). The sound waves travel through the earth and are received by detection devices at different locations. The path of the waves, as refracted by layers in the earth, can be calculated from the relative intensities and times of sound received. From knowledge of the paths, estimates can be made of the composition of the layers.

The acoustic properties of a hall filled with people are very different from those of the empty hall. Acoustical engineers sometimes fill the seats with felt-covered sandbags while making tests.

As mentioned, diffraction is a property of sound waves. Sound waves readily bend around corners and barriers to reach the listener within range. Sound waves reflect, as do rope or water waves, wherever they encounter a boundary between different media. The architectural accidents called "whispering galleries" show vividly how sound can be focused by reflection from curved surfaces. All these effects are of interest in the study of acoustics. Moreover, the proper acoustical design of public buildings is now recognized as an important function by most good architects.

In this chapter, you have studied the basic phenomena of mechanical waves, ending with the theory of sound propagation. The explanations of these phenomena were considered the final triumph of Newtonian mechanics as applied to the transfer of

An "anechoic chamber" being used for research in acoustics. Sound is almost completely absorbed during multiple reflections among the wedges of soft material that cover the walls.

The concert hall of the University of Illinois Krannert Center for the Performing Arts was acoustically designed for unamplified performances.

energy of particles in motion. Most of the general principles of acoustics were discovered in the 1870's. Since then, the study of acoustics has become involved with such fields as quantum physics. Perhaps its most important influence on modern physics has been its effect on the imagination of scientists. The successes of acoustics encouraged them to take seriously the power of the wave viewpoint, even in fields far from the original one—the mechanical motion of particles that move back and forth or up and down in a medium.

?

33. List five wave behaviors that can be demonstrated with sound waves.

34. Can sound waves be polarized? Explain.

study guide

1. The *Project Physics* materials particularly appropriate for Chapter 12 include:

Experiments
Sound

Activities
Standing Waves on a Drum and a Violin
Moiré Patterns
Music and Speech Activities
Measurement of the Speed of Sound
Mechanical Wave Machines

Film Loops
Superposition
Standing Waves in a String
Standing Waves in a Gas
Four Loops on Vibrations

2. Some waves propagate at such a high speed that you are usually not aware of any delay in energy transfer. Give an example of a compression wave in a solid, started by an action at one end, that propagates so quickly that you are not aware of any delay before an effect is exhibited at the other end.

3. Describe the differences in phase of oscillation of various parts of your body as you walk. What points are exactly in phase? Which points are exactly ½ cycle out of phase? Are there any points ¼ cycle out of phase?

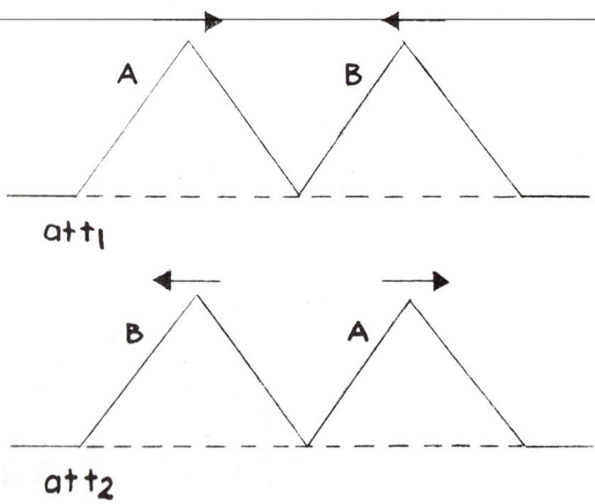

4. Pictured are two pulse waves (A and B) on a rope at the instants before and after they overlap (t_1 and t_2). Divide the elapsed time between t_1 and t_2 into four equal intervals and plot the shape of the rope at the end of each interval.

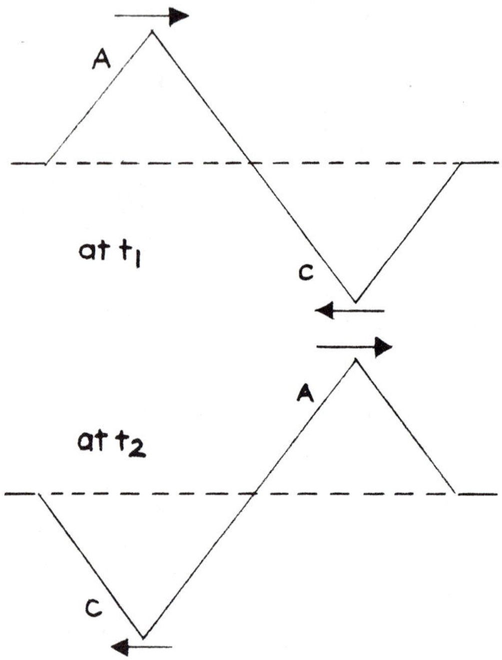

5. Repeat SG 4 for the two pulses (A and C) pictured above.

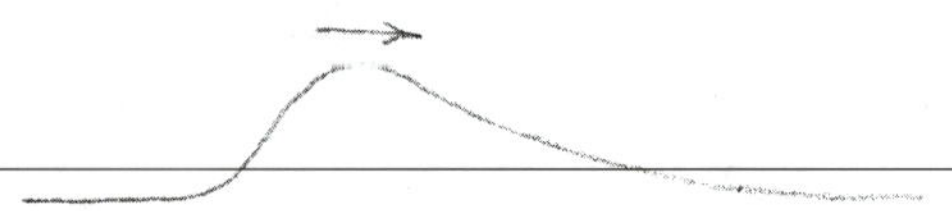

6. The wave above propagates to the right along a rope. What is the shape of the wave propagating to the left that could for an instant cancel this one completely?

7. The velocity of a portion of rope at some instant as transverse waves are passing through it is the superposition of the velocities of waves passing through that portion. Is the kinetic energy of a portion of the rope the superposition of the kinetic energies of waves passing through that region? Justify your answer.

8. Shown in the figure are two waves of slightly different frequency. Find their sum graphically. You will probably find that the sum has "beats." Can you find a way to determine the frequency of the beats if you know the frequencies of the waves?

9. What shape would the nodal regions have for sound waves from two loudspeakers?

10. Imagine a detection device for waves is moved slowly to either the right or left of the point labeled A_0 in the figure on page 366. Describe what the detection device would register.

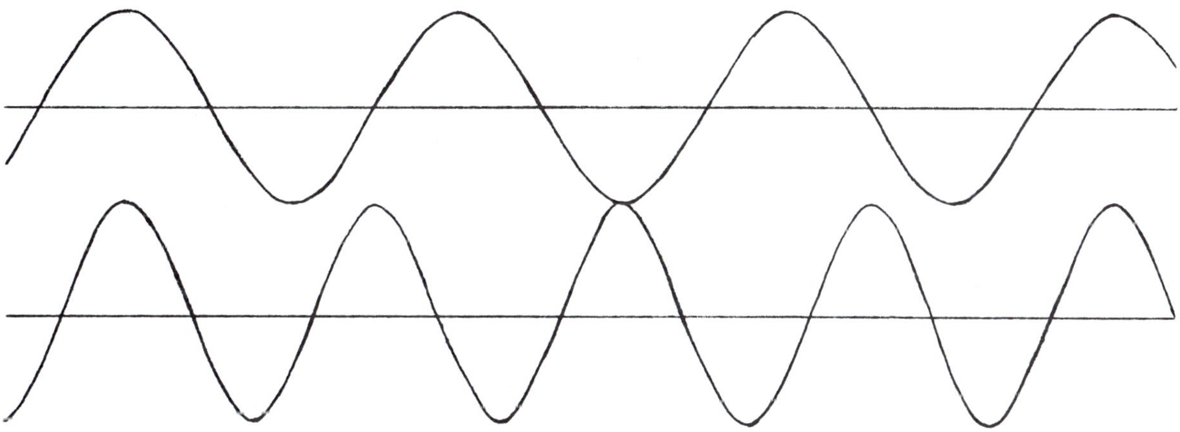

11. What kind of interference pattern would you expect to see if the separation between two in-phase sources were less than the wavelength λ? Where would the nodal and antinodal lines be if the two in-phase sources were separated by the distance λ? by $\lambda/2$? Convince yourself that one additional nodal line appears on each side of the central antinodal line whenever the separation between the two in-phase sources is increased by one wavelength.

12. Derive an equation, similar to $n\lambda\ell = dx_n$, for *nodal* points in a two-source interference pattern (where d is the separation of the sources, ℓ the distance from the sources, and x_n the distance of the nth node from the center line).

13. If you suddenly disturbed a stretched rubber hose or Slinky with a frequency that precisely matched a standing wave frequency, would standing waves appear immediately? If not, what factors would determine the time delay?

14. (a) What is the speed of sound in air if middle C (256 Hz) has a wavelength of 1.34 m?

(b) What is the wavelength in water of middle C if sound waves travel at 1,500 m/sec in water?

(c) What is the period of middle C in air and in water?

15. Different notes are sounded with the same guitar string by changing its vibrating length (that is, pressing the string against a brass ridge). If the full length of the string is L, what lengths must it be shortened to in order to sound (a) a "musical fourth," (b) a "musical fifth," (c) an "octave"?

16. An oscilloscope displays a picture of an electric wave so that its amplitude and frequency can be easily measured. What is the frequency of the incoming waves if eight complete cycles cover 10 cm of the oscilloscope screen and the electron beam is moving across the screen at 100 cm/sec?

17. Standing sound waves can be set up in the air in an enclosure (like a bottle or an organ pipe). In a pipe that is closed at one end, the air molecules at the closed end are not free to be displaced, so the standing wave must have a displacement node at the closed end. At the open end, however, the molecules are almost completely free to be displaced, so the standing waves must have an antinode near the open end.

(**a**) What will be the wavelength of the fundamental standing wave in a pipe of length L closed at one end? (Hint: What is the longest wave that has a node and an antinode a distance L apart?)

(**b**) What is a general expression for possible wavelengths of standing waves in a pipe closed at one end?

(**c**) Answer (a) and (b) for the case of a pipe open at *both* ends.

18. Imagine a spherical blob of Jello in which you can set up standing vibrations. What would be some of the possible modes of vibration? (Hint: What possible symmetrical nodal surfaces could there be?)

19. Suppose that straight-line ripple waves approach a thin straight barrier that is a few wavelengths long and that is oriented with its length parallel to the wave fronts. What do you predict about the nature of the diffraction pattern along a straight line behind the barrier that is perpendicular to the barrier and passes through the center of the barrier? Why do people who design breakwaters need to concern themselves with diffraction effects?

20. A megaphone directs sound along the megaphone axis if the wavelength of the sound is small compared to the diameter of the opening. Estimate the upper limit of frequencies that are diffracted at a cheerleader's megaphone opening. Can you hear what a cheerleader shouts even though you are far off the axis of the megaphone?

21. Explain why it is that the narrower a slit in a barrier is, the more nearly it can act like a point source of waves.

22. If light is also a wave, why have you not seen light being diffracted by the slits of a picket fence, or diffracted around the corner of a house?

23. Assuming that light is a wave phenomenon, what is the wavelength of green light if the first node in a diffraction pattern is found 10 cm from the center line at a distance of 4 m from the slits which have a separation distance of 2.5×10^{-3} cm?

24. By actual construction with a ruler and compass on a tracing of the photograph on page 377, show that rays for the reflected wave front appear to come from S′. Show also that this is consistent with $\theta_r = \theta_i$.

25. A straight-line wave approaches a right-angle reflecting barrier as shown in the figure. Find the shape, size, and direction of propagation of the wave after it has been completely reflected by the barrier.

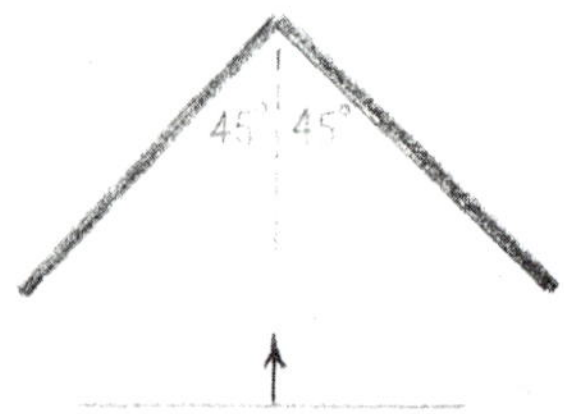

26. With ruler and compass reproduce part (b) of the figure at the bottom of page 377 and find the distance from the circle's center to the point P in terms of the radius of the circle r. Make the radius of your circle much larger than the one in the figure. (Hint: The dotted lines are along radii.)

27. Convince yourself that a parabolic reflector will actually bring parallel wave fronts to a sharp focus. Draw a parabola $y = kx^2$ (choosing any convenient value for k) and some parallel rays along the axis as in part (c) of the figure at the bottom of page 377. Construct line segments perpendicular to the parabola where the rays hit it, and draw the reflected rays at equal angles on the other side of these lines.

28. The *focal length* of a curved reflector is the distance from the reflector to the point where parallel rays are focused. Use the drawing in SG 24 to find the focal length of a parabola in terms of k.

29. Recalling that water surface waves travel slower in shallow water, what would you expect to happen to the shape of the following wave as it continues to the right? Pay particular attention to the region of varying depth. Can you use the line of reasoning above to give at least a partial explanation of the cause of breakers near a beach?

30. If the frequency of a wave traveling in a medium is increased, what will happen to its speed? What determines the speed of waves in a medium?

31. A straight-line wave in a ripple tank approaches a boundary between deep and shallow water as shown. Describe the shape of the wave as it passes through the boundary and then as it continues in the shallow water.

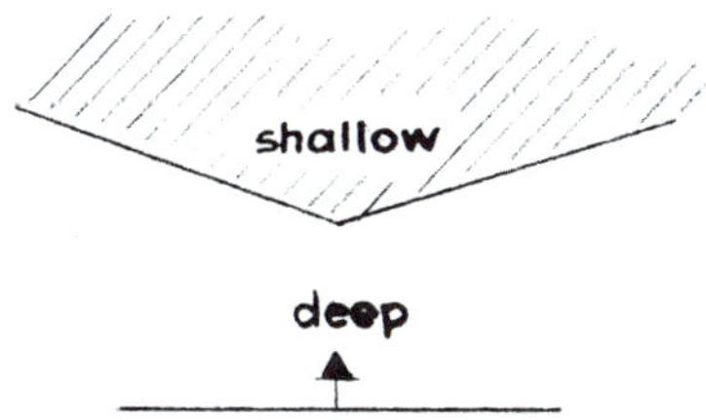

32. The diagram below shows two successive positions. AB and CD, of a wave train of sound or light, before and after crossing an air–glass boundary. The time taken to go from AB to DC is one period of the wave.

(**a**) Indicate and label an angle equal to angle of incidence θ_A.

(**b**) Indicate and label an angle equal to anglo of refraction θ_B.

(**c**) Label the wavelength in air λ_A.

(**d**) Label the wavelength in glass λ_B.

(**e**) Show that $v_A/v_B = \lambda_A/\lambda_B$.

(**f**) If you are familiar with trigonometry, show that $\sin \theta_A/\sin \theta_B = \lambda_A/\lambda_B$.

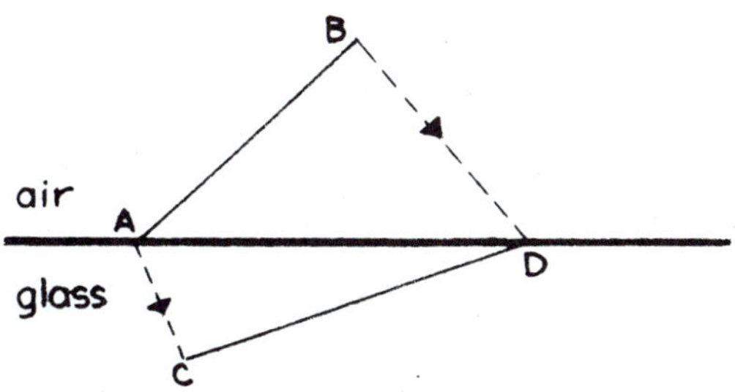

33. A periodic ripple-tank wave passes through a straight boundary between deep and shallow water. The angle of incidence at the boundary is 45°, and the angle of refraction is 30°. The propagation speed in the deep water is 0.35 m/sec, and the frequency of the wave is 10 Hz. Find the wavelengths in the deep and shallow water.

34. What is the speed of water waves in a ripple tank if waves generated at 10 Hz pass through slits 3 cm apart and create a diffraction pattern whose third node is 10 cm from the center line at a distance of 40 cm from the slits?

35. Look at figure (c) on page 379. Prove that if a wave were to approach the boundary between medium 1 and medium 2 from the right, along the same direction as the refracted wave in the figure, it would be refracted along the direction of the incident wave in the figure. This is another example of a general rule: If a wave follows a set of rays in one direction, then a wave can follow the same set of rays in the opposite direction. In other words, wave paths are reversible.

36. Suppose that in an extremely quiet room you can barely hear a buzzing mosquito at a distance of 1 m.

(**a**) What is the sound power output of the mosquito?

(**b**) How many mosquitoes would it take to supply the power for one 100-W reading lamp?

(**c**) If the swarm were at 10 m distance, what would the sound be like? (Sound intensity diminishes in proportion to the square of the distance from a point source.)

37. How can sound waves be used to map the floors of oceans?

38. Estimate the wavelength of a 1,000-Hz sound wave in air, in water, in steel (refer to data in text). Do the same if f = 10,000 Hz. Design the dimensions of an experiment to show two-source interference for 1,000-Hz sound waves.

39. Waves reflect from an object in a definite direction only when the wavelength is small compared to the dimensions of the object. This is true for sound waves as well as for any other. What does this tell you about the sound frequencies a bat must generate if it is to catch a moth or a fly? Actually, some bats can detect the presence of a wire about 0.12 mm in diameter. Approximately what frequency would that require?

EPILOGUE

Seventeenth-century scientists thought they could eventually explain all physical phenomena by reducing them to matter and motion. This mechanistic viewpoint became known as the Newtonian world view or Newtonian cosmology, since its most impressive success was Newton's theory of planetary motion. Newton and other scientists of his time proposed to apply similar methods to other problems, as mentioned in the Prologue to this unit.

The early enthusiasm for this new approach to science is vividly expressed by Henry Power in his book *Experimental Philosophy* (1664). Addressing his fellow natural philosophers (or scientists, as we would now call them), he wrote:

> You are the enlarged and elastical Souls of the world, who, removing all former rubbish, and prejudicial resistances, do make way for the Springy Intellect to flye out into its desired Expansion. . . .
>
> . . . This is the Age wherein (me-thinks) Philosophy comes in with a Spring-tide . . . I see how all the old Rubbish must be thrown away, and carried away with so powerful an Inundation. These are the days that must lay a new Foundation of a more magnificent Philosophy, never to be overthrown: that will Empirically and Sensibly canvass the *Phaenomena* of Nature, deducing the causes of things from such Originals in Nature, as we observe are producible by Art, and the infallible demonstration of Mechanicks; and certainly, this is the way, and no other, to build a true and permanent Philosophy.

In Power's day, there were many people who did not regard the old Aristotelian cosmology as rubbish. For them, it provided a comforting sense of unity and interrelation among natural phenomena. They feared that this unity would be lost if everything was reduced simply to atoms moving randomly through space. The poet John Donne, in 1611, complained bitterly of the change already taking place in cosmology:

> And new Philosophy calls all in doubt,
> The Element of fire is quite put out;
> The Sun is lost, and th' earth, and no man's wit
> Can well direct him where to looke for it.
> And freely men confesse that this world's spent,
> When in the Planets, and the Firmament
> They seeke so many new; then see that this
> Is crumbled out againe to his Atomies
> 'Tis all in peeces, all coherence gone;
> All just supply, and all Relation . . .

Newtonian physics provided powerful methods for analyzing the world and uncovering the basic principles of motion for individual pieces of matter. The richness and complexity of processes in the real world seemed infinite. Could Newtonian physics deal as successfully with these real events as with ideal processes in a hypothetical vacuum? Could the perceptions of colors, sounds, and smells really be reduced to "nothing but" matter and motion? In the seventeenth century, and even in the eighteenth century, it was too soon to expect Newtonian physics to answer these questions. There was still too much work to do in establishing the basic principles of mechanics and applying them to astronomical problems. A full-scale attack on the properties of matter and energy had to wait until the nineteenth century.

This unit covered several successful applications and extensions of Newtonian mechanics that were accomplished by the end of the nineteenth century. For example, we discussed the conservation laws, new explanations of the properties of heat and gases, and estimates of some properties of molecules. We introduced the concept of energy, linking mechanics to heat and to sound. Unit 4 will show similar links to light, electricity, and magnetism. We also noted that applying mechanics on a molecular level requires statistical ideas and presents questions about the direction of time.

Throughout most of this unit, we have emphasized the application of mechanics to separate pieces or molecules of matter. But scientists found that the molecular model was not the only way to understand the behavior of matter. Without departing from basic Newtonian cosmology, scientists could also interpret many phenomena (such a sound and light) in terms of wave motions in continuous matter. By the middle of the nineteenth century, it was generally believed that all physical phenomena could be explained by a theory that was built on the use of either particles or waves. In the next unit, you will discover how much or how little validity there was in this belief. You will begin to see the rise of a new viewpoint in physics, based on the concept of field.

The Newtonian world view, as extended to include electrical and statistical phenomena, was immensely successful, so successful that at the close of the nineteenth century the fundamental properties of the physical universe seemed to be well understood. All that remained for research were details and applications. However, this was not to be. As you will see in Unit 5, the advances in physics brought about by developments in the nineteenth century led to new problems that required revolutionary changes in our understanding of the world.

UNIT

Light and Electromagnetism

PROLOGUE

The conviction that the world and everything in it consists of *matter in motion* drove scientists to search for mechanical models for light and electromagnetism; that is, they tried to imagine how the effects of light, electricity, and magnetism could be explained in detail as the action of material objects. (For example, consider the way light bounces off a mirror. A model for this effect might picture light as consisting of particles of matter that behave somewhat like tiny ping-pong balls.) Such mechanical models were useful for a time, but in the long run proved far too limited. Still, the search for these models led to many new discoveries, which in turn brought about important changes in science, technology, and society. These discoveries and their effects form the subject of this unit. This Prologue covers the development of various models and briefly indicates their effect on present ideas of the physical world.

From the seventeenth century on there were two competing models for light. One model tried to explain light in terms of particles; the other, in terms of waves. In the first half of the nineteenth century, the wave model won general acceptance

It was inconceivable to many scientists that one body could directly affect another across empty space. They devised a variety of schemes to fill the space in between with something that would transmit the effect, first with material "ether," later with mathematical "fields." Some of these schemes are illustrated on this and the next page. This model is by Euler (eighteenth century).

because it was better able to account for newly discovered optical effects. Chapter 13 tells the story of the triumph of the wave theory of light. The wave theory remained supreme until the early part of the twentieth century, when it was found (as you will see in Unit 5) that neither waves nor particles alone could account for all the behavior of light.

As experiments established that electric and magnetic forces have some characteristics in common with gravitational forces, scientists developed new theories of electricity and magnetism. Modeled on Newton's treatment of gravitation, these new theories assumed that there are forces between electrified and magnetized bodies that vary inversely with the square of the distance. This assumption was found to account for many observations. Of course, the drafters of these theories also assumed that bodies can exert forces over a distance without having to touch one another.

Action-at-a-distance theories were remarkably successful in providing a quantitative explanation for some aspects of electromagnetism. However, these theories did not at the time provide a really complete explanation. Instead, another means of description, based on the idea of *fields,* became widely accepted by the end of the nineteenth century. It is now generally believed to be the best way to discuss *all* physical forces. The concept of field is introduced in Chapter 14 and developed further in the last chapter of the unit.

Many scientists felt that action-at-a-distance theories, however accurate in their predictions, failed to give a satisfactory physical explanation of how one body exerts a force on another. Newton himself was reluctant to assume that one body can act on another through empty space. In a letter to Richard Bentley he wrote:

Above: Maxwell's model of the ether (nineteenth century).

Right: This model of the ether is by Descartes (seventeenth century).

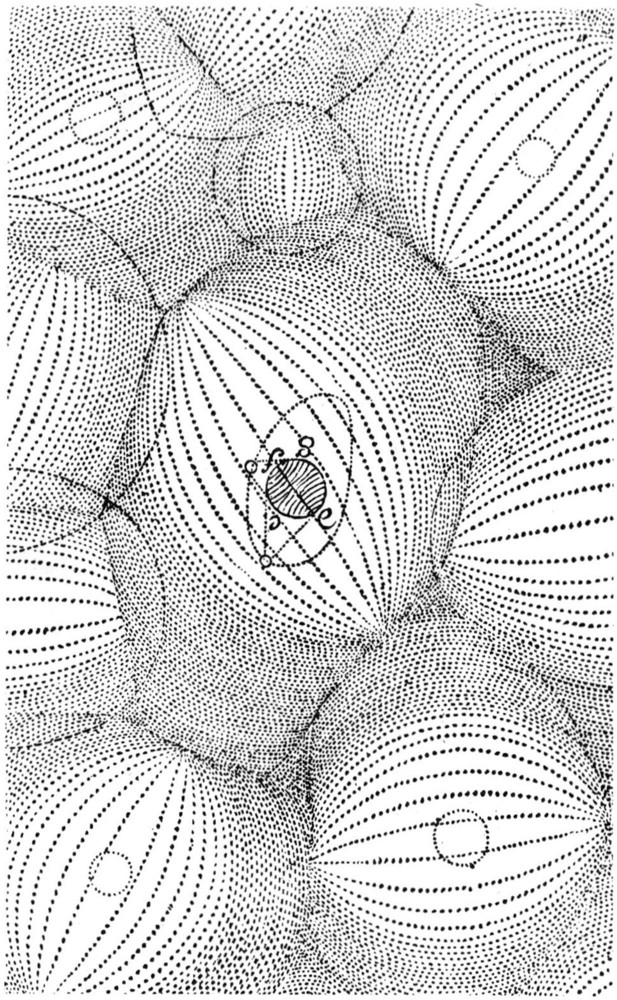

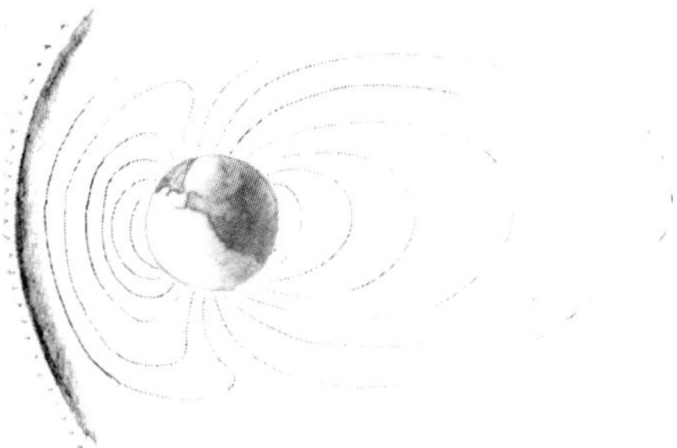

Above is a drawing representing the magnetic field around the earth. It is not the more symmetrical field the earth would have on its own, but as disturbed by streams of charged particles from the sun.

The text of this letter is reproduced exactly as Newton wrote it.

'Tis unconceivable to me that inanimate brute matter should (without the mediation of something else wch is not material) operate upon & affect other matter without mutual contact; . . . And this is one reason why I desire you would not ascribe innate gravity to me. That gravity should be innate inherent & essential to matter so yt one body may act upon another at a distance through a vacuum wthout the mediation of any thing else by & through wch their action or force may be conveyed from one point to another is to me so great an absurdity that I believe no man who has in philosophical matters any competent faculty of thinking can ever fall into it.

Some seventeenth-century scientists were less cautious than Newton. They proposed that objects are surrounded by atmospheres that extend to the most distant regions and transmit gravitational, electric, and magnetic forces from one body to another. The atmospheres proposed at this time were not made a part of a quantitative theory. But in the nineteenth century the atmosphere concept was revived. Numerous attempts were made to develop mathematically the properties of a medium that would transmit the waves of light. The name *luminiferous ether* was given to this hypothetical "light-bearing" substance.

The rapid discovery of new electrical and magnetic effects in the first half of the nineteenth century stimulated the scientific imagination. Michael Faraday (1791–1867), who made many of the important discoveries, developed a model with lines of force assigned to the space surrounding electrified and magnetized bodies. Faraday showed how these lines of force could account for many electromagnetic effects.

In a paper he wrote at age 17, William Thomson (1824–1907) showed how the equations used to formulate and solve a problem in electrostatics could also be used to solve a heat-flow problem. *Electrostatics* deals with the effects of forces between charges at rest. At the time, electrostatics was most simply and effectively treated by assuming that electrical forces can act at a distance. On the other hand, the flow of heat was generally held to result from the action of parts that touch. Thomson showed that the same mathematical formulation could be used for theories based on completely different physical assumptions. Perhaps, then, it was more important to find correct mathematical tools than to choose a particular mechanical model.

William Thomson (Lord Kelvin) was a Scottish mathematical physicist. He contributed to the fields of electricity, mechanics, and thermodynamics and to such practical developments as an improved ship's compass and the first trans-Atlantic cable. The Kelvin scale of absolute temperature is named for him.

James Clerk Maxwell (1831–1879), inspired by Faraday's physical models and by Thomson's mathematics, attempted to develop a mathematical theory of electromagnetism. Maxwell first assumed an imaginary ether consisting of gears and idler wheels. Then he gradually worked out a set of equations that described the properties of electric and magnetic fields. These equations were later found to be remarkably successful. They described quite accurately the electric and magnetic effects already known to occur. Moreover, they led Maxwell to predict new effects based on the idea of a propagating wave disturbance in electric and magnetic fields. The speed he predicted for such electromagnetic waves was nearly the same as the measured speed of light. This similarity suggested to Maxwell that light might *be* an electromagnetic wave.

Radio telescope at the National Radio Astronomy Observatory, Greenbank, West Virginia.

The concept of field, together with the concept of energy, provides a way of treating the influence of one body on another without speaking of action at a distance or of a material medium that transmits the action. The field concept has proved its usefulness over and over again during the twentieth century.

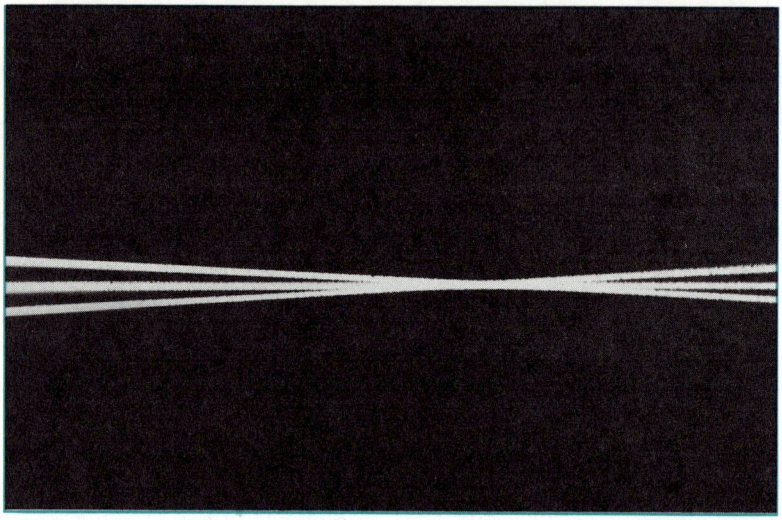

Light

13.1 Introduction: What is light?

Behold the Light emitted from the Sun,
What more familiar, and what more unknown;
While by its spreading Radiance it reveals
All Nature's Face, it still itself conceals . . .
[Richard Blackmore, *Creation II*, 1715]

Light is a form of energy. The physicist can describe a beam of light by stating measurable values of its speed, wavelength or frequency, and intensity. But to physicists, as to all people, "light" also means brightness and shade, the beauty of summer flowers and fall foliage, of red sunsets, and of the canvases painted by masters. These are simply different ways of appreciating light. One way concentrates on light's measurable aspects; this approach has been enormously fruitful in physics and technology. The other way concerns aesthetic responses to viewing light in nature or art. Still another way of considering light deals with the biophysical process of vision.

SG 1

These aspects of light are not easily separated. Thus, in the early history of science, light presented more subtle and more elusive problems than did most other aspects of physical experience. Early ideas on its nature often confused light with vision. This confusion is still evident in young children. When playing hide-and-go-seek, some children "hide" by covering their eyes with their hands; apparently they think that they cannot be seen when they cannot see. The association of vision with light persists in the language of the adult world. People often talk about the sun "peeping out of the clouds."

Some Greek philosophers believed that light travels in straight lines at high speed and contains particles that stimulate the sense of vision when they enter the eye. For centuries after the Greek era, limited attention was paid to the nature of light, and this particle model survived almost intact. Around 1500, Leonardo da Vinci, noting a similarity between sound echoes and the reflection of light, speculated that light might have a wave character.

A decided difference of opinion about the nature of light emerged among scientists of the seventeenth century. Some, including Newton, favored a model largely based on the idea of light as a stream of particles. Others, including Huygens, supported a wave model. By the late nineteenth century, there appeared to be overwhelming evidence in support of the wave model. This chapter will deal with the question: *How accurate is a wave model in explaining the observed behavior of light?* The wave model will be taken as a hypothesis, and the evidence that supports it examined. Remember that any scientific model, hypothesis, or theory has two chief functions: to explain what is known and to make predictions that can be tested experimentally. Both of these aspects of the wave model will be discussed. The result will be rather surprising. The wave model turns out to work splendidly for all properties of light known before the twentieth century. But in Chapter 18 you will find that for some purposes a particle model must be used. Then in Chapter 20, *both* models will be combined, merging two apparently conflicting theories.

The ancient opinion, later proved by experiment, that light travels in straight lines and at high speed has been mentioned. The daily use of mirrors shows that light can also be reflected. Light also can be refracted, and it shows the phenomena of interference and diffraction. You studied all of these properties in Chapter 12, which discussed the behavior of waves. If necessary, you should refresh your memory about the basic ideas of that chapter before going on to the study of light. You will also encounter other phenomena such as dispersion, polarization, and scattering. As you will see, these also fit into the wave model.

Model, analogy, hypothesis, and *theory* have similar but distinct meanings when applied to physics. An *analogy* is a corresponding situation which, though perhaps totally unrelated to the situation at hand, helps you understand it. Many electronic circuits have analogs in mechanical systems. A *model* is a corresponding situation that may well be what "is really going on" and therefore can be taken more seriously as an explanation. An electron rotating around a nucleus is one model for the atom. An *hypothesis* is a statement that can usually be directly or indirectly tested. To Franklin, the statement "lightning is caused by electricity" was at first an hypothesis. A *theory* is a more general construction, perhaps putting together several models and hypotheses to explain a collection of effects that previously seemed unrelated. Newton's explanation of Kepler's laws, Galileo's experiments in mechanics and, finally, the Cavendish experiment were all part of the theory of universal gravitation. This is a good example of a theory.

13.2 | Propagation of light

There is ample evidence that light travels in straight lines. The fact that one cannot see "around the corner" of an obstacle is an

Light beams travel in straight lines.

obvious example of such evidence. A shadow cast by the sun has the sharply defined outlines given by a large but very distant light source. Similarly, sharp shadows are cast by smaller sources closer by. The distant sun and the nearby small source are approximate *point* sources of light. Such point sources produce sharp shadows.

SG 2

Images as well as shadows can demonstrate that light travels in straight lines. Before the invention of the modern camera with its lens system, a light-tight box with a pinhole in the center of one face was widely used. As the *camera obscura,* the device was highly popular in the Middle Ages. Leonardo da Vinci probably used it as an aid in his sketching. In one of his manuscripts he says that "a small aperture in a window shutter projects on the inner wall of the room an image of the bodies which are beyond the aperture." He includes a sketch to show how the straight-line propagation of light explains the formation of an image.

Camera obscura **is a Latin phrase meaning "dark chamber."**

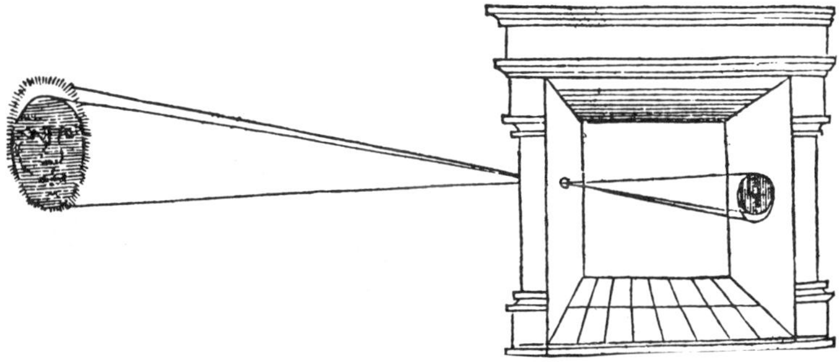

First published illustration of a camera obscura, *used to observe a solar eclipse in January 1544, from a book by the Dutch physician and mathematician Gemma Frisius.*

It is often convenient to use a straight line to represent the direction in which light travels. The pictorial device of an infinitely thin *ray* of light is useful for thinking about light, but no such rays actually exist. A light beam emerging from a good-sized hole in a screen is as wide as the hole. You might expect that if you made the hole extremely small, you would get a very narrow beam of light, ultimately just a single ray. This is not the case. Diffraction effects (such as you observed for water and sound waves in Chapter 12) appear when the beam of light passes

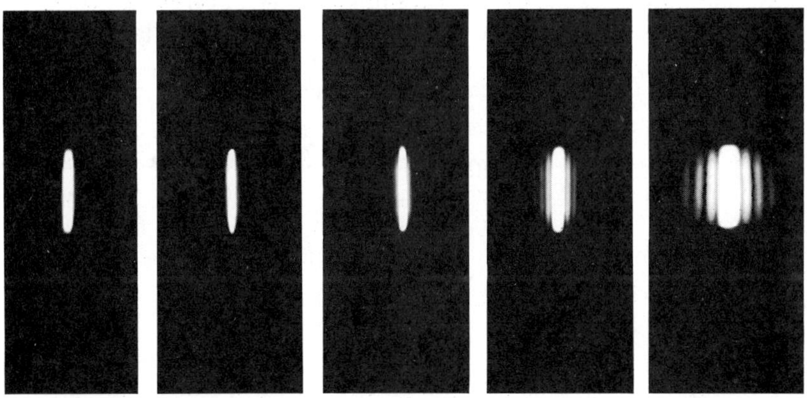

An attempt to produce a "ray" of light. To make the pictures at the left, a parallel beam of red light was directed through increasingly narrow slits to a photographic plate. (Of course, the narrower the slit, the less light gets through. This was compensated for by longer exposures in these photographs. The slit widths, from left to right, were 1.5 mm, 0.7 mm, 0.4 mm, 0.2 mm, and 0.1 mm.

through a small hole (see opposite). So an infinitely thin ray of light, although it is pictorially useful, cannot be produced in practice. But the idea can still be used in order to *represent the direction* in which a train of parallel waves in a beam of light is traveling.

Actually, the beam of light produced by a laser comes as close as possible to the ideal case of a thin, parallel bundle of rays. As you will learn later in this unit (Chapter 15), light is produced by the vibrations of electrons within the atoms of its source. In most light sources, ranging from incandescent and fluorescent bulbs to the sun and stars, these vibrations occur independently

SG 3 of one another. Each of the vibrating atoms produces an individual wavelet, and the sum of the wavelets from all the atoms makes up the total emerging light beam. As a result, light from such sources spreads out in all directions. A more or less parallel beam of light can be produced by using a set of pinholes or by using mirrors or lenses, as found, for instance, in flashlights, automobile headlights, and searchlights. However, as you can quickly determine for yourself, the beams of light they produce still diverge noticeably.

In contrast, lasers are designed in such a way that their atoms vibrate and produce light in unison with one another, rather than individually and at random. As a result, the atoms produce their wavelets simultaneously; this can yield a total beam of considerable intensity and much more nearly monochromatic (that is, of a single color) than the light from any conventional source. In addition, since the individual wavelets from the atoms of a laser are produced simultaneously, they are able to interfere with each other constructively to produce a beam of light that is narrow and very nearly parallel. In fact, such light spreads out so little that beams from lasers, when directed at the surface of the moon 400,000 km away, have been found to produce spots of light only a meter in diameter. (You will learn more about lasers later in this text.)

Given that light seems to travel in straight lines, can we tell how fast it goes? Galileo discussed this problem in his *Two New Sciences* (published in 1638). He pointed out that everyday experiences might lead one to conclude that light propagates instantaneously. But these experiences, when analyzed more closely, really show only that light travels much faster than sound. For example, "when we see a piece of artillery fired, at a great distance, the flash reaches our eyes without lapse of time; but the sound reaches the ear only after a noticeable interval." But how do you really know whether the light moved "without lapse of time" unless you have some accurate way of measuring the lapse of time?

SG 4 Galileo then described an experiment by which two people standing on distant hills flashing lanterns might measure the speed of light. (This experiment will be analyzed in SG 4.) He concluded that the speed of light is probably finite, not infinite. Galileo, however, was not able to estimate a definite value for it.

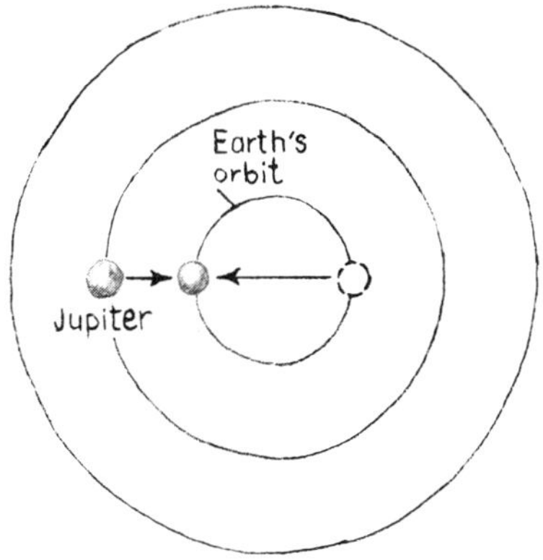

Experimental evidence was first successfully related to a finite speed for light by a Danish astronomer, Ole Römer. Detailed observations of Jupiter's satellites had shown an unexplained irregularity in the times recorded between successive eclipses of the satellites by the planet. Such an eclipse was expected to occur at 45 sec after 5:25 A.M. on November 9, 1676. In September of that year, Römer announced to the Academy of Sciences in Paris that the eclipse would be 10 min late. On November 9, astronomers at the Royal Observatory in Paris carefully observed the eclipse. Though skeptical of Römer's mysterious prediction, they reported that the eclipse did occur late, just as he had foreseen.

Later, Römer revealed the theoretical basis of his prediction to the baffled astronomers at the Academy of Sciences. He explained that the originally expected time of the eclipse had been calculated from observations made when Jupiter was near the earth. But now Jupiter had moved farther away. The delay in the eclipse occurred simply because light from Jupiter takes time to reach the earth. Obviously, this time interval must be greater when the relative distance between Jupiter and the earth in their orbits is greater. In fact, Römer estimated that it takes about 22 min for light to cross the earth's own orbit around the sun.

The importance of Römer's work was not so much that it led to a particular value of the speed of light, but rather that it established that the propagation of light is not instantaneous but takes a finite time and that he obtained a value of the right order of magnitude.

Shortly after this, the Dutch physicist Christian Huygens used Römer's data to make the first calculation of the speed of light. Huygens combined Römer's value of 22 min for light to cross the earth's orbit with his own estimate of the diameter of the earth's orbit. (This distance could be estimated for the first time in the seventeenth century, as a result of the advances in astronomy described in Unit 2.) Huygens obtained a value which, in modern units, is about 2×10^8 m/sec. This is about two-thirds of the presently accepted value (see below). The error in Huygens' value was due mainly to Römer's overestimate of the time interval. Scientists now know that it takes light only about 16 min to cross the earth's orbit.

The speed of light has been measured in many different ways since the seventeenth century. Since the speed is very great, it is necessary to use either a very long distance or a very short time interval or both. The earlier methods were based on measurements of astronomical distances. In the nineteenth century, rotating slotted wheels and mirrors made it possible to measure very short time intervals so that distances of a few kilometers could be used. The development of electronic devices in the twentieth century allowed measurement of even shorter time intervals. Today, the speed of light is one of the most accurately known physical constants. Because of the importance of the value of the speed of light in modern physical theories, physicists are continuing to improve their methods of measurement.

SG 5
SG 6

The most accurate recent measurements indicate that the speed of light in vacuum is 299,792,456.2 m/sec. The uncertainty of this value is thought to be about 1 m/sec, or 0.000001%. The speed of light is usually represented by the symbol c; for most purposes it is sufficient to use the approximate value $c = 3 \times 10^8$ m/sec.

Two narrow beams of light, coming from the upper left, strike a block of glass. Can you account for all the observed effects?

?

1. Can a beam of light be made increasingly narrow by passing it through narrower and narrower slits? What property of light does this experiment demonstrate?

2. What reason did Römer have for thinking that the eclipse of a particular satellite of Jupiter would be observed later than expected?

3. What was the most important outcome of Römer's work?

Katherine Burr Blodgett (1898–1979). Dr. Blodgett developed "invisible" glass by applying 44 layers of a one-molecule thick transparent liquid soap to glass to reduce reflections from its surface. Today, nearly all camera lenses and optical devices have nonreflective coatings on their surfaces. These coatings facilitate the efficient passage of light.

13.3 | Reflection and refraction

What does each model of light predict will happen when light traveling in one medium (e.g., air) hits the boundary of another medium (e.g., glass)? The answers to this question depend on whether a particle or a wave theory of light is used. Here is an opportunity to test which theory is better.

Reflection and refraction from the wave viewpoint were discussed in Chapter 12. Recall the results obtained there and apply them to light:

1. A ray may be taken as the line drawn perpendicular to a wave's crest lines. Such a ray represents the direction in which a train of parallel waves is traveling.

2. In reflection, the angle of incidence (θ_i) is equal to the angle of reflection (θ_r).

3. Refraction involves a change of wavelength and speed of the wave as it passes into another medium. When the speed decreases, the wavelength decreases, and the ray bends in a direction toward a line perpendicular to the boundary. This bending toward the perpendicular is observed when a ray of light passes from air to glass.

SG 7–12

The incident, reflected and refracted rays are all in the same plane, a plane perpendicular to the surface. See Sec. 12.10.

What about explaining the same observations by means of the particle model? To test this model, first consider the nature of the surface of glass. Though apparently smooth, it is actually a wrinkled surface. A powerful microscope would show it to have endless hills and valleys. If particles of light were at all similar

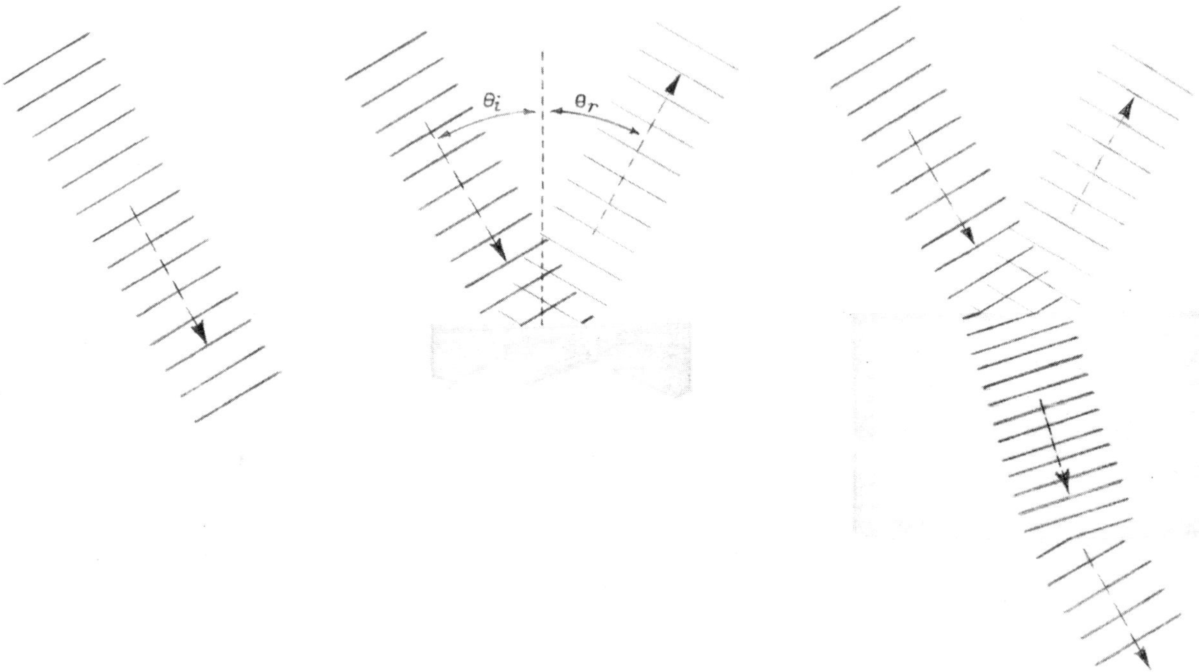

to little balls of matter, then on striking such a wrinkled surface they would scatter in all directions. They would not be reflected and refracted as shown above. Therefore, Newton argued, there must actually be "some feature of the body which is evenly diffused over its surface and by which it acts upon the ray without immediate contact." Obviously, in the case of reflection, the acting force would have to be one that repelled the particles of light. Similarly, a force that attracted light particles instead of repelling them could explain refraction. As a particle of light approached a boundary of another medium, it would first have to overcome the repelling force. If it did that, it would then meet an attractive force in the medium that would pull it into the medium. Since the attractive force would be a vector with a component in the direction of the particle's original motion, the particle's speed would increase. If the ray of particles were moving at an oblique angle to the boundary, it would change direction as it entered the medium, bending toward the line perpendicular to the boundary.

According to the *particle* model, therefore, you can make the following statements about reflection and refraction:

1. A ray represents the direction in which the particles are moving.

2. In reflection, the angles of incidence and reflection are equal. This prediction can be derived by applying the law of conservation of momentum (Chapter 9) to particles repelled by a

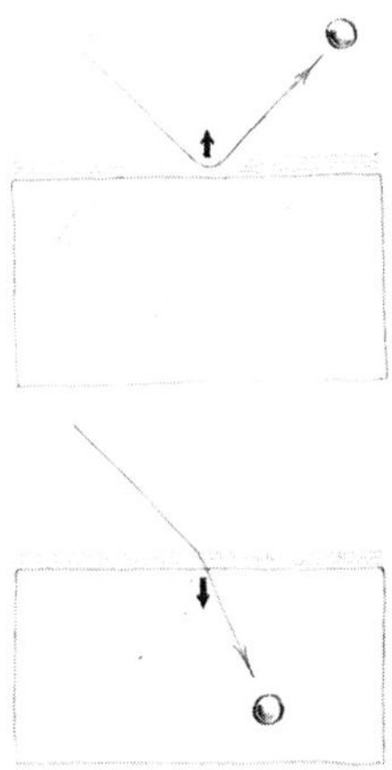

SG 13

Notice that to make this argument we have had to make an assumption about the size of Newton's light "particles." The particles must be at least as small as the irregularities in the surface of a mirror. Similarly, a concrete wall is quite rough, but a tennis ball rebounds from such a wall almost exactly as light reflects from a mirror.

The surface of a mirror as shown by a scanning electron microscope. The surface is a 3-micron thick aluminum film. The magnification here is nearly 26,000 times. (μ stands for micron; 1μ = 10^{-6} m.)

force as shown on the last sketch on the previous page (see also SG 7).

3. Refraction involves a change of speed of the particles as they enter another medium. In particular, when an attractive power acts, *the speed increases,* and the ray is bent into the medium.

Compare these features of the particle model with the corresponding features of the wave model. The only difference is in the predicted speed for a refracted ray. You *observe* that a ray is bent toward the perpendicular line when light passes from air into water. The particle theory *predicts* that light has a *greater* speed in the second medium. The wave theory *predicts* that light has a *lower* speed.

You might think that it would be fairly easy to devise an experiment to determine which prediction is correct. All one has to do is measure the speed of light after it has entered water and compare it with the speed of light in air. But in the late seventeenth and early eighteenth centuries, when Huygens was supportinig the wave model and Newton a particle model, no such experiment was possible. The only available way of measuring the speed of light was an astronomical one. Not until the middle of the nineteenth century did Armand H. L. Fizeau and Jean B. L. Foucault measure the speed of light in water. The results agreed with the predictions of the wave model: The speed of light is less in water than in air.

Actually, by the time these experiments were done, most physicists had already accepted the wave model for other reasons. The Foucault–Fizeau experiments of 1850 were widely regarded as driving the last nail in the coffin of the Newtonian particle theory.

?

4. What evidence showed conclusively that Newton's particle model for light could not explain all aspects of refraction?

5. If light has a wave nature, what changes take place in the speed, wavelength, and frequency of light on passing from air into water?

13.4 | Interference and diffraction

From the time of Newton until the early nineteenth century, most physicists favored the particle theory of light. Newton's own prestige contributed greatly to this support. Early in the nineteenth century, however, the wave theory was revived by Thomas Young. In experiments made between 1802 and 1804, Young found that light shows the phenomenon of *interference.* (Interference patterns were discussed in Sec. 12.6 in connection with water waves.) The particle theory of light could not easily explain such patterns. Young's famous "double-slit experiment" provided convincing evidence that light does have properties that are explainable only in terms of waves.

Young's experiment must be done in the lab, rather than just talked about, but we will describe it briefly here. Basically, it involves splitting a single beam of light into two beams. The split beams are then allowed to overlap, and the two wave trains interfere, constructively in some places and destructively in others. To simplify the interpretation of the experiment, assume that it is done with light that has a single definite wavelength λ.

Young used a black screen with a small hole punched in it to produce a narrow beam of sunlight in a dark room. In the beam he placed a second black screen with two narrow slits cut in it, close together. Beyond this screen he placed a white screen. The light coming through each slit was diffracted and spread out into the space beyond the screen. The light from each slit interfered with the light from the other, and the interference pattern showed on the white screen. Where interference was constructive, there was a bright band on the screen. Where interference was destructive, the screen remained dark.

It is remarkable that Young actually found, by experiment, numerical values for the very short wavelength of light. Here is his result:

> From a comparison of various experiments, it appears that the breadth of the undulations constituting the extreme red light must be supposed to be, in air, about one 36 thousandth of an inch [7×10^{-7} m], and those of the extreme violet about one 60 thousandth [4×10^{-7} m].

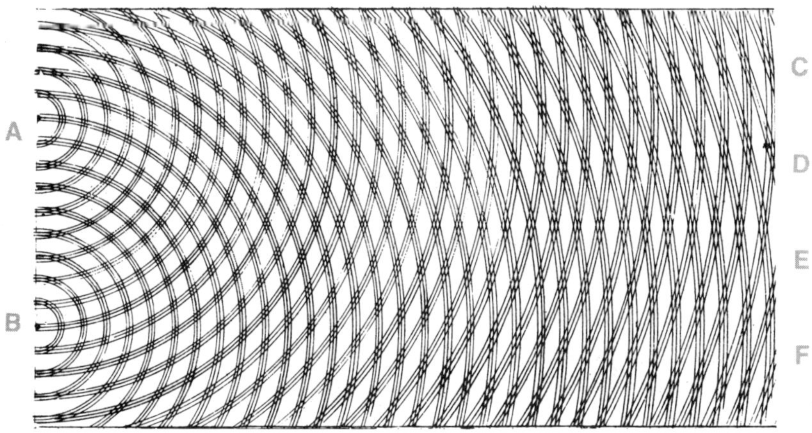

In announcing his result, Young took special pains to forestall criticism from followers of Newton, who was generally considered a supporter of the particle theory. He pointed out that Newton himself had made several statements favoring a theory of light that had some aspects of a wave theory.

Thomas Young (1773–1829) was an English linguist, physician, and expert in many fields of science. At the age of 14 he was familiar with Latin, Greek, Hebrew, Arabic, Persian, French, and Italian, and later was one of the first scholars successful at decoding Egyptian hieroglyphic inscriptions. He studied medicine in England, Scotland, and Germany. While still in medical school, he made original studies of the eye and later developed the first version of what is now known as the three-color theory of vision. Young also did research in physiology on the functions of the heart and arteries and studied the human voice mechanism, through which he became interested in the physics of sound and sound waves. Young then turned to optics and showed that many of Newton's experiments with light could be explained in terms of a simple wave theory of light. This conclusion was strongly attacked by some scientists in England and Scotland who were upset by the implication that Newton might have been wrong.

SG 14

Thomas Young's original drawing showing interference effects in overlapping waves. The alternate regions of reinforcement and cancellation in the drawing can be seen best by placing your eye near the right edge and sighting at a grazing angle along the diagram.

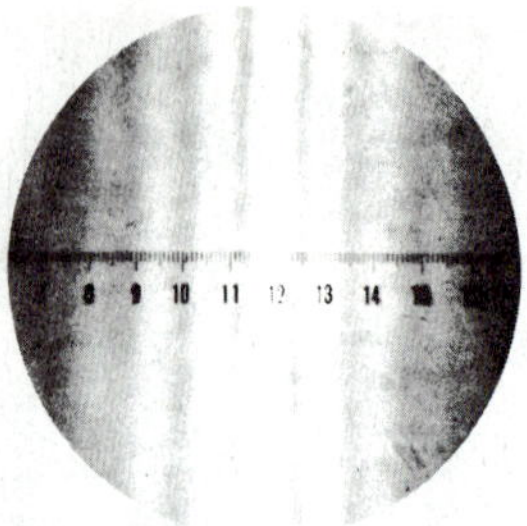

A Polaroid photograph taken through a Project Physics *magnifier placed about 30 cm behind a pair of closely spaced slits. The slits were illuminated with a narrow but bright light source.*

Augustin Jean Fresnel (1788–1827) was an engineer of bridges and roads for the French government. In his spare time, he carried out extensive experimental and theoretical work in optics. Fresnel developed a comprehensive wave model of light that successfully accounted for reflection, refraction, interference, and polarization. He also designed a lens system for lighthouses that is still used today.

SG 15

Nevertheless, Young was not taken seriously. It was not until 1818, when the French physicist Augustin Fresnel proposed his own mathematical wave theory, that Young's research began to get the credit it deserved. Fresnel also had to submit his work for approval to a group of physicists who were committed to the particle theory. One of them, the mathematician Simon Poisson, tried to refute Fresnel's wave equations. If these equations really did describe the behavior of light, Poisson said, a very peculiar thing ought to happen when a small solid disk is placed in a beam of light. Diffraction of the light waves all around the edge of the round disk should lead to constructive interference at the center. In turn, this constructive interference should produce a bright spot in the center of the disk's shadow on a white screen placed at certain distances behind the disk. But the particle theory of light allowed no room for ideas such as diffraction and constructive interference. In addition, such a bright spot had never been reported, and even the very idea of a bright spot in the center of a shadow seemed absurd. For all of these reasons, Poisson announced that he had refuted the wave theory.

Fresnel accepted the challenge, however, and immediately arranged for Poisson's prediction to be tested by experiment. The result was that a bright spot *did* appear in the center of the shadow, as predicted by Poisson on the basis of Fresnel's wave theory.

Gradually, scientists realized the significance of the Young double-slit experiment and the Poisson bright spot. Support for the particle theory of light began to crumble. By 1850, the wave model of light was generally accepted. Physicists had begun to concentrate on working out the mathematical consequences of this model and applying it to the different properties of light.

6. How did Young's experiments support the wave model of light?

7. In what way is diffraction involved in Young's experiments?

8. What phenomenon was predicted by Poisson on the basis of Fresnel's wave theory?

13.5 | Color

The coloring agents found in prehistoric painting and pottery show that humans have appreciated color since earliest times. But no scientific theory of color was developed before the time of Newton. Until then, most of the accepted ideas about color had come from artist–scientists like da Vinci, who based their ideas on experiences with mixing pigments.

Unfortunately, the lessons learned in mixing pigments rarely apply to the mixing of different-colored light beams. In ancient times, it was thought that light from the sun was "pure light." Color resulted from adding impurity, as when "pure light" was refracted in glass.

Newton became interested in colors when, as a student at Cambridge University, he set out to construct an astronomical telescope. One troublesome defect of the telescope was a fuzzy ring of color that always surrounded the image formed by the telescope lens. Perhaps in an attempt to understand this particular defect, Newton began his extensive study of color.

In 1672, at the age of 29, Newton published a theory of color in the *Philosophical Transactions* of The Royal Society of London. This was his first published scientific paper. He wrote:

> ... in the beginning of the Year 1666 (at which time I applyed myself to the grinding of Optick glasses of other figures than *Spherical,*) I procured me a Triangular glass-Prisme, to try therewith the celebrated *Phaenomena* of *Colours*. And in order thereto haveing darkened my chamber, and made a small hole in my window-shuts, to let in a convenient quantity of the Suns light, I placed my Prisme at his entrance, that it might be thereby refracted to the opposite wall. It was at first a very pleasing divertisement, to view the vivid and intense colours produced thereby....

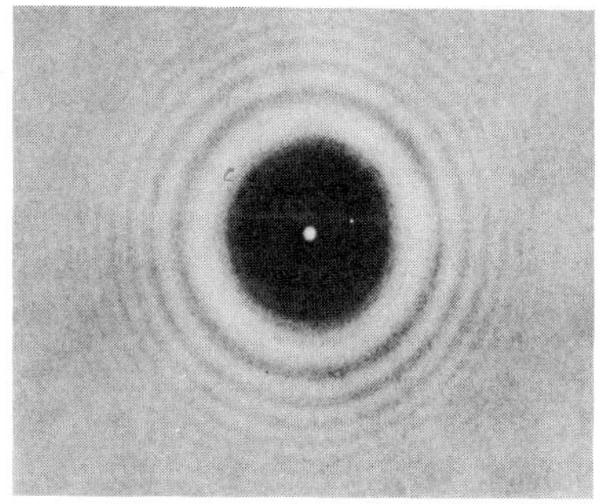

Diffraction pattern caused by an opaque circular disk, showing the Poisson bright spot in the center of the shadow. Note also the bright and dark fringes of constructive and destructive interference. (You can make similar photographs yourself; see the activity "Poisson's Spot" in the Handbook.*)*

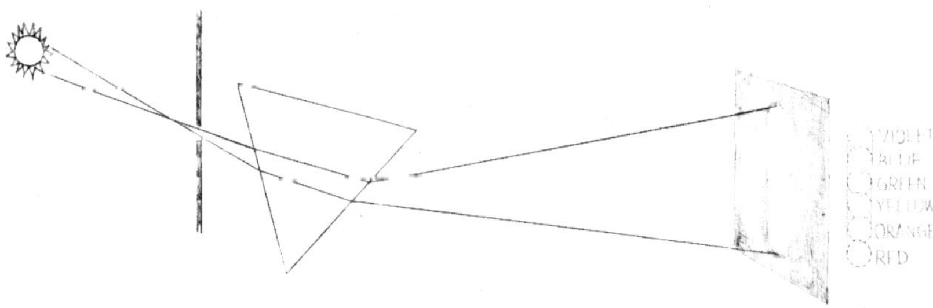

The drawing at the left is based on Newton's diagram of the refraction of sunlight by a prism.

The cylindrical beam of "white" sunlight from the circular opening passed through the prism and produced an elongated patch of colored light on the opposite wall. This patch was violet at one end, red at the other, and showed a continuous gradation of colors in between. For such a pattern of colors, Newton invented the name *spectrum.*

But, Newton wondered, where do the colors come from? And why is the image spread out in an elongated patch rather than circular? Newton passed the light through different thicknesses of the glass, changed the size of the hole in the window shutter, and even placed the prism outside the window. None of these changes had any effect on the spectrum. Perhaps some unevenness or irregularity in the glass produced the spectrum, Newton thought. To test this possibility, he passed the colored

Close Up

Diffraction and Detail

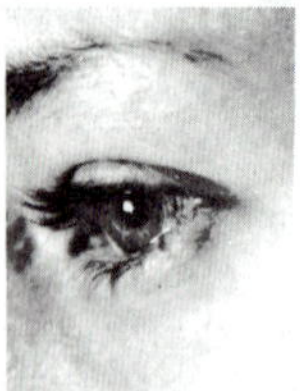   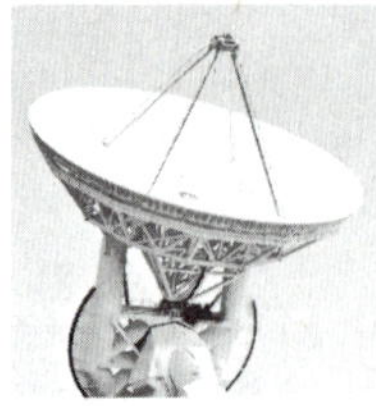 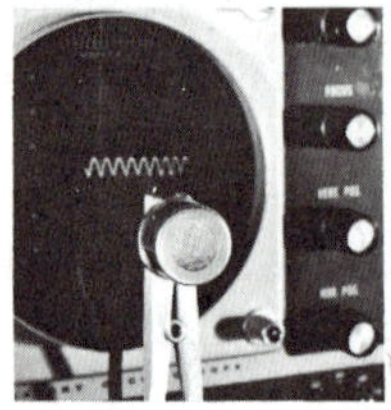 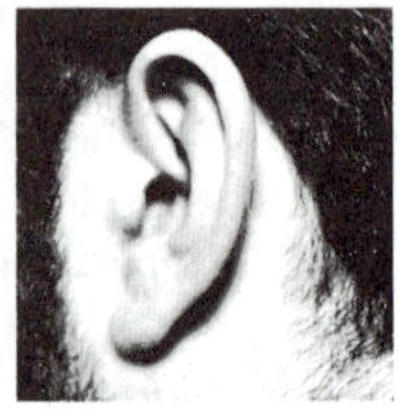

The photograph above shows the diffraction image of a point source of light. Diffraction by the camera lens opening has spread the light energy into a bright central disk surrounded by alternate dark and bright rings. The photographs below show an array of point sources, recorded through a progressively smaller and smaller hole. The array could represent a star cluster, surface detail on Mars, granules in living cells, or simply specific points on some object.

The diffraction of the waves from the edges of the hole limits the detail of information that it is possible to receive. As the hole through which we observe the array below becomes smaller, the diffraction image of each point spreads out and begins overlapping the diffraction images of other points. When the diffraction patterns for the points overlap sufficiently, it is impossible to distinguish between them.

This problem of diffraction has many practical consequences. We obtain most of the information about our environment by means of waves (light, sound, radio, etc.) which we receive through some sort of hole: the pupil of the eye, the entrance to the ear or a microphone, the aperture of an optical telescope or radio telescope, etc. In all these cases, then, diffraction places a limit on the detail with which the sources of waves can be discriminated.

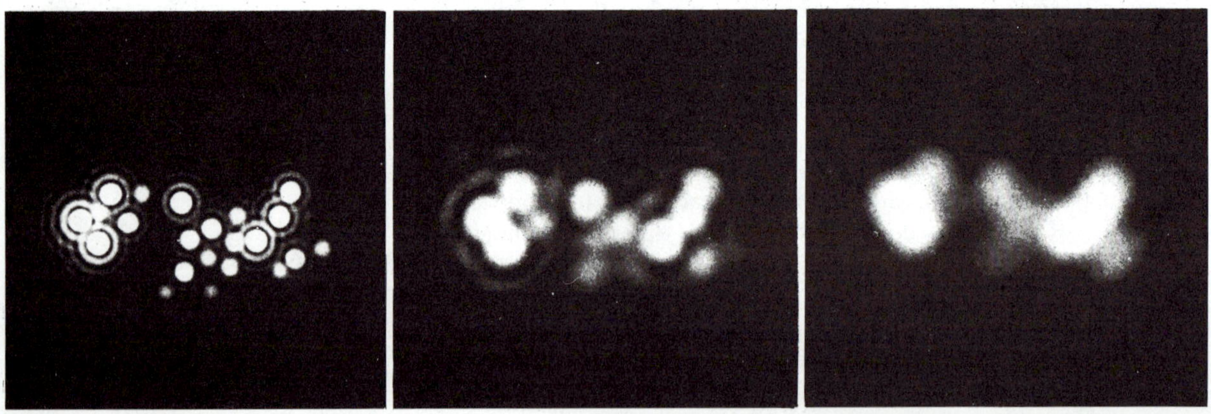

rays from one prism through a second similar prism turned upside down. If some irregularity in the glass caused the beam of light to spread out, then passing this beam through the second prism should spread it out even more. Instead, the second prism, when properly placed, brought the colors *back together* fairly well. A spot of *white* light was formed, as if the light had not passed through either prism.

By such a process of elimination, Newton convinced himself of a belief that he probably had held from the beginning: *White light is composed of colors.* The prism does not manufacture or add the colors; they are there all the time, but mixed up so that they cannot be distinguished. When white light passes through a prism, each of the component colors is refracted at a different angle. Thus, the beam is spread into a spectrum.

As a further test of this hypothesis, Newton cut a small hole in a screen on which a spectrum was projected. In this way, light of a single color could be separated out and passed through a second prism. He found that the second prism had no further effect on this single-color beam, aside from refracting it more. Once the first prism had done its job of separating the colored components of white light, the second prism could not change the color of the components.

Summarizing his conclusions, Newton wrote:

> Colors are not *Qualifications of Light* derived from Refraction or Reflection of natural Bodies (as 'tis generally believed) but Original and Connate Properties, which in divers Rays are divers. Some Rays are disposed to exhibit a Red Colour and no other; some a Yellow and no other, some a Green and no other, and so of the rest. Nor are there only Rays proper and particular to the more Eminent Colours, but even to all their intermediate gradations.

Most colors observed for real materials are "body" colors, produced by selective absorption of light which penetrates a little beyond the surface before being scattered back. This explains why the light transmitted by colored glass has the same color as the light reflected from it. Thin metallic films, however, have "surface" color, resulting from selective regular reflection. Thus, the transmitted light will be the complement of the reflected light. For example, the light transmitted by a thin film of gold is bluish-green, while that reflected is yellow.

Apparent colors of objects. So far, Newton had discussed only the colors of rays of light. In a later section of his paper he raised the important question: Why do objects appear to have different colors? Why is the sky blue, the grass green, a paint pigment yellow or red? Newton proposed a very simple answer:

> That the Colours of all Natural Bodies have no other Origin than this, that they . . . Reflect one sort of Light in greater plenty than another.

In other words, a red pigment looks red to us because when white sunlight falls on it, the pigment absorbs most of the rays of other colors of the spectrum and reflects mainly the red to our eyes.

According to Newton's theory, color is not a property of an object by itself. Rather, color depends on how the object reflects and absorbs the various colored rays that strike it. Newton backed up this hypothesis by pointing out that an object may appear to have a different color when a different kind of light shines on it. For example, consider a pigment that reflects much more red light than green or blue light. When illuminated by white light, it will reflect mostly the red component of the white light, and so will appear red. But if it is illuminated with blue light, there is no red for it to reflect, and it can reflect only a very little of the blue light. Thus, it will appear to be dark and slightly blue.

SG 16

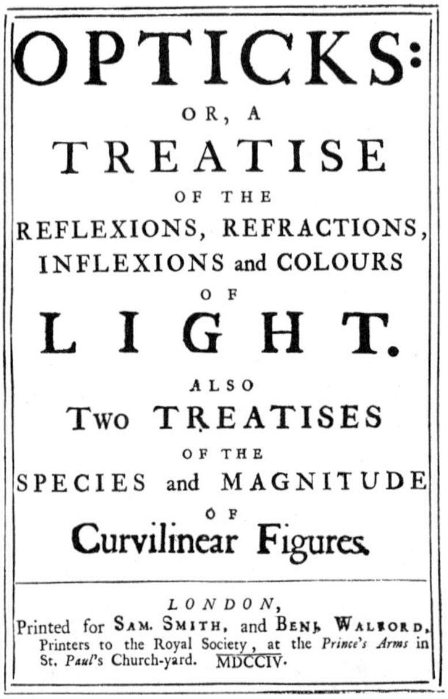

OPTICKS:
OR, A
TREATISE
OF THE
REFLEXIONS, REFRACTIONS, INFLEXIONS and COLOURS
OF
LIGHT.
ALSO
Two TREATISES
OF THE
SPECIES and MAGNITUDE
OF
Curvilinear Figures.

LONDON,
Printed for SAM. SMITH, and BENJ. WALFORD, Printers to the Royal Society, at the *Prince's Arms* in St. *Paul's* Church-yard. MDCCIV.

Title page from the first edition of Newton's Opticks *(1704), in which he described his theory of light.*

Reactions to Newton's theory. Newton's theory of color met with violent opposition at first. Other British scientists, especially Robert Hooke, objected on grounds that postulating a different kind of light for each color was unnecessary. It would be simpler to assume that the different colors were produced from pure white light by some kind of modification. For example, the wave front might be twisted so that it is no longer perpendicular to the direction of motion.

Newton was aware of the flaws in Hooke's theory, but he disliked public controversy. In fact, he waited until after Hooke's death in 1703 to publish his own book, *Opticks* (1704), in which he reviewed the properties of light.

Newton's *Principia* was a much more important work from a purely scientific viewpoint. But his *Opticks* had also considerable influence on the literary world. English poets gladly celebrated the discoveries of their country's greatest scientist. Most poets, of course, were only dimly aware of the significance of Newton's theory of gravity. The technical details of the geometric axioms and proofs of the *Principia* were beyond them. But Newton's theory of colors and light provided good material for poetic fancy, as in James Thomson's, "To the Memory of Sir Isaac Newton" (1727):

> . . . First the flaming red,
> Springs vivid forth; the tawny orange next;
> And next delicious yellow; by whose side
> Fell the kind beams of all-refreshing green.
> Then the pure blue, that swells autumnal skies,
> Ethereal played; and then, of sadder hue,
> Emerged the deepened indigo, as when
> The heavy-skirted evening droops with frost;
> While the last gleamings of refracted light
> Died in the fainting violet away.

SG 17

Leaders of the nineteenth-century Romantic movement in literature and the German "Nature Philosophers" did not think so highly of Newton's theory of color. The scientific procedure of dissecting and analyzing natural phenomena by experiments was distasteful to them. They preferred to speculate about the unifying principles of all natural forces, hoping somehow to grasp nature as a whole. The German philosopher Friedrich Schelling wrote in 1802:

> Newton's *Opticks* is the greatest illustration of a whole structure of fallacies which, in all its parts, is founded on observation and experiment.

To the nineteenth-century physicists who were trying to use Newton's theory to explain newly discovered color phenomena, Goethe addressed the following poem:

May ye chop the light in pieces
Till it hue on hue releases;
May ye other pranks deliver,
Polarize the tiny sliver
Till the listener, overtaken,
Feels his senses numbed and shaken—
Nay, persuade us shall ye never
Nor aside us shoulder ever,
Steadfast was our dedication—
We shall win the consummation.

The German poet and Nature Philosopher Goethe (mentioned in Chapter 11) spent many years trying to overthrow Newton's theory of colors. Using his own observations, as well as passionate arguments, Goethe insisted on the purity of white light in its natural state. He rejected Newton's hypothesis that white light is a mixture of colors. Instead, he suggested, colors may be produced by the interaction of white light and its opposite—darkness. Goethe's observations on color perception were of some value to science. But his theory of the physical nature of color could not stand up under detailed experiment. Newton's theory of the colors of the spectrum remained firmly established.

SG 18, 19

?

9. *How did Newton show that white light was not "pure"?*
10. *Why could Newton be confident that, say, green light was not itself composed of different colors of light?*
11. *How would Newton explain the color of a blue shirt?*
12. *Why was Newton's theory of color attacked by the Nature Philosophers?*

The nanometer is 10^{-9} m, a convenient unit for measuring wavelengths of visible and ultraviolet light, X rays, and sizes of atomic dimension. The Ångstrom, 10^{-10} m, was used previously.

13.6 Why is the sky blue?

Newton suggested that the apparent colors of natural objects depend on which color is most strongly reflected or scattered to the viewer by the object. In general, there is no simple way of predicting from the surface structure, chemical composition, etc., what colors a substance will reflect or scatter. However, the blue color of the clear sky can be explained by a fairly simple argument.

As Thomas Young showed experimentally (Sec. 13.4), different wavelengths of light correspond to different colors. The wavelength of light may be specified in units of *nanometers* (abbreviated nm; 1 nm = 10^{-9} m) or, alternatively, in Ångstroms (Å), equal to 10^{-10} m. The range of the spectrum visible to humans is from about 400 nm (4000 Å) for violet light to about 700 nm (7,000 Å) for red light.

The Ångstrom unit is named after Anders Jonas Ångstrom, a Swedish astronomer who, in 1862, used spectroscopic techniques to detect the presence of hydrogen in the sun.

Small obstacles can scatter the energy of an incident wave in all directions, and the amount of scattering depends on the wavelength. This fact can be demonstrated by simple

The amount of scattering of different waves by a tiny obstacle is indicated here for three wavelengths.

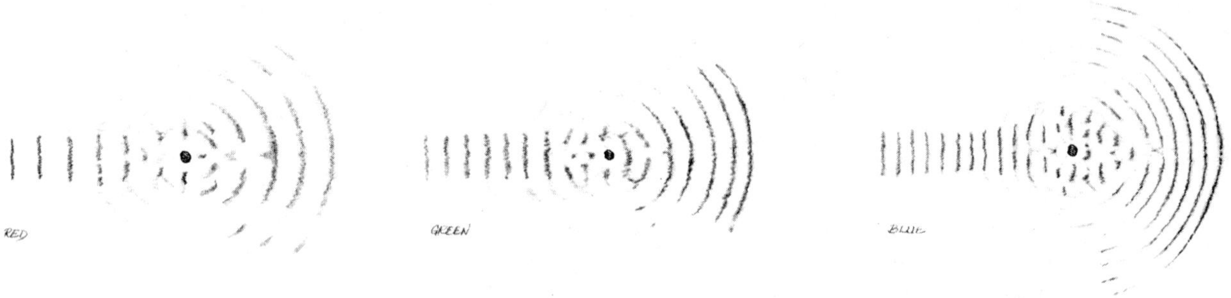

experiments with water waves in a ripple tank. As a general rule, *the larger the wavelength is compared to the size of the obstacle, the less the wave is scattered by the obstacle.* For particles smaller than one wavelength, the amount of scattering of light varies inversely with the fourth power of the wavelength. For example, the wavelength of red light is about twice the wavelength of blue light. Therefore, the scattering of red light is about one-sixteenth as much as the scattering of blue light.

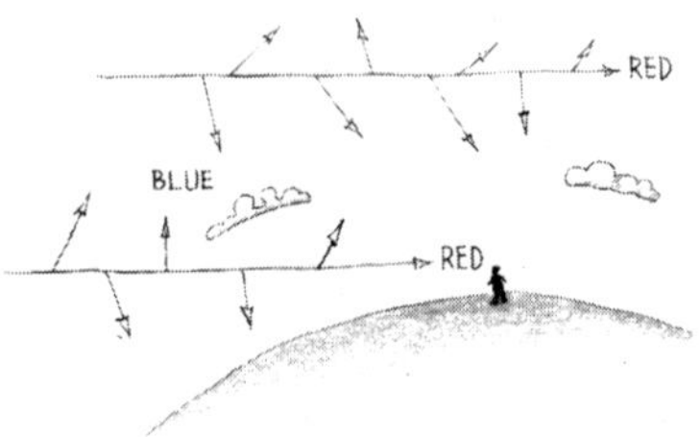

If you look at a sunset on a hazy day, you receive primarily unscattered colors, such as red; whereas if you look overhead, you will receive primarily scattered colors, the most dominant of which is blue.

Now you can understand why the sky is blue. Light from the sun is scattered by molecules and particles of dust in the sky, which are usually very small compared to the wavelengths of visible light. Thus, light of short wavelengths (blue light) is much more strongly scattered by the particles than is light of longer wavelengths. When you look up into a clear sky, it is mainly this scattered light that enters your eyes. The range of scattered short wavelengths (and the color sensitivity of the human eye) lead to the sensation of blue. On the other hand, suppose you look at a sunset on a very hazy day. You receive a beam that has had the blue light almost completely scattered out, while the longer wavelengths have *not* been scattered out. So you perceive the sun as reddish.

If the earth had no atmosphere, the sky would appear black, and stars would be visible by day. In fact, starting at altitudes of about 16 km, where the atmosphere becomes quite thin, the sky does look black, and stars can be seen during the day, as astronauts have found.

If light is scattered by particles considerably larger than one wavelength (such as water droplets in a cloud), there is not much difference in the scattering of different wavelengths. So we receive the mixture we perceive as white.

Sometimes the air contains dust particles or water droplets as large as the wavelength of visible light (about 10^{-6} m). If so, colors other than blue may be strongly scattered. For example, the quality of sky coloring changes with the water-vapor content of the atmosphere. On clear, dry days the sky is a much deeper blue than on clear days with high humidity. The intensely blue skies of Italy and Greece, which have inspired poets and painters for centuries, are a result of exceptionally dry air.

The blue-gray haze that often covers large cities is caused mainly by particles emitted by internal combustion engines (cars, trucks) and by industrial plants. Even when idling, a typical automobile engine emits more than 100 billion particles per second. Most of these particles are invisible, ranging in size from 10^{-6} m to 5×10^{-9} m. Such particles provide a framework to which gases, liquids, and other solids adhere. These larger particles then scatter light and produce haze. Gravity has little effect on the particles until they become very large by collecting more matter. They may remain in the atmosphere for months if not cleaned out by repeated rain, snow, or winds. The influences of such clouds of haze or smog on the climate and on human health are substantial.

?

13. How does the scattering of light waves by tiny obstacles depend on the wavelength?

14. How would you explain the color of the earth's sky? What do you expect the sky to look like on the moon? Why?

13.7 | Polarization

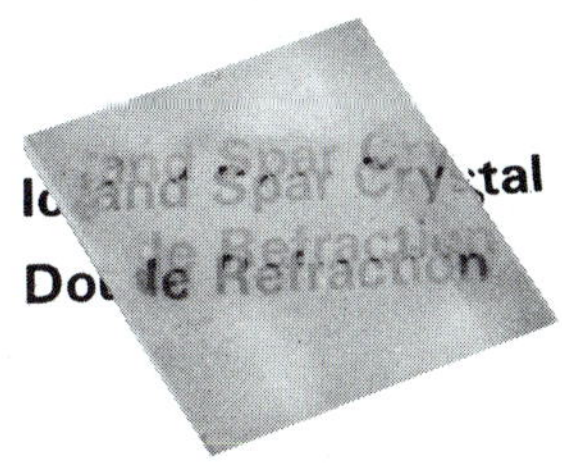

Hooke and Huygens proposed that light is in many ways like sound, that is, that light is a wave propagated through a medium. Newton could not accept this proposal and argued that light must also have some particle-like properties. He noted two properties of light that, he thought, could not be explained unless light had particle properties. First, a beam of light is propagated in straight lines, while waves such as sound spread out in all directions and go around corners. This objection could not be answered until early in the nineteenth century, when Young measured the wavelength of light and found it to be extremely small. Even the wavelength of red light, the longest wavelength of the visible spectrum, is less than a thousandth of a millimeter. As long as a beam of light shines on objects or through holes of ordinary size (a few millimeters or more in width), the light will appear to travel in straight lines. Diffraction and scattering effects do not become strikingly evident until a wave passes over an object or through a hole whose size is about equal to or smaller than the wavelength.

Newton's second objection was based on the phenomenon of "polarization" of light. In 1669, the Danish scientist Erasmus Bartholinus discovered that crystals of Iceland spar (calcite) could split a ray of light into two rays. Writing or small objects viewed through the crystal looked double.

Newton thought this behavior could be explained by assuming that light is made up of particles that have different "sides," for example, rectangular cross sections. The double images, he thought, represent a sorting out of light particles that had entered the medium with different orientations.

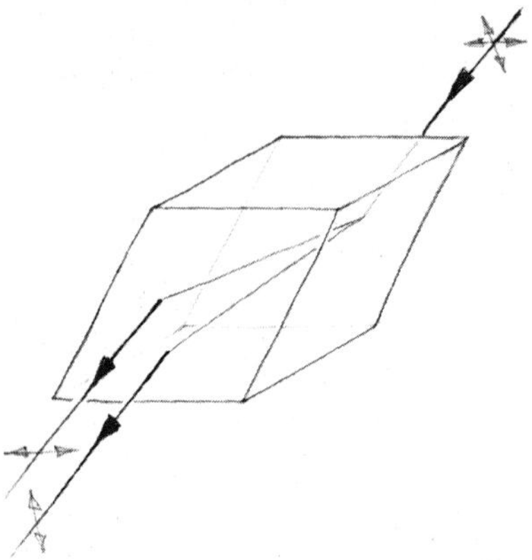

Double refraction by a crystal of Iceland spar. The "unpolarized" incident light can be thought of as consisting of two polarized components. The crystal separates these two components, transmitting them through the crystal in different directions and with different speeds.

The same short wave train on the rope approaches the slotted board in each of the three sketches. Depending on the orientation of the slot, the train of waves (a) goes entirely through the slot; (b) is partly reflected and partly transmitted with changed angles of rope vibration; or (c) is completely reflected.

Around 1820, Young and Fresnel gave a far more satisfactory explanation of polarization, using a modified wave theory of light. Before then, scientists had generally assumed that light waves, like sound waves, must be *longitudinal*. (And, as Newton believed, longitudinal waves could not have any directional property.) Young and Fresnel showed that if light waves are *transverse*, this could account for the phenomenon of polarization.

In a transverse wave, the motion of the medium itself is always perpendicular to the direction of propagation of the wave (see Chapter 12). This does not mean that the motion of the medium is always in the same direction. In fact, it could be in any direction in a plane perpendicular to the direction of propagation. However, if the motion of the medium *is* mainly in one direction (for example, vertical), the wave is *polarized.* (Thus, a polarized wave is really the *simplest* kind of transverse wave. An unpolarized transverse wave is more complicated, since it is a mixture of various transverse motions.) The way in which Iceland spar separates an unpolarized light beam into two polarized beams is sketched in the margin.

Scientific studies of polarization continued throughout the nineteenth century. Practical applications, however, were frustrated, mainly because polarizing substances like Iceland spar were scarce and fragile. One of the best polarizers was "herapathite," or sulfate of iodo-quinine, a synthetic crystal. The needle-like herapathite crystals absorb light that is polarized in the direction of the long crystal axis and absorb very little of the light polarized in a direction at 90° to the long axis.

Herapathite crystals were so fragile that there seemed to be no way of using them. But in 1928, Edwin H. Land, while still a freshman in college, invented a polarizing plastic sheet he called "Polaroid." His first polarizer was a plastic film in which many

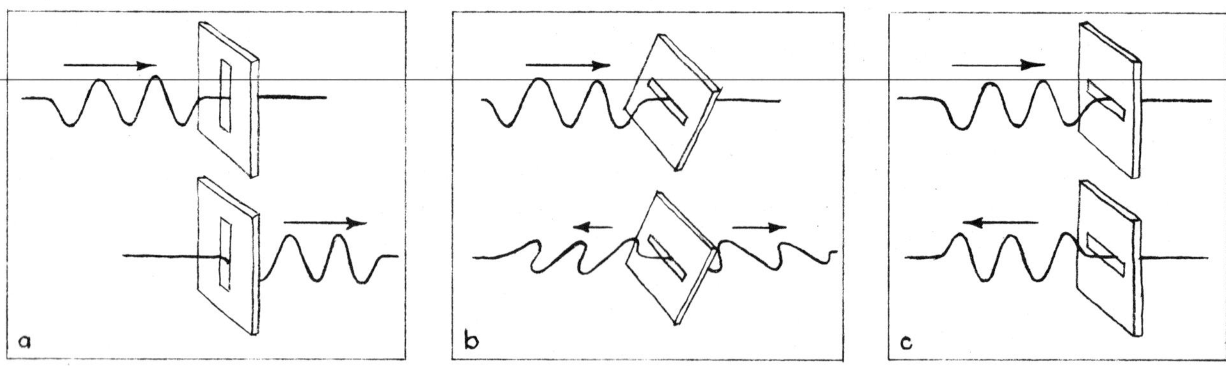

Later, Land improved Polaroid by using polymeric molecules composed mainly of iodine in place of herapathite crystals.

microscopic crystals of herapathite were imbedded. When the plastic was stretched, the needle-like crystals lined up in one direction. Thus, they all acted on incoming light in the same way.

Some properties of a polarizing material are easily demonstrated. Hold a polarizing sheet, for example, the lens of a pair of polarizing sunglasses, in front of a light source. Then look at the first polarizing sheet through a second one. Rotate the first sheet. You will notice that, as you do so, the light alternately brightens and dims. You must rotate the sheet through an angle of 90° to go from maximum brightness to maximum dimness.

How can this be explained? The light that strikes the first sheet is originally unpolarized, that is, a mixture of waves polarized in various directions. The first sheet transmits only those waves that are polarized in one direction and absorbs the rest. The transmitted wave going toward the second sheet is now polarized in one direction. Whenever this direction coincides with the direction of the long molecules in the second sheet, the wave will be absorbed by the second sheet (that is, the wave will set up vibrations within the molecules of the crystals and will lose most of its energy). However, if the direction is *perpendicular* to the long axis of the crystal molecules, the polarized light will go through the second sheet without much absorption.

Interference and diffraction effects required a wave model for light. To explain polarization phenomena, the wave model was made more specific. It was shown that polarization could be explained if light waves were transverse. This model for light explains well enough all the characteristics of light considered so far. But you will see in Unit 5 that the model turned out to require even further extension.

Ordinary light, when scattered by particles, shows polarization to different degrees, depending on the direction of scattering. The eyes of bees, ants, and other animals are sensitive to the polarization of scattered light from the clear sky. This enables a bee to navigate by the sun, even when the sun is low on the horizon or obscured. Following the bees' example, engineers have equipped airplanes with polarization indicators for use in Arctic regions.

SG 20, 21

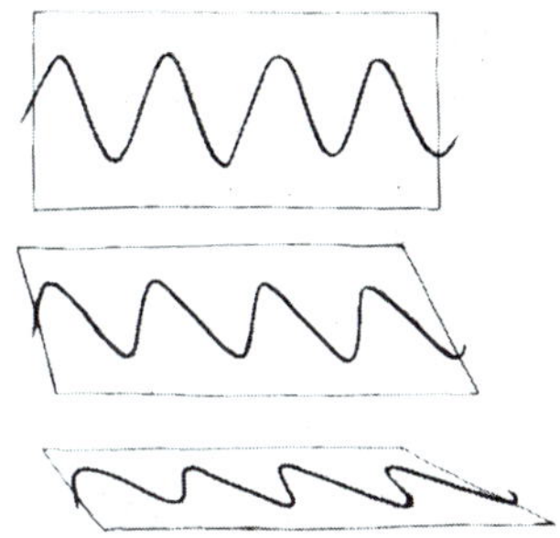

?

15. What two objections did Newton have to a wave model?

16. What phenomena have been discussed that agree with a wave model of light?

17. Can sound waves be polarized?

13.8 | The ether

One factor seems clearly to be missing from the wave model for light. Chapter 12 discussed waves as disturbances that propagate in some substance or "medium," such as a rope or water. What is the medium for the propagation of light waves?

Is air the medium for light waves? No, because light can pass through airless space, as it does between the sun or other stars and the earth. Even before it was definitely known that there is no air between the sun and the earth, Robert Boyle had tried the experiment of pumping almost all of the air out of a glass container. He found that the objects inside remained visible.

It was difficult to think of a disturbance without specifying what was being disturbed. So it was natural to propose that a

"Ether" was originally the name for Aristotle's fifth element, the pure transparent fluid that filled the heavenly sphere. It was later called "quintessence" (see Secs. 2.1 and 6.4).

medium for the propagation of light waves existed. This medium was called the *ether*. In the seventeenth and eighteenth centuries, the ether was imagined to be an invisible fluid of very low density. This fluid could penetrate all matter and fill all space. It might somehow be associated with the "effluvium" (something that "flows out") that was imagined to explain magnetic and electric forces. But light waves must be transverse in order to explain polarization, and transverse waves usually propagate only in a *solid* medium. A liquid or a gas cannot transmit transverse waves for any significant distance for the same reason that you cannot "twist" a liquid or a gas. So nineteenth-century physicists assumed that the ether must be a solid.

As stated in Chapter 12, the speed of propagation increases with the stiffness of the medium, and decreases with its density. The speed of propagation of light is very high compared to that of other kinds of waves, such as sound. Therefore, the ether was

"Entrance to the Harbor," a painting by Georges Seurat (1888). Art historians believe that Seurat's technique of pointillism, *the use of tiny dots of pure color to achieve all effects in a painting, was based on his understanding of the physical nature of light. The Museum of Modern Art; Lillie P. Bliss Collection.*

thought to be a very stiff solid with a very low density. Yet it seems absurd to say that a stiff, solid ether fills all space. The planets move without slowing down, so apparently they encounter no resistance from a stiff ether. And, of course, you feel no resistance when you move around in a space that transmits light freely.

Without ether, the wave theory seemed improbable. But the ether itself had absurd properties. Until early in this century, this problem remained unsolved, just as it had for Newton. You will see how, following Einstein's modification of the theory of light, the problem was solved.

In order to transmit transverse waves, the medium must have some tendency to return to its original shape when it has been deformed by a transverse pulse. As Thomas Young remarked on one such ether model, "It is at least very ingenious, and may lead us to some satisfactory computations; but it is attended by one circumstance which is perfectly *appalling* in its consequences. . . . It is only to solids that such a *lateral* resistance has ever been attributed: so that . . . it might be inferred that the luminiferous ether, pervading all space, and penetrating almost all substances, is not only high elastic, but absolutely solid!!!"

?

18. Why was it assumed that there existed an "ether" that transmitted light waves?

19. What remarkable property must the ether have if it is to be the mechanical medium for the propagation of light?

study guide

1. The *Project Physics* learning materials particularly appropriate for Chapter 13 include:

Experiments

Refraction of a Light Beam
Young's Experiment—the Wavelength of Light

Activities

Thin-Film Interference
Handkerchief Diffraction Grating
Photographic Diffraction Patterns
Poisson's Spot
Photographing Activities
Color
Polarized Light
Make an Ice Lens

In addition the following *Project Physics* materials can be used with Unit 4 in general:

Film

People and Particles

2. A square card, 3 cm on a side, is held 10 cm from a small penlight bulb, and its shadow falls on a wall 15 cm behind the card. What is the size of the shadow on the wall? (A diagram of the situation will be useful.)

3. The row of photographs on page 399 shows what happens to a beam of light that passes through a narrow slit. The row of photographs on page 375 of Chapter 12 shows what happens to a train of water wave that passes through a narrow opening. Both sets of photographs illustrate single-slit diffraction, but the photographs are not at all similar in appearance. Explain the difference in appearance of the photographs and also how they are similar.

4. An experiment to determine whether or not the propagation of light is instantaneous is described by Galileo as follows:

> Let each of two persons take a light contained in a lantern, or other receptacle, such that by the

interposition of the hand, the one can shut off or admit the light to the vision of the other. Next let them stand opposite each other at a distance of a few cubits [1 cubit = about 0.5 m] and practice until they acquire such skill in uncovering and occulting their lights that the instant one sees the light of his companion he will uncover his own. After a few trials the response will be so prompt that without sensible error [svario] the uncovering of one light is immediately followed by the uncovering of the other, so that as soon as one exposes his light he will instantly see that of the other. Having acquired skill at this short distance let the two experimenters, equipped as before take up positions separated by a distance of two or three miles and let them perform the same experiment at night, noting carefully whether the exposures and occultations occur in the same manner as at short distances; if they do, we may safely conclude that the propagation of light is instantaneous, but if time is required at a distance of three miles which, considering the going of one light and the coming of the other, really amounts to six, then the delay ought to be easily observable. . . .

But later he states:

In fact, I have tried the experiment only at a short distance, less than a mile, from which I have not been able to ascertain with certainty whether the appearance of the opposite light was instantaneous or not; but if not instantaneous, it is extraordinarily rapid. . . .

(a) Why was Galileo unsuccessful in the above experiment?

(b) How would the experiment have to be altered to be successful?

(c) What do you think is the longest time that light might have taken in getting from one observer to the other without the observers detecting the delay? Use this estimate to arrive at a lower limit for the speed of light that is consistent with Galileo's description of the result.

(d) Why do you suppose that the first proof of the finite speed of light was based on celestial observations rather than on terrestrial observations?

5. A convenient unit for measuring astronomical distances is the *light year*, defined as the distance that light travels in 1 yr. Calculate the number of meters in a light year to two significant figures.

6. Calculate how much farther than expected Jupiter must have been from earth when Römer predicted a 10-min delay for the eclipse of 1676.

7. Suppose a space vehicle had a speed 1/1000 that of light. How long woud it take to travel the 4.3 light years from the earth to the closest known star other than the sun, Proxima Centauri. Compare the speed given for the space vehicle with the speed of approximately 10 km/sec maximum speed (relative to the earth) that a space capsule reaches on an earth–moon trip.

8. Newton supposed that the reflection of light off shiny surfaces is due to "some feature of the body which is evenly diffused over its surface and by which it acts upon the ray without contact." The simplest model for such a feature would be a repulsive force that acts only in a direction perpendicular to the surface. In this problem you are to show how this model, together with the laws of mechanics, predicts that the angles of incidence and reflection must be equal. Proceed as follows:

(a) Draw a clear diagram showing the incident and reflected rays. Also show the angles of incidence and reflection (θ_1 and θ_2). Sketch a coordinate system on your diagram that has an *x*-axis parallel to the surface and a *y*-axis perpendicular to the surface. Note that the angles of incidence and reflection are defined to be the angles between the incident and reflected rays and the *y*-axis.

(b) Supposing that the incident light consists of particles of mass *m* and speed *v*, what is the kinetic energy of a single particle? Write mathematical expressions for the *x* and *y* components of the momentum of an incident light particle.

(c) If the repulsive force due to the reflecting surface does no work on the particle and acts only perpendicular to the surface, which of the quantities that you have described in part (b) is conserved?

(d) Show algebraically that the speed *u* of the reflected particle is the same as the speed *y* of the incident particle.

(**e**) Write mathematical expressions for the components of the momentum of the reflected particle.

(**f**) Show algebraically that θ_1 and θ_2 must be equal angles.

9. In the diagram below, find the shortest path from point A to any point on the surface M and then to point B. Solve this by trial and error, perhaps by experimenting with a short piece of string held at one end by a tack at point A. (A possible path is shown, but it is not necessarily the shortest one.) Notice that the shortest distance between A, M, and B is also the *least-time* path for a particle traveling at a constant speed from A to M to B. What path would light take from A to M to B? Can you make a statement of the law of reflection in terms of this principle instead of in terms of angles?

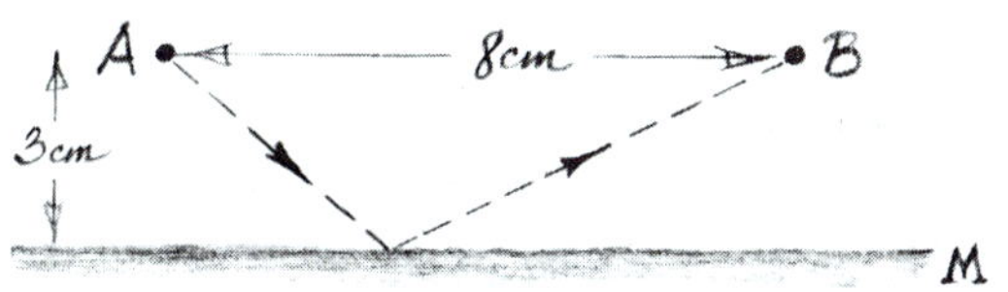

10. What is the shortest mirror in which a 180-cm-tall man can see himself entirely? (Assume that both he and the mirror are vertical and that he places the mirror in the most favorable position.) Does it matter how far away he is from the mirror? Do your answers to these questions depend on the distance from his eyes to the top of his head?

11. Suppose the reflecting surfaces of every visible object were somehow altered so that they completely absorbed any light falling on them; how would the world then appear to you?

12. Objects are visible as a whole if their surfaces reflect light, enabling your eyes to intercept cones of reflected light diverging from *each part* of the surface. The accompanying diagram shows such a cone of light (represented by two diverging rays) entering the eye from a book.

Draw clear, straight diagrams to show how a pair of diverging rays can be used to help explain the following phenomena.

(**a**) The mirror image of an object appears to be just as far behind the mirror as the object is in front of the mirror.

(**b**) A pond appears shallower than it actually is.

(**c**) A coin is placed in an empty coffee mug so that the coin cannot *quite* be seen. The coin becomes visible if the mug is filled with water.

13. Because of atmospheric refraction, you see the sun in the evening for some minutes after it is really below the horizon and also for some minutes before it is actually above the horizon in the morning.

(**a**) Draw a simple diagram to illustrate how this phenomenon occurs.

(**b**) What would sunset be like on a planet with a very thick and dense (but still transparent) atmosphere?

14. In a particle theory of light, refraction could be explained by assuming that the particle is accelerated by an attractive force as it passes from air or a vacuum toward a medium such as glass. Assume that this accelerating force can act on the particle *only* in a direction perpendicular to the surface. Use vector diagrams to show that the speed of the particle in the glass would have to be greater than in air.

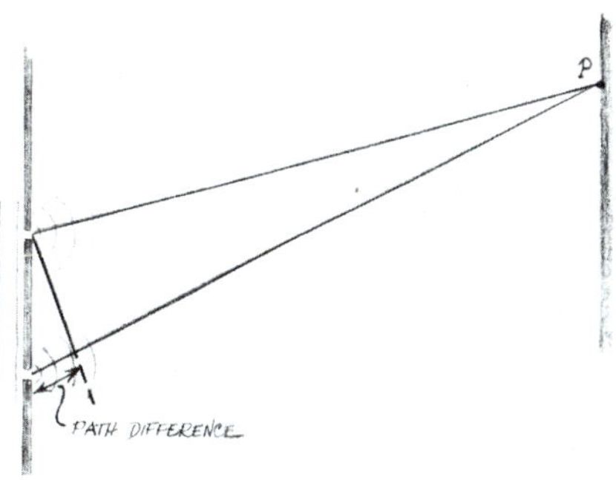

15. Plane parallel waves of single-wavelength light illuminate the two narrow slits, resulting in an interference pattern of alternate bright and dark fringes being formed on the screen. The bright fringes represent zones of constructive interference. Therefore, they appear at a point such as P in the diagram above only if the diffracted waves from the two slits arrive at P in phase. The diffracted waves will be in phase at point P only if the path difference is a whole number of wavelengths (that is, only if the path difference equals $m\lambda$, where $m = 0, 1, 2, 3, \ldots$).

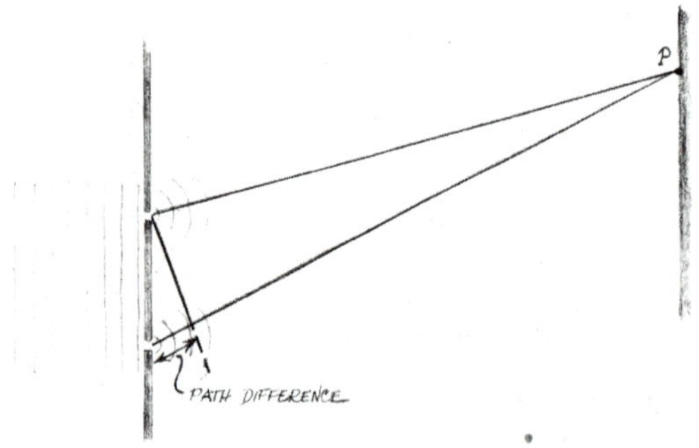

(a) What path difference results in destructive interference at the screen?

(b) The separation between two successive bright fringes depends on the wavelengths of the light used. Would the separation be greater for red light or for blue light?

(c) For a particular color of light, how would the pattern change if the distance of the screen from the slits is increased? (Hint: Make two diagrams.)

(d) What changes occur in the pattern if the slits are moved closer together? (Hint: Make two diagrams.)

(e) What happens to the pattern if the slits themselves are made narrower?

16. Use Young's estimate of the wavelengths of violet (400 nm) and red (700 nm) light to calculate their frequencies.

17. How far apart would the slits in Young's experiment have been if he had noticed the first node of the interference pattern 5 cm from the center line on a screen 15 m away while using green light?

18. The layers of atoms in a crystal act as slits in diffracting beams of X rays. How far from the center line of a viewing screen 1 m from the crystal would you expect to find the first node using X rays of 10^{-10} m wavelength if the crystalline layers are 10^{-7} m apart?

19. Recalling diffraction and interference phenomena from Chapter 12, show how the wave theory of light can explain the bright spot found in the center of the shadow of a disk illuminated by a point source.

20. It is now a familiar observation that clothing of certain colors appears different in artificial light and in sunlight. Explain why.

21. Another poem by James Thomson (1728):

Meantime, refracted from yon eastern cloud,
Bestriding earth, the grand ethereal bow
Shoots up immense; and every hue unfold,
In fair proportion running from the red
To where the violet fades into the sky.
Here, awe-ful Newton, the dissolving clouds
Form, fronting on the sun, thy showery prism;
And to the sage-instructed eye unfold
The various twine of light, by thee disclosed
From the white mingling blaze.

How do you think it compares with the poem on p. ■■

(a) as poetry?

(b) as physics?

22. Green light has a wavelength of approximately 5×10^{-7} m (500 nm). What frequency corresponds to this wavelength? Compare this frequency to the carrier frequency of the radio waves broadcast by a radio station you listen to.

23. One way to achieve privacy in apartments facing each other across a narrow courtyard while still allowing residents to enjoy the view of the courtyard and the sky above the courtyard is to use polarizing sheets placed over the windows. Explain how the sheets must be oriented for maximum effectiveness.

24. To prevent car drivers from being blinded by the lights of approaching autos, polarizing sheets could be placed over the headlights and windshields of every car. Explain why these sheets would have to be oriented the same way on every vehicle and must have their polarizing axis at 45° to the vertical.

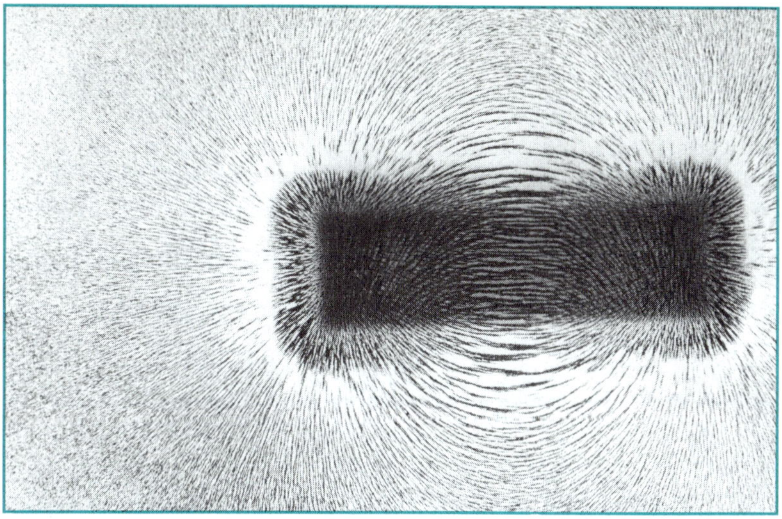

Electric and Magnetic Fields

14.1 | Introduction

SG 1

The subject "electricity and magnetism" makes up a large part of modern physics and has important connections with almost all other areas of physics, chemistry, and engineering. It would be impossible to cover this subject fully in the time available in an introductory course. This chapter will consider only a few main topics, which will serve as a foundation for later chapters. Major applications of the information in this chapter will appear later; the development of electrical technology (Chapter 15), the study of the nature of light and electromagnetic waves (Chapter 16), and the study of properties of atomic and subatomic particles (Unit 5).

This chapter will cover electric charges and the forces between them. The discussion will be brief, because the best way to learn about this subject is not by reading but by doing experiments in the laboratory (see Experiment 4-3 in the *Handbook*). Next, you will see how the idea of a "field" simplifies the description of electric and magnetic effects. Then, you will study electric currents, which are made up of moving charges. By combining the concept of field with the idea of potential energy, quantitative relations between current, voltage, and power can be established. These relations will be needed for the practical applications discussed in Chapter 15.

Finally, at the end of this chapter, you will come to the relation between electricity and magnetism. This relation has very important consequences for both technology and basic physical theory. You will begin by studying a simple physical phenomenon: the interaction between moving charges and magnetic fields.

An inside view of the heavy ion linear accelerator at Berkeley, California. In this device electric fields accelerate charged atoms to high energies.

14.2 | The curious properties of lodestone and amber: Gilbert's *De Magnete*

Two natural substances, amber and lodestone, have aroused interest since ancient times. Amber is sap that long ago oozed from certain softwood trees, such as pine. Over many centuries, it hardened into a semitransparent solid ranging in color from yellow to brown. It is a handsome ornamental stone when polished, and it sometimes contains the remains of insects that were caught in the sticky sap. Ancient Greeks recognized a strange property of amber. If rubbed vigorously against cloth, it can attract nearby objects such as bits of straw or grain seeds.

Lodestone is a mineral that also has unusual properties. It attracts iron. Also, when suspended or floated, a piece of lodestone always turns to one particular position, a north-south direction. The first known written description of the navigational use of lodestone as a compass in Western countries dates from the late twelfth century, but its properties were known even earlier in China. Today, lodestone would be called magnetized iron ore.

The histories of lodestone and amber are the early histories of magnetism and electricity. The modern developments in these subject areas began in 1600 with the publication in London of William Gilbert's book *De Magnete*. Gilbert (1544–1603) was an influential physician, who served as Queen Elizabeth's chief physician. During the last 20 years of his life, he studied what was already known of lodestone and amber. Gilbert made his own experiments to check the reports of other writers and summarized his conclusion in *De Magnete*. The book is a classic in scientific literature, primarily because it was a thorough and largely successful attempt to test complex speculation by detailed experiment.

Gilbert's first task in his book was to review and criticize what had previously been written about lodestone. He discussed various theories about the cause of magnetic attraction. One of the most popular theories was suggested by the Roman author Lucretius:

Lucretius was one of the early writers on atomic theory; see the Prologue to Unit 5.

> Lucretius ... deems the attraction to be due to this, that as there is from all things a flowing out ["efflux" or "effluvium"] of minutest bodies, so there is from iron an efflux of atoms into the space between the iron and the lodestone—a space emptied of air by the lodestone's atoms [seeds]; and when these begin to return to the lodestone, the iron follows, the corpuscles being entangled with each other.

Gilbert himself did not accept the effluvium theory as an explanation for magnetic attraction, although he thought it might apply to electrical attraction.

When it was discovered that lodestone and magnetized needles or iron bars tend to turn in a north–south direction, many authors offered explanations. But, says Gilbert,

> ... they wasted oil and labor, because, not being practical in the research of objects of nature, being acquainted only with books, being led astray by certain erroneous physical systems, and having made no magnetical experiments, they constructed certain explanations on a basis of mere opinions, and old-womanishly dreamt the things that were not. Marcilius Ficinus chews the cud of ancient opinions, and to give the reason for the magnetic direction seeks its cause in the constellation Ursa ... Paracelsus declares that there are stars which, gifted with the lodestone's power, do attract to themselves iron ... All these philosophers ... reckoning among the causes of the direction of the magnet, a region of the sky, celestial poles, stars ... mountains, cliffs, vacant space, atoms, attractional ... regions beyond the heavens, and other like unproved paradoxes, are world-wide astray from the truth and are blindly wandering.

Gilbert himself proposed the real cause of the lining-up of a suspended magnetic needle or lodestone: The earth itself is a lodestone. Gilbert also performed a clever experiment to show that his hypothesis was a likely one. Using a large piece of natural lodestone in the shape of a sphere, he showed that a small magnetized needle placed on the surface of the lodestone acts just as a compass needle does at different places on the earth's surface. (In fact, Gilbert called his lodestone the *terrella*, or "little earth.") If the directions along which the needle lines up are marked with chalk on the lodestone, they form meridian circles. Like the lines of equal longitude on a globe of the earth, these circles converge at two opposite ends that may be called "poles." At the poles, the needle points perpendicular to the surface of the lodestone. Halfway between, along the "equator," the needle lies along the surface. Small bits of iron wire placed on the spherical lodestone also line up in these same directions.

The immensely important idea of "field" was introduced into physics by Michael Faraday early in the nineteenth century, and developed further by Kelvin and Maxwell (see Secs. 14.4 and 16.2).

Discussion of the actions of magnets now generally involves the idea that magnets set up "fields" all around themselves. The field can act on other objects, near or distant. Gilbert's description of the force exerted on the needle by his spherical lodestone was a step toward the modern field concept:

> The terrella's force extends in all directions. . . . But whenever iron or other magnetic body of suitable size happens within its sphere of influence it is attracted; yet the nearer it is to the lodestone the greater the force with which it is borne toward it.

"Electric" comes from the Greek word *electron*, meaning "amber."

Gilbert also included a discussion of electricity in his book. He introduced the word *electric* as the general term for "bodies that attract in the same way as amber." Gilbert showed that electric and magnetic forces are different. For example, a lodestone always attracts iron or other magnetic bodies. An electric object exerts its attraction only when it has been recently rubbed. On the other hand, an electric object can attract small pieces of many different substances. But magnetic forces act only between a few types of substances. Objects are attracted to a rubbed electric object along lines directed toward one center region. But magnets always have *two* regions (poles) toward which other magnets are attracted.

Gilbert went beyond summarizing the known facts of electricity and magnets. He suggested new research problems that were pursued by others for many years. For example, he proposed that while the poles of two lodestones might either attract or repel each other, electric bodies could never exert repelling forces. However, in 1646, Sir Thomas Browne published the first account of electric repulsion. To systematize such observations a new concept, *electric charge*, was introduced. In the next section, you will see how this concept can be used to describe the forces between electrically charged bodies.

?

1. List separately the "curious properties of lodestone" and those of amber. Which properties are now called "electric effects" and which are called "magnetic effects"?

2. How did Gilbert demonstrate that the earth behaves like a spherical lodestone?

3. How does the attraction of objects by amber differ from the attraction by lodestone?

14.3 | Electric charges and electric forces

As Gilbert strongly argued, the facts of electrostatics must be learned in the laboratory rather than by just reading about them. This section, therefore, is only a brief outline to prepare you for your own experience with the phenomena.

The behavior of amber was discussed earlier. When rubbed, it mysteriously acquires the property of picking up small bits of grain, cork, paper, hair, etc. To some extent, all materials show this effect when rubbed, including rods made of glass or hard rubber, or strips of plastic. There are two other important basic observations: (1) When two rods of the same material are rubbed with something made of another material, the rods *repel* each other. Examples that were long ago found to work especially well are two glass rods rubbed with silk cloth, or two hard rubber rods rubbed with fur. (2) When two rods of *different* material are rubbed (for example, a glass rod rubbed with silk, and a rubber rod rubbed with fur, the two rods may *attract* each other.

Benjamin Franklin (1706–1790), American statesman, inventor, scientist, and writer, was greatly interested in the phenomena of electricity; his famous kite experiment and invention of the lightning rod gained him wide recognition. Franklin is shown here observing the behavior of a bell whose clapper is connected to a lightning rod.

These and thousands of similar experimentally observable facts can be summarized in a systematic way by adopting a very simple model. While describing a *model* for electrostatic attraction and repulsion, remember that this model is *not* an experimental fact which you can observe separately. It is, rather, a set of invented ideas which help describe and summarize observations. It is easy to forget this important difference between experimentally observable facts and invented explanations. Both are needed, but they are not the same thing! The model adopted consists of the concept of "charge" and three rules. An object that is rubbed and given the property of attracting small bits of matter is said to "be electrically charged" or to "have an electric charge." Also, imagine that there are two kinds of charge. All objects showing electrical behavior are imagined to have either one or the other of the two kinds of charge. The three rules are:

1. There are only two kinds of electric charge.
2. Two objects charged alike (that is, having the same kind of charge) repel each other.
3. Two objects charged oppositely attract each other.

When two different uncharged materials are rubbed together (for example, the glass rod and the silk cloth), they acquire opposite kinds of charge. Benjamin Franklin, who did many experiments with electric charges, proposed a mechanical model for such phenomena. In his model, charging an object electrically involved the transfer of an "electric fluid" that was present in all matter. When two objects were rubbed together, some electric fluid from one passed into the other. One body then had an extra amount of fluid and the other a lack of fluid. An excess of fluid produced one kind of electric charge, which

Franklin called "positive." A lack of the same fluid produced the other kind of electric charge, which he called "negative."

Previously, some theorists had proposed "two-fluid" models involving both a "positive fluid" and a "negative fluid." Normal matter contained equal amounts of these two fluids, so that they cancelled out each other's effects. When two different objects were rubbed together, a transfer of fluids occurred. One object received an excess of positive fluid, and the other received an excess of negative fluid.

There was some dispute between advocates of one-fluid and two-fluid models, but both sides agreed to speak of the two kinds of electrical charges as "+" or "−." It was not until the late 1890's that experimental evidence gave convincing support to any model for "electric charge." There were, as it turned out, elements of truth in both one-fluid and two-fluid models. The story will be told in some detail in Unit 5. For the present, we can say that there are in fact two different material "fluids." But the "negative fluid" moves around much more easily than the "positive fluid." So most of the electric phenomena discussed so far actually involve an excess or lack of the mobile "negative fluid," or, in modern terms, an excess or lack of electrons.

Franklin thought of the electric fluid as consisting of tiny particles, and that is the present view, too. Consequently, the word "charge" is often used in the plural. For example, we usually say "electric charges transfer from one body to another."

What is amazing in electricity, and indeed in other parts of physics, is that so few concepts are needed to deal with so many different observations. For example, a third or fourth kind of charge is not needed in addition to "+" and "−." That is to say, no observation of charged objects requires some additional type of charge that might have to be called "÷" or "×."

Even the behavior of an *un*charged body can be understood in terms of + and − charges. Any piece of matter large enough to be visible can be considered to contain a large amount of electric charge, both positive and negative. If the positive charge is equal to the negative charge, the piece of matter will appear to have zero charge, no charge at all. The effects of the positive and negative charges simply cancel each other when they are added together or are acting together. (This is one advantage of calling the two kinds of charge positive and negative rather than, say, *x* and *y*.) The electric charge on an object usually means the slight *excess* (or net) of either positive or negative charge that happens to be on that object.

Our experience with Newton's law of gravitation is affecting our question. We are assuming that the force depends only on a single property and on distance.

The electric force law. What is the "law of force" between electric charges? In other words, how does the force depend on the *amount of* charge and on the *distance* between the charged objects?

The first evidence of the nature of such a force law was obtained in an indirect way. About 1775, Benjamin Franklin noted that a small cork hanging near the outside of an electrically charged metal can was strongly attracted. But when he lowered the cork by a thread into the can, he found that no force was experienced by the cork no matter what its position inside the can.

Franklin did not understand why the walls of the can did not attract the cork when it was inside, but did when it was outside. He asked his friend Joseph Priestley to repeat the experiment.

Priestley verified Franklin's results and went on to reach a brilliant conclusion from them. He remembered from Newton's *Principia* that gravitational forces behave in a similar way. Inside a hollow planet, the net gravitational force on an object (the sum of all the forces exerted by all parts of the planet) would be exactly zero. This result also follows mathematically from the law that the gravitational force between any two individual pieces of matter is inversely proportional to the square of the distance between them. Priestley therefore proposed that forces exerted by charges vary inversely as the square of the distance, just as do forces exerted by massive bodies. (Zero force inside a hollow conductor is discussed on page 431.) The force exerted between bodies owing to the fact that they are charged is called "electric" force, just as the force between uncharged bodies is called "gravitational" force. (Remember that all forces are known by their mechanical effects, by the push or acceleration they cause on material objects!)

Priestley's proposal was based on reasoning by analogy, that is, by reasoning from a parallel, well-demonstrated case. Such reasoning alone could not *prove* that electrical forces are inversely proportional to the square of the distance between charges. But it strongly encouraged other physicists to test Priestley's hypothesis by experiment.

The French physicist Charles Coulomb provided direct experimental evidence for the inverse-square law for electric charges suggested by Priestley. Coulomb used a *torsion balance* which he had invented. A diagram of the balance appears on the following page. A horizontal, balanced insulating rod is suspended by a thin silver wire. The wire twists when a force is exerted on the end of the rod, and the twisting effect can be used as a measure of the force.

Coulomb attached a charged body, A, to one end of the rod and placed another charged body, B, near it. The electrical force exerted on A by B caused the wire to twist. By measuring the twisting effect for different separations between the centers of spheres A and B, Coulomb found that the force between spheres varied in proportion to $1/R^2$, where R represents the distance between the centers:

Joseph Priestley (1773–1804), a Unitarian minister and physical scientist, was persecuted in England for his radical political ideas. One of his books was burned, and a mob looted his house because of his sympathy with the French Revolution. He moved to America, the home of Benjamin Franklin, who had stimulated Priestley's interest in science. Primarily known for his identification of oxygen as a separate element that is involved in combustion and respiration, he also experimented with electricity. In addition, Priestley can claim to be the developer of carbonated drinks (soda-pop).

Charles Augustin Coulomb (1738–1806) was born into a family of high social position and grew up in an age of political unrest. He studied science and mathematics and began his career as a military engineer. While studying machines, Coulomb invented his torsion balance, with which he carried out intensive investigations on the mechanical forces caused by electrical charges. These investigations were analogous to the work of Cavendish on gravitation (see Unit 2).

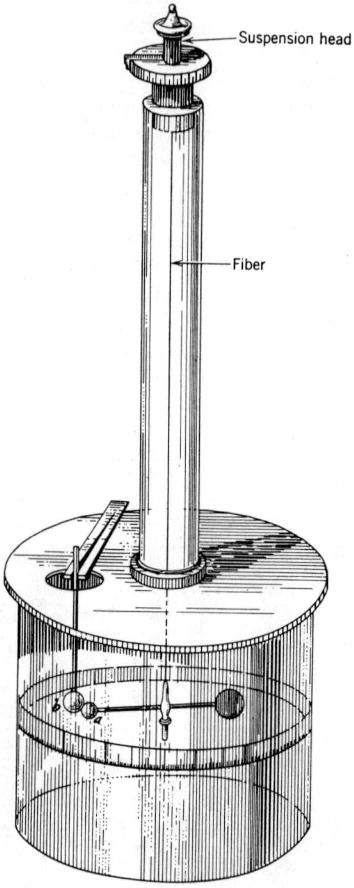

Coulomb's torsion balance.

$$F_{el} \propto \frac{1}{R^2}$$

Thus, he directly confirmed Priestley's suggestion. *The electric force of repulsion for like charges, or attraction for unlike charges, varies inversely as the square of the distance between charges.*

Coulomb also demonstrated how the magnitude of the electric force depends on the magnitudes of the charges. There was not yet any accepted method for measuring quantitatively the amount of charge on an object. (In fact, nothing said so far would suggest how to measure the magnitude of the charge on a body.) Yet Coulomb used a clever technique based on symmetry to compare the effects of different amounts of charge. He first showed that if a charged metal sphere touches an uncharged sphere of the same size, the second sphere becomes charged also. You might say that, at the moment of contact between the objects, some of the charge from the first "flows" or is "conducted" to the second. Moreover, after contact has been made, the two spheres are found to share the original charge *equally*. (This is demonstrated by the observable fact that they exert equal forces on some third charged body.) Using this principle, Coulomb started with a given amount of charge on one sphere. He then shared this charge by contact among several other identical but uncharged spheres. Thus, he could produce charges of one-half, one-quarter, one-eighth, etc., of the original amount. In this way, Coulomb varied the charges on the two original test spheres independently and then measured the change in force between them. He found that, for example, when the charges on the two spheres are both reduced by one-half, the force between the spheres is reduced to one-quarter its previous value. In general, he found that the magnitude of the electric force is proportional to the *product* of the charges. The symbols q_A and q_B can be used for the net charge on bodies A and B. The magnitude F_{el} of the electric force that each exerts on the other is proportional to $q_A \times q_B$, and may be written as $F_{el} \propto q_A q_B$.

Coulomb summarized his results in a single equation that describes the electric forces two small charged spheres A and B exert on each other:

$$F_{el} = k\frac{q_A q_B}{R^2}$$

R represents the distance between the centers, and k is a constant whose value depends on the units of charge and length that are used. This form of the law of force between two electric charges is now called Coulomb's law. The value of k is discussed below. For the moment, note one striking fact about Coulomb's law: It has exactly the same form as Newton's law of universal gravitation! Yet these two great laws arise from completely different sets of observations and apply to completely different kinds of phenomena. Why they should match so exactly is to this day a fascinating puzzle.

The unit of charge. Coulomb's law can be used to define a unit of charge. For example, assign k a value of exactly 1. Then define a unit charge so that two unit charges separated by a unit distance exert a unit force on each other. There actually is a set of units based on this choice. However, another system of electrical units, the "MKSA" system, is more convenient to use. In this system, the unit of charge is derived not from electrostatics, but from the unit of current—the *ampere* (A). (This will be discussed in Sec. 14.12.) The unit of charge is called the *coulomb* (C). It is defined as the amount of charge that flows past a point in a wire in 1 sec when the current is equal to 1 A. In Sec. 14.6, you will see that 1 C corresponds to the charge of $1/1.6 \times 10^{19}$ electrons.

The *Project Physics* documentary film "People and Particles" shows an experiment designed to demonstrate whether Coulomb's law applies to charges at distances as small as 10^{-15} cm. (It does.)

Meter–Kilogram–Second–Ampere

The ampere (A), or "amp," is a familiar unit frequently used to describe the current in electrical appliances. The effective amount of current in a common 100-watt light bulb is approximately 1 A. Therefore, the amount of charge that goes through the bulb in 1 sec is about 1 C. It might seem that a coulomb is a fairly small amount of charge. However, 1 C of *net* charge collected in one place is unmanageably large! In the light bulb, 1 C of negative charge moves through the filament each second. However, these negative charges are passing through a more or less stationary arrangement of *positive* charges in the filament. Thus, the *net* charge on the filament is zero.

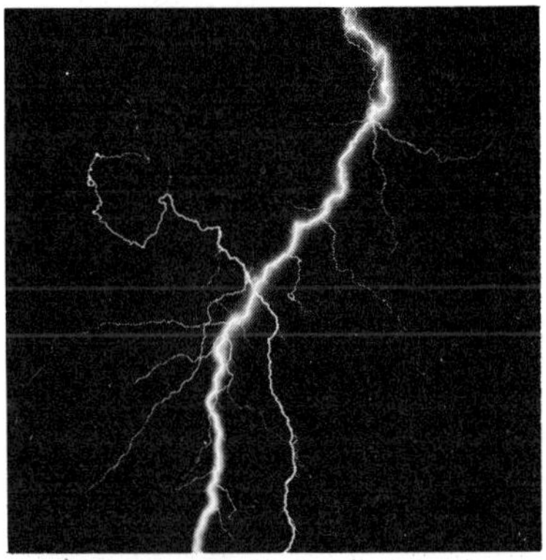

A stroke of lightning is, on the average, about 40,000 amperes and transfers about 50 coulombs of charge between the cloud and the ground.

Taking the coulomb (1 C) as the unit of charge, you can find the constant k in Coulomb's law experimentally. Simply measure the force between known charges separated by a known distance. The value of k turns out to equal about 9 billion newton-meters squared per coulomb squared (9×10^9 N-m^2/C^2). So two objects, each with a *net* charge of 1 C, separated by a distance of 1 m, would exert forces on each other of 9 billion N. This force is roughly the same as a weight of 1 million tons! We never observe such large forces because we cannot actually collect so much net charge in one place. Nor can we exert enough force to bring two such charges so close together. The mutual repulsion of like charges is so strong that it is difficult to keep a charge of more than a thousandth of a coulomb on an object of ordinary size. If you rub a pocket comb on your sleeve enough to produce a spark when the comb is brought near a conductor (such as a sink faucet), the net charge on the comb will be far less than 1 millionth of a coulomb. Lightning discharges usually take place when a cloud has accumulated a net charge of a few hundred coulombs distributed over its very large volume.

SG 3

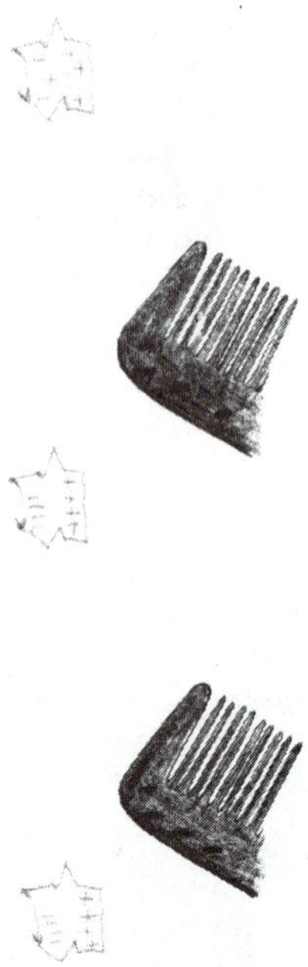

Electrostatic induction. As noted, and as you have probably observed, an electrically charged object can often attract small pieces of paper. But the paper itself has no net charge; it exerts no force on other pieces of paper. At first sight then, its attraction to the charged object might seem to contradict Coulomb's law. After all, the force ought to be zero if either q_A or q_B is zero. To explain the attraction, recall that uncharged objects contain equal amounts of positive and negative electric charges. When a charged body is brought near a neutral object, it may rearrange the positions of some of the charges in the neutral object. The negatively charged comb does this when held near a piece of paper. Some of the positive charges in the paper shift toward the side of the paper nearest the comb, and a corresponding amount of negative charge shifts toward the other side. The paper still has no *net* electric charge. But some of the positive charges are slightly *closer* to the comb than the corresponding negative charges are. So the attraction to the comb is greater than the repulsion. (Remember that the force gets weaker with the square of the distance, according to Coulomb's law. The force would be only one fourth as large if the distance were twice as large.) In short, there is a net attraction of the charged body for the neutral object. This explains the old observation of the effect rubbed amber had on bits of grain and the like.

A charged body *induces* a shift of charge in or on the nearby neutral body. Thus, the rearrangement of electric charges inside or on the surface of a neutral body caused by the influence of a nearby charged object is called *electrostatic induction.* In Chapter 16, you will see how the theory of electrostatic induction played an important role in the development of the theory of light.

?

4. In the following sentences, underline the words or phrases that do not simply describe observable facts, but that have been specifically "invented" to help understand such observations.

(a) Like charges repel each other. A body that has a net positive charge repels any body that also has a net positive charge; that is, two glass rods that have both been rubbed will tend to repel each other. A body that has a net negative charge repels any other body that also has a net negative charge.

(b) Unlike charges attract each other. A body that has a net positive charge attracts any body that has a net negative charge and vice versa.

5. What experimental fact led Priestley to propose that electrical forces and gravitational forces change with distance in a similar way?

Close Up

Electric Shielding

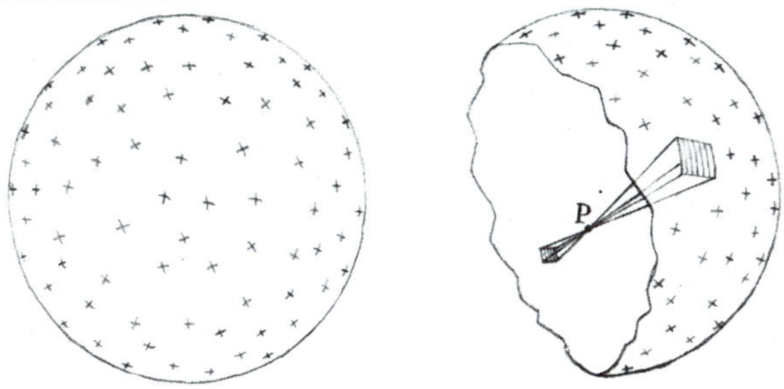

Consider any point charge P inside an even, spherical distribution of charges. For any small patch of charges with total charge Q_1 on the sphere there is a corresponding patch on the other side of P with total charge Q_2. But the areas of the patches are directly proportional to the squares of the distances from P. Therefore, the total charges Q_1 and Q_2 are also directly proportional to the squares of the distances from P. The electric field due to each patch of charge is proportional to the area of the patch, and inversely proportional to the square of the distance from P. So the distance and area factors cancel. The forces on P due to the two patches at P are exactly equal in magnitude. But the forces are also in opposite directions. So the net force on P is zero owing to Q_1 and Q_2. Since this is true for all pairs of charge patches, the net electric field at P is zero.

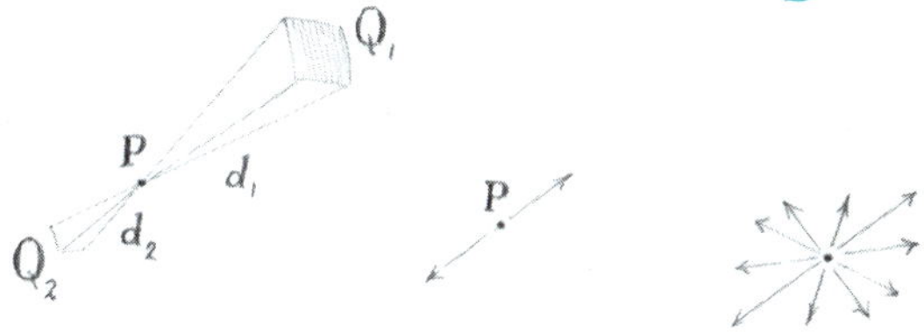

In general, charges on a closed conducting surface arrange themselves so that the electric force inside is zero just as they do on a sphere as shown in the diagrams above. Even if the conductor is placed in an electric field, the surface charges will rearrange themselves so as to keep the net force zero everywhere inside. Thus, the region inside any closed conductor is "shielded" from any *external* electric field. This is a very important practical principle.

Whenever stray electric fields might disturb the operation of some electric equipment, the equipment can be enclosed by a shell of conducting material. Some uses of electric shielding can be seen in the photographs of the back of a TV receiver, below.

Closeup of a tube in the tuning section of the TV set on the left. Surrounding the tube is a collapsible metal shield. Partly shielded tubes can be seen elsewhere in that photo.

A section of shielded cable.

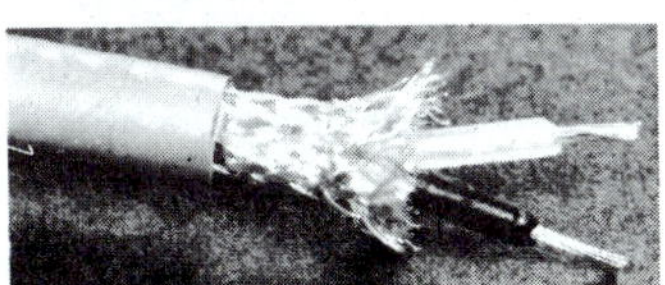

6. What two facts about the force between electric charges did Coulomb demonstrate?

SG 4 *7. If the distance between two charged objects is doubled, how is the electrical force between them affected? How is the force affected if the charge on one of them is cut to one-quarter its former size?*

SG 5 *8. Are the coulomb and ampere both units of charge?*

9. State the force law, tell the direction of the force between two charged particles, and give the value of k. Verify that a force of 9,000,000,000 N (9×10^9 N) exists between two particles with charges of 1 C each if the particles are 1 m apart.

14.4 | Forces and fields

Gilbert described the action of the lodestone by saying it had a "sphere of influence" surrounding it. He meant that any other magnetic body coming inside this sphere would be attracted. In addition, the strength of the attractive force would be greater at places closer to the lodestone. In modern language, you would say that the lodestone is surrounded by a *magnetic field.*

The word "field" is used in many ways. Here, some familiar kinds of fields will be discussed, and then the idea of physical fields as used in science will be gradually developed. This exercise should remind you that most terms in physics are really adaptations, with important changes, of commonly used words. Velocity, acceleration, force, energy, and work are examples you have already encountered in this course.

One ordinary use of the concept of field is illustrated by the "playing field" in various sports. The football field, for example, is a place where teams compete according to rules that confine the important action to the area of the field. "Field" in this case means a *region of interaction.*

In international politics, people speak of "spheres" or "fields" of influence. A field of political influence is also a region of interaction. But unlike a playing field, it has no sharp boundary line. A country usually has greater influence on some countries and less influence on others. So in the political sense, "field" refers also to an *amount* of influence, more in some places and less in others. Moreover, the field has a *source,* that is, the country that exerts the influence.

There are similarities here to the concept of field as used in physics. But there is also an important difference. To define a field in physics, it must be possible to assign a numerical value of field strength to every point in the field. This part of the field idea will become clearer if you consider some situations that are more directly related to the study of physics. First think about these situations in everyday language, then in terms of physics.

Situation	*Description of Your Experience*
(a) You are walking along the sidewalk toward a street lamp at night.	"The brightness of light is increasing."
(b) You stand on the sidewalk as an automobile moves down the street with its horn blaring.	"The sound gets louder and then softer."
(c) On a hot summer day, you walk barefoot out of the sunshine and into the shade on the sidewalk.	"The sidewalk is cooler here than in the sunshine."

You can also describe these experiences in terms of fields:

(a) The street lamp is surrounded by a field of illumination. The closer you move to the lamp, the stronger is the field of illumination as registered on your eye or on a light meter you might be carrying. For every point in the space around the street lamp, you could assign a number that represents the strength of the field of illumination at that place.

(b) The automobile horn is surrounded by a sound field. You are standing still in your frame of reference (the sidewalk). A pattern of field values goes past you with the same speed as the car. You can think of the sound field as steady but moving with the horn. At any instant, you could assign a number to each point in the field to represent the intensity of sound. At first the sound is faintly heard as the weakest part of the field reaches you. Then the more intense parts of the field go by, and the sound seems louder. Finally, the loudness diminishes as the sound field and its source (the horn) move away.

(c) In this case, you are walking in a temperature field. This field is intense where the sidewalk is in the sunshine and weaker where it is in the shade. Again, you could assign a number to each point in the field to represent the temperature at that point.

Notice that the first two fields are each produced by a single source. In (a) the source is a stationary street lamp; in (b) it is a moving horn. In both cases the field strength gradually increases as your distance from the source decreases. But in the third case (c) the field is produced by a complicated combination of influences: the sun, clouds in the sky, the shadow cast by nearby buildings, and other factors. Yet the description of the field itself is just as simple as that of a field produced by a single source. One numerical value is associated with each point in the field.

So far, all examples have been simple *scalar* fields. No direction has been involved in the value of the field at each point. On the next page are maps of two fields for the layer of air

Note that meteorologists have a convention for representing vectors different from the one we have been using. What are the advantages and disadvantages of each?

Close Up

Pressure and Velocity Fields

These maps, adapted from those of the U.S. Weather Bureau, depict two fields, air pressure at the earth's surface and high-altitude wind velocity, for two successive days. Locations at which the pressure is the same are connected by lines. The set of such pressure "contours" represents the overall field pattern. The wind velocity at a location is indicated by a line (showing direction) and feather lines—one for every 10 mph. (The wind velocity over the tip of Florida, for example, is a little to the east of due north and is approximately 30 mph.)

Air pressure at the earth's surface | High-altitude wind velocity

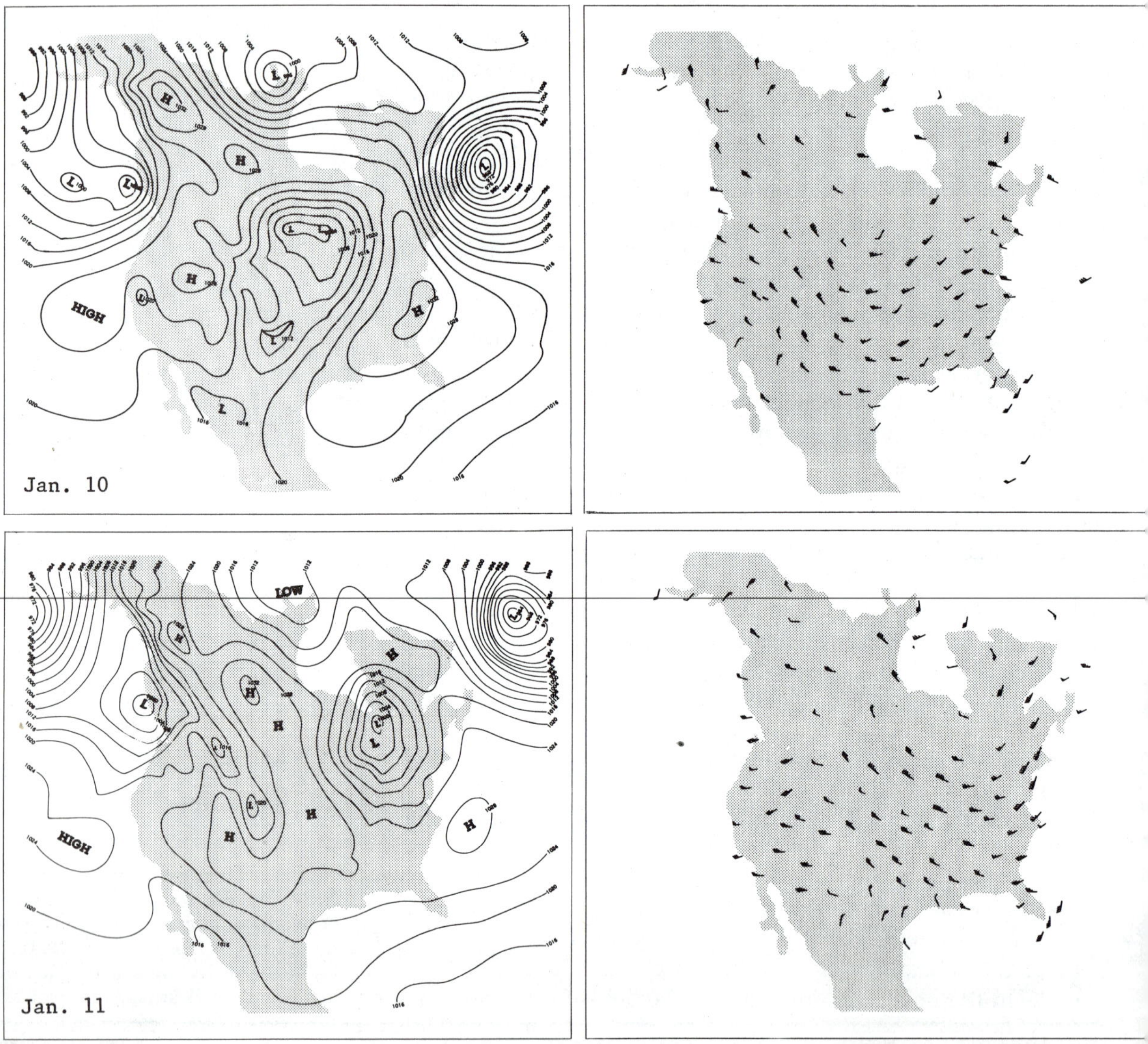

over North America on two consecutive days. There is a very important difference between the field mapped at the left and that mapped at the right. The air pressure field (on the left) is a scalar field; the wind velocity field (on the right) is a vector field. For each point in the pressure field, a single number (a scalar quantity) gives the value of the field at that point. But for each point in the wind velocity field, the value of the field is given by both a numerical value (magnitude) and a *direction,* that is, by a vector.

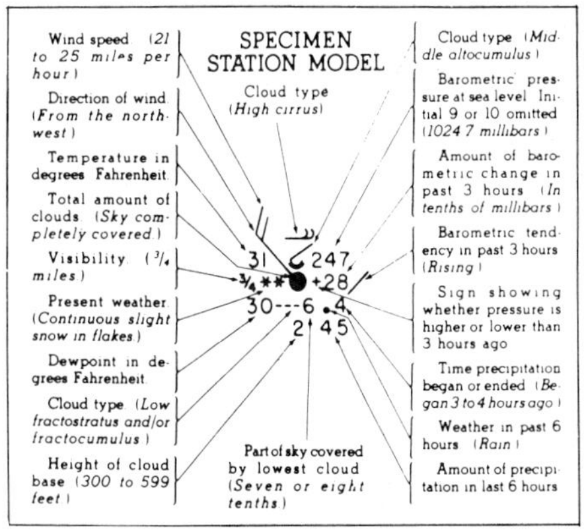

Key for a U.S. Weather Bureau map.

These field maps can help in more or less accurately predicting what conditions might prevail in the field on the next day. Also, by superimposing the maps for pressure and wind velocity, you can discover how these two kinds of fields are related to each other.

Physicists actually use the term "field" in three different senses: (1) the value of the field *at a point* in space; (2) the set or collection of all values everywhere in the space where that field exists; (3) the region of space in which the field has values other than zero. In reading the rest of this chapter, you will not find it difficult to decide which meaning applies each time the term is used.

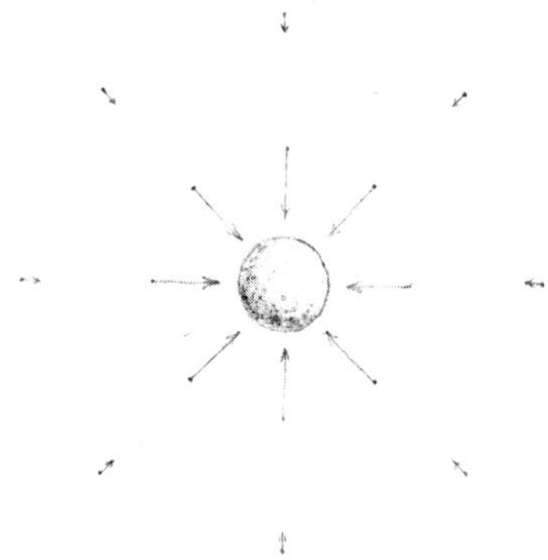

The gravitational force field. Before returning to electricity and magnetism, let us illustrate a bit further the idea of a field. A good example is the gravitational force field of the earth. Recall that the force $\vec{F}_{grav}$ exerted by the earth on any object above its surface acts in a direction toward the center of the earth. So the field of force of gravitational attraction is a *vector* field, which can be represented by arrows pointing toward the center of the earth. In the illustration in the margin, a few such arrows are shown, some near, some far from the earth.

The strength, or numerical magnitude, of the earth's gravitational force field at any chosen point depends on the distance of the point from the center of the earth. This follows from Newton's theory, which states that the magnitude of the gravitational attraction is inversely proportional to the square of the distance R:

$$F_{grav} = G \times \frac{Mm}{R^2}$$

where M is the mass of the earth, m is the mass of the test body, R is the distance between the centers of earth and the test body, and G is the universal gravitational constant.

In this equation, F_{grav} also depends on the mass of the test body. It would be more convenient to define a field that depends only on the properties of the source, whatever the mass of the test body. Then you could think of the field as existing in space and having a definite magnitude and direction at every point.

The mass of the test body would not matter. In fact, it would not matter whether there were any test body present at all. As it happens, such a field is easy to define. By slightly rearranging the equation for Newton's law of gravitation, you can write:

$$F_{grav} = m\left(\frac{GM}{R^2}\right)$$

Then, define the gravitational field strength $\vec{g}$ around a spherical body of mass M as having a magnitude GM/R^2 and a direction the same as the direction of $\vec{F}_{grav}$, so that:

$$\vec{F}_{grav} = m\vec{g}$$

where $g = GM/R^2$. Thus, note that $\vec{g}$ at a point in space is determined by the source mass M and the distance R from the source, and does *not* depend on the mass of any test object.

The total or net gravitational force at a point in space is usually determined by more than one source. For example, the moon is acted on by the sun as well as by the earth and to a smaller extent by the other planets. In order to define the field resulting from any configuration of massive bodies, take $\vec{F}_{grav}$ to be the *net* gravitational force due to *all* sources. Then *define* $\vec{g}$ in such a way that you can still write the simple relationship $\vec{F}_{grav} = m\vec{g}$; that is, define $\vec{g}$ by the equation:

SG 4

$$\vec{g} = \frac{\vec{F}_{grav}}{m}$$

Thus, the gravitational field strength at any point is the *ratio* of the net gravitational force $\vec{F}_{grav}$ acting on a test body at that point to the mass m of the test body.

Recall that F_{el} is *called* an "electric" force because it is caused by the presence of charges. But, as with all forces, we know it exists and can measure it only by its mechanical effects on bodies.

Electric fields. The strength of any force field can be defined in a similar way. According to Coulomb's law, the electric force exerted by one relatively small charged body on another depends on the product of the *charges* of the two bodies. Consider a charge q placed at any point in the electric field set up by a charge Q. Coulomb's law, describing the force F_{el} experienced by q, can be written as:

$$F_{el} = k\frac{Qq}{R^2} \quad \text{or} \quad F_{el} = q\frac{kQ}{R^2}$$

As in the discussion of the gravitational field, the expression for force here is divided into two parts. One part, kQ/R^2, depends only on the charge Q of the source and distance R from it. This part can be called "the electric field strength due to Q." The second part, q, is a property of the body being acted on. Thus, the electric field strength $\vec{E}$ due to charge Q is *defined* as having magnitude kQ/R^2 and the same direction as $\vec{F}_{el}$. The electric force is then the product of the test charge and the electric field strength:

$$\vec{F} = q\vec{E} \quad \text{and} \quad \vec{E} = \frac{\vec{F}_{el}}{q}$$

The equation *defines* $\vec{E}$ for an electric force field. Thus, the electric field strength $\vec{E}$ at a point in space is the *ratio* of the net electric force $\vec{F}_{el}$ acting on a test charge at that point to the magnitude q of the test charge. This definition applies whether the electric field results from a single point charge or from a complicated distribution of charges. The same kind of superposition principle holds which you have already seen many times. Fields set up by many sources superpose, forming a single net field. The vector specifying the magnitude of the net field at any point is simply the vector sum of the values of the fields due to each individual source. SG 7 SG 8

So far, we have passed over a complication not encountered in dealing with gravitation. There are *two* kinds of electric charge, positive (+) and negative (−). The forces they experience when placed in the same electric field are opposite in direction. By agreement, scientists define the direction of the vector $\vec{E}$ as the direction of the force exerted by the field on a *positive* test charge. Given the direction and magnitude of the field vector $\vec{E}$ at a point, then by definition the force vector $\vec{F}_{el}$ acting on a charge q is $F_{el} = q\vec{E}$. A positive charge, say +0.00001 C, placed at this point will experience a force $\vec{F}_{el}$ in the same direction as $\vec{E}$ at that point. A negative charge, say −0.00001 C, will experience a force of the same magnitude, but in the *opposite* direction. Changing the sign of q from + to − automatically changes the direction of $\vec{F}_{el}$ to the opposite direction. SG 9

?

10. *What is the difference between a scalar field and a vector field? Give examples of each.*

11. *Describe how one can find, by experiment, the magnitude and the directions of:*

(a) the gravitational field at a certain point in space.

(b) the electric field at a certain point in space.

12. *Why would the field strengths* $\vec{g}$ *and* $\vec{E}$ *for the test bodies be unchanged if* m *and* q *were doubled?*

13. *A negatively charged test body is placed in an electric field where the vector* $\vec{E}$ *is pointing downward. What is the direction of the force on the test body?* SG 10

14. *What is the electric field at a point if a test particle of* 3×10^{-5} *C experiences a force of* 10^{-2} *N upward?*

15. *What is the force on another particle of 5×10^{-3} N at the same point?*

16. *What useful simplification results when you use the field concept instead of the Coulomb force law?*

14.5 | The smallest charge

In Sec. 14.3, you read that an electrified comb can pick up a small piece of paper. Obviously, the electric force on the paper must exceed the gravitational force exerted on the paper by the earth. This observation indicates that electric forces generally are stronger than gravitational forces. Using the same principle, the gravitational force on a microscopically small object (which still contains several billion atoms) can be balanced against the electrical force on the same object when the object has a net electric charge of only a single electron! (The electron is one of the basic components of the atom. Other properties of atoms and electrons will be discussed in Unit 5.) This fact is the basis of a method for actually measuring the electron's charge. The method was first employed by the American physicist Robert A. Millikan in 1909. Millikan's experiment will be described in detail in Sec. 18.3. The basic principle is discussed here, since it provides such a clear connection between the ideas of force, field, and charge.

Millikan used fine droplets of oil from an atomizer. The droplets became charged as they formed a spray. The oil was convenient because of the low rate of evaporation of the droplet.

Suppose a small body of mass *m*, for example, a tiny drop of oil or a small plastic sphere, has a net negative electric charge of magnitude *q*. The negatively charged body is placed in an electric field *E* directed downward. A force $\vec{F}_{el}$ of magnitude qE is now exerted on the body in the *upward* direction. Of course, there is also a downward gravitational $F_{grav} = mg$ on the object. The body will accelerate upward or downward, depending on whether the electric force or the gravitational force is greater. By adjusting the magnitude of the electric field strength $\vec{E}$ (that is, by changing the source that sets up $\vec{E}$), the two forces can be balanced.

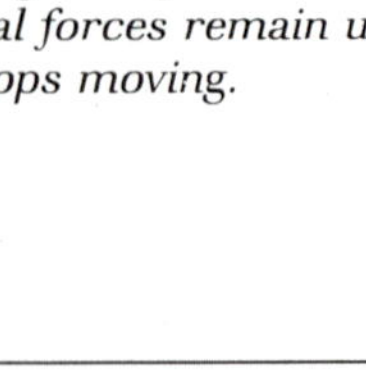

When $m\vec{g}$ and $q\vec{E}$ are balanced, frictional forces remain until the body stops moving.

What happens when the two forces are balanced? Remember that if a zero *net* force acts on a body, the body can have no acceleration; that is, it would be at rest or continue to move at some constant velocity. In this case, air resistance is also acting as long as the drop moves at all and will soon bring the drop or sphere to rest. The drop will then be in equilibrium. In fact, it will be suspended in mid-air. When this happens, the magnitude of the electric field strength $\vec{E}$ which was applied to produce this condition is recorded.

SG 11

Since now the electric force balances the gravitational force, the following must hold:

$$qE = mg$$

You can calculate the charge q from this equation if you know the quantities E, m, and g, since

$$q = \frac{mg}{E}$$

But how can m be determined?

Thus, you can find, in the laboratory, what values of charge q a very small test object can carry.

When you do this, you will discover a remarkable fact: *All possible charges in nature are made up of whole multiples of some smallest charge.* This smallest possible charge is called the *magnitude of the charge on one electron.* By repeating the experiment many times with a variety of small charges, you can find the value of the charge on one electron (q_e). In effect, this is what Millikan did. He obtained the value of $q_e = 1.6024 \times 10^{-19}$ C for the electron charge. (For most purposes you can use the value 1.6×10^{-19} C.) This value agrees with the results of many other experiments done since then. No experiment has yet revealed the existence of a smaller unit of charge. (Some physicists have speculated, however, that there might be $\frac{1}{3}q_e$ associated with a yet-to-be-found subatomic particle called the *quark.* Although new experimental evidence supports the existence of quarks, current theory predicts that quarks cannot be isolated singly but exist only in collections that have a total charge of either zero or q_e.)

The magnitude of the charge on the electron is symbolized by q_e, and its sign is negative. Any charge q is therefore given by $q = nq_e$ where n is the whole number of individual charges, each of magnitude q_e.

Therefore, 1 coulomb is the magnitude of the charge on $1/1.6 \times 10^{19}$ electrons.

SG 16–21

?

17. How can the small oil drops or plastic spheres used in the Millikan experiment experience an electric force upward *if the electric field is directed* downward*?*

18. What do the results of the Millikan experiment indicate about the nature of electric charge?

14.6 | The law of conservation of electric charge

For many centuries, the only way to charge objects electrically was to rub them. In 1663, Otto von Guericke made and described a machine that would aid in producing large amounts of charge by rubbing:

> ... take a sphere of glass which is called a phial, as large as a child's head; fill it with sulphur that has been pounded in a mortar and melt it sufficiently over a fire. When it is cooled again break the sphere and take out the globe and keep it in a dry place. If you think it best, bore a hole through it so that it can be turned around an iron rod or axle....

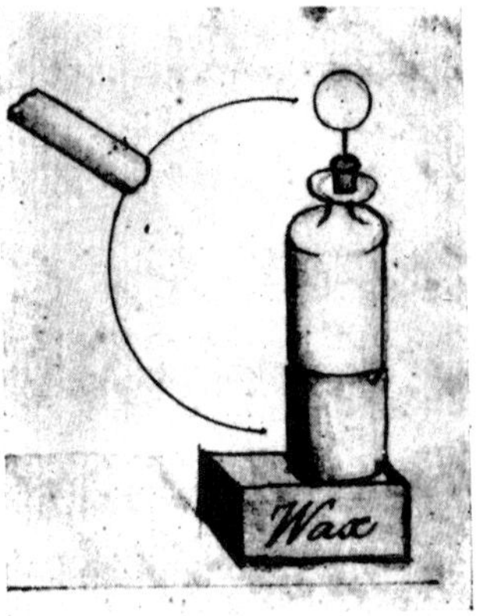

Franklin's drawing of a Leyden jar, standing on an insulating block of wax. The rod in the stopper was connected to a conducting liquid in the bottle. A charge given to the ball would hold, through the non-conducting glass wall, an equal amount of the opposite charge on the metal foil wrapped around the outside. A Leyden jar can hold a large charge because positive charges hold negative charges on the other side of a nonconducting wall.

When von Guericke rested his hand on the surface of the sulphur globe while rotating it rapidly, the globe acquired enough charge to attract small objects.

By 1750 electrical machines were far more powerful, and vigorous research on the nature of electricity was going on in many places. Large glass spheres or cylinders were whirled on axles supported by heavy wooden frames. A stuffed leather pad was sometimes substituted for the human hands. The charge on the globe was often transferred to a large metal object (such as a gun barrel) suspended nearby.

These machines were powerful enough to deliver strong electrical shocks and to produce frightening sparks. In 1746, Pieter van Musschenbroek, a physics professor at Leyden, reported on an accidental and nearly fatal discovery in a letter beginning, "I wish to communicate to you a new, but terrible, experiment that I would advise you never to attempt yourself." Musschenbroek was apparently trying to capture electricity in a bottle, for he had a brass wire leading from a charged gun barrel to a jar filled with water. A student was holding the jar in one hand while Musschenbroek cranked the machine. When the student touched the brass wire with his free hand, he received a tremendous shock. They repeated the experiment, this time with the student at the crank and Musschenbroek holding the jar. The jolt was even greater than before (the student must have been very energetic at the crank). Musschenbroek wrote later that he thought "... it was all up with me ..." and that he would not repeat the experience for the whole kingdom of France. Word of the experiment spread rapidly, and the jar came to be called a Leyden jar. In fact, Musschenbroek had inadvertently discovered that charge could be stored in a properly constructed solid object. Devices such as Leyden jars that have a capacity for storing electric charge are now called *capacitors*.

The Leyden jar came to Benjamin Franklin's attention. Franklin performed a series of experiments with it and published his analysis of its behavior in 1747. In these experiments, Franklin first showed that the effects of different kinds of charges (positive and negative) can cancel each other. Because of this cancellation, he concluded that positive and negative charges were not really different. As mentioned before, Franklin thought that *one* kind of electricity was enough to explain all phenomena. He believed that a positive charge resulted from an excess of "electric fluid" or "electric fire," and a negative charge from a shortage of it.

This view led Franklin to the powerful and correct idea that electric charge is neither created nor destroyed. Objects become positively or negatively charged by rearrangement of the electric charges already present in them. This is a matter of redistribution rather than creation. Similarly, positive and negative charges can cancel or neutralize each other's effect without being destroyed. This is the modern principle of *conservation of charge*. It is taken to be as basic a law of nature as are the conservation principles of momentum and of energy. The principle of the law of conservation of electric charge can be stated in this way: *The net amount of electric charge in a closed system remains constant, regardless of what reactions occur in the system. Net amount of charge* is defined as the difference between the amount of + and of − charge. (For example, a *net* charge of +1 C would describe 1 C of positive charge all by itself, or a combination of 11 C of positive charge and 10 C of negative charge.) If the + and − are taken as actual numerical signs, instead of only as convenient labels for two different kinds of charges, then the *net* charge can be called the *total* charge. Simply adding charges with + and − signs will then give the difference between the amounts of positive and negative charge.

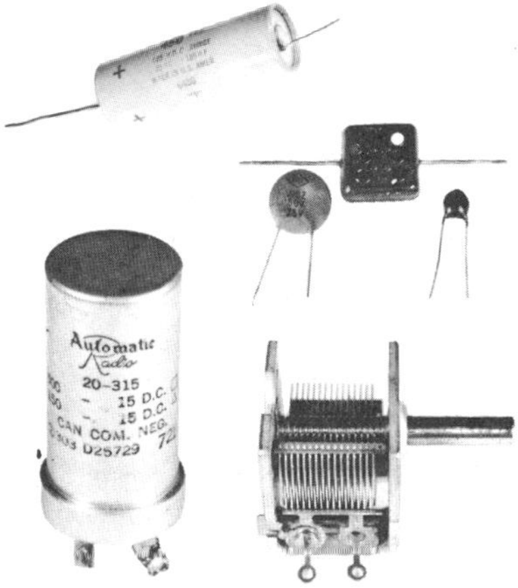

Capacitors, familiar to anyone who has looked inside a radio, are descendents of the Leyden jar. They have many different functions in modern electronics.

The law or principle of conservation of electric charge is widely useful. Its applications range from designing circuits to analyzing subatomic reactions (see the *Project Physics* Supplemental Unit "Elementary Particles"). One interesting possibility allowed by the electric charge conservation law is that charges can appear or disappear suddenly in a closed system, as long as the appearance or disappearance involves *equal* amounts of + and − charge. (An example of such a spontaneous appearance of + and − charges, in the form of a negative electron and a positron, is a central part of the experiment in the *Project Physics* film "People and Particles.")

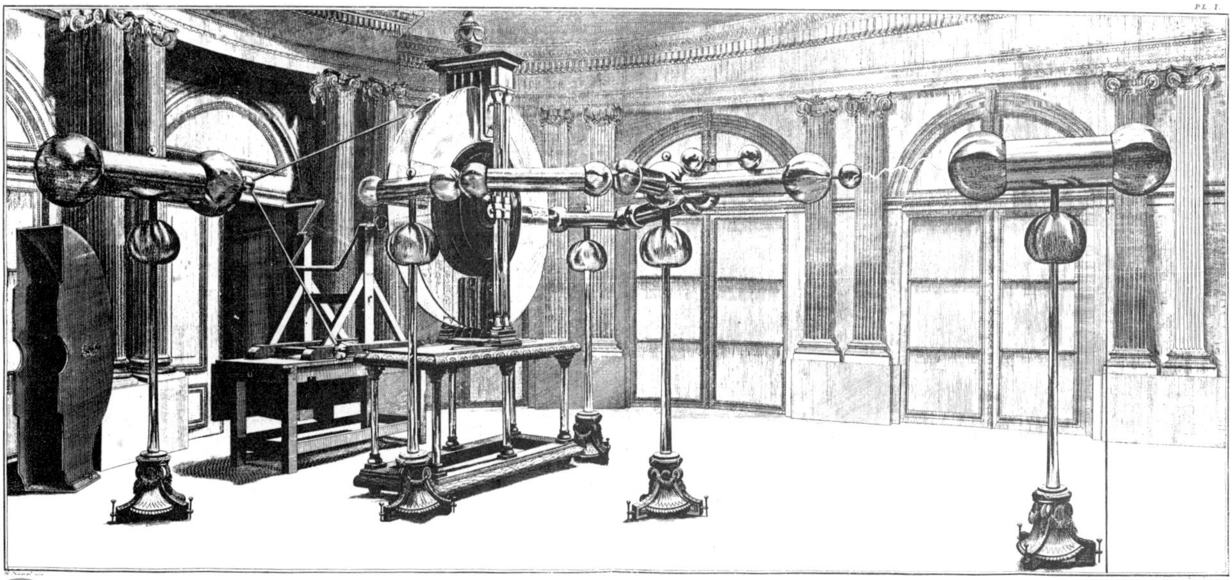

Electrostatic equipment of the 1700's.

19. What does the law of conservation of electric charge demand when, for example, a + charge appears inside a closed system?

14.7 | Electric currents

Count Alessandro Volta (1745–1827) was given his title by Napoleon in honor of his electrical experiments. He was Professor of Physics at the University of Pavia, Italy. Volta showed that the electric effects previously observed by Luigi Galvani, in experiments with frog legs, were due to the metals and not to any special kind of "animal electricity."

SG 15

Touching a charged object to one end of a chain or gun barrel will cause the entire chain or barrel to become charged. The obvious explanation is that the charges move through and spread over the object. Electric charges move easily through some materials, called *conductors*. Metal conductors were most commonly used by the early experimenters, but salt solutions and very hot gases also conduct charge easily. Other materials, such as glass and dry fibers, conduct charge hardly at all. Such materials are called nonconductors or *insulators*. Dry air is a fairly good insulator. (Damp air is not; you may have difficulty keeping charges on objects in electrostatic experiments on a humid day.) If the charge is great enough, however, even dry air suddenly will become a conductor, allowing a large amount of charge to shift through it. The heat and light caused by the sudden rush of charge produces a "spark." Sparks were the first obvious evidence of moving charges. Until late in the eighteenth century, a significant flow of charge, that is, an *electric current*, could be produced only by discharging a Leyden jar. Such currents lasted only for the brief time it took for the jar to discharge.

In 1800, Alessandro Volta discovered a much better way of producing electric currents. Volta's method involved two different metals, each held with an insulating handle. When put into contact and then separated, one metal took on a positive charge and the other a negative charge. Volta reasoned that a much larger charge could be produced by stacking up several pieces of metal in alternate layers. This idea led him to undertake a series of experiments that produced an amazing finding, reported in a letter to the Royal Society in England in March of 1800:

> Yes! the apparatus of which I speak, and which will doubtless astonish you, is only an assemblage of a number of good conductors of different sorts arranged in a certain way. 30, 40, 60 pieces or more of copper, or better of silver, each in contact with a piece of tin, or what is much better, of zinc, and an equal number of layers of water or some other liquid which is a better conductor than pure water, such as salt water or lye and so forth, or pieces of cardboard or of leather, etc., well soaked with these liquids....
>
> I place horizontally on a table or base one of the metallic plates, for example, one of the silver ones, and on this first plate I place a second plate of zinc; on this second plate I lay one of the moistened discs; then another plate of silver, followed immediately by another of zinc, on which I place again a moistened disc. I thus continue in the same way coupling a plate of silver with one of zinc, always in the same sense, that is to say, always silver below and zinc above or *vice versa*, according as I began, and inserting between these couples a

moistened disc; I continue, I say, to form from several of these steps a column as high as can hold itself up without falling.

Volta showed that one end, or "terminal," of the pile was charged positive, and the other charged negative. He then attached wires to the first and last disks of his apparatus, which he called a "battery." Through these wires, he obtained electricity with exactly the same effects as the electricity produced by rubbing amber, by friction in electrostatic machines, or by discharging a Leyden jar.

Most important of all, Volta's battery could produce a more or less *steady* electric current for a long period of time. Unlike the Leyden jar, it did not have to be charged from the outside after each use. Now the properties of electric currents as well as of static electric charges could be studied in a controlled manner. This was the device needed to start the series of inventions that have so greatly changed civilization.

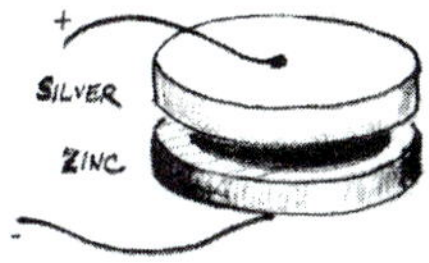

Voltaic "cell"

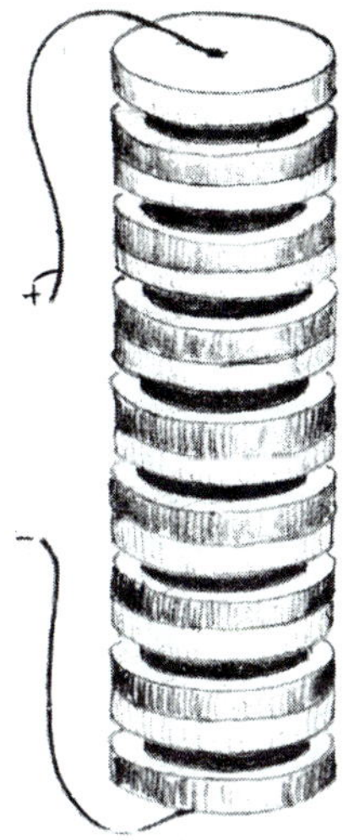

Voltaic "pile" or battery

?

20. In what ways was Volta's battery superior to a Leyden jar?

14.8 | Electric potential difference

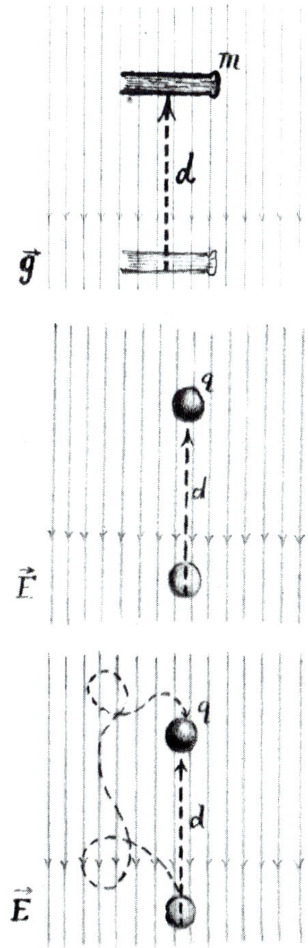

Sparks and heat are produced when the terminals of an electric battery are connected. These phenomena show that energy from the battery is being transformed into light, sound, and heat energy. The battery itself converts chemical energy to electrical energy. This, in turn, is changed to other forms of energy (such as heat) in the conducting path between the terminals. In order to understand electric currents and how they can be used to transport energy, a new concept, which has the common name *voltage*, is needed.

You learned in mechanics (Unit 3) that *change in potential energy* is equal to the work required to move an object frictionlessly from one position to another (Sec. 10.2). For example, a book's gravitational potential energy is greater when the book is on a shelf than when it is on the floor. The increase in potential energy is equal to the work done in raising the book from floor to shelf. This difference in potential energy depends on three factors: the mass m of the book, the magnitude of the gravitational field strength g, and the difference in height d between the floor and the shelf.

Similarly, the *electric* potential energy changes when work is done in moving an electric charge from one point to another in an electric field. Again, this change of potential energy $\Delta(PE)$ can be directly measured by the work that is done. The magnitude of this change in potential energy, of course, depends on the magnitude of the test charge q. Dividing $\Delta(PE)$ by q gives a quantity that does not depend on how large q is. Rather, it depends only on the intensity of the electric field and on the location of the beginning and end points. The new quantity is called *electric potential difference*. Electric potential difference is defined as *the ratio of the change in electrical potential energy $\Delta(PE)$ of a charge q to the magnitude of the charge.* In symbols,

$$V = \frac{\Delta(PE)}{q}$$

As is true for gravitational potential energy, there is no absolute zero level of electric potential energy. The *difference* in potential energy is the significant quantity. The symbol V is used both for "potential difference" as in the equation at the right, and as an abbreviation for volt, the unit of potential difference (as in 1 V = 1 J/Coul).

The units of electric potential difference are those of energy divided by charge, or joules per coulomb. The term used as the abbreviation for joules/coulomb is *volt* (V). The electrical potential difference (or *voltage*) between two points is 1 V if 1 J of work is done in moving 1 C of charge from one point to the other.

$$1 \text{ volt} = 1 \text{ joule/coulomb}$$

SG 16–21

The potential difference between two points in a steady electric field depends on the location of the points. It does *not* depend on the *path* followed by the test charge. Whether the path is short or long, direct or roundabout, the same work is done per unit charge. Similarly, a hiker does the same work per kilogram of mass in the pack against the gravitational field, whether climbing straight up or spiraling up along the slopes. Thus, the electrical potential difference between two points in a field is similar to the difference in gravitational potential energy between two points (Sec. 10.2).

A 1½-volt cell is one that has a potential difference of 1½ volts between its two terminals. (This type of cell is often called a "battery," although technically a battery is the name for a group of connected cells.)

A simple case will help you to see the great importance of this definition of potential difference. Calculate the potential difference between two points in a uniform electric field of magnitude E produced by oppositely charged parallel plates. Work must be done in moving a positive charge q from one point to the other directly against the lines of electric force. The amount of work required is the product of the force F_{el} exerted on the charge (where $F_{el} = qE$), and the distance d through which the charge is moved. Thus,

$$\Delta(PE) = qEd$$

Substituting this expression for $\Delta(PE)$ in the definition of electric potential difference gives for the simple case of a uniform field:

$$V = \frac{\Delta(PE)}{q} = \frac{qEd}{q} = Ed$$

In practice it is easier to measure electric potential difference V (with a voltmeter) than to measure electric field strength E. The relationship is often useful in the form $E = V/d$, which can be used to find the intensity of a uniform electric field.

Electric potential energy, like gravitational potential energy, can be converted into kinetic energy. A charged particle placed in an electric field, but free of other forces, will accelerate. In doing so, it will increase its kinetic energy at the expense of electric potential energy. (In other words, the electric force on the charge acts in such a way as to push it toward a region of lower potential energy.) A charge q "falling" through a potential difference V increases its kinetic energy by qV if nothing is lost by friction (as in a vacuum tube). The *increase* in kinetic energy is equal to the *decrease* in potential energy. So the sum of the two at any moment remains constant. This is just one particular case of the general principle of energy conservation, even though only electric forces are acting.

The conversion of electric potential energy to kinetic energy is used in *electron accelerators* (a common example is a television picture tube). An electron accelerator usually begins with an electron "gun." The "gun" has two basic parts: a wire and a metal can in an evacuated glass tube. The wire is heated red-hot, causing electrons to escape from its surface. The nearby can is charged positively, producing an electric field between the hot wire and the can. The electric field accelerates the electrons through the vacuum toward the can. Many electrons stick to the can, but some go shooting through a hole in one end of it. The stream of electrons emerging from the hole can be further accelerated or focused by additional cans. (You can make such an electron gun for yourself in the laboratory experiment. "Electron Beam Tube. I.") Such a beam of charged particles has a wide range of uses both in technology and in research. For example, it can make a fluorescent screen glow, as in a television picture tube or electron microscope. Or it can be used to break atoms apart, producing interesting particles for study, or X rays for medical purposes or research. When moving through a potential difference of 1 V, an electron with a charge of 1.6×10^{-19} C increases its kinetic energy by 1.6×10^{-19} J. This amount of energy is called an *electron volt*, abbreviated eV. Multiples are 1 KeV (= 1,000 eV), 1 MeV (= 10^6 eV), and 1 BeV (= 10^9 eV). Energies of particles in accelerators are commonly expressed in such multiples. In a television tube, the electrons in the beam are accelerated across an electric potential difference of about 20,000 V. Thus, each electron has an energy of about 20 KeV. The largest accelerator now operating gives (for research purposes) charged particles with kinetic energies of about 1,000 GeV.

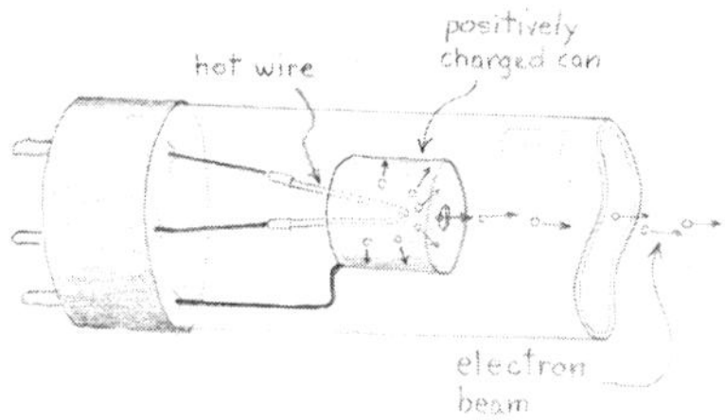

Electrically charged particles (electrons) are accelerated in an "electron gun" as they cross the potential difference between a hot wire (filament) and a can in an evacuated glass tube.

Particle accelerators come in a wide variety of shapes and sizes. They can be as common as a 1,000-volt tube in an oscilloscope or 20,000-volt TV "guns," or as spectacular as the one shown below. (Or see the Cambridge Electron Accelerator, which was the scene for two *Project Physics* films, "People and Particles" and "Synchrotron.")

---

k : kilo- (10^3)

M : mega- (10^6)

B : billion (10^9)

(B is often replaced by G : giga-)

? *21. How is the electric potential difference, or "voltage," between two points defined?*

SG 22

22. *Does the potential difference between two points depend on the path followed in taking a charge from one to the other? Does it depend on the magnitude of the charge moved?*

23. *Is the electron volt a unit of (a) charge, (b) potential difference (voltage), or (c) something else? If (c), what is it?*

Above left; A section of the evacuated tube through which the electrons travel. The electrons are accelerated in steps by electric fields in a long line of accelerating cavities, similar to those in the photograph on page 422.

Above right: The site of Stanford University's 3.2-km electron accelerator, in which electrons are given kinetic energies of up to 20 BeV.

14.9 | Electric potential difference and current

The acceleration of an electron in a vacuum by an electric field is the simplest example of a potential difference affecting a charged particle. A more familiar example is electric current in a metal wire. In this arrangement, the two ends of the wire are attached to the two terminals of a battery. Chemical changes inside a battery produce an electric field that continually drives charges to the terminals, one charged negatively, the other positively. The "voltage" of the battery tells how much energy per unit charge is available when the charges move in any *external* path from one terminal to the other along the wire, for example.

In metallic conductors, the moving charge is the negative electron, with the positive "mother" atom fixed. But all effects are the same as if positive charges were moving in the opposite direction. By an old convention, the latter is the direction usually chosen to describe the direction of current.

Electrons in a metal do not move freely as they do in an evacuated tube, but continually interact with the metal atoms. If the electrons were really completely free to move, a constant voltage would make them *accelerate* so that the current would increase with time. This does not happen. A simple relation between current and voltage first found by Georg Wilhelm Ohm is at least approximately valid for most metallic conductors: *The total current I in a conductor is proportional to the potential difference V applied between the two ends of the conductor.* Using the symbol I for the current and V for the potential difference,

$$I \propto V$$

or

$$I = constant \times V$$

This simple relation is called *Ohm's law.* It is usually written in the form

$$I = \frac{V}{R}$$

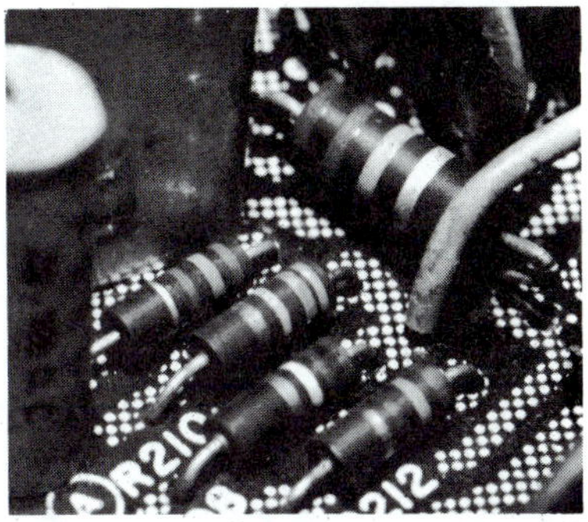

Close-up of part of the electric circuit in the TV set pictured on page 431. These "resistors" have a fairly constant voltage-to-current ratio. (The value of the ratio is indicated by colored stripes, each color standing for a number: 0, 1, 2, . . . 9.)

where R is a constant called the *resistance* of the conducting path. Thus, Ohm's law assumes that the resistance of a given conducting path does not depend on current or voltage. Resistance *does* depend on the material and dimensions of the path, such as the length and diameter of a wire. Resistance is not strictly constant for any conducting path; it varies with changes in temperature, for example.

SG 23

Ohm's law applies closely enough for practical technical work. But it does not have the general validity of the law of universal gravitation or Coulomb's law. In this course, you will use it mainly in lab work and in connection with electric light bulbs and power transmission in Chapter 15.

?

24. How does the current in a metallic conductor change if the potential difference between the ends of the conductor is doubled?

25. What does it mean to say a resistor has a resistance of 5 megohms *(5×10^6 ohms)?*

26. How would you test whether Ohm's law applies to a given piece of wire?

14.10 | Electric potential difference and power

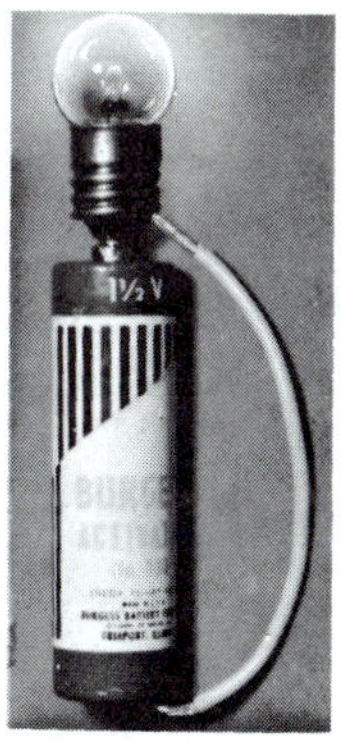

Suppose a charge could move freely from one terminal to the other in an evacuated tube. The work done on the charge would then simply increase the kinetic energy of the charge. However, a charge moving through some material such as a wire transfers energy to the material by colliding with atoms. Thus, at least some of the work goes into heat energy. A good example of this process is a flashlight bulb. A battery forces charges through the filament wire in the bulb. The electric energy carried by the charges is converted to heat energy in the filament. The hot

Close Up

Electrical Conduction in Metals

While charges cannot move freely through an insulator, they can move freely through a conductor. Yet when a conductor (say, a piece of copper wire or a steel knife blade) is connected between the two terminals of a battery, a steady current starts immediately and persists until the battery is discharged. This is puzzling. The battery sets up a potential difference between the two ends of the conductor and so there is an electric field along the conductor. This means that there is an electrical force on the charges. If this were the net force on the charges, they would be moving faster and faster. In that case, the current should increase with time, a situation not at all like what actually happens.

An acceptable model for a conductor must be a little more complex, then, than a substance "through which charge can move freely." One of the first useful models for a conductor (and one which is still used today) was constructed around 1900 by Drude and Lorentz. They pictured the atoms of a perfect crystal of metal locked into position in a regular array (called a lattice). Each atom has one or more electrons (depending on the metal) that are shared with all the other atoms in the metal. These mobile electrons are always in random motion at very high speeds (roughly 10^6 m/sec for copper), very much like the molecules of a gas studied in Unit 3. The electrons' motion is *much* faster, though, than that of the gas molecules at the same temperature (the reason for this was not discovered until about 1930 when quantum mechanics was applied to the problem).

An electric current exists where there is *net* flow of charge along the wire. As long as the electrons are moving at random, the net flow is zero on the average. The electrons are constantly experiencing collisions with any metal atom which gets "out of line," for example, impurities in the metal or imperfections in the lattice, and with vibrations of the atoms caused by their own random thermal motion. On the average, an electron travels freely for a time t between consecutive collisions (for copper, this time t is about 10^{-14} sec).

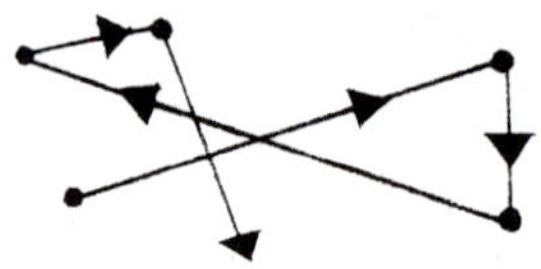

Path of an electron.

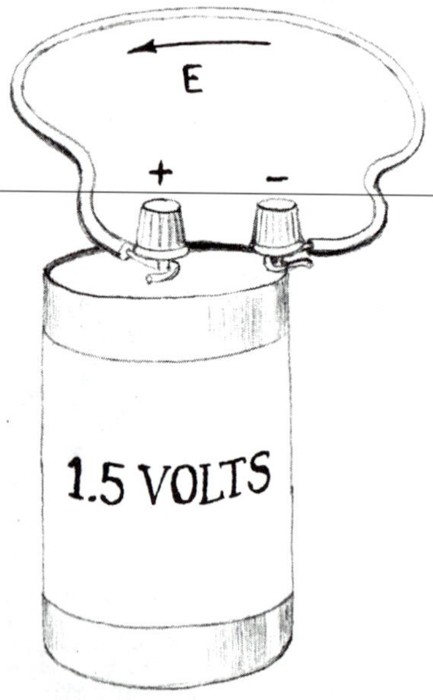

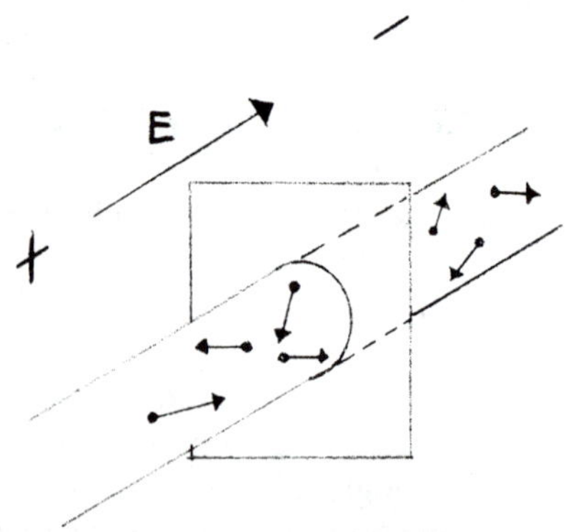

No net flow of electrons past the surface.

When a battery is connected to the metal, there is an electric field E created along the length of the conductor. This field does indeed accelerate the electrons, but since they move freely only for a time t, the change in their velocity caused by the field is just

$$\Delta\vec{v} = \vec{a}\,t = \vec{E}\frac{q_e}{m}t$$

This *additional* velocity imparted to the electrons is called the "drift velocity" and is responsible for the conduction of electricity. Since $\vec{E}$ is proportional to the battery's voltage, it is easy to see that the current will be proportional to the voltage (Ohm's law) so long as the average time between collisions, t, does not change. For example, when a metal is cooled, the thermal motion of the atoms is reduced and collisions with these thermal vibrations become less frequent. Therefore, cooling a metal makes it a better conductor. Similarly, a very pure sample of copper is a better conductor than a sample with many impurities from which electrons are scattered as they move. A more quantitative model can also be described (though that is not necessary to understanding the basic model). Picture a piece of wire of length L, cross-sectional area A, with an average of n electrons in each cubic centimeter.

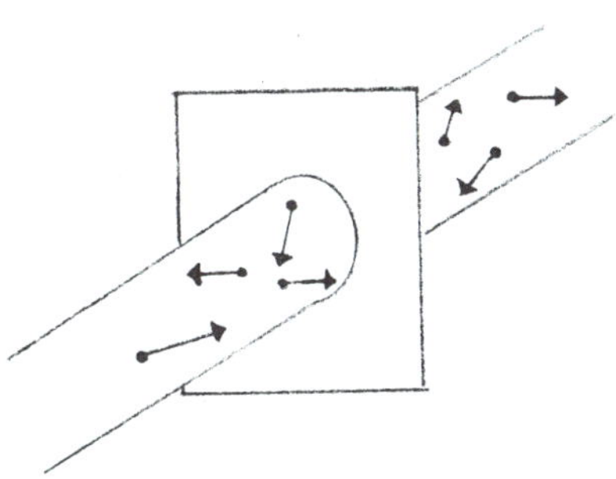

A net flow of electrons past the surface.

Ignore the *random* motion of the electrons, since this makes no contribution to the conduction, and picture all the electrons moving with the drift velocity.

$$\vec{v}_d = \Delta\vec{v} = \vec{E}\frac{q_e}{m}t$$

The current is just the amount of charge crossing the surface each second:

$$I = \begin{pmatrix}\text{(no. of electrons)}\\ \text{crossing surface}\\ \text{in 1 sec}\end{pmatrix} \times q_e$$

The number of electrons crossing the surface each second is nAv_d (just as you calculated in Chapter 11 for gas molecules). Thus,

$$I = \left(\frac{n\,q_e^2 t_A}{m}\right)E$$

But $E = V/L$ if the wire is uniform so that the field is a constant along its length, and

$$I = \left(\frac{n\,q_e^2 t}{m}\frac{A}{L}\right)V$$

V, i.e., $I \propto v$. But this is Ohm's law! Thus, this model determines the resistance of a wire as

$$R = \left(\frac{m\,L}{n\,q_e^2\,t_A}\right)$$

where $R = v/I$. It follows that, for a given material, doubling the length should double the resistance; doubling the cross-sectional area should halve the resistance. This is just what is found experimentally.

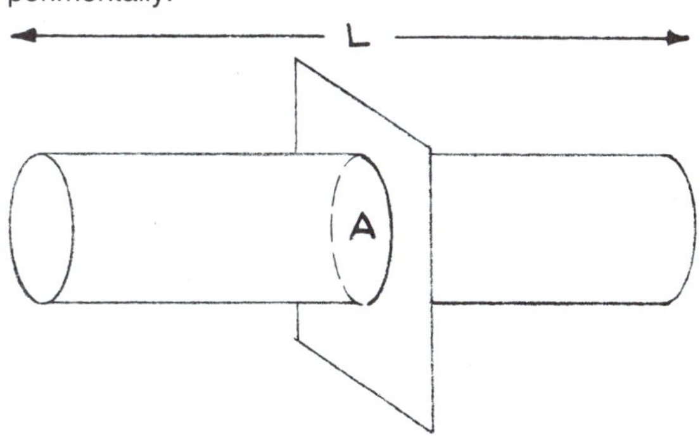

Example: A small flashlight bulb connected to a 1.5-V cell will have a current of about 0.1 A in its filament. At what rate is electric work being done to heat the filament in the bulb?

$$P = VI$$
$$= 1.5 \text{ V} \times 0.1 \text{ A}$$
$$= 0.15 \text{ W}$$

(Only a small fraction of this power goes into the visible light energy radiated from the filament.)

filament in turn radiates energy, a small fraction of which is in the form of visible light. Recall now that "voltage" (electric potential difference) is the amount of *work* done per unit of charge transferred. So the product of voltage and current gives the amount of *work* done per unit *time*:

$$V \text{ (joules/coulomb)} \times I \text{ (coulombs/sec)} = VI \text{ (joules/sec)}$$

Work done per unit time is called *power* (as defined in Sec. 10.6 of Unit 3). The unit of power, equal to 1 J/sec, is called a *watt* (W). Using the definition of ampere (1 C/sec) and volt (1 J/C), the equation for power P is

$$P \text{ (watts)} = V \text{ (volts)} \times I \text{ (amperes)}$$

What energy transformation does this work accomplish? As the positive charge moves to a lower potential level, it does work against material by colliding with atoms. The electric energy of the charge is converted to heat energy. If V is the voltage between the two ends of some material carrying a current I, the power converted to heat in the material is given by $P = VI$. This can be expressed equally well in terms of the resistance of the material, substituting IR for V:

SG 24–27

$$P = IR \times I$$
$$P = I^2R$$

Thus, *the heat produced by a current is proportional to the square of the current*. Joule was the first to find this relationship experimentally. The discovery was part of his series of researches on conversion of different forms of energy (Sec. 10.7). The fact that heat production is proportional to the *square* of the current is very important in making practical use of electric energy. You will learn more about this in the next chapter.

27. What happens to the electrical energy used to move charge through a conducting material?

28. How does the power converted to heat in a conductor change if the current in the conductor is doubled?

14.11 Currents act on magnets

Early in the eighteenth century, reports began to appear that lightning changed the magnetization of compass needles and made magnets of knives and spoons. Some researchers believed that they had magnetized steel needles by discharging a Leyden jar through them. These reports suggested that electricity and magnetism were closely related in some way. But the casual observations were not followed up with deliberate, planned experiments that might have led to useful concepts and theories.

None of these early reports surprised the nineteenth-century Nature Philosophers in Europe. They were convinced that all phenomena observed in nature were only different effects of a single "force." Their belief in the unity of physical forces naturally led them to expect that electrical and magnetic forces were associated or related in some way.

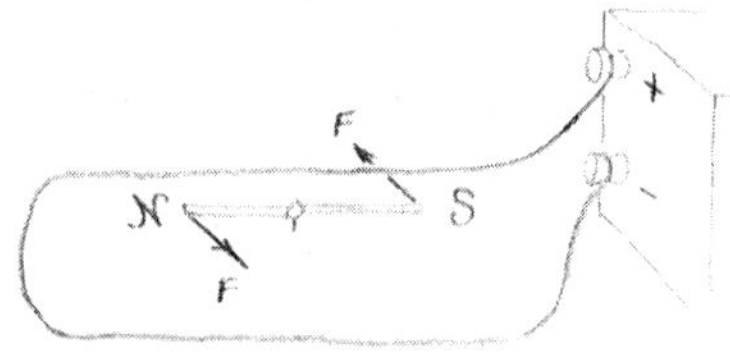

The first concrete evidence of a connection between electricity and magnetism came in 1820, when Oersted performed an extremely important series of experiments. (See illustrations on next page.) Oersted placed a magnetic compass needle directly beneath a long horizontal conducting wire. The wire lay along the earth's magnetic north–south line, so that the magnetic needle was naturally lined up parallel to the wire. When Oersted connected the wire to the terminals of a battery, the compass needle swung toward an east–west orientation, nearly perpendicular to the wire! Charge at rest does not affect a magnet. But charge in motion (a current) does exert an odd kind of force on a magnet.

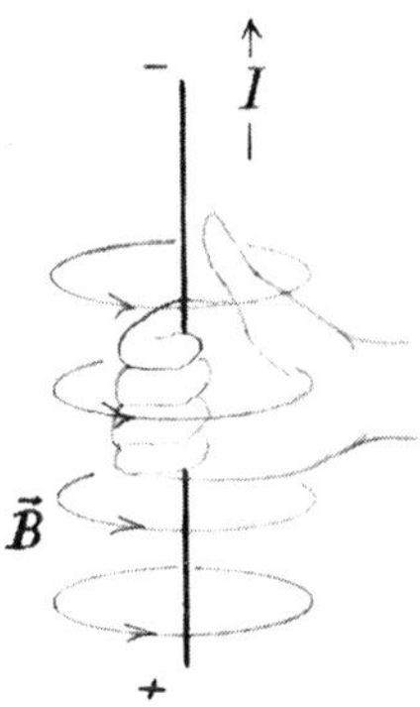

Remember this useful rule: If the thumb points in the direction of the flow of charge, the fingers curl in the direction of the lines of the magnetic field $\vec{B}$. The magnitude of $\vec{B}$ is discussed in Sec. 14.13. Use the right hand for positive charge flow, left hand for negative charge flow.

Oersted's results were the first ever found in which a force did *not* act along a line connecting the sources of the force. (Forces between planets, between electric charges, or between magnetic poles all act along such a line.) The force exerted between the current-carrying wire and each magnetic pole of the compass needle is not along the line from the wire to the pole. In fact, for the needle to twist as it does, the force must be acting *perpendicular* to such a line. The magnetic needle is *not* attracted or repelled by the wire, but is *twisted* sideways by forces on its poles.

This was a totally new kind of effect. No wonder it had taken so long before anyone found the connection between electricity and magnetism. Closer examination revealed more clearly what was happening in this experiment. The long, straight, current-carrying wire sets up a magnetic field. This field turns a small magnet so that the north–south line on the magnet is tangent to a circle whose center is at the wire and whose plane lies *perpendicular* to the wire. Thus, the current produces a *circular* magnetic field, not a centrally directed magnetic field as had been expected.

The direction of the magnetic field vector $\vec{B}$ at each point is defined as *the direction of the force on the north-seeking pole of a compass needle placed at that point.* The force on the south-seeking pole will be in a direction exactly opposite to the field direction. A compass needle will respond to the opposite forces on its ends by turning until it points as closely as possible in the direction of the field. You can get a clue to the "shape" of the magnetic field around a current by sprinkling tiny slivers of iron

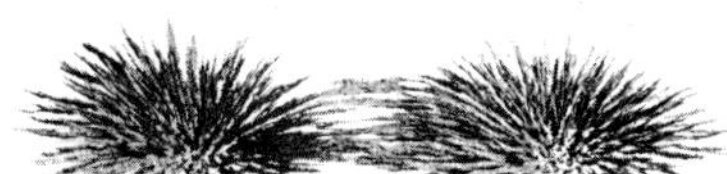

Needle-like iron oxide crystals in the magnetic field of a bar magnet. The bar magnet is under the paper on which the iron oxide crystals have been spread.

Close Up

Hans Christian Oersted

Hans Christian Oersted (1777–1851), a Danish physicist, studied the writings of the Nature Philosopher Schelling and wrote extensively on philosophical subjects himself. In an essay published in 1813, Oersted predicted that a connection between electricity and magnetism would be found. In 1820, he discovered that a magnetic field surrounds an electric current when he placed a compass under a current-carrying wire. In later years he vigorously denied the suggestion of other scientists that his discovery of electromagnetism had been accidental.

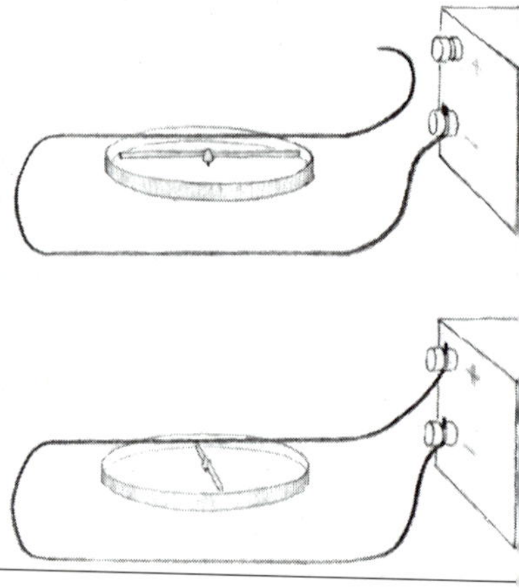

Oersted's experiment.

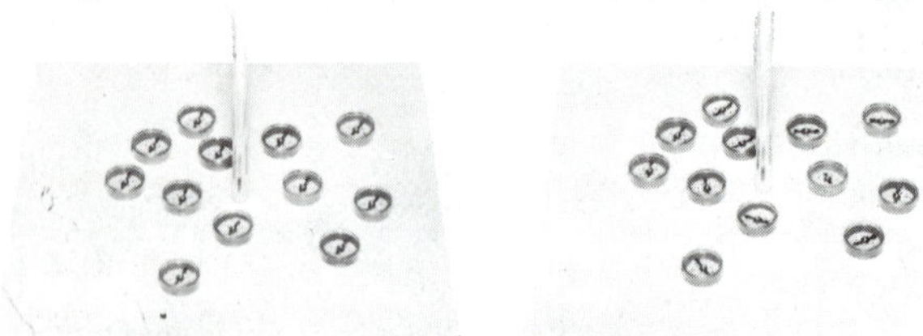

Left: An array of tiny compasses on a sheet of cardboard placed perpendicular to a brass rod. Right: When there is a strong current in the rod, the compass needles are deflected from their normal north–south line by the magnetic field set up by the current. This experiment, too, indicates that the lines of magnetic force due to the current are circular around the rod.

on a sheet of paper through which the current-carrying wire is passing. The slivers become magnetized and behave like tiny compass needles, indicating the direction of the field. The slivers also tend to link together end-to-end. Thus, the pattern of slivers indicates magnetic lines of force around any current-carrying conductor or bar magnet. These lines form a "picture" of the magnetic field.

You can use a similar argument to find the "shape" of a magnetic field produced by a current in a *coil* of wire, instead of a straight wire. To do this, bend the wire into a loop so that it goes through the paper in two places. The magnetic effects of the different parts of the wire on the iron slivers produce a field pattern similar to that of a bar magnet.

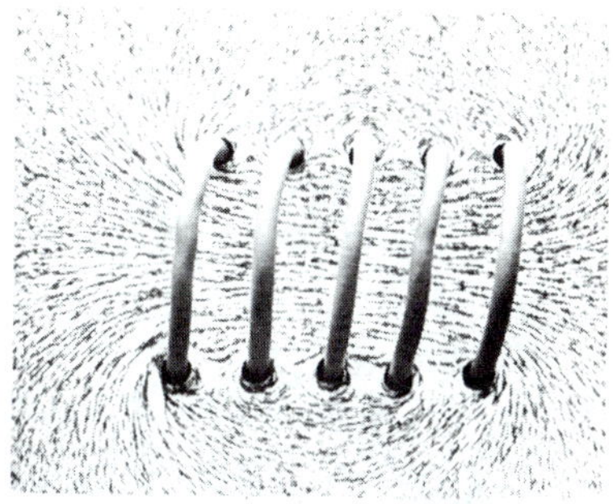

Iron filings in the magnetic field produced by current in a coil of wire.

?

29. Under what conditions can electric charges affect magnets?

30. What was surprising about the force a current exerted on a magnet?

31. How do we know that a current produces any magnetic field near it? What is the "shape" of the field anywhere near a straight conductor?

SG 28

14.12 | Currents act on currents

Oersted's experiment was one of those rare occasions when a discovery suddenly opens up an exciting new subject of research. In this case, no new equipment was needed. At once, dozens of scientists throughout Europe and America began intensive studies on the magnetic effects of electric currents. The work of André-Marie Ampère (1775–1836) stands out among all the rest. Ampère was called the "Newton of electricity" by James Clerk Maxwell, who decades later constructed a complete theory of electricity and magnetism. Ampère's work is filled with elegant mathematics. Without describing his theory in detail, we can trace some of his ideas and review some of his experiments.

Ampère's thoughts raced forward as soon as he heard Oersted's news. He began with a line of thought somewhat as follows: Magnets exert forces on each other, and magnets and currents exert forces on each other. Do currents then exert forces on other currents? The answer is not necessarily yes. Arguing from symmetry is inviting and often turns out to be right. But the conclusions to which such arguments lead are not logically or physically necessary. Ampère recognized the need to let experiment answer his question. He wrote:

André-Marie Ampère (1775–1836) was born in a village near Lyons, France. There was no school in the village, and Ampère was self-taught. His father was executed during the French Revolution, and Ampère's personal life was deeply affected by his father's death. Ampère became a professor of mathematics in Paris and made important contributions to physics, mathematics, and the philosophy of science. His self-portrait is reproduced above.

When M. Oersted discovered the action which a current exercises on a magnet, one might certainly have suspected the

Replica of Ampère's current balance. The essential part is a fixed horizontal wire (foreground), and just behind it, hanging from a hinged support, a shorter segment of wire. Current is produced in both wires, and the force between them is measured.

existence of a mutual action between two circuits carrying currents; but this was not a necessary consequence; for a bar of soft iron also acts on a magnetised needle, although there is not mutual action between two bars of soft iron.

Ampère put his hunch to the test. On September 30, 1820, within a week after word of Oersted's work reached France, Ampère reported to the French Academy of Sciences. He had indeed found that two parallel current-carrying wires exert forces on each other. They did so even though the wires showed no evidence of net electric charges.

Ampère made a thorough study of the forces between currents. He investigated how they depend on the distance between the wires, the relative positions of the wires, and the amount of current. In the laboratory, you can repeat these experiments and work out the "force law" between two currents. We need not go into the quantitative details here, except to note that the force between currents can be used to measure how much current is flowing. In fact, the magnetic force between currents is now the quantity preferred for *defining* the unit of current. This unit is called the *ampere,* as mentioned in Sec. 14.3. One ampere (1 A) is defined as the amount of current in each of two long, straight, parallel wires, set 1 m apart, that causes a force of exactly 2×10^{-7} N to act on each meter of each wire.

? **32.** *What was Ampère's hunch?*

SUMMARY OF ELECTRIC UNITS

Quantity	*Symbol*	*Unit*
Current	I	The *ampere* (A) is the fourth fundamental unit in the so-called MKSA system (meter, kilogram, second, ampere) which is now widely used by physicists. For definition, see last paragraph (Sec. 14.12).
Charge	Q	The *coulomb* (C) is defined as the amount of charge that flows in one second, when the current is one ampere.
Potential difference	V	The *volt* (V) is defined as the electric potential difference between two points such that one joule of work is done in moving one coulomb of charge between those points.
Electric power	P	The *watt* (W) is defined as the rate of energy flow (or work done per second, or "power") which corresponds to one joule per second. Thus, a current of one ampere due to a potential difference of one volt corresponds to one watt power. The *kilowatt* is equal to 1,000 watts.
Work	W	The *kilowatt-hour* (kWh) is the amount of energy expended (work done) when one kilowatt of power is used for one hour. It is equal to 3,600,000 joules (1,000 joules/sec × 3,600 sec).
Resistance	R	The *ohm* (Ω) is defined as the resistance of a material that allows a current of just one ampere to pass through if the potential difference across the material is one volt.
Electric field	$\vec{E}$	Electric field can be expressed either in terms of the force experienced by a unit charge (newtons per coulomb), or in terms of the rate at which the electric potential difference increases (volts per meter).
Magnetic field	$\vec{B}$	The magnitude of magnetic field is defined in terms of the force experienced per meter of length by a conductor carrying a current of one ampere. The units are thus newtons per ampere-meter. The name for this unit is the *tesla* (T).

14.13 | Magnetic fields and moving charges

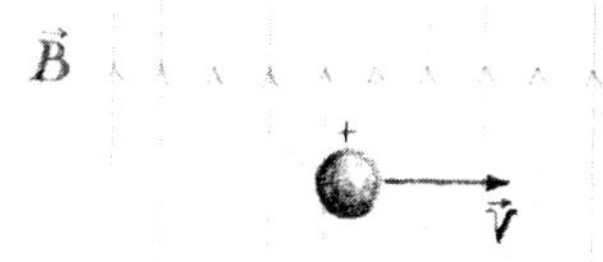

(a) When the charge q moves with velocity $\vec{v}$ perpendicular to $\vec{B}$,

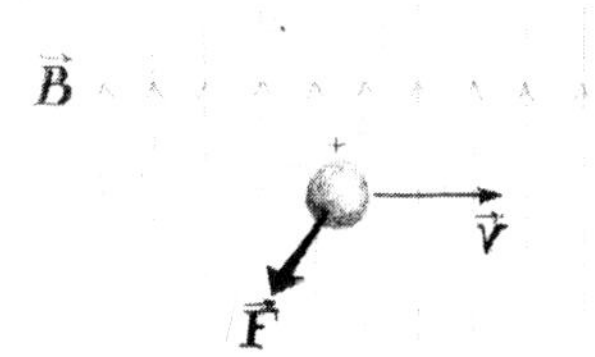

(b) there is a force $\vec{F}$ as shown, proportional to q, v, and B.

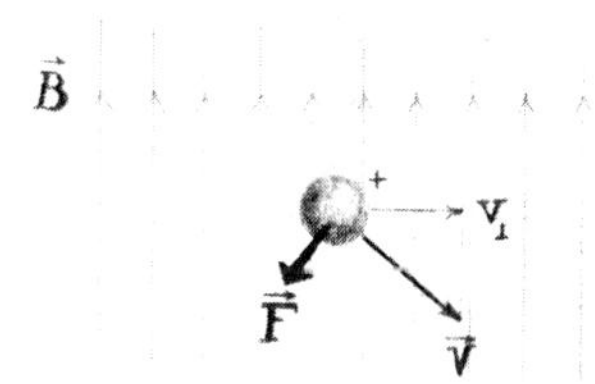

(c) If $\vec{v}$ is not perpendicular (⊥) to $\vec{B}$, there is a smaller force, proportional to $v_\perp$ instead of v.

In the last two sections, the interactions of currents with magnets and with each other were discussed. The concept of *magnetic field* greatly simplifies the description of these phenomena.

As you saw in studying Coulomb's law, electrically charged bodies exert forces on each other. When the charged bodies are at rest, the forces are "electric" forces, or Coulomb forces. "Electric fields" act as the sources of these forces. But when the charged bodies are moving (as when two parallel wires carry currents), new forces *in addition to* the electric forces are present. These new forces are called "magnetic" and are caused by "magnetic fields" set up by the moving charges.

Magnetic interaction of moving charged bodies is not as simple as electric interaction. Remember the description of Oersted's experiment. The direction of the force exerted by a current on a magnet is perpendicular both to the direction of the current and to the line between the magnet and current. For the moment, however, it is not necessary to examine the forces on current-carrying conductors. After all, the force on a wire is believed to be caused by forces on the individual electric charges moving in it. How do such individual charges behave when moving freely in an external magnetic field? Once some simple rules have been established for the behavior of free charged particles, wires will be discussed in the next chapter. There you will see how these simple rules are enough to explain the operation of electric generators and electric motors. (You will also see how these inventions have changed civilization.)

The rules summarized in the remainder of this section are best learned in the laboratory. All you need is a magnet and a device for producing a beam of charged particles, for example, the "electron gun" described in Sec. 14.8. (Recommended procedures are described in the experiment "Electron Beam Tube. I." in the *Handbook*.)

The force on a moving charged body. Suppose you have a fairly uniform magnetic field $\vec{B}$, produced either by a bar magnet or by a current in a coil. How does this external field act on a moving, charged body? You can find by experiment that the charge experiences a force and that the force depends on three quantities: (1) the charge q on a body, (2) the velocity $\vec{v}$ of the body, (3) the strength of the external field $\vec{B}$ through which the body is moving.

The force depends not only on the *magnitude* of the velocity, but also on its *direction*. If the body is moving in a direction

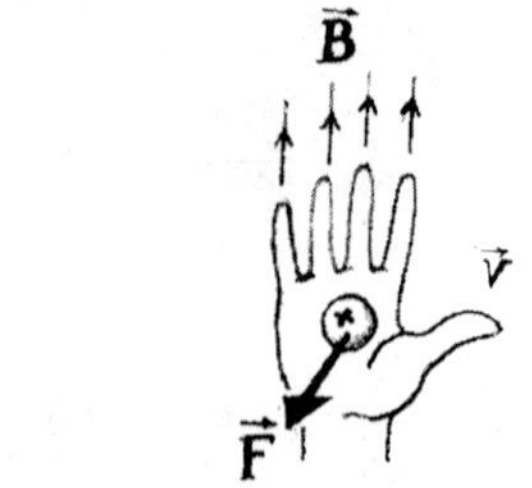

Remember this useful rule: If your fingers point along $\vec{B}$ *and your thumb along* $\vec{v}$, $\vec{F}$ *will be in the direction your palm would push. For positive charges use the right hand, and for negative use the left hand.*

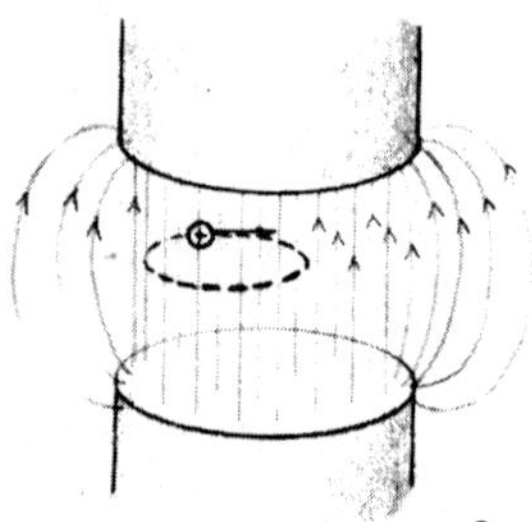

a

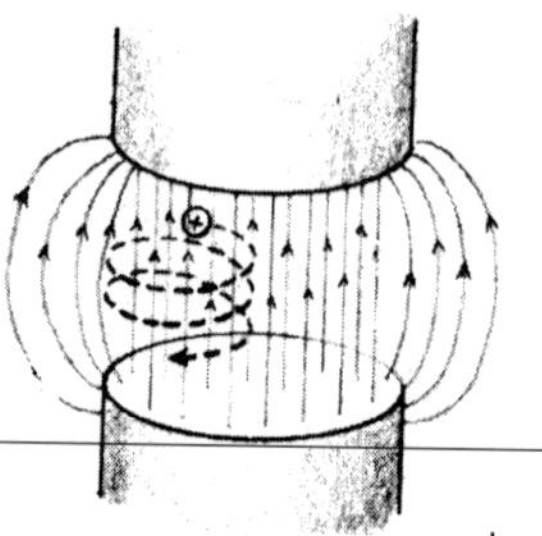

b

perpendicular to the field $\vec{B}$, the magnitude of the force is proportional to *both* of these quantities; that is,

$$F \propto qvB$$

which can also be written as

$$F = kqvB$$

where k is a proportionality constant that depends on the units chosen for F, q, v, and B.

But if the charge is moving in a direction *parallel* to $\vec{B}$, there is no force! For all other directions of motion, the magnitude of the force is somewhere between the full value and zero. In fact, the force is proportional to the *component* of the velocity that is perpendicular to the field direction, $v_{\perp}$. Therefore, a more general expression for the force is

$$F \propto qv_{\perp} B$$

or

$$F = kqv_{\perp} B$$

where k is the same constant as before. *The direction of the force is always perpendicular to the direction of the field. It is also perpendicular to the direction of motion of the charged body.*

The force exerted by an external magnetic field on a moving charged particle can be used to *define* the unit of magnetic field $\vec{B}$. This is done by taking the proportionality constant k as equal to one. This definition is convenient here, since we are dealing mainly with how magnetic fields act on moving charges (rather than with forces between bar magnets). So in the special case when $\vec{B}$ and $\vec{v}$ are *at right angles* to each other, the magnitude of the deflecting force becomes simply

$$F = qvB$$

The path of a charged body in a magnetic field. The force on a moving charged body in a magnetic field is always "off to the side"; that is, the force is perpendicular to the body's direction of motion at every moment. Therefore, the magnetic force does not change the *speed* of the charged body. Rather, it changes the *direction* of the velocity vector. If a charged body is moving exactly perpendicular to a uniform magnetic field, there will be a constant sideways push. The body will move along a circular path, in a plane perpendicular to the direction of the magnetic
SG 29 field. If B is strong enough, the particle will be trapped in a
SG 30 circular orbit [as in sketch (*a*) in the margin].

What if the charged body's velocity has some component along the direction of the field but not exactly parallel to it? The body will still be deflected into a curved path, but the component of its motion *along* the field will continue undisturbed. So the particle will trace out a coiled (helical) path [as in sketch (*b*) in the margin]. If the body is initially moving exactly parallel to the magnetic field, there is no deflecting force at all, since $v_{\perp}$ is zero.

SG 31

Some important examples of the deflection of charged particles by magnetic fields are discussed in Unit 5 and Unit 6. These examples include particle accelerators and bubble chambers. One example of "coiled" motion is found in the Van Allen radiation belts. A stream of charged particles, mainly from the sun, but also from outer space, continually sweeps past the earth. Many of these particles are deflected into spiral paths by the magnetic field of the earth and become "trapped" in the earth's field. The extensive zones containing these rapidly moving trapped particles are called the Van Allen belts. Particles from these zones sometimes work their way toward the earth's magnetic poles. When they hit the atmosphere, they excite the atoms of the gases to radiate light. This is the cause of the aurora ("northern lights" and "southern lights").

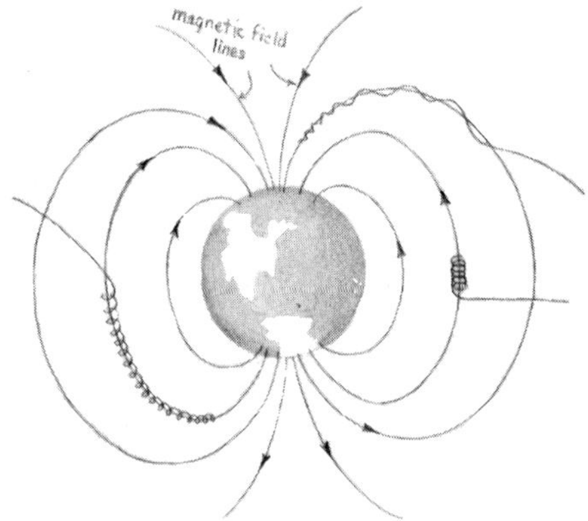

A simplified sketch of a variety of paths taken by charged particles in the earth's magnetic field. The Van Allen belts are regions of such trapped particles.

This chapter has dealt with the interaction between currents and magnets and between magnetic fields and charged particles. At first reading, many students consider this topic to be a very abstract part of pure physics. Yet the study of these interactions has had important social and practical effects on the whole civilized world. You will look at some of these effects in the next two chapters.

The American physicist James A. Van Allen directed the design of instruments carried by the first American satellite, Explorer I.

?

33. *Which of the following affect the magnitude of the deflecting force on a moving charged particle?*

(a) the component of the velocity parallel to the magnetic field

(b) the component of the velocity perpendicular to the field

(c) the magnetic field $\vec{B}$ *itself*

(d) the magnitude of the charge

(e) the sign of the charge

34. *Which of the items in the preceding question affect the direction of the deflecting force on the charged particle?*

SG 32, 33

35. *Why does the deflecting force on a moving charged particle not change the speed of the charged particle? Does it ever do any work on it?*

36. *What are differences between deflecting forces on a charged object due to*

(a) gravity?

(b) an electric field?

(c) a magnetic field?

The aurora photographed from Alaska. The glow is produced when the upper atmosphere is excited by charged particles trapped in the earth's magnetic field.

study guide

1. The *Project Physics* learning materials particularly appropriate for Chapter 14 include:

Experiments
Electric Force. I
Electric Forces. II
Forces on Currents
Currents, Magnets, and Forces
Electron Beam Tube. I
Electron Beam Tubes. II

Activities
Detecting Electric Fields
Voltaic Pile
An 11¢ Battery
Measuring Magnetic Field Intensity
More Perpetual Motion Machines
An Isolated North Magnetic Pole?

2. How much must you alter the distances between two charged objects in order to keep the force on them constant, if you also

(a) triple the net charge on each?

(b) halve the net charge on each?

(c) double the net charge on one and halve the net charge on the other?

3. How far apart in air must two charged spheres be placed, each having a net charge of 1 C, so that the force on them is 1 N?

4. If electrostatic induction does not involve the addition or subtraction of charged particles, but instead is just a redistribution of charged particles, how can attraction result from induction?

5. A carbon-coated (and therefore conducting) ping-pong ball hanging by a nylon (nonconducting) thread from a ring stand is touched with a finger to remove any slight charge it may have had. Then a negatively charged rod is brought up close to but *not touching* the ball. While the rod is held there, the ball is momentarily touched with a finger; then the rod is removed. Does the ball now have a net charge? How would you test whether it has? If you think it has, make a few simple sketches to show how it became charged, indicating clearly what kind of charge it has been left with.

6. **(a)** Calculate the strength of the gravitational field of the moon at a point on its surface. The mass of the moon is taken to be 7.3×10^{22} kg, and its radius is 1.74×10^{6} m.

(b) Calculate the gravitational field at a point near the surface of a small but extremely dense star whose radius is 1.5×10^{6} m and whose density is about 10^{22} kg/m^3.

(c) The gravitational field of any uniform spherical shell is zero inside the shell. Use this principle together with Newton's gravitational force law and the formula for the volume of a sphere ($\frac{4}{3}\pi r^3$) to find out how the gravitational field at a point *P* *inside* a solid spherical planet depends on the distance *r* from the center. (Assume the planet's density is uniform throughout.)

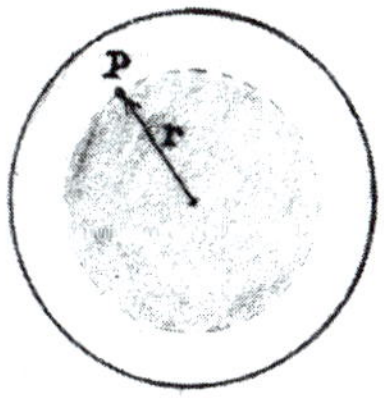

7. An electric field exerts a force on a charged particle placed in the field. What else can you say about this situation, considering the fact that Newton's third law holds in this case, too?

8. The three spheres A, B, and C are fixed in the positions shown. Determine the direction of the net electrical force on sphere C, which is positively charged, if

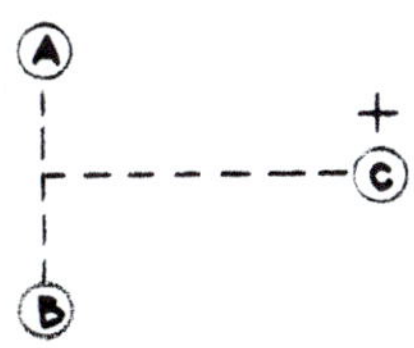

(a) A and B carry equal positive charges.

(b) A and B have charges of equal magnitude, but the charge on B is negative and that on A positive.

9. A sphere with a negative charge of 4×10^{-2} C is in the middle of a room. All test charges are placed at a distance of 2.5 m west of the center of the sphere.

(a) Use Coulomb's force law to calculate the force on particles with charges of $3q_e$, $6q_e$, $10q_e$, and $34q_e$.

(b) Find the electric field at the test point, then use $F = qE$ to determine the force on the particles in (a).

(c) What important physical principle is the basis of the electric field concept?

10. An electric field strength exists at the earth's surface of about 100 N/C, directed downward.

(a) What is the net charge on the earth? (As Newton had shown for gravitational forces, the field of a uniformly charged sphere can be calculated by assuming all of the charge is concentrated at its center.)

(b) Because the earth is a conductor, most of the net charge is on the surface. What, roughly, is the average amount of net charge per square meter of surface? Does this seem large or small, compared to familiar static charges like those that can be produced on combs?

11. In oscilloscope tubes, a beam of electrons is deflected as it is passed between two pairs of oppositely charged plates. Each pair of plates, as can be seen in the photograph at the top of the following page, is shaped something like the sketch to the right of the photograph. Sketch in roughly what you think the lines of force in the electric field between a pair of such oppositely charged plates would be like.

12. Is air friction acting on the moving oil drop a help or a hindrance in the experiment described for measurement of the charge of the electron? Explain your answer briefly.

13. The magnitude of the electron's charge is 1.6×10^{-19} C. How many electrons are required to make 1 C of charge?

14. Calculate the ratio of the electrostatic force to the gravitational force between two electrons a distance of 10^{-10} m apart. (The mass of the electron is approximately 10^{-30} kg; recall that $G = 6.7 \times 10^{-11}$ N-m^2/kg^2.)

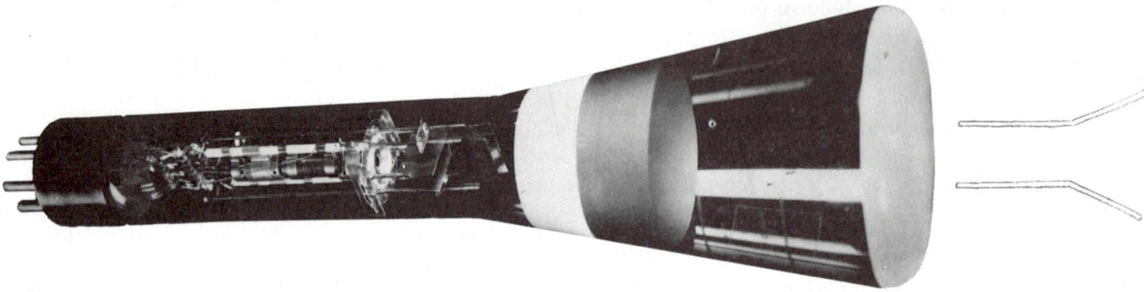

15. Electrical forces are similar in some respects to gravitational forces. So it is reasonable to imagine that charged particles such as the electron may move in stable orbits around other charged particles. Then, just as the earth is a "gravitational satellite" of the sun, the electron would be an "electric satellite" of some *positively* charged particle. If this particle has a very large mass compared to that of the electron, you can assume it will remain stationary at the center of the electron's orbit. Suppose the particle has a charge equal in magnitude to the charge of the electron and that the electron moves in a circular orbit.

(a) The centripetal force acting on the moving electron is provided by the electrical (Coulomb) force between the electron and the positively charged particle. Write an equation representing this statement. From this equation derive another equation showing how the kinetic energy of the electron is related to its distance from the positively charged particle.

(b) Calculate the kinetic energy of the electron if the radius of its orbit is 10^{-10} m.

(c) What will be the speed of the electron if it has the kinetic energy you calculated in part (b)? (The mass of the electron is approximately 10^{-30} kg.)

16. A hard rubber or plastic comb rubbed against wool can often be shown to be charged. Why does a metal comb not readily show a net charge produced by rubbing unless it is held by an insulating handle?

17. What is the potential difference between two points in an electric field if 6×10^{-4} J of work is done against the electric forces in moving 2×10^{-5} C of charge from one point to the other?

18. If there is no potential difference between any points in a region, what must be true of

(a) the electric potential energy in that region?

(b) the electric field?

19. Electric field intensity $\vec{E}$ can be measured in either of two equivalent units: newtons per coulomb and volts per meter. Using the definitions of volt and joule, show that newton/coulomb is actually the same as volt/meter. Can you give the reason for the equivalence in words?

20. By experiment, if the distance between the surfaces of two conducting spheres is about 1 cm, an electric potential difference of about 30,000 V between them is required to produce a spark in ordinary air. (The higher the voltage above 30,000 V, the "fatter" the spark for this gap distance.) What is the minimum electric field strength (in the gap between the surfaces) necessary to cause sparking?

21. The gap between the two electrodes in an automobile spark plug is about 1 mm. If the voltage produced between them by the ignition coil is about 10,000 V, what is the approximate electric field strength in the gap?

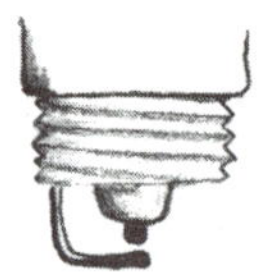

22. One can think of an electric battery as "pumping" charges into its terminals. This pumping continues until the electric potential difference between the terminals reaches a certain value where those charges already there repel newcomers from inside the battery. Usually this value is very close to the voltage marked on the battery.

(a)

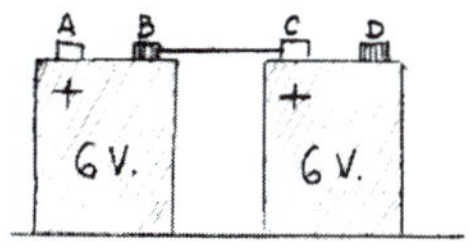

What would happen if you connected two or more batteries in a sequence? For example, the battery on the right, above, maintains terminal C at an electric potential 6 V higher than terminal D. This is what the + indicates under C; its electric potential is higher than that of the other terminal of the *same* battery. The battery on the left maintains terminal A at a potential 6 V higher than terminal B. If you connect B to C with a good conductor, so that B and C are at the same potential level, what is the potential difference between A and D?

What would the potential difference be between the extreme left and right terminals in the following set-ups?

(b)

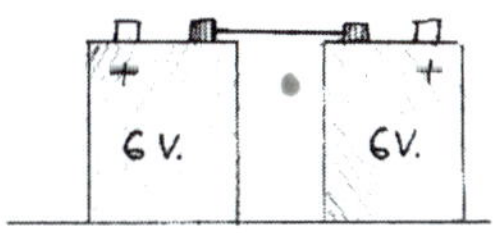

(c)

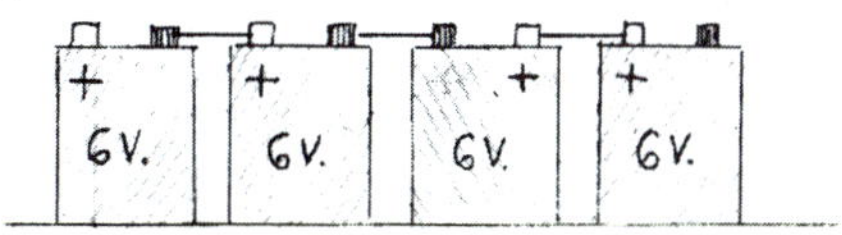

23. (a) What kinetic energy will an electron gain in an evacuated tube if it is accelerated through a potential difference of 100 V? State your answer in electron volts and also in joules. (The magnitude of the charge on the electron is 1.6×10^{-19} C.)

(b) What speed will it acquire because of the acceleration? (The mass of the electron is 10^{-30} kg.)

24. (a) A battery of 12 V is connected to a circuit with a resistance of 3 ohms. What is the current? If the voltage is doubled and the resistance is constant, what is the new current in the circuit?

(b) Why are the following terms used: "voltage *across*" and "current *through*"?

25. What is the resistance of a dc circuit that allows a current of 4 A to flow if 100 V is applied across it? Assuming that Ohm's law applies to the circuit, what is the voltage if the current is cut in half? If Ohm's law does not apply to this circuit, what is the relationship of current to voltage?

26. What is the power used by a circuit in which 3 A flow across a 50-V supply? What is the resistance of this circuit?

27. When power demand is high, an electric company will often lower the voltage to its customers. If your house is operating six 100-W light bulbs, one 200-W television, one 5,000-W clothes dryer, and one 25-W radio, what is your total power use before and after a 5% voltage reduction on 120-V service? What is your current use before and after the voltage cut? Are you being cheated by the electric power company?

28. A circuit breaker will cut off the current to a circuit when the current through the circuit exceeds the rating marked on the circuit breaker. How many 150-W light bulbs can you put on a circuit with a 10-A circuit breaker if a 500-W stereo is already on the circuit? (Assume that the house voltage is 120 V.)

29. Suppose three resistors are each connected to a battery and to a current meter. The following table gives two of three quantities related by Ohm's law for three separate cases. Complete the table.

	Voltage (volts)	*Current* (amperes)	*Resistance* (ohms)
(a)	2		0.5
(b)	10	2	
(c)		3	5

30. The electric field at the earth's surface can increase to about 10^4 V/m under thunderclouds.

(a) About how large a potential difference between ground and cloud does that imply?

(b) A set of lightning strokes can transfer as much as 50 C of charge. Roughly how much energy would be released in such a discharge?

31. "International Physics Co.'s Pulsed Radiation Facility is now producing electron beams (40,000 A at 4 MeV) as a routine operation. This beam can deposit precisely 5,000 J of energy in 30 nanoseconds." (From an advertisement in the journal *Physics Today*.)

The term "4MeV" means that the charges in the beam have an energy that would result from being accelerated across a potential difference of 4 million volts. A "nanosecond" is a billionth of a second. Are these published values consistent with one another? (Hint: Calculate the power of the beam in two different ways.)

32. An electron "gun" includes several electrodes, kept at different voltages, to accelerate and focus the electron beam. But the energy of electrons in the beam that emerges from the gun depends only on the potential difference between their source (the hot wire) and the final accelerating electrode. In a color TV picture tube, this potential difference is 20 to 30 kV. A triple gun assembly (one each for red, blue, and green) from a color TV set is shown in the photograph below.

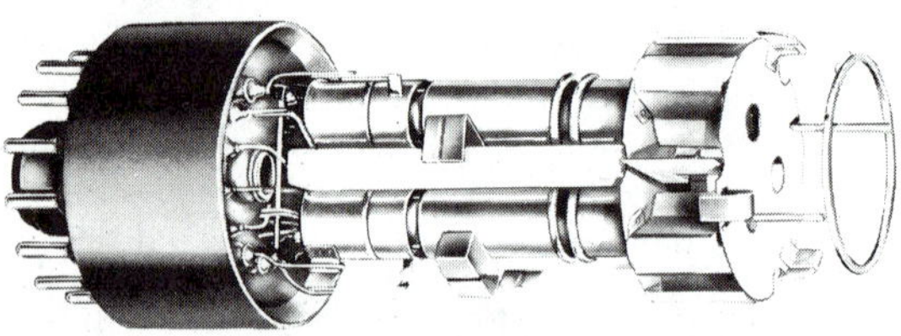

Suppose the beam in a TV tube is accelerated through 20,000 V and forms an average current on the order of 10^{-3} A. Roughly what is the power being dissipated against the screen of the tube?

33. Calculate the power dissipated in each of the three circuit elements of SG 23.

34. A student trying to show the magnetic effect of a current on a pocket compass slowly slid the compass along a tabletop toward a wire lying on the table and carrying a constant current. The student was surprised by the lack of any observed turning effect on the compass needle. How would you explain these observations?

35. The sketch shows two long, parallel wires, lying in a vertical north–south plane (the view here is toward the west). A horizontal compass is located midway between the two wires. With no current in the wires, the needle points N. With 1 A in the upper wire, the needle points NW.

(a) What is the direction of this 1-A current?

(b) What current (magnitude and direction) in the lower wire would restore the compass to its original position?

36. **(a)** What is the definition of the ampere (A) given in Sec. 14.12?

(b) The force between two wires carrying an electric current varies directly with the current in each of the wires and inversely with the distance separating them. What is the force between two wires 3 m long, 0.5 m apart, carrying 5 A and 8 A, respectively?

37. The deflecting force on a charged particle moving perpendicularly to a uniform magnetic field is always perpendicular to its velocity vector. Therefore, it is directed at every moment toward a single point—the center of the circular path the particle follows.

(a) The magnetic force (given by the expression qvB) therefore provides a centripetal force always given by mv^2/R). Show that the radius of the circle R is directly proportional to the momentum of the particle mv.

(b) What information would you need to determine the ratio of the particle's charge to its mass?

38. By referring to the information given in the last problem,

(a) find an equation for the period of the circular motion of a charged particle in a uniform magnetic field.

(b) show mathematically that the radius of the helical path will be smaller where the magnetic field strength is greater. (See sketch.)

(c) using the right-hand rule, show that the direction of the deflecting force on the particle *opposes* the movement of the particle into the region of stronger field.

39. If the energy of charged particles approaching the earth (from the sun) is very great, they will not

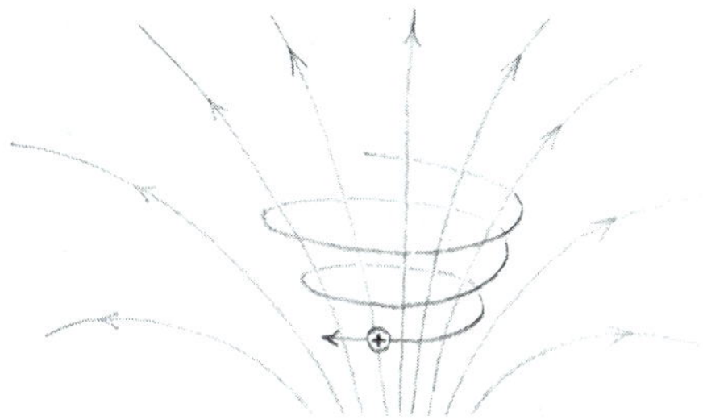

be trapped in the Van Allen belts. Rather, they will be somewhat deflected, continuing on past or into the earth. The direction of the lines of force of the earth's magnetic field is toward the earth's north end. If you set up a detector for positively charged particles on the earth, would you expect to detect more particles by directing it slightly toward the east or slightly toward the west?

40. William Gilbert, in *De Magnete,* recorded that a piece of amber that had been rubbed attracted smoke rising from a recently extinguished candle. The smoke particles had been charged by passing through the ionized gases of the flame. After the development of electrostatic machines, experiments were done on the discharges (called *corona discharges*) from sharp or pointed needlelike electrodes. As long ago as 1824 it was found that passing such a discharge through a jar filled with fog cleared the fog from the jar. A similar experiment was performed using tobacco smoke. The corona discharge in these experiments ionized the gas, which in turn charged and precipitated the water droplets of the fog or the smoke particles.

However, no successful industrial precipitator was built until Frederick Cottrell succeeded in using the

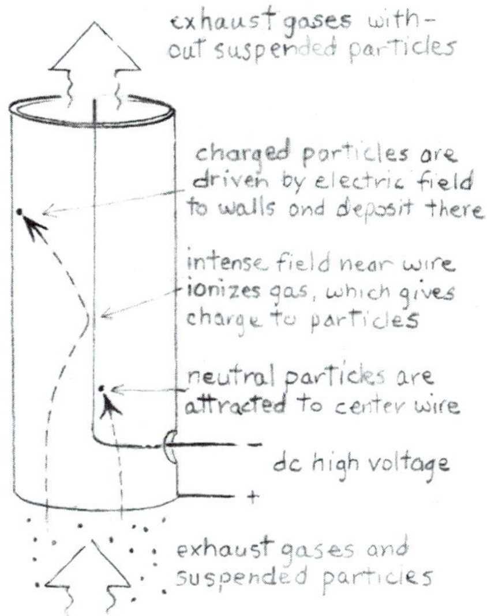

electric generator, high-voltage transformer, and mechanical rectifier developed late in the nineteenth century. Cottrell achieved both a strong corona discharge and a high potential difference between the discharge electrode and the collecting electrode. Since that time, many "electrostatic precipitators" have been built by electrical engineers to collect particulate matter. The most important such particles are fly ash from the burning of coal in the electrical power industry.

(**a**) Why are *neutral* particles attracted to the central wire?

(**b**) Would the precipitator work if the central wire were + and the casing −?

Faraday and the Electrical Age

15.1 The problem: Getting energy from one place to another

SG 1 In Chapter 10, the development of steam engines in the eighteenth and nineteenth centuries was discussed. These engines enabled Europe and America to make use of the vast stores of energy contained in coal, wood, and oil. By burning fuel, chemical energy can be converted to heat energy, which in turn can be used to make steam. By letting the steam expand against a piston or a turbine blade, heat energy can be converted to mechanical energy. In this way, a coal-fueled steam engine can power machinery.

Steam engines had two major defects. First, the mechanical energy was available only at the place where the steam engine

was located. Second, practical steam engines were big, hot, and dirty. As the use of machines run by steam engines increased, people were crowded together in factories, and their homes stood in the shadow of the smoke stacks. Even steam-powered locomotives, though useful for transportation, were limited by their size and weight. They also added further to polluting the air.

These defects could be partially overcome by using one central power plant for sending out energy for use at a distance. This energy could drive machines of any desired size and power at the most practical locations. After Volta's development of the battery, many scientists and inventors speculated that electricity might provide such a means of distributing energy and running machines. But the energy in batteries is quickly used up unless it is delivered at a low rate. A better way of generating electric currents was needed. When such a way was found, it changed the whole shape of life in homes, factories, farms, and offices. It even changed the very appearance of cities and landscapes.

SG 2

In this chapter you will see another example of how discoveries in basic physics have given rise to new technologies. These technologies have revolutionized and benefited modern civilization. But they have brought some new problems in their turn.

SG 3

Ampère also sensed that electricity might transmit not only energy but also *information* to distant places.

Two versions of Faraday's electromagnetic rotator. In each, the cup was filled with mercury so that a current could be passed between the base and overhead support.

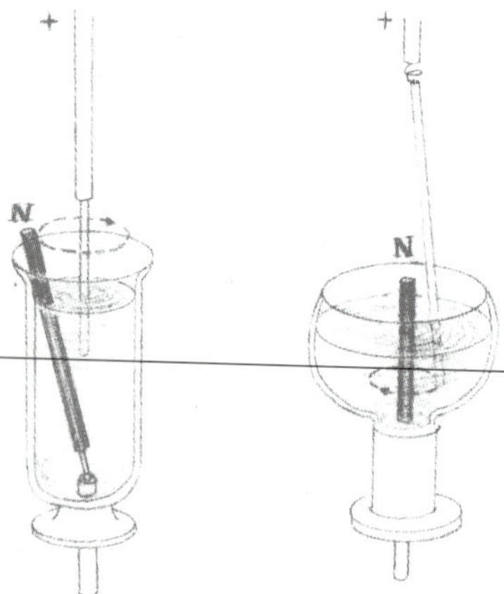

In one version (left) the north end of a bar magnet revolves along the circular electric lines of force surrounding the fixed current. In the other version (right), the rod carrying the current revolves around the fixed bar magnet, moving always perpendicular to the magnetic lines of force coming from the pole of the magnet.

The first clue to the broader use of electricity came from Oersted's discovery that a magnetic needle is deflected by a current from a battery. Since an electric current can exert a force on a magnet, many physicists naturally speculated that a magnet could somehow produce a current in a wire. (Such reasoning from symmetry is common in physics and often is useful.) Soon after the news of Oersted's discovery reached Paris, the French physicists Biot, Savart, and Ampère began research on the interactions of electricity and magnetism. (Some of their results were mentioned in Chapter 14.) A flood of other experiments and speculations on electromagnetism poured from all over the world into the scientific journals. Yet the one key discovery—how to generate an ample and continuous electric current—still eluded everyone.

15.2 | Faraday's first electric motor

Scientific journals regularly print brief announcements of the technical details of new discoveries. From time to time they also provide valuable in-depth surveys of recent broad advances in science. The need for such a review article is especially great after a burst of activity of the kind that followed Oersted's discovery of electromagnetism in 1820.

In 1821, the editor of the British journal *Annals of Philosophy* asked Michael Faraday to review the experiments and theories of electromagnetism that had appeared in the previous year. Faraday's first discovery in electromagnetism came on September 3, 1821. Repeating Oersted's experiment (described in Sec. 14.11), he put a compass needle at various places around a current-carrying wire. Faraday was particularly struck by one fact: The force exerted by the current on each pole of the magnet tended to carry the pole along a circular line around the wire. As he expressed it later, the wire is surrounded by *circular lines of force*: a circular magnetic field. Faraday then constructed an "electromagnetic rotator" based on this idea. It worked. Though very primitive, it was the first device for producing continuous motion by the action of a current: the first electric motor.

Faraday also designed an arrangement in which the magnet was fixed and the current-carrying wire rotated around it. (If a current exerts a force on a magnet, the magnet should exert an equal force on the current, according to Newton's third law.) As in many other cases, Faraday was guided by the idea that for every effect of electricity on magnetism, there must exist a corresponding effect of magnetism on electricity. Of course, it was not always so obvious what form the corresponding effect would take.

? *1. Why does the magnetic pole of Faraday's "electromagnetic rotator" move in a circle around a fixed wire?*

15.3 | The discovery of electromagnetic induction

Armed with his "lines of force" idea of electric and magnetic fields, Faraday joined the search for a way of producing currents by magnetism. Scattered through his diary in the years after 1824 are many descriptions of such experiments. Each report ended with a note: "exhibited no action" or "no effect."

Finally, in 1831, came the breakthrough. Like many discoveries that follow much research and discussion among scientists, this one was made almost at the same time by two scientists working independently in different countries. Faraday was not quite the first to produce electricity from magnetism. *Electromagnetic induction* (the production of a current by magnetism) was actually discovered first by the American scientist Joseph Henry. At the time Henry was teaching mathematics and philosophy at an academy in Albany, New York. Unfortunately for the reputation of American science, teachers at the Albany Academy were expected to spend all their time on teaching and related duties. There was little time left for research. Henry had hardly any opportunity to follow up his discovery, which he made during a one-month vacation. He was not able to publish his work until a year later. In the meantime, Faraday had made a similar discovery and published his results.

Faraday is known as the discoverer of electromagnetic induction not simply because he published his results first. More importantly, he conducted exhaustive investigations into all aspects of the subject. His earlier experiments and his ideas about lines of force had suggested that a current in one wire should somehow induce a current in a nearby wire. Oersted and Ampère had shown that a *steady* electric current produced a *steady* magnetic field around the circuit carrying the current. Perhaps a steady electric current could somehow be generated if a wire were placed near or around a very strong magnet. Or a steady current might be produced in one wire by a very large steady current in another wire nearby. Faraday tried all these possibilities, with no success.

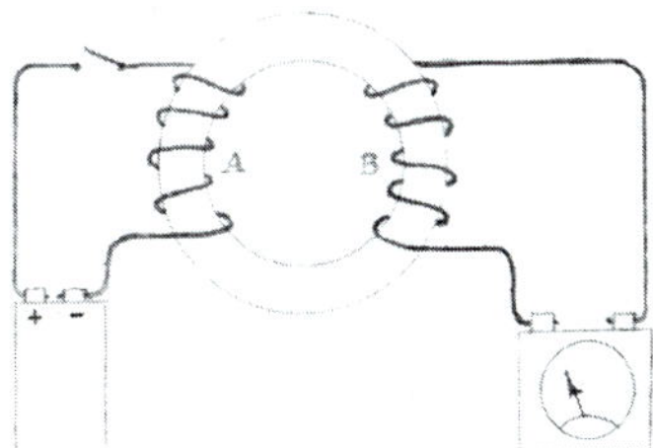

The solution Faraday found in 1831 came partly by accident. He was experimenting with two wire coils that had been wound around an iron ring (see illustration in the margin). He noted that a current appeared in one coil while the current in the other coil was being switched on or off. When a current was turned on in coil A, a current was induced in coil B, but it lasted

Close Up

Michael Faraday

Michael Faraday (1791–1867) was the son of an English blacksmith. In his own words:

> My education was of the most ordinary description, consisting of little more than the rudiments of reading, writing and arithmetic at a common day-school. My hours out of school were passed at home and in the streets.

At the age of 12 he went to work as an errand boy at a bookseller's store. Later he became a bookbinder's assistant. When Faraday was about 19 he was given a ticket to attend a series of lectures given by Sir Humphry Davy at the Royal Institution in London. The Royal Institution was an important center of research and education in science, and Davy was Superintendent of the Institution. Faraday became strongly interested in science and undertook the study of chemistry by himself. In 1813, he applied to Davy for a job at the Royal Institution and Davy hired him as a research assistant. Faraday soon showed his genius as an experimenter. He made important contributions to chemistry, magnetism, electricity and light, and eventually succeeded Davy as superintendent of the Royal Institution.

Because of his many discoveries, Faraday is generally regarded as one of the greatest of all experimental scientists. Faraday was also a fine lecturer and had an extraordinary gift for explaining the results of scientific research to non-scientists. His lectures to audiences of young people are still delightful to read. Two of them, "On the Various Forces of Nature" and "The Chemical History of a Candle," have been republished in paperback editions.

Faraday was a modest, gentle, and deeply religious man. Although he received many international scientific honors, he had no wish to be knighted, preferring to remain without title.

Faraday's laboratory at the Royal Institution.

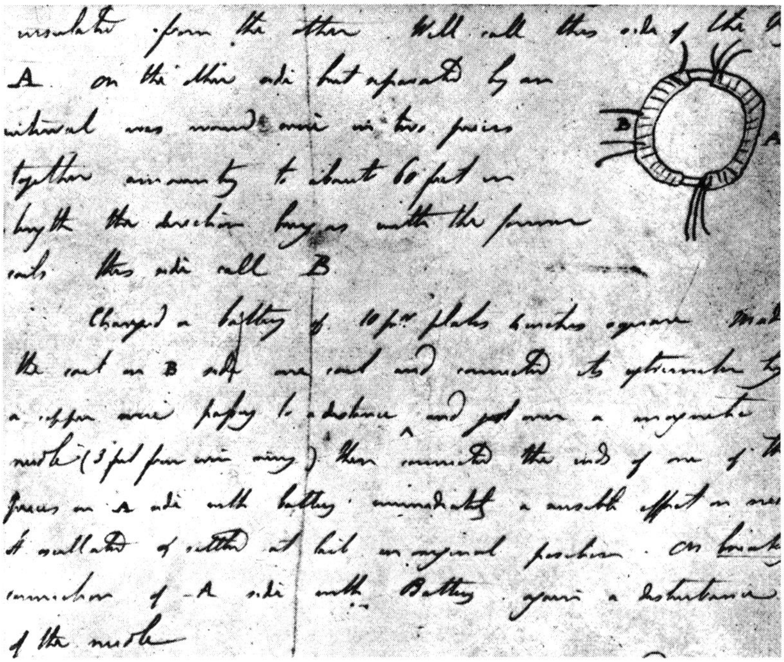

Part of a page in Faraday's diary where he recorded the first successful experiment in electromagnetic induction (about one-half the actual size).

only a moment. As soon as there was a steady current in coil A, the current in coil B disappeared. When the current in coil A was turned off, a current again appeared briefly in coil B.

To summarize Faraday's result: A current in a stationary wire can induce a current in another stationary wire *only while the current is changing*. A steady current in one wire cannot induce a current in another wire.

Faraday was not satisfied with merely observing and reporting his accidental arrangement and its important result. Guided by his concept of "lines of force," he tried to find out the basic principles involved in electromagnetic induction.

According to Faraday's theory, the changing current in coil A would change the lines of magnetic force in the whole iron ring. The change in lines of magnetic force in the part of the ring near coil B would then induce a current in B. But if this was really the correct explanation, Faraday asked himself, should it not be possible to produce the same effect in another way? In particular:

1. Is the iron ring really necessary to produce the induction effect? Or does the presence of iron merely strengthen an effect that would also occur without it?

2. Is coil A really necessary? Or could current be induced simply by changing the magnetic lines of force through coil B in some other way, such as by moving a simple magnet relative to the wire?

Faraday answered these questions almost immediately by performing further experiments. First, he showed that the iron ring was not necessary. Starting or stopping a current in one coil of wire would induce a momentary current in a nearby coil, with only air (or a vacuum) between the coils. (See top figure on this page. Note that there is no battery in the circuit at the right, only a meter to measure the induced current.) Second, he studied what happened when a bar magnet was inserted into or removed from a coil of wire. He found that a current was induced at the instant of insertion or removal. (See second figure at the left.) In Faraday's words,

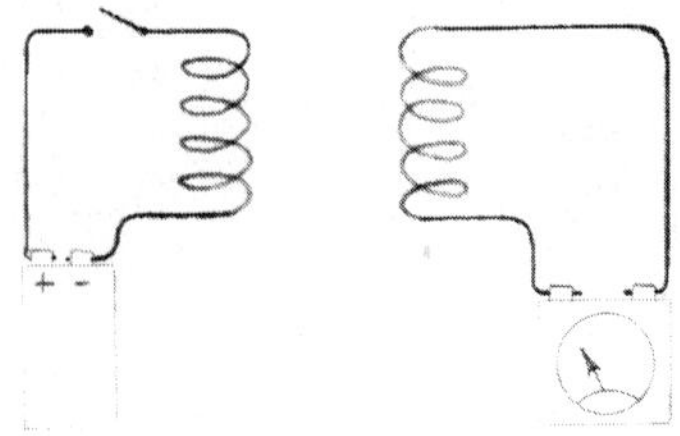

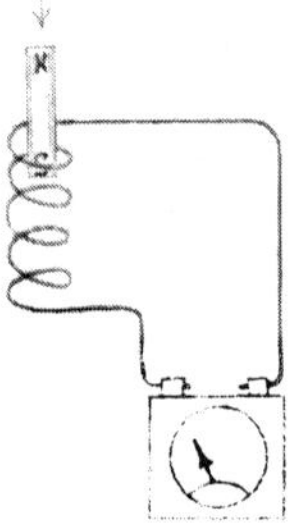

> A cylindrical bar magnet . . . had one end just inserted into the end of the helix cylinder; then it was quickly thrust in the whole length and the galvanometer needle moved; when pulled out again the needle moved, but in the opposite direction. The effect was repeated every time the magnet was put in or out. . . .

Note that this is a primitive *electric generator*; it provides electric current by having some mechanical agent move a magnet.

Having done these and many other experiments, Faraday stated his general principle of electromagnetic induction. Basically, it is that *changing lines of magnetic force can induce a current in a wire*. The needed "change" in lines of force can be produced either by a magnet moving relative to a wire or by a changing current. In the case of the moving magnet, Faraday described the wire as "cutting across" lines of force. In the case of changing current, the lines of force "cut across" the wire. He later used the word *field* to refer to the arrangement and intensity of lines of force in space. Therefore, a current can be induced in a circuit by changes in a magnetic field around the circuit. Such changes may result either from relative motion of
SG 4 wire and field or simply from a change in intensity of the field.

So far, Faraday had produced only momentary surges of current by induction. This was hardly an improvement over batteries as a source of current. Was it possible to produce a continual current by electromagnetic induction? To do this
SG 5 would require a situation in which magnetic lines of force were *continually changing* relative to the conductor. Using a simple magnet, the relative change could be produced either by moving the magnet or by moving the conductor. This is just what Faraday did. He turned a copper disk between the poles of a magnet. (See illustration in margin.) A steady current was produced in a circuit connected to the disk through brass contacts or "brushes." His device, called the "Faraday disk dynamo," was the first constant-current electric generator. This particular arrangement did not turn out to be very practical, but it showed that continuous generation of electricity was possible.

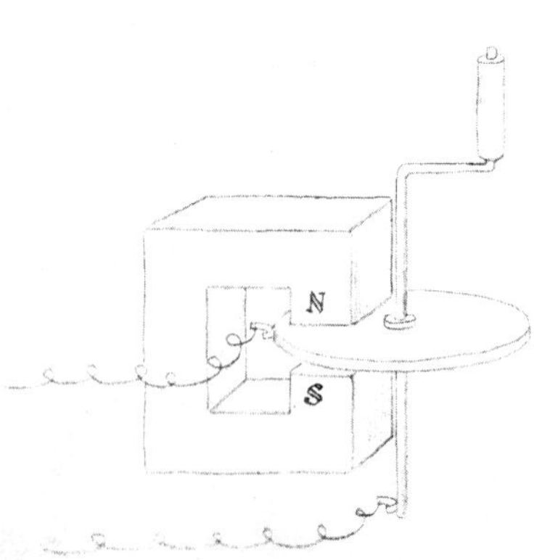

These first experimental means of producing a continuous current were important aids to understanding the connection between electricity and magnetism. Moreover, they suggested the possibility of eventually generating electricity on a large scale. The production of electrical current involves changing energy from one form to another. When electrical energy appears, it is at the cost of some other form of energy. In the electric battery, chemical energy (the energy of formation of chemical compounds) is converted to electrical energy. Batteries are useful for many portable applications (automobiles and flashlights, for example). But it is not practical to produce large amounts of electrical energy by this means. There is, however, a vast supply of mechanical energy available from many sources. Electrical energy could be produced on a large scale if some reasonably efficient means of converting mechanical energy to electrical energy were available. This mechanical energy might be in the form of wind, or falling water, or continuous mechanical motion produced by a steam engine. The discovery of electromagnetic induction showed that, at least in principle, it was possible to produce electricity by mechanical means. In this sense, Faraday can rightly be regarded as the founder of the modern electrical age.

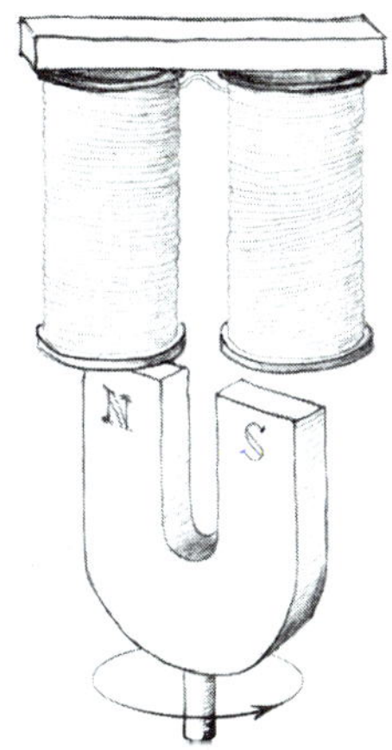

One generator of 1832 had a permanent horseshoe magnet rotated by hand beneath two stationary coils in which current was induced.

?

2. *Why is Faraday considered the discoverer of electromagnetic induction?*
3. *What is the general definition of electromagnetic induction?*

15.4 | Generating electricity by the use of magnetic fields: The dynamo

Faraday had shown that when a conducting wire moves relative to a magnetic field, a current is produced. Whether it is the wire or the magnetic field that moves does not matter. What counts is the relative motion of one with respect to the other. Once the principle of electromagnetic induction had been discovered, experimenters tested many combinations of wires and magnets in relative motion. One basic type of generator (or "dynamo," as it was often called) was widely used in the nineteenth century. In fact, it remains the basic model for many generators today.

This form of generator is basically a coil of wire that can be rotated in a magnetic field. The coil is connected to an external circuit by sliding contacts. On page 472, in the diagram above and on the left, the "coil" is shown for simplicity as a single rectangular loop of wire. This loop rotates around an axis XY between the north and south poles of a magnet. Two conducting rings *d* and *e* are permanently attached to the loop and, therefore, also rotate around the axis. Conducting brushes *f* and *g* complete a circuit through a meter at *h* that indicates the current produced. The complete circuit is *abdfhgea*. (Note that

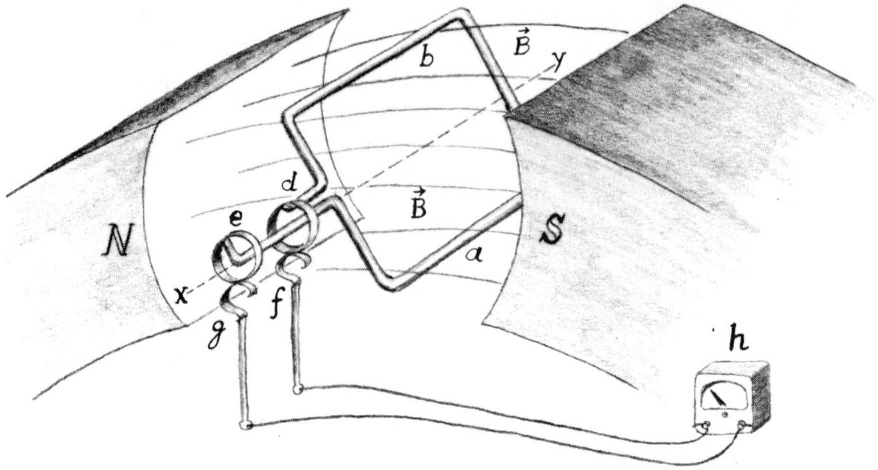

Alternating-current generator.

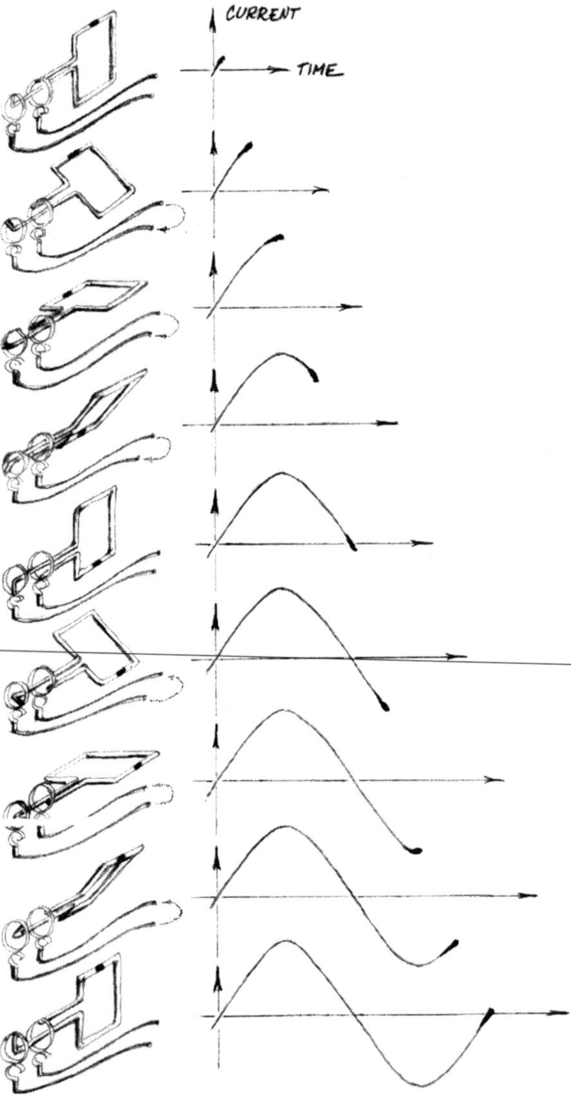

one part of the wire goes through ring *d* without touching it and connects to *e*.)

Initially, the loop is at rest between the magnetic poles and no charge flows through it. Now suppose the loop is rotated counter-clockwise. The wire's long sides *a* and *b* now have a component of motion perpendicular to the direction of the magnetic lines of force; that is, the wire "cuts across" lines of force. This is the condition for inducing an electric current in the loop. The greater the rate at which the lines are cut, the greater the induced current.

To understand better what is going on in the wire, you should understand its operation in terms of the force on the charges in the wire. It is the movement of these charges that forms the current. The charges in the part of the loop labeled *b* are being physically moved together with the loop across the magnetic field. Therefore, they experience a magnetic force given by qvB (as described in Sec. 14.13). This force pushes the charges in the wire "off to the side." In this situation, "off to the side" is *along the wire*.

What about side *a*? It is also moving through the field and "cutting" lines of force, but in the opposite direction. So the charges in *a* experience a push along the wire in the direction opposite to those in *b*. This is just what is needed; the two effects reinforce each other in

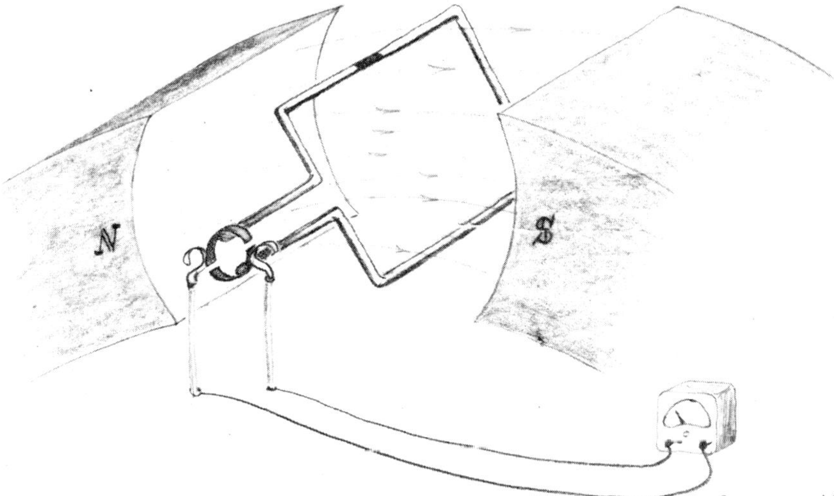

Direct-current generator.

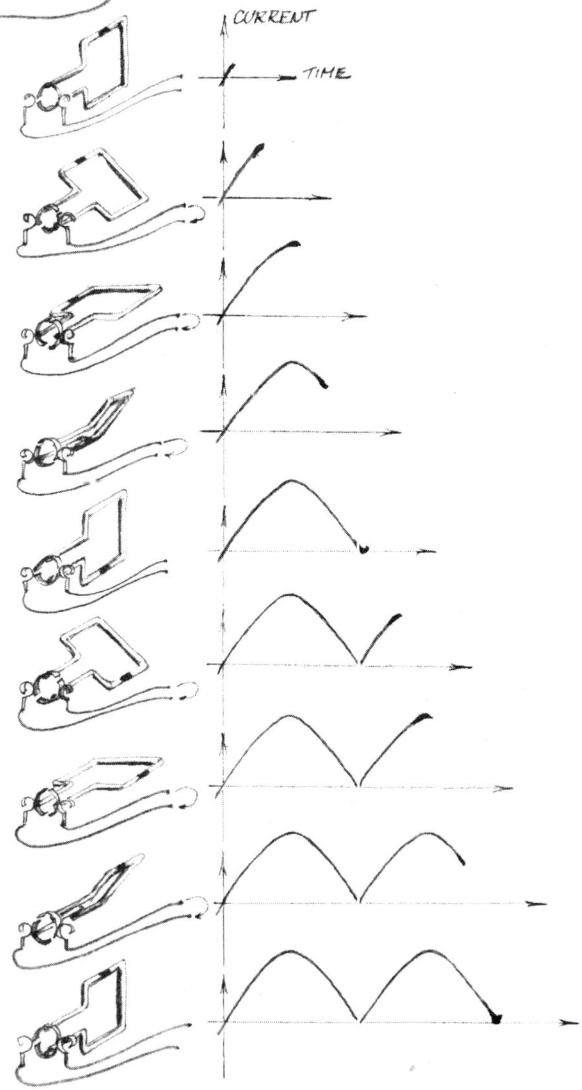

generating a current around the whole loop. The "push" that produces the current can also be regarded as resulting from a potential difference ("voltage") induced in the loop of wire. Therefore, a generator produces both "voltage" and current.

The generator just described produces *alternating current* (abbreviated ac). The current is called "alternating" because it regularly reverses (alternates) its direction. This is indicated in the margin on page 472. At the time this kind of generator was first developed, in the 1830's, alternating current could not be used to run machines. Instead *direct current* (dc) was needed.

In 1832, Ampère announced that his instrument maker, Hippolyte Pixii, had solved the problem of generating direct current. Pixii modified the ac generator by means of a device called the *commutator*. The name comes from the word *commute*, to interchange or to go back and forth. The commutator is a split cylinder (see top of this page) inserted in the circuit. In the ac generator (previous page), brushes *f* and g are always connected to the same part of the loop. But with the commutator, the brushes *reverse connections* each time the loop passes through the vertical position. Just as the direction of current induced in the loop is at the point of reversing, the contacts reverse. As a result, the current in the ouside circuit is always in the same direction.

Although the current in the outside circuit is always in the same direction, it is not constant. It rises and falls rapidly between zero and its maximum value, as shown in the drawings on page 473. In working generators, many sets of loops and commutators are connected together on the same shaft. In this way, their induced currents reach their maximum and zero values at different times. The *total* current from all of them together is then more uniform.

SG 6

Whether a generator delivers alternating or direct current, the electric power (energy per unit time) produced at every instant is given by the same equation developed in Sec. 14.10. For example, suppose that a wire (for example, the filament wire in a light bulb) with resistance R is substituted for the meter at h. If the current generated in the circuit at a given time is I, the electrical energy per unit time delivered to the wire is given by I^2R. For alternating current, the power output varies from instant to instant. But the *average* output power is simply $(I^2)_{av}R$. This electrical energy, of course, does not appear by itself, without any source. That would violate the laws of conservation of energy. In the generator, the "source" of energy is clearly the mechanical energy that keeps the coils rotating in the magnetic field. This mechanical energy is provided by a steam or gasoline engine, or by water power, wind power, etc. The generator is thus a device for converting mechanical energy to electrical energy.

SG 7–9

?

4. What is the position of a rotating loop when it generates maximum current? minimum? Why?

5. What is the purpose of the commutator?

6. Where does the energy delivered by the generator come from?

15.5 | The electric motor

The greatest obstacle to practical use of electric motors was the lack of cheap electric current to run them. The chemical energy in batteries was quickly exhausted. The dynamo, invented almost simultaneously by Faraday and Henry in 1832, was at first also not at all economical in producing electrical current when mechanical energy was expended on it. Generators that used mechanical power efficiently to produce electric power were needed. But to design such generators required an understanding of the details of operation, and this understanding took nearly 50 years.

In fact, a chance event marked the effective start of the electric power age. This event was an accidental discovery at the Vienna

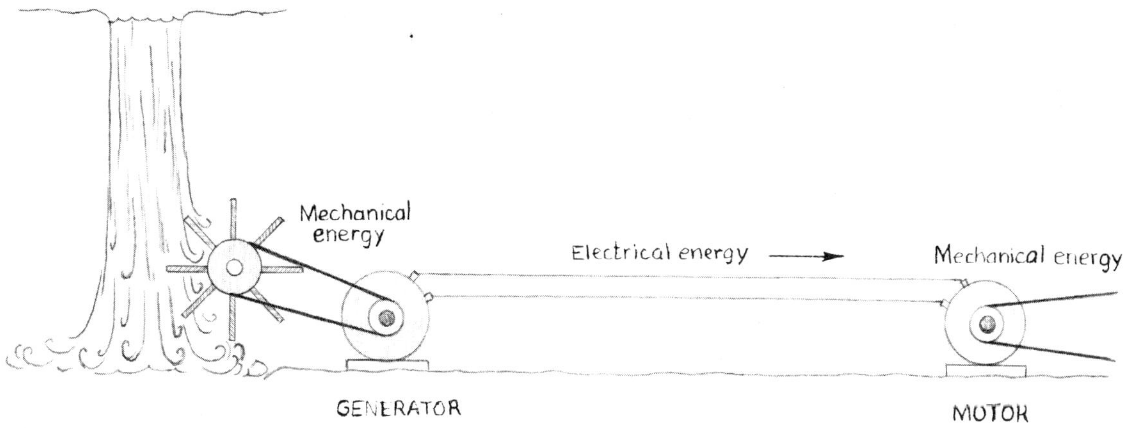

Exhibition of 1873. The story goes that an unknown worker at the Exhibition just happened to connect two dynamos together. The first dynamo, which was mechanically driven, generated current, and this current then passed through the coils of the second dynamo. Amazingly, the second dynamo then ran as an electric motor, driven by the electricity generated by the first dynamo.

This accidental discovery that a generator (dynamo) could function as a motor was immediately utilized at the Exhibition. A small artificial waterfall was used to drive the generator. Its current then drove the motor, which in turn operated a device that did mechanical work. This is the basic operation of a modern electrical transmission system. A turbine driven by steam or falling water drives a generator that converts the mechanical energy to electrical energy. Conducting wires transmit the electricity over long distances to motors, toasters, electric lights, etc. These devices in turn convert the electrical energy to mechanical energy, heat, or light.

SG 10

SG 11

The development of electrical generators shows an interaction of science and technology different from that of the development of steam engines. As was pointed out in Chapter 10, the early steam engines were developed by practical inventors. These inventors had no knowledge of what is now considered to be the correct theory of heat (thermodynamics). But their development of the steam engine, and attempts by Sadi Carnot and others to improve its efficiency through theoretical analysis, contributed greatly to the establishment of thermodynamics. In that case, the advance in technology came before the advance of science. In the case of electromagnetism, the reverse occurred. A large amount of scientific knowledge was built up by Ampère, Faraday, Kelvin, and Maxwell before any serious practical application succeeded. The scientists, who understood electricity better than anyone else, were not especially interested in commercial applications. And the inventors, who hoped to make huge profits from electricity, knew very little theory. After Faraday announced

A commercial generator. As in almost all large generators, the coils of wire in which current is induced are around the outside, and electromagnets are rotated on the inside.

his discovery of electromagnetic induction, people started making generators to produce electricity immediately. But it was not until 40 years later that inventors and engineers understood enough to work with such necessary concepts as lines of force and field vectors. With the introduction of the telegraph, telephone, radio, and alternating-current power systems, a much greater mathematical knowledge was needed to work with electricity. Universities and technical schools started to give courses in electrical engineering. Gradually, a group of specialists developed who were familiar with the physics of electricity and who also knew how to apply it.

Water-driven electric generators producing power at the Tennessee Valley Authority. The plant can generate electric energy at a rate of over 100,000,000 watts.

?

7. How would you make an electric motor out of a generator?

8. What prevented the electric motor from being an immediate economic success?

9. What chance event led to the beginning of the electric power age?

15.6 | The electric light bulb

The growth of the electric industry has resulted largely from the great public demand for electrical products. One of the first commercially successful electrical products in the United States was the electric light bulb. Its success is an interesting case of the relationship between physics, industry, and society.

At the beginning of the nineteenth century, buildings and homes were lit by candles and oil lamps. There was almost no street lighting in cities except for a few lights hung outside houses at night. The natural gas industry was just starting to change this situation. London got its first street lighting system in 1813, when gas lights were installed on Westminster Bridge. However, the social effects of gas lighting were not all beneficial. For example, gas lighting in factories enabled employers to extend an already long and difficult working day into one still longer.

In 1801, the British chemist Humphry Davy noted that a brilliant spark or arc appeared when he broke contact between carbon rods connected to the two terminals of a battery. This discovery led to the development of the *arc light*.

The arc light was not practical for general use until steam-driven electrical generators replaced expensive batteries as a source of current. In the 1860's and 1870's, arc lights began to be used for street lighting and lighthouses. However, they were too glaring and too expensive for use in the home. Also, the carbon rods burned up in a few hours because of the high temperatures produced by the arc. This need for frequent service and replacement made the system inconvenient. (Arc lights are still used for some high-intensity purposes, such as spotlights in theaters and motion picture projectors.)

As Davy and other scientists showed, light can be produced simply by heating a wire to a high temperature by passing a current through it. This method is known as *incandescent* lighting. The major technical drawback was that the wire filament gradually burned up. The obvious solution was to enclose the filament in a glass container from which all the air had been removed. But this was easier said than done. The vacuum pumps available in the early nineteenth century could not produce a strong enough vacuum for this purpose. It was not until 1865, when Hermann Sprengel in Germany invented an improved vacuum pump, that the electric light bulb in its modern form could be developed. (Sprengel's pump also greatly aided Crookes and others in scientific experiments leading to important discoveries in atomic physics. These discoveries will be discussed in Chapter 18.)

Thomas Edison was not the first to invent an incandescent light, nor did he discover any essentially new scientific

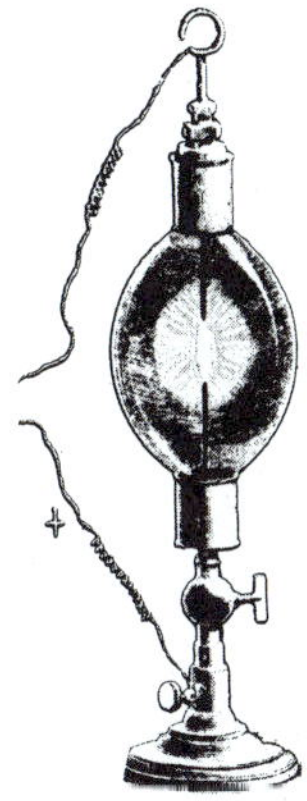

Davy's arc lamp.

Demonstrations of the new electric light during a visit of Queen Victoria and Prince Albert to Dublin, Ireland from Illustrated London News, *August 11, 1849.*

In the late 1800's, dynamo-powered arc lamps were used in some European cities.

Lewis Howard Latimer (1848–1928). The son of a runaway slave, Latimer became one of the original associates of Thomas Edison. Latimer was an inventor, patent authority, poet, draftsman, author, and musician.

principles. What he did was develop a practical light bulb for use in homes. Even more important, he worked out a distribution system for electricity. His system not only made the light bulb practical, but opened the way for mass consumption of electrical energy in the United States.

Edison started by making an important assumption about *how* people would want to use their light bulbs. He decided that each customer must be able to turn on and off any single bulb without affecting the other bulbs connected to the circuit. This meant that the bulbs must be connected "in parallel," like the rungs of a ladder, rather than "in series."

Edison in his laboratory.

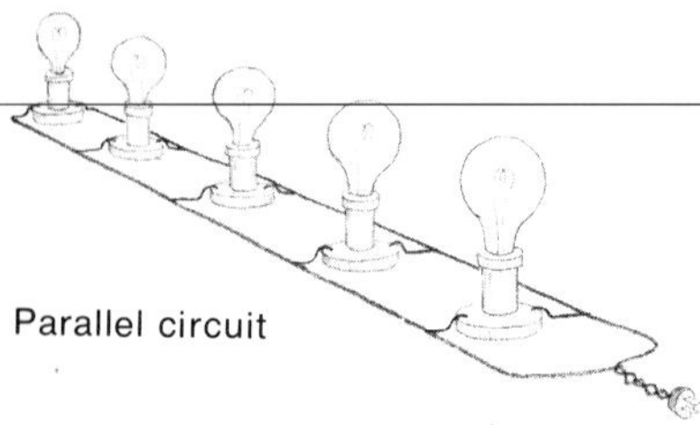

Parallel circuit

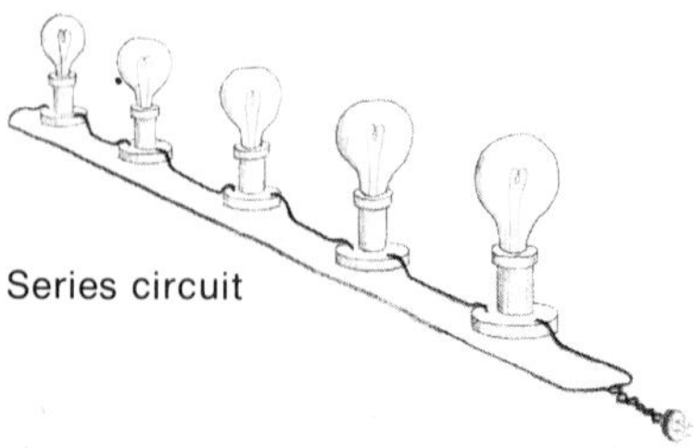

Series circuit

The choice of parallel rather than series circuits had important technical consequences. In a series circuit, the same current goes through each bulb. In a parallel circuit, only part of the total current available from the source goes through any one bulb. To keep the total current needed from being too large, the current in each bulb has to be small.

As noted in Chapter 14, the heating effect of a current depends on both the resistance of the wire and the amount of current. The rate at which heat energy is produced is I^2R; that is, the rate goes up directly as the resistance, but increases as the *square* of the current. Therefore, most inventors used high-current, low-resistance bulbs and assumed that parallel circuits would not be practical. Edison realized that a small current can have a large heating effect if the resistance is high enough.

So Edison began a search for a suitable high-resistance, nonmetallic substance for his filaments. To make such a filament, he first had to bake or "carbonise" a thin piece of a substance. Then he sealed it inside an evacuated glass bulb with wires leading out. His assistants tried more than 1,600 kinds of material: "paper and cloth, thread, fishline, fiber, celluloid, boxwood, coconut-shells, spruce, hickory, hay, maple shavings, rosewood, punk, cork, flax, bamboo, and the hair out of a redheaded Scotchman's beard." Edison's first successful high-resistance lamp was made with carbonized cotton thread in a high-vacuum sealed bulb. It burned continuously for two days before it fell apart. This was in October 1879. The following year, Edison produced lamps with filaments made from bamboo and paper.

SG 12, 13

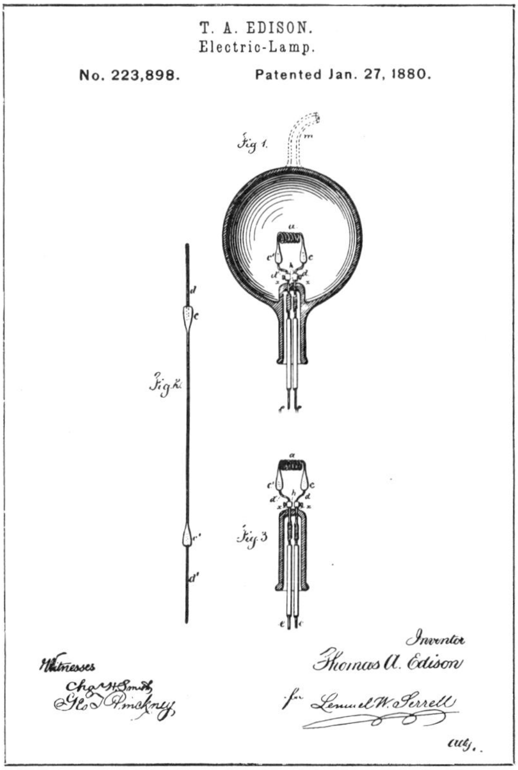

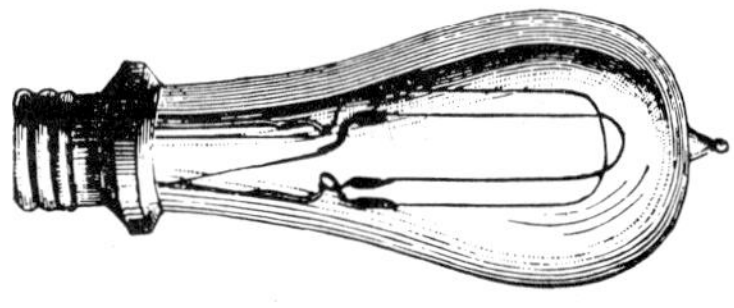

One type of Edison lamp. Note the familiar filament and screw-type base.

The Edison Electric Light Company began to install lighting systems in 1882. After only 3 years of operation, the Edison company had sold 200,000 lamps. It had a virtual monopoly of the field and paid big dividends to its stockholders.

The electric light bulb had changed somewhat since Edison's original invention. For example, the carbonized filaments of the older lamps had been replaced in newer bulbs by thin tungsten

EDISON'S LIGHT.

The Great Inventor's Triumph in Electric Illumination.

A SCRAP OF PAPER.

It Makes a Light, Without Gas or Flame, Cheaper Than Oil.

TRANSFORMED IN THE FURNACE.

Complete Details of the Perfected Carbon Lamp.

FIFTEEN MONTHS OF TOIL.

Story of His Tireless Experiments with Lamps, Burners and Generators.

SUCCESS IN A COTTON THREAD.

The Wizard's Byplay, with Bodily Pain and Gold "Tailings."

HISTORY OF ELECTRIC LIGHTING.

The near approach of the first public exhibition of Edison's long looked for electric light, announced to take place on New Year's Eve at Menlo Park, on which occasion that place will be illuminated with the new light, has revived public interest in the great inventor's work, and throughout the civilized world scientists and people generally are anxiously awaiting the result. From the beginning of his experiments in electric lighting to the present time Mr. Edison has kept his laboratory guardedly closed, and no authoritative account (except that published in the HERALD some months ago relating to his first patent) of any of the important steps of his progress has been made public—a course of procedure the inventor found absolutely necessary for his own protection. The HERALD is now, however, enabled to present to its readers a full and accurate account of his work from its inception to its completion.

A LIGHTED PAPER.

Edison's electric light, incredible as it may appear, is produced from a little piece of paper—a tiny strip of paper that a breath would blow away. Through

wires. Tungsten had the advantages of greater efficiency and longer life.

The widespread use of light bulbs confirmed the soundness of Edison's theory about what people would buy. It also led to the rapid development of systems of power generation and distribution. The need for more power for lighting spurred the invention of better generators, the harnessing of water power, and the invention of the steam turbine. Success in providing large quantities of cheap energy made other uses of electricity practical. Once homes were wired for electric lights, the current could be used to run sewing machines, vacuum cleaners, washing machines, toasters, and (later on) refrigerators, freezers, radios, and television sets. Once electric power was available for relatively clean public transportation, cities could grow rapidly in all dimensions. Electric elevators made high-rise buildings practical, while electric tramways and subways rapidly transported people from their homes to jobs and markets.

We are now so accustomed to more sophisticated applications of electricity that it is hard to realize the impact of something as simple as the light bulb. But most people who lived through the period of electrification, which was as late as the 1930's and 1940's in many rural areas of the United States, agreed that the electrical appliance that made the greatest difference in their daily lives was the electric light bulb.

?

10. Why were arc lights not used for illuminating homes?

11. What device was essential to the development of the incandescent lamp?

12. Why did Edison require a substance with a high resistance for his light bulb filaments?

13. What were some of the major effects the introduction of electric power had on everyday life?

15.7 Ac versus dc and the Niagara Falls power plant

Section 15.4 stated that the earliest electric generators produced alternating current, which could be changed into direct current by the use of a commutator. Throughout most of the nineteenth century, most engineers believed that only dc was useful in practical applications of electricity. However, as the demand for electric power increased, some disadvantages of dc became evident. One problem was that the commutator complicated the mechanical design of generators, especially if the ring had to rotate at high speed. This difficulty became even more serious after the introduction of steam turbines in the 1890's, since turbines work most effectively at high speeds. Another disadvantage was there was no convenient way of changing the generated voltage of a direct current supply.

Opposite page: First newspaper account of Edison's invention (New York **Herald**, *December 21, 1879).*

Why should it be necessary to change the voltage with which current is driven through a transmission system? One reason involves the amount of power lost in heating the transmission wires. The power output of a generator depends (as indicated in Sec. 14.10) on the output *voltage* of the generator and the *amount of current*:

SG 14

$$P_{\text{total}} = VI$$

The power made available by the generator is transmitted to the line and to the consumer. The *same* amount of power can be delivered at smaller I if V is somehow made larger. When there is a current I in a transmission wire of resistance R, the portion of the power lost as heat in transmission is proportional to the resistance and the square of the current:

$$P_{\text{heat loss}} = I^2R$$

The power finally available to consumers is $P_{\text{total}} - P_{\text{heat loss}}$. For transmission lines of a given resistance R, where the value of R is fixed by the wires themselves, the current I should be as small as possible in order to minimize the power loss. Obviously, therefore, electricity should be transmitted at low current and at high voltage.

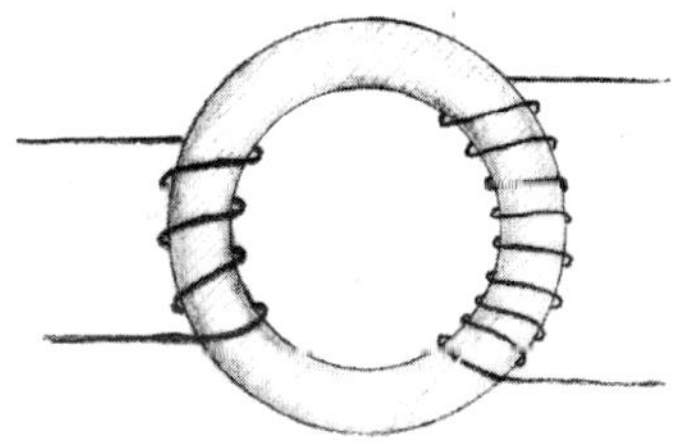

However, most generators cannot produce electricity at very high voltages. To do so would require excessively high speeds of the moving parts. Some way of "stepping up" the generated electricity to a high voltage for transmission is needed. But some way of "stepping down" voltage again at the other end, where the consumer uses the power, is also needed. For most applications of electricity, especially in homes, it is neither convenient nor safe to use high voltages. In short, *transformers* are needed at both ends of the transmission line.

A transformer can easily be made by a simple change in Faraday's induction coil (Sec. 15.4). Recall that Faraday wound a coil of wire called the *secondary* coil) around one side of an iron ring. He then induced a current in this secondary coil by *changing* a current in another coil (the *primary* coil) wound around the other side of the ring. A current is induced in the secondary coil whenever the primary current changes. If the primary current is changing all the time, then a current is continually induced in the secondary. An alternating current applied to the primary coil (for example, from a generator without a commutator) induces an alternating current in the secondary coil.

A steady current (dc) in the primary induces no current at all in the secondary; transformers work on ac.

By the law of conservation of energy, the output power from a transformer cannot exceed the input power. So if the output voltage is increased (by a greater coil ratio for the secondary coil), the output current will decrease proportionally.

SG 15–17

One more concept is important to understanding a simple electric transformer. If the secondary has *more* turns than the primary, the alternating voltage produced across the secondary coil will be *greater* than that across the primary. If the secondary has *fewer* turns than the primary, the alternating voltage produced across the secondary will be *lower* than the voltage across the primary. This fact was discovered by Joseph Henry, who built the first transformer in 1838.

The first ac system was demonstrated in Paris in 1883. An experimental line that powered arc and incandescent lighting, through transformers, was installed in a railway line in London in 1884. Another one was exhibited shortly afterward in Italy. An American engineer, George Westinghouse, saw the Italian system and bought the American patent rights for it. Westinghouse had already gained a reputation from his invention of the railway air brake. He also had set up a small electrical engineering company in Pittsburgh. After improving the design and construction of transformers, the Westinghouse Electric Company set up its first commercial installation in 1886. Its purpose was to distribute alternating current for incandescent lighting in Buffalo, New York.

SG 18

When Westinghouse introduced its ac system in the United States, the Edison Electric Light Company held an almost complete monopoly of the incandescent lighting business. The Edison Company had invested a great deal of money in dc generating plants and distribution systems for most of the large cities. Naturally, Edison was alarmed by a new company that claimed to produce electric power for illumination with a much cheaper system. A bitter public controversy followed. Edison attempted to show that ac was unsafe because of the high voltage used for transmission. In the middle of the dispute, the New York State Legislature passed a law establishing electrocution as a means of capital punishment. This event seems to have added to the popular fear of high voltage.

Nevertheless, the Westinghouse system continued to grow. There were no spectacular accidents, and the public began to accept ac as reasonably safe. The invention of the "rotary converter" (essentially an ac motor driving a dc generator) also helped to end the dispute. It could change ac into dc for use in local systems already set up with dc equipment, or it could power individual dc motors. So the Edison company (later merged into General Electric) did not have to go out of business when ac was generally adopted.

The final victory of the ac system was assured in 1893, when ac was chosen for the new hydroelectric plant at Niagara Falls. In 1887, businessmen in Buffalo had dangled a $100,000 prize before "the Inventors of the World." The prize would go to the inventor who designed a system for utilizing the power of the

Niagara River "at or near Buffalo, so that such power may be made practically available for various purposes throughout the city." The contest attracted worldwide attention. Large quantities of electrical power had never before been transmitted over such a distance. It was 32 km from Niagara Falls to Buffalo. The success or failure of this venture would influence the future development of electrical distribution systems for other large cities.

It was a close decision whether to use ac or dc for the Niagara Falls system. Ac could be generated and transmitted more efficiently. But the demand for electricity in 1890 was mainly for lighting. This meant that there would be a peak demand in the evening. The system would have to operate at less than full capacity during the day and late at night. Because of this variation in the demand for electricity, some engineers believed that a dc system would be cheaper to operate. This was because *batteries* could be used to back up the generators in periods of peak demand. Thomas Edison was consulted, and without hesitation he recommended dc. But the Cataract Construction Company, which had been formed to administer the project, delayed making a decision.

The issue was still in doubt in 1891, when the International Electrical Exhibition opened in Frankfurt, Germany. There, a fairly high-voltage ac line carrying sizable quantities of power 176 km from Frankfurt to Lauffen was demonstrated. Tests of the line showed an efficiency of transmission of 77%; that is, for every 100 W fed in at one end of the line, only 23 were wasted by heating effects in the line. The other 77 W were delivered as useful power. The success of this demonstration reinforced the gradual change in expert opinion in favor of ac over dc. Finally, the Cataract Company decided to construct an ac system.

After the ac system had been established, it turned out that the critics had been wrong about the variation of demand for electricity throughout the day. Electricity found many uses besides lighting. In the 1890's, electric motors were already being used in street railway cars, sewing machines, and elevators. Because of these diverse uses, the demand for electricity was spread out more evenly during each 24-hour period. In the particular case of the Niagara Falls power plant, the source of energy was the steady flow of water down the Niagara River. This made it possible to produce energy continuously without much extra cost. (The boiler for a steam turbine would either have to be kept supplied with fuel late at night, or shut down and started up again in the morning.) Since hydroelectric power was available at night at low cost, new uses for it became possible. The Niagara Falls plant attracted electric furnace industries, continually producing aluminum, abrasives, silicon, and graphite. Previously, the electrochemical processes involved in these

Wilson Dam (Tennessee Valley Authority), Alabama.

The general principle of hydroelectric power generation is shown in this sketch. Water flowing from a higher to a lower level turns turbine blades attached to a generator shaft. The details of construction vary widely.

Niagara Power Plant (above right).

SG 19

industries had been too expensive for large-scale use. Cheap power now made them practical. These new industries in turn provided a constant demand for power, making the Niagara project even more profitable than had been expected.

The first transmission of power to Buffalo took place in November 1896. By 1899, there were 5,000 horsepower units in operation at Niagara. The stockholders of the Cataract Construction Company already had earned a profit of better than 50% on their investment. The electrochemical industries, which had not figured in the original plans at all, were using more power than lighting and motors together.

There is a postscript to the story of ac versus dc. Dc is now coming back into favor for long-distance transmission of electric power at high voltages.

?

14. Give one reason why it is more economical to transmit electric power at high voltage and low current than at low voltage and high current.

15. Why will transformers not operate if steady dc is furnished for the primary coil?

15.8 Electricity and society

An optimistic view. Many times during the last 100 years, enthusiastic promoters have predicted a marvelous future for us all, based on the application of electricity to all phases of life. Machinery run by electricity will do all the backbreaking physical labor that has been the lot of 99% of the human race throughout the ages, and still is for most of humanity today. The average citizen will have nothing to do except supervise machinery for a few hours a day and then go home to enjoy a life of leisure. Electric machines will also do all the household chores, such as cleaning, laundering, ironing, cooking, and dishwashing.

Others, including President Franklin D. Roosevelt, who believed that country life is more natural and healthy than city life, conceived of electrical technology as having a social purpose. In the nineteenth century, the steam engine had provided a source of power that could take over most work done by humans and animals, but to use this power people had had to crowd into the cities, close to the power generating plant. Now that electrical transmission of power at a distance was possible, people could go back to the countryside without sacrificing the comforts of city life. Heating, lighting, and refrigeration by electricity would make life easier and more sanitary in difficult climates. One of the major achievements of Roosevelt's administration in the 1930's was the rural electrification program. This program gave loans to rural cooperatives for installing electrical generating and distribution systems in areas where private power companies had found it unprofitable to operate. Federal power projects such as the Tennessee Valley Authority also assisted in the campaign to make electricity available to everyone. Electricity made country life a bit easier, reducing the physical labor involved in farming and lengthening the day for leisure and education. In this way, electrification should have helped to reverse the migration of people from rural to urban areas.

An effect of electricity unforeseen by its original promoters has been its tendency to unite a large country into a single social unit by providing rapid transportation and even more rapid communication between the different parts. In transportation, electricity-operated devices play essential roles both in the mass manufacture and in the operation of cars, trucks, and buses. As to communication, remember that human society evolves much as do biological organisms: All parts develop in step and increase their interdependence. It follows that telecommunication and modern civilization *had* to develop together. The telephone is

SG 20

Close Up

Commercial Distribution of Electric Power

The commercial distribution of ac electric power requires elaborate transmission facilities. Generator output voltages of about 10^4 volts are stepped up to about 10^5 volts for transmission, stepped down to about 10^4 volts for local distribution, and further stepped down to about 10^2 volts by neighborhood power-pole transformers. Within the home, they may be stepped down further (often to 6 volts for doorbells and electric trains, as with the transformer shown below) and stepped up by transformers in radio and TV sets for operating high-voltage tubes.

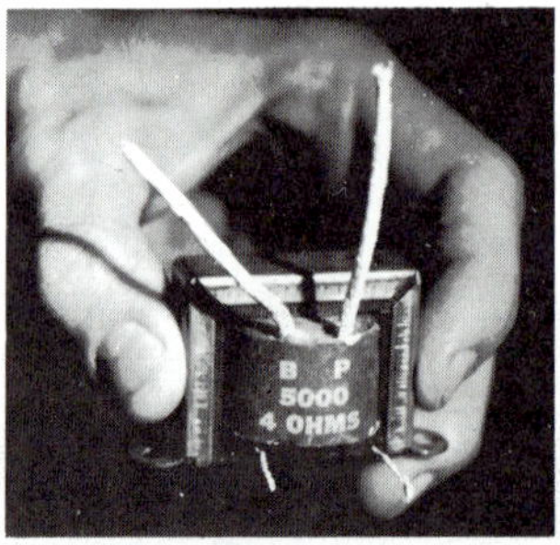

Major electric transmission lines in the United States. In many cases several lines are represented by a single line on the map. Not shown are the small-capacity lines serving widely scattered populations in the mountainous and desert areas. In the densely populated areas, only the high-voltage lines are shown.

Top: New York City before the blackout of 1977.

Bottom: After the blackout.

most valuable in a complicated, cosmopolitan society. In fact, many of the basic institutions of society, for example, a free Press, could not operate today without rapid, two-way electronic communication.

Optimists envision electricity doing more and more things for a larger and larger part of the population. Electric appliances such as refrigerators and air conditioners will contribute to healthier, more comfortable lives the world over. Electronic communications will continue to spread, allowing an ever-greater exchange of facts, opinions, and cultures. Electric machines will do more and more of our difficult work for us. Thanks to advances in science and related technology, many people no longer have to spend almost all of their time working for the bare necessities of life. Whatever it is that we really want to do, the optimists say, electricity can help us do it better.

A less optimistic opinion. Wonderful as all this seems, many people take a much dimmer view of the "progress" cited above. They point, for example, to dwindling resources of fossil fuel (coal, oil, and gas). They argue that industries in the more advanced countries have used up in only 200 years most of the reserves of chemical energy accumulated over the last 200 million years. Moreover, these industries have generally polluted the air and water, except when confronted by outraged public opinion. Other skeptics claim that a social system has been created in which the virtues of "honest toil and pride of workmanship" are endangered. Instead, most people work at dull, trivial jobs, while many others suffer chronic unemployment. And while it is true that many people in wealthy, industrial countries enjoy high standards of living and can purchase many gadgets and luxuries, these things have not fulfilled real human and social needs. Even the "demand" for them is not real, but is artificially created by advertising campaigns and planned obsolescence. Therefore, such items do not bring happiness and peace of mind, but only a growing, mindless clamor for more and more material possessions. Meanwhile, the materially less fortunate people are separated by a wider and wider gap from the richer ones and look on in growing envy and anger.

What about "labor-saving" household appliances? Do they really make life easier for the upper- and middle-income families who own them? Not much, say the critics, because the work done by such machines was previously done by servants, anyway. Of course, industrialization and electrification *have* created some jobs for people with little training. These jobs may be more attractive than work as a servant, but they still do not pay very well. As a result, many low-income families cannot afford more than one major electrical appliance. The one

What it costs in a metropolitan area to run some electric home appliances (approximate, in cents):

refrigerator, frostless, 16 cu. ft.	50	a day
freezer, 16 cu. ft.	50	a day
range	15	a meal
oven	5	an hour
oven, self-cleaning	30	extra
dishwasher	10	a load
toaster	0.5	a slice
toaster oven	15	an hour
coffee maker	3	a pot
blender	0.2	a blend
mixer	0.5	a mix
TV, color	5	an hour
TV, black & white	2.5	an hour
radio-phonograph	1	an hour
light bulb, 100-watt	1	an hour
air conditioner, 24°–25° C	10	an hour
blanket, king-size	10	a night
clock	15	a month
vacuum cleaner	8	an hour
washing machine	5	a load
dryer	30	a load
iron	10	an hour
hair dryer	5	an hour
shaver	0.01	a shave

appliance usually purchased by such families is a television set, which, the skeptics say, is not likely to contribute much to improving the quality of life.

The decentralization of population which electricity was supposed to help produce has come about, but in an unexpected way. The upper- and middle-income inhabitants of cities have indeed been able to escape to the suburbs where they can enjoy the convenience and pleasures of the electrical age. But they have left behind them urban ghettos crowded with minority groups. These people are naturally angry at being deprived of the benefits of the "affluent society" and the suburban life presented to them on television. As for the farmer, modern technology has made agriculture into a giant industry with no place for the small landholder.

Electrical communications and rapid transportation are binding us more and more into a close-knit, interdependent social system. But this has its disadvantages, too. For example, an electronic computer may be used by an employer or a state to dredge up all a person's past mistakes. The threat of war has become ever more terrifying because of modern weapons. In the same way, the threat of an authoritarian state is much greater when government adopts the tools of rapid communication and information processing to its own purposes.

Electricity: Good or bad? Such criticisms illustrate the other half of the total story. Electrical devices, like all other technological improvements based on scientific discovery, are

Electric power lines in New York State.

SG 21

neither good nor bad by themselves. Electricity increases enormously the whole range of possibilities open to us. But choices among these possibilities still have to be made on the basis of value systems outside the framework of science or technology. Important decisions lie ahead concerning electrification, the use of nuclear power, automation and other uses of computers, and many other applications of electricity. These decisions cannot be left to the experts in physics or engineering, to the public or private utilities, or to government agencies. They must be made by citizens who have taken the trouble to learn something about the physical forces that play such an important role in modern civilization.

15.9 | Alternate energy sources

There are other ways to obtain energy that can be used to produce electricity, for example, harnessing the tides, wind power, heat from the earth's depths, nuclear reactions, and the energy that comes directly from the sun. This last, solar power, is receiving intense study because of the vast amount of energy potentially available.

Just outside the earth's atmosphere the sun's radiation provides 1,360 W of power to each square meter of surface. By the time the radiation reaches the earth's surface, much of this is lost, because of atmospheric absorption and clouds. Since the earth rotates, direct sunlight is available in any particular spot on the earth's surface for only about 8 hr (on the average) each day. Depending upon the location, then, between 150 and 450 W/m^2 are delivered to ground level when averaged over a 24-hr period. This power can be used to boil water with which to power an ordinary steam turbine generating plant or to heat water for household use.

An experimental solar house. Glass panels trap heat through the greenhouse effect. Solar cells beneath the panels convert light to electricity, which can then be stored in batteries. Heat from the cells is transferred to a coolant and stored for later use.

Sunlight can also be converted directly into electricity in photocells or "solar" cells by a process called *photovoltaic conversion*. The basic operation of the photocell is explained on page 492. Even the best photocell is unable to turn all the energy that strikes it into electrical energy, for much of the light is reflected, transmitted, or turned into heat within the cell. The ratio of energy striking the cell to electrical energy produced is called the *efficiency*. While the best efficiency obtainable at this writing is 23% for very specially prepared cells, a more typical value for commercial cells is 12%. Moreover, the cost of a solar cell is high, currently around $6 per watt of output. This is many times the cost per watt of either conventional or nuclear power plants.

While solar power could solve many of the problems resulting from fuel shortages and the environmental pollution of other energy sources, the generation of even a significant fraction of our nation's electric power directly from the sun is dependent on finding ways of building far less expensive collection systems for accepting the sun's energy and storage systems for providing the electricity when it is needed (at night, for example, or on cloudy days). Probably the most practical use of solar energy during the next decade or two will be to assist in the heating of houses and in the production of hot water for home use.

15.10 | The efficiency of an electric power plant

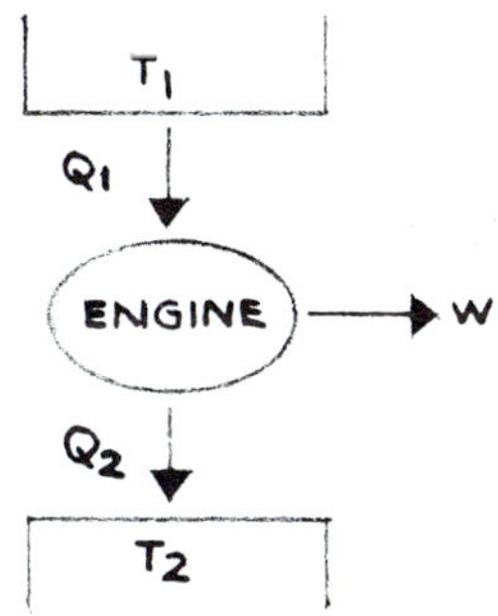

An electric power plant, whether powered by fossil fuels (coal, oil, or gas) or nuclear fuel, needs both a heat engine and a generator in order to produce electricity. The thermodynamic limit of the efficiency of a heat engine sets very severe constraints on how much of the energy released from burning the fuel is ultimately available as electrical energy. (This limit does not, of course, apply to hydroelectric plants.)

You saw in Unit 3 (Sec. 10.7) that any engine that converts heat into mechanical work must also release heat into the environment. A diagram of this process, which can be applied to Watt's steam engine or to a large steam turbine, has been sketched in the margin. T_1 and T_2 are the temperatures of the hot and cold reservoirs, respectively; Q_1 is the heat fed into the engine; Q_2 is the waste heat released into the environment; and W is the mechanical work obtained from the engine.

The second law of thermodynamics states that in the best possible circumstances the efficiency (η) of the heat engine can be no greater than

$$\eta = 1 - \frac{T_2}{T_2}$$

This law was discussed in detail on page 296 in Sec. 10.7.

What does this mean for a power plant? Fuel is burned in a combustion chamber; the chemical (or nuclear) energy is converted into thermal energy that keeps the combustion chamber at the temperature T_1. Water, heated by the combustion chamber in the boiler, circulates through the plant as steam during parts of its route and as liquid during others. In most plants, very high-pressure steam is created in the boiler. This steam pushes against the blades of a turbine, doing work on the turbine, and leaves the turbine as steam at a much lower pressure and temperature. The electric generator converts the mechanical work done on the turbine into electric energy; this process is not restricted by the second law because no thermal energy is involved. Finally, the steam must be condensed so that the water can retrace its route through the plant. This is done

Close Up

The Photocell

As you will see in Unit 5 when you study the photoelectric effect, light can give up energy to the electrons in a metal, causing the electrons to leave the metal with kinetic energy. Since the metal is

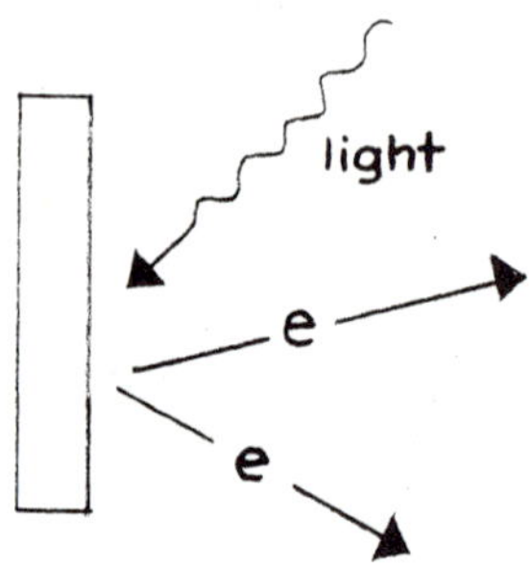

originally uncharged, the loss of electrons leaves it with a positive charge. If a conductor is placed where it can collect the electrons, it will become negatively charged. Such a device is called a phototube, and it can be used to generate an electric

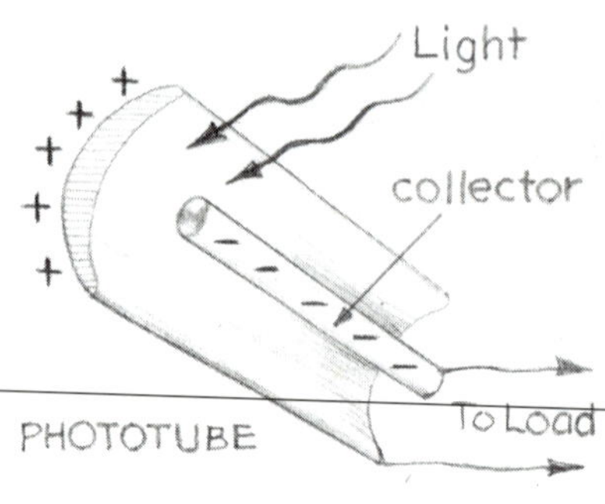

current from sunlight. Notice that as the charge builds up, there will be a tendency for the electrons to turn around and go the "wrong" way because the negative charge already on the collector will repel them (and the positive charge on the metal will make it more difficult for electrons to escape):

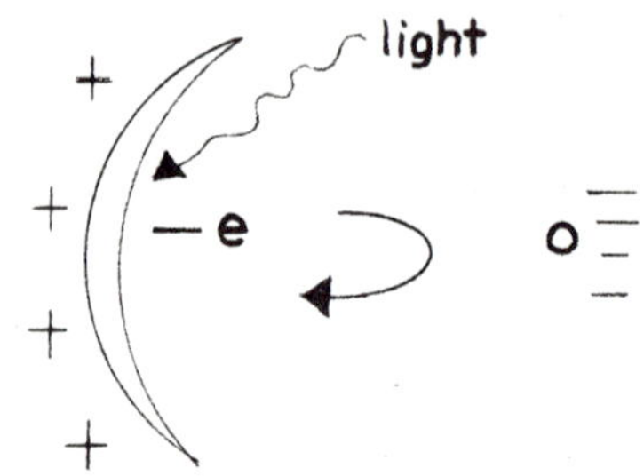

What is needed here is some sort of "one-way gate" that lets the electrons flow in one direction only. To a limited extent, this can be done by constructing the light-sensitive plate from a metal that gives up its electrons easily, and the collector from a metal that does not. No phototube has yet been built that serves as a useful energy source.

A better "one-way gate" can be constructed as a sandwich of two thin layers of materials that absorb light and that conduct electricity far better than an insulator but not as well as a metal. Silicon and germanium are examples of such materials (called "semiconductors"). A property of a junction between two suitable semiconductors is that an electric current can flow in only one direction. (The reason for this will not be explained here. You can learn more about this effect, which makes transistors work as well, by reading, for example, the supplementary unit on *Electronics*.)

Light striking the junction causes a separation of charge, just as it does in a phototube. The separation of charge creates a potential difference between the two layers. The process is called the *photovoltaic effect* because light creates a voltage. The voltage generates an electric current if a circuit joins them. The "sandwich" is called a photocell or a solar cell.

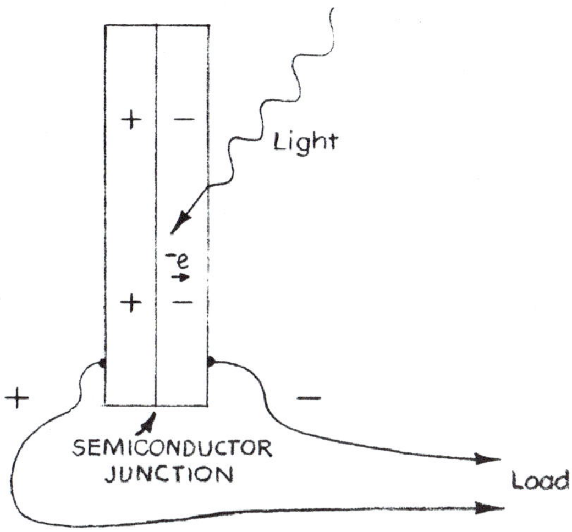

Several things keep a solar cell from turning all the energy of light into electricity. First, some of the light is reflected and some passes right through the cell; this light is simply never put to use at all. Second, though the junction is a good "one-way gate" it is not perfect. A certain fraction of the + and − charges that are created do, in fact, recombine within the cell (a process called "recombination"). When this happens, the net effect is that some of the light's energy is converted not into electricity, but just into heating up the cell.

Current research is aimed at improving several aspects of solar cells: (1) reducing the recombination of the + and − charges to make the cell as efficient as possible; (2) making cells of a sort that can be mass produced inexpensively.

These solar cells absorb light energy and convert it directly to electrical energy.

Solar cells in the panels shown in the photograph were used on the Skylab orbiting space station.

by allowing heat Q_2 to escape to the environment at temperature T_2. The whole process is shown schematically in the sketch below:

For the best efficiency, T_2 should be as low as possible, and T_1 as high as possible. However, T_2 is fixed by the environment, since cooling air or water must be used at whatever temperature is available outside. This is generally about 20–25°C (about 300°K). T_1 is limited by technology and chemistry. Metals weaken and melt when they get too hot. For a modern fossil plant, T_1 may be as high as 500°C (about 773°K). In a nuclear power plant, caution suggests more conservative limits, and therefore T_1 is typically 400°C (about 673°K). The lower temperature is necessary in particular to avoid damaging the fuel rods.

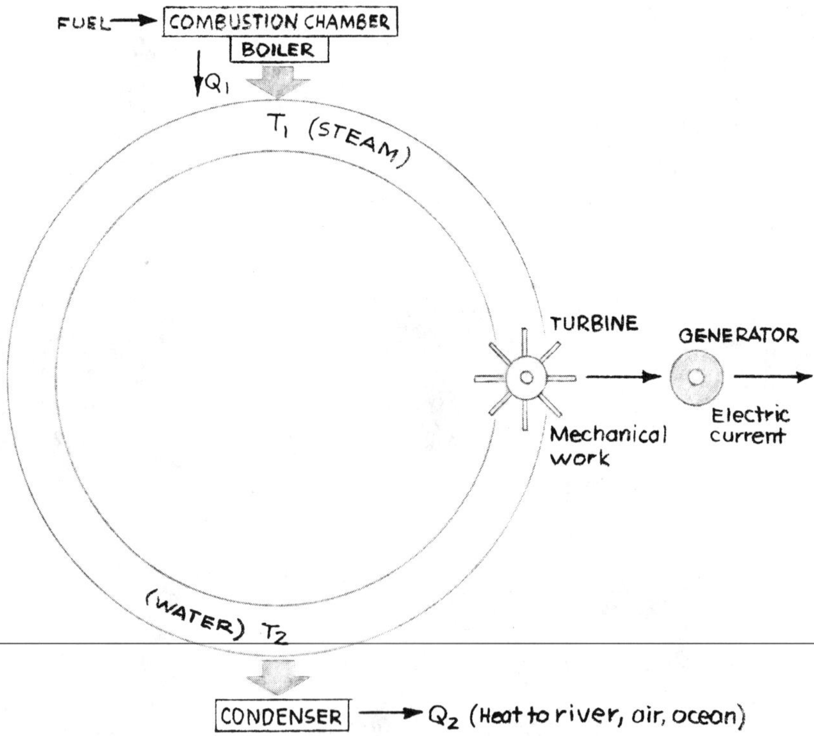

The maximum possible efficiency is thus

$$\eta = 1 - \frac{T_2}{T_1} = 1 - \frac{300}{773} \sim 0.6 \qquad \text{(fossil)}$$

$$= 1 - \frac{300}{673} \approx 0.5 \qquad \text{(nuclear)}$$

Therefore, *even if there were no losses of any kind whatsoever*, a power plant could only turn about half of the thermal energy into electrical energy. For each joule of electrical energy produced, two joules of energy will have to be provided originally by the fuel. The remaining joule will be released to the environment (into a river, the ocean, or the air) as thermal pollution.

The preceding paragraph describes the maximum possible efficiency of a perfect Carnot engine. Real power plants are significantly less efficient. Modern fossil fuel plants can achieve about 38% or 40% in practice; nuclear plants, because of the lower value of T_1, can manage about 30%. Older fossil plants have efficiencies of 30% or less. These additional losses are due to the fact that turbines are not ideal Carnot engines (they have friction; some heat simply leaks through them without doing any work at all) and the fact that there are minor losses in generators, transformers, and power lines. A useful rule of thumb is that the overall efficiency of a power plant is about 33%.

A typical fossil-fuel power plant that produces electricity by burning coal.

What this analysis shows is, very roughly, that any time you use 1 J of electrical energy, about 3 J of thermal energy were produced at the power plant, and 2 J were released into the environment, mostly near the plant. For example, if you heat a room with a small electric heater, about three times as much fuel has to be burned to produce the needed energy when the same fuel is burned directly within the room itself (in a gas stove, for example). This is the trade-off for the fact that electricity is such a convenient source of power.

study guide

1. The *Project Physics* learning materials particularly appropriate for Chapter 15 include:

Activities
Faraday Disk Dynamo
Generator Jump Rope
Simple Meters and Motors
Simple Motor–Generator Demonstration
Physics Collage
Bicycle Generator
Lapis Polaris, Magnes

2. What sources of energy were there for industry before the electrical age? How was the energy transported to where it was needed?

3. Oersted discovered that a magnetic needle was affected by a current. Would you expect a magnetic needle to exert a force on a current? Why? How would you detect this force?

4. In which of these cases will electromagnetic induction occur?

(a) A battery is connected to a loop of wire that is being held near another closed loop of wire.

(b) A battery is disconnected from a loop of wire held near another loop of wire.

(c) A magnet is moved through a loop of wire.

(**d**) A loop of wire is held in a steady magnetic field.

(**e**) A loop of wire is moved across a magnetic field.

5. It was stated on page 474 that the output of a dc generator can be made smoother by using multiple windings. If each of two loops were connected to commutators as shown what would the output current of the generator be like?

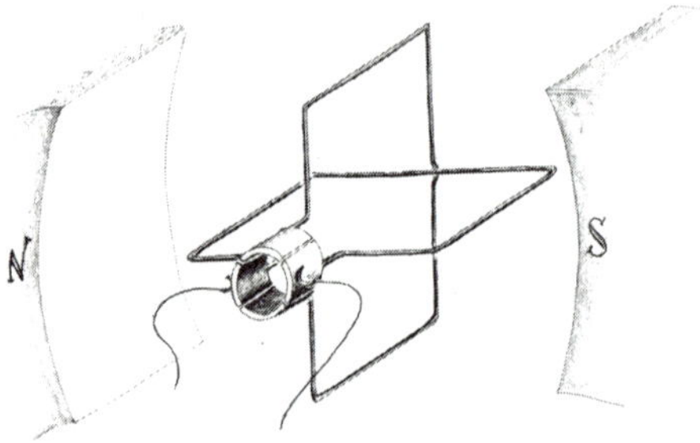

6. Refer to the simple ac generator shown on page 472. Suppose the loop is being rotated counter-clockwise by some externally applied mechanical force. Consider the segment *b* as it is pictured in the third drawing, moving *down* across the magnetic field. (Remember the useful rule: If your fingers point along $\vec{B}$, and your thumb along $\vec{v}$, $\vec{F}$ will be in the direction your palm would push. For positive charges use the right hand, and for negative, use the left hand.)

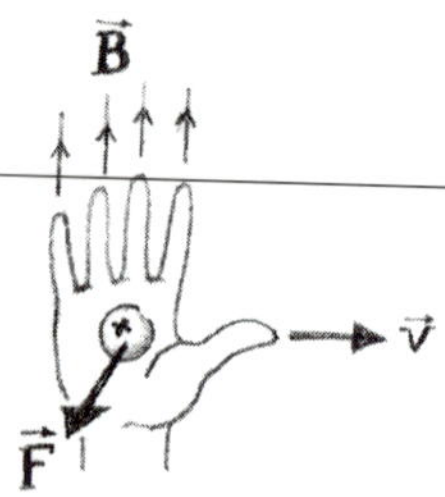

(**a**) Use the hand rule to determine the direction of the current induced in *b*.

(**b**) The induced current is an additional motion of the charges, which also move across the external magnetic field. Thus, an additional magnetic force acts on segment *b*. Use the hand rule to determine the direction of the additional force. *Before* doing so try to guess the direction of the force.)

(**c**) Determine the direction of the additional force on the charges in the segment labeled *a*, which is moving upward across the field

7. Why is a generator coil harder to rotate when it is connected to an appliance to which it provides current, such as a lamp, than when it is disconnected from any load?

8. Suppose two bar magnets, each held by one end at the same level but a few feet apart, are dropped simultaneously. One of them passes through a closed loop of wire. Which magnet reaches the ground first? Why?

9. Sketch a situation in which a wire is perpendicular to a magnetic field, and use the hand rule to find the direction of the force on the current. Imagine the wire moves sideways in response to the force. This sideways motion is an additional motion across the field, and so each charge in the wire experiences an additional force. In what direction is the additional force on the charges?

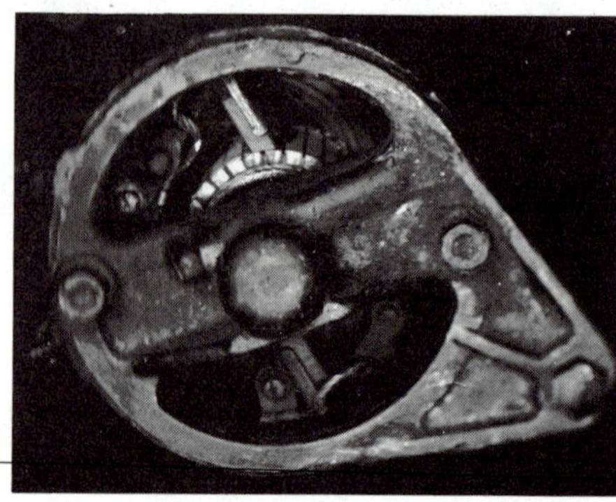

Right: Multiple commutator segments of an electric generator for use in an automobile.

10. Connect a small dc motor to a battery through a current meter. By squeezing on the motor shaft, vary the speed of the motor. On the basis of your answer to 9, can you explain the effect that the speed of the motor has on the current?

11. There are two ways to connect resistors in a circuit. In a series circuit, the resistors receive the same total input current and share the voltage in direct proportion to their resistance. In a parallel circuit, the resistors receive the same total input voltage and share the current in inverse proportion to their resistances.

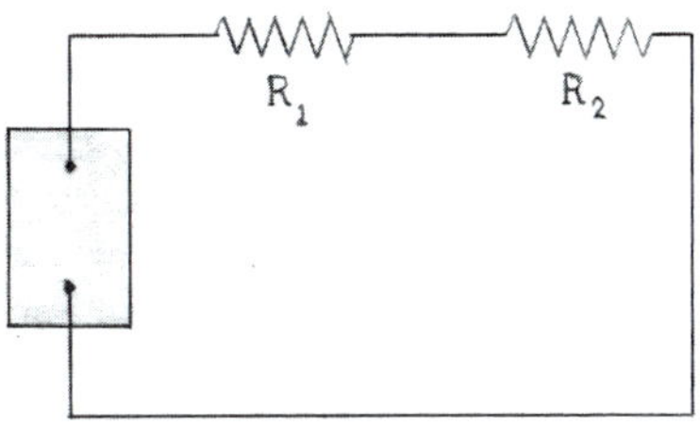

SERIES

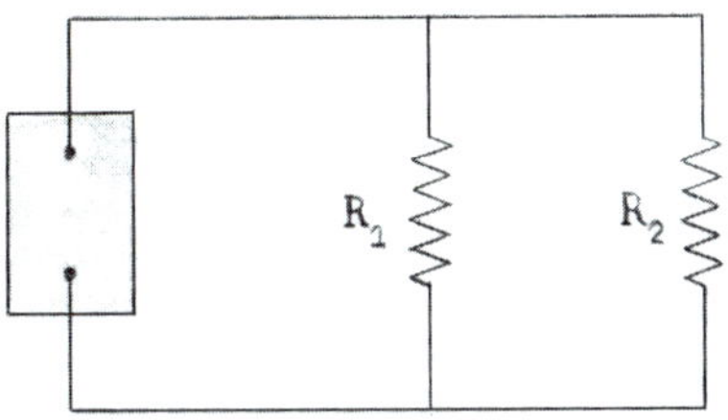

PARALLEL

(a) Which circuit stops functioning completely if one resistor is removed?

(b) In which circuit does the total current decrease if more resistors are added while the input voltage remains the same?

(c) Which circuit represents the way your home is wired?

(d) Using this model, explain why the current in your home increases when more resistors (appliances, light bulbs, etc.) are added to the circuit.

12. Suppose that the resistors in the diagrams above are each 4 ohms and the batteries supply 12 V. Find the total resistance of each circuit and the total current flowing in the circuit. Also find the voltage across and the current through each resistor.

13. Find the total voltages and currents, the voltage across and the current through each resistor, if $R_1 = 5$ ohms and $R_2 = 10$ ohms with a 50-V battery connected to the circuit.

14. A dozen Christmas-tree lights are connected in series and plugged into a 120-volt wall outlet.

(a) If each lamp dissipates 10 W of heat and light energy, what is the current in the circuit?

(b) What is the resistance of each lamp?

(c) What would happen to these lamps if they were connected in parallel across the 120-volt line? Why?

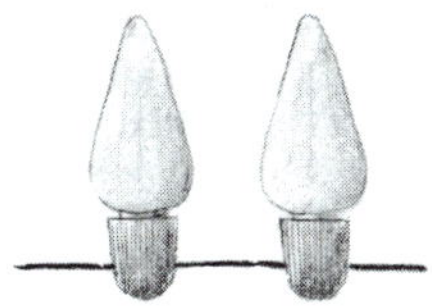

15. Suppose you wanted to connect a dozen 10-W lamps in *parallel* across a 120-V line. What resistance must each lamp have in this case? To determine the resistance, proceed by answering the following questions.

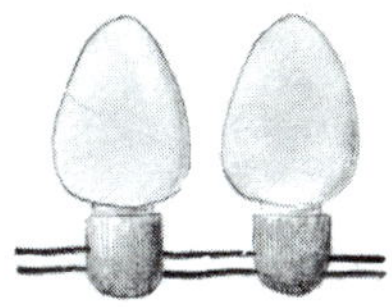

(a) What current will there be in each lamp?

(b) What is the resistance of each lamp?

Compare the total current for this string of 10-W lamps with the total current in the string of lamps in the previous question.

16. A man who built his own boat wanted to equip it with running lights and an interior light using a connecting wire with a resistance of ⅕ ohm. But he was puzzled about whether a 6-V system or a 12-V system would have less heating loss in the connecting wires. Suppose that his interior lamp is to be a 6-W lamp. (A 6-W lamp designed for use in 6-V systems has a resistance of 6 ohms.)

(a) If it were to operate at its full 6-V, 6-W rating, what current would the lamp require?

(b) If the current calculated in (a) were the actual current, what power loss would there be in the connecting wires?

(c) What would be the answers to (a) and (b) if a 12-V battery and a 12-V, 6-W bulb were used?

(d) Because of the resistance of the connecting wires, the lamps described will not actually operate at full capacity. Recalculate parts (a) and (b) to determine what would be the actual currents, power losses, and power consumptions of the lamps.

17. A transformer for an electric toy train is used to "step down" the voltage from 120 V to 6 V. As in most transformers, the output power from the secondary coil is only a little less than the input power to the primary coil. If the current in the primary coil is 0.25 A, what is the current in the secondary coil?

18. For a transformer, the ratio of the secondary voltage to the primary voltage is the same as the ratio of the number of turns of wire on the secondary coil to the number of turns of wire on the primary coil. If a transformer were 100% efficient, the output power would equal the input power. Assume such is the case, and derive an expression for the ratio of the secondary current to the primary current in terms of the turn ratio.

19. In a transformer, there is no connection between the input and output coils. Use the principle of electromagnetic induction to explain why there is a current in the output coils.

20. On many transformers, thicker wire (having lower resistance) is used for one of the coils than for the other. Which would you expect has the thicker wire, the low-voltage coil or the high-voltage coil?

21. Comment on the advisability and possible methods of getting out of a car over which a high-voltage power line has fallen.

22. What factors made Edison's recommendation for the use of dc for the Niagara Falls system in error?

23. Write a report comparing the earliest electric automobiles with those being developed now.

24. What were some of the major effects (both good and bad) of electricity on society?

25. What limits the efficiency of electric power plants? How efficient are they at best?

Electromagnetic Radiation

Radio telescope in Alaska, framed by Northern Lights.

16.1 | Introduction

On April 11, 1846, the distinguished physicist Sir Charles Wheatstone was scheduled to give a lecture at the Royal Institution in London. Michael Faraday was to introduce Wheatstone to the audience. At the last minute, just as Faraday and Wheatstone were about to enter the lecture hall, Wheatstone got stage fright, turned around, and ran out into the street. Faraday had to give the lecture himself. Normally, Faraday discussed only his actual experiments in public. But on this occasion he revealed certain speculations which, as he later admitted, he would never have made public had he not suddenly been forced to speak for an hour.

SG 1

Faraday's speculations dealt with the nature of light. Faraday, like Oersted before him, believed that all the forces of nature are somehow connected. Electricity and magnetism, for example, could not be separate forces that just happen to exist in the same universe. Rather, they must be different forms of one basic

Nature Philosophy was discussed in the Epilogue to Unit 2, in Sec. 10.9, and its effect on Oersted in Sec. 14.11.

phenomenon. This belief paralleled that of Schelling and other German Nature Philosophers at the beginning of the nineteenth century. It had inspired Oersted to search in the laboratory for a connection between electricity and magnetism. Eventually he found such a connection in his discovery that an electric current in a conductor can turn a nearby magnet.

Faraday, too, had been guided by a belief in the unity of natural forces. Could *light* also be another form of this basic "force"? Or rather, to use more modern terms, is light a form of *energy*? If so, scientists should be able to demonstrate experimentally its connection with other forms of energy such as electricity and magnetism. Faraday did succeed in doing just this. In 1845, he showed that light traveling through heavy glass had its plane of polarization rotated by a magnetic field applied
SG 2 to the glass.

This experiment convinced Faraday that there is a definite connection between light and magnetism. But he could not resist going one step further in his unrehearsed lecture the following year. Perhaps, he suggested, light itself is a vibration of magnetic lines of force. Suppose, for example, that two charged or magnetized objects are connected by an electric or magnetic line of force. If one of them moved, Faraday reasoned, a disturbance would be transmitted along the line of force. Furthermore, if light waves were vibrations of lines of force, then an elastic substance such as "ether" would not be needed in order to explain the propagation of light. The concept of the ether could be replaced if it could be shown that lines of force themselves have the elastic properties needed for wave transmission.

Faraday could not make his idea more precise. He lacked the mathematical skill needed to prove that waves could propagate along lines of electric or magnetic force. Other physicists in Britain and Europe might have been able to develop a mathematical theory of electromagnetic waves. But at the time these scientists either did not understand Faraday's concept of lines of force or did not consider them a good basis for a mathematical theory. Ten years passed before James Clerk Maxwell, a Scottish mathematical physicist, saw the value of the idea of lines of force and started using mathematics to express Faraday's concepts.

16.2 Maxwell's formulation of the principles of electromagnetism

The work of Oersted, Ampère, Henry, and Faraday had established two basic principles of electromagnetism:

1. *An electric current in a conductor produces magnetic lines of force that circle the conductor.*

2. *When a conductor moves across externally set up magnetic lines of force, a current is induced in the conductor.*

In the 1860's, James Clerk Maxwell developed a mathematical theory of electromagnetism. In it, he added to and generalized these principles so that they applied to electric and magnetic fields in conductors, in insulators, and even in space free of matter.

Maxwell began by putting Faraday's theory of electricity and magnetism into mathematical form. In 1855, less than 2 years after completing his undergraduate studies at Cambridge University, Maxwell presented to the Cambridge Philosophical Society a long paper. Entitled "On Faraday's Lines of Force," it described how these lines are constructed:

> ... if we commence at any point and draw a line so that, as we go along it, its direction at any point shall always coincide with that of the resultant force at that point, this curve will indicate the direction of that force for every point through which it passes, and might be called on that account a *line of force*. We might in the same way draw other lines of force, till we had filled all space with curves indicating by their direction that of the force at any assigned point.

Maxwell stated that his paper was designed to "show how, by a strict application of the ideas and methods of Faraday, the connection of the very different orders of phenomena which he has discovered may be clearly placed before the mathematical mind." During the next 10 years, Maxwell created his own models of electric and magnetic induction. In developing his theory, he first proposed a mechanical model for the electrical and magnetic quantities observed experimentally by Faraday and others. Maxwell then expressed the operation of the model in a group of equations that gave the relations between the electric and magnetic fields. He soon found these equations to be the most useful way to represent the theory. Their power allowed him eventually to discard the mechanical model altogether. Maxwell's mathematical view is still considered by physicists to be the proper approach to the theory of electromagnetic phenomena. If you go on to take another physics course after this introductory one, you will find that the development of Maxwell's mathematical model (Maxwell's equations) is one of the high points of the course. However, it will require vector calculus.

Maxwell's work contained an entirely new idea of far-reaching consequences: *An electric field that is changing with time must be accompanied by a magnetic field.* Not only do *currents in conductors* produce fields around them, but *changing electric fields in insulators* such as glass, air, or empty space also produce magnetic fields.

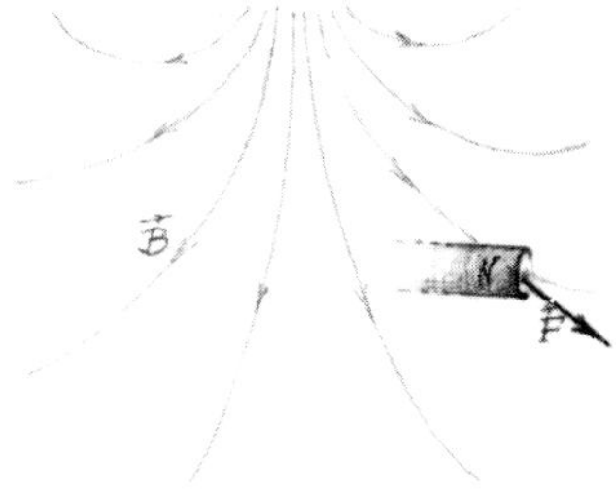

Magnetic lines of force indicate the direction of magnetic force on a north magnetic pole. (The force on a south pole is in the opposite direction.)

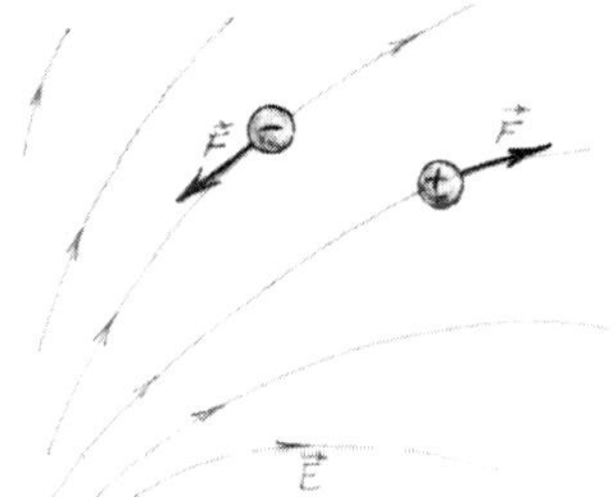

Electric lines of force indicate the direction of electric force on a positive test charge. (The force on a negative charge is in the opposite direction.)

James Clerk Maxwell (1831–1879).

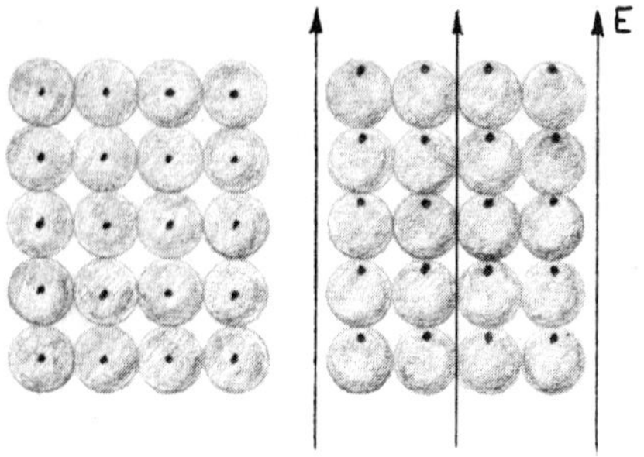

When an electric field is set up in an insulating material (as in the diagram at the right, above), the + and − charges, which are bound to one another by attraction, are displaced. This displacement forms a current. (The + charges are represented by dots, and − charges by shaded circles.)

It is one thing to accept this newly stated connection between electric and magnetic fields. But it is harder, and more interesting, to *understand* the physical necessity for such a connection. The paragraphs below are intended to make it clearer.

An uncharged insulator (such as glass, wood, paper, or rubber) contains equal amounts of negative and positive charges. In the normal state, these charges are distributed evenly. Thus, the *net* charge is zero in every region of the material. But when the insulator is placed in an electric field, these charges are subjected to electrical forces. The positive charges are pushed in one direction, the negative in the opposite direction. Unlike the charges in a conductor, the charges in an insulating material are *not* free to move far through the material. The charges can be displaced only a small distance before restoring forces in the insulator balance the force of the electric field. If the strength of the field is increased, the charges will be displaced further. The changing displacement of charges that accompanies a changing electric field in an insulator forms a current. Maxwell called this current a *displacement current*. He assumed that this momentary displacement current in an insulator surrounds itself with a magnetic field just as a conduction current of the same magnitude does.

SG 3

A changing electric field produces a magnetic field (see page 503 left): When the electric field $\vec{E}$ between a pair of charged plates starts to increase in intensity, a magnetic field $\vec{B}$ is induced. The faster $\vec{E}$ changes, the more intense $\vec{B}$ is. When $\vec{E}$ momentarily has reached its maximum value, $\vec{B}$ has decreased to zero momentarily. When $\vec{E}$ diminishes, a $\vec{B}$ field is again induced, in the opposite direction, falling to zero as $\vec{E}$ returns to its original strength.

In an insulator, the displacement current is defined as *the rate at which the charge displacement changes*. This rate is directly proportional to the rate at which the electric field is changing in time. Thus, the magnetic field that circles the displacement current can be considered a consequence of the time-varying electric field. Maxwell assumed that this model, developed for matter, also applies to *space free of matter* (though at first glance this seems absurd). Therefore, under all circumstances, *an electric field that is changing with time surrounds itself with a magnetic field*. Previously, it was thought that the only current that produced a magnetic field was the current in a conductor. Now Maxwell predicted that a magnetic field would also arise from a changing electric field, even in empty space. Unfortunately, this field was very small in comparison to the magnetic field produced by the current in the conductors of the apparatus. So it was not at that time possible to measure it directly. But as you will see, Maxwell predicted consequences that soon *could* be tested.

A changing magnetic field produces an electric field (see page 503 right): When the magnet field $\vec{B}$ between the poles of an electromagnet starts to increase, an electric field $\vec{E}$ is induced. The faster $\vec{B}$ changes, the more intense $\vec{E}$ is. When $\vec{B}$ momentarily has reached its maximum value, $\vec{E}$ has decreased to zero momentarily. When $\vec{B}$ diminishes, an $\vec{E}$ field is again induced, in the opposite direction, falling to zero as $\vec{B}$ returns to its original strength.

According to Maxwell's theory, then, the two basic principles of electromagnetism should be expanded by adding a third:

3. *A changing electric field in space produces a magnetic field.* The induced magnetic field vector $\vec{B}$ is in a plane perpendicular to the changing electric field vector $\vec{E}$. The magnitude of $\vec{B}$ depends on the rate at which $\vec{E}$ is changing—not on $\vec{E}$ itself, but on $\Delta\vec{E}/\Delta t$. Therefore, the higher the frequency of alteration of $\vec{E}$, the greater the field $\vec{B}$ so induced.

Consider a pair of conducting plates connected to a source of current, as shown at the right. Charges are moved onto or away from plates through the conductors connecting them to the source. Thus, the strength of the electric field $\vec{E}$ in the space between the plates changes with time. This changing electric field produces a magnetic field $\vec{B}$ as shown. (Of course, only a few of the infinitely many lines for $\vec{E}$ and $\vec{B}$ are shown.)

An additional principle, known before Maxwell, assumed new significance in Maxwell's work because it is so symmetrical to statement 3 above:

4. *A changing magnetic field in space produces an electric field.* The induced electric field vector $\vec{E}$ is in a plane perpendicular to the changing magnetic field vector $\vec{B}$. The magnitude of $\vec{E}$ depends on the *rate* at which $\vec{B}$ is changing—not on $\vec{B}$ itself, but on $\Delta\vec{B}/\Delta t$. Consider the changing magnetic field produced by, say, temporarily increasing the current in an electromagnet. (See the illustration in the right margin of this page.) This changing magnetic field induces an electric field in the region around the magnet. If a conductor happens to be lined up in the direction of the induced electric field, the free charges in the conductor will move under the field's influence. Thus, a current in the direction of the induced field will arise in the conductor. This electromagnetic induction had been discovered experimentally by Faraday (Sec. 15.3).

Maxwell's theories of the total set of relations between electric and magnetic fields were not at once directly testable. When the test finally came, it concerned his prediction of the existence of waves traveling as interrelating electric and magnetic fields, that is, electromagnetic waves.

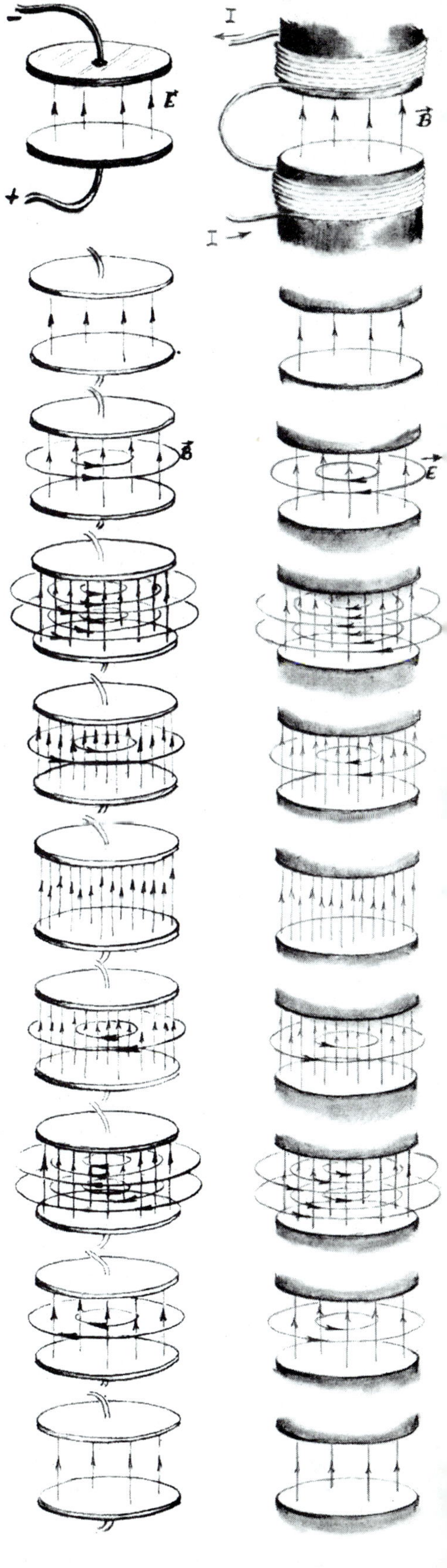

?

1. *When there is a changing electric field, what else occurs (according to Maxwell)?*
2. *What is a displacement current?*
3. *What are the four principles of electromagnetism?*

16.3 | The propagation of electromagnetic waves

Suppose in a certain region of space, an electric field that changes with time is created. According to Maxwell's theory, an electric field $\vec{E}$ that varies in time simultaneously induces a magnetic field $\vec{B}$ that also varies with time. (The magnetic field also varies with the distance from the region where the changing electric field was created.) Similarly, a magnetic field that is changing with time simultaneously induces an electric field that

The electric and magnetic field changes occur together, much like the "action" and "reaction" of Newton's third law.

changes with time. (Here, too, the electric field also changes with distance from the region where the changing magnetic field was created.)

As Maxwell realized and correctly predicted, mutual induction of time- and space-changing electric and magnetic fields should set up an unending sequence of events. First, a time-varying electric field in one region produces a time- and space-varying magnetic field at points near this region. But this *magnetic* field produces a time and space-varying *electric* field in the space surrounding it. And *this* electric field produces time- and space-varying magnetic fields in *its* neighborhood, and so on. Thus, suppose that an electromagnetic disturbance is started at one location, say by vibrating charges in a hot gas or in the transmitter wire of a radio or television station. This disturbance can travel to distant points through the mutual generation of the electric and magnetic fields. The fluctuating, interlocked electric and magnetic fields propagate through space in the form of an *electromagnetic wave*, a disturbance in the electric and magnetic field intensities in space.

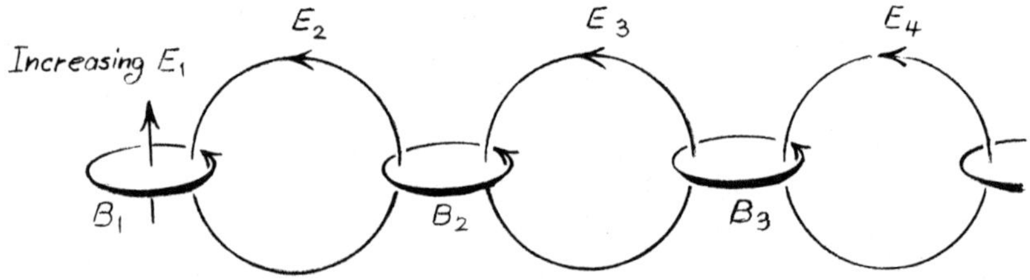

Electric and magnetic fields linked by induction, from Max Born, Einstein's Theory of Special Relativity *(1924). An increasing electric field* E_1 *at the left (or a current) surrounds itself with a magnetic field* B_1*. As* B_1 *changes, it induces an interlinking electric field* E_2*, etc. The chainlike process continues with finite velocity. This is only a symbolic picture of the process, which propagates itself in all directions.*

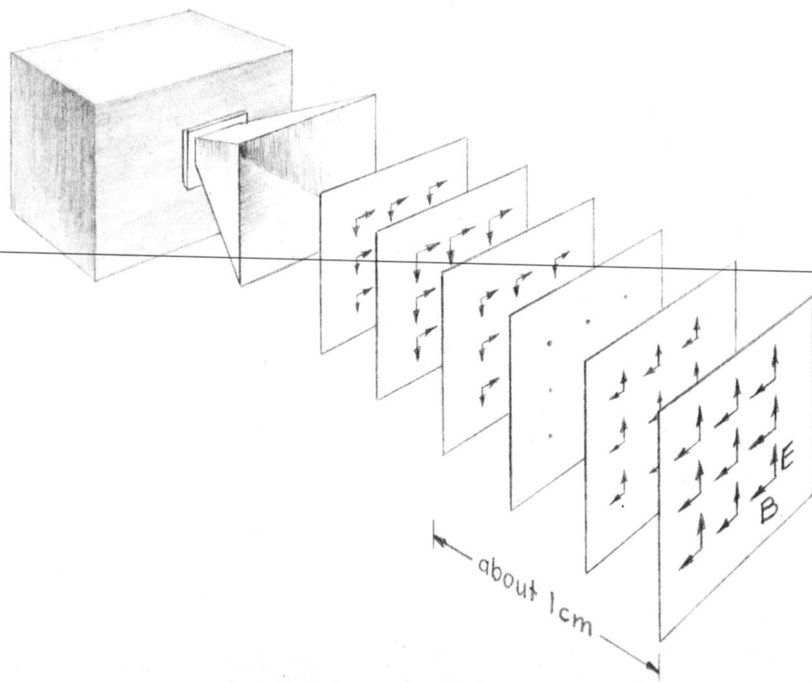

In a microwave oscillator, which you may see in your laboratory work, electric oscillations in a circuit are led onto a rod in a metal "horn." In the horn they generate a variation in electric and magnetic fields that radiates away into space. This drawing represents an instantaneous "snapshot" of almost plane wave fronts directly in front of such a horn.

In Chapter 12 it was shown that waves occur when a disturbance created in one region produces at a later time a disturbance in adjacent regions. Snapping one end of a rope produces, through the action of one part of the rope on the other, a displacement at points farther along the rope and at a later time. Dropping a pebble into a pond produces a disturbance that moves away from the source as one part of the water acts on neighboring parts. Time-varying electric and magnetic fields produce a disturbance that moves away from the source as the varying fields in one region create varying fields in neighboring regions.

SG 4, 5

What determines the speed with which electromagnetic waves travel? Recall first that for mechanical waves the speed of propagation is determined by the stiffness and density of the medium. Speed increases with increasing stiffness, but decreases with increasing density. This relation between wave speed, stiffness, and density holds for mechanical wave motions and for many other types of waves. Only the barest outline of how Maxwell proceeded beyond this point is given here. First, he assumed that a similar "stiffness and density" relation would hold for electromagnetic waves. Then he computed what he thought to be the "stiffness" and "density" of electric and magnetic fields propagating through the hypothetical ether. In finding values for these two properties of the electric and magnetic fields, Maxwell was guided by his mechanical model representing the ether. In this model, stiffness was related to the electric field, and density to the magnetic field. Next, he proved mathematically that the *ratio* of these two factors, which should determine the wave speed, is the same for all strengths of the fields. Finally, Maxwell demonstrated that the speed of the waves (if they exist!) is a definite quantity that can be deduced from measurements in the laboratory.

As was stated in Chapter 12, page 359, the speed of propagation depends on both the stiffness and density of the medium; the relation can be written

$$\text{speed} = \sqrt{\frac{\text{stiffness}}{\text{density}}}$$

The necessary measurements of the factors involved actually had been made 5 years earlier by the German scientists Weber and Kohlrausch. Using their published values, Maxwell calculated that the speed of the supposed electromagnetic waves should be about 311,000,000 m/sec. He was immediately struck by the fact that this large number was very close to a measured speed already well known in physics. In 1849, Armand Fizeau had measured the speed of *light* and had obtained a value of about 315,000,000 m/sec. The close similarity could have been a chance occurrence. But Maxwell believed that there must be a deep underlying reason for these two numbers being so nearly the same. The significance for physics seemed obvious to him. Making an enormous leap of the imagination, he wrote:

With better measurements, scientists now know that both Maxwell's predicted speed and Fizeau's measured speed should have come to just under 3×10^8 m/sec or 2.99793×10^8 m/sec.

> The velocity of the transverse undulations in our hypothetical medium, calculated from the electromagnetic experiments of MM. Kohlrausch and Weber, agrees so exactly with the velocity

Maxwell had shown that in an electromagnetic disturbance $\vec{E}$ and $\vec{B}$ should be perpendicular to each other and to the direction of propagation of the wave. Therefore, in the language of Chapter 12, electromagnetic waves are *transverse*. And as was noted in Chapter 13, it was long known that light waves are transverse.

Recall from page 503: The magnitude of B depends on the rate at which E changes (or $\Delta E/\Delta t$). Therefore, an electric field oscillating at a very high frequency induces magnetic fields that are large compared to the ordinary magnetic field surrounding the conductor for the current. But circuits to produce such high-frequency oscillations were not available in Maxwell's time.

of light calculated from the optical experiments of M. Fizeau, that we can scarcely avoid the inference that *light consists in the transverse undulations of the same medium which is the cause of electric and magnetic phenomena.*

Here then was an explanation of light waves and at the same time a joining of the previously separate sciences of electricity, magnetism, and optics. Maxwell realized the importance of his discovery. Now he set to work making the theory mathematically sound and freeing it from his admittedly artificial model.

Maxwell's synthesis of electromagnetism and optics, after it had been experimentally confirmed (see Sec. 16.4), was seen as a great event in physics. In fact, physics had known no greater time since the 1680's, when Newton was writing his monumental work on mechanics. Of course, Maxwell's electromagnetic theory had arisen in Maxwell's mind in a Newtonian, mechanical framework. But it had grown out of that framework, becoming another great general physical theory, independent of its mechanical origins. Like Newtonian mechanics, Maxwell's electromagnetic field theory succeeded spectacularly. You will see something of that success in the next few sections. The success occurred on two different levels: the practical and the theoretical. Practically, it led to a host of modern developments, such as radio and television. On the theoretical level, it led to a whole new way of viewing phenomena. The universe was not only a Newtonian machine of whirling and colliding parts; it included fields and energies that no machine could duplicate. As you will see later, Maxwell's work formed a basis of the special theory of relativity. Other physical theories were nourished by it also. Eventually, however, results accumulated that did not fit Maxwell's theory; something more was needed. Starting about 1925, after a quarter-century of discovery, the development of quantum mechanics led to a larger synthesis, which included Maxwell's electromagnetism.

?

4. What discovery did Maxwell make upon calculating the speed with which electromagnetic disturbances should travel?

5. What is Maxwell's synthesis?

16.4 | Hertz's experiments

Did Maxwell establish without doubt that light actually does consist of electromagnetic waves, or even that electromagnetic waves exist at all? No. Most physicists remained skeptical for several years. The fact that the ratio of two quantities determined by electrical experiments came out equal to the speed of light certainly suggested *some* connection between electricity and light. No one would seriously argue that this was only a coincidence. But stronger evidence was needed before the rest of Maxwell's theory, with its displacement current, could be accepted.

What further evidence was needed to persuade physicists that Maxwell's theory was correct? Maxwell showed that his theory could explain all the known facts about electricity, magnetism, and light. But so could other theories, although with less sweeping connections between their separate parts. To a modern physicist, the other theories proposed in the nineteenth century seem much more complicated and artificial than Maxwell's. But at the time, Maxwell's theory seemed strange to physicists who were not accustomed to thinking in terms of fields. It could be accepted over other theories only if it could be used to predict some *new* property of electromagnetism or light.

Maxwell himself made two such predictions from his theory. He did not live to see them verified experimentally in 1888, for he died in 1879 at the age of 48. Maxwell's most important prediction was that electromagnetic waves of many different frequencies could exist. All such waves would propagate through space at the speed of light. Light itself would correspond to waves of only a small range of high frequencies (from 4×10^{14} Hz to 7×10^{14} Hz). These are frequencies detectable by the human eye.

The unit "cycles/sec" is called the *hertz*, after Heinrich Hertz. It is abbreviated Hz.

To test this prediction required inventing apparatus that could both produce and detect electromagnetic waves, preferably of frequencies other than light frequencies. This was first done by the German physicist Heinrich Hertz, whose contribution was triggered by a chance observation. In 1886, Hertz noticed a peculiar effect produced during the sparking of an induction coil. As was well known, sparks sometimes jump the air gap between the terminals of an induction coil (see drawing). You will recall (Chapter 15) that an induction coil can be used to produce high voltages if there are many more turns of wire on one side than on the other. Ordinarily, air does not conduct electricity. But when there is a very large potential difference between two wires a short distance apart, a conducting pathway may form briefly as air molecules are ionized. A short burst of electricity then may pass through, attended by a visible spark. Each visible spark produced is actually a series of many small sparks, jumping rapidly back and forth (oscillating) between the terminals. Hertz found that he could control the spark's frequency of oscillation by changing the size and shape of metal plates attached to the spark gap of the induction coil.

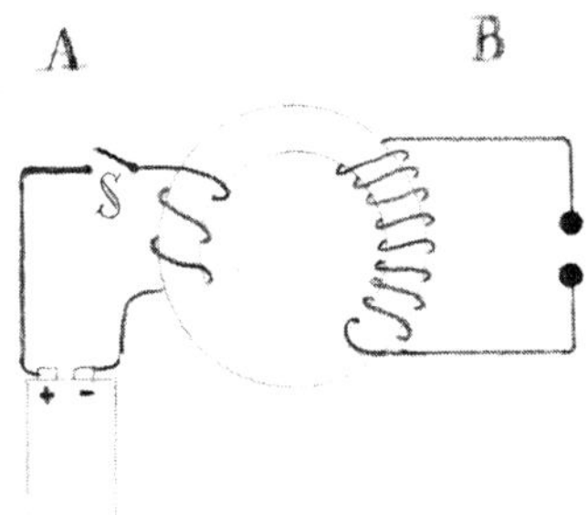

Operation of the induction coil: Starting and stopping the current in coil A with a vibrating switch S produces a rapidly changing magnetic field in the iron core. This rapidly changing field induces high-voltage peaks in the many-turn coil B and can cause a spark to jump across the air gap. Spark coils for use in car engines operate in this way.

Hertz then bent a simple piece of wire so that there was a short gap between its two ends. When it was held near an induction coil, *a spark jumped across the air gap in the wire just*

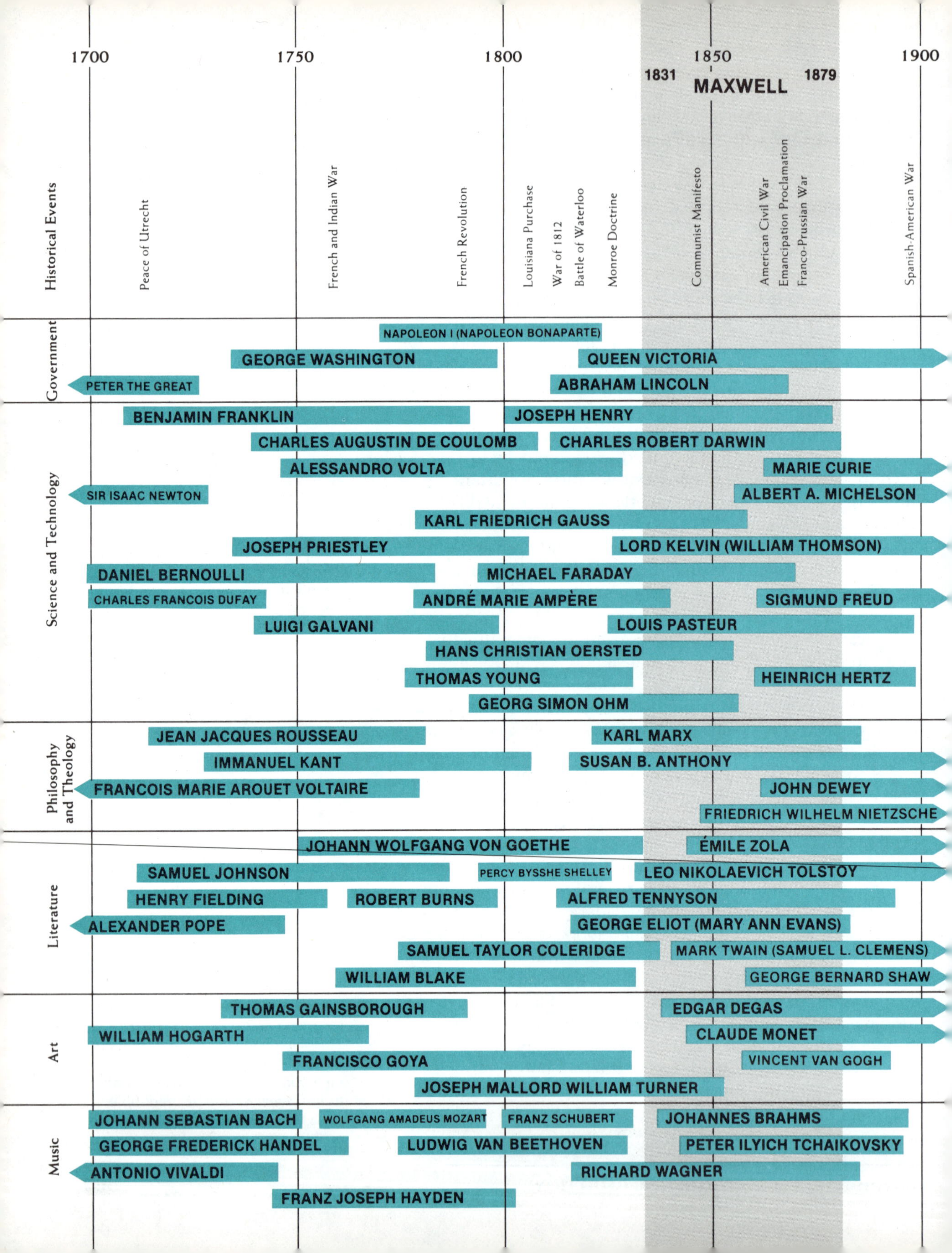
1700
1750
1800
1850
1900
1831
MAXWELL
1879
Historical Events
Peace of Utrecht
French and Indian War
French Revolution
Louisiana Purchase
War of 1812
Battle of Waterloo
Monroe Doctrine
Communist Manifesto
American Civil War
Emancipation Proclamation
Franco-Prussian War
Spanish-American War
Government
NAPOLEON I (NAPOLEON BONAPARTE)
GEORGE WASHINGTON
QUEEN VICTORIA
PETER THE GREAT
ABRAHAM LINCOLN
Science and Technology
BENJAMIN FRANKLIN
JOSEPH HENRY
CHARLES AUGUSTIN DE COULOMB
CHARLES ROBERT DARWIN
ALESSANDRO VOLTA
MARIE CURIE
SIR ISAAC NEWTON
ALBERT A. MICHELSON
KARL FRIEDRICH GAUSS
JOSEPH PRIESTLEY
LORD KELVIN (WILLIAM THOMSON)
DANIEL BERNOULLI
MICHAEL FARADAY
CHARLES FRANCOIS DUFAY
ANDRÉ MARIE AMPÈRE
SIGMUND FREUD
LUIGI GALVANI
LOUIS PASTEUR
HANS CHRISTIAN OERSTED
THOMAS YOUNG
HEINRICH HERTZ
GEORG SIMON OHM
Philosophy and Theology
JEAN JACQUES ROUSSEAU
KARL MARX
IMMANUEL KANT
SUSAN B. ANTHONY
FRANCOIS MARIE AROUET VOLTAIRE
JOHN DEWEY
FRIEDRICH WILHELM NIETZSCHE
Literature
JOHANN WOLFGANG VON GOETHE
ÉMILE ZOLA
SAMUEL JOHNSON
PERCY BYSSHE SHELLEY
LEO NIKOLAEVICH TOLSTOY
HENRY FIELDING
ROBERT BURNS
ALFRED TENNYSON
ALEXANDER POPE
GEORGE ELIOT (MARY ANN EVANS)
SAMUEL TAYLOR COLERIDGE
MARK TWAIN (SAMUEL L. CLEMENS)
WILLIAM BLAKE
GEORGE BERNARD SHAW
Art
THOMAS GAINSBOROUGH
EDGAR DEGAS
WILLIAM HOGARTH
CLAUDE MONET
FRANCISCO GOYA
VINCENT VAN GOGH
JOSEPH MALLORD WILLIAM TURNER
Music
JOHANN SEBASTIAN BACH
WOLFGANG AMADEUS MOZART
FRANZ SCHUBERT
JOHANNES BRAHMS
GEORGE FREDERICK HANDEL
LUDWIG VAN BEETHOVEN
PETER ILYICH TCHAIKOVSKY
ANTONIO VIVALDI
RICHARD WAGNER
FRANZ JOSEPH HAYDEN

when a spark jumped across the terminals of the induction coil. This was a surprising new phenomenon. Hertz reasoned that as the spark jumps back and forth across the gap of the induction coil, it must set up rapidly changing electric and magnetic fields. According to Maxwell's theory, these changes propagate through space as electromagnetic waves. (The frequency of the waves is the same as the frequency of oscillations of the sparks.) When the electromagnetic waves pass over the bent wire, they set up rapidly changing electric and magnetic fields there, too. A strong electric field produces a spark in the air gap, just as the transmitter field did between the terminals of the induction coil. Since the field is rapidly changing, sparks can jump back and forth between the two ends of the wire. This wire, therefore, serves as a detector of the electromagnetic waves generated by the induction coil. Hertz's observation of the induced spark was the first solid clue that electromagnetic waves do exist.

Radio and television "static" is often produced by sparking in electrical appliances and in the ignition of passing cars. This fact shows that high-frequency oscillations occur in sparks.

SG 6

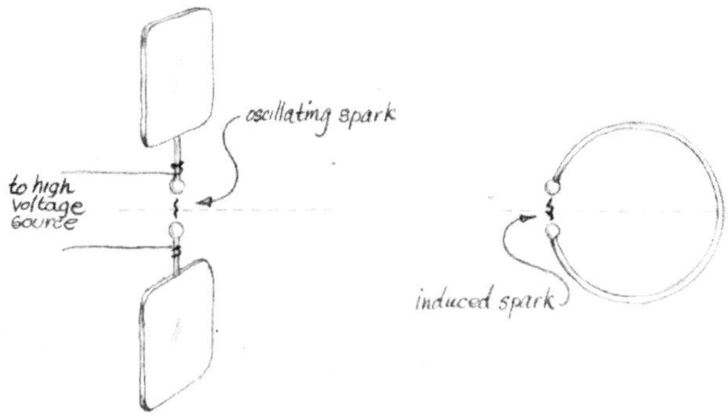

Heinrich Hertz (1857–1894) was born in Hamburg, Germany. During his youth, Hertz was mainly interested in languages and the humanities, but was attracted to science after his grandfather gave him some apparatus. Hertz did simple experiments in a small laboratory which he had fitted out in his home. After completing secondary school (and a year of military service) he undertook the serious study of mathematics and physics at the University of Berlin in 1878. In 1882, Hertz devoted himself to the study of electromagnetism, including the recent and still generally unappreciated work of Maxwell. Two years later he started his famous experiments on electromagnetic waves. During the course of this work, Hertz discovered the photoelectric effect, which has had a profound influence on modern physics. You will study this effect in Chapter 18 (Unit 5).

Hertz showed that the electromagnetic radiation from his induction coil has all the usual properties of light waves. It can be reflected at the surface of solid bodies, including metallic conductors. In addition, the angle of reflection is equal to the angle of incidence. The electromagnetic radiation can be focused by concave metallic mirrors. It shows diffraction effects when it passes through an opening in a screen. All interference phenomena can be shown, including standing waves. Also, electromagnetic waves are refracted by prisms made of glass, wood, plastic, and other nonconducting material. (All these experiments, with more modern apparatus, can be done in your laboratory.) By setting up a standing-wave pattern with a large metal reflector, Hertz was also able to determine the distance between consecutive nodes and thus measure the wavelength. He determined the frequency of the oscillating electric current through an analysis of his circuits. Thus, he was able to determine the speed of his waves and found it to be the same value that Maxwell had predicted: the speed of light.

Hertz's experiments dramatically confirmed Maxwell's electromagnetic theory. They showed that electromagnetic waves

Instead of relying on oscillating sparks, modern electronic circuits use the wires of a transmitting antenna. Through the wires move the oscillating currents that radiate electromagnetic waves.

SG 7
SG 8

actually exist, that they travel with the speed of light, and that they have the familiar characteristics of light. Now mathematical physicists rapidly accepted Maxwell's theory and applied it with great success to the detailed analysis of a wide range of phenomena.

Thus, at the end of the nineteenth century, Maxwell's electromagnetic theory stood with Newton's laws of mechanics as an established part of the foundations of physics.

?

6. What predictions of Maxwell's were verified by Hertz?

7. What did Hertz use as a detector of electromagnetic waves?

16.5 | The electromagnetic spectrum

Radio stations regularly announce their frequencies in megahertz (MHz) for the FM band and kilohertz (kHz) for the AM band.

Hertz's induction coil produced electromagnetic radiation with a wavelength of about 1 m. This is about 1 million times the wavelength of visible light. Later experiments showed that a very wide and continuous range of electromagnetic wavelengths (and frequencies) is possible. The entire possible range is called the *electromagnetic spectrum.* A range of frequencies from about 1 Hz to 10^{25} Hz, corresponding to a wavelength range from 10^{8} m to 10^{-17} m, has been studied. Many of these frequency regions have been put to practical use.

Light, heat, radio waves, and X rays are names given to radiations in certain regions of the electromagnetic spectrum. In each of these regions radiation is produced or observed in a particular way. For example, light may be perceived directly through its effect on the retina of the eye. But to detect radio waves requires electronic equipment. The named regions overlap. For example, some radiation is called "ultraviolet" or "X ray," depending on how it is produced.

All waves in the electromagnetic spectrum, although produced and detected in various ways, behave as predicted by Maxwell's theory. All electromagnetic waves travel through empty space at the same speed—the speed of light. They all carry energy; when they are absorbed, the absorber is heated, as is food in a microwave oven. Electromagnetic radiation, whatever its frequency, can be emitted only if energy is supplied to the source of radiation, which is, ultimately, a charge that is undergoing acceleration. This charge acceleration can be produced in many ways. For example, heating a material will increase the vibrational energy of charged particles. Also, one can vary the motion of charges on an electric conductor (an antenna) or cause a charged particle to change its direction. In these and other processes, work is done by the force that is applied to

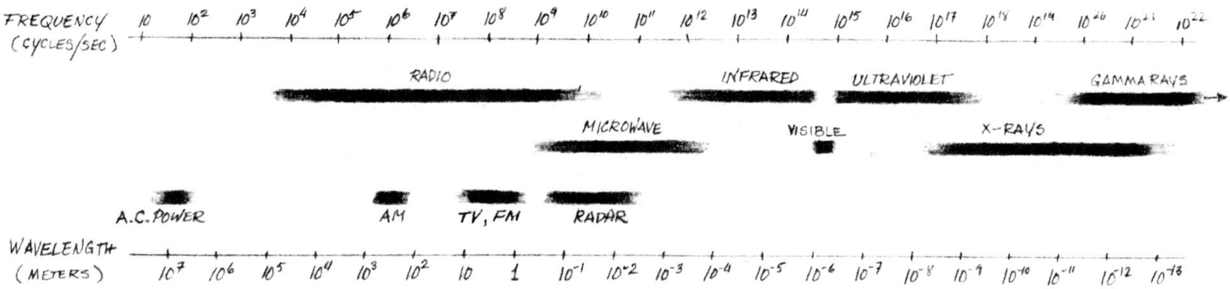

A chart of the electromagnetic spectrum.

accelerate the electric charge. Some of the energy supplied to the antenna in doing this work is "radiated" away; that is, it propagates away from the source as an electromagnetic wave.

The work of Maxwell and Hertz opened up a new scientific view of nature. It also prepared for a rapid blooming of new technologies, such as radio, TV, radar, etc. As was done before, for example, in the chapter on electric motors and generators, a brief description of these indirect consequences of a scientific advance is given below.

SG 9

Radio. Electromagnetic waves having frequencies of 10^4 to 10^7 Hz are reflected quite well by electrically charged layers that exist in the upper atmosphere. This reflection makes it possible to detect radio waves at great distances from the source. Radio signals have wavelengths from tens to thousands of meters. Such waves can easily diffract around relatively small obstacles such as trees or buildings. But large hills and mountains may cast "dark" shadows.

SG 10
SG 11

In December 1901, Guglielmo Marconi successfully detected radio waves sent from Newfoundland to Ireland. Marconi's work showed that long-distance radio communication was possible and revealed the previously unsuspected layers of ionized particles in the upper atmosphere.

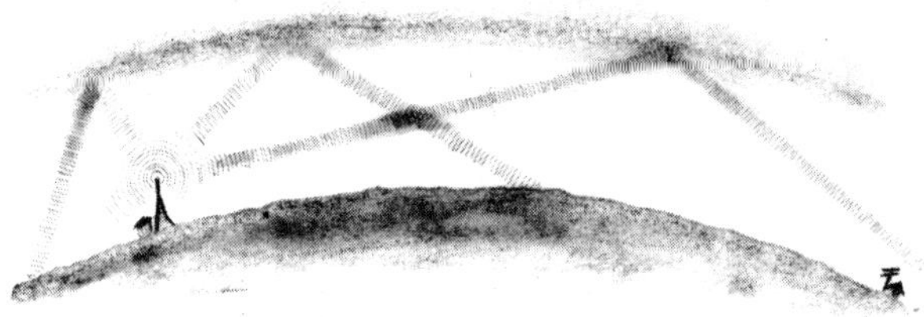

Radio waves that can cross large distances, either directly or by relay, are very useful for carrying information. Communication is accomplished by changing the signal according to an agreed code that can be deciphered at the receiving end. The first radio communication was achieved by turning the signal on and off in an agreed pattern, such as Morse code. Later, sounds were coded by continuous variations in the *amplitude* (that is, the *intensity*) of the broadcast wave (AM). Later still, the information was coded as *frequency* variations in the broadcast wave (FM). In broadcast radio and television, the "decoding" is done in the receiver serving the loudspeaker or TV picture tube. The output message from the receiver takes the same form that it had at the transmitter.

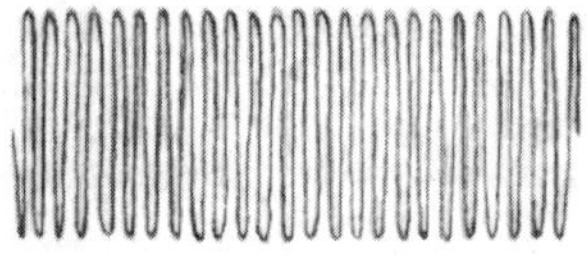

A "carrier" radio wave.

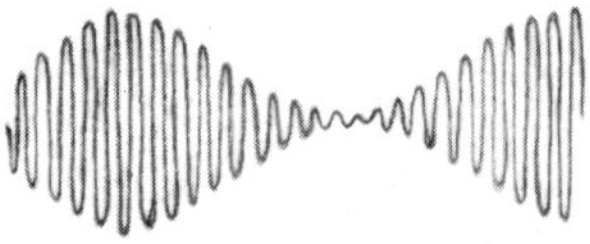

AM (amplitude modulation): Information is coded as variations in the amplitude (or intensity) of the carrier.

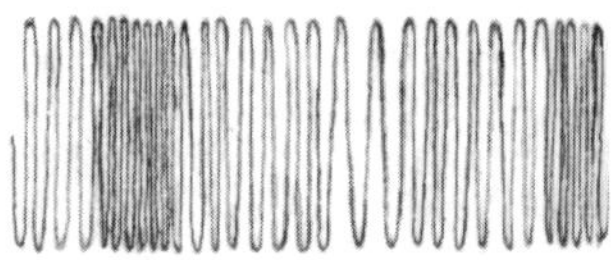

FM (frequency modulation): Information is coded as variations in the frequency of the carrier.

Because signals from different stations should not be received at the same spot on the dial, it is necessary to restrict their transmission. The International Telecommunication Union (ITU) controls radio transmission and other means of international communication. Within the United States, the Federal Communications Commission (FCC) regulates radio transmission. In order to reduce the interference of one station's signal with another, the FCC assigns suitable frequencies to radio stations. It also limits their power or the power radiated in particular directions, and may restrict the hours of transmission.

Television and radar. Television and FM broadcasting stations operate at frequencies of about 10^8 Hz. Waves at these frequencies are not reflected by the layers of electric charge in the upper atmosphere. Rather, the signals travel in nearly straight lines and pass into space instead of following the curvature of the earth. Thus, they can be used in communication between the earth and the moon, for example. But on earth, coaxial cables or relay stations are necessary to transmit signals between points more than about 80 km apart, even if there are no mountains in the way. Signals can be transmitted from one distant place to another, including from one continent to another, by relay satellites.

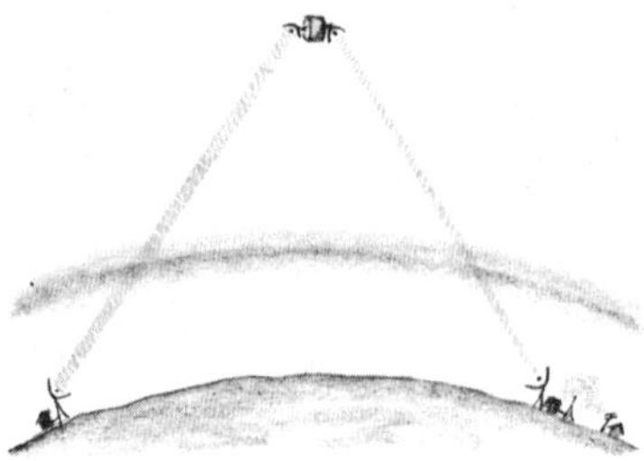

Satellites are used to relay microwaves all over the world. The microwaves can carry radio or TV information.

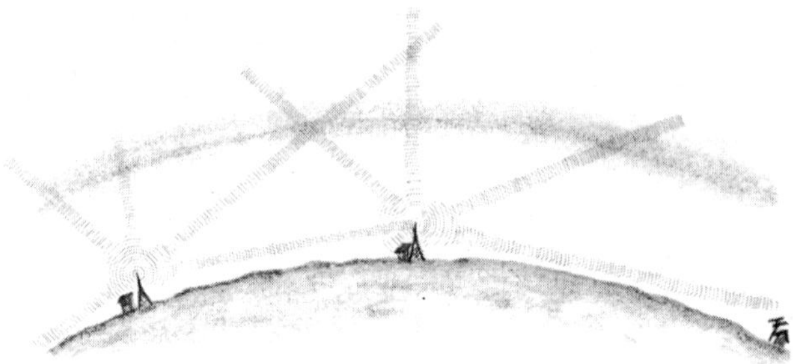

Since these signals have wavelengths of only about 1 m, they are not diffracted much around objects that have dimensions of several meters, such as cars, ships, or aircraft. Thus, the reflected portion of signals of wavelengths from 1 m down to 1 mm is used to detect such objects. The interference between the direct waves and reflection of these waves by passing airplanes can distort a television picture considerably. The signal also may be radiated in the form of pulses. If so, the time from the emission of a pulse to the reception of its echo measures the distance of the reflecting object. This technique is called radio detection and ranging, or *radar*. By means of the reflection of a beam that is pulsed, both the direction and distance of an object can be measured.

An important principle in radio transmission and detection is that of the resonant or "tuned" circuit. Reference to this can be found in the *Radio Amateur's Handbook* or in basic texts listed in most radio supply catalogues, such as *Allied Radio*.

SG 12–18

Microwave radiation. Electromagnetic waves with wavelengths of 10^{-1} to 10^{-4} m are often called *microwaves*. This radiation interacts strongly with matter and thus has uses other than communication. Water, for example, readily absorbs radiation with a wavelength on the order of 10 cm. Thus, any moist substance placed in a region of intense microwave radiation of this wavelength (meat, soup, or a cake batter, for example) will become hot very quickly. Because the heat is generated within the substance itself, rather than conducted inward from the outside, foods can be cooked very rapidly in a microwave oven. It is, however, important to keep the radiation confined to the oven because microwave radiation can damage living tissue.

A photograph made with film sensitive only to infrared radiation.

Infrared radiation. In heated bodies, the atoms themselves give off electromagnetic waves shorter than about 10^{-4} m. This "radiant heat" is usually called *infrared* rays, because most of the energy is of wavelengths slightly longer than the red end of the visible spectrum. While associated mainly with heat radiation, infrared rays do have some properties that are the same as those of visible light. The shorter infrared waves affect specially treated photographic film, and photographs taken with infrared radiation show some interesting effects. Also, scattering by small particles in the atmosphere is very much less for long wavelengths (Sec. 13.6). Thus, infrared rays can penetrate smoky haze dense enough to block visible light.

SG 19
SG 20

Visible light. The visual receptors in the human eye are sensitive to electromagnetic radiation with wavelengths between about 7×10^{-7} and 4×10^{-7} m. Radiation of these wavelengths is usually called light, or visible light. The eye is most sensitive to the green and yellow parts of the spectrum. This peak sensitivity corresponds roughly to the peak of solar radiation that reaches the earth's surface.

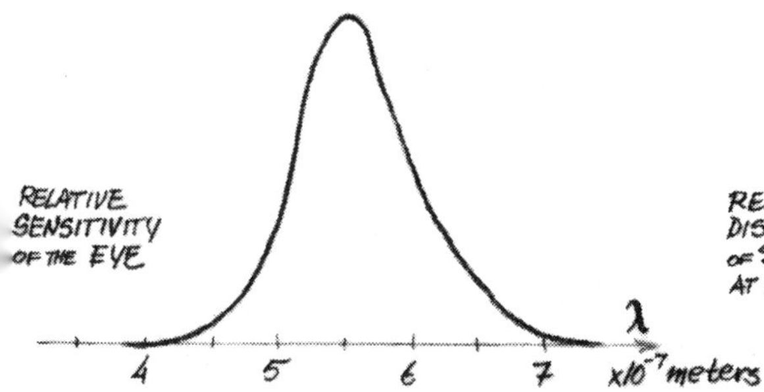

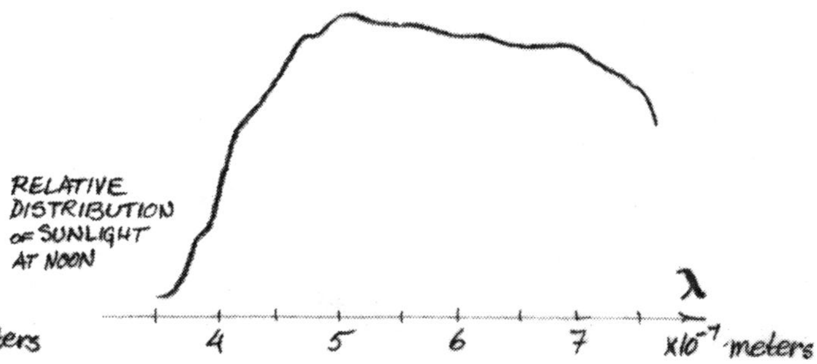

SG 21

Ultraviolet light. Electromagnetic waves shorter than the visible violet (4×10^{-7} to 10^{-8} m) are called *ultraviolet*. The ultraviolet region of the spectrum is of just as much interest in spectrum study as the visible and infrared. The atoms of many elements emit ultraviolet radiations that are characteristic of those

Close Up

Astronomy Across the Spectrum

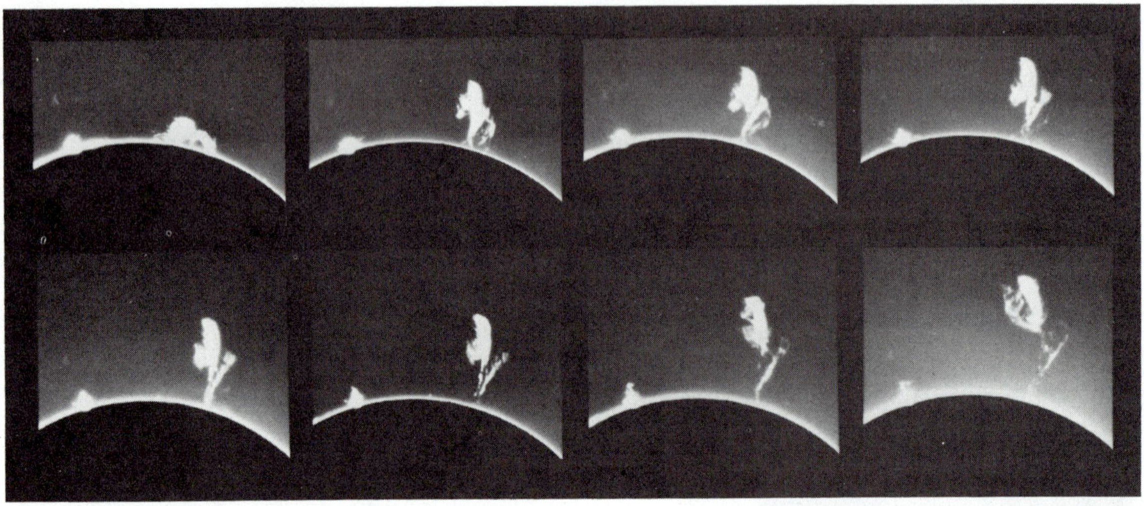

The electromagnetic spectrum comprises more than the rainbow effect produced by passing white light through a prism. Electromagnetic radiation of different wavelengths provides different kinds of information. You are familiar with the effects of various parts of the spectrum: sunburn (ultraviolet rays), visible light, heat (infrared), sound, and vibrations (for example, earth tremors). Scientists make use of electromagnetic radiations in such fields as astronomy, earth and life sciences, and communications. Several astronomical applications of the electromagnetic spectrum are shown here.

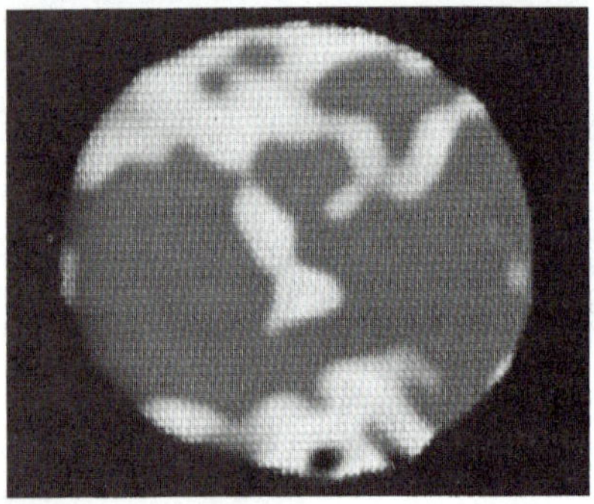

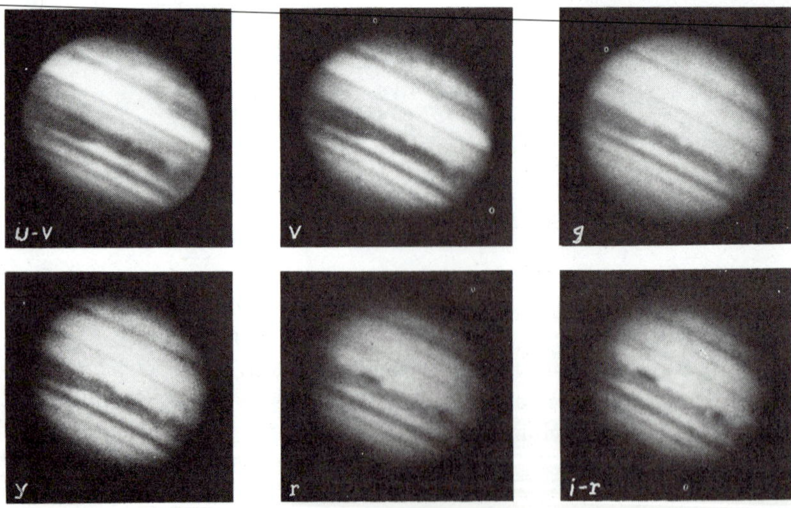

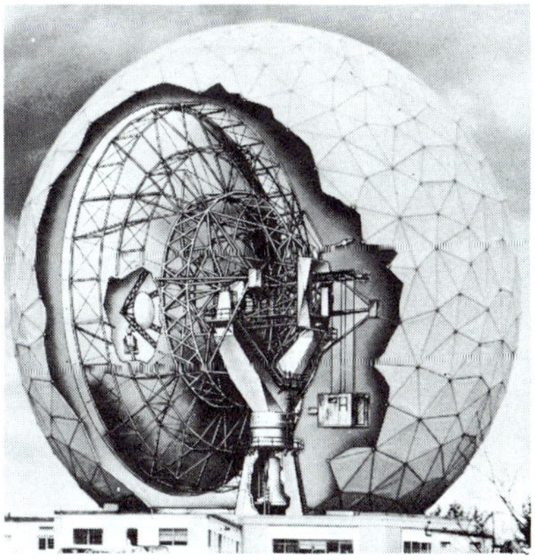

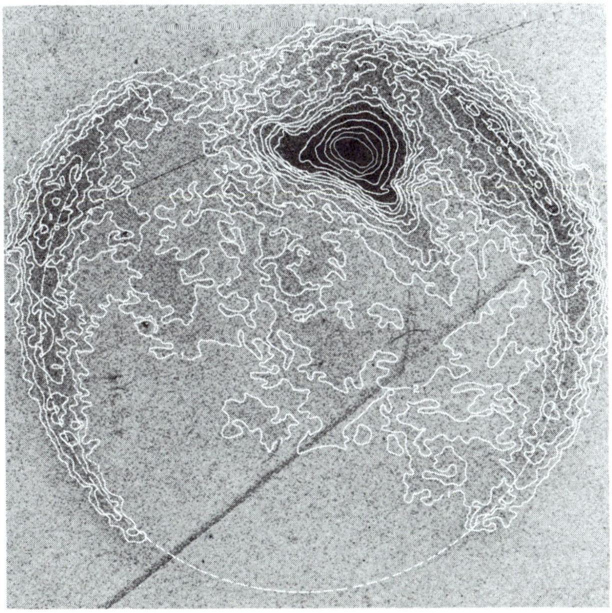

The photos on page 514 show: (top) a solar flare photographed in visible light; (middle) a computer-enhanced image of the star Betelguese taken with the 4-m Mayall telescope; (bottom) Jupiter photographed in ultraviolet, violet, green, yellow, red, and infrared. The photos on this page show: (top left) the huge radio telescope in Arecibo, Puerto Rico; (top right) the steerable Haystack antenna in Massachusetts; (bottom left) a contour map of the infrared brightness of a portion of the sky.

Electromagnetic waves generally are produced in the *acceleration* of charged particles.

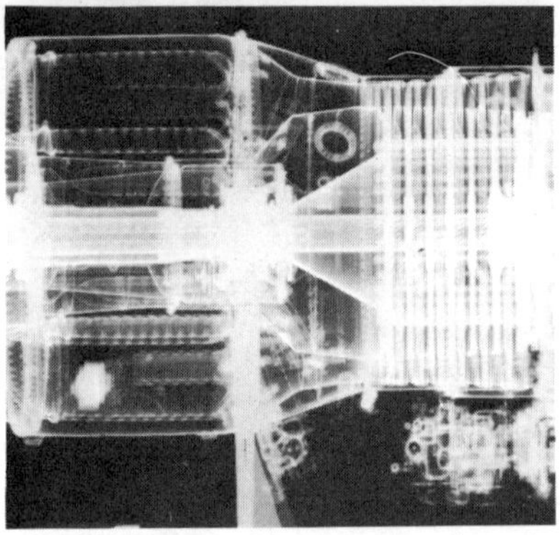

X-ray photos of (top) a chambered nautilus sea shell and (bottom) a jet engine.

elements. Ultraviolet light, like visible light, can cause photochemical reactions in which radiant energy is converted directly into chemical energy. Typical of these reactions are those that occur in silver bromide in the photographic process, in the production of ozone in the upper atmosphere, and in the production of a dark pigment, known as melanin, in the skin.

X rays. This radiation involves wavelengths from about 10^{-8} m to 10^{-17} m. Usually, it is produced by the sudden deflection or stopping of electrons when they strike a metal target. The maximum frequency of the radiation generated is determined by the energy with which the electrons strike the target. In turn, this energy is determined by the voltage through which the electrons are accelerated (Sec. 14.8). So the maximum frequency increases with the accelerating voltage. The higher the frequency of the X rays, the greater is their power to penetrate matter. But the distance of penetration also depends on the nature of the material being penetrated. X rays are readily absorbed by bone, which contains calcium; they pass much more easily through less dense organic matter (such as flesh) containing mainly the light atoms hydrogen, carbon, and oxygen. This fact, combined with the ability of X rays to affect a photographic plate, has led to some of the medical uses of X-ray photography. X rays can damage living cells and should be used with great caution and only by trained technicians. But some kinds of diseased cells are injured more easily by X rays than are healthy cells. Thus, a carefully controlled X-ray beam is sometimes used to destroy cancerous growths or other harmful cells.

X rays produce interference effects when they fall on a crystal in which atoms and molecules are arranged in a regular pattern. Different portions of the incident beam of X rays are reflected from different planes of atoms in the crystal structure. These reflected rays can interfere constructively, and this fact can be used in either of two ways. If the spacing of the atoms in the crystal is known, the wavelength of the X rays can be calculated. If the X-ray wavelength is known, the distance between crystal planes, and thus the structure of the crystal, can be determined. X rays are now widely used by chemists, mineralogists, and biologists in studying the structure of crystals and complex molecules. You will use these ideas in Chapter 18.

The glow in the photograph is caused when gamma rays emitted by radioactive cobalt cylinders interact with the surrounding pool of water.

Gamma rays. The gamma-ray region of the electromagnetic spectrum overlaps the X-ray region (see page 511). Gamma radiation is emitted mainly by the unstable nuclei of natural or artificial radioactive materials. You will study gamma rays further in Unit 6.

8. *Why do radio waves not cast noticeable "shadows" behind such obstacles as trees or small buildings?*

9. *Why are relay stations often needed in transmitting television signals?*

10. *How is the frequency of X rays related to their penetration of matter?*

11. *How do the wavelengths used in radar compare with the wavelengths of visible light?*

16.6 | What about the ether now?

The "luminiferous ether" had been proposed specifically as a medium for the propagation of light waves. Maxwell found that the ether could also be thought of as a medium for transmitting electric and magnetic forces. Later, he realized that he could drop his specific model of the ether entirely if he focused on the mathematical form of the theory. Yet, just before his death in 1879, Maxwell wrote an article in which he still supported the ether concept:

> Whatever difficulties we may have in forming a consistent idea of the constitution of the aether, there can be no doubt that the interplanetary and interstellar spaces are not empty, but are occupied by a material substance or body, which is certainly the largest, and probably the most uniform body of which we have any knowledge...

Maxwell was aware of the failures of earlier ether theories. Near the beginning of the same article he said:

> Aethers were invented for the planets to swim in, to constitute electric atmospheres and magnetic effluvia, to convey sensations from one part of our bodies to another, and so on, till all space had been filled three or four times over with aethers. It is only when we remember the extensive and mischievous influence on science which hypotheses about aethers used formerly to exercise, that we can appreciate the horror of aethers which sober-minded men had during the 18th century....

Maxwell had formulated his electromagnetic theory mathematically, independent of any particular model of the ether. Why, then, did he continue to speak of the "great ocean of aether" filling all space? It seemed unthinkable to Maxwell that there could be vibrations without something that vibrates, or waves without a medium. Also, to many nineteenth-century physicists the idea of "action at a distance" seemed absurd. How

James Clerk Maxwell (1831–1879) was born in Edinburgh, Scotland, in the same year Faraday discovered electromagnetic induction. Unlike Faraday, Maxwell came from a well-off family. He was educated at the Edinburgh Academy and the University of Edinburgh. He showed a lively interest in how things happened when he was scarcely 3 years old. As a child he constantly asked, "What's the go of that?" He studied mechanisms, from a toy top to a commercial steam engine, until he had satisfied his curiosity about how they worked. On the abstract side, his formal studies, begun at the Academy in Edinburgh and continued through his work as an undergraduate at Cambridge, gave Maxwell experience in using mathematics to develop useful parallels among apparently unrelated occurrences. His first publication appeared in the proceedings of the Royal Society of Edinburgh when he was only 14 years old. By the time he was 17, he had published three papers on the results of his original research. In the 1870's he organized the Cavendish Laboratory at Cambridge University, which became a world center for physics research for the next several decades.

He was one of the main contributors to the kinetic theory of gases, to statistical mechanics and thermodynamics, and also the theory of color vision. His greatest achievement was his electromagnetic theory. Maxwell is generally regarded as the most profound and productive physicist between the time of Newton and Einstein.

could one object exert a force on another body far away if something did not transmit the force? One body is said to act *on* another, and the word *on* gives the idea of contact. Thus, according to accepted ways of describing the world in common language, the ether seemed somehow necessary.

Yet 25 years after Maxwell's death the ether concept had lost much of its support. Within another decade, it had vanished from the collection of useful concepts. In part, the success of Maxwell's theory itself helped to undermine the general belief in the existence of an ether, simply because his equations did not depend on details of the ether's structure. In fact, they could be taken to describe the relations between changes of electric and magnetic fields in space without any reference to the ether at all.

SG 23

A comparable effect *is* observed with sound waves: They go faster with respect to the ground when traveling *with* the wind than when traveling *against* the wind.

Another difficulty with belief in the ether was that all attempts to detect the motion of the earth relative to the ether failed. If light is a kind of vibration of an ether that fills all space, then light should travel at a definite speed relative to the ether. But the earth must also be moving through the ether in its annual orbit around the sun. Thus, the earth should be moving like a ship, against an "ether wind" at some times, and with it at other times. Under these conditions, the apparent speed of light should be observed to differ. When the earth and a beam of light are moving in the same direction through the ether, the observed speed of light should not be the same as when the earth and the light are moving in opposite directions.

Michelson first tried the experiment in 1881, stimulated by speculations on how to measure the effect of ether on light propagation, given in a letter of Maxwell's published just after Maxwell's death.

Theorists computed the time required for light to make a round trip with and against the ether wind. They compared this interval with the time calculated for a round trip in the absence of an ether wind. The expected time difference was found to be very small: only 10^{-15} sec for a round trip of 30 m. This is too short a time difference to measure directly, but it is of the same order as the time for one vibration of visible light. Therefore, the difference might be detected from observations of a properly produced interference pattern. In 1887, the American scientists Albert A. Michelson and Edward Morley used a device sensitive enough to detect an effect only 1% as great as that predicted by the ether theory. Neither this experiment nor the many similar experiments done since then have revealed the existence or expected effects of an ether wind.

Supporters of the ether concept offered various explanations for this unexpected result. For example, they suggested that objects moving at high speeds relative to the ether might change their size in just such a way as to make this relative speed undetectable. But even those who made such attempts to rescue the ether concept felt their proposals to be forced and artificial. Finally, a decisive development led scientists to abandon the ether concept. This breakthrough was not a specific experiment, but a brilliant proposal by a young man of 26 years. The man was Albert Einstein, who in 1905 suggested that a new and deep union of mechanics and electromagnetism could be achieved without the ether model. A few brief remarks here will provide a setting for your further study of relativity at a later time.

More on relativity theory appears in Chapter 20.

Einstein showed that the equations of electromagnetism can be written to fit the same principle of relativity that holds for mechanics. In Sec. 4.4, the Galilean principle of relativity was discussed. It states that *the same laws of mechanics apply in each of two frames of reference that have a constant velocity relative to each other*. Thus, it is impossible to tell by any mechanical experiment whether or not a laboratory (reference frame) is at rest or is moving with constant velocity. The principle is illustrated by common experience within a ship, car, plane, or train moving at a constant speed in a straight line. The observer finds that objects move, remain at rest, fall, or respond to applied force in just the same way they do when the ship, or whatever, is at rest. Galileo, a convinced Copernican, applied this principle to the motion of objects with respect to the earth. You will recall the example of a stone falling straight down alongside a tower. Galileo argued that this event gives no indication whether the earth is fixed and the sun in motion, or the sun fixed and the earth in motion.

Einstein in 1908.

Einstein extended this principle of relativity beyond mechanics. It applied, he proposed, to *all* of physics, including electromagnetism. A main reason for this assumption appears to have been his feeling that nature could not be lopsided; relativity could not apply only to *part of* physics. Einstein then added a second basic conjecture. He stated that *the speed of any light beam moving through free space is the same for all observers,* even when they are moving relative to each other or relative to the light source! This bold statement resolved the question of why the motion of observers with respect to the ether did not show up in experiments on the speed of light. In fact, Einstein rejected the ether and all other attempts to provide a "preferred frame of reference" for light propagation. The price of making these assumptions of his was, Einstein showed, the necessity of revising some common-sense notions of space and time. Einstein showed that Maxwell's equations are fully consistent with extending the principle of relativity to all physics. This was yet another great synthesis of previously separate ideas, like the syntheses forged by Copernicus, Newton, and Maxwell.

Some of the other important consequences of Einstein's theory of relativity will be discussed in Unit 5.

SG 24–28

?

13. Why did Maxwell (and others) cling to the concept of an ether?

14. Whose argument finally showed that the ether was an unnecessary hypothesis?

study guide

1. The *Project Physics* learning materials particularly appropriate for Chapter 16 include:

Activities
Microwave Transmission System

Film Loop
Standing Electromagnetic Waves

2. What inspired Oersted to look for a connection between electricity and magnetism?

3. A current in a conductor can be caused by a steady electric field. Can a *displacement* current in an insulator be similarly caused? Explain your answer briefly.

4. What causes an electromagnetic wave to be initiated? to be propagated?

5. What is the "disturbance" that travels in each of the following waves?

(**a**) water waves

(**b**) sound waves

(**c**) electromagnetic waves

6. In Hertz's detector, it is the electric field strength in the neighborhood of the wire that makes the sparks jump. How was Hertz able to show that the waves from the induction coil spark gap were polarized?

7. What evidence did Hertz obtain that his induction-coil-generated waves have many properties similar to visible light waves?

8. Give several factors that contributed to the 25-year delay in the general acceptance by scientists of Maxwell's electromagnetic wave theory.

9. What evidence is there for believing that electromagnetic waves carry energy? Does this suggest why the early particle theory of light had some success?

10. What is the wavelength of an electromagnetic wave generated by the 60-Hz alternating current in power lines? by radio broadcasts at the standard AM radio frequencies (between 500 and 1,500 $\times$ 10^3 Hz)? by broadcasts on the AM Hz "Citizen Band" (26.225 $\times$ 10^6 Hz)?

11. How short are "short-wave" radio waves? (Look at the frequencies indicated on the dial of a short-wave radio.)

12. Electric discharges in sparks, neon signs, lighting, and some atmospheric disturbances produce radio waves. The result is "static" or noise in AM radio receivers. Give other likely sources of such static.

13. Why must there be some federal control of the broadcast power and direction of radio and TV stations, but no such controls of the distribution of newspapers and magazines?

14. Many different kinds of radiation are broadcast from the Empire State Building. Antennas broadcast television and radio signals. Red warning lights protect airplanes. Infrared radiation leaks from all warm areas of the building. Even the electric wires broadcast a faint radiation of their own.

(**a**) Give an approximate frequency and wavelength for each of these four kinds of radiation.

(**b**) How would each kind of radiation behave if it encountered a 2-m space between two buildings?

15. If there are extraterrestrial beings of advanced civilizations, what method for gathering information about earth-people might they have?

16. Why can AM radio waves be detected at greater distances than the waves used for television and FM broadcasting?

17. Some relay satellites have a 24-hr orbit. Thus, they stay above the same point as the earth turns below them. What would the radius and location of the orbit of such a "synchronous" orbit be? (Refer to Unit 2 for whatever principles or constants you need.)

18. Explain why airplanes passing overhead cause "flutter" of a TV picture.

19. How much time would elapse between the sending of a radar signal to the moon and the return of the echo?

20. Refer to the black-and-white photograph on page 513 that was taken using film sensitive only to infrared. How do you explain the appearance of the trees, clouds, and sky?

21. Why do you think the eye is sensitive to the range of light wavelengths to which it *is* sensitive?

22. A sensitive thermometer placed in different parts of the visible light spectrum formed by a quartz prism will show a rise in temperature. This proves that all colors of light produce heat when absorbed. But the thermometer also shows an increase in temperature when its bulb is placed in either of the two dark regions to either side of the end of the visible spectrum. Why is this?

23. For each part of the electromagnetic spectrum discussed in Sec. 16.5, list the ways in which you have been affected by it. Give examples of things you have done with radiation in that frequency range, or of effects it has had on you.

24. What is the principal reason for the loss of support for the ether concept?

25. At many points in the history of science, the "natural" or common-sense way of looking at things has changed greatly. Attitudes toward action-at-a-distance are a case in point. What are some other examples?

26. Can intuition be educated; that is, can feelings about the fundamental aspects of reality be changed? Use attitudes toward action-at-a-distance of the ether as one example and give others.

27. Explain the "cat-less" grin shown below.

28. Write a brief essay on any two of the pictures on page 522, explaining in some detail what principles of physics they illustrate. (Select first the main principle at work in each of the situations shown. Also, you need not limit yourself to the principles discussed in this unit.)

29. In a couple of pages, summarize how this unit built up the story (and physical details) of the wave theory of light and the particle model of light. Include the model of light as a material wave in a material ether. Go on to the joining of the initially separate disciplines of electricity and magnetism, first with each other and then with the theory of light in Maxwell's general electromagnetic theory.

In this chapter, you have read about how mechanical models of light and electromagnetism faded away, leaving a model-less, mathematical *(and therefore abstract) field theory. The situation has been likened to that of the Cheshire Cat, in a story written by the Reverend Charles Dodgson, a mathematics teacher at Oxford, in 1862. An illustration is reproduced here.*

EPILOGUE

This unit has presented the story of how light and electromagnetism became understandable, first separately and then together. The particle model of light explained the behavior of light in terms of moving particles. On experiencing strong forces at a boundary, these particles were thought of as bouncing back or swerving in just the direction that light is observed to be reflected and refracted. The wave model accounted for these and other effects by treating light as transverse waves in a continuous medium. These rival theories of light provided helpful mechanical models for light viewed either as particles or as waves.

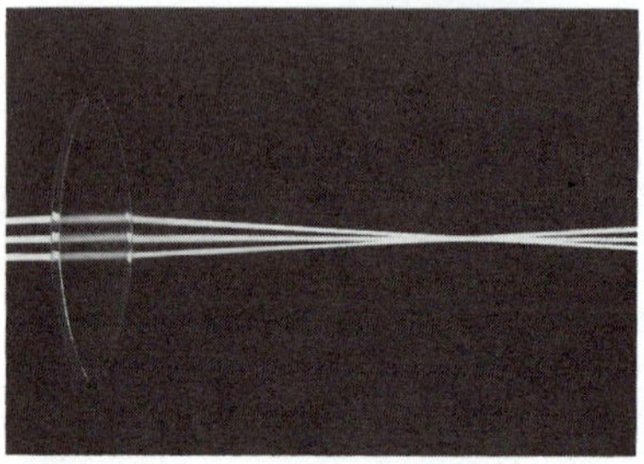

Mechanical models also worked, up to a point, in explaining electricity and magnetism. Both Faraday and Maxwell made use of mechanical models for electric and magnetic lines of force. Maxwell used these models as guides in developing a mathematical theory of electromagnetism that, when completed, went well beyond the models. This theory also explained light as an electromagnetic wave phenomenon.

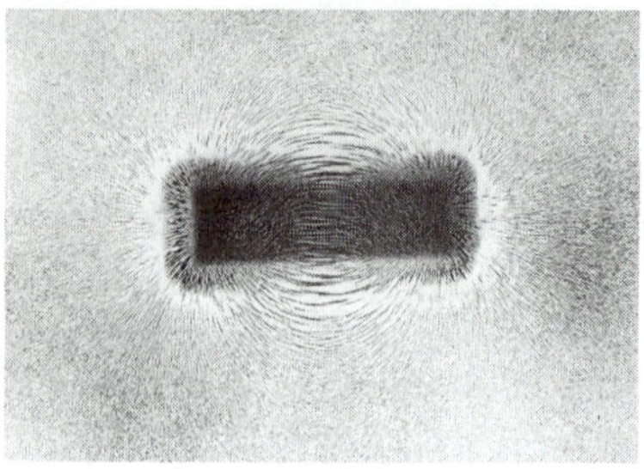

However, it proved essential, ultimately, to keep the equations describing the electric and magnetic fields, but to dispose of any specific mechanical model used to help derive the equations. But is there any way you can picture in your mind what a field "looks like"? Here is the response of the Nobel Prize-winning American physicist Richard Feynman to this question:

> I have asked you to imagine these electric and magnetic fields. What do you do? Do you know how? How do I imagine the electric and magnetic field? What do I actually see? What are the demands of scientific imagination? Is it any different from trying to imagine that the room is full of invisible angels? No, it is not like imagining invisible angels. It requires a much higher degree of imagination to understand the electromagnetic field than to understand invisible angels. Why? Because to make invisible angels understandable, all I have to do is to alter their properties a *little bit*—I make them slightly visible, and then I can see the shapes of their wings and bodies, and halos. Once I succeed in imagining a visible angel, the abstraction required—which is to take almost invisible angels and imagine them completely invisible—is relatively easy. So you say, "Professor, please give me an approximate description of the electromagnetic waves, even though it may be slightly inaccurate, so that I too can see them as well as I can see almost-invisible angels. Then I will modify the picture to the necessary abstraction."
>
> I'm sorry that I can't do that for you. I don't know how. I have no picture of this electromagnetic field that is in any sense accurate. I have known about the electromagnetic field a long time—I was in the same position 25 years ago that you are now, and I have had 25 years of experience thinking about these wiggling waves. When I start describing the magnetic field

moving through space, I speak of the E- and B-fields and wave my arms and you may imagine that I can see them. I'll tell you what I *can* see. I see some kind of vague shadowy wiggling lines—here and there is an E and B written on them somehow, and perhaps some of the lines have arrows on them—an arrow here or there which disappears when I look too closely at it. When I talk about the fields swishing through space, I have a terrible confusion between the symbols I use to describe the objects and the objects themselves. I cannot really make a picture that is even nearly like the true waves. So if you have some difficulty in making such a picture, you should not be worried that your difficulty is unusual.

The general trend in modern mechanics and electromagnetism can be summarized by saying that physical theories have become increasingly abstract and mathematical. Newton replaced the celestial machinery of early theories with a mathematical theory using the laws of motion and the inverse-square law. Maxwell developed a mathematical theory of electromagnetism that, as Einstein showed, did not require any material medium such as "ether." You are seeing here a growing but quite natural gap between common-sense ideas developed from direct human experiences and the subtle mathematical abstractions describing effects that you cannot sense directly.

Yet, in the end, even these highly abstract theories must make sense when put into ordinary language. And they *do* tell about the things you can see and touch and feel. They use abstract language, but have concrete tests and by-products. They have made it possible to devise the equipment that guides space probes to other planets and to design and operate the instruments that enable scientists to communicate with these probes. Not only are these theories at the base of all practical developments in electronics and optics, but they now also contribute to the understanding of vision and the nervous system.

Maxwell's electromagnetic theory and the interpretation given to electromagnetism and mechanics by Einstein in the special theory of relativity produced a profound change in the basic philosophical viewpoint of the Newtonian cosmology. (In this sense, Unit 4 marks a kind of watershed between the "old" and "new" ways of doing physics.) While it is too early to hope for a comprehensive statement of these changes, some aspects of a new cosmology can already be detected. For this purpose, you will now proceed to study the behavior of matter and the atomic theories developed to account for this behavior.

UNIT

Models of the Atom

PROLOGUE

In the early units of this text, you studied the motion of bodies of ordinary size, such as you deal with in ordinary life. You saw how the laws of mechanics and of electricity and magnetism can be used to describe and predict the behavior of these bodies. More important, you saw that these laws can also predict the behavior of bodies very different from those you are familiar with from everyday experience, for example, planets and the solar system at one extreme, and molecules of a gas or electrons in a metal at the other. In this unit, you will learn how to apply basic theories to the problem of the nature of matter. The phrase "the nature of matter" may seem simple now, but its meaning has been changing and growing over the centuries. The kinds of questions asked about matter and the methods used to answer these questions are continually changing. For example, during the nineteenth century, the study of the nature of matter consisted mainly of chemistry; in the twentieth century, the study of matter also involves atomic, nuclear, and elementary particle physics.

Since 1800, progress has been so rapid that it is easy to forget that people have theorized about matter for more than 2,500 years. In fact, some questions which have been answered only during the last hundred years were first asked more than 2,000 years ago. Some ideas which are considered new and exciting, such as the atomic structure of matter, were debated in Greece in the fifth and fourth centuries B.C. This prologue will review briefly the development of ideas concerning the nature of matter up to about 1800. This review will set the stage for the four chapters of Unit 5, which cover in greater detail the progress made since 1800. You will see that regardless of the form it takes (large or small, stable or shifting, solid, liquid, or gaseous) all matter is made up of separate particles called atoms. You will find that the atoms themselves have structure.

Theories based on the motion of bodies of ordinary size are useful only up to a point when dealing with atoms and their structure. Beyond that point, these theories break down. Nor are they able to deal with other extremes, for example, bodies that move very rapidly or bodies that are extraordinarily dense. Quantum mechanics and relativity were developed to describe the behavior of bodies that are far outside direct, everyday experience. These new theories do not prove mechanics to be "wrong." Rather, they show the existence of a boundary beyond which everyday experience cannot be extrapolated without clashing with new phenomena and new laws.

Finally, it will become apparent that quantum mechanics and relativity have their limits, too. There is a realm of behavior in the elementary particles that make up the nucleus, of which even quantum mechanics or relativity do not provide an adequate description. At present, no new theories have yet been shown to be fully satisfactory, either.

Early science had to develop out of ideas that were available before science started. These ideas came from experience with snow, wind, rain, mist, and clouds; heat and cold, salt and fresh water; wine, milk, blood, and honey; ripe and unripe fruit; fertile and infertile seeds. The most obvious and most puzzling facts were that plants, animals, and people were born, grew and matured, then aged and died. The world was continually changing, and yet, on the whole, it seemed to remain much the same. The causes of these changes and of the apparent continuity of nature were unknown. Often they were assigned to the actions of gods and demons. Myths concerning the creation of the world and the changes of the seasons were among the earliest creative productions of primitive people everywhere. Such myths helped people to come to terms with events they could see happening but could not rationally understand.

This gold earring, made in Greece about 600 B.C., shows the great skill with which ancient artisans worked metals. [Museum of Fine Arts, Boston]

Over a long period of time, humans developed some control over nature and materials. They learned how to keep warm and dry, to smelt ores, to make weapons and tools, to produce gold ornaments, glass, perfumes, and medicines. Eventually, in Greece, by the year 600 B.C., philosophers (literally "lovers of wisdom") had started to look for rational explanations of natural events. They wanted explanations that did not depend on the actions or the whims of gods or demons. They sought to discover the enduring, unchanging things out of which the world is made. How did these things give rise to the changes they perceived and to the great variety of material things? This search was the beginning of human attempts to understand the material world rationally, and it led to a theory of the nature of matter.

The earliest Greek philosophers thought that all the different things in the world were made out of a single basic substance. Some thought that water was the fundamental substance and that all other substances were derived from it. Others thought that air was the basic substance; still others favored fire. But neither water, air, nor fire was satisfactory. No one substance seemed to have enough different properties to give rise to the enormous variety of substances in the world. According to another view, introduced by Empedocles around 450 B.C., there were four basic types of matter: earth, air, fire, and water. All material things were made out of them. These four basic materials could mingle and separate and reunite in different proportions. In doing so, they could produce the variety of familiar objects as well as the changes in such objects. The basic four materials, called *elements*, were supposed to persist through all these changes. This theory was the first appearance of a *model of matter* explaining all material things as just different arrangements of a few elements.

The first atomic theory of matter was introduced by the Greek philosopher Leucippus, born about 500 B.C., and his pupil Democritus (460–370 B.C.). Only scattered fragments of the writings of these philosophers remain, but their ideas were discussed in considerable detail by the Greek philosophers Aristotle (389–321 B.C.) and Epicurus (341–270 B.C.) and by the Latin poet Lucretius (100–55 B.C.). To these men we owe most of our knowledge of ancient atomism.

The theory of the atomists was based on a number of assumptions:

1. Matter is eternal; no material thing can come from nothing, nor can any material thing pass into nothing.
2. Material things consist of very small indivisible particles. The word "atom" meant "uncuttable" in Greek. In discussing the ideas of the early atomists, the word "indivisibles" could be used instead of the word "atoms."
3. Atoms differ in their sizes and shapes.

4. Atoms exist in otherwise empty space (the void), which separates them and allows them to move from one place to another.

5. Atoms are continually in motion, although the nature and cause of the motion are not clear.

6. In the course of their motions, atoms come together and form combinations which are the material substances. When the atoms forming these combinations separate, the substances decay or break up. Thus, the combinations and separations of atoms give rise to the changes that take place in the world.

7. The combinations and separations take place according to natural laws that are not yet clear, but do not require the action of gods, demons, or other supernatural powers. In fact, one of the chief aims of the atomists was to liberate people from superstition and fear. As Lucretius put it, "fear in sooth takes hold of all mortals because they see many operations go on in earth and heaven, the causes of which they can in no way understand, believing them therefore to be done by divine power." By explaining natural events by the motion of atoms, Lucretius hoped to show "the manner in which all things are done without the hand of the gods."

With the above assumptions, the ancient atomists worked out a consistent story of change, which they sometimes called "coming-to-be" and "passing away." They could not demonstrate experimentally that their theory was correct. It was simply an explanation derived from assumptions that seemed reasonable to them. The theory was a "likely story." It was not useful for predicting new phenomena; but prediction became an important value for a theory only later. To the atomists, it was more significant that the theory also helped to allay an unreasonable belief in capricious supernatural beings.

The atomic theory was criticized severely by Aristotle. He argued logically, from his own assumptions, that no vacuum or void could exist. Therefore, the idea of atoms in continual motion must be rejected. (Aristotle was also probably sensitive to the fact that in his time belief in atomism was identified with atheism.) For a long time Aristotle's argument against the void was widely held to be convincing. Not until the seventeenth century did Torricelli's experiments (described in Chapter 11) show that a vacuum could indeed exist. Furthermore, Aristotle argued that matter is continuous and infinitely divisible, so that there can be no atoms.

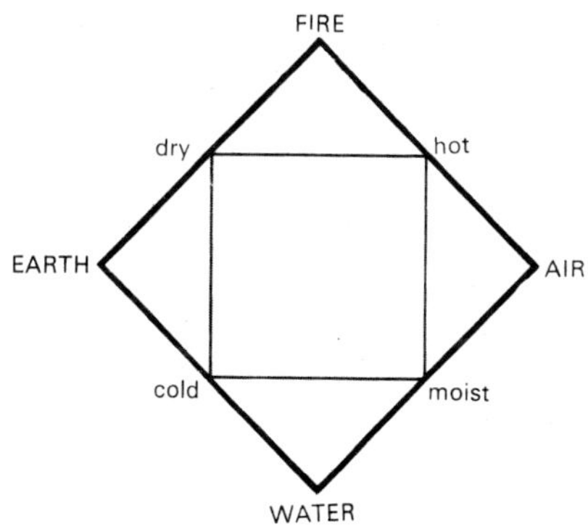

Aristotle developed a theory of matter as part of his grand scheme of the universe. This theory, with some modifications, was considered satisfactory by most philosophers of nature for nearly 2,000 years. Aristotle's theory was based on the four basic elements, Earth, Air, Fire, and Water, and four "qualities," Cold, Hot, Moist, and Dry. Each element was characterized by two

According to Aristotle in his *Metaphysics*, "There is no consensus concerning the number or nature of these fundamental substances. Thales, the first to think about such matters, held that the elementary substance is clear liquid. . . . He may have gotten this idea from the observation that only moist matter can be wholly integrated into an object—so that all growth depends on moisture. . . .

"Anaximenes and Diogenes held that colorless gas is more elementary than clear liquid, and that indeed, it is the most elementary of all simple substances. On the other hand, Hippasus of Metpontum and Heraclitus of Ephesus said that the most elementary substance is heat. Empedocles spoke of four elementary substances, adding dry dust to the three already mentioned. . . . Anaxagoras of Clazomenae said that there are an infinite number of elementary constituents of matter. . . ." [From a translation by D. E. Gershenson and D. A. Greenberg]

Laboratory of a sixteenth-century alchemist.

qualities (the nearer two to each side, as shown in the diagram at the left). Thus, the element

Earth is	Dry and	Cold,		
Water is		Cold and	Moist,	
Air is			Moist and	Hot,
Fire is				Hot and Dry.

According to Aristotle, it is always the first of the two qualities that dominates. In his theory, the elements are not unchangeable. Any one of them may be transformed into any other if one or both of its qualities change to their opposites. The transformation takes place most easily between two elements having one quality in common. Thus, Earth (dry and cold) is transformed into Water when dryness changes to moistness. Earth can be transformed into Air only if both of the qualities of Earth (dry and cold) are changed to their opposites (moist and hot).

As mentioned in Chapter 2, Aristotle was able to explain many natural phenomena by means of his ideas. Like the atomic theory, Aristotle's theory of coming-to-be and passing away was a consistent model of the nature of matter. It also had certain advantages over the atomic theory. For example, it was based on elements and qualities that were familiar to people; it did not involve atoms, which could not be seen or otherwise perceived, or a void, which was difficult to imagine. In addition, Aristotle's theory provided some basis for further experimentation: It supplied what seemed like a rational basis for the fascinating possibility of changing any material into any other.

Although the atomistic view was not completely abandoned, it found few supporters between 300 B.C. and about 1600 A.D. The atoms of Leucippus and Democritus moved through empty space, which contained no "spirit" and had no definite plan or purpose. Such an idea remained contrary to the beliefs of the major religions. Like the Athenians in the time of Plato and Aristotle, the later Christian, Hebrew, and Moslem theologians considered atomists atheistic and "materialistic" for claiming that everything in the universe could be explained in terms of matter and motion.

About 300 or 400 years after Aristotle, a type of research called *alchemy* appeared in the Near and Far East. Alchemy in the Near East combined Aristotle's ideas about matter with methods of treating ores and metals. One aim of the alchemists was to change or "transmute" ordinary metals into precious metals. Although they failed to do this, the alchemists found and studied many properties that are now classified as chemical properties. They invented some pieces of chemical apparatus, such as reaction vessels and distillation flasks, that (in modern form) are still common in chemical laboratories. They studied such processes as calcination, distillation, fermentation, and sublimation. In this sense, alchemy may be regarded as the chemistry of the Middle Ages. But alchemy left unsolved the fundamental questions. At the opening of the eighteenth century the most important of these questions were: (1) What is a chemical element? (2) What specifically is the chemical nature of the so-called elements, Earth, Air, Fire, and Water? (3) What is the nature of chemical composition and chemical change, especially burning? Until these questions were answered, it was impossible to make real progress in finding out the structure of matter. Real progress was delayed until about a century after the "scientific revolution" of the seventeenth century, which clarified the chief problems of astronomy and mechanics, but not of chemistry.

During the seventeenth century, however, some forward steps were made, supplying a basis for future progress in determining the nature of matter. The Copernican and Newtonian revolutions greatly undermined the authority of Aristotle. Now his ideas about matter were also more easily questioned. Atomic concepts were revived, offering a way of looking at things that was very different from Aristotle's ideas. As a result, theories involving atoms (or "particles" or "corpuscles") were again considered seriously. Boyle based his models on the idea of "gas particles." Newton also discussed the behavior of a gas (and even of light) by supposing it to consist of particles. In addition, there was now a successful science of mechanics. Through mechanics, scientists could hope to describe how the atoms interacted with each other. Thus, the stage was set for a general revival of atomic theory.

One of those who contributed greatly to the revival of atomism was Pierre Gassendi (1592–1655), a French priest and philosopher. He avoided the criticism of atomism as atheistic by saying that God created the atoms and bestowed motion upon them. Gassendi accepted the physical explanations of the atomists, but rejected their disbelief in the immortality of the soul and in Divine Providence. He was thus able to provide a philosophical justification of atomism that met some of the serious religious objections.

In the eighteenth century, chemistry became more quantitative. Weighing, in particular, was done more frequently and more carefully. New substances were isolated and their properties examined. The attitude that grew up in the second half of the century was apparent in the work of Henry Cavendish (1731–1810). According to a biographer, Cavendish regarded the universe as consisting

It was Cavendish, remember, who designed the sensitive torsional balance that made it possible to find a value for the gravitational constant *G* (Sec. 8.8).

> . . . solely of a multitude of objects which could be weighed, numbered, and measured; and the vocation to which he considered himself called was to weigh, number, and measure as many of those objects as his alloted threescore years and ten would permit. . . . He weighed the Earth; he analysed the Air; he discovered the compound nature of Water; he noted with numerical precision the obscure actions of the ancient element Fire.

Eighteenth-century chemistry reached its peak in the work of Antoine Lavoisier (1743–1794). Lavoisier worked out the modern views of combustion, established the law of conservation of mass and explained the elementary nature of hydrogen and oxygen and the composition of water. Above all, he emphasized the

Lavoisier's work on the conservation of mass was described in Chapter 9.

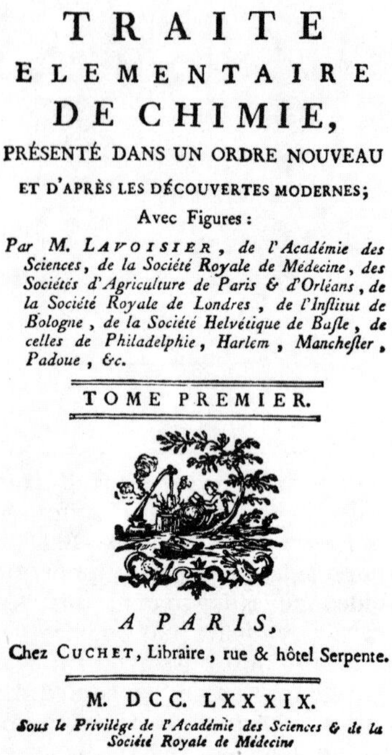
TRAITE
ELEMENTAIRE
DE CHIMIE,
PRÉSENTÉ DANS UN ORDRE NOUVEAU
ET D'APRÈS LES DÉCOUVERTES MODERNES;
Avec Figures:
Par M. Lavoisier, de l'Académie des Sciences, de la Société Royale de Médecine, des Sociétés d'Agriculture de Paris & d'Orléans, de la Société Royale de Londres, de l'Inftitut de Bologne, de la Société Helvétique de Bafle, de celles de Philadelphie, Harlem, Manchefter, Padoue, &c.
TOME PREMIER.
A PARIS,
Chez Cuchet, Libraire, rue & hôtel Serpente.
M. DCC. LXXXIX.
Sous le Privilège de l'Académie des Sciences & de la Société Royale de Médecine

Title page of Lavoisier's Traité Elémentaire de Chimie *(1789).*

quantitative aspects of chemistry. His famous book, *Traité Elémentaire de Chimie (Elements of Chemistry)*, published in 1789, established chemistry as a modern science. In it, Lavoisier analyzed the idea of an element in a way which is very close to modern views:

> . . . if, by the term *elements* we mean to express those simple and indivisible atoms of which matter is composed, it is extremely probable that we know nothing at all about them; but if we apply the term *elements*, or *principles of bodies*, to express our idea of the last point which analysis is capable of reaching, we must admit as elements all the substances into which we are capable, by any means, to reduce bodies by decomposition. Not that we are entitled to affirm that these substances we consider as simple may not be compounded of two, or even of a greater number of principles; but since these principles cannot be separated, or rather since we have not hitherto discovered the means of separating them, they act with regard to us as simple substances, and we ought never to suppose them compounded until experiment and observation have proved them to be so.

During the second half of the eighteenth century and the early years of the nineteenth century great progress was made in chemistry. This progress resulted largely from the increasing use of quantitative methods. Chemists found out more and more about the composition of substances. They separated many elements and showed that nearly all substances are *compounds*, that is, combinations of a fairly small number of chemical elements. They learned a great deal about how elements combine and form compounds and how compounds can be broken down into the elements of which they consist. This information allowed chemists to establish many empirical laws of chemical combination. Then chemists sought an explanation for these laws.

During the first 10 years of the nineteenth century, the English chemist John Dalton introduced a modified form of the old Greek atomic theory. Dalton's theory was an attempt to account for the laws of chemical combination. It is here that the modern story of the atom begins. Dalton's atomic theory was an improvement over that of the Greeks because it opened the way for quantitative study of the atom. Today, the existence of the atom is no longer a topic of speculation. There are many kinds of experimental evidence, not only for the existence of atoms but also for their inner structure. This unit will trace the discoveries and ideas that provided this evidence.

Between the nineteenth and twentieth centuries, convincing evidence was developed for the modern conception of atoms. Some of this evidence came from chemistry. Chapter 17 will review these ideas quite briefly. (The review assumes that you have already studied some chemistry.)

Chemical laboratory of the eighteenth century.

However, chemistry could not answer all questions about atoms; these questions could only be answered by physics. Physical evidence accumulated in the nineteenth century and in the early years of the twentieth century made it possible to propose models not only for the atomic structure of matter but for the interior structure of the atom itself. This evidence will be discussed in Chapters 18 and 19. You will see how this physical evidence required a revision of the laws upon which all physical explanations were based thus far, when these laws were applied to atomic phenomena.

Chapter 20 deals with the triumphs of two theories belonging to modern physics. These theories grew logically from the attempts to understand the structure of the atom. At the same time, you will begin to see that these new theories, too, are still evolving in order to deal with the newest discoveries.

A Summary of Some Ideas from Chemistry

17.1 | Elements, atoms, and compounds

This brief chapter is not meant to teach you chemistry but to refresh your memory about some ideas you have learned in previous chemistry courses. Unlike other chapters in *Project Physics*, there will be little history. You may also find many of the explanations incomplete.

Substances can be divided into two classes: compounds and elements. Compounds, by far the larger class, are substances that can be decomposed into other substances by chemical means (for example, heating or passing an electric current through them). Elements are unique in that they can not be decomposed chemically.

Electrolysis is discussed in Sec. 17.2.

Sometimes this definition was difficult to apply because chemists did not know all the ways to decompose substances. Soda (sodium bicarbonate) and potash (potassium carbonate) were considered elements before Humphry Davy, by using a strong electric current, showed that they could be decomposed.

Most elements can be combined with other elements to make compounds. When elements combine in this way, *precisely* the same ratio of masses of the constituents is required to make a particular compound. For example, 7.94 g of oxygen always

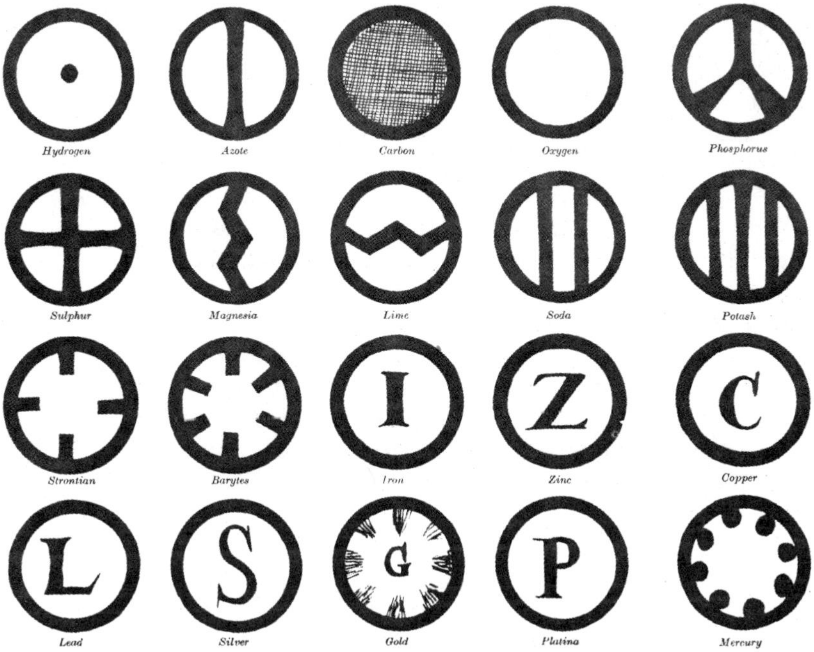

Dalton's symbols for "elements" (1808).

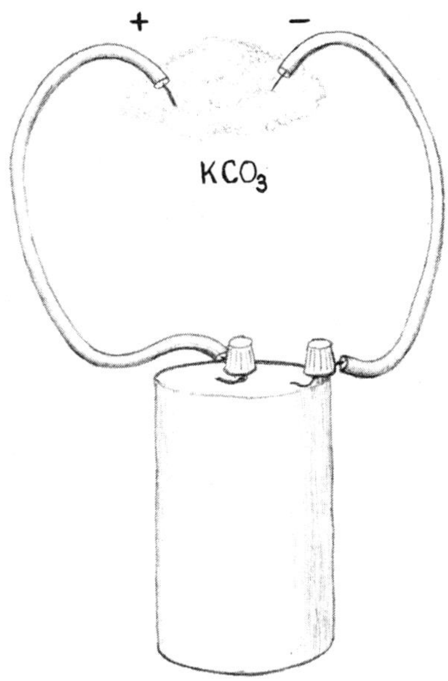

combine with 1.00 g of hydrogen to produce 8.94 g of water. If you start with 10.00 g of oxygen, you still get only 8.94 g of water, but there will be 2.06 g of oxygen left over. This rule applies to all compounds and is called *the law of fixed proportions*.

An alloy is a mixture of metals and is *not* a compound. Thus, $Cu_{.73}Sn_{.27}$ is a reasonable formula for an alloy.

A model that explains the law of fixed proportions asserts that any element consists of a collection of identical, indestructible atoms, a notion carefully investigated by John Dalton (1766–1844). When two or more atoms link together, they form a molecule. The molecule may be an element itself (if both atoms are the same), or it may be a compound (if the atoms are different). Since atoms are not divisible, the idea of joining two atoms of hydrogen to, say, 1¼ atoms instead of exactly one atom of oxygen is meaningless. The law of fixed proportions follows quite naturally from the hypothesis that elements are made up of identical, indestructible atoms.

It is useful to give the elements symbols, for example, "O" for oxygen, "H" for hydrogen, "Fe" for iron. With these symbols, formulas for the compounds can be written quite simply. The familiar formula H_2O indicates that a water molecule has two hydrogen atoms (2 H) and one oxygen atom (O). Other examples of simple formulas are NaCl, Fe_2O_3, and H_2SO_4. The subscript indicates the number of atoms of the particular element in one molecule of the compound.

Berzelius (1779–1848) introduced these symbols in the 1820's.

These formulas suggest another very useful concept that is called *combining capacity*. This concept is probably easiest to illustrate by example. Common table salt has the formula NaCl. This formula indicates that one atom of sodium (Na) combines

Cl: 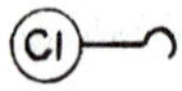

NaCl: 

$CaCl_2$: 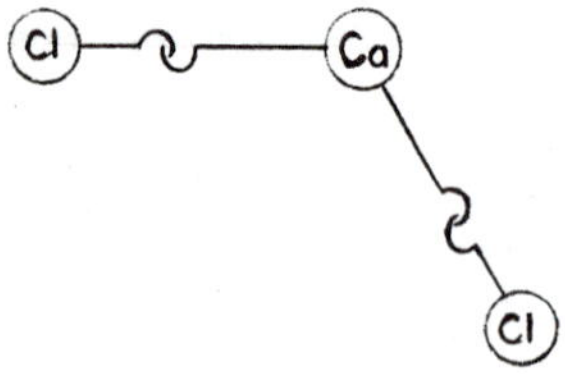

C: 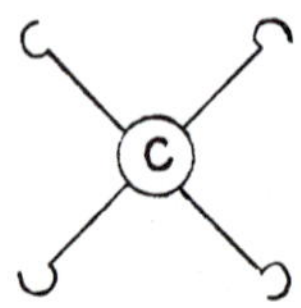

He: 

with one atom of chlorine (Cl) to form one molecule of table salt (NaCl). Another salt, calcium chloride (often used to melt ice on roads) has the formula $CaCl_2$; one atom of calcium (Ca) combines with two atoms of chlorine to form this compound. Carbon tetrachloride (CCl_4) is a liquid used in dry cleaning in which one carbon atom (C) is combined with four chlorine atoms.

The atoms Na, Ca, and C apparently have different capacities for combining with Cl. You could imagine that molecules are held together with little hooks. Then you could picture a Cl atom, for example, as having one hook. In this case, Na would need one hook also; Ca would need two; and C would need four. The number of hooks required for each atom in this model is called the combining capacity or *valence*. Actually, you would begin by assigning a combining capacity of 1 to hydrogen and arranging all other combining capacities to be mutually consistent. Since the compound HCl is stable, for example, Cl also has a combining capacity of 1. Ammonia has the formula NH_3, so nitrogen (N) has a combining capacity of 3. Proceeding this way, you could assign the combining capacity to each element in turn. For example, since water is represented by H_2O, oxygen's combining capacity is 2. There are a few complications, however. Some elements behave as though they have different combining capacities in different situations. At another extreme are elements like helium which do not form compounds at all (or only under very exotic conditions). A combining capacity of 0 is assigned to these elements.

Since the combining capacity, or *valence*, of each element is known, you can calculate the *equivalent mass* of each element. "Equivalent" means the mass of an equal number of atoms. For convenience, the mass of a hydrogen atom is taken as the standard unit. For example, we know by experiment that water (H_2O) contains 7.94 g of oxygen for each gram of hydrogen. Two hydrogen atoms (H^+) are needed to balance the valance of one oxygen atom (O^{-2}). Therefore, the equivalent mass of one oxygen atom is $2 \times 7.94 = 15.88$ compared to each hydrogen atom. Similarly, in the compound HCl, one atom of chlorine (atomic weight 35.4) combines with one atom of hydrogen (or 35.4 g of Cl combines with 1 g of H). Each chlorine atom has a mass equivalent to 35.4 that of a hydrogen atom. This equivalent weight for the same number of atoms is called the *gram atomic weight.*

You might try to imagine an SO_3 molecule like this:

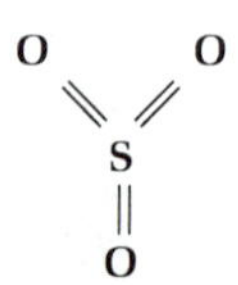

?

1. *What is the combining capacity of sulfur (S) in the compound H_2S? in the compound SO_3?*

2. *What is the combining capacity of aluminum in the compound Al_2O_3?*

3. *Methane (CH_4) has roughly 3 g carbon for 1 g of hydrogen. Use this information to find the approximate gram atomic weight of carbon.*

Physicists and chemists have now decided to use the most common isotope of carbon, C^{12}, to define atomic mass. With the atomic mass of C^{12} set at 12.000, hydrogen has an atomic mass of 1.008.

17.2 | Electricity and chemistry

While chemists were applying Dalton's atomic theory in the first decade of the nineteenth century, another development made possible experiments showing that electricity and the structure of matter were closely related. As you saw in Unit 4, Alessandro Volta invented the electric cell in 1800 (Sec. 14.7). Soon, batteries of these cells were providing the first available large, steady sources of electric current in several laboratories. A few weeks after Volta announced his discovery, it was found that an electric current could decompose water into oxygen and hydrogen. The decomposition of a compound by passing an electric current through it is called *electrolysis*. When the hydrogen and oxygen gases are collected and weighed, they form in the proportion of 7.94 g oxygen to 1.00 g hydrogen. This is exactly the proportion in which these elements combine to form water.

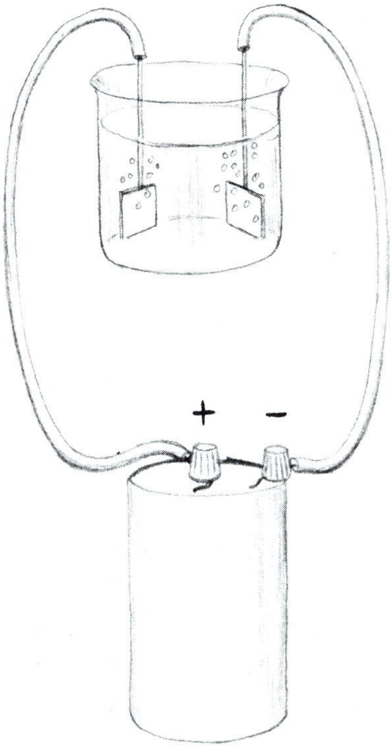

As mentioned in Sec. 17.1, when Humphry Davy (1778–1829) passed a current through soda and potash, he discovered the metals sodium and potassium. Michael Faraday, whose enormous contribution to electromagnetism was discussed in Unit 4, carried on with Davy's initial experiments and discovered two fundamental empirical laws of electrolysis:

1. The mass of an element released at an electrode during electrolysis is proportional to the amount of charge (= current × time) that has passed through the electrode.

Remember that current is defined as charge/time.

2. If A is the atomic mass of an element and v its combining capacity, a transfer of 96,540 C of charge (for example, 100 A for 965 sec) releases A/v grams of the element.

The quantity A/v has significance beyond electrolysis experiments. For example, the values of A/v are 8.00 for oxygen and 1.008 for hydrogen. The ratio $8.00/1.008 = 7.94$—just the ratio of the masses of oxygen and hydrogen that combine to produce water. In general, when two elements combine, the ratio of their masses is equal to the ratio of their values for A/v.

There are two profound implications in Faraday's work on electrolysis. First, it showed that electricity was somehow involved in holding atoms together to form molecules and hinted that a certain amount of electricity was connected with each atom. (In other words, electricity may also be atomic in nature.) As you will see in Chapter 18, this is a very useful idea. Second, electrolysis provided precisely the tool needed to help chemists determine atomic weights and combining capacities for many elements which are confusing if dealt with using other chemical techniques.

? 4. *Complete the following table:*

MASSES OF ELEMENTS RELEASED IN ELECTROLYSIS WHEN 96,540 C OF CHARGE HAVE PASSED

Element	*Atomic Mass (A)*	*Combining Capacity (v)*	*Mass Liberated (grams)*
Hydrogen	1.008	1	1.008
Chlorine	35.45	1	
Oxygen	16.00	2	
Copper		2	31.77
Zinc	65.37		32.69
Aluminum	26.98		8.99

17.3 | The Periodic Table

When all the elements are arranged in order of increasing atomic weight, a curious regularity becomes apparent. The physical properties of the elements (melting and boiling points, density, heat and electrical conductivity, hardness, etc.) all change together in the same cyclic fashion. Specifically, when the elements are arranged in a table as shown below, the elements below one another in each column or group share physical properties to a remarkable degree. Therefore, these elements can be considered to belong to the same "family" of elements.

Group→ Period ↓	I	II											III	IV	V	VI	VII	0
1	1.0080 H 1																	4.0026 He 2
2	6.939 Li 3	9.012 Be 4											10.811 B 5	12.011 C 6	14.007 N 7	15.999 O 8	18.998 F 9	20.183 Ne 10
3	22.990 Na 11	24.32 Mg 12											26.98 Al 13	28.09 Si 14	30.97 P 15	32.06 S 16	35.45 Cl 17	39.95 Ar 18
4	39.10 K 19	40.08 Ca 20	44.96 Sc 21	47.90 Ti 22	50.94 V 23	52.00 Cr 24	54.94 Mn 25	55.85 Fe 26	58.93 Co 27	58.71 Ni 28	63.54 Cu 29	65.37 Zn 30	69.72 Ga 31	72.59 Ge 32	74.92 As 33	78.96 Se 34	79.91 Br 35	83.80 Kr 36
5	85.47 Rb 37	87.62 Sr 38	88.91 Y 39	91.22 Zr 40	92.91 Nb 41	95.94 Mo 42	(99) Tc 43	101.07 Ru 44	102.91 Rh 45	106.4 Pd 46	107.87 Ag 47	112.40 Cd 48	114.82 In 49	118.69 Sn 50	121.75 Sb 51	127.60 Te 52	126.9 I 53	131.30 Xe 54
6	132.91 Cs 55	137.34 Ba 56	* 57–71	178.49 Hf 72	180.95 Ta 73	183.85 W 74	186.2 Re 75	190.2 Os 76	192.2 Ir 77	195.09 Pt 78	196.97 Au 79	200.59 Hg 80	204.37 Tl 81	207.19 Pb 82	208.98 Bi 83	(210) Po 84	(210) At 85	222 Rn 86
7	(223) Fr 87	226.05 Ra 88	† 89-103	(261) Rf 104	(260) Ha 105	(263) 106	etc.											

Atomic mass (A) — 4.0026; Abbreviation for element — He; Atomic number (Z) — 2

*Rare-earth metals	138.91 La 57	140.12 Ce 58	140.91 Pr 59	144.27 Nd 60	(147) Pm 61	150.35 Sm 62	151.96 Eu 63	157.25 Gd 64	158.92 Tb 65	162.50 Dy 66	164.93 Ho 67	167.26 Er 68	168.93 Tm 69	173.04 Yb 70	174.97 Lu 71
† Actinide metals	(227) Ac 89	232.04 Th 90	(231) Pa 91	238.03 U 92	(237) Np 93	(242) Pu 94	(243) Am 95	(245) Cm 96	(249) Bk 97	(249) Cf 98	(253) E 99	(255) Fm 100	(256) Mv 101	(253) No 102	(257) Lr 103

TABLE 17.1 LIST OF THE ELEMENTS

Element	*Symbol*	*Atomic Number (Z)*	*Element*	*Symbol*	*Atomic Number (Z)*
Actinium	Ac	89	Mercury	Hg	80
Aluminum	Al	13	Molybdenum	Mo	42
Americium	Am	95	Neodymium	Nd	60
Antimony	Sb	51	Neon	Ne	10
Argon	Ar	18	Neptunium	Np	93
Arsenic	As	33	Nickel	Ni	28
Astatine	At	85	Niobium	Nb	41
Barium	Ba	56	Nitrogen	N	7
Berkelium	Bk	97	Nobelium	No	102
Beryllium	Be	4	Osmium	Os	76
Bismuth	Bi	83	Oxygen	O	8
Boron	B	5	Palladium	Pd	46
Bromine	Br	35	Phosphorus	P	15
Cadmium	Cd	48	Platinum	Pt	78
Calcium	Ca	20	Plutonium	Pu	94
Californium	Cf	98	Polonium	Po	84
Carbon	C	6	Potassium	K	19
Cerium	Ce	58	Praseodymium	Pr	59
Cesium	Cs	55	Promethium	Pm	61
Chlorine	Cl	17	Protactinium	Pa	91
Chromium	Cr	24	Radium	Ra	88
Cobalt	Co	27	Radon	Rn	86
Copper	Cu	29	Rhenium	Re	75
Curium	Cm	96	Rhodium	Rh	45
Dysprosium	Dy	66	Rubidium	Rb	37
Einsteinium	Es	99	Ruthenium	Ru	44
Erbium	Er	68	Rutherfordium	Rf	104
Europium	Eu	63	Samarium	Sm	62
Fermium	Fm	100	Scandium	Sc	21
Fluorine	F	9	Selenium	Se	34
Francium	Fr	87	Silicon	Si	14
Gadolinium	Gd	64	Silver	Ag	47
Gallium	Ga	31	Sodium	Na	11
Germanium	Ge	32	Strontium	Sr	38
Gold	Au	79	Sulfur	S	16
Hafnium	Hf	72	Tantalum	Ta	73
Hahnium	Ha	105	Technetium	Tc	43
Helium	He	2	Tellurium	Te	52
Holmium	Ho	67	Terbium	Tb	65
Hydrogen	H	1	Thallium	Tl	81
Indium	In	49	Thorium	Th	90
Iodine	I	53	Thulium	Tm	69
Iridium	Ir	77	Tin	Sn	50
Iron	Fe	26	Titanium	Ti	22
Krypton	Kr	36	Tungsten	W	74
Lanthanum	La	57	Uranium	U	92
Lawrencium	Lr	103	Vanadium	V	23
Lead	Pb	82	Xenon	Xe	54
Lithium	Li	3	Ytterbium	Yb	70
Lutetium	Lu	71	Yttrium	Y	39
Magnesium	Mg	12	Zinc	Zn	30
Mendelevium	Md	101	Zirconium	Zr	40

On the left, Group I contains the family of *alkali* metals: lithium, sodium, potassium, rubidium, and cesium. This is a group of soft metals with extremely low densities, low melting points, and similar chemical behavior. Another family of elements, called the *halogens*, is found in Group VII: fluorine, chlorine, bromine, and iodine. These elements combine violently with many metals and form white, crystalline salts (*halogen* means "salt former"). These salts have similar formulas, such as NaF, NaCl, NaBr, NaI, or MgF_2, $MgCl_2$, $MgBr_2$, and MgI_2.

Dmitri Mendeleev (1834–1907) published the first periodic table in 1869. It contained the 63 elements known at the time.

Occasionally, it was necessary to depart from the overall scheme of ordering the elements by increasing atomic weight in the periodic table. For example, the chemical properties of argon (Ar) and potassium (K) demand that they be placed in the eighteenth and nineteenth positions in order to fall into groups characteristic of their families. On the basis of their atomic masses alone (39.948 for argon; 39.102 for potassium) their positions would have been reversed.

The physical reason for this reversal is now well understood and will be discussed later.

The number that designates the place an element has in the periodic table is called the *atomic number* of the element. The atomic number is usually represented by the symbol Z; thus, for hydrogen $Z = 1$, for chlorine $Z = 17$, and so on. In Chapter 19, you will see that the atomic number has a fundamental physical meaning that is fixed by the structure of the particular atom. Any successful model of atoms must explain why the elements arrange themselves in a periodic table and predict why an element of $Z = 2$ is chemically quite similar to one of $Z = 10$ and very distinct from one of $Z = 3$. You will see that by combining the ideas of several physicists, including Rutherford, Bohr, and Pauli, such a model has been constructed and is one of the great successes of physics during the first third of the twentieth century.

study guide

1. The *Project Physics* learning materials particularly appropriate for Chapter 17 include the following:

Experiment
Electrolysis

Activities
Dalton's Puzzle
Electrolysis of Water
Periodic Table
Single-Electrode Plating
Activities from *Scientific American*

Film Loops
Production of Sodium by Electrolysis

2. The chemical compound zinc oxide (molecular formula ZnO) contains equal numbers of atoms of zinc and oxygen. Using values of atomic masses from the periodic table (on page 536), find the percentage by mass of zinc in zinc oxide. What is the percentage of oxygen in zinc oxide?

3. The chemical compound zinc chloride (molecular formula $ZnCl_2$) contains two atoms of chlorine for each atom of zinc. Using values of atomic masses from the periodic table, find the percentage by mass of zinc in zinc chloride.

4. When a 5.00-g sample of ammonia gas is completely decomposed into its elements, nitrogen and hydrogen, 4.11 g of nitrogen are obtained. The molecular formula of ammonia is NH_3. Find the mass of a nitrogen atom relative to that of a hydrogen atom. Compare your result with the one you would get by using the values of the atomic masses in the periodic table. If the two results are different, how do you account for the difference?

5. From the information in SG 4, calculate how much nitrogen and hydrogen are needed to make 1.2 kg of ammonia.

6. A sample of nitric oxide gas, weighing 1.00 g, after separation into its components, is found to have contained 0.47 g of nitrogen. Taking the atomic mass of oxygen to be 16.00, find the corresponding numbers that express the atomic mass of nitrogen relative to oxygen if the molecular formula of nitric oxide is (a) NO; (b) NO_2; (c) N_2O.

7. Given the molecular formulas HCl, NaCl, $CaCl_2$, $AlCl_3$, $SnCl_4$, PCl_5, find possible combining capacities of sodium, calcium, aluminum, tin, and phosphorus.

8. In recent editions of the *Handbook of Chemistry and Physics,* the valence numbers of the elements are printed in or below one of the periodic tables. Ignore the negative valence numbers and plot (to element 65) a graph of maximum valences observed versus atomic mass. What periodicity do you find? Is there any physical or chemical significance to this periodicity?

9. According to the table in Question 4 (p. 536), when about 96,500 C of charge pass through a water solution, how much oxygen will be released at the same time when (on the other electrode) 1.008 g of hydrogen are released? How much oxygen will be produced when a current of 3 A is passed through water for 60 min (3,600 sec)?

10. If a current of 0.5 A is passed through molten zinc chloride in an electrolytic apparatus, what mass of zinc will be deposited in

(a) 5 min (300 sec)?

(b) 30 min?

(c) 120 min?

11. (a) For 20 min (1,200 sec), a current of 2.0 A is passed through molten zinc chloride in an electrolytic apparatus. What mass of chlorine will be released at the anode?

(b) If the current is passed through molten zinc iodide rather than molten zinc chloride, what mass of iodine will be released at the anode?

(c) Would the quantity of zinc deposited in part (b) differ from that in part (a)? Why?

(d) How would you set up a device for plating a copper spoon with silver?

12. 96,540 C in electrolysis frees A grams of a monovalent element ($v = 1$) of atomic mass A, for example, hydrogen when hydrochloric acid is used as electrolyte. How much chlorine will be released on the other electrode?

13. 96,540 C in electrolysis always frees A grams of a monovalent element, $A/2$ grams of a divalent element ($v = 2$), etc. What relation does this suggest between valence and "atoms" of electricity?

14. The idea of chemical elements composed of identical atoms makes it easier to understand the phenomena discussed in this chapter. Could the phenomena be explained without using the idea of atoms? Chemical phenomena usually involve a fairly large quantity of material (in terms of the number of "atoms"). Do such phenomena provide sufficient evidence for Dalton's belief that an element consists of atoms, all of which are exactly identical with each other?

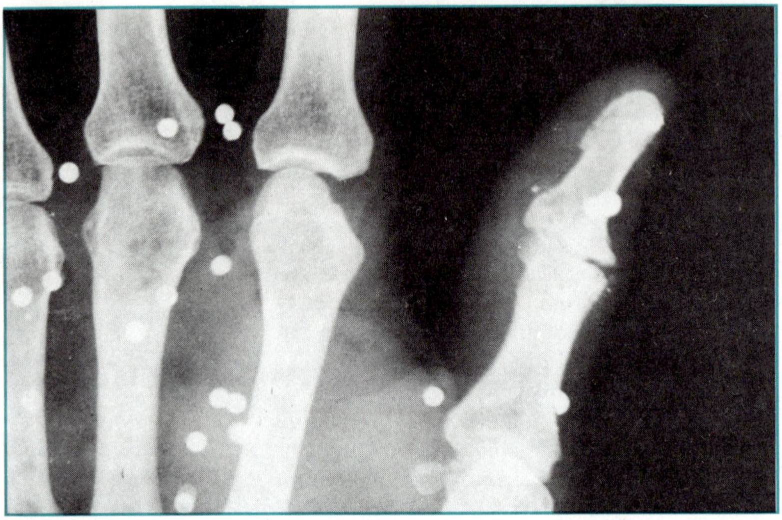

CHAPTER 18

Electrons and Quanta

18.1 The idea of atomic structure

SG 1

Chemistry in the nineteenth century had succeeded remarkably in accounting for combining proportions and in predicting chemical reactions. This success had convinced most scientists that matter is indeed composed of atoms. But there remained a related question: Are atoms really indivisible, or do they consist of still smaller particles?

You can see how this question arose by thinking a little more about the periodic table. Mendeleev had arranged the elements in the order of increasing atomic mass. But the atomic masses of the elements cannot explain the *periodic* features of Mendeleev's table.

Why, for example, do (a) the 3rd, 11th, 19th, 37th, 55th, and 87th elements, with quite different atomic masses, have similar chemical properties?

Why are these properties somewhat different from those of (b) the 4th, 12th, 20th, 38th, 56th, and 88th elements in the list, but

a. These elements burn when exposed to air; they decompose water, often explosively.
b. These elements react slowly with air or water.
c. These elements rarely combine with any other element.

J. J. Thomson and F. B. Jewett (later president of Bell Laboratories) inspecting vacuum tubes at the New Jersey laboratories in 1923.

greatly different from the properties of (c) the 2nd, 10th, 18th, 36th, 54th, and 86th elements?

The periodicity in the properties of the elements led to speculation that atoms might have structure, that they might be made up of smaller pieces. The properties changed gradually from group to group. This fact suggested that some unit of atomic structure might be added from one element to the next, until a certain portion of the structure is completed. The completed condition would occur in the atom of a noble gas. In an atom of the next heavier element, a new portion of the structure would be started, and so on. The methods and techniques of classical chemistry could not supply experimental evidence for such structure. In the nineteenth century, however, discoveries and new techniques in physics opened the way to proof that atoms actually do consist of smaller pieces. Evidence piled up to support the conclusion that the atoms of different elements differ in the number and arrangement of these pieces.

In this chapter, you will study the discovery of one structural unit that all atoms contain, the electron. Then you will see how experiments with light and electrons led to the revolutionary idea that *light* energy is transmitted in separate "chunks." Chapter 19 will describe the discovery of another part of the atom, the nucleus. Finally, you will see how Niels Bohr combined these pieces to create a workable model of the atom. The story starts with the discovery of cathode rays.

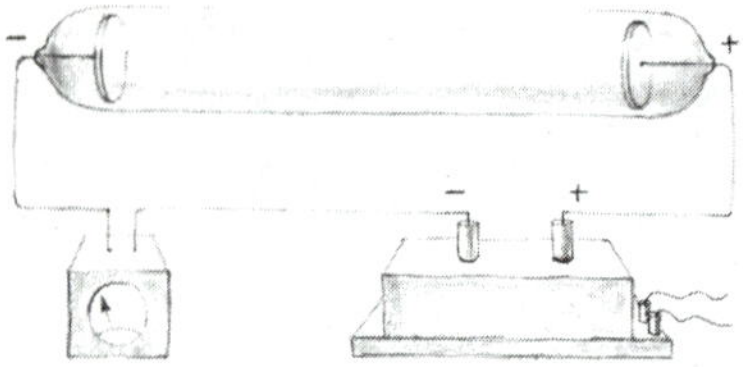

Cathode-ray apparatus.

Substances that glow when exposed to light, particularly ultraviolet, are called fluorescent. "Fluorescent lights" are essentially Geissler tubes with an inner coating of fluorescent powder.

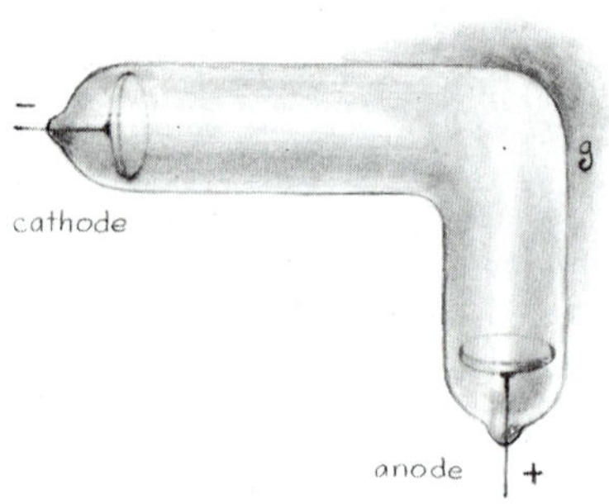

Bent Geissler tube. The most intense green glow appeared at g.

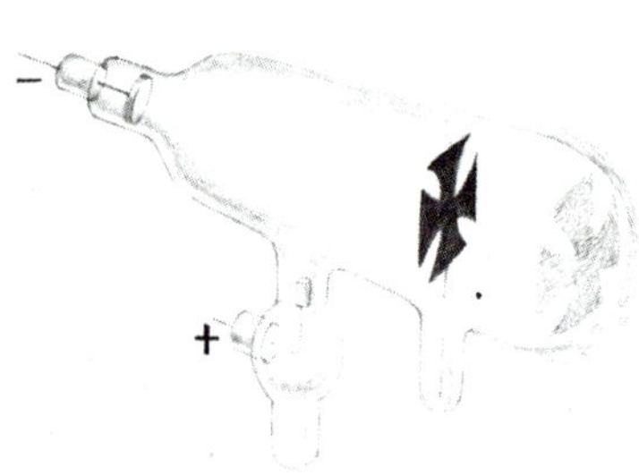

A Crookes tube.

18.2 | Cathode rays

In 1855, the German physicist Heinrich Geissler invented a powerful vacuum pump. This pump could remove enough gas from a strong glass tube to reduce the pressure to 0.01% of normal air pressure. It was the first major improvement in vacuum pumps after Guericke's invention of the air pump, two centuries earlier. This new pump made possible the electric light bulb, the electron tube, and other technologically valuable inventions over the next 50 years. It also opened new fields to pure scientific research. Geissler's friend Julius Plücker connected one of Geissler's evacuated tubes to a battery. He was surprised to find that, at the very low pressure obtained with Geissler's pump, electricity flowed through the tube. Plücker used apparatus similar to that sketched in the margin. He sealed a wire into each end of a strong glass tube. Inside the tube, each wire ended in a metal plate, called an electrode. Outside the tube, each wire ran to a source of high voltage. (The negative plate is called the *cathode*, and the positive plate is called the *anode*.) A meter indicated the current going through the tube.

Plücker and his student Johann Hittorf noticed that when an electric current passed through the low-pressure gas in a tube, the tube itself glowed with a pale green color. Several other scientists observed this effect, but two decades passed before anyone undertook a thorough study of the glowing tubes. By 1875, Sir William Crookes had designed new tubes for studying the glow. When he used a bent tube (see figure at the left) the most intense green glow appeared on the part of the tube that was directly opposite the cathode (at g). This suggested that the green glow is produced by something that comes out of the cathode and travels down the tube until it hits the glass. Another physicist, Eugen Goldstein, was also studying the effects of passing an electric current through a gas at low pressure. Goldstein coined a name for whatever it was that appeared to be coming from the cathode: *cathode rays*. For the time being, the nature of these cathode rays was a mystery.

To study the nature of the rays, Crookes did some clever experiments. He reasoned that if cathode rays could be stopped before they reached the end of the tube, the intense green glow would disappear. He therefore introduced barriers like the Maltese cross (made of metal) as in the sketch in the margin. A shadow of the barrier appeared in the midst of the green glow at the end of the tube. The cathode seemed to act like a source that radiates a kind of light; the cross acted like a barrier blocking the light. The shadow, cross, and cathode appeared along one straight line. Therefore, Crookes concluded, cathode rays, like light rays, travel in straight lines. Next, Crookes moved a magnet near the tube, and the shadow moved. Thus, he found that magnetic fields deflect cathode rays (which does not happen with light).

In the course of many experiments, Crookes found the following properties of cathode rays:

1. No matter what material the cathode is made of, it produces rays with the same properties.
2. In the absence of a magnetic field, the rays travel in straight lines perpendicular to the surface that emits them.
3. A magnetic field deflects the path of the cathode rays.
4. The rays can produce some chemical reactions similar to the reactions produced by light. For example, certain silver salts change color when hit by the rays.
5. In addition, Crookes suspected (but did not succeed in showing) that charged objects deflect the path of cathode rays.

J. J. Thomson later observed this to be possible.

Physicists were fascinated by the cathode rays. Some thought that the rays must be a form of light. After all, they have many of the properties of light: they travel in straight lines and produce chemical changes and fluorescent glows just as light does. According to Maxwell's theory of electricity and magnetism, light consists of electromagnetic waves. So the cathode rays might, for example, be electromagnetic waves of frequency much higher than that of visible light.

However, while magnetic fields do not bend light, they do bend the path of cathode rays. Chapter 14 described how magnetic fields exert forces on currents, that is, on moving electric charges. A magnetic field deflects cathode rays in the same way that it deflects negative charges. Therefore, some physicists believed that cathode rays consisted of negatively charged particles.

The debate over whether cathode rays are a form of electromagnetic waves or a stream of charged particles continued for 25 years. Finally, in 1897, J. J. Thomson made a series of experiments that convinced physicists that cathode rays are negatively charged particles. A detailed account of the discovery of the electron is given in Chapter 2 of the Supplemental Unit B, "*Discoveries in Physics*."

Sir Joseph John Thomson (1856–1940), one of the greatest British physicists, attended Owens College in Manchester, England, and then Cambridge University. He worked on the conduction of electricity through gases, on the relation between electricity and matter, and on atomic models. His greatest single contribution was the discovery of the electron. Thomson was the head of the famous Cavendish Laboratory at Cambridge University, where one of his students was Ernest Rutherford.

It was then well known that the paths of charged particles are affected by both magnetic and electric fields. By assuming that cathode rays were negatively charged particles, Thomson could predict what should happen when they passed through such fields. For example, an electric field of just the right magnitude and direction should exactly balance the deflection of a beam of cathode rays by a magnetic field. As Thomson discovered, the predictions were verified. Thomson could therefore conclude that cathode rays were indeed made up of negatively charged particles. He was then able to calculate, from the experimental data, the ratio of the charge of a particle to its mass. This ratio is represented by q/m, where q is the charge and m is the mass of the particle. Thomson found that the rays coming from cathodes made of different materials all had the same value of q/m: 1.76×10^{11} C/kg.

Thus, it was clear that cathode rays must be made of something all materials have in common. Thomson's negatively charged particles were later called *electrons.* The value of q/m for the cathode-ray particles was about 1,800 times larger than the values of q/m for charged hydrogen atoms (ions), which can be shown to be 9.6×10^7 C/kg, as measured in electrolysis experiments of the kind discussed in Sec. 17.2 (see table on page 536). Thomson concluded from these results that either the *charge* of the cathode-ray particles is much *greater* than that of the hydrogen ion, or the *mass* of the cathode-ray particles is much *less* than the mass of the hydrogen ion.

SG 2

The data in the table in Question 4 of Chapter 17 show that 1.008 g of hydrogen is freed when 96,540 C of electric charge are transferred; this implies q/m = 96,540 C/1.008 g, or about 9.6×10^7 C/kg.

Thomson also measured the charge q on these negatively charged particles with methods other than deflection by electric and magnetic fields. His experiments were not very accurate. But they were good enough to indicate that the charge of a cathode-ray particle was the same or not much different from that of the hydrogen ion in electrolysis. In view of the large value of q/m, Thomson concluded that the mass of cathode-ray particles is much less than the mass of hydrogen ions.

In short, the cathode-ray particles, or electrons, were found to have two important properties: (1) they were emitted by a wide variety of cathode materials, and (2) they were much smaller in mass than the hydrogen atom, which has the smallest known mass. Thomson therefore concluded that the cathode-ray particles form a part of all kinds of matter. He suggested that the atom is not the ultimate limit to the subdivision of matter; rather, the electron is part of an atom and is perhaps even a basic building block of atoms. Scientists now know that this is correct. The electron, whose existence Thomson had first proved by quantitative experiment, is one of the fundamental or "elementary" particles of which matter is made.

In the article in which he published his discovery, Thomson also speculated about how electrons might be arranged in atoms of different elements. He thought that such arrangements might account for the periodicity of the chemical properties of the elements. As you will see in the next chapter, Thomson did not say the *last* word about the arrangement and number of electrons in the atom. But he did say the *first* word about it.

?

1. What was the most convincing evidence to support the fact that cathode rays were not electromagnetic radiation?

2. What was the reason given for the ratio q/m *for electrons being 1,800 times larger than* q/m *for hydrogen ions?*

3. What were two main reasons for Thomson's belief that electrons may be "building blocks" from which all atoms are made?

18.3 The measurement of the charge of the electron: Millikan's experiment

After the ratio of charge to the mass (q/m) of the electron had been determined, physicists tried to measure the value of the charge q itself. If the charge could be determined, the mass of the electron could be found from the known value of q/m. In the years between 1909 and 1916, the American physicist Robert A. Millikan succeeded in measuring the charge of the electron. This quantity is one of the fundamental constants of physics; it comes up again and again in atomic and nuclear physics as well as in electricity and electromagnetism.

Millikan's "oil-drop experiment" is still an informative experiment that students can do. It is described in general outline on page 547. Millikan found that the electric charge that a small object such as an oil drop can pick up is always a simple multiple of a certain *minimum value*. For example, the charge may have the value -4.8×10^{-19} C, -1.6×10^{-19} C, -6.4×10^{-19} C, or -1.6×10^{-18} C. But it never has a charge of, say, -2.4×10^{-19} C, and it never has a value smaller than -1.6×10^{-19} C. In other words, electric charges always come in multiples (1, 2, 3, . . .) of 1.6×10^{-19} C, a quantity often symbolized by q_e. Millikan correctly took this minimum charge to be the amount of charge of a single electron.

From now on, the magnitude of the charge of the electron is represented by q_e

$$q_e = 1.6 \times 10^{-19} \text{ C}$$

The sign of the charge is negative for the electron.

The magnitude of the charges of nuclei or atomic and molecular ions is also always a multiple of the electron charge q_e. For example, a chemist may refer to a "doubly charged oxygen ion." This means that the magnitude of the charge of the ion is $2q_e$, or 3.2×10^{-19} C.

SG 3

Note that Millikan's experiments did not prove that no charges smaller than q_e can exist. However, no experiment has yet proved the existence of smaller charges. Recent theoretical advances suggest that in some very high-energy experiments, an elementary particle of charge ⅓ q_e may eventually be discovered. But no such "fractional" charge is expected to be found on nuclei, ions, or droplets.

In 1964, an American physicist, Murray Gell-Mann, suggested that particles with charge equal to ⅓ or ⅔ of q_e might exist. He named these particles "quarks." The word comes from James Joyce's novel *Finnegan's Wake*. However, current theory suggests that quarks are always bound together in groups whose charge is equal to q_e.

In everyday life, the electric charges are huge compared to that on one electron. Thus, you usually think of such charges or currents as being continuous, just as you think of the flow of water in a river as continuous rather than as a flow of individual molecules. A current of 1A, for example, is equivalent to the flow of 6.25×10^{18} electrons per second. The "static" electric charge you accumulate by shuffling over a rug on a dry day consists of something like 10^{12} electron charges.

Thomson found that

$$q_e/m = 1.76 \times 10^{11} \text{ C/kg}$$

According to Millikan's experiment, the magnitude of q_e is 1.6×10^{-19} C.

Therefore, the mass of an electron is:

$$m = \frac{1.6 \times 10^{-19} \text{ C}}{1.76 \times 10^{11} \text{ C/kg}}$$
$$= 0.91 \times 10^{-30} \text{ kg}$$

(Mass of a hydrogen ion is 1.66×10^{-27} kg. This is approximately the value of one "atomic mass unit.")

Since the work of Millikan, other experiments involving many different fields within physics have all pointed to the charge q_e as being fundamental in the structue and behavior of atoms, nuclei, and smaller particles. For example, it has been shown directly that cathode-ray particles carry this basic unit of charge, that they are, in other words, electrons.

By combining Millikan's value for the electron charge q_e with Thomson's value for the ratio of charge to mass (q_e/m), you can calculate the mass of a single electron (see margin). The mass found for the electron is about 10^{-30} kg. From electrolysis experiments (see Sec. 17.2), the charge-to-mass ratio of a hydrogen ion is known to be 1,836 times smaller than the charge-to-mass ratio of an electron. But an electron and a hydrogen ion form a neutral hydrogen atom when they combine. Therefore, it is reasonable to expect that they have equal and opposite charges. If so, the *mass* of the hydrogen ion is 1,836 times as great as the mass of the electron; that is, the mass of the hydrogen ion is $1{,}836 \times 0.91 \times 10^{-30}$ kg $= 1.66 \times 10^{-27}$ kg. This is approximately the value of one *atomic mass unit*.

? *4. Oil drops pick up different amounts of electric charge. On what basis did Millikan decide that the lowest charge he found was actually just one electron charge?*

18.4 | The photoelectric effect

In 1887, the German physicist Heinrich Hertz was testing Maxwell's theory of electromagnetic waves (see Sec. 16.4). He noticed that a metallic surface can emit electric charges when light of very short wavelength falls on it. Because light and electricity are both involved, the name *photoelectric effect* was given to this phenomenon. When the electric charges so produced passed through electric and magnetic fields, their paths were changed in the same ways as the paths of cathode rays. It was therefore deduced that the electric charges consist of negatively charged particles. In 1898, J. J. Thomson measured the value of the ratio *q/m* for these photoelectrically emitted particles. Using the same method that he had used for cathode-ray particles, Thomson got the same value for the photoelectric particles as he had for the cathode-ray particles. These experiments (and others) demonstrated that photoelectric particles had the same properties as electrons. In fact, physicists consider them to be ordinary electrons, although they are often referred to as *photoelectrons* to indicate their origin. Later work showed that all substances (solids, liquids, and gases) exhibit the photoelectric effect under appropriate conditions. However, it is convenient to study the effect with metallic surfaces.

Close Up

Millikan's Oil-drop Experiment

R. A. Millikan's own apparatus (about 1910) for measuring the charge of the electron is seen in the photograph below.

In principle, Millikan's experiment is simple; the essential part of the apparatus is sketched above. When oil is sprayed into the chamber containing two horizontal plates, the tiny droplets formed are electrically charged as they emerge from the spray nozzle. The charge of a droplet is what must be measured. Consider a small oil drop of mass m carrying an electric charge q. It is situated between the two horizontal plates that are separated by a distance d and at an electrical potential difference V. There will be a uniform electric field $\vec{E}$ between the plates, of strength V/d (see Section 14.8). This field can be adjusted so that the electrical force $q\vec{E}$ exerted upward on the drop's charge will balance the force ma_g exerted downward by gravity. In this balanced situation,

$$F_{el} = F_{grav}$$

Therefore,

$$qE = ma_g$$

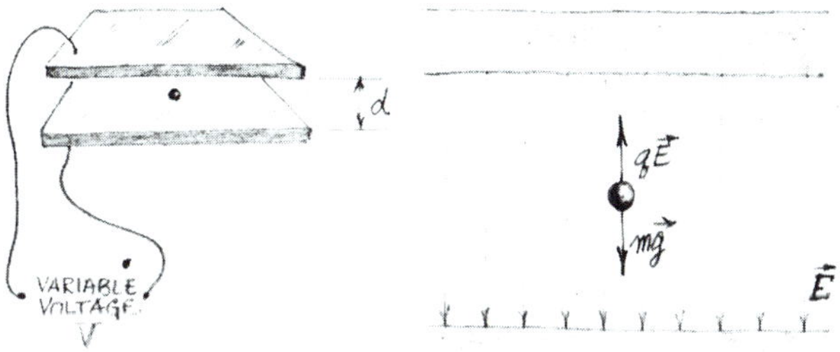

or

$$q = ma_g/E$$

The mass of the drop can, in principle, be determined from its radius and the density of the oil from which it was made. Millikan had to measure these quantities by an indirect method. (Today it is possible to do the experiment with small manufactured polystyrene spheres instead of oil drops. Their mass is known, so that some of the complications of the original experiment can be avoided.) Millikan's remarkable result was that the charge q on objects such as an oil drop is always a multiple (1, 2, 3 . . .) of a smallest charge, which he identified with the magnitude of the charge of one electron (q_e).

The photoelectric effect, which you will study in greater detail, has had an important place in the development of atomic physics. The effect could not be explained in terms of the ideas of physics you have studied so far. New ideas had to be introduced to account for the experimental results. In particular, a revolutionary concept was introduced: *quanta*. A new branch of physics called *quantum theory* developed, at least in part, because of the explanation provided for the photoelectric effect.

The basic information for studying the photoelectric effect comes from two kinds of measurements: (1) measurements of the *photoelectric current* (the number of photoelectrons emitted per unit time), and (2) measurements of the *kinetic energies* with which the photoelectrons are emitted.

The best way to study this and most other parts of physics is by actually doing the experiments discussed.

The *photoelectric current* can be studied with an apparatus like that sketched in (a) below. Two metal plates, C and A, are sealed inside a well-evacuated quartz tube. (Quartz glass is transparent to ultraviolet light as well as visible light.) The two plates are connected to a source of potential difference (for example, a battery). In the circuit is also an ammeter. As long as light strikes plate C, as in (b), electrons are emitted from it. If the potential of plate A is positive relative to plate C, these emitted photoelectrons will accelerate to plate A. (Some emitted electrons will reach plate A even if it is not positive relative to C.) The resulting "photoelectric" current is indicated by the ammeter. The result of the experiment is that the stronger the beam of light of a given color (frequency), the greater the photoelectric current.

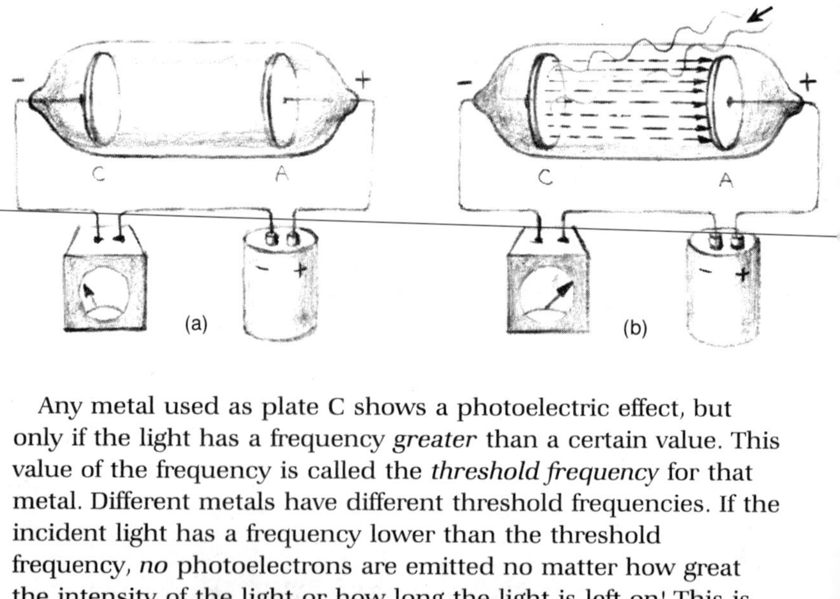

Schematic diagram of apparatus for photoelectric experiments.

Any metal used as plate C shows a photoelectric effect, but only if the light has a frequency *greater* than a certain value. This value of the frequency is called the *threshold frequency* for that metal. Different metals have different threshold frequencies. If the incident light has a frequency lower than the threshold frequency, *no* photoelectrons are emitted no matter how great the intensity of the light or how long the light is left on! This is the first of a set of surprising discoveries.

The *kinetic energies of the electrons* can be measured in a slightly modified version of the apparatus, sketched in (c) below. The battery is reversed so that plate A now tends to repel the photoelectrons. The voltage can be changed from zero to a value just large enough to keep any electrons from reaching plate A, as indicated in (d).

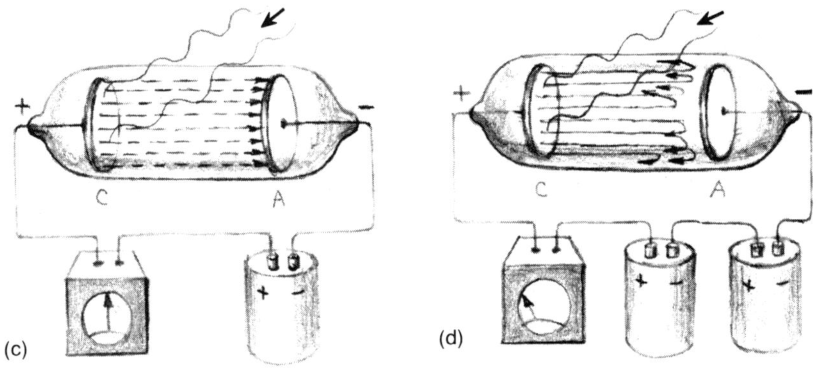

(c) (d)

When the voltage across the plates is zero, the meter will indicate a current. This reading shows that the photoelectrons, emerging with kinetic energy from the metallic surface, can reach plate A. As the repelling voltage is increased, the photoelectric current decreases. Eventually a certain voltage is reached at which the current becomes zero, as indicated in (d) above. This voltage, which is called the *stopping voltage,* is a measure of the maximum kinetic energy of the emitted photoelectrons (KE_{max}). If the stopping voltage is called V_{stop}, this maximum kinetic energy is given by the relation

$$KE_{max} = V_{stop}\, q_e$$

SG 4

In Sec. 14.8, you saw that the change in potential energy of a charge is given by Vq. In Unit 3, you saw that (in the absence of friction) the decrease in kinetic energy in a system is equal to the increase in its potential energy.

The results can be stated more precisely. Only the important experimental results are listed here. Their theoretical interpretation will be discussed later.

1. A substance shows a photoelectric effect only if the incident light has a frequency above a certain threshold frequency (symbol f_o).

2. If light of a given frequency does produce a photoelectric effect, the photoelectric current from the surface is proportional to the intensity of the light falling on it.

3. If light of a given frequency releases photoelectrons, the emission of these electrons is *immediate*. The time interval between the instant the light strikes the metallic surface and the appearance of electrons is at most 3×10^{-9} sec and probably much less. In some experiments, the light intensity used was extremely low. According to the classical theory, it should take several hundred seconds for an electron to accumulate enough energy from such light to be emitted. But even in these cases,

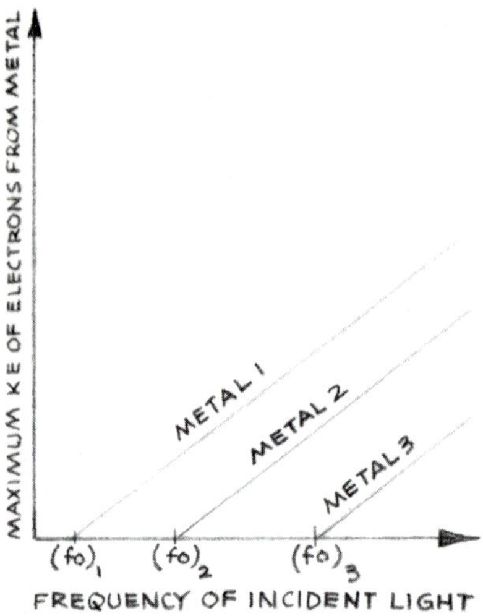

Photoelectric effect: Maximum kinetic energy of the electrons as a function of the frequency of the incident light; different metals yield lines that are parallel, but have different threshold frequencies.

electrons are sometimes emitted about a billionth of a second after the light strikes the surface.

4. The maximum kinetic energy of the photoelectrons increases in direct proportion to the *frequency* of the light that causes their emission. (Maximum *KE* is *not* dependent on the *intensity* of the incident light.) The way in which the maximum kinetic energy of the electrons varies with the frequency of the incident light is shown in the margin. The symbols $(f_o)_1$, $(f_o)_2$, and $(f_o)_3$ stand for the different threshold frequencies of three different substances. For each substance, the experimental data points fall on a straight line. All the lines have the same slope.

What is most astonishing about the results is that photoelectrons are emitted if the light frequencies are *a little above* the threshold frequency, no matter how weak the beam of light. But if the light frequencies are just *a bit below* the threshold frequency, no electrons are emitted *no matter how great the intensity of the light beam is*.

Findings (1), (3), and (4) could not be explained on the basis of the classical electromagnetic theory of light. How could a low-intensity train of light waves, spread out over a large number of atoms, concentrate, in a very short time interval, enough energy on one electron to knock the electron out of the metal?

Furthermore, the classical wave theory could not account for the existence of a threshold frequency. There seemed to be no reason why a sufficiently intense beam of low-frequency radiation should not produce photoelectricity if low-intensity radiation of higher frequency could produce it. Neither could classical theory explain why the maximum kinetic energy of the photoelectrons increases directly with the frequency of the light but is independent of the intensity. Thus, the photoelectric effect posed a challenge that the classical wave theory of light was not able to meet.

?

5. Light falling on a certain metal surface causes electrons to be emitted. What happens to the photoelectric current as the intensity of the light is decreased?

6. What happens as the frequency of the light is decreased?

7. Sketch a rough diagram of the equipment and circuit used to demonstrate the main facts of photoelectricity.

18.5 | Einstein's theory of the photoelectric effect

The explanation of the photoelectric effect was the major work cited in the award to Albert Einstein of the Nobel Prize in physics for the year 1921. Einstein's theory, proposed in 1905, played a major role in the development of atomic physics. The theory was based on a daring proposal, for few of the experimental details were known in 1905. Moreover, the key point of Einstein's explanation contradicted the classical ideas of the time.

Einstein assumed that energy of light is not distributed evenly over the whole expanding wave front (as the classical theory assumed). Instead, the light energy is concentrated in separate "lumps." In addition, the amount of energy in each of these regions is not just any amount, but a definite amount of energy that is proportional to the frequency f of the wave. The proportionality factor is a constant (symbol h); it is called Planck's constant for reasons which will be discussed later. Thus, in this model, the light energy in a beam of frequency f comes in pieces, each of amount hf. The amount of radiant energy in each piece is called a *quantum* of energy. It represents the smallest possible quantity of energy for light of that frequency. The quantum of light energy was later called a *photon*.

$h = 6.6 \times 10^{-34}$ J-sec

There is no explanation clearer or more direct than Einstein's. A quote from his first paper (1905) on this subject is given here. Only the notation is changed, in order to agree with modern practice (including the notation used in this text):

SG 5

> . . . According to the idea that the incident light consists of quanta with energy hf, the ejection of cathode rays by light can be understood in the following way. Energy quanta penetrate the surface layer of the body, and their energy is converted, at least in part, into kinetic energy of electrons. The simplest picture is that a light quantum gives up all its energy to a single electron; we shall assume that this happens. The possibility is not to be excluded, however, that electrons receive their energy only in part from the light quantum. An electron provided with kinetic energy inside the body may have lost part of its kinetic energy by the time it reaches the surface. In addition, it is to be assumed that each electron, in leaving the body, has to do an amount of work W (which is characteristic of the body). The electrons ejected directly from the surface and at right angles to it will have the greatest velocities perpendicular to the surface. The maximum kinetic energy of such an electron is
>
> $$KE_{max} = hf - W$$
>
> If the body plate C is charged to a positive potential, V_{stop}, just large enough to keep the body from losing electric charge, we must have
>
> $$KE_{max} = hf - W = V_{stop}\, q_e$$
>
> where q_e is the magnitude of the electronic charge . . .

Each electron must be given a minimum energy to emerge from the surface, because it must do work against the forces of attraction as it leaves the rest of the atoms.

This equation is usually called Einstein's photoelectric equation.

SG 6–8

If the derived formula is correct, then V_{stop}, when plotted as a function of the frequency of the incident light, should yield a

How Einstein's theory explains the photoelectric effect:

1. No photoelectric emission below threshold frequency. Reason: Low-frequency photons do not have enough energy to provide electrons with *KE* sufficient to leave the metal.

2. Current $\propto$ light intensity. Reason: One photon ejects one electron.

3. Immediate emission. Reason: A single photon provides the energy concentrated in one place.

4. KE_{max} increases directly with frequency above f_0. Reason: The work needed to remove the electron is $W = hf_0$; any energy left over from the original photon is now available for kinetic energy of the electron.

Student apparatus for photoelectric experiments often includes a vacuum phototube, like the one shown at the left. The collecting wire corresponds to A in (a) on page 548, and is at the center of a cylindrical photosensitive surface that corresponds to C. The frequency of the light entering the tube is selected by placing colored filters between the tube and a white light source, as shown at the right.

straight line whose slope should be independent of the nature of the substance illuminated.

Compare Einstein's photoelectric equation with the experimental results to test whether or not his theory accounts for those results. According to the equation, the kinetic energy is greater than zero only when hf is greater than W. Therefore, an electron can be emitted only when the frequency of the incident light is greater than a certain lowest value f_o (where $hf_o = W$.)

Next, according to Einstein's photon model, it is an individual photon that ejects an electron. Now, the intensity of the light is proportional to the number of the photons in the light beam. In addition, the number of photoelectrons ejected is proportional to the number of photons incident on the surface. Therefore, the number of electrons ejected (and with it the photoelectric current) is proportional to the intensity of the incident light.

According to Einstein's model, the light energy is concentrated in the quanta (photons). So no time is needed for collecting light energy. Instead, the quanta transfer their energy immediately to the photoelectrons, which emerge after the very short time required for them to escape from the surface. (See SG 9 and 10.)

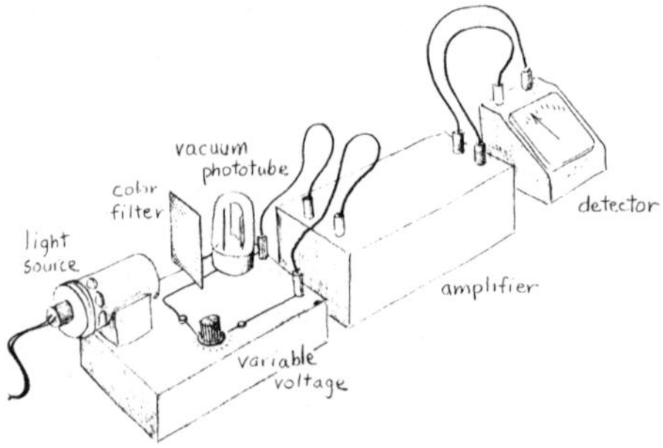

Finally, the photoelectric equation predicts that the greater the frequency of the incident light, the greater the maximum kinetic energy of the ejected electrons. According to the photon model, the photon's energy is directly proportional to the light frequency. The minimum energy needed to eject an electron is the energy required for the electron to escape from the metal surface. This explains why light of frequency less than some frequency f_o cannot eject any electrons. The kinetic energy of the escaping electron is the difference between the energy of the absorbed photon and the energy lost by the electron in escaping the surface.

Thus, Einstein's photoelectric equation agreed qualitatively with the experimental results. There remained two quantitative tests to be made: (1) Does the maximum energy vary in direct proportion to the light frequency? (2) Is the proportionality factor h really the same for all substances? For 10 years, experimental physicists attempted these quantitative tests. One experimental

Close Up

Albert Einstein

Albert Einstein (1879–1955) was born in the city of Ulm, in Germany. Like Newton, he showed no particular intellectual promise as a youngster. He received his early education in Germany, but at the age of 15, dissatisfied with the discipline in school and militarism in the nation, he left and went to Switzerland. After graduation from the Polytechnic Institute in Zurich, Einstein (in 1902) found work in the Swiss Patent Office in Berne. This job gave Einstein a salary to live on and an opportunity to use his spare time for working in physics on his own. In 1905, he published three papers of immense importance. One dealt with quantum theory and included his theory of the photoelectric effect. Another treated the problem of molecular motions and sizes, and worked out a mathematical analysis of the phenomenon of "Brownian motion." Einstein's analysis and subsequent experimental work by Jean Perrin, a French physicist, provided a strong argument for the molecular motions assumed in the kinetic theory. Einstein's third 1905 paper provided the theory of special relativity which revolutionized modern thought about the nature of space and time, and of physical theory itself.

In 1915, Einstein published a paper on the theory of general relativity. In it he provided a new theory of gravitation that included Newton's theory as a special case.

When Hitler and the Nazis came to power in Germany in 1933, Einstein went to the United States and became a member of the Institute for Advanced Study at Princeton. He spent the rest of his working life seeking a unified theory which would include gravitation and electromagnetism. Near the beginning of World War II, Einstein wrote a letter to President Roosevelt, warning of the war potential of an "atomic bomb," for which the Germans had all necessary knowledge and motivation to work. After World War II, Einstein devoted much of his time to organizations advocating world agreements to end the threat of atomic warfare.

The equation $KE_{max} = hf - W$ can be said to have led to two Nobel Prizes: one to Einstein, who derived it theoretically, and one to Millikan, who verified it experimentally. This equation is the subject of a *Project Physics* laboratory experiment.

difficulty was that the value of W for a metal changes greatly if there are impurities (for example, a layer of oxidized metal) on the surface. Finally, in 1916, Robert A Millikan established that there is indeed a straight-line relationship between the frequency of the absorbed light and the maximum kinetic energy of the photoelectrons. (See the graph on page 550.) To obtain his data, Millikan designed an apparatus in which the metal photoelectric surface was cut clean while in a vacuum. A knife inside the evacuated volume was manipulated by an electromagnet outside the vacuum to make the cuts. This rather intricate arrangement was required to achieve a pure metal surface.

SG 11

The straight-line graphs Millikan obtained for different metals all had the same slope, even though the threshold frequencies were different. The value of h could be obtained from Millikan's measurements, and it was the same for each metal surface. Also, it agreed very well with a value obtained by means of other, independent methods. So Einstein's theory of the photoelectric effect was verified quantitatively.

Max Planck (1858–1947), a German physicist, was the originator of the quantum theory, one of the two great revolutionary physical theories of the twentieth century. (The other is Einstein's relativity theory.) Planck won the Nobel Prize in 1918 for his quantum theory. He tried for many years to show that this theory could be understood in terms of the classical physics of Newton and Maxwell, but this attempt did not succeed. Quantum physics is fundamentally different, through its postulate that energy in light and matter is not continuously divisible, but exists in quanta of definite amount.

Actually, the first suggestion that the energy in electromagnetic radiation is "quantized" (comes in definite quanta) did not come from the photoelectric effect. Rather, it came from studies of the heat and light radiated by hot solids. The concept of *quanta of energy* was introduced by Max Planck, a German physicist, in 1900, five years before Einstein's theory. (Thus, the constant h is known as *Planck's constant*.) Planck was trying to explain how the heat (and light) energy radiated by a hot body is related to the frequency of the radiation. Classical physics (nineteenth-century thermodynamics and electromagnetism) could not account for the experimental facts. Planck found that the facts could be interpreted only by assuming that atoms, on radiating, change their energy in separate, quantized amounts. Einstein's theory of the photoelectric effect was actually an extension and application of Planck's quantum theory of thermal radiation. Einstein postulated that the change in the atom's energy is carried off as a distinct photon rather than being spread continuously over the light wave.

The experiments and the theory of radiation are much more difficult to describe than the experiments and theory of the photoelectric effect. That is why the concept of quanta of energy was introduced here by means of the photoelectric effect. By now, many tests have been made of both Planck's and Einstein's conceptions. In all cases, Planck's constant h is found to have the same basic position in quantum physics that Newton's universal constant G has in the physics of gravitation.

The photoelectric effect presented physicists with a real dilemma. According to the classical wave theory, light consists of electromagnetic waves extending continuously throughout space. This theory was highly successful in explaining optical

Robert Andrews Millikan (1868–1953), an American physicist, attended Oberlin College, where his interest in physics was only mild. After his graduation, he became more interested in physics, taught at Oberlin while taking his master's degree, and then obtained his doctor's degree from Columbia University in 1895. After more study in Germany, Millikan went to the University of Chicago, where he became a professor of physics in 1910. His work on the determination of the electronic charge took place from 1906 to 1913. He was awarded the Nobel Prize in physics in 1923 for this research and for the very careful experiments which resulted in the verification of the Einstein photoelectric equation (Sec. 18.4). In 1921, Millikan moved to the California Institute of Technology, eventually to become its president.

phenomena (reflection, refraction, polarization, interference). But it could not account for the photoelectric effect. Einstein's theory, which postulated the existence of separate quanta of light energy, accounted for the photoelectric effect. But it could not account for the other properties of light. The result was that there were two models whose basic concepts seemed to contradict each other. According to one, light is a wave phenomenon; according to the other, light has particle-like properties. Each model had its limits, successes, and failures. What, if anything, could be done about the contradictions between the two models? You will see later that this problem and its treatment have a central position in modern physics.

?

8. Einstein's idea of a quantum of light had a definite relation to the wave model of light. What was it?

9. Why does the photoelectron not have as much energy as the quantum of light that causes it to be ejected?

10. What does a stopping voltage of 2.0 V indicate about the photoelectrons emerging from a metal surface?

18.6 | X rays

Wilhelm Konrad Röntgen (1845–1923)

In 1895, a surprising discovery was made. Like the photoelectric effect, it did not fit in with accepted ideas about electromagnetic waves and eventually needed quanta for its explanation. The discovery was that of X rays by the German physicist Wilhelm

Röntgen. Its consequences for atomic physics and technology were dramatic and important.

On November 8, 1895, Röntgen was experimenting with the newly found cathode rays, as were many physicists all over the world. According to a biographer,

> . . . he had covered the all-glass pear-shaped tube [Crookes tube; see Sec. 18.2] with pieces of black cardboard, and had darkened the room in order to test the opacity of the black paper cover. Suddenly, about a yard from the tube, he saw a weak light that shimmered on a little bench he knew was nearby. Highly excited, Röntgen lit a match and, to his great surprise, discovered that the source of the mysterious light was a little barium platinocyanide screen lying on the bench.

Barium platinocyanide, a mineral, is one of the many chemicals known to *fluoresce* (emit visible light when illuminated with ultraviolet light). But no source of ultraviolet light was present in Röntgen's experiment. Cathode rays had not been observed to travel more than a few centimeters in air. So, neither ultraviolet light nor the cathode rays themselves could have caused the fluorescence. Röntgen therefore deduced that the fluorescence involved rays of a new kind. He named them *X rays,* that is, rays of an unknown nature. In an intensive series of experiments over the next 7 weeks, he determined the properties of this new radiation. Röntgen reported his results on December 28, 1895, in a paper whose title (translated) is "On a New Kind of Rays."

The discovery of X rays was narrowly missed by several physicists, including Hertz and Lenard (another well-known German physicist). An English physicist, Frederick Smith, found that photographic plates kept in a box near a cathode-ray tube became fogged, so he told his assistant to keep the plates in another place!

X rays were often referred to as Röntgen rays, after their discoverer.

Röntgen's paper described nearly all of the properties of X rays that are known even now. It described the method of producing the rays and proved that they originated in the glass wall of the tube, where the cathode rays struck it. Röntgen showed that X rays travel in straight lines from their place of origin and that they darken a photographic plate. He reported in detail the ability of X rays to penetrate various substances, such as paper, wood, aluminum, platinum, and lead. Their penetrating power was greater through light materials (paper, wood, flesh) than through dense materials (platinum, lead, bone). He described photographs showing "the shadows of bones of the hand, of a set of weights inside a small box, and of a piece of metal whose inhomogeneity becomes apparent with X rays." He gave a clear description of the shadows cast by the bones of the hand on the fluorescent screen. Röntgen also reported that the X rays were not deflected by a magnetic field, nor did they show reflection, refraction, or interference effects in ordinary optical apparatus.

J. J. Thomson discovered one of the most important properties of X rays a month or two after the rays themselves had become known. He found that when the rays pass through a gas, they make it a conductor of electricity. Thomson attributed this effect to "a kind of electrolysis, the molecule being split up, or nearly split up by the Röntgen rays." The X rays, in passing through the gas, knock electrons loose from some of the atoms or molecules of the gas. The atoms or molecules that lose these electrons become positively charged. They are called *ions* because they resemble the positive ions in electrolysis, and the gas is said to be *ionized.* Also, the freed electrons may attach themselves to previously neutral atoms or molecules, giving them negative charges.

It is easy to see why a charged electroscope will be discharged when the air around it is ionized. It attracts the ions of the opposite charge from the air.

Röntgen and Thomson found, independently, that electrified bodies lose their charges when the air around them is ionized by X rays. The rate of discharge depends on the intensity of the rays. This property was therefore used as a convenient quantitative means of measuring the intensity of an X-ray beam. As a result, careful quantitative measurements of the properties and effects of X rays could be made.

One problem that aroused keen interest following the discovery of X rays concerned the nature of the mysterious rays. Unlike charged particles (electrons, for example) they were not deflected by magnetic or electric fields. Therefore, it seemed that they had to be either neutral particles or electromagnetic waves. It was difficult to choose between these two possibilities. On the one hand, no neutral particles of atomic size (or smaller) that had the penetrating power of X rays were then known. The existence of such particles would be extremely hard to prove, because there was no way of getting at them. On the other hand, if the X rays were electromagnetic waves, they would have to have *extremely short wavelengths* because only in this case, according to theory, could they have high penetrating power and show no refraction or interference effects with ordinary optical apparatus.

Such a particle, called a neutron, was discovered in 1932. But the neutron has nothing to do with X rays.

As discussed in Chapters 12 and 13, distinctly wave-like properties become apparent only when waves interact with objects (like slits in a barrier) that are smaller than several wavelengths across. The wavelength hypothesized for X rays would be on the order of 10^{-10} m. So a demonstration of their wave behavior would require a diffraction grating with slits spaced about 10^{-10} m apart. Evidence from kinetic theory and from chemistry indicated that atoms were about 10^{-10} m in diameter. It was suggested, therefore, that X rays might be diffracted measurably by crystals in which the atoms form orderly layers about 10^{-10} m apart.

In 1912, such experiments succeeded. The layers of atoms did act like diffraction gratings, and X rays did, indeed, act like electromagnetic radiations of very short wavelength (like *ultra*-ultraviolet light). These experiments are more complicated to interpret than diffraction of a beam of light by a single, two-dimensional optical grating. The diffraction effect occurs in three

SG 12

SG 13

dimensions instead of two. Therefore, the diffraction patterns are far more elaborate (see the illustration below).

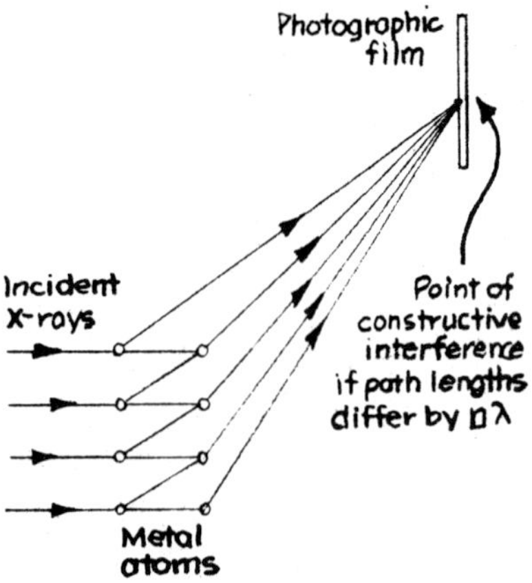

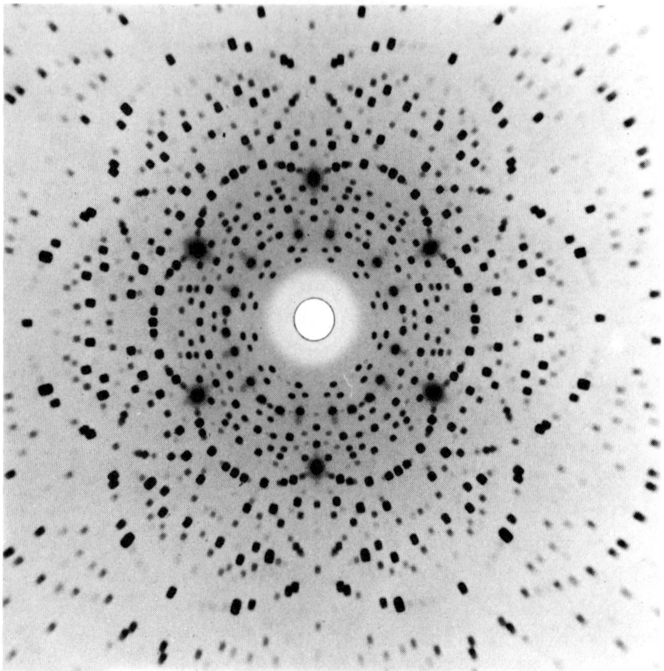

X-ray diffraction patterns from a metal crystal. The black spots are produced by constructive interference of X rays scattered from atoms. The drawing indicates how one of the spots is formed on the photographic film.

In addition to wave properties, X rays were found to have quantum properties. For example, they can cause the emission of electrons from metals. These electrons have greater kinetic energies than those produced by ultraviolet light. (The ionization of gases by X rays is also an example of the photoelectric effect. In this case, the electrons are freed from the atoms and molecules of the gas.) Thus, X rays also require quantum theory for the explanation of some of their behavior. So, like light, X rays were shown to have both wave and particle properties.

SG 14
SG 15
SG 16

Röntgen's discovery of X rays excited scientists throughout the world. His experiments were immediately repeated and extended in many laboratories in both Europe and America. Scientific journals during the year 1896 were filled with letters and articles describing new experiments or confirming the results of earlier experiments. This widespread experimentation was made easier since, during the years before Röntgen's discovery, the passage of electricity through gases had been a popular topic for study by physicists. Many physics laboratories therefore had cathode-ray tubes and could produce X rays easily.

Intense interest in X rays was generated by the spectacular use of these rays in medicine. Within three months of Röntgen's discovery, X rays were put to practical use in surgical operations in a hospital in Vienna. The use of this new aid to surgery spread rapidly. Since Röntgen's time, X rays have revolutionized some phases of medical practice, especially the diagnosis of some diseases and treatment of some forms of cancer. Extremely important uses of X rays also occur in other fields of applied science, both physical and biological. Among these are the study of the crystal structure of materials; "industrial diagnosis," such as the search for possible defects in materials and engineering structures; the study of old paintings and sculptures; and many others.

?

11. X rays were the first "ionizing" radiation discovered. What does "ionizing" mean?

12. What are three properties of X rays that led to the conclusion that X rays were electromagnetic waves?

13. What was the experimental and theoretical evidence to support that X rays had a very short wavelength?

18.7 | Electrons, quanta, and the atom

By the beginning of the twentieth century, enough chemical and physical information was available to allow many physicists to devise models of atoms. It was known that negative particles with identical properties (electrons) could be obtained from many different substances and in different ways. This suggested that electrons are parts of all atoms. Electrons are negatively charged. But samples of an element are ordinarily electrically *neutral*. Therefore, the atoms making up such samples are also presumably neutral. If so, the presence of negative electrons in an atom would seem to require the presence of an equal amount of positive charge.

Section 18.2 discussed a comparison of the values of q/m for the electron and for charged hydrogen atoms. As mentioned, hydrogen atoms are nearly 2,000 times more massive than electrons. Experiments (which will be discussed in some detail in Chapter 22 of Unit 6) showed that electrons make up only a very small part of the atomic mass in any atom. Consequently, any model of an atom must take into account the following information: (a) an electrically neutral atom contains equal amounts of positive and negative charge; (b) the negative charge is associated with only a small part of the mass of the atom. In addition, any atomic model should answer at least two questions: (1) How *many* electrons are there in an atom? (2) How are the electrons and the positive charge *arranged* in an atom?

During the first 10 years of the twentieth century, several atomic models were proposed, but none was satisfactory. The early models were all based entirely upon classical physics, that

Close Up

X rays

Originally, X rays were produced in Röntgen's laboratory when cathode rays (electrons) struck a target (the glass wall of the tube). Today, X rays usually are produced by directing a beam of high-energy electrons onto a metal target. As the electrons are deflected and stopped, X rays of various energies are produced. The maximum energy a single ray can have is the total kinetic energy the incident electron gives up on being stopped. So the greater the voltage across which the electron beam is accelerated, the more energetic and penetrating are the X rays. One type of X-ray tube is shown in the sketch below. A stream of electrons is emitted from a cathode C and accelerated to a tungsten target T by a strong electric field (high potential difference).

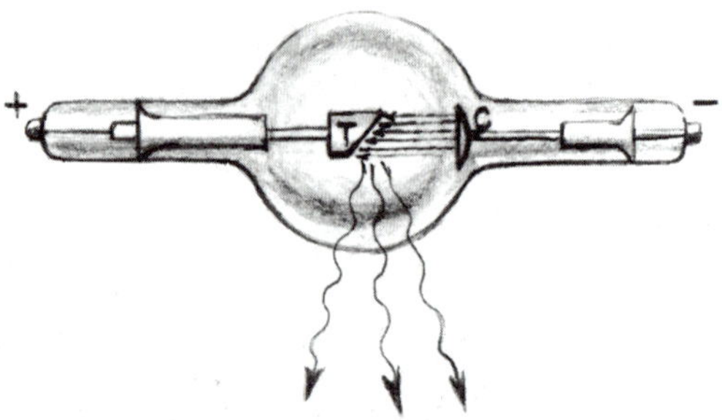

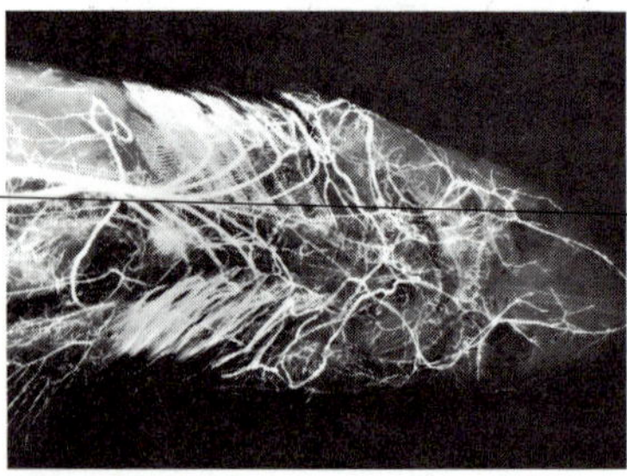

Above is the head of a dogfish shark. The blood vessels have been injected with a fluid that absorbs X rays so that the vessels can be studied.

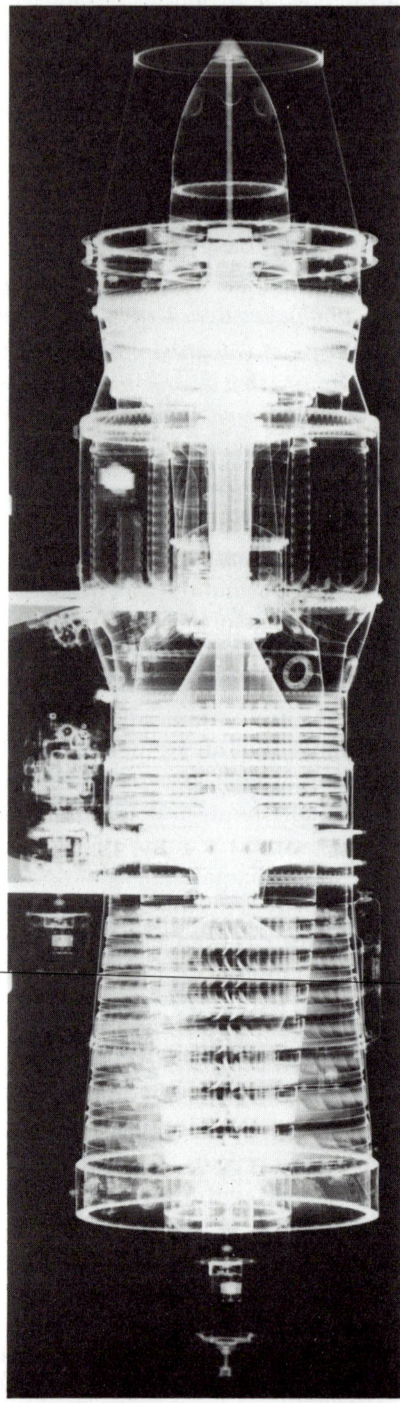

An X ray of a jet engine. X-ray photographs are often used to discover internal structural damage and flaws in pieces of complex machinery like this engine and nuclear reactor components, as well as in more mundane objects such as bowling pins and golf balls.

Above is a rose, photographed with X rays. The potential difference between the electron-emitting cathode and the target in the X-ray tube was 30,000 volts.

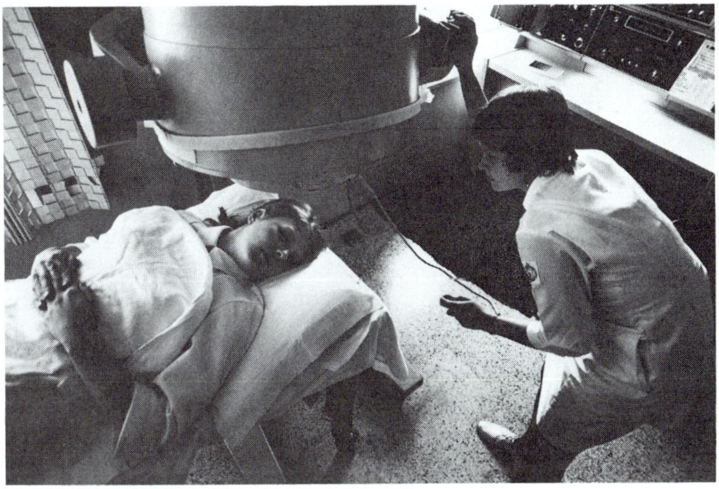

Immediately above is illustrated the familiar use of X rays in dentistry. Because X rays can injure tissues, a great deal of caution is required in using them. The shortest possible pulse of X rays is used, lead shielding is provided for the body, and the technician stands behind a wall of lead and lead glass.

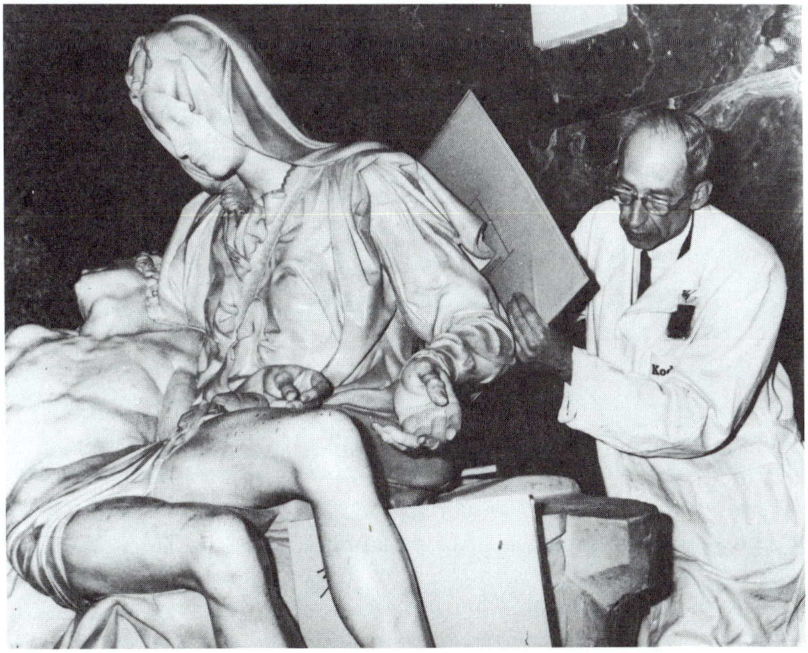

This photograph illustrates another use of X rays to discover internal damage, this time in a piece of art. Here, a technician is preparing to take X-ray photographs of Michelangelo's "Pieta."

is, upon the physics of Newton and Maxwell. No one knew how to invent a model that took account of Planck's theory of quantization of energy. More detailed experimental facts were also needed. For example, this was the period during which the charge on the electron and the main facts of photoelectricity were still being found. Nevertheless, scientists cannot and should not wait until every last fact is in, for that will never happen. It is impossible even to know what the missing facts *are* unless you have some sort of model. Even an incomplete or a partly wrong model will provide clues on which to build a better one.

See the *Project Physics* film loop "*Thomson Model of the Atom.*"

Until 1911, the most popular model for the atom was one proposed by J. J. Thomson in 1904. Thomson suggested that an atom consisted of a sphere of positive electricity in which an equal amount of negative charge was distributed in the form of electrons. Under this assumption, the atom was like a pudding of positive electricity, with the negative electricity scattered in it like raisins. The positive "fluid" was assumed to act on the negative charges, holding them in the atom by electric forces only. Thomson did not specify how the positive "fluid" was held together. The radius of the atom was taken to be of the order of 10^{-10} m. This value was based on information from the kinetic theory of gases and other considerations (see SG 13). With this model, Thomson was able to calculate that certain arrangements of electrons would be stable. This was the first requirement for explaining the existence of stable atoms. Thomson's theory also suggested that chemical properties might be associated with particular groupings of electrons. A systematic repetition of chemical properties might then occur among groups of elements. But it was not possible to deduce the detailed structure of the atoms of particular elements, nor could any detailed comparison with the actual periodic table be made. The problem of precisely how the electrons would arrange themselves was simply too difficult for Thomson to solve quantitatively.

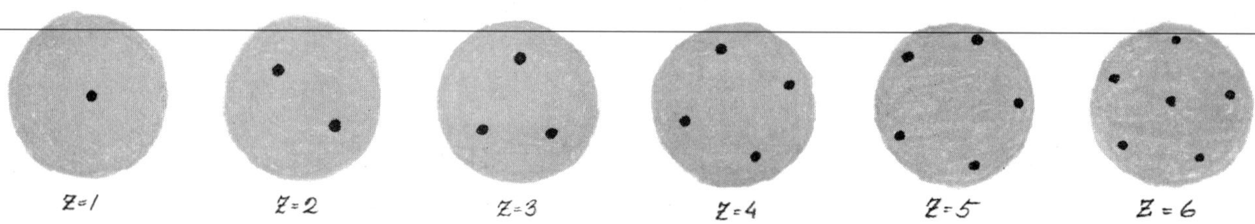

Some stable (hypothetical) arrangements of electrons in Thomson atoms. The atomic number Z is interpreted as equal to the number of electrons.

Chapter 19 will discuss some additional experimental information that provided valuable clues to improved models of the structure of atoms. You will also see how one of the greatest physicists of our time, Niels Bohr, combined the experimental evidence then available with the new concept of quanta in a successful theory of atomic structure. Bohr's model was eventually replaced by more sophisticated ones. But it led to the presently accepted theory of the atom and to this day is quite adequate for explaining most of the main facts of concern in this course.

14. Why was most of the mass of an atom believed to be associated with positive electric charge?

study guide

1. The *Project Physics* learning materials particularly appropriate for Chapter 18 include the following:

Experiments
The Charge-to-Mass Ratio for an Electron
The Measurement of Elementary Charge
The Photoelectric Effect

Activities
Measuring q/m for the Electron
Cathode Rays in a Crookes Tube
X Rays from a Crookes Tube
Lighting a Bulb Photoelectrically

Transparencies
Photoelectric Experiment
Photoelectric Equation

2. In Thomson's experiment on the ratio of charge to mass of cathode-ray particles (page 543), the following might have been typical values for B, V, and d: With a magnetic field B alone, the deflection of the beam indicated a radius of curvature of the beam within the field of 0.114 m for $B = 1.0 \times 10^{-3}$ T.* With the same magnetic field, the addition of an electric field in the same region ($V = 200$ volts, plate separation $d = 0.01$ m) made the beam go on straight through.

(a) Find the speed of the cathode-ray particles in the beam.

(b) Find q/m for the cathode-ray particles.

*The SI unit for B is N/A-m and is now called the *tesla*, symbol T (after the electrical engineer Nikola Tesla).

3. Given the value for the charge on the electron, show that a current of 1A is equivalent to the movement of 6.25×10^{18} electrons per second past a given point.

4. In the apparatus shown in (d) in Sec. 18.4, an electron is turned back before reaching plate A and eventually arrives at electrode C from which it was ejected. It arrives with some kinetic energy. How does this final energy of the electron compare with the energy it had as it left the electrode C?

5. At light frequencies below the threshold frequency no photoelectrons are emitted. What might happen to the light energy?

6. For most metals, the work function W is about 10^{-18} J. Light of what frequency will cause photoelectrons to leave the metal with virtually no kinetic energy? In what region of the spectrum is this frequency?

7. What is the energy of a light photon that corresponds to a wavelength of 5×10^{-7} m? 5×10^{-8} m?

8. The minimum or threshold frequency of light from emission of photoelectrons for copper is 1.1×10^{15} Hz. When ultraviolet light of frequency 1.5×10^{15} Hz shines on a copper surface, what is the maximum energy of the photoelectrons emitted, in joules? in electron volts?

9. What is the lowest-frequency light that will cause the emission of photoelectrons from a surface whose work function is 2.0 eV (that is, an energy of at least 2.0 eV is needed to eject an electron)?

10. Monochromatic light of wavelength 5×10^{-7} m falls on a metal cathode to produce photoelectrons. The light intensity at the surface of the metal is 10^2 J (m^2 = sec).

(a) What is the frequency of the light?

(b) What is the energy (in joules) of a single photon of the light?

(c) How many photons fall on 1 m^2 in 1 sec?

(d) If the diameter of an atom is about 10^{-10} m, how many photons fall on one atom in 1 sec, on the average?

(e) How often would one photon fall on one atom, on the average?

(f) How many photons fall on one atom in 10^{-10} sec, on the average?

(g) Suppose the cathode is a square 0.05 m on a side. How many electrons are released per second, assuming every photon releases a photoelectron? (In fact, only about one in 50 photons does so.) How big a current would this be in amperes?

11. Roughly how many photons of visible light are given off per second by a 1-W flashlight? (Only about 5% of the electric energy input to a tungsten-filament bulb is given off as visible light.) Hint: First find the energy, in joules, of an average photon of visible light.

12. Recall from Sec. 17.2 that 96,540 C of charge will deposit 31.77 g of copper in the electrolysis of copper sulfate. In Sec. 18.3, the charge of a single electron was reported to be 1.6×10^{-19} C.

(a) How many electrons must be transferred to deposit 31.77 g of copper?

(b) The density of copper is 8.92 g/cm^3. How many copper atoms would there be in the 1 cm^3? (Actually copper has a combining number of 2, which suggests that two electrons are required to deposit a single copper atom.)

(c) What is the approximate volume of each copper atom?

(d) What is the approximate diameter of a copper atom? (For this rough approximation, assume that the atoms are cubes.)

13. The approximate size of atoms can be calculated in a simple way from X-ray scattering experiments. The diagram below represents a beam of X rays reaching a crystal and the paths of two portions of the X-ray wave front leaving the crystal. Part of the front was scattered from the first layer (or plane) of atoms and part by the second layer, the third layer, and so on. The distance between layers is *d*. Each layer of atoms may be pictured as a partly transparent, plane mirror. Thus, each plane reflects some of the X rays specularly (just as light is reflected from a plane surface of water). The planes of atoms in the crystal are commonly called "Bragg planes," after W. H. Bragg who, with his son, W. L. Bragg, developed this part of the theory of X-ray diffraction in 1913. This scattering is shown schematically in the diagram.

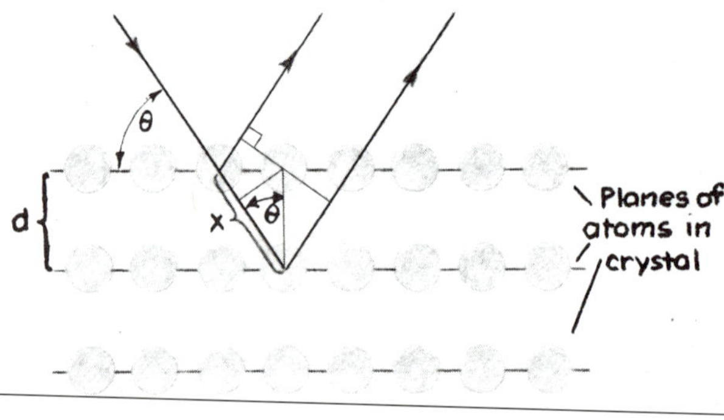

Note that the wave front of the X ray reflected from the second plane travels a distance $2x$ further than that reflected from the first plane.

(a) Show that $x = d \sin \theta$, where d is the distance between consecutive planes.

(b) At what angle θ will the wave scattered from the second plane interfere *constructively* with that scattered from the first plane? destructively?

(c) What will be the effect on your answers to (b) if you take into account the third and subsequent planes?

14. The highest frequency, f_{max}, of the X rays produced by an X-ray tube is given by the relation

$$hf_{max} = q_e V$$

where h is Planck's constant, q_e is the charge of an electron, and V is the potential difference at which the tube operates. If V is 50,000 volts, what is f_{max}?

15. The equation giving the maximum energy of the X rays in the preceding problem looks like one of the equations in Einstein's theory of the photoelectric effect. How would you account for this similarity? for the difference?

16. What potential difference must be applied across an X-ray tube for it to emit X rays with a minimum wavelength of 10^{-11} m? What is the energy of these X rays in joules? in electron volts?

17. A *glossary* is a collection of terms applicable to a special field of knowledge. Make a glossary of terms that appeared for the first time in this course in Chapter 18. Write an informative statement or definition for each term.

18. In his *Opticks,* Newton proposed a set of hypotheses about light which, taken together, formed a fairly successful model of light. The hypotheses were stated as questions. Three of the hypotheses are given below:

> Are not all hypotheses erroneous, in which light is supposed to consist in pression or motion waves . . . ? [Quest. 28]
>
> Are not the rays of light very small bodies emitted from shining substances? [Quest. 29]
>
> Are not gross bodies and light convertible into one another, and may not bodies receive much of their activity from the particles of light which enter their composition? [Quest. 30]

(a) In what respect is Newton's model similar to and different from the photon model of light?

(b) Why would Newton's model be insufficient to explain the photoelectric effect? What predictions can be made with the photon model that cannot be made with Newton's?

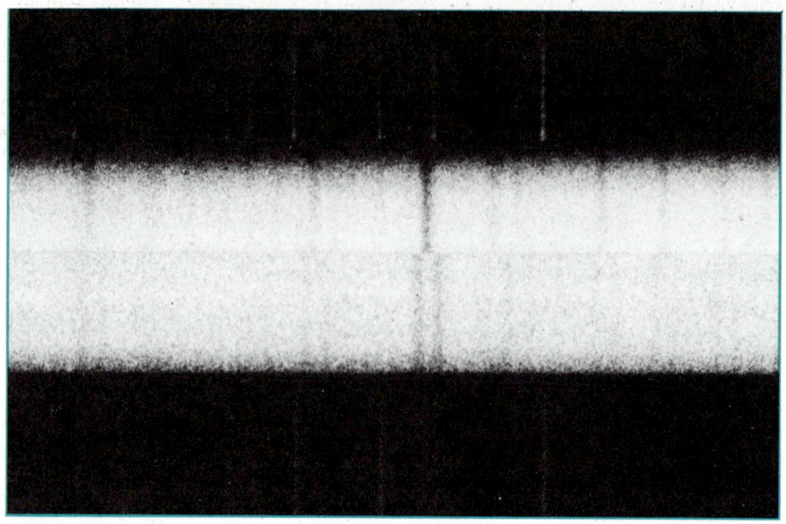

The Rutherford-Bohr Model of the Atom

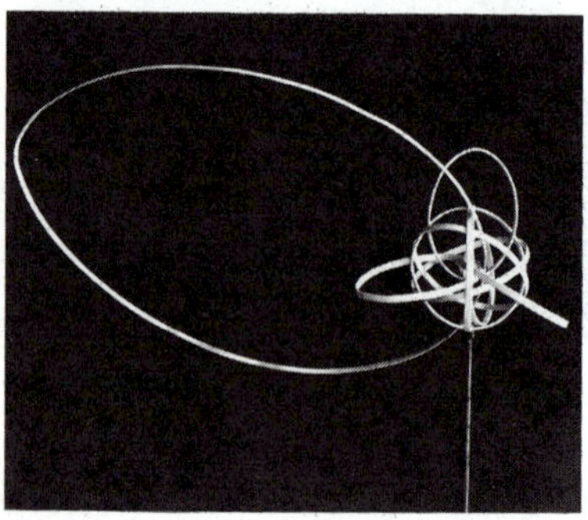

A sculptor's construction representing the Bohr model of a sodium atom.

19.1 | Spectra of gases

SG 1 One of the first real clues to understanding atomic structure appeared in the study of emission and absorption of light by different elements. The results of this study are so important to the story that their development will be reviewed in some detail.

It has long been known that light is emitted by gases or vapors when they are excited in any one of several ways: (1) by heating the gas to a high temperature, as when a volatile substance is put into a flame; (2) by an electric discharge through gas in the space between the terminals of an electric arc; (3) by a continuous electric current in a gas at low pressure (as in the now familiar "neon sign").

Close Up

Three Classes of Spectra

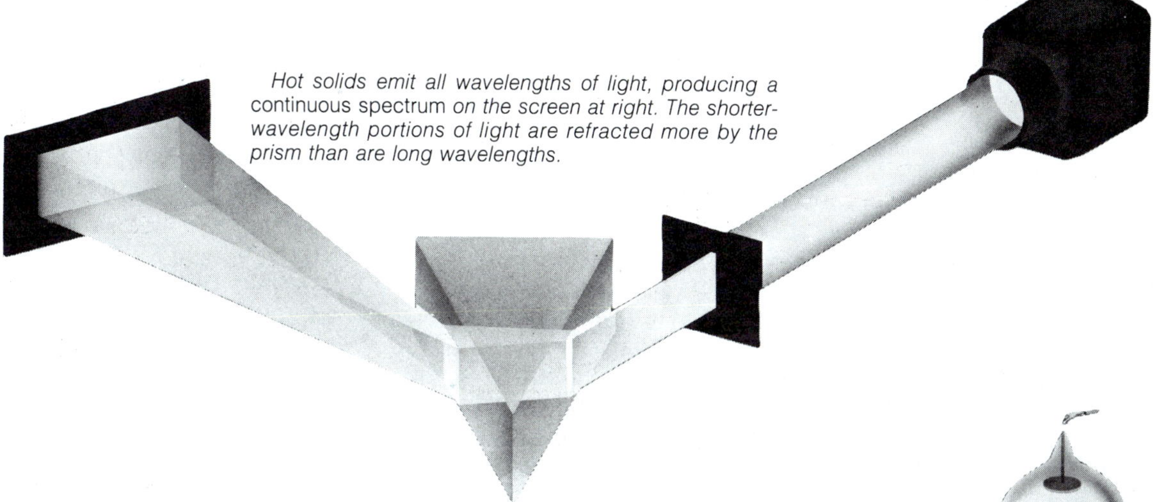

Hot solids emit all wavelengths of light, producing a continuous spectrum on the screen at right. The shorter-wavelength portions of light are refracted more by the prism than are long wavelengths.

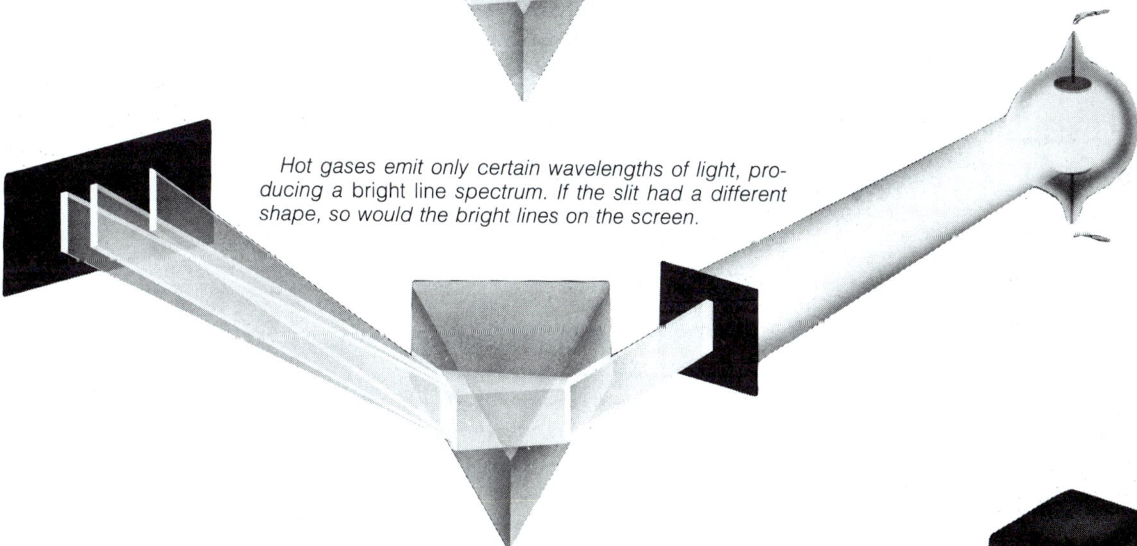

Hot gases emit only certain wavelengths of light, producing a bright line spectrum. If the slit had a different shape, so would the bright lines on the screen.

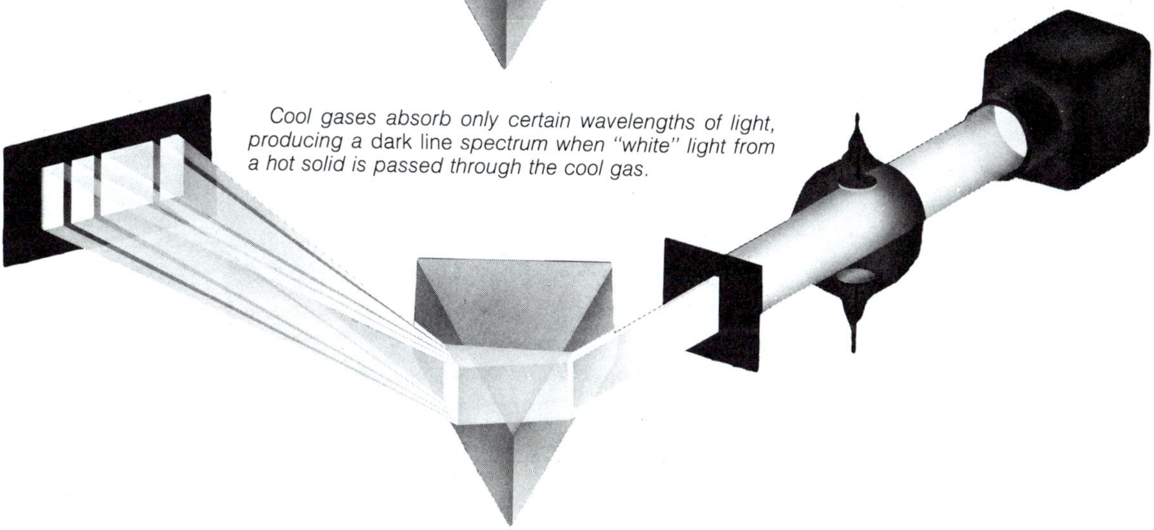

Cool gases absorb only certain wavelengths of light, producing a dark line spectrum when "white" light from a hot solid is passed through the cool gas.

The pioneer experiments on light emitted by various excited gases were made in 1752 by the Scottish physicist Thomas Melvill. He put one substance after another in a flame; "having placed a pasteboard with a circular hole in it between my eye and the flame . . ., I examined the constitution of these different lights with a prism." Melvill found that the spectrum of light from a hot gas was different from the well-known rainbow-colored spectrum of a glowing solid or liquid. Melvill's spectrum was not an unbroken stretch of color continuously graded from violet to red. Rather, it consisted of individual patches, each having the color of that part of the spectrum in which it was located. There were dark gaps (missing colors) between the patches. Later, more general use was made of a narrow slit through which to pass the light. Now the emission spectrum of a gas was seen as a set of bright lines (see the figure in the margin on this page). The bright lines are in fact colored images of the slit. Such spectra show that light from a gas is a mixture of only a few definite colors or narrow wavelength regions of light.

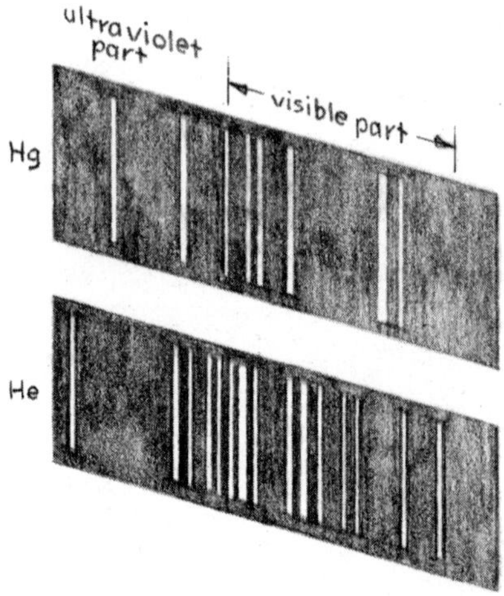

Parts of the line emission spectra of mercury (Hg) and helium (He), redrawn from photographic records, containing both visible and ultraviolet rays.

Melvill also noted that the colors and locations of the bright spots were different when different substances were put in the flame. For example, with ordinary table salt in the flame, the dominant color was "bright yellow" (now known to be characteristic of the element sodium). In fact, the line emission spectrum is markedly different for each chemically different gas because each chemical element emits its own characteristic set of wavelengths. (See the figure in the margin.) In looking at a gaseous source without the aid of a prism or a grating, the eye combines the separate colors. It perceives the mixture as reddish for glowing neon, pale blue for nitrogen, yellow for sodium vapor, and so on.

Some gases have relatively simple spectra. Thus, the most prominent part of the visible spectrum of sodium vapor is a pair of bright yellow lines. (This is why, for example, the flame in a gas stove turns yellow when soup, or any liquid containing salt, boils over.) Some gases or vapors have very complex spectra. Iron vapor, for example, has some 6,000 bright lines in the visible range alone.

In 1823, the British astronomer John Herschel suggested that each gas could be identified by its unique line spectrum. By the early 1860's, the physicist Gustav R. Kirchhoff and the chemist Robert W. Bunsen, in Germany, had jointly discovered two new elements (rubidium and cesium) by noting previously unreported emission lines in the spectrum of the vapor of a mineral water. This was the first of a series of such discoveries. It started the development of a technique for speedy chemical analysis of small amounts of materials by *spectrum analysis*. The "flame test" you may have performed in chemistry class is a simple application of this analysis.

In 1802, the English scientist William Wollaston saw in the spectrum of sunlight something that had been overlooked before. Wollaston noticed a set of seven sharp, irregularly spaced *dark* lines across the continuous solar spectrum. He did not understand why they were there and did not investigate further. A dozen years later, the German physicist Joseph von Fraunhofer, using better instruments, detected many hundreds of such dark lines. To the most prominent dark lines, von Fraunhofer assigned the letters A, B, C, etc. These dark lines can be easily seen in the sun's spectrum with even quite simple modern spectroscopes. The letters A, B, C, . . . are still used to identify them.

Spectroscope: **A device for examining the spectrum by eye.**

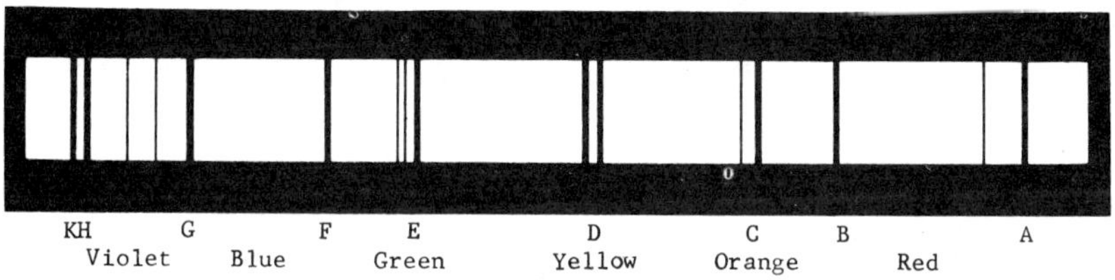

The "Fraunhofer dark lines" in the visible part of the solar spectrum. Only a few of the most prominent lines are represented.

In the spectra of several other bright stars, von Fraunhofer found similar dark lines. Many, but not all, of these lines were in the same positions as those in the solar spectrum.

Spectrometer **or** ***spectrograph:*** **A device for measuring the wave length of the spectrum and for recording the spectra (for example, on film).**

In 1859, Kirchhoff made some key observations that led to better understanding of both dark-line and bright-line spectra of gases. It was already known that the two prominent yellow lines in the emission spectrum of heated sodium vapor had the same wavelengths as two neighboring prominent dark lines in the solar spectrum. (The solar spectrum lines were the ones to which von Fraunhofer had assigned the letter D.) It was also known that the light emitted by a glowing solid forms a perfectly continuous spectrum that shows no dark lines. Kirchhoff now experimented with light from a glowing solid, as shown on page 567. The light was first passed through cooler sodium vapor and then dispersed by a prism. The spectrum produced had two prominent dark lines at the same place in the spectrum as the D-lines of the sun's spectrum. It was therefore reasonable to conclude that the light from the sun, too, was passing through a mass of sodium gas. This was the first evidence of the chemical composition of the gas envelope around the sun.

Kirchhoff's experiment was repeated with various other relatively cool gases placed between a glowing solid and the prism. Each gas produced its own characteristic set of dark lines. Evidently each gas in some way absorbs light of certain wavelengths from the passing "white" light. In addition, Kirchhoff showed that the wavelength of each absorption line matches the wavelength of a bright line in the emission spectrum of the same gas. The conclusion is that a gas can absorb *only* light of those wavelengths which, when excited, it can emit. (Note that not *every* emission line is represented in the absorption spectrum. Soon you will see why.)

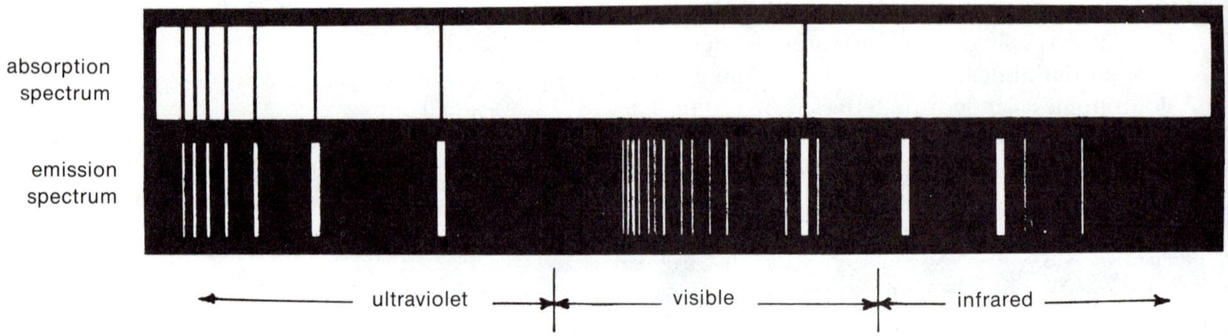

Comparison of the line absorption spectrum and line emission spectrum of sodium vapor.

Each of the various von Fraunhofer lines across the spectra of the sun and other stars has now been identified with the action of some gas as tested in the laboratory. In this way, the whole chemical composition of the outer region of the sun and other stars has been determined. This is really quite breathtaking from several points of view. First, it is surprising that scientists can learn the chemical composition of immensely distant objects. It is even more surprising that chemical materials out there are the same as those on earth. (That this is true is clearly shown by the fact that even very complex absorption spectra are reproduced exactly in star spectra.) Finally, this result leads to a striking conclusion: The physical processes that cause light absorption in the atom must be the same among the distant stars as on earth.

In these facts you can see a hint of how *universal* physical law really is. Even at the farthest edges of the cosmos from which the earth receives light, the laws of physics appear to be the same as for common materials close at hand in the laboratory! This is just what Galileo and Newton had intuited when they proposed that there is no difference between terrestrial and celestial physics.

SG 2

?

1. What can you conclude about a source if its light gives a bright-line spectrum?

2. What can you conclude about the source if its light gives a dark-line spectrum?

3. What evidence is there that the physics and chemistry of materials at great distances from earth are the same as those of matter close at hand?

19.2 | Regularities in the hydrogen spectrum

Of all the spectra, the line emission spectrum of hydrogen is especially interesting for both historical and theoretical reasons. In the visible and near-ultraviolet regions, the emission spectrum consists of an apparently systematic series of lines. (See the illustration at the right.) In 1885, Johann Jakob Balmer found a simple formula (an empirical relation) which gave the wavelengths of the lines known at the time. The formula is

$$\lambda = b\left(\frac{n^2}{n^2 - 2^2}\right)$$

The quantity b is a constant which Balmer determined empirically and found to be equal to 364.56×10^{-9} m); n is a whole number, different for each line. Specifically, to give the observed value for the wavelength, n must be 3 for the first (red) line of the hydrogen emission spectrum (named H_α); $n = 4$ for the second (green) line (H_β); $n = 5$ for the third (blue) line (H_γ); and $n = 6$ for the fourth (violet) line (H_δ). The table below shows excellent agreement (within 0.02%) between Balmer's calculations from his empirical formula and previously measured values.

Johann Jakob Balmer (1825–1898) was a teacher at a girl's school in Switzerland. His interest in mathematical puzzles and numerology led him to study wavelengths of spectra listed in tables.

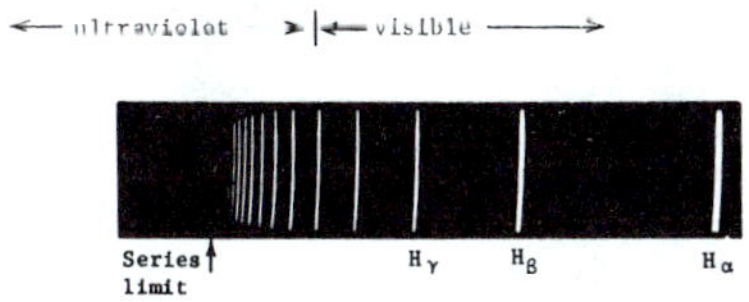

The Balmer lines of hydrogen, redrawn from a photograph made with a film sensitive to ultraviolet as well as visible light. The lines get more crowded as they approach the series limit in the ultraviolet.

WAVELENTH λ, IN NANOMETERS (10^{-9} m), FOR HYDROGEN EMISSION SPECTRUM*

Name of Line	*n*	*From Balmer's Formula*	*By Ångström's Measurement*	*Difference*
H_α	3	656.208	656.210	+0.002
H_β	4	486.08	486.074	−0.006
H_γ	5	434.00	434.01	+0.01
H_δ	6	410.13	410.12	−0.01

*Data for hydrogen spectrum (Balmer, 1885)

It took nearly 30 years before anyone understood why Balmer's empirical formula worked so well; why the hydrogen atom emitted light whose wavelength made such a simple sequence. But this did not keep Balmer from speculating that there might be other series of unsuspected lines in the hydrogen spectrum. Their wavelengths, he suggested, could be found by replacing the 2^2 in his equation with numbers such as 1^2, 3^2, 4^2, and so on. This suggestion stimulated many scientists to search for such additional spectral series. The search turned out to be fruitful, as you will see shortly.

To use modern notation, first rewrite Balmer's formula in a form that will be more useful:

$$\frac{1}{\lambda} = R_H\left(\frac{1}{2^2} - \frac{1}{n^2}\right)$$

In this equation, which can be derived from the first one, R_H is a constant, equal to $4/b$. (It is called the *Rydberg constant for hydrogen*, in honor of the Swedish spectroscopist J. R. Rydberg. Following Balmer, Rydberg made great progress in the search for various spectral series.) The series of lines described by Balmer's formula are called the *Balmer series*. Balmer constructed his formula from known λ of only four lines. The formula could be used to predict that there should be many more lines in the same series (indeed, infinitely many such lines as n takes on values such as $n = 3, 4, 5, 6, 7, 8, \ldots \infty$). The figure in the margin indicates that this has indeed been observed. Moreover, every one of the lines is correctly predicted by Balmer's formula with considerable accuracy.

Following Balmer's speculative suggestion of replacing 2^2 by other numbers gives the following possibilities:

$$\frac{1}{\lambda} = R_H\left(\frac{1}{1^2} - \frac{1}{n^2}\right), \quad \frac{1}{\lambda} = R_H\left(\frac{1}{3^2} - \frac{1}{n^2}\right), \quad \frac{1}{\lambda} = R_H\left(\frac{1}{4^2} - \frac{1}{n^2}\right)$$

and so on. Each of these equations describes a possible series. All these hypothetical series of lines can then be summarized by one overall formula:

$$\frac{1}{\lambda} = R_H\left(\frac{1}{n_f^2} - \frac{1}{n_i^2}\right)$$

where n_f is a whole number that is fixed for any one series for which wavelengths are to be found. (For example, it is 2 for all lines in the Balmer series.) The letter n_i stands for integers that take on the values $n_f + 1, n_f + 2, n_f + 3, \ldots$ for the successive individual lines in a given series. (Thus, for the first two lines of the Balmer series, n_i is 3 and 4.) The constant R_H should have the same value for all of these hydrogen series.

So far, this discussion has been merely speculative. No series, no single line fitting the general formula, *need* exist (except for the observed Balmer series, where $n_f = 2$). But when physicists began to look for these hypothetical lines with good spectrometers, they found that they do exist!

In 1908, F. Paschen in Germany found two hydrogen lines in the infrared. Their wavelengths were correctly given by setting $n_f = 3$ and $n_i = 4$ and 5 in the general formula. Many other lines in this "Paschen series" have since been identified. With improved experimental apparatus and techniques, new regions of the spectrum could be explored. Thus, other series gradually were added to the Balmer and Paschen series. In the table on page 573, the name of each series listed is that of the discoverer.

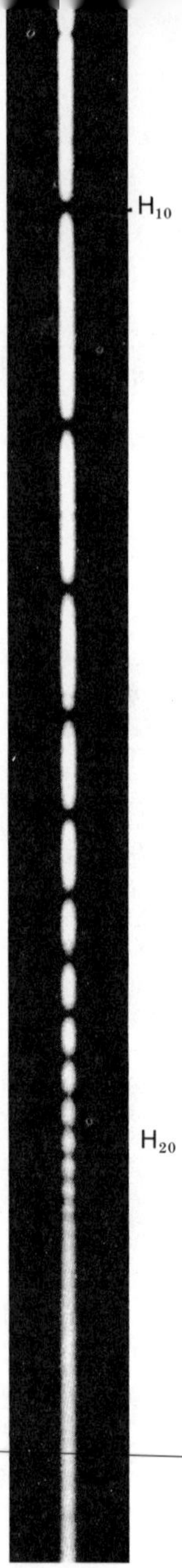

Part of the absorption spectrum observed in the light from the star Rigel (β Orion). The dark lines are at the same location as the lines caused by absorption by hydrogen gas in the ultraviolet region as produced in a laboratory; they match the lines of the Balmer series as indicated by the H numbers (where H_1 would be H_α, H_2 would be H_β, etc.). This photograph thus indicates the presence of hydrogen in the star.

SERIES OF LINES IN THE HYDROGEN SPECTRUM

Name of Series	*Date of Discovery*	*Region of Spectrum*	*Values in Balmer Equation*
Lyman	1906–1914	ultraviolet	$n_f = 1,\ n_i = 2, 3, 4, \ldots$
Balmer	1885	untraviolet–visible	$n_f = 2,\ n_i = 3, 4, 5, \ldots$
Paschen	1908	infrared	$n_f = 3,\ n_i = 4, 5, 6, \ldots$
Brackett	1922	infrared	$n_f = 4,\ n_i = 5, 6, 7, \ldots$
Pfund	1924	infrared	$n_f = 5,\ n_i = 6, 7, 8, \ldots$

Balmer hoped that his formula for hydrogen spectra might be a pattern for finding series relationships in the spectra of other gases. This suggestion bore fruit also. Balmer's formula itself did not work directly in describing spectra of gases other than hydrogen. But it did inspire formulas of similar mathematical form that successfully described order in portions of many complex spectra. The Rydberg constant R_H also reappeared in such empirical formulas. SG 3

No model based on classical mechanics and electromagnetism could be constructed that would explain the spectra described by these formulas. What you have already learned in Chapter 18 about quantum theory suggests one line of attack. Obviously, the emission and absorption of light from an atom must correspond to a decrease and an increase of the atom's energy. If atoms of an element emit light of only certain frequencies, then the energy of the atoms must be able to change only by certain amounts. These changes of energy must involve some rearrangement of the parts of the atom. SG 4 SG 5

?

4. What evidence did Balmer have that there were other series of lines in the hydrogen spectrum, with terms 3^2, 4^2, etc., instead of 2^2?

5. Often discoveries result from grand theories (like Newton's) or from a good intuitive grasp of phenomena (like Faraday's). What led Balmer to his relation for spectra?

6. From the Balmer formula in the last form given above, is there any upper limit to the wavelengths of light emitted by the hydrogen atom?

19.3 | Rutherford's nuclear model of the atom

A new basis for atomic models was provided during the period 1909–1911 by Ernest Rutherford. Rutherford was interested in the rays emitted by radioactive substances, especially α (alpha) rays. As you will see in the first chapter of Unit 6, α rays consist of

SG 6

In somewhat the same way, you could, in principle, use a scattering experiment to discover the size and shape of an object hidden in a cloud or fog. You could do so by directing a series of projectiles at the unseen object and tracing their paths back after deflection.

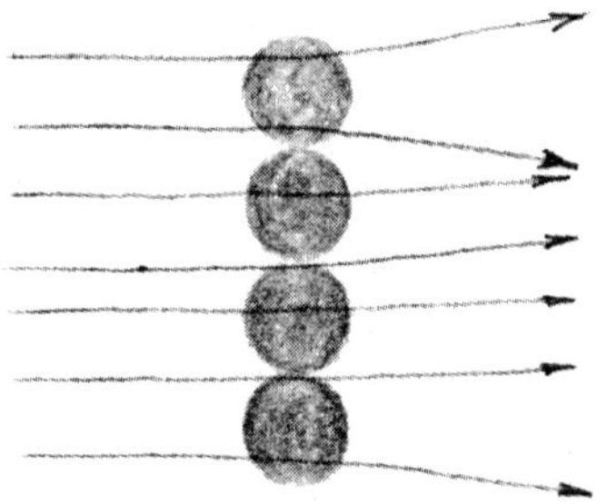

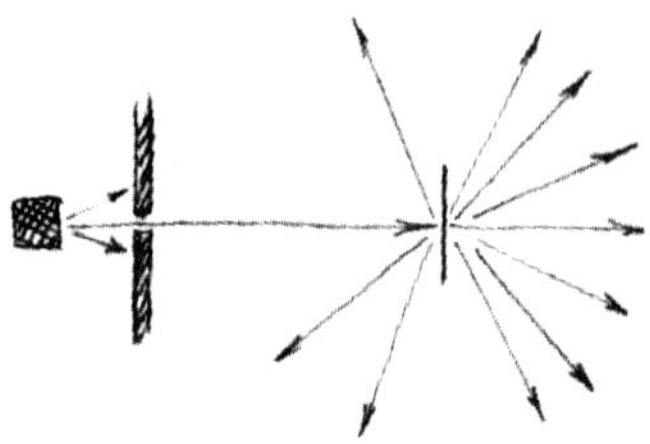

positively charged particles. These particles are positively charged helium ions with masses about 7,500 times greater than the electron mass. Some radioactive substances emit α particles at very high rates and energies. Such particles are often used as projectiles in bombarding samples of elements. The experiments that Rutherford and his colleagues did with α particles are examples of a highly important kind of experiment in atomic and nuclear physics: the scattering experiment.

In a scattering experiment, a narrow, parallel beam of projectiles (for example, α particles, electrons, X rays) is aimed at a target. The target is usually a thin foil or film of some material. As the beam strikes the target, some of the projectiles are deflected, or scattered, from their original direction. The scattering is the result of the interaction between the particles in the beam and the atoms of the material. A careful study of the projectiles after scattering can yield information about the projectiles, the atoms, and the interaction between them. If you know the mass, energy, and direction of the projectiles and see how they are scattered, you can deduce properties of the atoms that scattered the projectiles.

Rutherford noticed that when a beam of α particles passed through a thin metal foil, the beam spread out. This scattering may be thought of as caused by electrostatic forces between the positively charged α particles and the charges that make up atoms. Atoms contain both positive and negative charges. Therefore, an α particle undergoes both repelling and attracting forces as it passes through matter. The magnitude and direction of these forces depend on how closely the particle approaches the centers of the atoms among which it moves. When a particular atomic model is proposed, the extent of the expected scattering can be calculated and compared with experiment. The Thomson model of the atom predicted almost no chance that an α particle would be deflected by an angle of more than a few degrees.

The breakthrough which led to the modern model of the atom followed a discovery by one of Rutherford's assistants, Hans Geiger. Geiger found that the number of particles scattered through angles of 10° or more was much greater than the number predicted by the Thomson model. In fact, 1 out of about every 8,000 α particles was scattered through an angle greater than 90°. Thus, a significant number of α particles virtually bounced right back from the foil. This result was entirely unexpected. According to Thomson's model, the atom should have acted only slightly on the projectile, rather like a cloud in which fine dust is suspended. Some years later, Rutherford wrote:

> . . . I had observed the scattering of α-particles, and Dr. Geiger in my laboratory had examined it in detail. He found, in thin

pieces of heavy metal, that the scattering was usually small, of the order of one degree. One day Geiger came to me and said, "Don't you think that young Marsden, whom I am training in radioactive methods, ought to begin a small research?" Now I had thought that, too, so I said, "Why not let him see if any α-particles can be scattered through a large angle?" I may tell you in confidence that I did not believe that they would be, since we knew that the α-particle was a very fast, massive particle, with a great deal of [kinetic] energy, and you could show that if the scattering was due to the accumulated effect of a number of small scatterings, the chance of an α-particle's being scattered backward was very small. Then I remember two or three days later Geiger coming to me in great excitement and saying, "We have been able to get some of the α-particles coming backward . . ." It was quite the most incredible event that has ever happened to me in my life. It was almost as incredible as if you fired a 15-inch shell at a piece of tissue paper and it came back and hit you. On consideration, I realized that this scattering backward must be the result of a single collision, and when I made calculations I saw that it was impossible to get anything of that order of magnitude unless you took a system in which the greater part of the mass of the atom was concentrated in a minute nucleus. It was then that I had the idea of an atom with a minute massive centre, carrying a charge.

Ernest Rutherford (1871–1937) was born, grew up, and received most of his education in New Zealand. At age 24 he went to Cambridge, England, to work at the Cavendish Laboratory under J. J. Thomson. From there he went to McGill University in Canada, then home to be married, and back to England again, to Manchester University. At these universities, and later at the Cavendish Laboratory where he succeeded J. J. Thomson as director, Rutherford performed important experiments on radioactivity, the nuclear nature of the atom, and the structure of the nucleus. Rutherford introduced the concepts "alpha," "beta," and "gamma" rays, "protons," and "half-life." His contributions will be further discussed in Unit 6, "The Nucleus." *For his scientific work, Rutherford was knighted and received a Nobel Prize.*

These experiments and Rutherford's interpretation marked the origin of the modern concept of the *nuclear atom*. Look at the experiments and Rutherford's conclusion more closely. Why must the atom have its mass and positive charge concentrated in a tiny nucleus at the center about which the electrons are clustered?

SG 6
SG 7

A possible explanation of the observed scattering is that the foil contains concentrations of mass and charge, that is, positively charged nuclei. These nuclei are much more dense than in Thomson's atoms. An α particle heading directly toward one of them is stopped and turned back. In the same way, a ball would bounce back from a rock but not from a cloud of dust particles. The figure on page 576 is based on one of Rutherford's diagrams in his paper of 1911, which laid the foundation for the modern theory of atomic structure. It shows two positively charged α particles, A and A′. The α particle A is heading directly toward a massive nucleus N. If the nucleus has a positive electric charge, it will repel the positive α particle. Because of this electrical repulsive force, A will slow to a stop at some distance r from N and then move directly back. A′ is an α particle that is *not* headed directly toward the nucleus N. It is repelled by N along a path which calculation shows must be a hyperbola. The deflection of A′ from its original path is indicated by the angle ϕ.

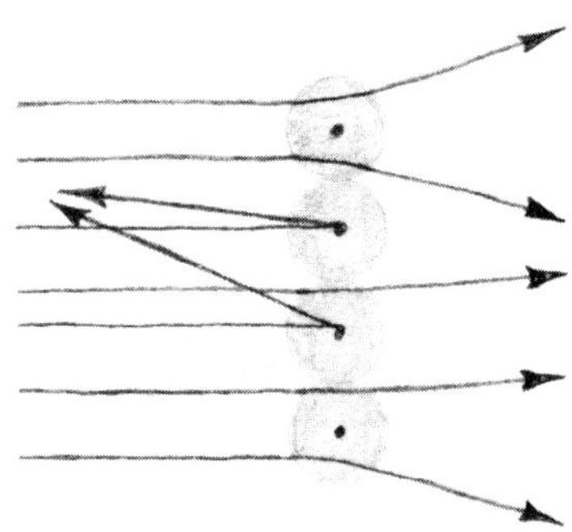

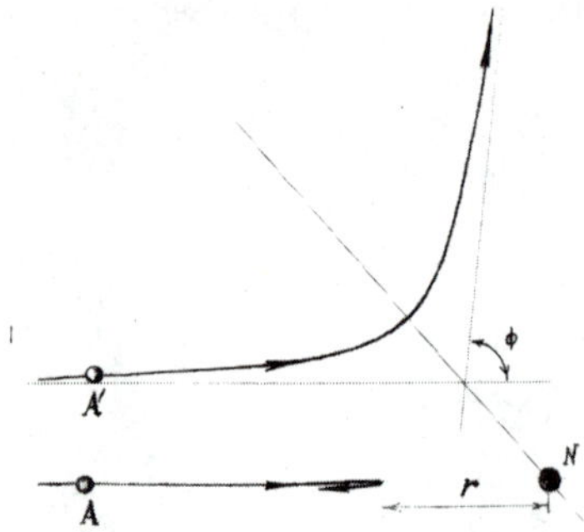

Paths of two α particles A and A′ approaching a nucleus N. (Based on Rutherford, Philosophical Magazine, *Vol. 21, 1911, p. 669.)*

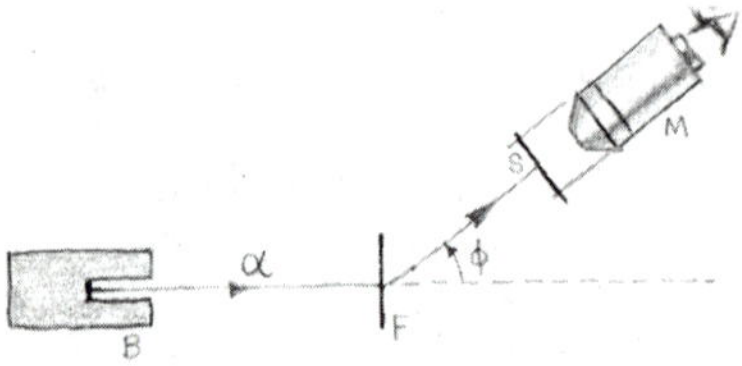

Rutherford's scintillation apparatus was placed in an evacuated chamber so that the α particles would not be slowed down by collisions with air molecules.

Rutherford considered the effects on the α particle's path of the important variables: the particle's speed, the foil thickness, and the quantity of charge Q on each nucleus. According to the model, *most* of the α particles should be scattered through small angles, because the chance of approaching a very small nucleus nearly head-on is so small. But a significant number of α particles should be scattered through large angles.

Geiger and Marsden tested these predictions with the apparatus sketched in the margin. The lead box B contains a radioactive substance (radon) that emits α particles. The particles emerging from the small hole in the box are deflected through various angles ϕ in passing through the thin metal foil F. The number of particles deflected through each angle ϕ is found by letting the particles strike a zinc sulfide screen S. Each α particle that strikes the screen produces a scintillation (a momentary pinpoint of fluorescence). These scintillations can be observed and counted by looking through the microscope M. The microscope and screen can be moved together along the arc of a circle. In later experiments, the number of α particles at any angle ϕ was counted more conveniently by a counter invented by Geiger (see sketch in the margin of page 577). The Geiger counter, in its more recent versions, is now a standard laboratory item.

Geiger and Marsden found that the number of α particles counted depended on the scattering angle, the speed of the particles, and the thickness of the foil. These findings agreed with Rutherford's predictions and supported an atomic model in which most of the mass and all positive charge occupy a very small region at the center of the atom.

?

7. Why are α particles scattered by atoms? Why is the angle of scattering mostly small but sometimes large?

8. What was the basic difference between the Rutherford and the Thomson models of the atom?

19.4 | Nuclear charge and size

Despite the success of Rutherford's model, a problem remained. There still was no way to measure independently the nucleus charge Q. However, the scattering experiments had confirmed Rutherford's predictions about the effect of the speed of the α particle and the thickness of the foil on the angle of scattering. As often happens when part of a theory is confirmed, it is reasonable to proceed temporarily as if the whole theory were justified; that is, pending further proof, one could assume that

q_e = magnitude of charge on one electron.

the value of Q needed to explain the observed scattering data was the correct value of Q for the actual nucleus. On this basis, scattering data were compiled for several different elements, among them carbon, aluminum, and gold. The following nuclear charges yielded the best agreement with experiments: for carbon, $Q = 6\ q_e$; for aluminum, $Q = 13$ or $14\ q_e$; and for gold, $Q = 78$ or $79\ q_e$. Similarly, values were found for other elements.

The magnitude of the positive charge of the nucleus was an important and welcome piece of information about the atom. The atom as a whole is, of course, electrically neutral. So if the nucleus has a positive charge of $6\ q_e$, 13 or $14\ q_e$, etc., the number of electrons surrounding the nucleus must be 6 for carbon, 13 or 14 for aluminum, etc. Thus, for the first time, scientists had a good idea of just how many electrons an atom may have. But an even more important fact was soon noticed. For each element, the value for the nuclear charge, in multiples of q_e, was close to the atomic number Z, the place number of that element in the periodic table! The results of scattering experiments with α particles were not yet precise enough to make this conclusion with certainty. But the data indicated that *each nucleus has a positive charge Q numerically equal to Zq_e.*

This suggestion made the picture of the nuclear atom much clearer and simpler. On this basis, the hydrogen atom ($Z = 1$) has one electron outside the nucleus. A helium atom ($Z = 2$) has in its neutral state two electrons outside the nucleus. A uranium atom ($Z = 92$) has 92 electrons. Additional experiments further supported this simple scheme. The experiments showed that it was possible to produce singly ionized hydrogen atoms, H^+, and doubly ionized helium atoms, He^{++}, but not H^{++} or He^{+++}. Evidently, a hydrogen atom has only one electron to lose, and a helium atom only two. Unexpectedly, the concept of the nuclear atom thus provided new insight into the periodic table of the elements. The nuclear concept suggested that the periodic table is really a listing of the elements according to *the number of electrons around the nucleus or according to the number of positive units of charge on the nucleus.*

These results cleared up some of the difficulties in Mendeleev's periodic table. For example, the elements tellurium and iodine had been assigned positions $Z = 52$ and $Z = 53$ on the basis of their chemical properties. This positioning contradicted the order of their atomic weights. But now Z was seen to correspond to a fundamental fact about the nucleus. Thus, the reversed order of atomic weights was understood to be only an accident rather than a basic fault in the scheme.

As an important additional result of these scattering experiments, the size of the nucleus could be estimated. Suppose an α particle is moving directly toward a nucleus. Its kinetic energy on approach is transformed to electrical potential energy.

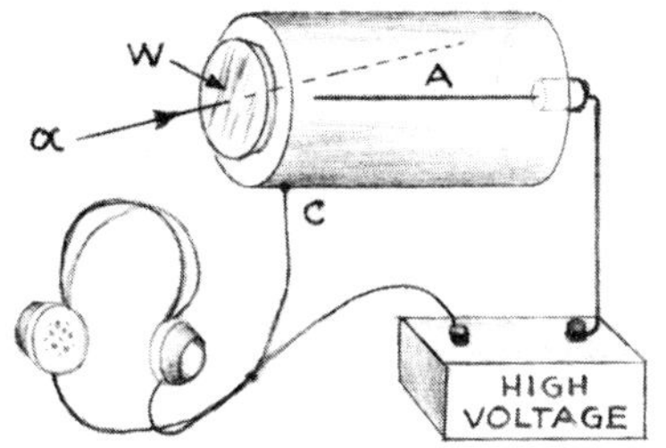

A Geiger counter (1928). It consists of a metal cylinder C containing a gas and a thin wire A that is insulated from the cylinder. A potential difference slightly less than that needed to produce a discharge through the gas is maintained between the wire (anode A) and cylinder (cathode C). When an α particle enters through the thin mica window (W), it frees a few electrons from the gas molecules. The electrons are accelerated toward the anode, freeing more electrons along the way by collisions with gas molecules. The avalanche of electrons constitutes a sudden surge of current that can be amplified to produce a click in the headphones or to operate a register (as in the Project Physics *scaler, used in experiments in Unit 6).*

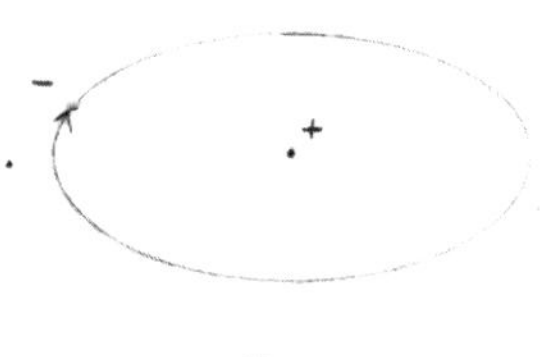

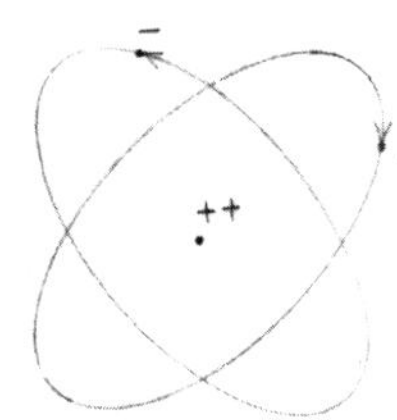

SG 8

It slows down and eventually stops, like a ball rolling up a hill. The distance of closest approach can be computed from the original kinetic energy of the α particle and the charges of α particle and nucleus. The value calculated for the closest approach is approximately 3×10^{-14} m. If the α particle does not penetrate the nucleus, this distance must be at least as great as the sum of the radii of α particles and nucleus; thus, the radius of the nucleus could not be larger than about 10^{-14} m. But 10^{-14} m is only about 1/1,000 of the known radius of an atom. Furthermore, the total volume of the atom is proportional to the cube of its radius. So it is clear that the atom is mostly empty, with the nucleus occupying only 1 billionth of the space! This explains how α particles or electrons can penetrate thousands of layers of atoms in metal foils or in gases, with only an occasional large deflection backward.

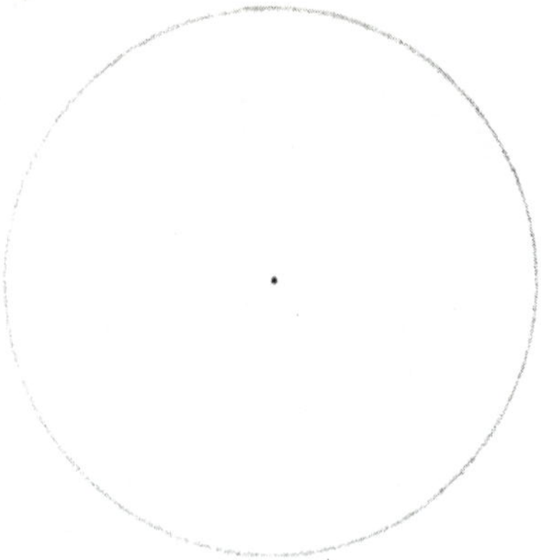

The central dot representing the nucleus in relation to the size of the atom as a whole is about 100 times too large. Popular diagrams of atoms often greatly exaggerate the relative size of the nucleus, perhaps in order to suggest the greater mass.

Successful as this model of the nuclear atom was in explaining scattering phenomena, it raised many new questions: What is the arrangement of electrons about the nucleus? What keeps the negative electron from falling into a positive nucleus by electrical attraction? Of what is the nucleus composed? What keeps it from exploding on account of the repulsion of its positive charges?
SG 9 Rutherford realized the problems raised by these questions and the failure of his model to answer them. But he rightly said that one should not expect one model, made on the basis of one set of puzzling results which it explains well, also to handle all other puzzles. Additional assumptions were needed to complete the model and answer the additional questions about the details of atomic structure. The remainder of this chapter will deal with the theory proposed by Niels Bohr, a young Danish physicist who joined Rutherford's group just as the nuclear model was being announced.

?

9. What does the "atomic number" of an element refer to, according to the Rutherford model of the atom?

10. What is the greatest positive charge that an ion of lithium (the next heaviest element after helium) could have?

19.5 | The Bohr theory: the postulates

Assume that an atom consists of a positively charged nucleus surrounded by a number of negatively charged electrons. What, then, keeps the electrons from falling into the nucleus, pulled in by the electric force of attraction? One possible answer is that an atom may be like a planetary system with the electrons revolving in orbits around the nucleus. Instead of the gravitational force, the electric attractive force between the nucleus and an electron would supply a centripetal force. This centripetal force would tend to keep the moving electron in orbit.

This idea seems to be a good start toward a theory of atomic structure. But a serious problem arises concerning the stability of a "planetary" atom. According to Maxwell's theory of electromagnetism, a charged particle radiates energy when it is accelerated. An electron moving in an orbit around a nucleus continually changes its velocity vector. In other words, it is *always being accelerated* by the centripetal electric force. The electron, therefore, should lose energy by emitting radiation. A detailed analysis of the electron's motion shows that the electron should be drawn steadily closer to the nucleus. (Somewhat similarly, an artificial satellite loses energy because of friction in the upper atmosphere and gradually spirals toward the earth.) Within a very short time, the energy-radiating electron should actually be pulled into the nucleus. According to classical physics, mechanics and electromagnetism, a planetary atom would not be stable for more than a very small fraction of a second.

"Classical physics" refers to those areas of physics firmly established before about 1900 and based on Newton's mechanics, Maxwell's electromagnetism, and Carnot's thermodynamics.

The idea of a planetary atom was nevertheless appealing. Physicists continued to look for a theory that would include a stable planetary structure and predict separate line spectra for the elements. Niels Bohr, then an unknown young physicist who had just received his PhD, succeeded in constructing such a theory in 1912–1913. This theory was widely recognized as a major victory. Although it had to be modified later to account for many more phenomena, it showed how to attack atomic problems by using quantum theory. Today, it seems a rather naive way of thinking about the atom, compared with more recent quantum-mechanical theories. But in fact, considering what it was designed to do, Bohr's theory is an impressive example of a successful physical model.

Since Bohr incorporated Rutherford's idea of the nucleus, the model that Bohr's theory discusses is often called the Rutherford–Bohr model.

Bohr introduced two new postulates specifically to account for the existence of stable electron orbits and separate emission spectra. These postulates may be stated as follows:

1. Contrary to the predictions of classical physics, there are states for an atomic system in which electromagnetic radiation simply does not occur, despite any acceleration of the charges. These states are called the *stationary states* of the atom.

2. Any emission or absorption of radiation, either as visible light or other electromagnetic radiation, corresponds to a sudden transition between two such stationary states. The radiation emitted or absorbed has a frequency f, determined by the relation $hf = E_i - E_f$. (In this equation, h is Planck's constant, and E_i and E_f are the energies of the atom in the initial and final stationary states, respectively.)

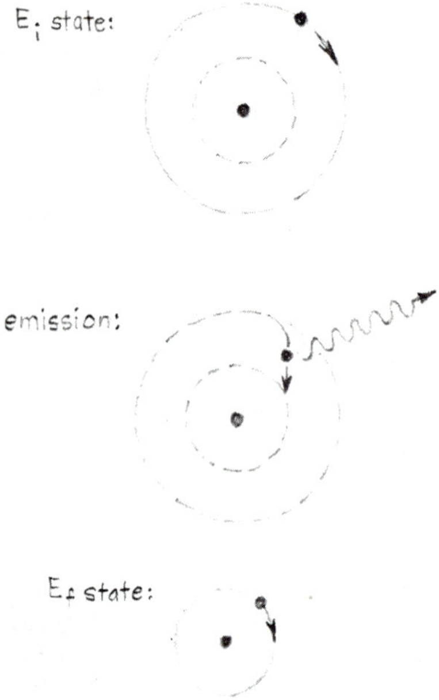

Quantum theory had begun with Planck's idea that *atoms emit light* only in definite amounts of energy. This concept was extended by Einstein's idea that *light travels* only as definite parcels of energy. Now it was extended further by Bohr's idea that *atoms exist* only in definite energy states. But Bohr also used the quantum concept in deciding which of all the *conceivable* stationary states were actually *possible*. An example of how Bohr did this is given in the next section.

For simplicity, the hydrogen atom, with a single electron revolving around the nucleus, is used. Following Bohr, it is assumed that the possible electron orbits are simply circular. Light is emitted by the atom when it changes from one state to another (see marginal sketches). The details of some additional assumptions and calculations are worked out on page 582. Bohr's result for the possible stable orbit radii r_n was $r_n = an^2$, where a is a constant $(h^2/4\pi^2 mkq_e^2)$ that can be calculated from known physical values, and n stands for any whole number, 1, 2, 3. . . .

?

11. What was the main evidence to support the fact that an atom could exist only in certain energy states?

12. How did Bohr deal with the fact that as long as an electron is steadily orbiting a nucleus, it does not radiate electromagnetic energy?

19.6 | The size of the hydrogen atom

Bohr's result is remarkable. In hydrogen atoms, the possible orbital radii of the electrons are whole multiples of a constant which can at once be evaluated; that is, n^2 takes on values of 1^2, 2^2, 3^2, . . . , and all factors to the left of n^2 are quantities known previously by independent measurement! Calculating the value $(h^2/4\pi^2 mkq_e^2)$ gives 5.3×10^{-11} m. Therefore, *according to Bohr's model,* the radii of stable electron orbits should be $r_n = 5.3 \times 10^{-11}\ \text{m} \times n^2$, that is, 5.3×10^{-11} m when $n = 1$ (first allowed orbit), $4 \times 5.3 \times 10^{-11}$ m when $n = 2$ (second allowed orbit), $9 \times 5.3 \times 10^{-11}$ m when $n = 3$, etc. In between these values, there are no allowed radii. In short, the separate allowed electron orbits are spaced around the nucleus in a regular way, with the allowed radii quantized in a regular manner. (See the drawing at the top of page 586.) Emission and absorption of light should therefore correspond to the transition of the electron from one allowed orbit to another.

SG 10

This is just the kind of result hoped for. It tells which radii are possible and where they lie. But so far, it had all been model building. Do the orbits in a real hydrogen atom actually correspond to this model? In his first paper of 1913, Bohr was able to give at least a partial yes as an answer. It had long been known that the normal "unexcited" hydrogen atom has a radius of about 5×10^{-11} m (that is the size of the atom obtained, for example, by interpreting measured characteristics of gases in light of the kinetic theory). This known value of about 5×10^{-11} m corresponds excellently to the prediction from the equation for orbital radius r if n has the lowest value, namely 1. Now there was a way to understand the size of the neutral, unexcited hydrogen atom. For every such atom, the size corresponds to the size of the innermost allowed electron orbit. That orbit, fixed by nature, is described by the quantization rule.

? 13. *Why do all unexcited hydrogen atoms have the same size?*

14. *Why does the hydrogen atom have just the size it has?*

19.7 | Other consequences of the Bohr model

With his two postulates, Bohr could calculate the radius of each permitted orbit. In addition, he could calculate the total energy of the electron in each orbit, the energy of the stationary state.

The results that Bohr obtained may be summarized in two simple formulas. As you saw, the radius of an orbit with quantum number n is given by the expression

$$r_n = n^2 r_1$$

where r_1 is the radius of the first orbit (the orbit for $n = 1$) and has the value 5.3×10^{-9} cm or 5.3×10^{-11} m.

The energy (the sum of kinetic and electric potential energy) of the electron in the orbit with quantum number n can also be computed from Bohr's postulate (see SG 11). As pointed out in Chapter 10, it makes no sense to assign an absolute value to potential energy. Only *changes* in energy have physical meaning. Therefore, any convenient zero level can be chosen. For an electron orbiting in an electric field, the mathematics is particularly simple if, as a zero level for energy, the state $n = \infty$ is chosen. At this level, the electron would be infinitely far from the nucleus (and therefore free of it). The energy for any other state E_n is then the *difference* from this free state. The possible energy states for the hydrogen atom are therefore

Note: Do not confuse this use of E for energy with the earlier use of E for electric field.

$$E_n = \frac{1}{n^2}E_1$$

where E_1 is the total energy of the atom when the electron is in the first orbit. E_1 is the lowest energy possible for an electron in a

Close Up

Bohr's Quantization Rule and the Size of Orbits

The magnitude of the charge on the electron is q_e; the charge on a nucleus is Zq_e, and for hydrogen ($Z = 1$) it is just q_e. The electric force with which the hydrogen nucleus attracts its electron is therefore

$$F_{el} = k\frac{q_e q_e}{r^2}$$

where k is the coulomb constant, and r is the center-to-center distance. If the electron is in a stable circular orbit of radius r around the nucleus, moving at a constant speed v, then the centripetal force is equal to mv^2/r. Since the centripetal force is provided by the electric attraction,

$$\frac{mv^2}{r} = k\frac{q_e^2}{r^2}$$

In the last equation, m, q_e, and k are constants; r and v are variables, whose values are related by the equation. What are the possible values of v and r for stationary states of the atom?

You can begin to get an answer if you write the last equation in slightly different form. Multiplying both sides by r^2 and dividing both sides by v, you get

$$mvr = \frac{kq_e^2}{v}$$

The quantity on the left side of this equation is the product of the momentum of the electron and the radius of the orbit. You can use this quantity to characterize the stable orbits. According to classical mechanics, the radius of the orbit could have any value, so the quantity mvr could also have any value. Of course, classical physics also seemed to deny that there could be *any* stable orbits in the hydrogen atom. But Bohr's first postulate implies that certain stable orbits (and only those) are permitted. So Bohr needed to find the rule that decides *which* stable orbits are possible. Here Bohr appears to have been largely guided by his intuition. He found that what was needed was the recognition that the quantity mvr does not take on just any value, but only certain *allowed values*. These values are defined by the relation

$$mvr = n\frac{h}{2\pi}$$

where h is Planck's constant, and n is a positive integer; that is, $n = 1, 2, 3, 4, \ldots$ (but not zero). When the possible values of mvr are restricted in this way, the quantity mvr is said to be *quantized*. The integer n that appears in the formula is called the *quantum number*. The main point is that each quantum number ($n = 1, 2, 3 \ldots$) corresponds to one allowed, stable orbit of the electron.

If you accept this rule, you can at once describe the "allowed" states of the atom, for example, in terms of the radii r of the possible orbits. You can combine the last expression above with the classical centripetal force relation as follows. The quantization rule is

$$mvr = n\frac{h}{2\pi}$$

so

$$r = \frac{nh}{2\pi mv}$$

and

$$r^2 = \frac{n^2h^2}{4\pi^2m^2v^2}$$

From classical mechanics,

$$\frac{mv^2}{r} = k\frac{q_e^2}{r^2}$$

so

$$v^2 = \frac{kq_e^2}{mr}$$

Substituting this "classical" value for v^2 into the quantization expression for r^2 gives

$$r^2 = \frac{n^2h^2}{4\pi^2m^2\left(\dfrac{kq_e^2}{mr}\right)}$$

Simplifying, you get the expression for the allowed radii, r_n:

$$r_n = \frac{n^2h^2}{4\pi^2kmq_e^2} = \left(\frac{h^2}{4\pi^2kmq_e^2}\right)n^2$$

hydrogen atom. Its value is -13.6 eV (the negative value means only that the energy is 13.6 eV *less* than the free state value E_∞). This is called the *ground state*. In that state, the electron is most tightly "bound" to the nucleus. The value of E_2, the first "excited" state above the ground state, is $1/2^2 \times -13.6 \text{ eV} = -3.4$ eV (only 3.4 eV less than in the free state).

According to the formula for r_n, the first Bohr orbit has the smallest radius, with $n = 1$. Higher values of n correspond to orbits that have larger radii. The higher orbits are spaced further and further apart, and the force field of the nucleus falls off even more rapidly. So the work required to move out to the next larger orbit actually becomes smaller and smaller. Also, the jumps in energy from one level of allowed energy E to the next become smaller and smaller.

19.8 | The Bohr theory: the spectral series of hydrogen

The different radii and energies a hydrogen atom can have are shown schematically in the diagram on page 586.

The most spectacular success of Bohr's model was that it could be used to explain all emission (and absorption) lines in the hydrogen spectrum; that is, Bohr could use his model to derive, and so to explain, the Balmer formula! By Bohr's second postulate, the radiation emitted or absorbed in a transition in an atom should have a frequency f determined by

$$hf = E_i - E_f$$

If n_f is the quantum number of the final state and n_i is the quantum number of the initial state, then according to the result for E_n,

$$E_f = \frac{1}{n_f^2} E_1 \quad \text{and} \quad E_i = \frac{1}{n_i^2} E_1$$

The frequency of radiation emitted or absorbed when the atom goes from the initial state to the final state is therefore determined by the equation

$$hf = \frac{E_1}{n_i^2} - \frac{E_1}{n_f^2} \quad \text{or} \quad hf = E_1 \left(\frac{1}{n_i^2} - \frac{1}{n_f^2} \right)$$

To deal with wavelength λ (as in Balmer's original formula) rather than frequency f, use the relation between frequency and wavelength given in Unit 3. The frequency is equal to the speed of the light wave divided by its wavelength: $f = c/\lambda$. Substituting c/λ for f in this equation and then dividing both sides by the constant hc (Planck's constant times the speed of light), gives

$$\frac{1}{\lambda} = \frac{E_1}{hc} \left(\frac{1}{n_i^2} - \frac{1}{n_f^2} \right)$$

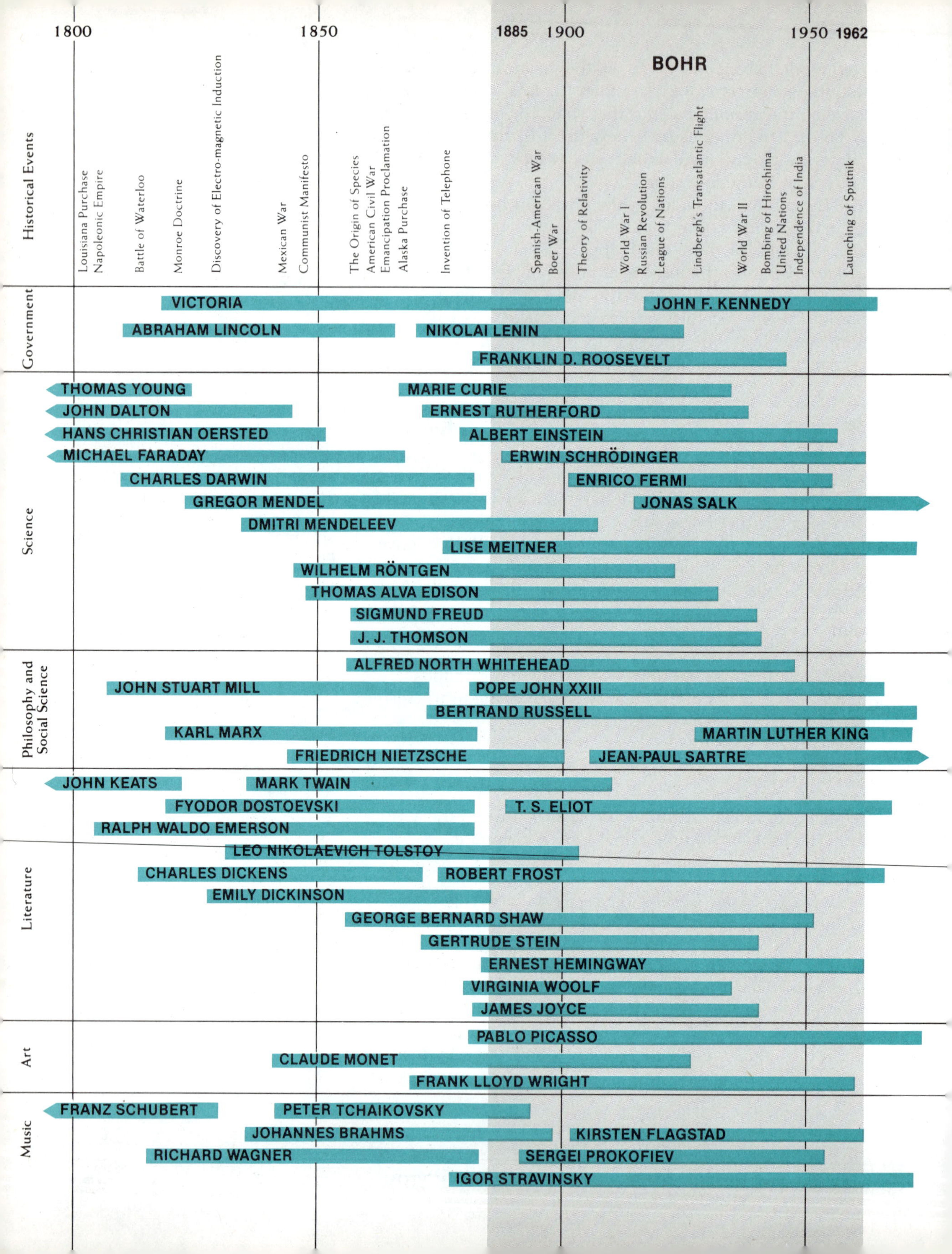

1800
1850
1885
1900
1950
1962
BOHR
Historical Events
Louisiana Purchase
Napoleonic Empire
Battle of Waterloo
Monroe Doctrine
Discovery of Electro-magnetic Induction
Mexican War
Communist Manifesto
The Origin of Species
American Civil War
Emancipation Proclamation
Alaska Purchase
Invention of Telephone
Spanish-American War
Boer War
Theory of Relativity
World War I
Russian Revolution
League of Nations
Lindbergh's Transatlantic Flight
World War II
Bombing of Hiroshima
United Nations
Independence of India
Launching of Sputnik
Government
VICTORIA
JOHN F. KENNEDY
ABRAHAM LINCOLN
NIKOLAI LENIN
FRANKLIN D. ROOSEVELT
Science
THOMAS YOUNG
MARIE CURIE
JOHN DALTON
ERNEST RUTHERFORD
HANS CHRISTIAN OERSTED
ALBERT EINSTEIN
MICHAEL FARADAY
ERWIN SCHRÖDINGER
CHARLES DARWIN
ENRICO FERMI
GREGOR MENDEL
JONAS SALK
DMITRI MENDELEEV
LISE MEITNER
WILHELM RÖNTGEN
THOMAS ALVA EDISON
SIGMUND FREUD
J. J. THOMSON
Philosophy and Social Science
ALFRED NORTH WHITEHEAD
JOHN STUART MILL
POPE JOHN XXIII
BERTRAND RUSSELL
KARL MARX
MARTIN LUTHER KING
FRIEDRICH NIETZSCHE
JEAN-PAUL SARTRE
Literature
JOHN KEATS
MARK TWAIN
FYODOR DOSTOEVSKI
T. S. ELIOT
RALPH WALDO EMERSON
LEO NIKOLAEVICH TOLSTOY
CHARLES DICKENS
ROBERT FROST
EMILY DICKINSON
GEORGE BERNARD SHAW
GERTRUDE STEIN
ERNEST HEMINGWAY
VIRGINIA WOOLF
JAMES JOYCE
Art
PABLO PICASSO
CLAUDE MONET
FRANK LLOYD WRIGHT
Music
FRANZ SCHUBERT
PETER TCHAIKOVSKY
JOHANNES BRAHMS
KIRSTEN FLAGSTAD
RICHARD WAGNER
SERGEI PROKOFIEV
IGOR STRAVINSKY

According to Bohr's model, then, this equation gives the wavelength λ of the radiation emitted or absorbed when a hydrogen atom changes from one stationary state with quantum number n_i to another with n_f.

How does this prediction from Bohr's model compare with the long-established *empirical* Balmer formula for the Balmer series? This, of course, is the crucial question. The Balmer formula was given on page 571:

$$\frac{1}{\lambda} = R_H\left(\frac{1}{2^2} - \frac{1}{n^2}\right)$$

You can see at once that the equation for λ of emitted (or absorbed) light derived from the Bohr model is exactly the same as Balmer's formula, if $R_H = -E_1/hc$ and $n_f = 2$.

The Rydberg constant R_H was long known from spectroscopic measurements to have the value of $1.097 \times 10^7 \text{ m}^{-1}$. Now it could be compared with the value for $-(E_1/hc)$. Remarkably, there was fine agreement, as can be shown without much difficulty. R_H, previously regarded as just an experimentally determined constant, was now shown to be a number that could be calculated from known constants, namely, the mass and charge of the electron, Planck's constant, and the speed of light.

SG 11

Niels Bohr (1885–1962) at the time he did the work described in this section.

More important, you can now see the *meaning*, in physical terms, of the old empirical formula for the lines (H_α, H_β, . . .) in the Balmer series. All the lines in the Balmer series simply correspond to transitions from various initial states (various values of n_i) to the same final state, for which $n_f = 2$. Thus, photons having the frequency or wavelength of the line H_α are emitted when electrons in a gas of hydrogen atoms "jump" from their $n = 3$-state to their $n = 2$-state (see diagram, page 586). The H_β line corresponds to "jumps" from $n = 4$ to $n = 2$, and so forth.

When the Bohr theory was proposed in 1913, emission lines in only the Balmer and Paschen series for hydrogen were known definitely. Balmer had suggested, and the Bohr model agreed, that additional series should exist. Further experiments revealed the Lyman series in the ultraviolet portion of the spectrum (1916), the Brackett series (1922), and the Pfund series (1924). In each series, the measured frequencies of the lines were found to be those predicted by Bohr's theory. Similarly, the general formula that Balmer guessed might apply for all spectral lines of hydrogen was explained. The lines of the Lyman series correspond to transitions from various initial states to the final state $n_f = 1$; the lines of the Paschen series correspond to transitions from various initial states to the final state $n_f = 3$; etc. (See table on page 573.) The general scheme of possible transitions among the first six orbits is shown in the figure on page 586. Thus, the theory not only related known information

SG 12
SG 13

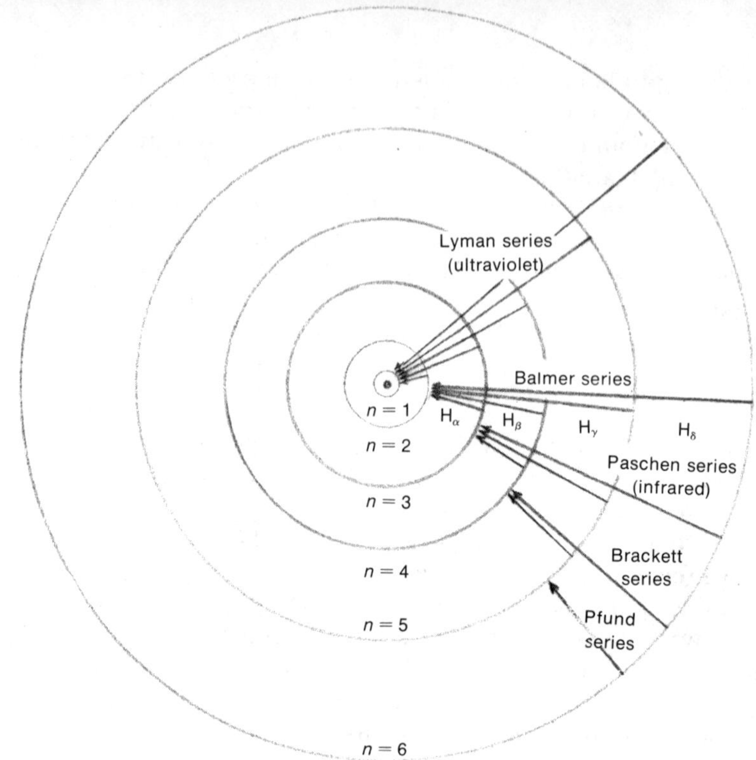

A schematic diagram of the relative energy levels of electron states in an atom of hydrogen.

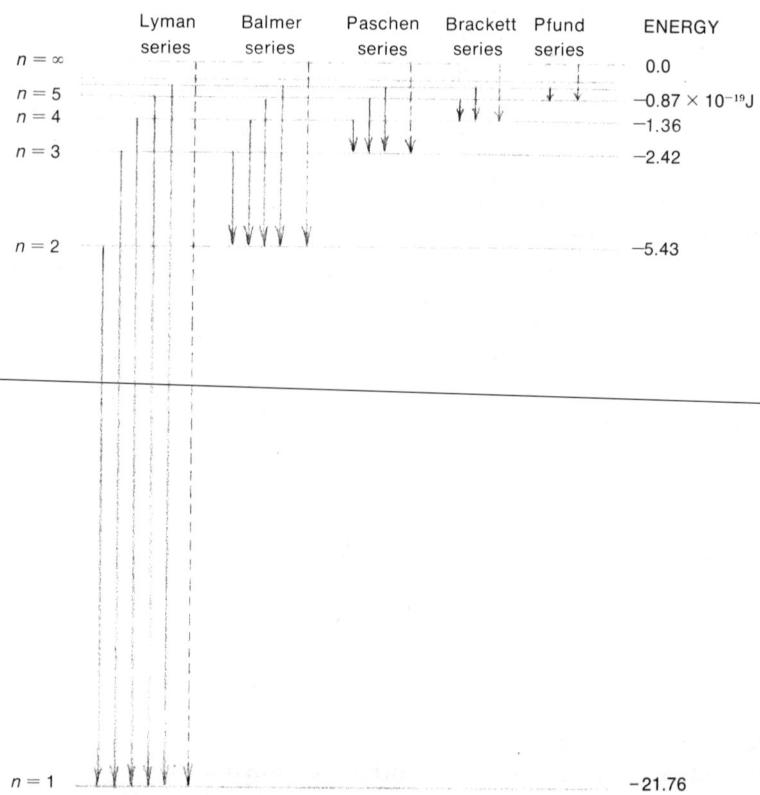

Energy-level diagram for the hydrogen atom. Possible transitions between energy states are shown for the first few levels (from n = 2 *to* n = 1*, or from* n = 3 *to* n = 2 *or* n = 1*, etc.). The dotted arrow for each series indicates the* series limit*, a transition from* n = ∞*, the state where the electron is completely free (infinitely far) from the nucleus.*

about the hydrogen spectrum, but also predicted correctly the wavelengths of previously unknown series of lines in the spectrum. Moreover, it provided a reasonable physical model; Balmer's general formula had offered no physical reason. The schematic diagram shown on page 586 is useful as an aid for the imagination. But it has the danger of being too specific. For instance, it leads you to think of the emission of radiation in terms of "jumps" of electrons between orbits. Although this is a useful idea, in Chapter 20 you will see why it is impossible to detect an electron moving in such orbits. Nor can you watch an electron "jump" from one orbit to another. A second way of presenting the results of Bohr's theory yields the same facts but does not adhere as closely to a picture of orbits. This scheme is shown in the bottom figure on page 586. It focuses not on orbits but on the corresponding possible energy states. These energy states are all given by the formula $E_n = 1/n^2 \times E_1$. In terms of this *mathematical model*, the atom is normally unexcited, with an *energy* E_1 about -22×10^{-19} J (-13.6 eV). Absorption of energy can place the atoms in an excited state, with a correspondingly higher energy. The excited atom is then ready to emit light, with a consequent reduction in energy. The energy absorbed or emitted always shifts the total energy of the atom to one of the values specified by the formula for E_n. Thus, the hydrogen atom may be represented by means of the energy-level diagram.

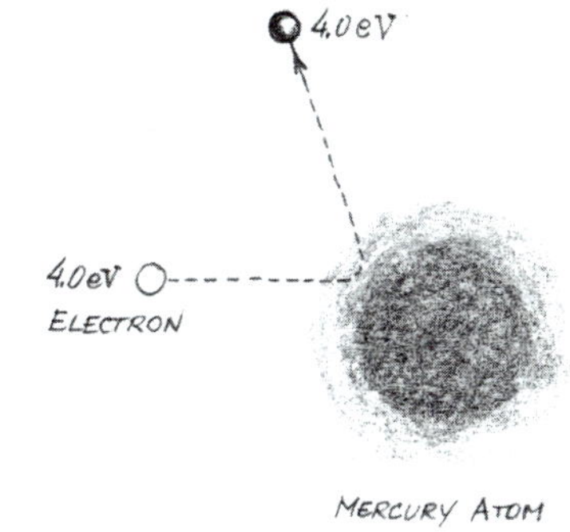

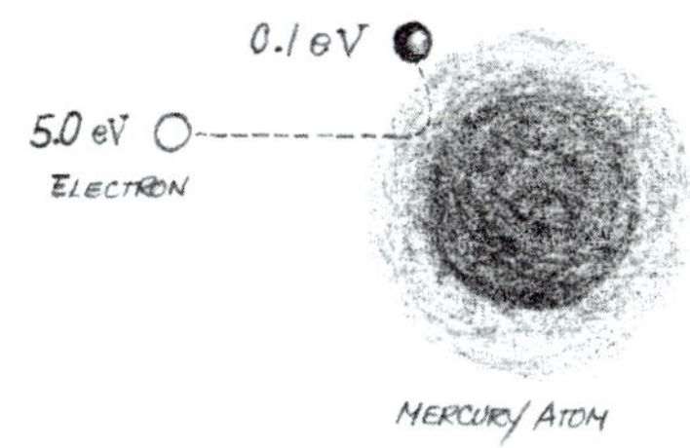

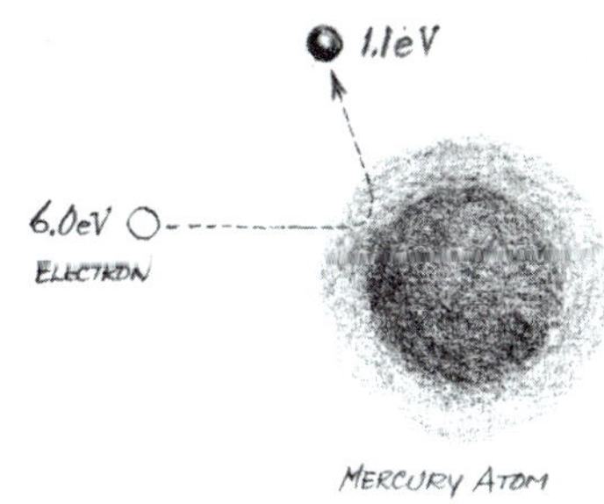

?

15. Balmer had predicted accurately the other spectral series of hydrogen 30 years before Bohr did. Why is Bohr's prediction considered more significant?

16. How does Bohr's model explain line absorption spectra? On page 570, you saw that an absorption spectrum did not contain all the lines of the emission spectrum. Based on the Bohr model, why is this so?

19.9 | Stationary states of atoms: the Franck–Hertz experiment

The success of the Bohr theory in accounting for the spectrum of hydrogen left this question: Could experiments show directly that atoms have only certain, separate energy states? In other words, were there really gaps between the energies that an atom can have? A famous experiment in 1914, by the German physicists James Franck and Gustav Hertz, showed that these separate energy states do indeed exist.

Franck and Hertz bombarded atoms with electrons from an electron gun. They were able to measure the energy lost by electrons in collisions with atoms. They could also determine the energy gained by atoms in these collisions. In their first experiment, Franck and Hertz bombarded mercury vapor contained in a chamber at very low pressure. The procedure was equivalent to measuring the kinetic energy of electrons on leaving the electron gun, and again after they had passed through the mercury vapor. The only way electrons could lose energy was in collisions with mercury atoms. Franck and Hertz found that when the kinetic energy of the electrons leaving the gun was small (up to several electron volts) the electrons still had almost exactly the same energy after passage through the mercury vapor as they had on leaving the gun. This result could be explained in the following way. A mercury atom is several hundred thousand times more massive than an electron. When it has low kinetic energy, the electron just bounces off a mercury atom, much as a golf ball thrown at a bowling ball would bounce off. A collision of this kind is called an "elastic" collision (discussed in Sec. 9.6). In an elastic collision, the mercury atom (bowling ball) takes up only a negligible part of the kinetic energy of the electron (golf ball) so that the electron loses practically none of its kinetic energy.

Physicists now know two ways of "exciting" an atom: by absorption and by collision. In absorption, an atom absorbs a photon with just the right energy to cause a transition from the lowest energy level to a higher one. Collision may involve an electron from an electron gun or collisions among agitated atoms (as in a heated enclosure or a discharge tube).

But when the kinetic energy of the electrons was raised to 5 eV, the experimental results changed dramatically. When an electron collided with a mercury atom, the electron lost almost exactly 4.9 eV of energy. When the energy was increased to 6 eV, the electron still lost just 4.9 eV of energy in collision, being left with 1.1 eV of energy. These results indicated that a mercury atom cannot accept less than 4.9 eV of energy. Furthermore, when the mercury atom is offered somewhat more energy, for example, 5 or 6 eV, it still accepts only 4.9 eV. The accepted amount of energy cannot go into kinetic energy of the mercury because the atom is so much more massive than the electron. Therefore, Franck and Hertz concluded that the 4.9 eV is added to the internal energy of the mercury atom; that is, the mercury atom enters a stationary state with energy 4.9 eV greater than that of the lowest energy state, with no allowed energy level in between.

What happens to this extra 4.9 eV of internal energy? According to the Bohr model, this amount of energy should be emitted as electromagnetic radiation when the atom returns to its lowest state. Franck and Hertz looked for this radiation and found it. They observed that the mercury vapor emitted light at a wavelength of 253.5 nanometers. This wavelength was known to exist in the emission spectrum of hot mercury vapor. The wavelength corresponds to a frequency f for which the photon's energy, hf, is just 4.9 eV (as you can calculate). This result showed that mercury atoms had indeed gained (and then radiated away) 4.9 eV of energy in collisions with electrons.

SG 14, 15

Later experiments showed that mercury atoms bombarded by electrons could also gain other sharply defined amounts of energy, for example, 6.7 eV and 10.4 eV. In each case, the radiation emitted corresponded to known lines in the emission spectrum of mercury. In each case, similar results were obtained; the electrons always lost energy, and the atoms always gained energy, both in sharply defined amounts. Each type of atom studied was found to have separate energy states. The amounts of energy gained by the atoms in collisions with electrons always corresponded to known spectrum lines. Thus, direct experiment confirmed the existence of separate stationary states of atoms predicted by the Bohr theory of atomic spectra. This result was considered to provide strong evidence of the validity of the Bohr theory.

SG 16

? *17. How much kinetic energy will an electron have after a collision with a mercury atom if its kinetic energy before collision is (a) 4.0 eV? (b) 5.0 eV? (c) 7.0 eV?*

19.10 | The periodic table of the elements

In the Rutherford–Bohr model, atoms of the different elements differ in the charge and mass of their nuclei and in the number and arrangement of the electrons. Bohr came to picture the electronic orbits as shown on page 592, though not as a series of concentric rings in one plane, but as patterns in three dimensions.

These two pages will be easier to follow if you refer to the table of the elements and the periodic table in Chapter 17.

How does the Bohr model of atoms help to explain chemical properties? Recall that the elements hydrogen (atomic number $Z = 1$) and lithium ($Z = 3$) are somewhat alike chemically. Both have valences of 1. Both enter into compounds of similar types, for example, hydrogen chloride (HCl), and lithium chloride (LiCl). There are also some similarities in their spectra. All this suggests that the lithium atom resembles the hydrogen atom in some important respects. Bohr speculated that two of the three electrons of the lithium atom are relatively close to the nucleus, in orbits resembling those of the helium atom. But the third electron is in a circular or elliptical orbit outside the inner system. Since this inner system consists of a nucleus of charge $(+)\ 3q_e$ and two electrons each of the charge $(-)\ q_e$, its *net* charge is $(+)\ q_e$. Thus, the lithium atom may be roughly pictured as having a central core of charge $(+)\ q_e$. Around this core one electron revolves, somewhat as for a hydrogen atom. This similar physical structure, then, is the reason for the similar chemical behavior.

Close Up

Lasers

An atom in an excited state gives off energy by emitting a photon, a quantum of electromagnetic radiation, according to Bohr's second postulate (Section 19.5). Although Bohr's specific model of the atom has been vastly extended and incorporated into models based on a different approach (see Chapter 20), this postulate is still valid.

As you have seen, atoms can acquire internal energy, that is, be brought to an excited state, in many ways. In the Franck–Hertz experiment, inelastic collisions provided the energy; in a cool gas displaying a dark-line spectrum, it is the absorption of photons; in a spark or discharge tube, it is collisions between electrons and atoms. There are other mechanisms as well.

Once an atom has acquired internal energy, it can also get rid of it in several ways. An atom can give up energy in inelastic collisions, or (as discussed above) it can emit energy as electromagnetic radiation. There are many different kinds of inelastic collisions; which one an atom undergoes depends as much on its surroundings as on the atom itself.

There are also two different ways an atom can emit radiation. Spontaneous radiation is the kind considered elsewhere in this chapter. At some random (unpredictable) moment, the previously excited atom emits a photon (of frequency ν) and changes its state to one of lower energy (by an amount ΔE). If, however, there are other photons of the appropriate frequency ($\nu = \Delta E/h$) in the vicinity, the atom may be *stimulated* to emit its energy. The radiation emitted is at exactly the same frequency, polarization, and phase as the stimulating radiation. That is, it is exactly in step with the existing radiation; in the wave model of light, you can think of the emission simply increasing the amplitude of the oscillations of the existing electromagnetic field within which the emitting atom finds itself.

Stimulated emission behaves very much like the classical emission of radiation discussed in Chapter 16. A collection of atoms stimulating one another to emit radiation behaves much like an antenna. You can think of the electrons in the different atoms as simply vibrating in step just as they do in an ordinary radio antenna, although much, much faster.

Usually atoms emit their energy spontaneously long before another photon comes along to stimulate them. Most light sources therefore emit incoherent light, that is, light made up of many different contributions, differing slightly in frequency, out of step with each other, and randomly polarized.

Usually, most of the atoms in a group are in the ground state. Light that illuminates the group is more likely to be absorbed than to stimulate any emission, since it is more likely to encounter an atom in the ground state than in the appropriate excited state. But suppose conditions are arranged so that more atoms are in one of the excited states than are in the ground state. (Such a group of atoms is said to be inverted.) In that case, light of the appropriate frequency is more likely to stimulate emission than to be absorbed. Then an interesting phenomenon takes over. Stimulated emission becomes more probable the more light there is around. The stimulated emission from some atoms therefore leads to a chain reaction, as more and more atoms give up some of their internal energy to the energy of the radiation. The incident light pulse has been amplified. Such an arrangement is called a *laser* (*l*ight *a*mplifier using *s*timulated *e*mission of *r*adiation).

Physicists and engineers have developed many tricks for producing "inverted" groups of atoms, on which laser operation depends. Exactly what the tricks are is not important for the action of the laser itself, although without them the laser would be impossible. Sometimes it is possible to maintain the inversion even while the laser is working; that is, it is possible to supply enough energy by the mechanisms that excite the atoms (inelastic collisions with other kinds of atoms, for example) to compensate for the energy emitted as radiation. These lasers can therefore operate continuously.

There are two reasons laser light is very desirable for certain applications. First, it can be extremely *intense*; some lasers can emit millions of joules in minute fractions of a second, as all their atoms emit their stored energy at once. Second, it is *coherent*; the light waves are all in step with each other. Incoherent light waves are somewhat like the waves crisscrossing the surface of a pond in a gale. But coherent waves are like those in a ripple tank, or at a beach where tall breakers arrive rhythmically.

The high intensity of some lasers can be used for applications in which a large amount of energy must be focused on a small spot. Such lasers are used in industries for cutting and welding delicate parts. In medicine, they are used to re-attach the retina (essentially by searing a very small spot) in the eye.

The coherence of lasers is used in applications that require a stable light source emitting light of a precisely given frequency and polarization in one precise direction. Surveyors can use lasers to lay out straight lines, since the coherent beam spreads out very little with distance. Telephone companies can use them to carry signals in the same way they now use radio and microwaves.

These powerful lasers are part of the experimental Shiva fusion device. The energy supplied by the lasers is directed onto a tiny fuel pellet, causing the pellet to implode and thus release energy.

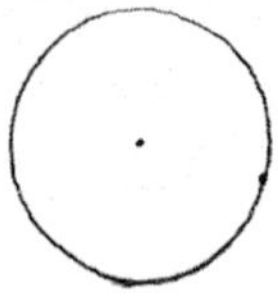

HYDROGEN (Z=1)

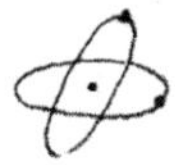

HELIUM (Z=2)

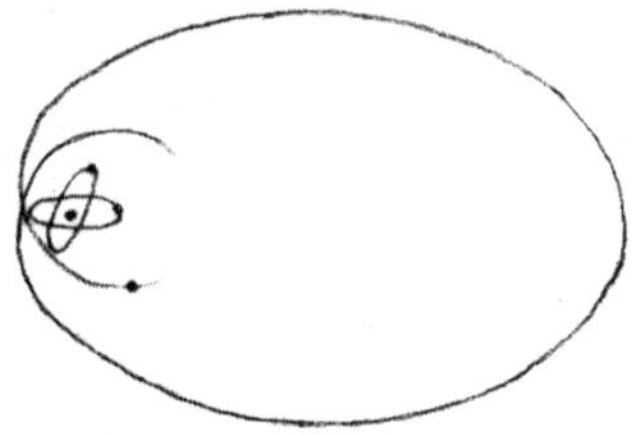

LITHIUM (Z=3)

NEON (Z=10)

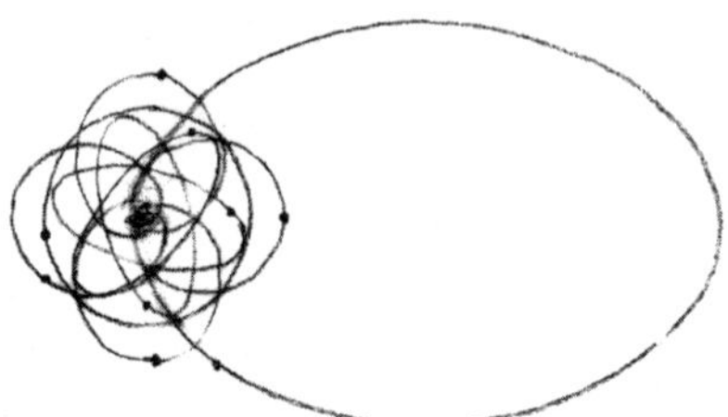

SODIUM (Z=11)

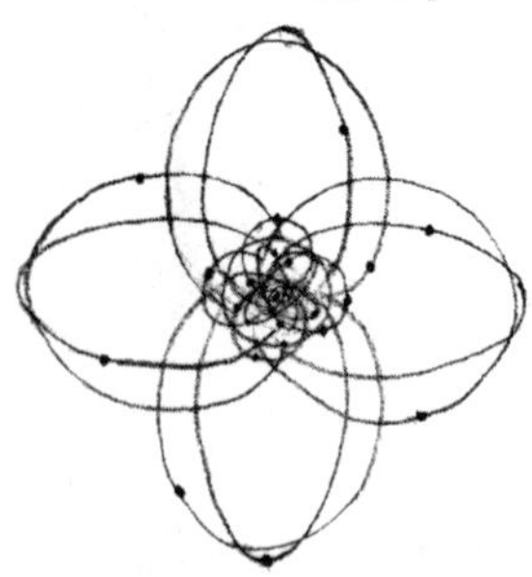

ARGON (Z=18)

The sketches above are based on diagrams Bohr used in his lectures.

Helium ($Z = 2$) is a chemically inert noble gas. These properties indicate that the helium atom is highly stable, having both of its electrons closely bound to the nucleus. It seems sensible, then, to regard both electrons as moving in the same *innermost shell* around the nucleus when the atom is unexcited. Moreover, because the helium atom is so stable and chemically inert, you may reasonably assume that this shell cannot hold more than two electrons. This shell is called the K-shell. The single electron of hydrogen is also said to be in the K-shell when the atom is unexcited. Lithium has two electrons in the K-shell, filling it to capacity; the third electron starts a new shell, called the L-shell. This single outlying and loosely bound electron is the reason why lithium combines so readily with oxygen, chlorine, and many other elements.

Sodium ($Z = 11$) is the next element in the periodic table that has chemical properties similar to those of hydrogen and lithium. This similarity suggests that the sodium atom also is hydrogen-like in having a central core about which one electron revolves. Moreover, just as lithium follows helium in the periodic table, sodium follows the noble gas neon ($Z = 10$). You may assume that two of neon's 10 electrons are in the first (K) shell, while the remaining eight electrons are in the second (L) shell. Because of the chemical inertness and stability of neon, you may further assume that these eight electrons fill the L-shell to capacity. For sodium, then, the 11th electron must be in a third shell, called the M-shell. Passing on to potassium ($Z = 19$), the next alkali metal in the periodic table, you may again picture an inner core and a single electron outside it. The core consists of a nucleus with charge (+) $19q_e$. There are two, eight, and eight electrons occupying the K-, L-, and M-shells, respectively. The 19th electron revolves around the core in a fourth shell, called the N-shell. The atom of the noble gas argon, with $Z = 18$, comes just before potassium in the periodic table. Argon again represents a tight and stable electron pattern, with two in the K-, eight in the L-, and eight in the M-shell.

These qualitative considerations have led to a consistent picture of electrons distributed in groups, or shells, around the nucleus. The arrangement of electrons in the noble gases may be considered particularly stable. For each new alkali metal in Group I of the periodic table, a new shell is started. Each alkali metal atom has a single electron around a core that resembles the pattern for the preceding noble gas. You may expect this outlying electron to be easily "loosened" by the action of neighboring atoms, and this agrees with the facts. The elements lithium, sodium, and potassium are alkali metals. In compounds or in solution (as in electrolysis), they may be considered to be in the form of ions such as Li^+, Na^+, and K^+. Each ion lacks one electron and so has one positive net charge (+) q_e. In the neutral atoms of these elements, the outer electron is relatively free to move about. This property has been used as the basis of a theory of electrical conductivity. According to this theory, a good conductor has many "free" electrons that can form a current under appropriate conditions. A poor conductor has relatively few "free" electrons. The alkali metals are all good conductors. Elements whose electron shells are filled are very poor conductors; they have no "free" electrons. In Chapter 14, you saw how electrical conduction takes place in metals. It is because metals have many "free" electrons that they are conductors.

In Group II of the periodic table, you would expect those elements that follow immediately after the alkali metals to have atoms with two outlying electrons. For example, beryllium ($Z = 4$) should have two electrons in the K-shell, thus filling it, and two in the L-shell. If the atoms of all these elements have two outlying electrons, they should be chemically similar, as indeed they are. Thus, calcium and magnesium, which belong to this group, should easily form ions such as Ca^{++} and Mg^{++}, each with a positive net charge of $(+)\ 2q_e$. This is also found to be true.

As a final example, consider those elements that immediately *precede* the noble gases in the periodic table. For example, fluorine atoms ($Z = 9$) should have two electrons filling the K-shell but only seven electrons in the L-shell, one less than enough to fill it. If a fluorine atom captures an additional

Shell Name	*Number of Electrons in Filled Shell*
K	2
L	8
M	18

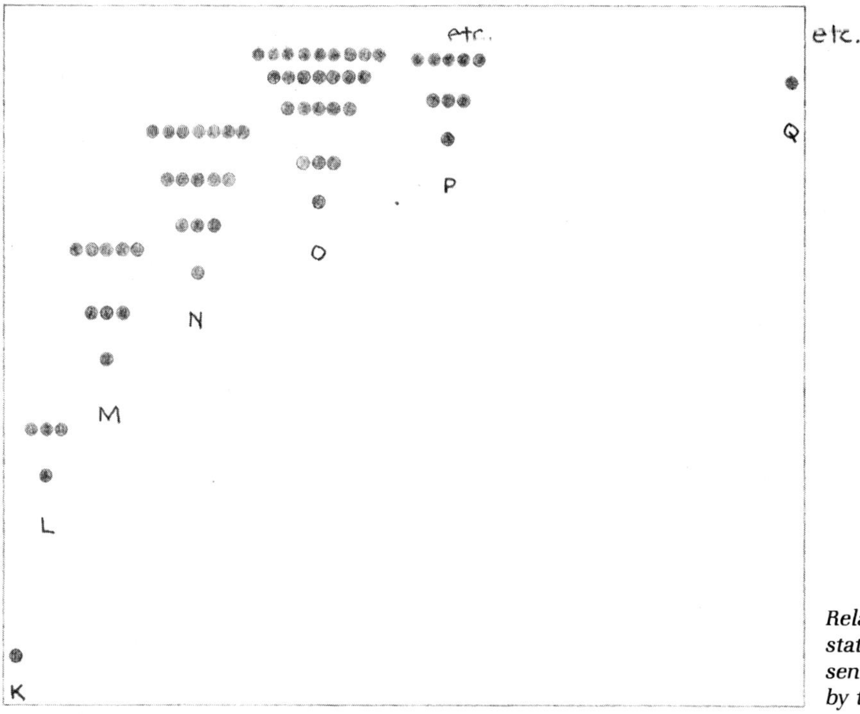

Relative energy levels of electron states in atoms. Each circle represents a state that can be occupied by two electrons.

electron, it should become an F^- ion with one negative net charge. The L-shell would then be filled, as it is for neutral neon ($Z = 10$), and you would expect the F^- ion to be relatively stable. This prediction agrees with observation. Indeed, all the elements immediately preceding the inert gases tend to form stable, singly charged negative ions in solution. In the solid state, you would expect these elements to lack free electrons. In fact, all of them are poor conductors of electricity.

As indicated in the diagram below, the seven main shells, K, L, M, . . . , Q, divide naturally into subshells. The shells fill with electrons so that the total energy of the atom is minimized.

Bohr carried through a complete analysis along these lines. Finally, in 1921, he proposed the form of the periodic table shown on page 593. The periodicity results from the completion of subshells. This phenomenon is complicated even beyond the shell overlap in the figure below by the interaction of electrons in the same subshell. Bohr's table, still useful, was the result of *physical* theory and offered a fundamental *physical* basis for understanding chemistry. For example, it showed how the structure of the periodic table follows from the shell structure of atoms. This was another triumph of the Bohr theory.

SG 17
SG 18

? *18. Why do the next heavier elements after the noble gases easily become positively charged?*

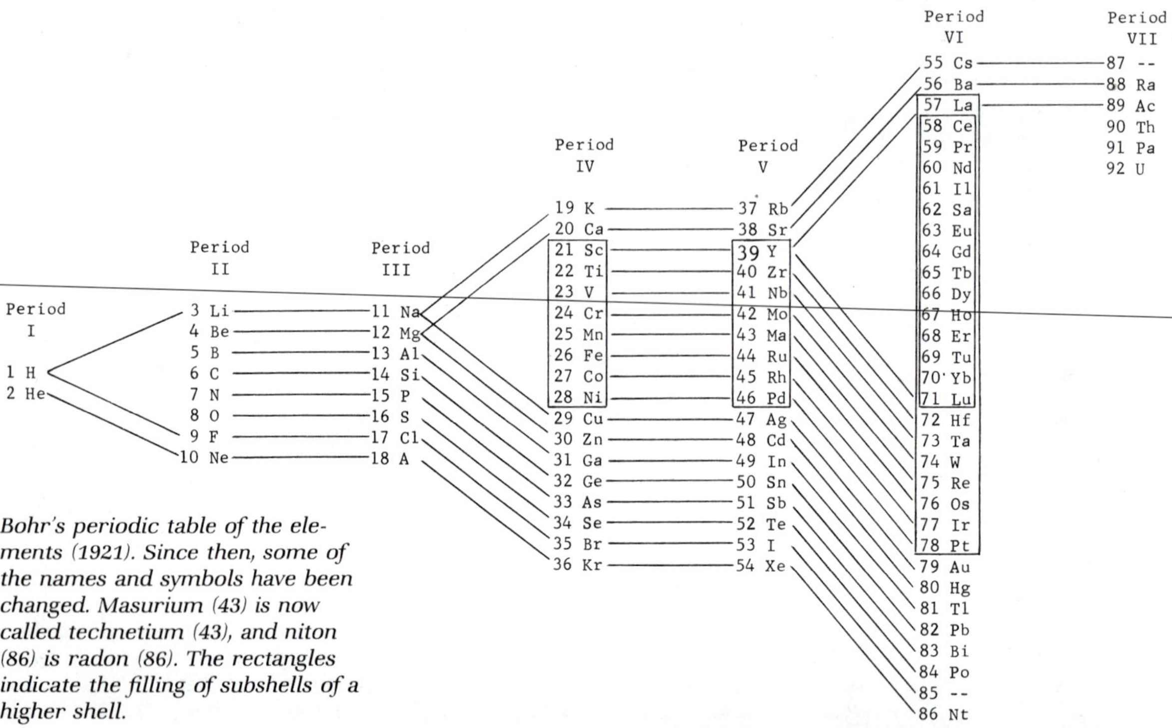

Bohr's periodic table of the elements (1921). Since then, some of the names and symbols have been changed. Masurium (43) is now called technetium (43), and niton (86) is radon (86). The rectangles indicate the filling of subshells of a higher shell.

19. Why are there only two elements in Period I, eight in Period II, eight in Period III, etc.?

19.11 | The inadequacy of the Bohr theory and the state of atomic theory in the early 1920's

Every model and every theory has its limits. The Bohr theory achieved great successes in the years between 1913 and 1924. But problems arose for which the theory proved inadequate. Bohr's theory accounted very well for the spectra of atoms with a single electron in the outermost shell. However, serious differences between theory and experiment appeared in the spectra of atoms with two or more electrons in the outermost shell. Experiments also revealed that when a sample of an element is placed in an electric or magnetic field, its emission spectrum shows additional lines. For example, in a magnetic field each line is split into several lines. The Bohr theory could not account in a quantitative way for some of the observed splittings. Furthermore, the theory supplied no method for predicting the relative brightness of spectral lines. These relative intensities depend on the probabilities with which atoms in a sample undergo transitions among the stationary states. Physicists wanted to be able to calculate the probability of a transition from one stationary state to another. They could not make such calculations with the Bohr theory.

By the early 1920's it was clear that the Bohr theory, despite its great successes, was limited. To form a theory that would solve more problems, Bohr's theory would have to be revised or replaced. But the successes of Bohr's theory showed that a better theory of atomic structure would still have to account for the existence of stationary states, which are separate, distinct atomic energy levels. Therefore, such a theory would have to be based on quantum concepts.

Besides the inability to predict certain properties of atoms at all, the Bohr theory had two additional shortcomings. First, it predicted some results that did not agree with experiment (such as incorrect spectra for elements with two or three electrons in the outermost electron shells). Second, it predicted results that could not be tested in any known way (such as the details of electron orbits). Although orbits were easy to draw on paper, they could not be observed directly. Nor could they be related to any observable properties of atoms. Planetary theory has very different significance when applied to a planet in an observable orbit than when applied to an electron in an atom. The precise position of a planet is important, especially in experiments such

In March 1913, Bohr wrote to Rutherford, enclosing a draft of his first paper on the quantum theory of atomic constitution. On March 20, 1913, Rutherford replied in a letter, the first part of which is quoted here.

"Dear Dr. Bohr:

I have received your paper and read it with great interest, but I want to look it over again carefully when I have more leisure. Your ideas as to the mode of origin of spectra in hydrogen are very ingenious and seem to work out well: but the mixture of Planck's ideas with the old mechanics make it very difficult to form a physical idea of what is the basis of it. There appears to me one grave difficulty in your hypothesis, which I have no doubt you fully realize, namely, how does an electron decide what frequency it is going to vibrate at when it passes from one stationary state to the other. It seems to me that you would have to assume that the electron knows beforehand where it is going to stop. . . ."

as photographing an eclipse or a portion of the surface of Mars from a satellite. But the moment-to-moment position of an electron in an orbit has no such meaning because it has no relation to any experiment physicists have been able to devise. It thus became evident that the Bohr theory led to some questions that could not be answered experimentally.

In the early 1920's, physicists, especially Bohr himself, began to work seriously on revising the basic ideas of the theory. One fact that stood out was that the theory started with a *mixture* of classical and quantum ideas. An atom was assumed to act according to the laws of classical physics up to the point where these laws did not work. Beyond this point, quantum ideas were introduced. The picture of the atom that emerged was an inconsistent mixture. It combined ideas from classical physics with concepts for which there was no place in classical physics. The orbits of the electrons were determined by the classical, Newtonian laws of motion. But of the many theoretical orbits, only a small portion were regarded as possible. Even these few orbits were selected by rules that contradicted classical mechanics. Again, the frequency calculated for the orbital revolution of electrons was quite different from the frequency of light emitted or absorbed when the electron moved from or to this orbit. Also, the decision that n could never be zero was necessary to prevent the model from collapsing by letting the electron fall on the nucleus. It became evident that a better theory of atomic structure would need a more consistent foundation in quantum concepts.

The contribution of the Bohr theory may be summarized as follows. It provided some excellent answers to the questions raised about atomic structure in Chapters 17 and 18. Although the theory turned out to be inadequate, it drew attention to how quantum concepts can be used. It indicated the path that a new theory would have to take. A new theory would have to supply the right answers that the Bohr theory gave. But it would also have to supply the right answers for the problems the Bohr theory could not solve. One of the most fascinating aspects of Bohr's work was the proof that physical and chemical properties of matter can be traced back to the fundamental role of *integers* (quantum numbers such as $n = 1, 2, 3, \ldots$). As Bohr said, "The solution of one of the boldest dreams of natural science is to build up an understanding of the regularities of nature upon the consideration of pure number." You can catch here an echo of the hope of Pythagoras and Plato, of Kepler and Galileo.

Remember, for example (Unit 1), how proudly Galileo pointed out, when announcing that all falling bodies are equally and constantly accelerated: "So far as I know, no one has yet pointed out that the distances traversed, during equal intervals of time, by a body falling from rest, stand to one another in the same ratio as the odd numbers beginning with unity [namely 1:3:5:7: . . .]."

Since the 1920's, a successful theory of atomic structure has been developed and generally accepted by physicists. It is part of *quantum mechanics*, so called because it is built directly on quantum concepts. It goes far beyond understanding atomic structure. In fact, it is the basis of the modern conception of events on a submicroscopic scale. Some aspects of this theory will be discussed in the next chapter. Significantly, Bohr himself was again a leading contributor.

? *20. The Bohr model of atoms is widely presented in science books. What is wrong with it? What is good about it?*

SG 19–23

study guide

1. The *Project Physics* materials particularly appropriate for Chapter 19 include:

Experiment
Spectroscopy

Activities
Measuring Ionization—A Quantum Effect
"Black Box" Atoms

Film Loop
Rutherford Scattering

Transparencies
Alpha Scattering
Energy Levels—Bohr Theory

2. (a) Suggest experiments to show which of the von Fraunhofer lines in the spectrum of sunlight result from abosrption in the sun's atmosphere rather than from absorption by gases in the earth's atmosphere.

(b) How might one decide from spectroscopic observations whether the moon and the planets shine by their own light or by reflected light from the sun?

3. Theoretically, how many series of lines are there in the emission spectrum of hydrogen? In all these series, how many lines are in the visible region?

4. The Rydberg constant for hydrogen, R_H, has the value 1.097×10^7/m. Calculate the wavelengths of the lines in the Balmer series corresponding to $n = 8$, $n = 10$, $n = 12$. Compare the values you get with the wavelengths listed in the table on page 571. Do you see any pattern in the values?

5. (a) As indicated in the figure on page 571, the lines in one of hydrogen's spectral series are bunched very closely at one end. Does the formula

$$\frac{1}{\lambda} = R_H \left(\frac{1}{n_f^2} - \frac{1}{n_i^2} \right)$$

suggest that such bunching will occur?

(b) The "series limit" corresponds to the last possible line(s) of the series. What value should be taken for n_i in the above equation to compute the wavelength of the series limit?

(c) Compute the series limit for the Lyman, Balmer, or Paschen series of hydrogen.

(d) Consider a photon with a wavelength corresponding to the series limit of the Lyman series. What energy could it carry? Express the answer in joules and in electron volts ($1 \text{ eV} = 1.6 \times 10^{-19}$ J).

6. In what ways are Thomson's and Rutherford's atomic models similar? In what ways do they differ?

7. In 1903, the German physicist Philipp Lenard (1864–1947) proposed an atomic model different from those of Thomson and Rutherford. He observed that, since cathode-ray particles can penetrate matter, most of the atomic volume must offer no obstacle to their penetration. In Lenard's model there were no electrons and no positive charges separate from the electrons. His atom was made up of particles called *dynamides*, each of which was an electric dipole possessing mass. (An electric dipole is a combination of a positive charge and a negative charge close together.) All dynamides were identical, and an atom contained as many of them as were needed to make up its mass. They were distributed throughout the volume of the atom. But their radius was so small compared with that of the atom that most of the atom was empty.

(a) In what ways does Lenard's model resemble those of Thomson and Rutherford? In what ways does it differ from those models?

(b) Why would you not expect α particles to be scattered through large angles if Lenard's model were valid?

(c) In view of the scattering of α particles that is observed, is Lenard's model valid?

8. Determine a likely upper limit for the effective size of a gold nucleus from the following facts and hypotheses:

i. A beam of α particles of known velocity $v = 2 \times 10^7$ m/sec is scattered from a gold foil. The manner of this scattering makes sense only if the α particles are repelled by nuclear charges that exert a Coulomb's law repulsion on the α particles.

ii. Some of these α particles come straight back after scattering. They therefore approach the nuclei up to a distance r from the nucleus' center. At this point, the initial kinetic energy $\frac{1}{2}m_\alpha v_\alpha^2$ is completely changed to the potential energy of the system.

iii. The potential energy of a system made up of an α particle of charge $2q_e$ at a distance r from a nucleus of charge Zq_e is given by the product of the "potential" (Zq_e/r) set up by the nucleus at distance r and the charge $(2q_e)$ of the α particle.

iv. The distance r can now be computed. You know v_α, m_α (7×10^{-27} kg, from other evidence), Z for gold atoms (see periodic table), and q_e (see Sec. 14.5).

v. The nuclear radius must be equal to or less than r. Thus, you have a reasonable upper limit for the size of this nucleus.

9. Physicists generally suppose that the atom and the nucleus are each spherical. They assume that the diameter of the atom is of the order of 10^{-10} m and that the diameter of the nucleus is of the order of 10^{-14} m.

(a) What are the evidences that these are reasonable suppositions?

(b) What is the ratio of the diameter of the nucleus to that of the atom?

10. The nucleus of the hydrogen atom is thought to have a radius of about 1.5×10^{-15}m. Imagine this atom magnified so that the nucleus is 0.1 mm across (the size of a grain of dust). How far away from it would the electron be in the Bohr orbit closest to it?

11. Show that the total energy of a neutral hydrogen atom made up of a positively charged nucleus and an electron is given by

$$E_n = \frac{1}{n^2} E_1$$

where E_1 is the energy when the electron is in the first orbit ($n = 1$) and where the value of $E_1 = -13.6$ eV. (You may consult other texts, for example, *Foundation of Modern Physical Science* by Holton and Roller, Sec. 34.4 and 34.7.) Program and hints:

i. The total energy E of the system is the kinetic and potential energy $KE + PE$ of the electron in its orbit. Since $mv^2/r = kq_e^2/r^2$ (see page 582), $KE = \frac{1}{2}mv^2$ can be quickly calculated.

ii. The electrical potential energy PE of a charged point object (electron) is given by the electrical potential V of the region in which it finds itself, times its own charge. The value of V set up by the (positive) nucleus at distance r is given by kq_e/r. The charge on the electron is $-q_e$. Therefore, $PE = -kq_e^2/r$. The meaning of the negative sign is simply that PE is taken to be zero if the electron is infinitely distant. The system radiates energy as the electron is placed closer to the nucleus. On the other hand, energy must be supplied to move the electron away from the nucleus.

iii. Now you can show that the total energy E is

$$E = KE + PE = -k\frac{q_e^2}{2r}$$

iv. Using the equation derived on page 582, namely $r = \dfrac{n^2h^2}{4\pi^2 mq_e^2}$, show that

$$E_n = -\frac{k^2 2\pi^2 m q_e^2}{n^2 h^2} = \frac{1}{n^2} E_1$$

where $E_1 = k^2 2\pi^2 m q_e^2 / h^2$.

The numerical value for this can be computed by using the known values (in consistent units) for k, m, q_e and h.

v. Find the numerical value of the energy of the hydrogen atom for each of the first four allowed orbits ($n = 1, 2, 3, 4$).

vi. As a final point, show that the quantity $-E_1/hc$ has the same value as the constant R_H, as claimed in Sec. 19.8.

12. Using the Bohr theory, how would you account for the existence of the dark lines in the absorption spectrum of hydrogen? Why are all the possible lines not seen?

13. A group of hydrogen atoms is excited (by collision or by absorption of a photon of proper frequency). They all reach the stationary state for which $n = 5$. Refer to the top figure on page 586 and list all possible lines emitted by this sample of hydrogen gas.

14. Make an energy-level diagram to represent the results of the Franck–Hertz experiment.

15. Many substances emit visible radiation when illuminated with ultraviolet light. This phenomenon is an example of fluorescence. Stokes, a British physicist of the nineteenth century, found that in fluorescence, the wavelength of the emitted light usually was the same or longer than the illuminating light. How would you account for this phenomenon on the basis of the Bohr theory?

16. Bohr's model of the hydrogen atom draws ideas from many branches of physics. Make a list of the ideas involved, and indicate whether each idea is in accord with classical physics or at odds with it.

17. Use the chart on page 593 to explain why atoms of potassium ($Z = 19$) have electrons in the N shell even though the M shell is not filled.

18. Use the chart on page 593 to predict the atomic number of the next inert gas after argon; that is, imagine filling the elecron levels with pairs of electrons until you reach an apparently stable, or complete, pattern. Do the same for the next inert gas following.

19. Make up a glossary, with definitions, of terms that appeared for the first time in this chapter.

20. The philosopher John Locke (1632–1704) proposed a science of human nature that was strongly influenced by Newton's physics. In Locke's atomistic view, elementary ideas ("atoms") are produced by elementary sensory experiences and then drift, collide, and interact in the mind. Thus, the formation of ideas was only a special case of the universal interactions of particles.

Does such an approach to the subject of human nature seem reasonable to you? What argument for and against this sort of theory can you think of?

21. In a textbook of physics, the following statement is made:

> Arbitrary though Bohr's new postulate may seem, it was just one more step in the process by which the apparently continuous macroscopic world was being analyzed in terms of a discontinuous, quantized, microscopic world. Although the Greeks had speculated about quantized matter (atoms), it remained for the chemists and physicists of the nineteenth century to give them reality. In 1900 Planck found it necessary to quantize the energy of electromagnetic waves. Also, in the early 1900's a series of experiments culminating in Millikan's oil-drop experiment conclusively showed that electric charge was quantized. To this list of quantized entities, Bohr added angular momentum (the product mvr).

(a) What other properties or things in physics can you think of that are "quantized"?

(b) What properties or things can you think of outside physics that might be said to be "quantized"?

22. Write an essay on the successes and failures of the Bohr model. Can it be called a good model? a simple model? a beautiful model?

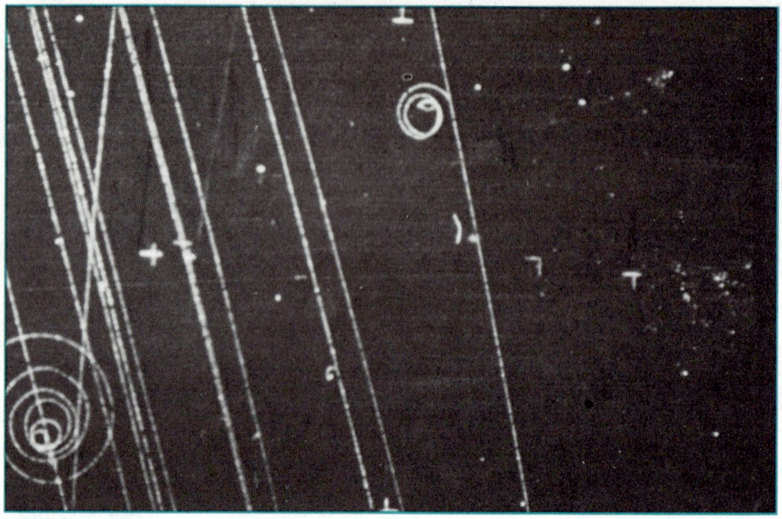

Some Ideas from Modern Physical Theories

20.1 | Some results of relativity theory

SG 1

Progress in atomic and nuclear physics has been based on two great advances in physical thought: quantum theory and relativity. A single chapter could not even begin to give a full account of the actual development of physical and mathematical ideas in these fields. But this chapter can offer you some notion of what kind of problems led to their development, suggest some of the unexpected conclusions, and prepare for material in later chapters. Chapter 20 can also introduce you to the beautiful ideas on relativity theory and quantum mechanics, and to nuclear physics, which is presented in more detail in Unit 6, "*The Nucleus*" and in the supplemental units.

In Chapter 18 and 19 you saw how quantum theory entered into atomic physics. To follow its further development into quantum mechanics, you need to know some of the results of the relativity theory. These results will also be essential to the treatment of nuclear physics. Therefore, this section is devoted to a brief discussion of one essential result of the theory of relativity introduced by Einstein in 1905, the same year in which he published the theory of the photoelectric effect.

Unit 1 discussed the basic idea of relativity. You saw that certain aspects of physical events appear the same from different frames of reference, even if the reference frames are moving with respect to one another. Mass, acceleration, and force seem to be such *invariant* quantities. Thus, Newton's laws relating these quantities should be equally valid in all reference frames.

By 1905, it had become clear that this is true enough for all ordinary cases of motion. But problems arise if the bodies involved move with respect to the observer at a speed more than a few percent of that of light. Einstein wondered whether the relativity principle could be extended to the mechanics of rapidly moving bodies and even to the description of electromagnetic waves. He found that this could be done only by replacing Newton's intuitive definitions of length and time with definitions that produce a more consistent physics. His work resulted in a new viewpoint, and this viewpoint is the most interesting part of Einstein's thinking. But here you will deal with high-speed phenomena from an essentially Newtonian viewpoint. The focus will be on the *corrections* required to make Newtonian mechanics better fit a new range of phenomena.

The diffraction pattern on the left was made by a beam of X rays passing through thin aluminum foil. The diffraction pattern on the right was made by a beam of electrons passing through the same foil. (The center of the pattern on the left is black because the center of the film was blocked off to prevent overexposing the film.)

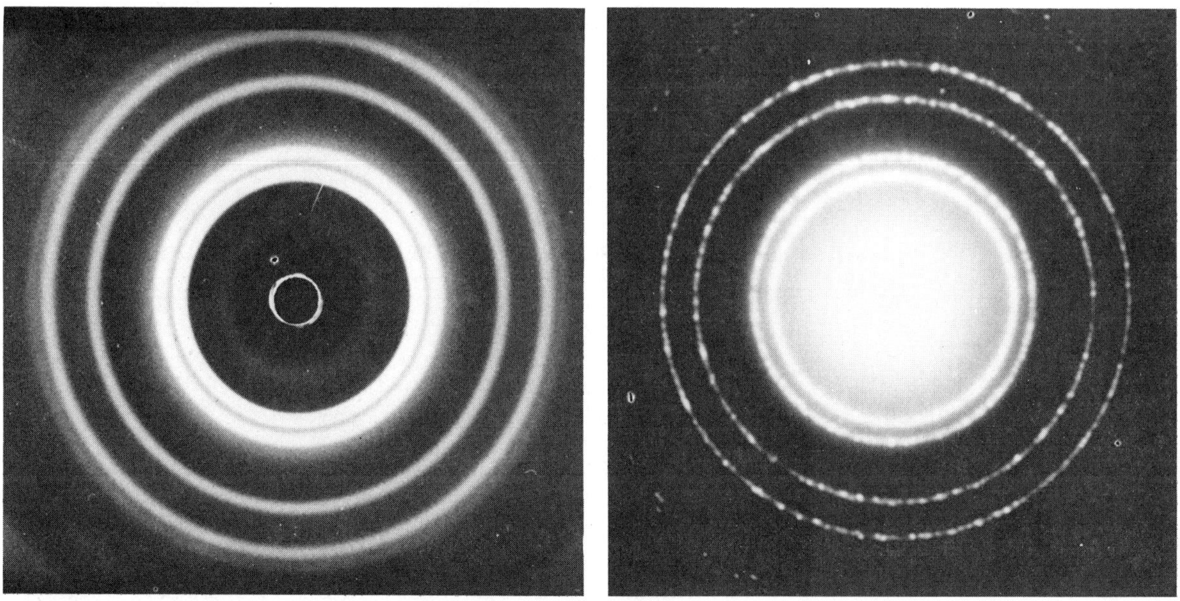

For bodies moving at speeds that are small compared to the speed of light, measurements predicted by relativity theory differ only very slightly from measurements predicted by Newtonian mechanics. You know that this is true because Newton's laws account very well for the motion of bodies with which you are familiar in ordinary life. Differences between relativistic mechanics and Newtonian mechanics become apparent in experiments involving high-speed particles.

You saw in Sec. 18.2 that J. J. Thomson devised a method for determining the speed v and the ratio of charge to mass q/m for electrons. Not long after Thomson discovered the electron, it was found that the value of q_e/m seemed to vary with the speed of the electrons. Between 1900 and 1910, several physicists found that electrons have the value $q_e/m = 1.76 \times 10^{11}$ C/kg only for speeds that are very small compared to the speed of light. As electrons were given greater speeds, the ratio became smaller. Relativity theory offered an explanation: The electron charge is invariant; it does not depend on the speed of the electrons. But the *mass* of an electron, as measured by an observer in a laboratory, should vary with speed. The mass should *increase*, according to the formula:

$$m = \frac{m_0}{\sqrt{1 - v^2/c^2}}$$

In this formula, v is the speed the electron has relative to the observer, while c is the speed of light in a vacuum. The quantity m_0 is the *rest mass*, that is, the electron's mass measured by an observer when the electron is at rest with respect to the observer; m is the mass of an electron measured while it moves with speed v relative to the observer; m may be called the *relativistic* mass. It is the mass determined, for example, by J. J. Thomson's method.

The ratio of relativistic mass to rest mass, m/m_0, is equal to $1/\sqrt{1 - v^2/c^2}$. The table below shows how this ratio varies as values of v/c approach 1. The value of m/m_0 becomes very large as v approaches c.

THE RELATIVISTIC INCREASE OF MASS WITH SPEED

v/c	m/m_0	v/c	m/m_0
0.0	1.000	0.95	3.203
0.01	1.000	0.98	5.025
0.10	1.005	0.99	7.089
0.50	1.155	0.998	15.82
0.75	1.538	0.999	22.37
0.80	1.667	0.9999	70.72
0.90	2.294	0.99999	223.6

The formula for the relativistic mass was derived by Einstein from fundamental ideas of space and time. It has been tested experimentally. Some of the results, for electrons with speeds so high that the value of v reaches about 0.8 c, are graphed below. At $v = 0.8\,c$, the relativistic mass m is about 1.7 times the rest mass m_0. The curve shows the theoretical variation of m as the value of v increases. The agreement of experiment and theory is excellent. The increase in mass with speed precisely accounts for the shrinking of the ratio q_e/m with speed, which was mentioned earlier.

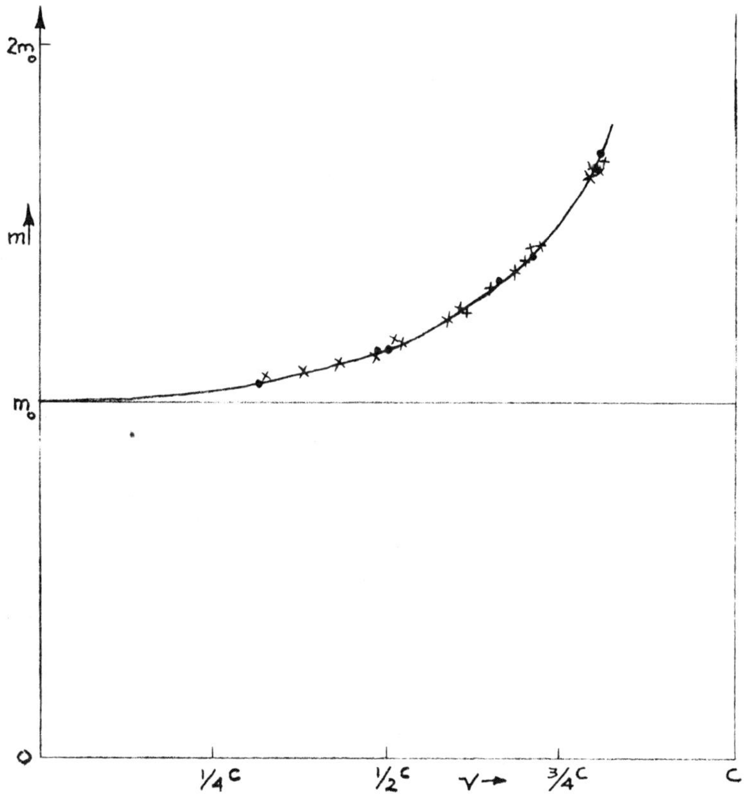

Variation of relativistic mass with speed (expressed as a fraction of the speed of light). The dots and crosses indicate the results of two different experiments.

SG 2
SG 3
SG 4

The formula for variation of mass with speed is valid for *all* moving bodies, not just for electrons and other atomic particles. But the large bodies that you encounter in everyday life move at very small speeds compared to the speed of light. Thus, for such bodies, the value of v/c is very small. The value of v^2/c^2 in the denominator is also very small, and the values of m and m_0 are so nearly the same that you cannot tell the difference. In other words, the relativistic increase in mass can be detected in practice only for particles of atomic or subatomic size. For it is only these particles to which accelerators can give speeds higher than a small fraction of c.

The effects discussed so far are mainly of historical interest because they eventually helped to convince physicists of the correctness of relativity theory. More recent experiments provide more striking evidence of the inadequacy of Newtonian physics

Unit 6 deals further with accelerators. The operation of one of these, the Cambridge Electron Accelerator (CEA) apparatus, is also the subject of the *Project Physics* film "Synchrotron."

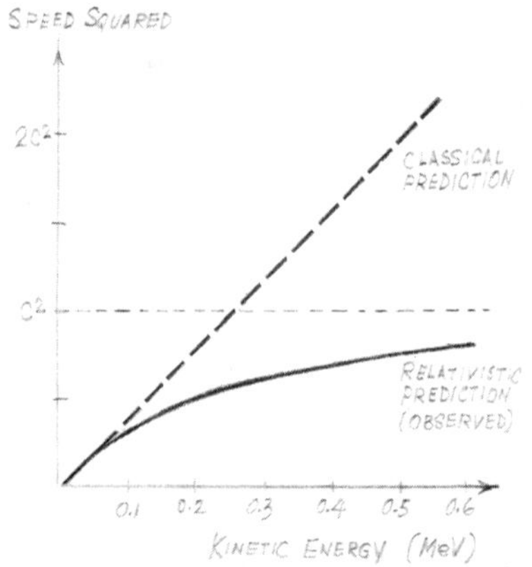

for particles with very high speeds. Electrons can be given very high energies by accelerating them in a vacuum by means of a high voltage V. Since the electron charge q_e is known, the energy increase, q_eV, is known. The rest mass m_0 of an electron is also known (see Sec. 18.3), and the speed v can be measured by timing the travel over a known distance. It is therefore possible to compare the value of the energy supplied, q_eV, with the expression for kinetic energy in classical mechanics, $\frac{1}{2}m_0v^2$. Experiments of this kind have shown that when electrons have speeds that are small compared to the speed of light, it is correct to write $\frac{1}{2}m_0v^2 = q_eV$. This relation was used in Sec. 18.5 in discussing the photoelectric effect. This could be done correctly because photoelectrons do indeed have small speeds, and m and m_0 have nearly the same value. But when the speed of the electron becomes so large that v/c is no longer a small fraction, the quantity $\frac{1}{2}m_0v^2$ no longer increases in proportion to q_eV. This disagreement increases as q_eV increases. The increase in kinetic energy still is equal to the amount of work done by the electrical field, q_eV. But the mass is no longer m_0, and so kinetic energy cannot be measured by $\frac{1}{2}m_0v^2$ or even by $\frac{1}{2}mv^2$. The value of v^2, instead of steadily increasing with energy supplied, approaches a limiting value: c^2.

One of several accelerators of its kind is the Stanford Linear Accelerator (SLAC), operated in California by Stanford University. In it electrons are accelerated to an energy that is equivalent to what they would gain in being accelerated by a potential difference of 10^{10} V. This is an enormous energy for electrons. The speed attained by the electrons is 0.999999999 c. At this speed, the relativistic mass m (both by calculation and by experiment) is over 10,000 times greater than the rest mass m_0!

Another way of saying mass increases with speed is this: *Any increase in kinetic energy is accompanied by an increase in mass.* If the kinetic energy measured from a given frame of reference is KE, the increase in mass Δm (above the rest mass) measured in that frame turns out to be proportional to KE:

$$\Delta m \propto KE$$

It takes a great deal of kinetic energy to give a measurable increase in mass. The proportionality constant is very small. In fact, Einstein showed that the constant would be $1/c^2$, where c is the speed of light in a vacuum:

$$\Delta m = \frac{KE}{c^2}$$

Thus, the total mass m of a body is its rest mass, m_0 plus KE/c^2:

$$m = m_0 + \frac{KE}{c^2}$$

To increase the mass of a body by 1 g, it would have to be given a kinetic energy of 10^{14} J.

Einstein proposed that the "mass equivalent" of kinetic energy is only a special case. In general, there should be a precise equivalence between mass and energy. Thus, the rest mass m_0 should correspond to an equivalent amount of "rest energy" E_0 such that $m_0 = E_0/c^2$; that is,

The rest energy m_0c^2 includes the potential energy, if there is any. Thus, a compressed spring has a somewhat larger rest mass and rest energy than the same spring when relaxed.

$$m = \frac{E_0}{c^2} + \frac{KE}{c^2}$$

Using the symbol E for the *total* energy of a body, $E = E_0 + KE$,

Do not confuse E with the symbol for electric field.

$$m = \frac{E}{c^2}$$

This is just what Einstein concluded in 1905: "The mass of a body is a measure of its energy content." This relation can be written in a more familiar form, which is probably the most famous equation in physics:

$$E = mc^2$$

SG 5, 6

The last four equations all represent the same idea: Mass and energy are different expressions for the *same* characteristic of a system. You should not think of mass as being "converted" to energy, or energy to mass. Rather, a body with a measured mass m *has* an energy E equal to mc^2. And a body of total energy E *has* a mass equal to E/c^2.

This equivalence has exciting significance. First, two great conservation laws become alternate statements of a single law: In any system whose total mass is conserved, the total energy is conserved also. Second, the idea arises that some of the rest energy might be transformed into a more familiar form of energy. Since the energy equivalent of mass is so great, a very small reduction in rest mass would release a tremendous amount of energy, for example, kinetic energy or electromagnetic radiation.

In Unit 6, you will see how such changes come about experimentally. Unit 6 will discuss additional experimental evidence that supports this relationship.

?

1. What happens to the measurable mass of a particle as its kinetic energy is increased?

2. What happens to the speed of a particle as its kinetic energy is increased?

20.2 | Particle-like behavior of radiation

One of these mass–energy relations can be used to look at light quanta and their interaction with atoms from a somewhat different point of view than that used in discussing the photoelectric effect on Bohr's model. Study of the photoelectric effect showed that a light quantum has energy hf, where h is Planck's constant and f is the frequency of the light. This concept also applies to X rays which, like visible light, are electromagnetic radiation, but of higher frequency than visible light. The photoelectric effect, however, did not tell anything about the *momentum* of a quantum. If a light quantum has energy, does it also have momentum?

SG 7 The magnitude of the momentum $\vec{p}$ of a body is defined as the product of its mass m and speed v: $p = mv$. Replacing m with its energy equivalent E/c^2 gives

$$p = \frac{Ev}{c^2}$$

Note that this equation is an expression for momentum, but that it contains no direct reference to mass. Now suppose this same equation is applied to the momentum of a photon of energy E. Since a photon moves at the speed of light, v would be replaced by the speed of light c to give

$$p = \frac{Ec}{c^2} = \frac{E}{c}$$

Remember, $E = hf$ for a light quantum. If you substitute this expression for E in $p = E/c$, you get the momentum of a light quantum:

$$p = \frac{hf}{c}$$

Or, using the wave relation that the speed equals the frequency times the wavelength, $c = f\lambda$, you can express the momentum as

$$p = \frac{h}{\lambda}$$

SG 8 Does it make sense to define the momentum of a photon in this way? It does, if the definition helps in understanding experimental results. The first successful use of this definition was in the analysis of an effect discovered by Arthur H. Compton. A review of Compton's work is given below.

Consider a beam of light (or X rays) striking the atoms in a target (such as a thin sheet of metal). According to classical electromagnetic theory, the light will be scattered in various directions, but its frequency will not change. The absorption of light of a certain frequency by an atom may be followed by reemission of light of a different frequency. But if the light wave is simply *scattered*, then according to classical theory the frequency should not change.

According to quantum theory, however, light is made up of photons. According to relativity theory, photons have momentum. Therefore, Compton reasoned, in a collision between a photon and an atom, the law of conservation of momentum should apply. According to this law (see Chapter 9), when a body of small mass collides with a massive object at rest, it simply bounces back or glances off. It experiences very little loss in speed and so very little change in energy. But if the masses of the two colliding objects are not very different, a significant amount of energy can be transferred in the collision. Compton calculated how much energy a photon should lose in a collision with an atom, if the photon's momentum is hf/c. He concluded that the change in energy is too small to observe if a photon simply bounces off an entire atom. But if a photon strikes an *electron*, which has a small mass, the photon should transfer a significant amount of energy to the electron.

In 1923, Compton was able to show that X rays did in fact behave like particles with momentum $p = h\lambda$ when they collided with electrons. Compton measured the frequency of the incident and scattered X rays and thus was able to determine the X-ray photon's change in momentum. By measuring the momentum of the scattered electron, he was able to verify that $p = h\lambda$ by using the law of conservation of momentum (see page 257 in Sec. 9.3). For this work, Compton received the Nobel Prize in 1927.

Arthur H. Compton (1892–1962)

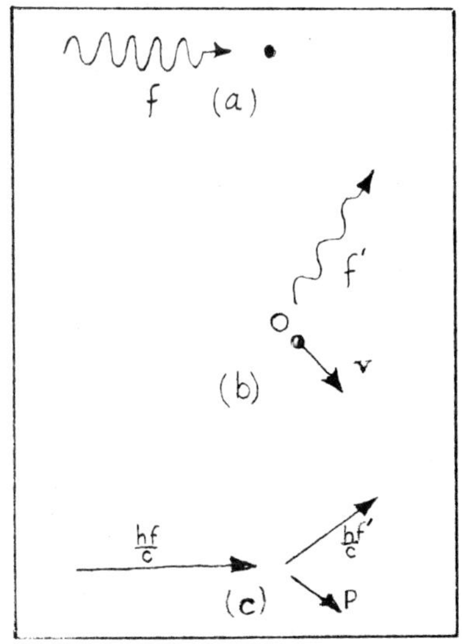

Compton's experiment: (a) X ray of frequency f *approaches an electron; (b) X ray is scattered, leaving at lower frequency* f′, *and electron recoils at velocity* $\vec{v}$. *(c) The momentum before "collision" (hf/c) is equal to the vector sum of the momentum afterwards.*

Compton's experiment showed that a photon can be regarded as a particle with a definite momentum as well as energy. It also showed that collisions between photons and electrons obey the laws of conservation of momentum and energy.

As noted in Sec. 18.5 in the discussion of the photoelectric effect, light has particle-like properties. The Compton effect gave additional evidence for this fact. To be sure, photons are not like ordinary particles, if only because photons do not exist at speeds other than that of light. (There can be no resting photons and, therefore, no rest mass for photons.) But in other ways, as in their scattering behavior, photons act much like particles of matter. For example, they have momentum as well as energy. Yet they also act like waves, having frequency and wavelength. In other words, electromagnetic radiation in some experiments exhibits behavior similar to what is thought of as particle behavior. In other experiments, its behavior is similar to what is thought of as wave behavior. This pattern of behavior is often referred to as the *wave–particle dualism of radiation*. Is a photon a wave or a particle? The only answer is that it can *act* like either, depending on what is being done with it.

SG 9
SG 10

? *3. How does the momentum of a photon depend on the frequency of the light? Why was momentum conservation not* *considered in the discussion of the photoelectric effect on Bohr's model?*

4. What did Compton do, and what did the experiment prove?

20.3 | Wave-like behavior of particles

In 1923, the French physicist Louis de Broglie suggested that the wave–particle dualism that applies to radiation might also apply to electrons and other atomic particles. Perhaps, he said, the wave–particle dualism is a fundamental property of all quantum processes. If so, particles that were always thought of as material particles can, in some circumstances, act like waves. De Broglie sought an expression for the wavelength that might be associated with wave-like behavior of an electron. He found such an expression by means of a simple argument.

The momentum of a photon of wavelength λ is $p = h/\lambda$. De Broglie thought that this relation might also apply to electrons with the momentum $p = mv$. He therefore boldly suggested that the wavelength of an electron is

The "de Broglie wavelength" of a material particle does not refer to anything having to do with light, but to some new wave property associated with the motion of matter itself. Sometimes these waves are called "pilot waves."

$$\lambda = \frac{h}{mv}$$

where m is the electron's mass and v its speed.

What does it mean to say that an electron has a wavelength equal to Planck's constant divided by its momentum? If this statement is to have any physical meaning, it must be possible to test it by some kind of experiment. Some wave property of the electron must be measured. The first such property to be measured was *diffraction*.

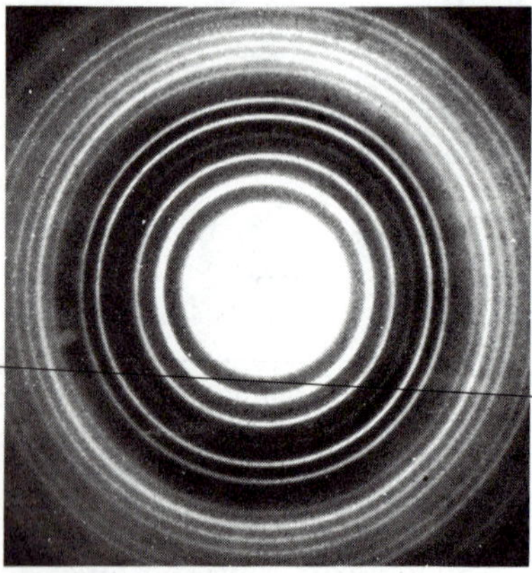

Diffraction pattern produced by directing a beam of electrons through polycrystalline aluminum (that is, many small crystals of aluminum oriented at random). With a similar pattern, G. P. Thomson demonstrated the wave properties of electrons 28 years after their particle properties were first demonstrated by J. J. Thomson, his father.

The relationship $\lambda = h/mv$ indicates that the wavelengths associated with electrons will be very short, even for fairly slow electrons. An electron accelerated across a potential difference of only 100 V would have a wavelength of only 10^{-10} m. So small a wavelength would not give measurable diffraction effects on encountering even a microscopically small object (say, 10^{-5} m).

By 1920, it was known that crystals have a regular lattice structure. The distance between rows of planes of atoms in a crystal is about 10^{-10} m. After de Broglie proposed that electrons have wave properties, several physicists suggested that the existence of electron waves might be shown by using crystals as diffraction gratings. Experiments begun in 1923 by C. J. Davisson and L. H. Germer in the United States yielded diffraction patterns similar to those obtained for X rays (see Sec. 18.6). Their method is illustrated in the drawing at the top of page 610. The experiment showed two things. First, electrons *do* have wave properties. One may say that an electron moves along the path taken by the de Broglie wave that is associated with the electron.

Close Up

The de Broglie Wavelength: Examples

A body of mass 1 kg moves with a speed of 1 m/sec. What is the de Broglie wavelength?

$$\lambda = \frac{h}{mv}$$

$$h = 6.6 \times 10^{-34} \text{ J-sec}$$

$$mv = 1 \text{ kg-m/sec}$$

$$\lambda = \frac{6.6 \times 10^{-34} \text{ J-sec}}{1 \text{ kg-m/sec}}$$

so

$$\lambda = 6.6 \times 10^{-34} \text{ m}$$

The de Broglie wavelength is many orders of magnitude smaller than an atom. Thus, it is much too small to be detected. There are, for example, no slits or obstacles small enough to show diffraction effects. You would expect to detect no wave aspects in the motion of this body.

An electron mass 9.1×10^{-31} kg moves with a speed of 2×10^{6} m/sec. What is its de Broglie wavelength?

$$\lambda = \frac{h}{mv}$$

$$h = 6.6 \times 10^{-34} \text{ J-sec}$$

$$mv = 1.82 \times 10^{-24} \text{ kg-m/sec}$$

$$\lambda = \frac{6.6 \times 10^{-34} \text{ J-sec}}{1.82 \times 10^{-24} \text{ kg-m/sec}}$$

so

$$\lambda = 3.6 \times 10^{-10} \text{ m}$$

The de Broglie wavelength is of atomic dimensions. For example, it is of the same order of magnitude as the distances between atoms in a crystal. So you can expect to see wave aspects in the interaction of electrons with crystals.

Prince Louis Victor de Broglie (1892–), whose ancestors served the French kings as far back as the time of Louis XIV, was educated at the Sorbonne in Paris. He proposed the idea of wave properties of electrons in his PhD thesis.

(a) (b) (c)

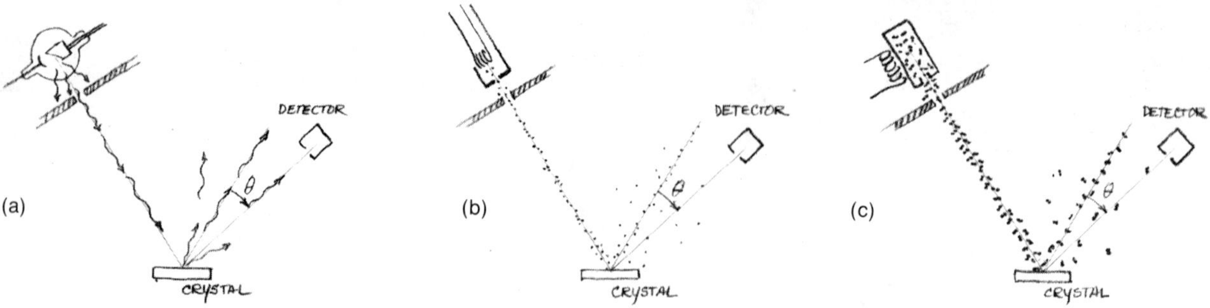

(d)

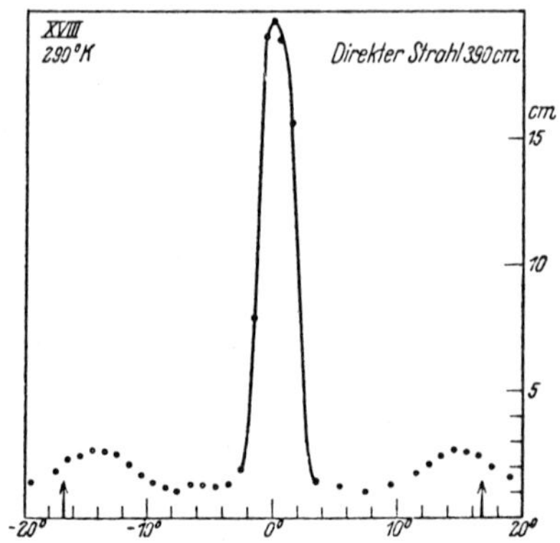

a. One way to demonstrate the wave behavior of X rays is to direct a beam at the surface of a crystal. The reflections from different planes of atoms in the crystal interfere to produce reflected beams at angles other than the ordinary angle of reflection.
b. A very similar effect can be demonstrated for a beam of electrons. The electrons must be accelerated to an energy that corresponds to a de Broglie wavelength at about 10^{-10} m. This would require an accelerating voltage of only about 100 volts.
c. Like any other beam of particles, a beam of molecules directed at a crystal will show a diffraction pattern. Diagram (c) shows how a beam of hydrogen molecules (H_2) can be formed by slits at the opening of a heated chamber. The average energy of the molecules is controlled by adjusting the temperature of the oven.
d. Diffraction pattern for H_2 molecules glancing off a crystal of lithium fluoride. The graph, reproduced from Zeitschrift für Physik *(1930), shows results obtained by I. Estermann and O. Stern in Germany. The detector reading is plotted against the deviation to either side of the angle of ordinary reflection. A low but distinct peak owing to diffraction is seen to each side of the ordinary reflection beams.*

SG 11
SG 12
SG 13

Second, electron wavelengths are correctly given by de Broglie's relation, $\lambda = h/mv$. These results were confirmed in 1927 when G. P. Thomson directed an electron beam through thin gold foil. Thomson found a pattern like the one on page 608. It resembles diffraction patterns produced by light beams going through thin slices of materials. By 1930, diffraction from crystals had been used to demonstrate the wave-like behavior of helium *atoms* and hydrogen *molecules*. (See the drawings above.)

According to de Broglie's hypothesis, wave–particle dualism is a general property not only of radiation but also of matter. This has been confirmed by all experiments. Scientists now customarily refer to electrons and photons as "particles" while recognizing that both have properties of waves as well. (Of course, there are also important differences between them.)

You will recall Bohr's postulate that the quantity *mvr* (called the angular momentum) of the electron in the hydrogen atom can have only certain values. De Broglie's relation, $\lambda = h/mv$, has an interesting yet simple application that supports this postulate. Bohr assumed that *mvr* can have only the values

$$mvr = n\frac{h}{2\pi} \qquad \text{where } n = 1, 2, 3, \ldots$$

Now, suppose that an electron wave is somehow spread over an orbit of radius *r* so that, in some sense, it "occupies" the orbit. Can *standing waves* be set up as indicated, for example, in the sketch in the margin? If so, the circumference of the orbit must be equal in length to a whole number of wavelengths, that is, to $n\lambda$. The mathematical expression for this condition of "fit" is:

$$2\pi r = n\lambda$$

Replacing λ by h/mv according to de Broglie's relation gives

$$2\pi r = n\frac{h}{mv}$$

or

$$mvr = n\frac{h}{2\pi}$$

This is just Bohr's quantization condition! The de Broglie relation for electron waves, and the idea that electrons have orbits that allow standing waves, allows you to *derive* the quantization that Bohr had to *assume*.

SG 14

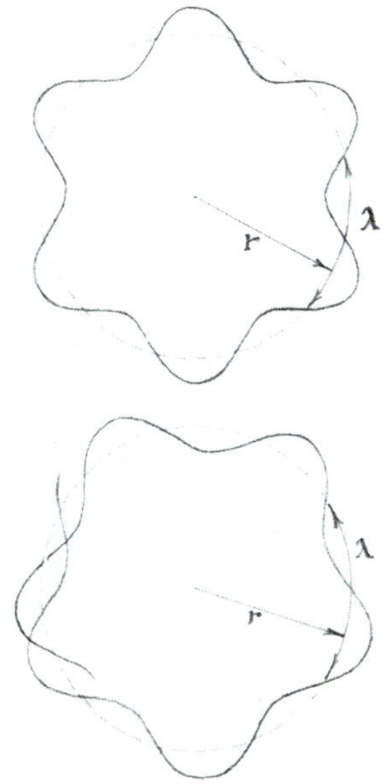

Only certain wavelengths will "fit" around a circle.

The result obtained indicates that you may picture the electron in the hydrogen atom in two ways. You may think of it as a particle moving in an orbit with a certain quantized value of mvr; or you may picture it as a standing de Broglie-type wave occupying a certain region around the nucleus.

Either way is incomplete by itself.

?

5. Where did de Broglie get the relation $\lambda = h/mv$ *for electrons?*

6. Why were crystals used to get diffraction patterns of electrons?

20.4 | Mathematical versus visualizable atoms

P. A. M. Dirac (1902–), an English physicist, was one of the developers of modern quantum mechanics. In 1932, at the age of 30, Dirac was appointed Lucasian Professor of Mathematics at Cambridge University, the post held by Newton.

It was now clear that "things" (electrons, atoms, molecules) long regarded as particles also show properties of waves. This fact is the basis for the presently accepted theory of atomic structure. This theory, *quantum mechanics*, was introduced in 1925. Its foundations were developed very rapidly during the next few years, primarily by Heisenberg, Born, Schrödinger, Bohr, and Dirac. At first, the theory appeared in two different mathematical forms, proposed independently by Heisenberg and Schrödinger. A few months later, these two forms were shown by Schrödinger to be equivalent, different ways of expressing the same relationships. Schrödinger's form of the theory is closer to the ideas of de Broglie (discussed in the last section). It is often referred to as *wave mechanics*.

Schrödinger sought to express the dual wave–particle nature of matter mathematically. Maxwell had formulated the

Max Born (1882–1969) was born in Germany, but left that country for England in 1933 when Hitler and the Nazis gained control. Born was largely responsible for introducing the statistical interpretation of wave mechanics.

Erwin Schrödinger (1887–1961) was born in Austria. He developed wave mechanics in 1926 and then fled from Germany in 1933 when Hitler and the Nazis came to power. From 1940 to 1956, when he retired, he was professor of physics at the Dublin Institute for Advanced Studies.

electromagnetic theory of light in terms of a wave equation. Physicists were familiar with this theory and its applications. Schrödinger reasoned that the de Broglie waves associated with electrons could be described in a way analogous to the classical waves of light. Thus, there should be a wave equation that holds for matter waves, just as there is a wave equation for electromagnetic waves. This mathematical part of wave mechanics cannot be discussed adequately without using advanced mathematics, but the physical ideas involved require only a little mathematics and are essential to understanding modern physics. Therefore, the rest of this chapter will discuss some of the physical ideas of the theory so as to make them seem reasonable. Some of the results of the theory and some of the significance of these results will also be considered.

Schrödinger successfully derived an equation for the "matter waves" that are associated with moving electrons. This equation, which has been named after him, defines the wave properties of electrons and also predicts particle-like behavior. The Schrödinger equation for an electron bound in an atom has a solution only when a constant in the equation has the whole-number values 1, 2, 3. . . . These numbers correspond to different energies. Thus, the Schrödinger equation predicts that only certain electron energies are possible in an atom. In the hydrogen atom, for example, the single electron can be in *only* those states for which the energy of the electron has the numerical values:

$$E_n = \frac{k^2 2\pi^2 m q_e^2}{n^2 h^2}$$

with n having only whole-number values. These are just the energy values that are found experimentally and just the ones given by the Bohr theory! In Schrödinger's theory, this result follows directly from the mathematical formulation of the wave and particle nature of the electron. The existence of these stationary states is not assumed, and no assumptions are made about orbits. The new theory yields all the results of the Bohr theory, with none of the Bohr theory's inconsistent hypotheses. The new theory also accounts for the experimental information for which the Bohr theory failed to account. For instance, it deals with the probability of an electron changing from one energy state to another.

On the other hand, quantum mechanics does not supply a physical model or visualizable "picture" of the atom. The planetary model of the atom has been given up, but has not been replaced by another simple picture. There is now a highly successful *mathematical* model, but no easily visualized *physical* model. The concepts used to build quantum mechanics are more abstract than those of the Bohr theory. Thus, it is difficult to get a "feeling" for atomic structure without training in the field. But the mathematical theory of quantum mechanics is much more powerful than the Bohr theory in predicting and explaining phenomena. Many problems that were previously unsolvable have been solved with quantum mechanics. Physicists have learned that the world of atoms, electrons, and photons cannot be thought of in the same mechanical terms as the world of everyday experience. Instead, the study of atoms presents some fascinating new concepts, which will be discussed in the next two sections. What has been lost in easy visualizability is made up for by an increase in fundamental understanding.

What does it mean to "visualize" or "picture" something? One answer is that it means relating an abstract idea to something that you are familiar with from everyday life; for example, a particle is like a baseball or a marble. But why should there be anything from everyday life that is exactly like an electron or an atom?

?

7. The set of energy states of hydrogen could be derived from Bohr's postulate that mvr = nh/2π. *In what respect was the derivation from Schrödinger's equation better?*

8. Quantum (or wave) mechanics has had great success. What is its drawback for those trained on physical models?

20.5 | The uncertainty principle

Up to this point, it has been assumed that any physical property can be measured as accurately as necessary. To reach any desired degree of accuracy would require only a sufficiently precise instrument. Wave mechanics showed, however, that even in thought experiments with ideal instruments there are limits to the accuracy that can be achieved.

Werner Heisenberg (1901–1976), a German physicist, was one of the developers of modern quantum mechanics (at the age of 23). He was the first to state the uncertainty principle. After the discovery of the neutron in 1932, he proposed the proton–neutron theory of nuclear structure.

Think how you would go about measuring the positions and velocity of a car moving slowly along a driveway. You could mark the position of the front end of the car at a given instant by making a scratch on the ground. At the same time, you could start a stopwatch. Then you could run to the end of the driveway, where you have previously placed another mark. At the instant when the front of the car reaches this point, you stop the watch. You then measure the distance between the marks and get the average speed of the car by dividing the distance traveled by the time elapsed. Since you know the direction of the car's motion, you know the average velocity. Thus, you know that at the moment the car reached the second mark it was at a certain distance from its starting point and had traveled at a certain average velocity. By going to smaller and smaller intervals, you could also get the instantaneous velocity at any point along its path.

How did you get the needed information? You located the car by sunlight bounced off the front end into your eyes. The light permitted you to see when the car reached a mark on the ground. To get the average speed, you had to locate the front end twice.

$$\lambda = \frac{c}{f} = \frac{3 \times 10^8 \text{ m/sec}}{10^6 \text{ sec}} = 300 \text{ m}$$

The extreme smallness of the atomic scale is indicated by these pictures made with techniques that are near the very limits of magnification, about 10,000,000 times in these reproductions. Below right: Pattern produced by charged particles repelled from the tip of a microscopically thin tungsten crystal. The entire section shown is only about 10 nanometers across. The finest detail that can be revealed by this "field–ion microscope" is about 0.1 nanometer. The bright spots indicate the locations of atoms along edges of the crystal, but should not be thought of as pictures of the atoms.

But suppose that you had decided to use reflected radio waves instead of light of visible wavelength. At 1,000 kHz, a typical value for radio signals, the wavelength is 300 m. This wavelength is very much greater than the dimensions of the car. Thus, it would be impossible to locate the position of the car with any accuracy. The wave would reflect from the car ("scatter" is a better term) in all directions. It would also sweep around any human-sized device you may wish to use to detect the wave direction. The wavelength has to be comparable with or smaller than the dimensions of the object before the object can be located well. *Radar* uses wavelengths from about 0.1 cm to about 3 cm, so a radar apparatus could be used instead of sunlight. But even radar would leave uncertainties as large as several centimeters in the two measurements of position. The wavelength of visible light is less than 10^{-6} m. For visible light, then, you could design instruments that would locate the position of the car to an accuracy of a few thousandths of a millimeter.

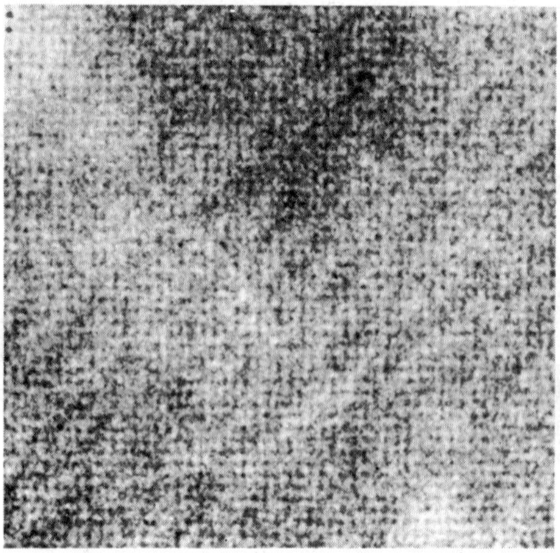

Above: Pattern produced by electron beam scattered from a section of a single gold crystal. The entire section of crystal shown is only 10 nanometers across. This is smaller than the shortest wavelength of ultraviolet light that could be used in a light microscope. The finest detail that can be resolved with this "electron microscope" is just under 0.2 nanometer. So the layers of gold atoms (spaced slightly more than 0.2 nanometer) show as a checked pattern; individual atoms are beyond the resolving power.

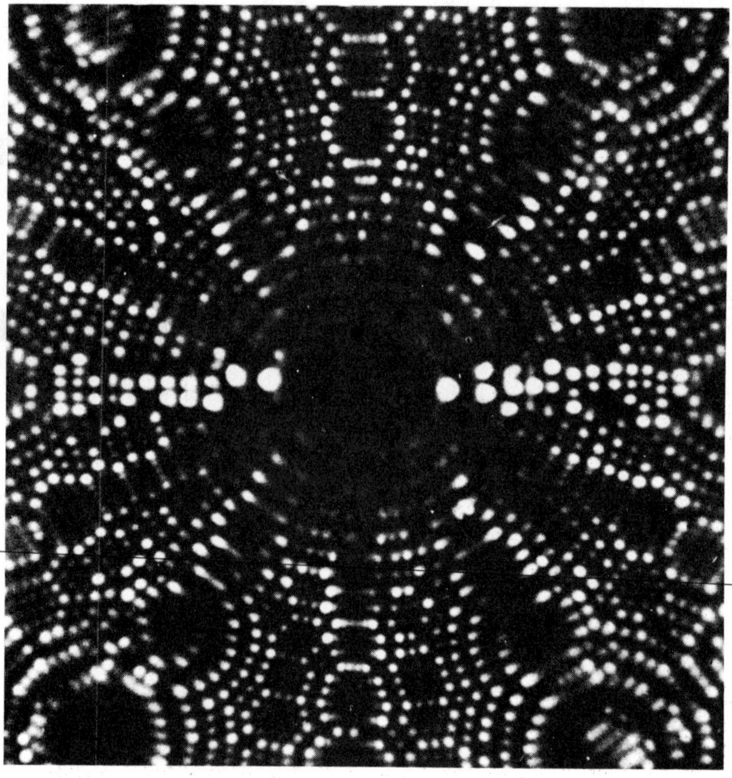

For more discussion of this problem of resolutions, refer to Unit 4.

Now think of an electron moving across an evacuated tube. You will try to measure the position and speed of the electron. But you must change your method of measurement. The electron is so small that you cannot locate its position by using visible light. (The wavelength of visible light, small as it is, is still at least 10^4 times greater than the diameter of an atom.)

You are attempting to locate an electron within a region the size of an atom (about 10^{-10} m across). So you need a light beam whose wavelength is about 10^{-10} or smaller. But a photon of such a short wavelength λ (and high frequency f) has very great momentum (h/λ) and energy (hf). Recalling Compton's work (Sec. 20.2), you know that such a photon will give the electron a strong kick when it is scattered by the electron. As a result, the velocity of the electron will be greatly changed, into a new and unknown direction. (This is a new problem, one you did not even think about when measuring the position of the car!) Therefore, when you receive the scattered photon, you can deduce from its direction where the electron *once was*; in this sense you can "locate" the electron. But in the process you have changed the velocity of the electron (in both magnitude and direction). In short, the more accurately you locate the electron (by using photons of shorter wavelength), the less accurately you can know its velocity. You could try to disturb the electron less by using less energetic photons. But because light exists in quanta of energy hf, a *lower*-energy photon will have a *longer* wavelength. This would create greater uncertainty about the electron's *position*!

SG 15

In other words, *it is impossible to measure both the position and velocity of an electron to unlimited accuracy*. This conclusion is expressed in the *uncertainty principle*, first stated by Werner Heisenberg. The uncertainty principle can be expressed quantitatively in a simple formula, derived from Schrödinger's wave equation for the motion of particles. Let Δx represent the uncertainty in position, and Δp the uncertainty in momentum. The product of these two uncertainties must be equal to, or greater than, Planck's constant divided by 2π:

The chief use made of the uncertainty principle is in general arguments in atomic theory rather than in particular numerical problems. Physicists do not really need to know exactly where an electron is. But they sometimes want to know if it *could* be in some *region* of space.

$$\Delta x \Delta p \geq \frac{h}{2\pi}$$

The same reasoning (and equation) holds for the experiment on the car. But the limitation has no practical consequence with such a massive object. (See the worked-out example on page 616.) It is only on the atomic scale that the limitation becomes evident and important.

Sg 16–18

?

9. If photons used in finding the velocity of an electron disturb the electron too much, why cannot the observation be improved by using less energetic photons?

10. If the wavelength of light used to locate a particle is too long, why cannot the location be found more precisely by using light of shorter wavelength?

Close Up

The Uncertainty Principle: Examples

Applied to a large mass.

Consider a car, with a mass of 1,000 kg, moving with a speed of about 1 m/sec. Suppose that in this experiment the inherent uncertainty Δv in the measured speed is 0.1 m/sec (10% of the speed). What is the minimum uncertainty in the position of the car?

$$\Delta x \Delta p \geq \frac{h}{2\pi}$$

$$\Delta p = m\Delta v = 100 \text{ kg-m/sec}$$

$$h = 6.63 \times 10^{-34} \text{ J-sec}$$

$$\Delta x \geq \frac{6.63}{6.28} \times \frac{10^{-34} \text{ J-sec}}{10^{2} \text{ kg-m/sec}}$$

$$\Delta x \geq 1 \times 10^{-36} \text{ m}$$

This uncertainty in position, which is many orders smaller than the size of an atom, is much too small to be observable. In this case, you can determine the position of the body with as high an accuracy as you would ever need.

Applied to a small mass.

Consider an electron, with a mass of 9.1×10^{-31} kg, moving with a speed of about 2×10^{6} m/sec. Suppose that the uncertainty Δv in the speed is 0.2×10^{6} m/sec (10% of the speed). What is the minimum uncertainty in the position of the electron?

$$\Delta x \Delta p \geq \frac{h}{2\pi}$$

$$\Delta p = m\Delta v = 1.82 \times 10^{-25} \text{ kg-m/sec}$$

$$h = 6.63 \times 10^{-34} \text{ J-sec}$$

$$\Delta x \geq \frac{6.63}{6.28} \times \frac{10^{-34} \text{ J-sec}}{1.82 \times 10^{-25} \text{ kg-m/sec}}$$

$$\Delta x \geq 5 \times 10^{-10} \text{ m}$$

The uncertainty in position is of the order of atomic dimensions and is significant in atomic problems. It is impossible to specify exactly where an electron is in an atom.

The reason for the difference between these two results is that Planck's constant h is very small, so small that the uncertainty principle becomes important only on the atomic scale. For ordinary-sized objects, the equations give the same result as if h had the value zero.

20.6 | Probability interpretation

To explore dualism further, it is necessary to review some ideas of probability. In some situations, no single event can be predicted with certainty. But it may still be possible to predict the *statistical probabilities* of certain events. On a holiday weekend during which perhaps 25 million cars are on the road, statisticians predict that about 600 people will be killed in accidents. It is not known which cars in which of the 50 states will be involved in the accidents. But, on the basis of past experience, the *average* behavior is still quite accurately predictable.

It is in this way that physicists think about the behavior of photons and material particles. As you have seen, there are basic limitations on the ability to describe the behavior of an individual particle. But the laws of physics often make it possible to describe the behavior of large collections of particles with good accuracy. Schrödinger's equations for the behavior of waves associated with particles give the *probabilities* for finding the particles at a given place at a given time.

To see how probability works, consider the situation of a star being photographed through a telescope. As you have already seen (for example, on the page on "Diffraction and Detail" in Chapter 13), the image of a point source is not a precise point. Rather , it is a *diffraction pattern*, a central spot with a series of progressively fainter circular rings.

The image of a star on the photographic film in the telescope would be a similar pattern. Imagine now that you wish to photograph a very faint star. If the energy in light rays were not quantized, it would spread continuously over ever-expanding wave fronts. Thus, you would expect the image of a very faint star to be exactly the same as that of a much brighter star, except that the intensity of light would be less over the whole pattern. However, the energy of light *is* quantized; it exists in separate quanta, "photons," of a definite energy. A photon striking a photographic emulsion produces a chemical change in the film at a single location, not all over the image area. If the star is very remote, only a few photons per second may arrive at the film. The effect on the film after a very short period of exposure would be something like the diffraction pattern in drawing A in the margin. As the exposure continued, the effect on the film would begin to look like B. Each successive photon falls on the photographic plate as if its location were decided by some wheel fixed to yield eventually not a completely random pattern but one with the radial symmetry shown in C. Finally, a pattern like C would be produced, just like the image produced by a much brighter star with a much shorter exposure.

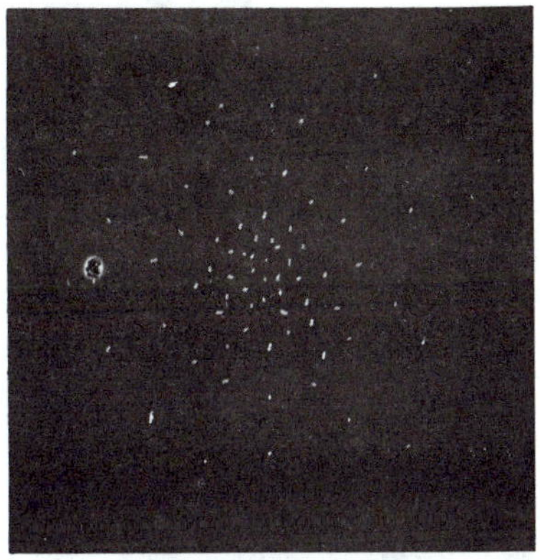

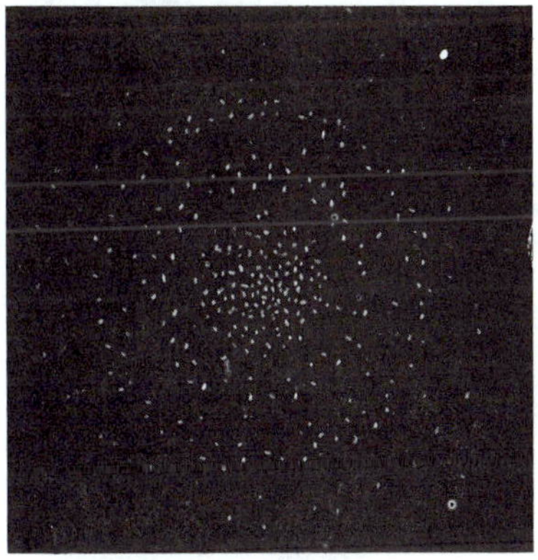

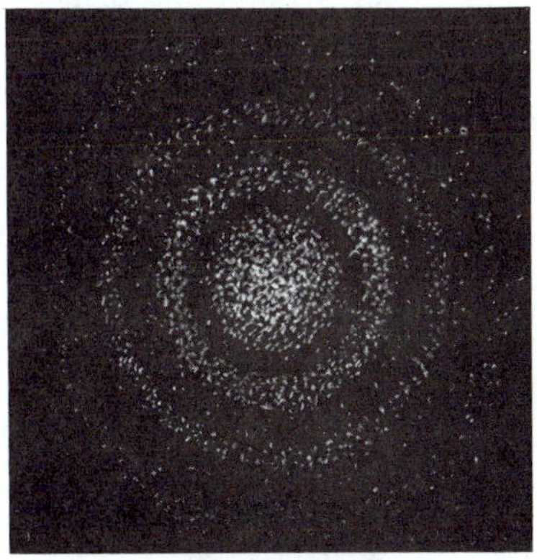

These sketches represent successive stages of a greatly enlarged image of a distant star on a photographic plate.

As discussed in connection with kinetic theory and disorder, it is easy to predict the average behavior of very large numbers of particles, even though nothing is known about the behavior of any single one of them. Unlike kinetic theory, however, the use of probabilities in quantum mechanics is not for convenience, but seems to be an intrinsic necessity. There is no other way to deal with quantum mechanics.

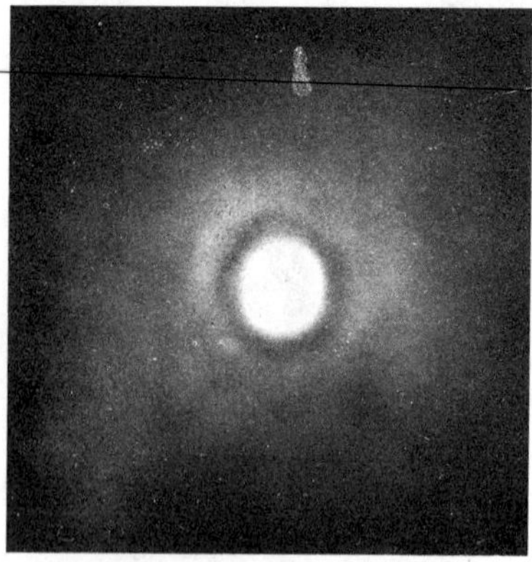

For tremendous numbers of quanta, the overall distribution is very well described by the distribution of wave intensity. For small numbers of quanta, the wave intensity is not very useful for predicting where they will go. You might expect them to go mostly to the "high-intensity" parts of the image, but you cannot predict exactly where. These facts fit together beautifully if you consider the wave intensity at a location to indicate the *probability* of a photon going there!

A similar connection can be made for de Broglie waves and particles of matter. For this purpose, rather than considering a diffraction pattern formed by an electron beam, consider an electron wave that is confined to a particular region in space. An example is the de Broglie wave associated with the electron in a hydrogen atom, which is spread out all over the atom. (Another example is the de Broglie wave of an electron in a good conductor of electricity.) In neither case would you think of the *electron* itself as spread out over the entire region; it is far more useful to picture the electron as a particle moving around the nucleus (or wandering throughout the conductor). The wave's amplitude at some location represents the *probability* of the electron being there, if a measurement of the electron's precise location were to be performed.

According to modern quantum theory, the hydrogen atom does not consist of a localized negative particle moving around a nucleus as in the Bohr model. Indeed, the theory does not provide any picture of the hydrogen atom. A description of the probability distribution is the closest thing to a picture that the theory provides. The probability distribution for the lowest energy state of the hydrogen atom is represented in the drawing above. The probability distribution for a higher energy state, still for a single electron, is represented in the drawing at the left below. In each case, whiter shading at a point indicates greater probability.

However, quantum theory is not really concerned with the position of any individual electron in any individual atom. Instead, the theory gives a mathematical representation that can be used to predict interaction with particles, fields, and radiation. For example, it can be used to calculate the probability that hydrogen will emit light of a particular wavelength. The intensity and wavelength of light emitted by a large number of hydrogen atoms can then be compared with these calculations. Comparisons such as these have shown that the theory agrees with experiment.

To understand atomic physics, you must deal with the average behavior of many atomic particles. The laws governing this average behavior are those of wave mechanics. The waves, it seems, are waves whose amplitudes are a measure of probability. The information (concerning the probability with which a particle will reach some position at a given time) travels through space in waves. These waves can interfere with each other in exactly the same way that water waves do. Now, for example, think of a beam of electrons passing through two slits. You can consider the electrons to be waves and compute their interference patterns. These patterns determine the directions in which there are high wave amplitudes (high probability of electrons going there). If there are no more slits or other interactions of the waves with matter, you can continue the description in terms of particles. You can say that the electrons are likely to (and on the average will) end up going in particular directions with particular speeds.

The success of wave mechanics emphasized the importance of the dual wave–particle nature of radiation and matter. How can a particle be thought of as "really" having wave properties? The answer is that matter, particularly on the atomic scale, need *not* be thought of as being either "really" particles or "really" waves. Ideas of waves and particles, taken from the world of visible things, just do not apply on the atomic scale.

In describing something that no one has ever seen or ever can see directly, it would be surprising if you could use the concepts of the visible world unchanged. It appeared natural before 1925 to talk about the transfer of energy in either wave terms or particle terms. Indeed, such terms were all physicists needed or knew at the time. Almost no one suspected that *both* wave and particle descriptions could apply to light and to matter. Even today, imagination and language have only these two ideas of waves and particles to stumble along on. Until new concepts appear, dualism cannot be wished away, but will remain the best way to handle experimental results.

Max Born, one of the founders of quantum mechanics, has written:

> The ultimate origin of the difficulty lies in the fact (or philosophical principle) that we are compelled to use the words of common language when we wish to describe a phenomenon, not by logical or mathematical analysis, but by a picture appealing to the imagination. Common language has grown by everyday experience and can never surpass these limits. Classical physics has restricted itself to the use of concepts of this kind; by analyzing visible motions it has developed two ways of representing them by elementary processes: moving particles and waves. There is no other way of giving a pictorial description of motions—we have to apply it even in the region of atomic processes, where classical physics breaks down.

The idea that the wave represents the probability of finding its associated particle in some specific condition of motion has had great success. Yet many scientists found it hard to accept the idea that it is impossible to know exactly what any one particle is doing. The most prominent of such disbelievers was Einstein. In a letter to Born written in 1926, he remarked:

> The quantum mechanics is very imposing. But an inner voice tells me that it is still not the final truth. The theory yields much, but it hardly brings us nearer to the secret of the Old One. In any case, I am convinced that He does not play dice.

"Deterministic" means here that if all the conditions of an isolated system are known and the laws describing interaction are known, then it is possible to predict precisely, according to "strict causality," what will happen next, without any need for probability.

Thus, Einstein agreed with the usefulness and success of wave mechanics, but he refused to accept probability-based laws as the final level of explanation in physics. The remark about not believing that God played dice (an expression he used many times later) expressed Einstein's faith that more basic, deterministic laws are yet to be found. Like Einstein, some other scientists refused to accept the probability laws in quantum mechanics. However, no one has yet succeeded in replacing Born's probability interpretation of quantum mechanics.

SG 19–23

Scientists agree that quantum mechanics works. It gives the right answers to many questions in physics; it unifies ideas and occurrences that were once unconnected; and it has produced many new experiments and new concepts. On the other hand, there is still vigorous argument about its basic significance. It yields probability functions, not precise trajectories. Some scientists see in this aspect of the theory an important indication of the nature of the world. For other scientists, the same fact indicates that quantum theory is incomplete. Some in this second group are trying to develop a more basic, nonstatistical theory. For such a theory, the present quantum theory is only a special, extreme case. As in other fields of physics, the greatest discoveries here may be those yet to be made.

?

11. *In wave terms, the bright lines of a diffraction pattern are regions where there is a high field intensity produced by constructive interference. What is the probability interpretation of quantum mechanics for the bright lines of a diffraction pattern?*

12. *Quantum mechanics can predict only probabilities for the behavior of any one particle. How, then, can it predict many phenomena, for example, half-lives and diffraction patterns, with great certainty?*

"Sea and Sky," by M. C. Escher.
(See SG 24.)

study guide

1. The *Project Physics* materials particularly appropriate for Chapter 20 include:

Activities

Standing Waves on a Band-Saw Blade

Turntable Oscillator Patterns Resembling de Broglie Waves

Standing Waves in a Wire Ring

Film Loops

Standing Waves on a Wire

2. How fast would you have to move to increase your mass by 1%?

3. The centripetal force on a mass moving with relativistic speed v around a circular orbit of radius R is $F = mv^2/R$, where m is the relativistic mass. Electrons moving at a speed 0.60 c are to be deflected in a circle of radius 1.0 m. What must be the magnitude of the force applied? ($m_0 = 9.1 \times 10^{-31}$ kg)

4. The formulas ($p = m_0v$, $KE = \frac{1}{2}m_0v^2$) used in Newtonian physics are convenient approximations to the more general relativistic formulas. The factor $1/\sqrt{1 - v^2/c^2}$ can be expressed as an infinite series of steadily decreasing terms by using a binomial series expansion. When this is done, it is found that

$$\frac{1}{\sqrt{1 - \frac{v^2}{c^3}}} = 1 + \frac{1}{2}\frac{v^2}{c^2} + \frac{3}{8}\frac{v^4}{c^4} + \frac{5}{16}\frac{v^6}{c^6} + \frac{35}{128}\frac{v^8}{c^8} + \ldots$$

(a) Show, by simple substitution, that when v/c is less than 0.1, the values of the terms drop off so rapidly that only the first few terms need be considered.

(b) You rarely observe familiar objects moving faster than about 3,000 m/sec; the speed of light is 3×10^8 m/sec, so the value of v/c for familiar objects is rarely greater than about 10^{-5}. What error is caused by using only the first two terms of the series?

(c) Substitute the first two terms of the series into the relativistic expression for kinetic energy and show that $KE = \frac{1}{2}m_0v^2$ is a good approximation for familiar objects.

5. According to relativity theory, changing the energy of a system by ΔE also changes the mass of the system by $\Delta m = \Delta E/c^2$. Something like 10^5 J per kilogram of substance are usually released as heat energy in chemical reactions.

(a) Calculate the mass change associated with a change of energy of 10^5 J.

(b) Why then are mass changes not detected in chemical reactions?

6. The speed of the earth in its orbit is about 3×10^4 m/sec. Its "rest" mass is 6.0×10^{24} kg.

(a) What is the kinetic energy of the earth in its orbit?

(b) What is the mass equivalent of that kinetic energy?

(c) By what percentage is the earth's "rest" mass increased at orbital speed?

(d) Refer back to Unit 2 to recall how the mass of the earth was found; was it the rest mass or the mass at orbital speed?

7. In relativistic mechanics, the formula $\vec{p} = m\vec{v}$ still holds, but the mass m is given by $m = m_0/\sqrt{1 - v^2/c^2}$. The rest mass of an electron is 9.1×10^{-31} kg.

(a) What is the electron's momentum when it is moving down the axis of a linear accelerator at a speed of 0.4 c with respect to the accelerator tube?

(b) What would Newton have calculated for the momentum of the electron?

(c) By how much would the relativistic momentum increase if the speed of the electron were doubled?

(d) What would Newton have calculated its change in momentum to be?

8. Calculate the momentum of a photon of wavelength 400×10^{-9} m. How fast would an electron have to move in order to have the same momentum?

9. Describe Compton's work. What did it prove?

10. What explanation would you offer for the fact that the wave aspect of light was shown to be valid before the particle aspect was demonstrated?

11. The electrons that produced the diffraction photograph on page 608 had de Broglie wavelengths of 10^{-10} m. To what speed must they have been accelerated? (Assume that the speed is small compared to c, so that the electron mass is about 10^{-30} kg.)

12. A ball of mass 0.2 kg moves with a speed of 1 m/sec. What is its de Broglie wavelength?

13. Show that the de Broglie wavelength of a classical particle of mass m and kinetic energy KE is given by

$$\lambda = \frac{h}{\sqrt{2m(KE)}}$$

14. A particle confined in a box cannot have a kinetic energy less than a certain amount; this least amount corresponds to the longest de Broglie wavelength that produces standing waves in the box; that is, the box size is one-half wavelength. For each of the following situations, find the longest de Broglie wavelength that would fit in the box. Then use $p = h/\lambda$ to find the momentum p, and use $p = mv$ to find the speed v.

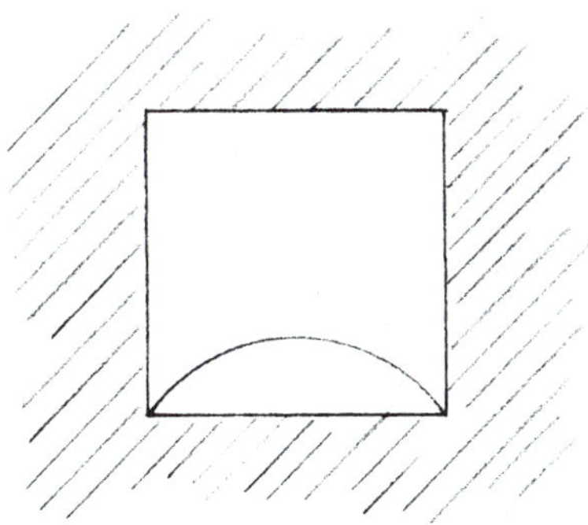

(a) a dust particle (about 10^{-9} kg) in a display case (about 1 m across)

(b) an argon atom (6.6×10^{-26} kg) in a light bulb (about 10^{-1} m across)

(c) a protein molecule (about 10^{-22} kg) in a bacterium (about 10^{-6} m across)

(d) an electron (about 10^{-30} kg) in an atom (about 10^{-10} m across)

15. Suppose that the only way you could obtain information about the world was by throwing rubber balls at the objects around you and measuring their speeds and directions of rebound. What kind of objects would you be unable to learn about?

16. A bullet can be considered a particle having dimensions in each direction of approximately 1 cm. It has a mass of about 10 g and a speed of about 3×10^4 cm/sec. Suppose you can measure its speed to an accuracy of ± 1 cm/sec. What is the corresponding uncertainty in its position according to Heisenberg's principle?

17. Show that if Planck's constant were equal to zero, quantum effects would disappear and even atomic particles would behave according to Newtonian physics. What effect would this have on the properties of light?

18. Some writers have claimed that the uncertainty principle proves that there is free will. Do you think this extrapolation from atomic phenomena to the world of living beings is valid?

19. A physicist has written:

> It is enough that quantum mechanics predicts the average value of observable quantities correctly. It is not really essential that the mathematical symbols and processes correspond to some intelligible physical picture of the atomic world.

Do you regard such a statement as acceptable? Give your reasons.

20. In Chapters 19 and 20, you saw that it is impossible to avoid the wave–particle dualism of light and matter. Bohr coined the word *complementarity* for the situation in which two opposite views seem valid and the correct choice depends only on which aspect of a phenomenon one chooses to consider. Can you think of situations in other fields (outside of atomic physics) to which this idea might apply?

21. Units 1–4 discussed the behavior of large-scale "classical particles" (for example, tennis balls) and "classical waves" (for example, sound waves). Such particles and waves in most cases can be described without any use of ideas such as the quantum of energy or the de Broglie matter–wave. Does this mean that there is one sort of physics ("classical physics") for the phenomena of the large-scale world and quite a different physics ("quantum physics") for the phenomena of the atomic world? Or does it mean that quantum physics really applies to all phenomena but is no different from classical physics when applied to large-scale particles and waves? What arguments or examples would you use to defend your answer?

22. If there are laws that describe precisely the behavior of atoms, one can reason that the future is completely determined by the present (and the present was determined in the ancient past). This idea of complete *determinism* was uncomfortable to many philosophers during the centuries following the great success of Newtonian mechanics. The great French physicist Pierre Laplace (1748–1827) wrote:

> Given for one instant an intelligence which could comprehend all the forces by which nature is animated and the respective situation of the beings who compose it—an intelligence sufficiently vast to submit these data to analysis—it would embrace in the same formula the movements of the greatest bodies of the universe and those of the lightest atom; for it, nothing would be uncertain and the future, as the past, would be present to its eyes. [*A Philosophical Essay on Probabilities*]

(The later statistical view of kinetic theory may have emphasized the difficulty of actually predicting the future. But it did not weaken the idea of an underlying chain of cause and effect.)

(a) Is Laplace's statement consistent with modern physical theory?

(b) What implications do you see in relativity theory for the idea of determinism?

(c) What implications do you see for determinism in quantum theory?

23. Those ancient Greeks who believed in natural law were also troubled by the idea of determinism. Compare the ideas expressed in the following passage from Lucretius' *On the Nature of Things* (about 80 B.C.) with somewhat analogous ideas of modern physics.

> If cause forever follows after cause
> In infinite, undeviating sequence
> And a new motion always has to come
> Out of an old one, by fixed law; if atoms
> Do not, by swerving, cause new moves which break
> The laws of fate; if cause forever follows,
> In infinite sequence, cause—where would we get
> This free will that we have, wrested from fate . . .
> What keeps the mind from having inside itself
> Some such compulsiveness in all its doings,
> What keeps it from being matter's absolute slave?
> The answer is that our free will derives
> From just that ever-so-slight atomic swerve
> At no fixed time, at no fixed place whatever.

24. Many scientists like the drawings of M. C. Escher such as that on page 621 because one can read into them some (not-too-farfetched) likeness to notions that are prominent in science. Do you see such a likeness between the drawing on page 621 and the dual nature of light? When does the likeness (analogy) break down?

EPILOGUE

This unit has traced the concept of the atom from the early ideas of the Greeks to the quantum mechanics now generally accepted by physicists. The search for the atom started with the qualitative assumptions of Leucippus and Democritus who thought that atoms offered a rational explanation of the behavior of matter. However, for many centuries most natural philosophers thought that other explanations, not involving atoms, were more reasonable. Atomism was pushed aside and received only occasional consideration until the seventeenth century.

With the growth of the mechanical philosophy of nature in the seventeenth and eighteenth centuries, particles (corpuscles) became important. Atomism was reexamined, mostly in connection with physical properties of matter. Galileo, Boyle, Newton, and others speculated on the role of particles for explaining the expansion and contraction of gases. Chemists speculated about atoms in connection with chemical change. Finally, Dalton began the modern development of atomic theory, introducing a quantitative conception that had been lacking: the relative atomic mass.

Chemists, in the nineteenth century, found that they could correlate the results of many chemical experiments in terms of atoms and molecules. They also found that there are relations between the properties of different chemical elements. Quantitative information about atomic masses provided a framework for the system organizing these relations: the periodic table of Mendeleev. During the nineteenth century, physicists developed the kinetic theory of gases. This theory, based on the assumption of very small corpuscles, particles, molecules, or whatever else they might be called, helped strengthen the position of the atomists. Other work of nineteenth-century physicists helped pave the way to the study of the structure of atoms, through the study of the spectra of the elements and of the conduction of electricity in gases, and through the discovery of cathode rays, electrons, and X rays.

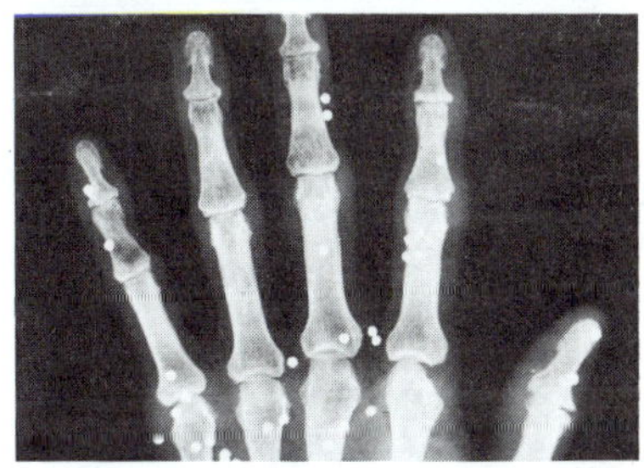

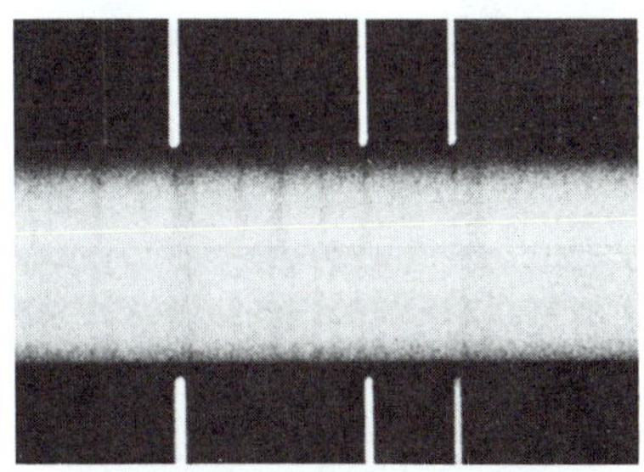

Nineteenth-century chemistry and physics converged, at the beginning of the twentieth century, on the problem of atomic structure. It became clear that the uncuttable, infinitely hard atom was too simple a model; that the atom itself is made up of small particles. The search for a model with structure began. Of the early models, that of Thomson gave way to Rutherford's nuclear atom, with its small, heavy, positively charged nucleus, surrounded somehow by negative charges. Then came the atom of Bohr, with its electrons thought to be moving in orbits like planets in a miniature solar system. The Bohr theory had many successes and linked chemistry and spectra to the physics of atomic structure. Beyond that, it could not advance substantially

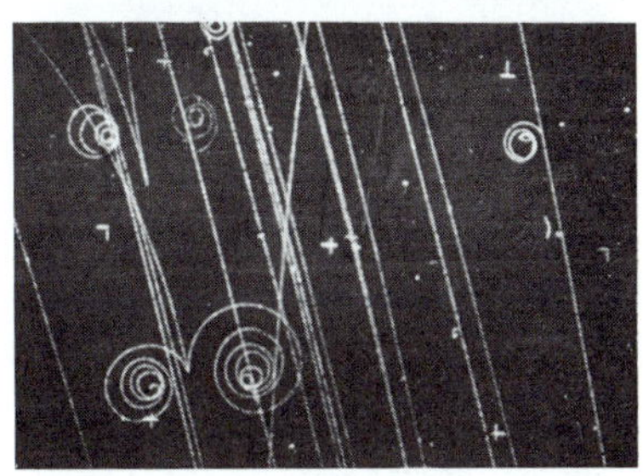

without giving up an easily grasped picture of the atom. The tool needed is the mathematical model, not pictures. Quantum mechanics makes it possible to calculate how atoms behave; it helps explain the physical and chemical properties of the elements. But at the most basic level, nature still has secrets. The next stage in the story of physics in Unit 6 is the nucleus at the center of the atom. Is the nucleus made up of smaller components? Does it have laws of physics all its own?

The study of the nucleus has been one of the most exciting branches of physics in the twentieth century. Progress in nuclear physics has advanced not only basic science but also technology, which both supplies tools for research and applies some of the results of research in practical ways. These applications, including the production of electricity from nuclear energy, the many clinical and industrial uses of radiation, and, of course, the military weapons, have had economic, social, and political consequences. The use and control of nuclear technology, therefore, are often front-page news, and citizens find it necessary to inform themselves about these problems in order to participate effectively in decisions that affect their lives.

UNIT

The Nucleus

PROLOGUE

In Unit 5, you learned that the atom consists of a very small, positively charged nucleus surrounded by electrons. Experiments on the scattering of α particles showed that the nucleus has dimensions of the order of 10^{-14} m. Since the diameter of an atom is of the order of 10^{-10} m, the nucleus takes up only a minute fraction of the volume of an atom. The nucleus, however, contains nearly all of the mass of the atom, as was also shown by the scattering experiments. The existence of the atomic nucleus and its properties raised new questions. Is the nucleus itself made up of still smaller units? If so, what are these units, and how are they arranged in the nucleus? What methods can be used to get answers to these questions? What experimental evidence can be used as a guide?

You saw in Unit 5 that the study of the properties and structure of atoms needed new physical methods. The methods that could be used to study the properties of bodies of ordinary size, that is, those with dimensions of the order of centimeters or meters, could not yield information about the structure of atoms.

It is reasonable to expect that it is still more difficult to get informaion about what goes on inside the nucleus, which is such a small part of the atom. New kinds of experimental data must be obtained. New theories must be devised to help correlate and explain the data. In these respects, the study of the nucleus is still another step on the long road from the very large to the very small along which you have traveled in this course. In this unit, you will further explore the problem of the constitution of matter by studying the atomic nucleus.

One of the first and most important steps to an understanding of the atomic nucleus was the discovery of radioactivity in 1896. This discussion of nuclear physics will, therefore, start with radioactivity. You will see how the study of radioactivity led to additional discoveries, to the development of methods for probing the nucleus, and to ideas about the constitution of the nucleus. In fact, the discovery that the atom has a nucleus was a consequence of the study of radioactivity. You will examine the interaction between experiment and theory, and the step-by-step development of ideas about the nucleus. You will see how particular experimental results led to new ideas, and how the latter, in turn, led to new experiments. This historical study is especially useful and interesting because nuclear physics is a new branch of physics, which has developed over a relatively short period of time. The reports and papers through which discoveries have been made known are readily available. The research is still going on and at an ever-increasing rate. Progress in nuclear physics is closely related to modern technology, which both supplies tools for further research and applies some of the research in practical ways. Some of these practical applications have serious economic and political consequences. Newspapers report about these applications almost daily, and it is the citizens' duty to inform themselves as well as they can in order to participate effectively in decisions that affect their lives.

Now that the use and control of nuclear technology is often front-page news, it may be difficult to realize that the study of the atomic nucleus is connected with a chance discovery made in 1896. But it was that discovery that touched off the whole enterprise called nuclear physics.

The Yankee Atomic Electric nuclear power station in Rowe, Massachusetts, which has been producing electricity with a capacity of 175,000 kW since 1961.

CHAPTER 21

Radioactivity

21.1 | Becquerel's discovery

SG 1

A legendary chapter in physics began with the discovery of the phenomenon known as "radioactivity" early in 1896 by the French physicist Henri Becquerel. It was another of those "accidents" that illustrate how the trained and prepared mind is able to respond to an unexpected observation.

Only two months before, in November 1895, Röntgen had discovered X rays. In doing so, he had unwittingly set the stage for the discovery of radioactivity. Röntgen had found that X rays came from the glowing spot on a glass tube where a beam of cathode rays (high-speed electrons) was hitting. (See Secs. 18.2 and 18.6 in Unit 5.) When the cathode-ray beam was turned off, the spot of light on the face of the glass tube disappeared; the X rays coming from that spot also stopped.

The emission of light by the glass tube when it is excited by the cathode-ray beam is an example of the phenomenon called

Right: *Uranium is mined in one of the thousands of tunnels dug into the mesa walls of the Colorado Plateau.*

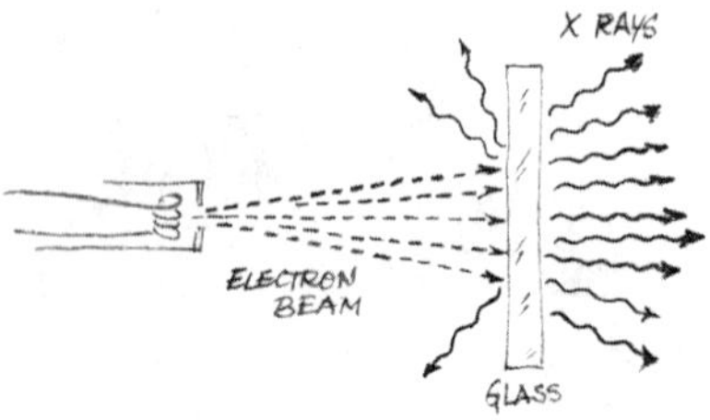

X-ray production by bombardment of electrons (cathode rays) on glass.

Röntgen showed that one method of detecting the presence of X rays is to let them expose a well-wrapped photographic plate. (See Sec. 18.6 in Unit 5.)

Henri Becquerel (1852–1908) received the 1903 Nobel Prize in physics (for the discovery of natural radioactivity) along with Pierre and Marie Curie (for the discovery of the radioactive elements radium and polonium).

fluorescence, which was well known before Röntgen's work. A considerable amount of research had been done on fluorescence during the latter part of the nineteenth century. A substance is said to be fluorescent if it immediately emits visible light when struck by (1) visible light of shorter wavelength, (2) invisible radiations, such as ultraviolet light, or (3) the beam of electrons that make up cathode rays. Fluorescence stops when the exciting light is turned off. (The term *phosphorescence* is generally applied to a related phenomenon, the emission of visible light that continues *after* the exciting light is turned off.)

Röntgen's observation that the X rays also came from the spot that showed fluorescence raised the suspicion that there was a close connection between X rays and fluorescence or phosphorescence. Becquerel was fortunate in having the necessary materials and training to study this problem. In addition, he was the son and grandson of physicists who had made important contributions to the field of fluorescence and phosphorescence. In his Paris laboratory, Becquerel had devised an instrument for examining materials in complete darkness a small fraction of a second after they had been exposed to a brilliant light. The question occurred to Becquerel: When bodies are made to fluoresce (or phosphoresce) in the visible region with sufficient intensity, do they also emit X rays in addition to the light rays? He tested a number of substances by exposing them to sunlight; his method of checking whether they also emitted invisible X rays followed Röntgen's idea: Is a well-wrapped photographic plate exposed by such invisible rays? One of the samples Becquerel used happened to be a salt of the metal uranium, a sample of potassium–uranyl sulfate. In his words:

I wrapped a . . . photographic plate . . . with two sheets of thick black paper, so thick that the plate did not become clouded by exposure to the sun for a whole day. I placed on the paper a crust of the phosphorescent substance, and exposed the whole thing to the sun for several hours. When I developed the photographic plate I saw the silhouette of the phosphorescent substance in black on the negative. If I placed between the phosphorescent substance and the paper a coin or a metallic screen pierced with an open-work design, the image of these objects appeared on the negative. The same experiment can be tried with a thin sheet of glass placed between the phosphorescent substance and the paper, which excludes the possibility of a chemical action resulting from vapors which might emanate from the substance when heated by the sun's rays.

We may therefore conclude from these experiments that the phosphorescent substance in question emits radiations which penetrate paper that is opaque to light. . . .

In his published paper, Becquerel was careful to conclude from his experiment only that "penetrating radiations" were emitted from the phosphorescent substance. He did not write that the substance emitted X rays while it phosphoresced, because he had not fully verified that the radiations were X rays (though the radiations were transmitted through the black paper, just as X rays are) or that they were actually related to the phosphorescence (though he strongly suspected that they were). Before he could investigate these possibilities, he made this discovery:

As it turned out, and will be shown in Sec. 21.3, the Becquerel rays are not X rays.

. . . among the preceding experiments some had been made ready on Wednesday the 26th and Thursday the 27th of February [1896]; and as on those days the sun only showed itself intermittently, I kept my arrangements all prepared and put back the holders in the dark in the drawer of the case, and left in place the crusts of uranium salt. Since the sun did not show itself again for several days, I developed the photographic plates on the 1st of March, expecting to find the images very feeble. On the contrary, the silhouettes appeared with great intensity. I at once thought that the action might be able to go on in the dark. . . .

Further experiments verified this surprising thought. Even when the uranium compound was not being excited by sunlight to phosphoresce, it continually emitted something that could penetrate black paper and other substances opaque to light, such as thin plates of aluminum or copper. Becquerel found that all the compounds of uranium, many of which were not phosphorescent at all, and metallic uranium itself had the same property. The amount of action on the photographic plate did not depend on what the particular compound of uranium was, but only on the amount of uranium present in it!

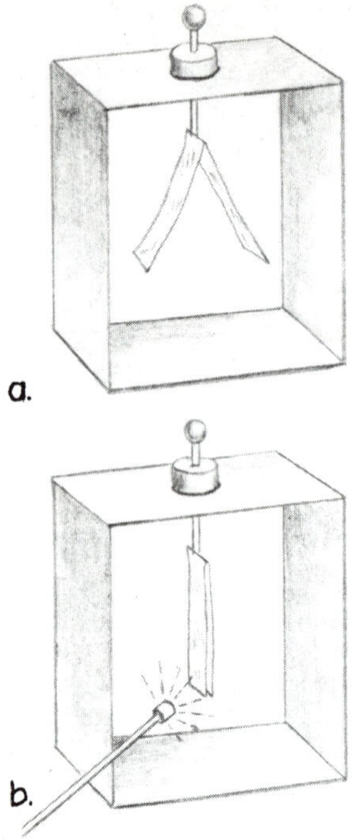

The ionizing effect of the Becquerel rays could be demonstrated with a charged electroscope (a). When a sample of uranium is held near the electroscope leaves (b), the rays cause gas molecules in the air to ionize, that is, to become electrically charged. Ions, with a charge opposite to that on the leaves, drift to the leaves and neutralize their charge. The time taken for the leaves to fall is a measure of the rate of ionization of the gas and, therefore, of the activity of the uranium source.

Becquerel also found that the persistent radiation from a sample of uranium did not appear to change, either in intensity or character, with the passing of time. Nor was a change in the activity observed when the sample of uranium or of one of its compounds was exposed to ultraviolet light, infrared light, or X rays. Moreover, the intensity of the uranium radiation (or "Becquerel rays," as they came to be known) was the same at room temperature (20°C), at 200°C, and at the temperature at which oxygen and nitrogen (air) liquefy, about −190°C. Thus, these rays seemed unaffected by physical (and chemical) changes.

Becquerel also showed that the radiations from uranium produced ionization in the surrounding air. They could discharge a positively or negatively charged body such as an electroscope. So the uranium rays resemble X rays in two important respects: their penetrating power and their ionization power. Both kinds of rays were invisible to the unaided eye, but both affected photographic plates. Still, X rays and Becquerel rays differed in at least two important ways: Compared to X rays, these newly discovered rays from uranium needed no cathode-ray tube or even light to start them, and they could not be turned off. Becquerel showed that even after a period of 3 years a given piece of uranium and its compounds continued to emit radiations spontaneously.

The years 1896 and 1897 were years of great excitement in physics, to a large extent because of the interest in the recently discovered X rays and in cathode rays. It quickly became evident that X rays could be used in medicine, and they were the subject of much research. In comparison, the properties of the Becquerel rays were less spectacular, and little work was done on them in the period from the end of May 1896 until the end of 1897. In any case, it seemed that somehow Becquerel rays were special cases of X-ray emission. Even Becquerel himself turned his attention to other work. But attention began to be attracted by the fact that the invisible rays from the uranium and its compounds appeared spontaneously.

Two questions were asked. First, what was the source of the energy creating the uranium rays and making it possible for them to penetrate opaque substances? Second, did any other of the 70 or more elements known then have properties similar to those of uranium? The first question was not answered for some time, although it was considered seriously. The second question was answered early in 1898 by the Curies, who, by doing so, opened a whole new field of research in physical science.

?

1. *Why was Becquerel experimenting with a uranium compound? Describe his experiment.*

2. *How did uranium compounds have to be treated in order to emit the "Becquerel rays"?*

3. *What were the properties of the "Becquerel rays"? In what ways were they similar to those of X rays?*

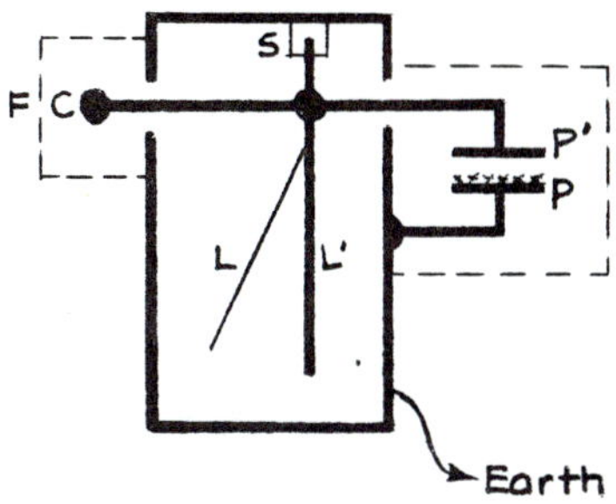

Sketch of an electroscope used by Pierre and Marie Curie in many of their early experiments. The active material is placed on a plate laid on top of the fixed circular plate P, which is connected with the case of the instrument and with the earth. The upper insulated plate P′ is connected with the insulated gold-leaf system LL′. S is an insulating support and L the gold leaf. The system is first charged to a suitable potential by means of the rod C. The rate of movement of the gold leaf is observed by means of a microscope.

21.2 Other radioactive elements are discovered

One of Becquerel's colleagues in Paris was the physicist Pierre Curie, who had recently married a Polish-born physicist, Marie Sklodowska. Marie Curie undertook a systematic study of the Becquerel rays and looked for other elements and minerals that might emit them. Using a sensitive type of electrometer that her husband had recently invented, she measured the small electric current produced when the rays ionized the air. This current was assumed to be (and actually is) proportional to the intensity of the rays. With this new technique, Curie could give a numerical value to the ionizing effect produced by the rays. These values were reproducible within a few percent from one experiment to the next with the same sample.

One of Marie Curie's first results was the discovery that the element thorium (Th) and its compounds emitted radiations with properties similar to those of the uranium rays. (The same finding was made independently in Germany by Gerhardt C. Schmidt, at about the same time.) The fact that thorium emits rays like those of uranium was of great importance; it showed that the mysterious rays were not a property peculiar just to one element. The discovery spurred the search for still other elements that might emit similar rays. The fact that uranium and thorium were the elements with the greatest known atomic masses indicated that the very heavy elements might have special properties different from those of the lighter elements.

The evident importance of the problems raised by the discovery of the uranium and thorium rays led Pierre Curie to lay aside his researches in other fields of physics and to work with his wife on these new problems. They began a herculean task. First, they found that the intensity of the emission from any thorium compound was directly proportional to the fraction by weight of the metallic element thorium present. (Recall that Becquerel had found a similar result for uranium compounds.) Moreover, the amount of radiation was independent of the physical conditions or the chemical combination of the active elements. These results led the Curies to the conclusion that the emission of the rays depended only on the presence of *atoms* of either of the two elements uranium or thorium. Atoms of other elements that were present were simply inactive or absorbed some of the radiation.

These ideas were especially important because they helped the Curies interpret their later experiments. For example, in their studies of the radioactivity of minerals they examined pitchblende, an ore containing about 80% uranium oxide (U_3O_8). They found that the emission from pitchblende, as measured by its effect in ionizing air, was about four or five times as great as that to be expected from the amount of uranium in the ore. The other elements known at the time to be associated with uranium in pitchblende, such as bismuth and barium, had been shown to be not radioactive. If emission of rays is an atomic phenomenon, the unexpected activity of pitchblende could be explained only by the presence of another, hitherto undiscovered, element in pitchblende, an element more active than uranium itself.

To explore this hypothesis, the Curies applied chemical separation processes to a large sample of pitchblende to try to isolate this hypothetical active substance. After each separation process, the products were tested, the inactive part discarded, and the active part analyzed further. Finally, the Curies obtained a highly active product that presumably consisted mainly of the unknown element. In a note titled "On a New Radioactive Substance Contained in Pitchblende," which they submitted to the French Academy of Sciences in July of 1898, they reported:

In this note, the term "radioactivity" was used for the first time.

> By carrying on these different operations . . . finally we obtained a substance whose activity is about 400 times greater than that of uranium. . . .
>
> We believe, therefore, that the substance which we removed from pitchblende contains a metal which has not yet been known, similar to bismuth in its chemical properties. If the existence of this new metal is confirmed, we propose to call it *polonium*, after the name of the native country of one of us.

Compare the positions of polonium (Po) and bismuth (Bi) in the periodic table on page 654.

Six months after the discovery of polonium, the Curies chemically separated another substance from pitchblende. They found the emission from it so intense as to indicate the presence of still another new element, even more radioactive than polonium! This substance had an activity per unit mass 900 times that of uranium and was chemically entirely different from uranium, thorium, or polonium. Spectroscopic analysis of this substance revealed spectral lines characteristic of the inactive element barium, but also a line in the ultraviolet region that did not seem to belong to any known element. The Curies reported their belief that the substance, "although for the most part consisting of barium, contains in addition a new element which produced radioactivity and, furthermore, is very near barium in its chemical properties." For this new element, so extraordinarily radioactive, they proposed the name *radium*.

Compare the positions of barium (Ba) and radium (Ra) in the periodic table.

The next step in making the evidence for the newly discovered elements more convincing was to determine their properties, especially their atomic masses. The Curies had made it clear that they had not yet been able to isolate either polonium or radium in pure metallic form or even to obtain a pure sample of a compound of either element. From the substance containing the strongly radioactive substance that they called radium, they had separated a part consisting of barium chloride mixed with a presumably very small quantity of radium chloride. Additional separations yielded an increasing proportion of radium chloride. The difficulty of this task is indicated by the Curies' remark that radium "is very near barium in its chemical properties," for it is very difficult to separate elements whose chemical properties are similar. Moreover, to obtain their highly radioactive substances in usable amounts, they had to start with a very large amount of pitchblende.

With an initial 100-kg shipment of pitchblende (from which the uranium salt had been removed to be used in the manufacture of glass) the Curies went to work in an abandoned woodshed at the School of Physics where Pierre Curie taught. Having failed to obtain financial support, the Curies made their preparations without technical help in this "laboratory." Marie Curie wrote later:

> I came to treat as many as twenty kilograms of matter at a time, which had the effect of filling the shed with great jars full of precipitates and liquids. It was killing work to carry the receivers, to pour off the liquids and to stir, for hours at a stretch, the boiling material in a smelting basin.

From the mixture of radium chloride and barium chloride they produced, only the average atomic mass of the barium and radium could be computed. At first an average value of 146 was obtained, as compared to 137 for the atomic mass of barium. After many additional purifications that increased the proportion of radium chloride, the average value for atomic mass rose to 174. Continuing the tedious purification proess for 4 years, during which she treated several tons of pitchblende residue, Marie Curie was able to report in July 1902 that she had isolated 0.1 g of radium chloride, so pure that spectroscopic examination showed no evidence of any remaining barium. She calculated the atomic mass of radium to be 225 (the present-day value is 226.03). The activity of radium is more than a million times that of the same mass of uranium.

The present yield of radium from 1 ton of high-grade uranium ore is about 0.2 g.

SG 2

4. How is the radioactive emission of an element affected by being combined into different chemical compounds?

5. Why did the Curies suspect the existence of another radioactive material in uranium ore, in addition to uranium itself?

Close Up

Marie and Pierre Curie

Marie Curie.

Marie and Pierre in their laboratory.

Pierre Curie (1859–1906) studied at the Sorbonne in Paris. In 1878, he became an assistant teacher in the physical laboratory there, and some years later, professor of physics. He was well known for his research on crystals and magnetism. Pierre married Marie Sklodowska in 1895 (she was 28 years old). After their marriage, Marie undertook her doctoral research on radioactivity. In 1898, Pierre joined his wife in this work. Their collaboration was so successful that in 1903 they were awarded the Nobel Prize in physics, which they shared with Becquerel. Pierre Curie was run over and killed by a horse-drawn vehicle in 1906. Marie Curie was appointed to his professorship at the Sorbonne, the first woman to have this post.

In 1911, Marie Curie was awarded the Nobel Prize in chemistry for the discovery of the two new elements, radium and polonium. She was the first person to win two Nobel prizes in science. The rest of her career was spent in the supervision of the Paris Institute of Radium, a center for research on radioactivity and the use of radium in the treatment of cancer.

Marie Curie died in 1934 of leukemia, a form of cancer of the leukocyte-forming cells of the body, probably caused by overexposure to the radiations from radioactive substances.

Marie and Pierre on a bicycling holiday.

Marie, Irène, and Pierre; all three won Nobel prizes.

6. What was the main difficulty in producing a pure sample of the element radium?

21.3 The penetrating power of the radiation. α, β, and γ rays

Once the extraordinary properties of radium became known, they excited interest both inside and outside the scientific world, and the number of people studying radioactivity increased rapidly. The main question that attracted attention was: What are the mysterious radiations emitted by radioactive bodies?

In 1899, Ernest Rutherford, whose theory of the nuclear atom was discussed in Chapter 19, started to seek answers to this question. Rutherford found that a sample of uranium emits at least two distinct kinds of rays: one that is very readily absorbed, which he called for convenience α rays (alpha rays), and the other more penetrating, which he called β rays (beta rays). In 1900, the French physicist P. Villard observed that the emission from radium contained rays much more penetrating than even the β rays; this type of emission was given the name γ (gamma) rays. The penetrating power of the three types of rays, as known at the time, is compared in the table below, first published by Rutherford in 1903.

APPROPRIATE THICKNESS OF ALUMINUM REQUIRED TO REDUCE THE RADIATION INTENSITY TO ONE-HALF ITS INITIAL VALUE

Radiation Type	*Thickness of Aluminum*
α	0.0005 cm
β	0.05
γ	8

SG 2

Thus, the Becquerel rays were more complex than had been thought even before the nature of α, β, and γ rays was ascertained. Of the three kinds of rays, the α rays are the most strongly ionizing and the γ rays the least. The power of penetration is inversely proportional to the power of ionization. This is to be expected; the penetrating power of the α rays from uranium is low because they expend their energy very rapidly in causing intense ionization. Alpha rays can be stopped, that is, almost all are absorbed, by about 0.006 cm of aluminum, by a sheet of ordinary writing paper, or by a few centimeters of air. Beta rays are completely stopped only after traveling many meters in air, or a centimeter in aluminum. Gamma rays can pass through many centimeters of lead, or through a meter of concrete, before being almost completely absorbed. One consequence of these properties of the rays is that heavy and expensive shielding is sometimes needed in the study or use of

The rays ionize and, consequently, break down molecules in living cells.

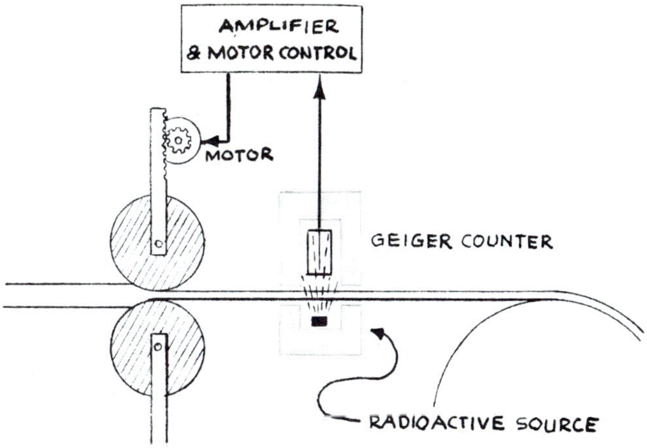

The absorption of β rays gives rise to many modern practical applications of radioactivity. One example is the thickness gauge illustrated in the photograph and drawing above. Sheet metal or plastic is reduced in thickness by rolling. The thickness is measured continuously and accurately by determining the intensity of the β rays that pass through the sheet. The rollers are adjusted so that the desired sheet thickness is obtained.

radiations, especially of γ rays, to protect people from harmful effects of the rays. In some cases, these "radiation shields" are as much as 3 m thick. Such shielding is required around a target at the output of an electron accelerator (where γ rays are created by a method different from radioactivity, as you will find later in this unit).

? *7. List α, β, and γ rays in order of penetrating ability. Why is penetrating power inversely related to ionizing power?*

21.4 | The charge and mass of α, β, and γ rays

Another method used to study the rays was to direct them through a magnetic field to see if they were deflected or deviated from their initial directions by the action of the field. This method came to provide one of the most widely used tools for the study of atomic and nuclear events. It is based on the now familiar fact that a force acts on a charged particle when it moves across a magnetic field. As was discussed in Sec. 14.13, this force acts always at right angles to the direction of motion of the charged particle. The particle experiences a continual deflection and, if sent into a uniform field at right angles, moves along the arc of a circle. (It might be wise to review Sec. 14.13 now.)

This property had been used in the 1890's by J. J. Thomson in his studies of cathode rays. He showed that these rays consist of very small negatively charged particles, or electrons (Chapter 18). Becquerel, the Curies, and others found that the α, β, and γ rays behaved differently from one another in a magnetic field. The behavior of the rays is illustrated in the diagram on page 640.

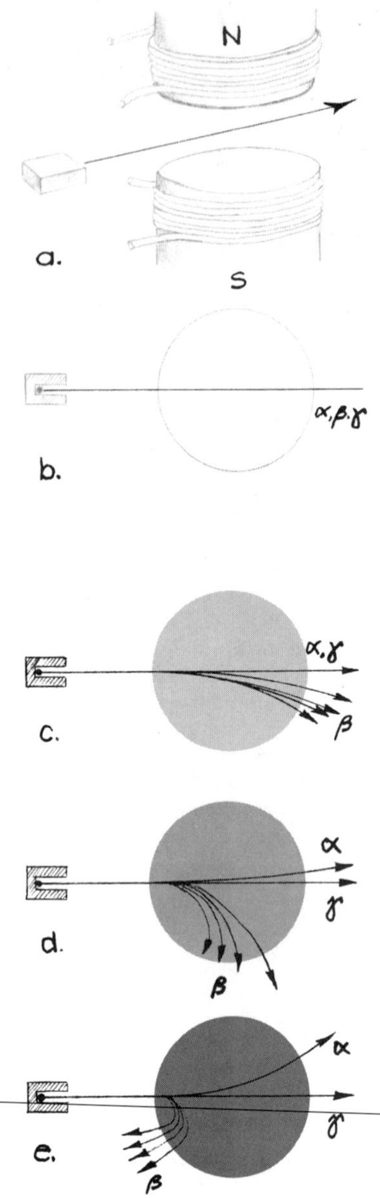

(a) α, β, and γ rays are separated from a sample of radioactive material by their passage through a magnetic field. (b) No magnetic field. (c) Weak magnetic field. (d) Stronger magnetic field. (e) Very strong magnetic field.

Suppose that some radioactive material, such as a sample of uranium, is placed at the end of a narrow hole in a lead block and that a narrow beam consisting of α, β, and γ rays escapes from the opening. If the beam enters a strong, uniform magnetic field (as in the last two drawings in the margin), the three types of rays will go along paths separated from one another. The γ rays continue in a straight line without any deviation. The β rays will be deflected to one side, moving in circular arcs of differing radii. The α rays will be deflected slightly to the other side, moving in a circular arc of large radius; they are rapidly absorbed in the air.

The direction of the deflection of the β rays in such a magnetic field is the same as that observed earlier in Thomson's studies of the properties of cathode rays. It was concluded, therefore, that the β rays, like cathode rays, consist of *negatively charged particles*. (The negative charge on the β particles was confirmed by the Curies in 1900; they caused the beam of the particles to enter an electroscope that became negatively charged.) Since the direction of the deflection of the α rays was opposite that of the β rays, it was concluded that the α rays consist of *positively charged particles*. Since the γ rays were not deflected, it was concluded that they are neutral, that is, they have no electric charge. No conclusion could be drawn from this type of experiment as to whether the γ rays are, or are not, particles.

The deflection of a charged particle in electric and magnetic fields depends on both its charge and mass. Therefore, the ratio of charge to mass for β particles can be calculated from measured deflections in fields of known intensity. Becquerel, investigating β particles in 1900, used a procedure that was essentially the same as that used by J. J. Thomson in 1897 to obtain a reliable value for the ratio of charge q_e to mass m_e for the particles in cathode rays. [At that time, the fact that a consistent single value of q_e/m had been found established quantitatively the *existence* of the electron (see Sec. 18.2).] By sending β rays through electric and magnetic fields, Becquerel was able to calculate the speed of the β particles. He obtained a value of q/m for β particles which was close to that found by Thomson for the electron, and so permitted the deduction that the β particles are electrons.

SG 4–6

The nature of the α radiation was more difficult to establish. It was necessary to use a very strong magnetic field to produce measurable deflections of α rays. The value of q/m found for α particles (4.8×10^7 C/kg) was about 4,000 times smaller than the q/m for β particles. The reason for the small q/m value could be a small value of q or a large value of m. Other evidence available at the time indicated that q for an α particle was not likely to be smaller than that for a β particle. It was therefore concluded that

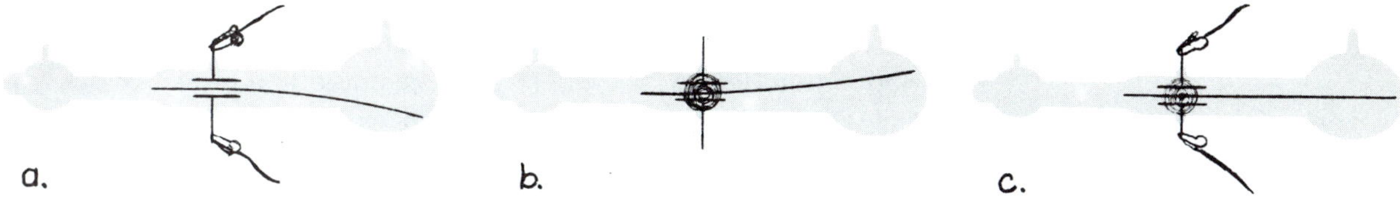

Electric and magnetic fields can be set up perpendicularly so that the deflections they cause in a beam of charged particles will be in opposite directions. Particles moving at a certain speed will not be deflected, because the electric and magnetic forces balance. (a) Electric field only. (b) Magnetic field only. (c) Both electric and magnetic field.

m would have to be much larger for the α particle than for the β particle.

The value of q/m given above for α particles is just one-half that of q/m found earlier for a hydrogen ion. The value would be explained in a reasonable way if the α particle were like a hydrogen molecule minus one electron (H_2^+), or if it were a helium atom (whose mass was known to be about four times that of a hydrogen atom) without its two electrons (He^{++}). Other possibilities might have been entertained, for example, bare nuclei of carbon, nitrogen, or oxygen would have about the same q/m ratio. In fact, however, the right identification turned out to be that of α particles with He^{++}. The clever experiment described in the following section provided the final proof.

The α particle was found to be the same as a helium nucleus and, therefore, has a mass of about 4 atomic mass units.

?

8. What was the evidence to support the theory that β particles are electrons?

9. What observation led to the suggestion that α particles are much more massive than β particles?

21.5 | The identity of α rays: Rutherford's "mousetrap"

It was known that the gas helium can be found imprisoned in radioactive minerals. In addition, Sir William Ramsey and Frederick Soddy had discovered, in 1903, that helium is given off from a radioactive compound, radium bromide. This led Rutherford to advance the hypothesis that the α particle is a doubly ionized helium atom, that is, a He atom minus its two electrons, or, as we would now say, the nucleus of a helium atom. In a series of experiments conducted from 1906 to 1909, Rutherford succeeded in proving the correctness of his hypothesis in several different ways. The last and most convincing of these experiments was made in 1909, with T. D. Royds, by constructing what Sir James Jeans later called "a sort of mousetrap for α particles."

The experiment used the radioactive element radon (Rn). Radon had been discovered by Pierre Curie and André Debierne in 1901; they had found that a gas is given off from radium. A small amount of the gas collected in this way was found to be a

Rutherford's "mousetrap" for identifying particles.

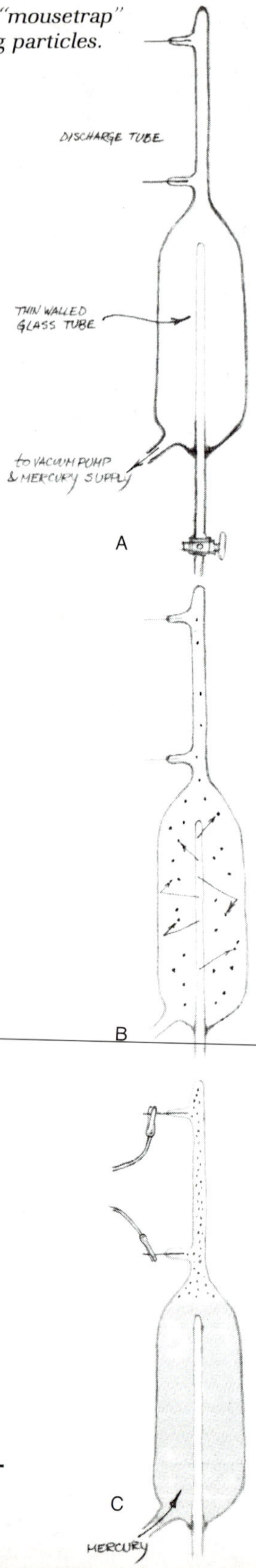

strong α emitter. The gas was shown to be a new element and was called "radium emanation" and later "radon." Ramsey and Soddy then found that when radon is stored in a closed vessel, helium always appears in the vessel also. Thus, helium is given off not only by radium by also by radon.

Rutherford and Royds put a small amount of radon in a fine glass tube with a wall only 0.01 mm thick. This wall was thin enough so that α particles could pass through it, but radon itself could not. The tube was sealed into a thick-walled, outer glass tube that had an electric discharge section at the top. (See sketch A in the margin.) The air was pumped out of the outer tube, and the apparatus was allowed to stand for about a week. During this time, while α particles from the radon passed through the thin walls of the inner tube, a gas was found gradually to collect in the previously evacuated space (sketch B). Mercury was then pumped in at the bottom to compress the very small quantity of gas and confine it in the discharge tube (sketch C). When a potential difference was applied to the electrodes of the discharge tube, an electric discharge was produced in the gas. The resulting light was examined with a spectroscope, and the spectral lines seen were characteristic of helium. (In a separate control experiment, helium gas itself was put in the inner, thin-walled tube and did not leak through the wall of the inner tube.)

Now it was clear to Rutherford how to interpret his results. He could safely conclude that the helium gas that collected in the outer tube was formed from α particles that had passed into the outer tube. Rutherford's result implied conclusions more important than just the identity of α particles. Apparently, an atom of an element (radon) can spontaneously emit a fragment (an α particle) that is the nucleus of *another* element (helium). A startling idea, but only the beginning of more startling things to come.

?

10. How did Rutherford know that the gas that appeared in the tube was helium?

21.6 | Radioactive transformations

The emission of α and β particles raised difficult questions with respect to existing ideas of matter and its structure. The rapid development of chemistry in the nineteenth century had made the atomic–molecular theory of matter highly convincing. According to this theory, a pure element consists of identical atoms, which are indestructible and unchangeable. But if a radioactive atom emits as substantial a fragment as an α particle (shown to be an ionized helium atom), can the radioactive atom remain unchanged? That did not seem plausible. Rather, it seemed that there must be a transformation in which the radioactive atom is changed to an atom of a different chemical element.

If an atom emits an α particle, a substantial part of its mass will be carried away by the α particle. What about the atoms that emit β particles? The β particle (shown to be an electron) is far less massive than the α particle. However, its mass is not zero; so a radioactive atom must also undergo some change when it emits a β particle. It was again difficult to escape the conclusion that radioactive atoms are, in fact, subject to division (into two parts of markedly unequal mass), a conclusion contrary to the basic concept that the atom is indivisible.

Another fundamental question arose in connection with the energy carried by the rays emitted by radioactive substances. As early as 1903, Rutherford and Soddy, and Pierre Curie and a young co-worker, A. Laborde, noted that a sample of radium kept itself at a higher temperature than its surroundings merely by reabsorbing some of the energy of the α particles emitted by atoms inside the sample. (Curie and Laborde found that 1 g of radium can produce about 0.1 kcal of heat per hour.) A sample of radium thus can continue to release energy year after year, and evidently for a very long time.

The continuing release of such a quantity of heat could not be explained by treating radioactivity as an ordinary chemical reaction. It was clear that radioactivity did not involve chemical changes in the usual sense. Energy was emitted by samples of pure elements; energy emission by compounds containing radioactive elements did not depend on the type of molecule in which the radioactive element was present. The origin of the production of heat had to be sought in some deep changes *within* the atoms of radioactive elements, rather than in chemical reactions among atoms.

Rutherford and Soddy proposed a bold theory of *radioactive transformation* to explain the nature of these changes. They suggested that when a radioactive atom emits an α or a β particle, it really breaks into two parts: the α or β particle that is emitted and a heavy, leftover part that is physically and chemically different from the "parent" atom. There was a good deal of evidence for the last part of the assumption. For example, the formation of radon gas from radium was known, as mentioned earlier. When the atomic mass of radon was determined, it turned out to be smaller than that of radium by just 4 atomic mass units, the mass of an α particle.

The idea of radioactive transformation can be represented by an "equation" similar to the kind used to represent chemical reactions. For example, using the symbols Ra and Rn to

The water is being boiled by the heat given off by a small capsule of cobalt-60. This capsule, the first ever made to produce heat from radioactive cobalt, was generating heat at the rate of 360 watts when this photo was taken.

Here He stands for the helium atom formed by the doubly charged α particle after it has picked up two electrons.

Rutherford and Soddy received Nobel prizes in chemistry for their work on the radioactive transformation of one element into another.

represent atoms of radium and radon, the transformation of radium into radon can be expressed as:

$$\text{Ra} \rightarrow \text{Rn} + \text{He}$$

The process of transformation can be described as the transformation ("disintegration," "decay," or "transmutation") of radium into radon, with the emission of an α particle.

Many decay processes in addition to the example just cited had been found and studied by the Curies, by Rutherford and his co-workers, and by others, and these processes fitted easily into the kind of scheme proposed by Rutherford and Soddy. For example, radon is radioactive also, emitting another α particle and thereby decaying into an atom of an element that was called "radium A" at the time. Radium A was later shown to be polonium (Po):

$$\text{Rn} \rightarrow \text{Po} + \text{He}$$

Polonium is also a radioactive solid. In fact, the original "parent" radium atoms undergo a series or chain of transformations into generation after generation of new, radioactive, "daughter" elements, ending finally with a "daughter" element that is nonradioactive or, in other words, stable.

SG 7

?

11. Why was radioactive decay believed not to be an ordinary chemical reaction?

12. Give an example of a radioactive transformation. Why is it contrary to the ideas of nineteenth-century chemistry?

21.7 | Radioactive decay series

The decay of radium and its daughters was found eventually to lead to a stable end product that was identified by its chemical behavior as *lead*. The chain beginning with radium has 10 members, some emitting α particles and others emitting β particles. Some gamma rays are emitted during the decay series, but gamma rays do not appear alone; they are emitted only together with an α particle or a β particle. Rutherford and Soddy also suggested that, since radium is always found in uranium ores, radium itself may be a member of a series starting with uranium as the ancestor of all the members. Research showed that this is indeed the case. Each uranium atom may in time give rise to successive daughter atoms, radium being the sixth generation and stable lead the fifteenth.

Table 21-1 shows all the members of the so-called *uranium–radium series*. The meaning of some of the symbols will be discussed in later sections. The number following the name of an element, as in uranium-238, indicates the atomic mass. Notice that there are heavier and lighter varieties of the element, such as uranium-238 and 235, polonium-218, 214, and 210. Much more will be said about these varieties in the next chapter.

Two other naturally occurring radioactive series have been found; one starts with thorium 232 and the other with uranium 235. (See SG 7 and 8, Chapter 22.)

TABLE 21-1. URANIUM–RADIUM DECAY SERIES

Old Name	*Present Name and Symbol*		*Mode of Decay*	*Half-Life*
Uranium I	Uranium-238	$_{92}U^{238}$	α,γ	4.51×10^9 years
Uranium X_1	Thorium-234	$_{90}Th^{234}$	β,γ	24.1 days
Uranium X_2	Protactinium-234	$_{91}Pa^{234}$	β,γ	1.18 minutes
Uranium II	Uranium-234	$_{92}U^{234}$	α,γ	2.48×10^5 years
Ionium	Thorium-230	$_{90}Th^{230}$	α,γ	8.0×10^4 years
Radium	Radium-226	$_{88}Ra^{226}$	α,γ	1620 years
Radium Emanation	Radon-222	$_{86}Rn^{222}$	α,γ	3.82 days
Radium A	Polonium-218	$_{84}Po^{218}$	α	3.05 minutes
Radium B	Lead-214	$_{82}Pb^{214}$	β,γ	26.8 minutes
Radium C	Bismuth-214	$_{83}Bi^{214}$	α,β,γ	19.7 minutes
Radium C′	Polonium-214	$_{84}Po^{214}$	α	1.64×10^{-4} seconds
Radium D	Lead-210	$_{82}Pb^{210}$	β,γ	21.4 years
Radium E	Bismuth-210	$_{83}Bi^{210}$	β,γ	5.0 days
Radium F	Polonium-210	$_{84}Po^{210}$	α,γ	138.4 days
Radium G	Lead-206	$_{82}Pb^{206}$	stable	

Each member of the series differs physically and chemically from its immediate parent above or daughter below; it should, therefore, be possible to separate the different members in any radioactive sample. This is by no means impossible to do, but the separation problem is made difficult by the fact that the different radioactive species decay at different rates, some very slowly, some rapidly, others at intermediate rates. These rates and their meaning will be discussed in the next section, but the fact that the rates differ gives rise to important effects that can be discussed now.

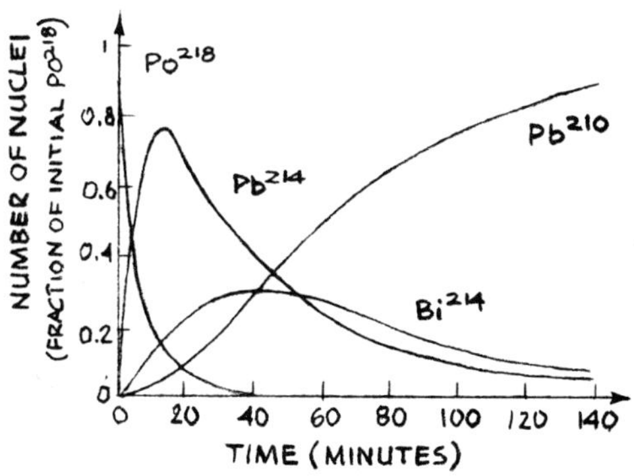

An interesting example is supplied by that portion of the uranium series that starts with the substance called polonium-218. A pure sample of polonium-218 may be collected by exposing to the gas radon a piece of ordinary material such as a thin foil of aluminum. Some of the radon atoms decay into polonium-218 atoms, which then stick to the surface of the foil. The graph at the right shows what becomes of the polonium-218. Polonium-218 (Po^{218}) decays into lead-214 (Pb^{214}), which decays into bismuth-214 (Bi^{214}), which decays into polonium-214 (not shown), then lead-210, etc. If the original sample contains 1,000,000 atoms of polonium-218 when it is formed, after 20 min it will contain about 10,000 Po^{218} atoms, about 660,0000 Pb^{214} atoms, about 240,000 Bi^{214} atoms, and about 90,000 Pb^{210} atoms. The number of Po^{214} atoms is negligibly small because most of the Po^{214} changes into Pb^{210} in a small fraction of a second. (Later, there would also begin to accumulate a signficant number of atoms from the list in Table 21-1.)

A sample of pure, freshly separated radium (Ra^{226}) would also change in composition in a complicated way, but much more slowly. Eventually it would consist of a mixture of some remaining radium-226, plus radon-222, polonium-218, lead-214, and all the rest of the members of the chain down to, and

SG 8 including, stable "radium G" (lead-206).

Similarly, a sample of pure uranium may contain, after a time, 14 other elements of which 13 (all but the last, stable portion) contribute to the radioactive emission, each in its own way. In all such cases, a complicated mixture of elements results. After starting as a pure α emitter, a sample eventually emits many α particles, β particles, and γ rays, apparently continuously and simultaneously.

It is evident that the separation of the different members of a radioactive chain from one another would be difficult, especially if some members of the chain decay rapidly. The determination of the chemical nature and the radioactive properties of each member required great experimental ingenuity. One successful method depended on the skillful chemical purification of a particular radioactive substance, as the Curies had done with radium and polonium. For example, suppose that a sample has been obtained from which all the radioactive atoms except those of radium-226 have been removed. The sample immediately starts to give off radon gas. The latter can be drawn off and its properties examined before it becomes seriously contaminated by the disintegration of many of its atoms into polonium-218. If this is done, it is found that radon decays (through several transformations) into lead much more quickly than radium decays into radon.

?

13. Give at least three reasons for the difficulty of separating decay products.

14. If you start with a sample made entirely of pure uranium-238 atoms, what emission is observed at the start? How will the emission change as time goes on?

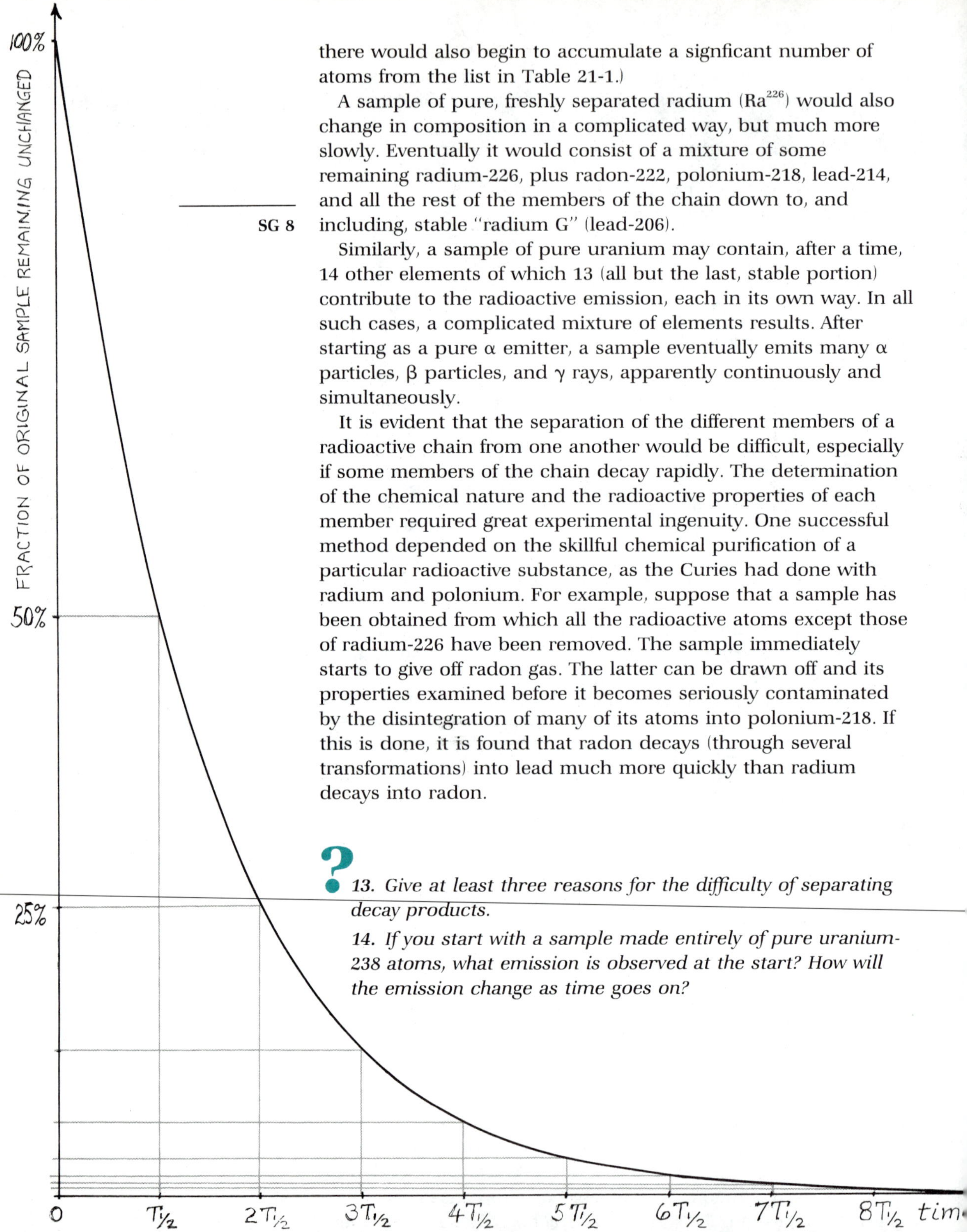

21.8 | Decay rate and half-life

In the last section, you saw that of 1,000,000 polonium-218 atoms present in a freshly prepared sample of that radioactive substance, only about 10,000 would remain after 20 min, the rest having decayed into atoms of lead-214 and its daughter products. It would take only 3 min following the preparation of the pure sample of Po^{218} for 50% of the atoms originally present in the sample to have decayed. In the case of radium (Ra^{226}), it would take 1,620 years for half of the radium atoms in a freshly prepared sample of radium to be transformed into radon atoms.

In 1898, the Curies obtained a total of about 200 g of radium. Seventy years later (1968), 194 g of this remained as radium. The other 6 g corresponded to 16×10^{24} radium atoms that had decayed into radon and subsequently into other elements during those 70 years.

These two examples illustrate the experimental fact that samples of radioactive elements show great difference in their rates of decay. These different rates are the result of *averages* of many individual, different decay events going on at random in a sample. Looking at *one* atom of any radioactive element, one never can tell when it will decay; some may decay as soon as they are produced, while others may never decay. Still, it has been found experimentally that for a large group of atoms of one kind, *the fraction of these atoms that decay per second* is unchangeable and aways the same for any large group of atoms of that kind. This fraction is almost completely independent of all physical and chemical conditions, such as temperature, pressure, and form of chemical combination. These remarkable properties of radioactivity deserve special attention, and the meaning of the italicized statement above now will be discussed in detail because it is basic to an understanding of radioactivity.

In a few cases, pressure and chemical combination have been found to make slight (and now well-understood) differences in the rate of decay.

Say, for example, that 1/1,000 of the atoms in a freshly prepared pure sample decay during the first second. Then you would expect that 1/1,000 of the remaining atoms will decay during the next second. But also, 1/1,000 of the atoms remaining after 10 sec will decay during the eleventh second, and so on. In fact, during any subsequent second of time, 1/1,000 of the atoms remaining at the beginning of that second will decay, at least until the number of remaining atoms becomes so small that predictions become very uncertain.

SG 9, 10

Since the *fraction* of the atoms that decay per unit time is a constant for each element, the *number* of atoms that decay per unit time will decrease in proportion to the diminishing number of atoms that have not yet changed. Consequently, if the percentage of surviving, unchanged atoms is plotted as a function of time, a curve like the one on the opposite page is obtained. The number of atoms in a sample that decay per unit time is the *activity* of the sample. Thus, the graph on the opposite page also represents the way in which the measured activity of a sample would decrease with time.

If the daughter atoms were also radioactive, then the change of measured activity would, of course, be complicated and not have so simple a form of graph.

The curve that shows the number of atoms that have not decayed as a function of time approaches the time axis asymptotically; that is, the number of survivors becomes small, but it never becomes zero. This is another way of saying that a definite "lifetime" in which all of the original atoms for a sample will have decayed cannot be assigned.

SG 11 However, it is possible to specify the time required for any particular *fraction* of a sample to decay, ½, ⅓, or 37%, for instance. For convenience in making comparisons, the fraction ½ has been chosen. Rutherford called the time required for the decay of one-half of the original atoms of a pure sample the *half-life*. Each kind of radioactive atom has a unique half-life, and thus the half-life of an element can be used to identify a radioactive element. As Table 21-1 shows, a wide variety of half-lives have been found.

For uranium-238, the parent of the uranium series, the half-life is 4.5 billion years. This means that after 4.5×10^9 years, half of the uranium-238 atoms will have decayed. For polonium-214, the half-life is of the order of 10^{-4} sec; that is, in only 1/10,000 of a second, half of an original sample of Po^{214} atoms will have decayed. If pure samples of each, containing the same number of atoms, were available, the initial activity (atoms decaying per second) of polonium-214 would be very strong and that of uranium-238 very feeble. If left for even 1 min though, the polonium would have decayed so thoroughly and, therefore, the number of its surviving atoms would be so small, that at this point the activity due to polonium would now be less than the activity of the uranium. Perhaps some radioactive elements, present in great quantities long ago, decayed so rapidly that no measurable traces are now left. On the other hand, many radioactive elements decay so slowly that during any ordinary experimentation time the counting rates that indicate decay seem to remain constant.

The principal advantage of the concept of half-life lies in the experimental result implied in the graph in the margin: For any element of half-life $T_{1/2}$, no matter how old a sample is, half of the atoms will still have survived after an additional time interval $T_{1/2}$. Thus, the half-life is not to be thought of as an abbreviation for "half a life." If one-half the original atoms remain unchanged after a time $T_{1/2}$, one-fourth (½ × ½) will remain after two consecutive half-life intervals $2T_{1/2}$, one-eighth after $3T_{1/2}$, and so on. Note how different the situation is for a population of, say, human beings instead of radioactive atoms. In a group of N_0 babies, half the number may survive to their 70th birthday; of these $N_0/2$ senior citizens, none is likely to celebrate a 140th birthday. But of N_0 radioactive atoms with a half-life of 70 years,
SG 12–15 $N_0/4$ will have remained intact after 140 years, $N_0/8$ after 210 years, etc. To put it differently, *the statistical probability of survival for atoms is unchanged by the age they have already reached.* In humans, of course, the probability of survival (say, for

Close Up

The Mathematics of Decay

The *activity* of a sample, the number of disintegrations per second, the decay rate are alternative expressions for the same quantity. Using the letter N to represent generally the number of atoms of a given kind present in a radioactive sample, the activity is $\Delta N/\Delta t$, where ΔN is the number of atoms disintegrating in the same interval Δt. But $\Delta N/\Delta t$ depends both on the type of atom involved, and how many happen to be in the sample. Therefore, a more useful quantity is needed. If, in a time interval Δt, ΔN atoms disintegrate out of a total number N, the *fraction* of atoms disintegrating is $\Delta N/N$. The *fraction* of atoms *disintegrating per unit time is* $\Delta N/N/\Delta t$. (this quantity can be thought of as the ratio of the activity $\Delta N/\Delta t$ to the total number, N.) This quantity, usually called λ or the decay constant, will be important, as you will see at once below. It is analogous to the death rate in a human population. In the United States, for example, about 5,000 persons die each day out of a population of about 200,000,000. The death rate is therefore one person per 40,000 per day (or one person per day per 40,000).

The beautifully simple mathematical aspect of radioactive decay is that the fraction of atoms decaying per second does not change with time. If initially there are N_0 atoms, and a certain fraction λ decay in 1 sec, the actual number of atoms decaying in 1 sec is λN_0. Then, at any later time t, when there are only N_t atoms remaining, the *fraction* that decay in 1 sec will still be λ, but the *number* of atoms decaying in one second is now λN_t, a smaller number than before.

The constant fraction λ of atoms decaying per unit time is called the *decay constant*. The value of this constant λ can be found for each radioactive species. For example, λ for radium is 1.36×10^{-11} per second, which means that on the average 0.0000000000136th of the total number of atoms in any sample of radium will decay in 1 sec.

The fact that λ is a constant can be represented by the expression

$$\lambda = \frac{\Delta N/\Delta t}{N} = \text{constant}$$

which can be rewritten as

$$\Delta N/\Delta t = \text{constant} \times N \quad \text{or} \quad \Delta N/\Delta t \propto N$$

This form of the relation expresses clearly the fact that the decay rate depends directly on the number of atoms left.

By using calculus, a relation of this type can be turned into an expression for N as a function of elapsed time t:

$$\frac{N_t}{N_o} = e^{-\lambda t} \quad \text{or} \quad N_t = N_o e^{-\lambda t}$$

where N_o is the number of atoms at $t = 0$, N_t is the number remaining unchanged at time t, and e is a mathematical constant that is approximately equal to 2.718. The factor $e^{-\lambda t}$ has the value 1 when $t = 0$, and decreases toward 0 as t increases. Since the decay constant appears as an exponent, the decay is called "exponential" and takes the form shown by the graph on page 646.

The relationship between the half-life $T_{1/2}$ and the decay constant λ can be derived as follows. Write the exponential decay equation in logarithmic form by taking the logarithm of both sides of the equation

$$\log \frac{N_t}{N_o} = \log e^{-\lambda t} = -\lambda t \log e$$

After a time equal to the half-life $T_{1/2}$, the ratio $N_t/N_o = 1/2$. So you can substitute $1/2$ for N_t/N_o if you substitute $T_{1/2}$ for t in the above equation, and get

$$\log (1/2) = -\lambda T_{1/2} \log e$$

The value of $\log (1/2)$ is -0.301 and the value of $\log e = 0.4343$; therefore,

$$-0.301 = -\lambda T_{1/2} (0.4343)$$
$$\text{and} \quad \lambda T_{1/2} = 0.693$$

So the product of the decay constant and the half-life is always equal to 0.693. Knowing either one allows you to compute the other.

For example, radium-226 has a decay constant $\lambda = 1.36 \times 10^{-11}$ per second; so

$$(1.36 \times 10^{-11}\ \text{sec}^{-1})\ T_{1/2} = 0.693$$
$$T_{1/2} = \frac{0.693}{1.36 \times 10^{-11}\ \text{sec}^{-1}}$$
$$T_{1/2} = 5.10 \times 10^{10}\ \text{sec}$$

Thus, the half-life of radium-226 is 5.10×10^{10} sec (about 1620 years).

another year) depends strongly on age, and so the concept "half-life" is not usable in this case.

The use of this statistical law, in practice, is justified because even a minute sample of a radioactive element contains very many atoms. For example, one-millionth of a gram of uranium contains 3×10^{15} atoms.

You have been considering the behavior not of individual atoms, but of a very large number of them. As you saw in considering the behavior of gases in Chapter 11, this method allows you to use laws of statistics to describe the average behavior of the group. If a hundred thousand people were to flip coins simultaneously just once, you could predict with good accuracy that about one-half of them would get heads. But you could not accurately predict that one particular person in this crowd would obtain heads on a single flip. If the total number of coins tossed is small (10) the observed count is likely to differ considerably from the prediction of 50% heads. From experiments in radioactivity, you can predict that a certain fraction of a relatively large number of atoms in a sample will survive in any given time interval (for example, one-half will survive to reach the age *T*), but not whether a particular atom will be among the survivors. As the sample of survivors decreases in size owing to disintegrations, predictions become less precise. Eventually, when only a few unchanged atoms are left, you could no longer make useful predictions at all. In short, the disintegration law is a *statistical* law and is thus applicable only to large populations of the radioactive atoms. Moreover, it makes no assumptions as to *why* the atoms disintegrate.

In the discussion of the kinetic theory of matter, you saw that it is a hopeless and meaningless task to try to describe the motions of each individual molecule, but you could calculate the *average* pressure of a gas containing a very large number of molecules. Similarly, in dealing with radioactivity, the inability to specify when each of the tremendous number of atoms in a normal sample will disintegrate makes a statistical treatment necessary and useful.

Radioactive decay pattern for strontium-90.

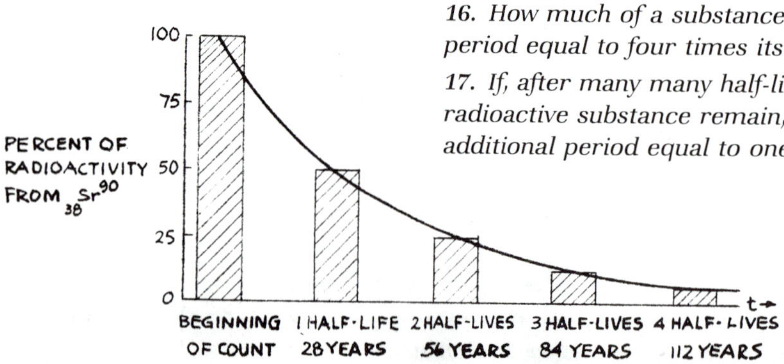

?

15. Why can one not specify the lifetime of a sample of radioactive atoms? of a single radioactive atom?

16. How much of a substance will be left unchanged after a period equal to four times its half-life?

17. If, after many many half-lives, only two atoms of a radioactive substance remain, what will happen during an additional period equal to one half-life?

study guide

1. The *Project Physics* learning materials particularly appropriate for Chapter 21 include:

Experiments
Random Events
Range of α and β Particles
Half-Life I
Half-Life II
Radioactive Tracers
Measuring the Energy of β Radiation

Activities
Magnetic Deflectionof β Rays
A Sweet Demonstration
Ionization of Radioactivity
Exponential Decay in Concentrations

Transparencies
Separation of α, β, γ Rays
Rutherford's α-Particle "Mousetrap"
Radioactive Disintegration Series

2. Which of the Curies' discoveries would have been unlikely if they had used Becquerel's photographic technique for detecting radioactivity?

3. A spectroscopic examination of the γ rays from bismuth-214 shows that rays of several discrete but different energies are present. The rays are said to show a "line spectrum." The measured wavelength corresponding to one of the lines is 0.0016 nanometer.

(**a**) Show that the energy of each of the γ-ray photons responsible for that line is 1.2×10^{-13} J. (Hint: See Chapter 20.)

(**b**) What is the equivalent energy in electron volts?

4. Suppose that in the figure on page 640 the magnetic field strength is 1.0×10^{-3} N/A-m.

(**a**) What would be the radius of curvature of the path of electrons entering the magnetic field with a speed of 1.0×10^{7} m/sec? (The charge and mass of the electron are 1.6×10^{-19} C and 9.1×10^{-31} kg, respectively.)

(**b**) If α particles entered the magnetic field with the same speed as the electrons in part (**a**), what would be the radius of curvature of their orbit? (The mass of an α particle is 6.7×10^{-27} kg.)

(**c**) Compare your answers to parts (**a**) and (**b**).

5. The electric field in the figure on page 641 is produced by a + charge at the top plate and a − charge at the bottom. What is the sign of charges in the beam going through the tube? What is the direction of the magnetic field (into or out of the page)?

6. If the electrons described in part (a) of SG 4 pass through crossed electric and magnetic fields as shown in part (c) of the figure on page 641,

(**a**) what must be the strength of the electric field to just balance the effect of a magnetic field of strength 1.0×10^{-3}N/A-m?

(**b**) what voltage must be supplied to the electric field deflecting plates to produce the electric field strength of part (**a**) of this problem if the plates are 0.10 m apart?

(**c**) what will happen to the α particles of SG 4(b) moving through the crossed magnetic and electric fields of this problem?

7. For each part below, select the most appropriate radiation(s): α, β, or γ.

(**a**) most penetrating radiation

(**b**) most easily absorbed by aluminum foil

(**c**) most strongly ionizing radiation

(**d**) may require thick "radiation shields" for protection

(**e**) cannot be deflected by a magnet

(**f**) largest radius of curvature when traveling across a magnetic field

(**g**) highest *q/m* value

(**h**) important in Rutherford's and Royd's "mousetrap" experiment

(**i**) involved in the transmutation of radium to radon

(**j**) involved in the transmutation of bismuth-210 to polonium-210

8. Suggest an explanation for the following observations:

The English physicist Sir William Crookes discovered in 1900 that immediately after a strongly radioactive uranium-containing compound was purified chemically, the uranium compound itself showed much smaller activity, and the separate residue that contained none of the uranium was strongly radioactive.

In 1901, Becquerel found that in such a case the uranium compound regained its original activity after several months, while the residue gradually lost most of its activity during the same time.

9. A Geiger counter shows that the rate of emission of β particles from an initially pure sample of a radioactive element decreases to one-half the initial rate in 25 hours.

(**a**) What fraction of the original number of radioactive atoms is still unchanged at that time?

(**b**) What fraction of the original number will have disintegrated in 50 hours?

(**c**) What assumptions have you made in giving these answers? How might you check them?

10. It took 10 years for 10% of the atoms of a certain freshly prepared sample of radioactive substance to decay. How much of the material that is left unchanged will decay in the *next* 10 years?

11. Suppose at time t_o a sample of pure radium consisted of 2.66×10^{21} atoms. (The half-life of radium is approximately 1,600 years.)

(**a**) If N_t is the number of radium atoms in the sample at a time t, make a graph of N_t versus time covering a period of 8,000 years.

(**b**) Show that at the end of 8,000 years, 8.3×10^{19} radium atoms still remain in the sample.

(**c**) From your graph, estimate the number of radium atoms in the sample after 4,000 years.

12. The capsule containing cobalt-60, shown and described on page 643, is reported to have an activity of 17,000 curies. One curie is defined as 3.70×10^{10} disintegrations per second.

(**a**) How much energy is released per disintegration in the cobalt-60?

(**b**) What would be the rate of heat production of that sample after 15 years? (The half-life of cobalt-60 is approximately 5 years.) Assume all radioactive emission is converted into heat in the water jacket around the sample.

13. Radioactive isotopes in quantities of 10 micro-curies or less can be purchased for instructional purposes from the Department of Energy. How many disintegrations per second occur in a 10 micro-curie sample? (For data, see SG 12.)

14. Below are the observed disintegration rates (counting rates) as a function of time for a radioactive sample.

Time (hr)	*Counting Rate (counts/min)*	*Time (hr)*	*Counting Rate (counts/min)*
0.0	....	6.0	1800
0.5	9535	7.0	1330
1.0	8190	8.0	980
1.5	7040	9.0	720
2.0	6050	10.0	535
3.0	4465	11.0	395
4.0	3300	12.0	290
5.0	2430		

(**a**) Plot the data, and determine the approximate half-life of this substance.

(**b**) How many atoms decay each minute for each 10^6 atoms in this sample? (Use the relationship between λ and T derived on page 649.) Does this number remain constant?

15. Rutherford and Soddy, working with samples of compounds of thorium, obtained results similar to those described in SG 8. Their results, published in 1903, are shown below.

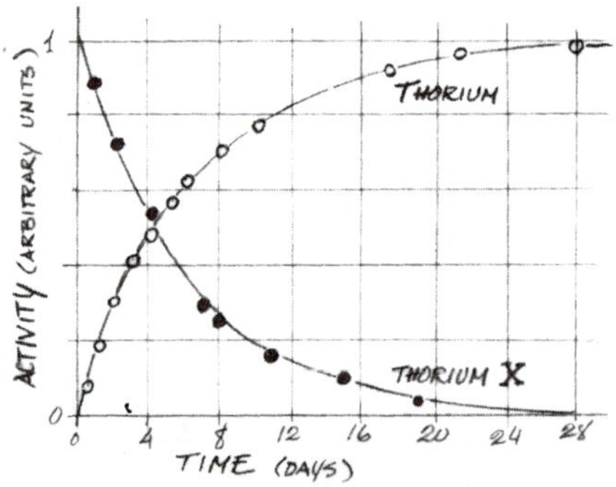

(**a**) What is the half-life of the material they called thorium X?

(**b**) In 1931, Rutherford was elevated to the British peerage, becoming "Baron Rutherford of Nelson." It is claimed that there is a connection between Rutherford's design of his coat of arms (shown below) and his work. What might the connection be?

RUTHERFORD OF NELSON.

16. Activity: Magnetic Deflection of β Rays. Clamp a radioactive β source securely a distance of about 30 cm from a Geiger tube. Place a sheet of lead at least 1 mm thick between source and counter to reduce the count to background level. Hold one end (pole) of a strong magnet above or to the side of the sheet, and change its position until the count rate increases appreciably. By what path do the β rays reach the counter? Try keeping the magnet in the same position but reversing the two poles; does the radiation still reach the counter? Determine the polarity of the magnet by using a compass needle. If β rays are particles, what is the sign of their charge? (See Experiment 4–6 for hints.)

17. Activity: A Sweet Demonstration. In Experiment 2, "Half-Life I," it is difficult to show that the number of dice "decaying" is directly proportional to the initial number of dice, because statistical fluctuations are fairly large with only 120 dice. An inexpensive way to show that ΔN is directly proportional to N is to use at least 400 sugar cubes (there are 198 in the commonly available 0.45-kg packages). Mark one face with edible food coloring. Then shake the cubes and record how many decayed as described in Experiment 6-2.

18. Activity: Ionization by Radioactivity. Place a different radioactive sample inside each of several identical electroscopes. Charge the electroscopes negatively (as by rubbing a hard, rubber comb on wool and touching the comb to the electroscope knob). Compare the times taken for the electroscopes to completely lose their charges, and interpret your observations.

Place no sample in one electroscope so that you can check how fast it discharges without a sample present. What causes this type of discharge?

19. Activity: Exponential Decay in Concentration. Stir 10 drops of food coloring into 1,000 cm^3 of water. Pour off 100 cm^3 into a beaker. Add 100 cm^3 of water, stir up the mixture, and pour off a second 100 cm^3 sample. Keep repeating until you have collected 10–15 samples.

The original concentration was 10 drops/1,000 cm^3 or 1 drop/100 cm^3. What is the concentration after one removal and the addition of pure water (one dilution cycle)? What is the concentration after two cycles? after three cycles? after n cycles? [Answer: $(0.9)^n$ drops/100 cm^3.]

What is the number of cycles required to reduce the concentration of approximately one-half of its original concentration?

How many times would you have to repeat the process to get rid of the dye completely?

PERIODIC TABLE OF ELEMENTS

Group→ Period↓	I	II											III	IV	V	VI	VII	0
1	1.0080 H 1																	4.0026 He 2
2	6.939 Li 3	9.012 Be 4											10.811 B 5	12.011 C 6	14.007 N 7	15.999 O 8	18.998 F 9	20.183 Ne 10
3	22.990 Na 11	24.32 Mg 12											26.98 Al 13	28.09 Si 14	30.97 P 15	32.06 S 16	35.45 Cl 17	39.95 Ar 18
4	39.10 K 19	40.08 Ca 20	44.96 Sc 21	47.90 Ti 22	50.94 V 23	52.00 Cr 24	54.94 Mn 25	55.85 Fe 26	58.93 Co 27	58.71 Ni 28	63.54 Cu 29	65.37 Zn 30	69.72 Ga 31	72.59 Ge 32	74.92 As 33	78.96 Se 34	79.91 Br 35	83.80 Kr 36
5	85.47 Rb 37	87.62 Sr 38	88.91 Y 39	91.22 Zr 40	92.91 Nb 41	95.94 Mo 42	(99) Tc 43	101.07 Ru 44	102.91 Rh 45	106.4 Pd 46	107.87 Ag 47	112.40 Cd 48	114.82 In 49	118.69 Sn 50	121.75 Sb 51	127.60 Te 52	126.9 I 53	131.30 Xe 54
6	132.91 Cs 55	137.34 Ba 56	* 57–71	178.49 Hf 72	180.95 Ta 73	183.85 W 74	186.2 Re 75	190.2 Os 76	192.2 Ir 77	195.09 Pt 78	196.97 Au 79	200.59 Hg 80	204.37 Tl 81	207.19 Pb 82	208.98 Bi 83	(210) Po 84	(210) At 85	222 Rn 86
7	(223) Fr 87	226.05 Ra 88	† 89-103	(261) Rf 104	(260) Ha 105	(263) 106	etc.											

*Rare-earth metals	138.91 La 57	140.12 Ce 58	140.91 Pr 59	144.27 Nd 60	(147) Pm 61	150.35 Sm 62	151.96 Eu 63	157.25 Gd 64	158.92 Tb 65	162.50 Dy 66	164.93 Ho 67	167.26 Er 68	168.93 Tm 69	173.04 Yb 70	174.97 Lu 71
† Actinide metals	(227) Ac 89	232.04 Th 90	(231) Pa 91	238.03 U 92	(237) Np 93	(242) Pu 94	(243) Am 95	(245) Cm 96	(249) Bk 97	(249) Cf 98	(253) E 99	(255) Fm 100	(256) Mv 101	(253) No 102	(257) Lr 103

Key (He cell): 4.0026 — Atomic mass (A); He — Abbreviation for element; 2 — Atomic number (Z)

Numbers above the symbols are the average atomic masses (relative to carbon-12). Parentheses enclose the mass number of the longest-lived isotope when there are no stable isotopes.

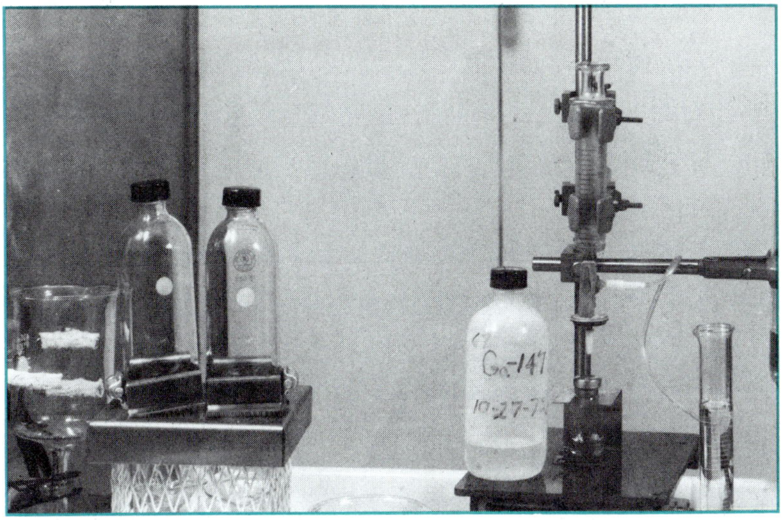

Isotopes

A mass spectrograph used to separate isotopes.

22.1 | The concept of isotopes

SG 1

The discovery that there are three radioactive series, each containing apparently new substances, created a serious problem. In 1910, there were still some empty spaces in the periodic table of the elements, but not enough spaces for the many new substances. The periodic table represents an arrangement of the elements according to their chemical properties, and, if it could not include the radioactive elements, it would have to be revised, perhaps in some drastic and fundamental way.

The clue to the puzzle lay in the observation that some of the newly found materials that were members of a radioactive series have *chemical* properties identical to those of well-known elements, although some of their *physical* properties are different.

For example, Uranium II, the "great-granddaughter" of Uranium I, was found to have the same chemical properties as Uranium I itself. When both were mixed together, the two could not be separated by chemical means. No chemist has detected, by chemical analysis, any difference between these two substances. But the two substances, now known as uranium-238 and uranium-234, do differ from each other in certain physical properties. As Table 21-1 shows, uranium-238 and 234 have quite different radioactive half-lives: 4.5×10^9 years and 2.5×10^5 years, respectively. The mass of a uranium-234 atom must be smaller than that of a uranium-238 atom by the mass of one α particle and two β particles. Another pair of radioactive substances, radium B and radium G, were found to have the same chemical properties as lead; when mixed with lead they could not be separated from it by chemical means. These substances are now known as lead-214 and lead-206, respectively. Lead-214 is radioactive, and lead-206 is stable. Table 21-1 indicates that the atoms must differ from each other in mass by the mass of two α particles and four β particles. There are many other examples of such physical differences among two or more radioactive substances with the same chemical behavior.

Soddy suggested a solution that threw a flood of light on the nature of matter and on the relationship of the elements in the periodic table. He proposed that a chemical element could be regarded as a pure substance only in the sense that all of its atoms have the same chemical properties; that is, a chemical element may in fact be a *mixture of atoms* having different radioactive behavior and different atomic masses, but all having the same chemical properties. This idea meant that one of the basic postulates of Dalton's atomic theory would have to be changed, namely, the postulate that the atoms of a pure element are alike in *all* respects. According to Soddy, it is only in chemical properties that the atoms of a given element are identical. The several physically different species of atoms making up a particular element occupy the same place in the periodic table, that is, have the same atomic number Z. Soddy called them *isotopes* of the element, from the Greek *isos* and *topos,* meaning *same* and *place* (same place in the periodic table). Thus, uranium-238 and uranium-234 are isotopes of uranium; lead-214 and lead-206 are isotopes of lead.

With this idea in mind, the many species of radioactive atoms in the three radioactive series were soon shown by chemical analysis to be isotopes of one or another of the last 11 elements in the periodic table, from lead to uranium. For example, the second and fifth members of the uranium series (see Table 21-1) were shown to be isotopes of thorium, with $Z = 90$; the 8th, 11th, and 14th members turned out to be isotopes of polonium ($Z = 84$). The old names and symbols given to the members of

This shorthand notation is explained further on page 665.

radioactive series upon their discovery were therefore rewritten to represent both the chemical similarity and physical difference among isotopes. The present names for uranium X_1 and ionium, for example, are thorium-234 and thorium-230 (as shown in Table 21-1). A modern "shorthand" form for symbolizing any species of atom, or *nuclide,* is also given in the same table (for example, $_{90}Th^{234}$ and $_{90}Th^{230}$ for two of the isotopes of thorium). The subscript (90 in both cases for thorium) is the atomic number *Z*, the place number in the periodic table; the superscript (234 or 230) is the mass number *A*, the approximate atomic mass in atomic mass units. Note that the chemical symbol (such as Th) adds nothing to the information given by the subscript.

Frederick Soddy (1877–1956), an English chemist, studied at Oxford and went to Canada in 1899 to work under Rutherford at McGill University in Montreal. There the two worked out their explanation of radioactive decay. Soddy returned to England in 1902 to work with Sir William Ramsay, the discoverer of the rare gases argon, neon, krypton, and xenon. Ramsay and Soddy showed, in 1903, that helium was continuously produced by naturally radioactive substances. In 1921, Soddy was awarded the Nobel Prize in chemistry for his discovery of isotopes. He was a professor of chemistry at Oxford from 1919 to 1936.

?

1. Why was it not necessary, after all, to expand the periodic table to fit in the newly discovered radioactive substances?

2. The symbol for the carbon-12 nuclide is $_6C^{12}$. What is the approximate atomic mass of carbon-12 in atomic mass units? What is its position in the list of elements (the atomic number Z)?

22.2 | Transformation rules

Two questions then arose: How do changes in chemical nature come about as an atom undergoes radioactive decay? More specifically, what determines whether the atomic number *Z* increases or decreases in a given radioactive transformation?

In 1913, the answers to these questions were given independently by Soddy in England and by A. Fajans in Germany. They each proposed two rules that systematized all the relevant observations for natural radioactivity. They are called the *transformation rules of radioactivity*. Recall that by 1913 Rutherford's nuclear model of the atom was generally accepted. Using this model, one could consider a radioactive atom to have an unstable nucleus that emits an α or β particle (sometimes with emission of a γ ray). Every nucleus has a positive charge Zq_e, where *Z* is the atomic number and q_e is the magnitude of the charge of an electron. The nucleus is surrounded by *Z* electrons that make the atom as a whole electrically neutral and determine the chemical behavior of the atom. An α particle has an atomic mass of about four units and a positive charge of two units, $+2q_e$. A β particle has a negative charge of one unit, $-q_e$, and very little mass compared to an α particle.

The transformation rules may now be stated as follows:

1. When a nucleus emits an α particle, the mass of the atom decreases by four atomic mass units, and the atomic number *Z*

of the nucleus decreases by two units; the resulting atom belongs to an element two spaces back in the periodic table.

2. When a nucleus emits a β particle, the mass of the atom is changed very little, but the atomic number Z increases by one unit; the resulting atom belongs to an element one place forward in the periodic table. When only a γ ray is emitted, there is no change in the number corresponding to the atomic mass, and none in the atomic number. Table 21-1 shows how these rules apply to the uranium–radium series, at least as far as the atomic number is concerned.

The Rutherford–Bohr model of the atom helps to explain why a change in chemical nature occurs as a result of α or β emission. Emission of an α particle takes two positive charges from the nucleus. The resulting new atom with its less positive nucleus can hold in its outer shells two fewer electrons than before, so two excess electrons drift away. The chemical behavior of atoms is controlled by the number of electrons; therefore, the new atom acts chemically like an atom of an element with an atomic number two units *less* than that of the parent atom. On the other hand, in the case of β emission, the nucleus, and with it the whole atom, becomes *more* positively charged, by one unit. The number of electrons that the atom can hold around the nucleus has increased by one, and after it has picked up an extra electron to become neutral again, the atom acts chemically as an atom with an atomic number one unit greater than that of the atom before the radioactive change occurred.

Example of α and β decay:

$$_{84}Po^{218} \rightarrow {}_{82}Pb^{214} + \alpha$$
$$_{90}Th^{234} \rightarrow {}_{91}Pa^{234} + \beta$$

By using the transformation rules, Soddy and Fajans were able to determine the place in the periodic table for every one of the substances (or nuclides) in the radioactive series; no revision of the existing periodic table was needed. Many of the nuclides between $Z = 82$ (lead) and $Z = 92$ (uranium) are now known to contain several isotopes each. These results were expected from the hypothesis of the existence of isotopes, but direct, independent evidence was also sought and obtained in 1914.

SG 2, 3

?

3. By how many units does the mass of an atom change during α decay? during β decay?

4. By how many units does the charge of a nucleus change during α decay? during β decay?

5. What are the transformation rules of radioactivity? Give an actual example of how they apply. How do these rules follow from the Rutherford–Bohr model of the atom?

22.3 | Direct evidence for isotopes of lead

Soddy knew that the stable end product of the uranium–radium series had the chemical properties of lead and that the end product of the thorium series also had the chemical properties of lead. But he realized that these end products should have atomic masses different from that of ordinary lead (that is, lead that was not produced from a radioactive series). This result follows from a simple calculation of the change in mass as an atom decays from the starting point of a radioactive series to the end point. The calculation may be simplified by ignoring beta decays in which no appreciable change in mass is involved.

In the uranium series, eight α particles, each with atomic mass of four units, are emitted. Therefore, the end product of the series starting from U^{238} is expected to have an atomic mass close to $238 - (8 \times 4) = 206$ units. In the thorium series, the end product derives from thorium-232, with an atomic mass of about 232 units, and six α particles are emitted along the way. It should therefore have an atomic mass close to $232 - (6 \times 4) = 208$ units. The average atomic mass of ordinary lead, found where there is no radioactive material evident, was known from chemical analysis to be 207.2 units.

There are four naturally occurring lead isotopes:

$${}_{82}Pb^{204}$$
$${}_{82}Pb^{206}$$
$${}_{82}Pb^{207}$$
$${}_{82}Pb^{208}$$

The first is found only as one of the isotopes of "ordinary" lead. Pb^{207} is also found as the end product of a decay chain starting with actinium.

The lead extracted from the mineral thorite, which consists mainly of thorium and contains only 1% or 2% by mass of uranium, may be presumed to be the final product of the thorium series. The atomic mass of lead extracted from thorite should therefore be significantly different both from the atomic mass of lead extracted from a uranium mineral, such as pitchblende, and different from the atomic mass of ordinary lead.

SG 13 involves the fact that the decay of U^{238} yields a distinct isotope of lead.

Here was a quantitative prediction, built on the transformation rules, that could be checked, and a number of chemists in Scotland, France, Germany, Austria, and the United States attacked the problem. The U.S. chemist T. W. Richards (later recipient of a Noble Prize in chemistry) found atomic masses as low as 206.08 for samples of lead from ores rich in uranium. Chemists in Austria found samples of lead, in the ore uraninite, with an atomic mass of 206.04. Soddy and others found samples of lead from thorite with atomic masses as high as 207.08 and 207.9. The results left no doubt that uranium was transformed into a light isotope of lead, and thorium into a heavier isotope of lead.

? *6. On what grounds was the existence of different atomic masses of lead predicted?*

22.4 | Positive rays

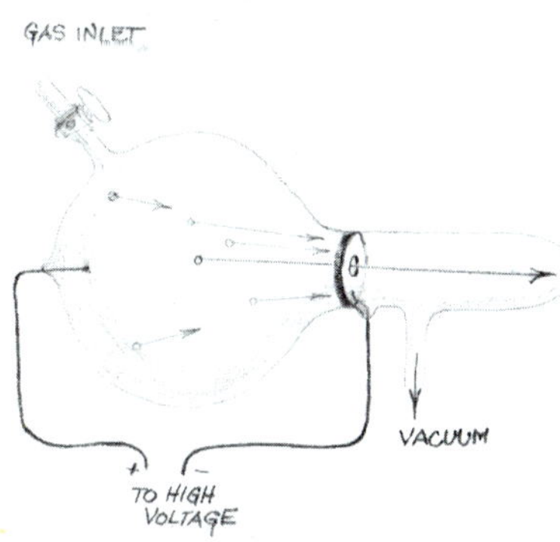

Discharge tube for producing a beam of positive ions. The low-pressure gas between anode and cathode is ionized by the action of the electric field. The positive ions are accelerated by the electric field toward the cathode; some of them do not fall on the plate but pass through a small hole and enter the well-evacuated region beyond, on the right side. Here, an external electric or magnetic field can be applied.

It was difficult to prove by independent, direct evidence that stable elements may be mixtures of isotopes. By definition, isotopes cannot be separated by ordinary chemical methods. Any attempt to separate a pair of isotopes must depend on a difference in some behavior that depends, in turn, on the difference between their atomic masses. Moreover, except for the very lightest elements, the difference in atomic mass is small compared to the atomic masses themselves. For the lead isotopes discussed in the last section, the difference was only two units in about 200 units, or about 1%. Any difference in a physical property between two isotopes having such a small mass difference would be expected to be very small, making separation difficult to achieve. Fortunately, when the question of the possible existence of isotopes of stable elements arose, a method was available that could be adapted to answer the question. This device, developed by J. J. Thomson in Great Britain and extended by A. J. Dempster in the United States and others, depended on the behavior of positively charged ions when these are traveling in electric and magnetic fields.

In a cathode-ray tube, electrons emitted from the cathode can knock electrons out of neutral atoms of gas with which they collide. It was thought that the positive ions produced in this way would accelerate toward the cathode and be neutralized there. In 1886, the German physicist Goldstein found that if a hole is made in the cathode, a ray passes through the cathode and emerges beyond it. The sketch in the margin (page 659) is a schematic diagram of a discharge tube for producing such rays. If the cathode fits the tube tightly, so that no gas can enter the region behind it, and if the holes are so small that very little gas can get through them, a high vacuum can be produced on the right side, where the ray emerges. The ray then has quite a long range and can be deflected by externally applied electric and magnetic fields. From the direction of the deflection, it could be concluded that the rays consist of positively charged particles. The rays were therefore called "positive rays" and were thought (correctly) to consist of positively charged ions of the atoms or molecules of the gas in the left side of the discharge tube.

J. J. Thomson (1856–1940) at work in the Cavendish laboratory.

In this manner, Thomson prepared positive rays from different gases and used the observed deflection produced by external fields to determine the relative masses of the atoms of the gases. It was a crucially important method, as you will see. Instead of the details of Thomson's early apparatus, an improved type of instrument based on the early form, and one that is still in common use will be described in this section.

The modern instrument typically consists of two main parts: The first part accelerates and then selects a beam of ions all moving with the same speed; in the second part, these ions pass through a magnetic field that deflects them from a straight path into several different curved paths determined by their relative masses. Ions of different mass are thus separated to such an extent that they can be detected separately. By analogy with the instrument that separates light of different wavelengths, the instrument that separates ions of different masses was called a *mass spectrograph*. Its operation (including how it can be used to measure q/m of ions) is explained on page 663. The details show what an ingenious and useful piece of equipment this is.

Thomson obtained results in his measurements of q/m for positive rays that were quite different from those that had been obtained for q/m of cathode-ray particles or β particles. Both the speeds of the ions and values of q/m were found to be smaller for gases with heavier molecules. These results were consistent with the idea that the positive rays are streams of positively ionized atoms or molecules.

Of course it would be very desirable if the values of q and m could be determined separately. The *magnitude* of q must be a multiple of the electron charge q_e; that is, it can only be q_e, or $2q_e$, or $3q_e$, $4q_e$, etc. The greater the charge on an ion, the greater the magnetic force will be and, therefore, the more curved the path of the ions. In the apparatus shown on page 663, a doubly ionized atom (an ion with charge $+2q_e$) will follow a path with half the radius of that of a singly ionized atom of similar type; a triply ionized atom will trace out a semi-circular path with one-third the radius, etc. Thus, for each type of atom analyzed, the path with the largest radius will be that taken by the ions with the single charge q_e. Since q is therefore known for each of the paths, the mass of the ions can be determined from the values of q/m found for each path.

Thus, study of positive rays with the mass spectrograph made it possible to measure for the first time the massesof individual atoms. [With the electrolysis methods that had been available befor (described in Sec. 17.7), it was possible to obtain only *average* masses for very large numbers of atoms.] The uncertainty of mass determinations made with modern mass spectrographs can be less than one part in a hundred thousand, that is, less than 0.001%. The difference in the masses of the isotopes of an element is thus easily detected, because in no case is it less than about 0.3%.

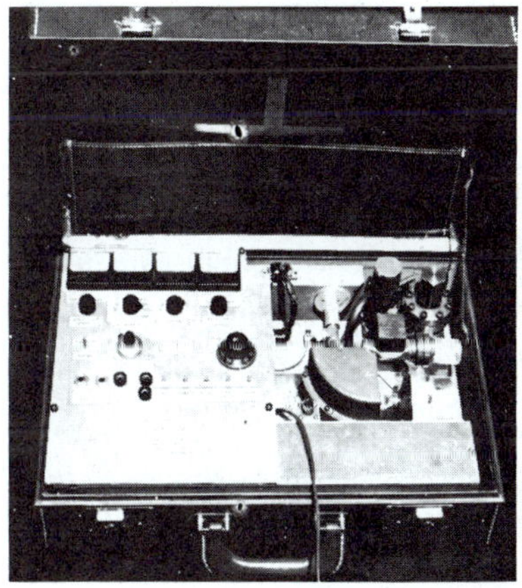

Some mass spectrometers are portable; small ones similar to this are carried aloft by rocket or balloon for the analysis of the gases in the upper atmosphere.

? *7. The radius of curvature of the path of an ion beam in a magnetic field depends on both the mass and speed of the ions. How must a mass spectrograph be constructed to allow separation of the ions in a beam by mass?*

22.5 | Separating isotopes

In Thomson's original instrument, the uncertainty in measured mass of ions was about 1%, but this was small enough to permit

Francis William Aston (1877–1945) studied chemistry at the University of Birmingham. In 1910, he went to Cambridge to work under J. J. Thomson. Aston was awarded the Nobel Prize in chemistry, in 1922, for his work on isotopic mass determinations with the mass spectrograph. In disagreement with Rutherford, Aston pictured a future in which the energy of the atom would be tapped by humanity. In his Nobel acceptance speech, he also spoke of the dangers involved in such a possibility. (Aston lived just long enough, by three months, to learn of the explosion of the nuclear bombs.)

Thomson to make the first observation of separated isotopes. He introduced a beam of neon ions from a discharge tube containing chemically pure neon into his mass spectrograph. The atomic mass of neon had been determined as 20.2 amu by means of the usual chemical methods for determining the atomic (or molecular) mass of a gas. At about the position on the photographic plate where ions of atomic mass 20 were expected to arrive, a dark line was observed when the plate was developed. But, in addition, there was also present nearby a faint line such as would indicate the presence of particles with atomic mass 22. No chemical element of gaseous compound was known which had this atomic or molecular mass. The presence of this line suggested, therefore, that neon might be a mixture of two isotopes, one with atomic mass 20, and the other with atomic mass 22. The average chemical mass 20.2 would result if neon contained about 10 times as many atoms of atomic mass 20 as those of atomic mass 22.

The tentative evidence from this physical experiment indicating that neon has two isotopes was so intriguing that Thomson's associate, F. W. Aston, looked for ways to strengthen the case for the existence of isotopes. It was well known from kinetic theory (see Sec. 11.5) that in a mixture of two gases with different molecular masses, the average molecular kinetic energy is the same for both. Therefore, the lighter molecules have a higher average speed than the heavier molecules and collide more often with the walls of a container. If the mixture is allowed to diffuse through a porous wall from one container into another, the slower, heavier molecules are less likely to reach and pass through the wall. The portion of the gas sample that does not get through the wall will, therefore, contain more of the heavier molecules than will the portion that does pass through the wall.

Aston allowed part of a sample of chemically pure neon gas to pass through such a wall. One pass accomplished only a slight separation of the lighter and heavier molecules, so a portion of

Two views of one of Aston's earlier mass spectrographs. The electromagnet was used to deflect a beam of charged atoms. Compare the photo with the sketch on the opposite page.

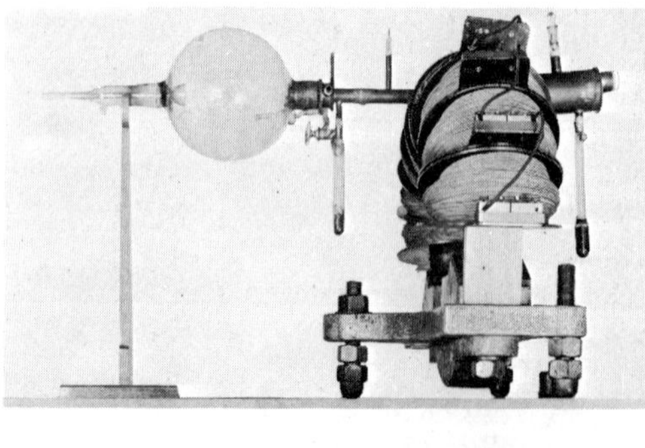

Close Up

The Mass Spectrograph

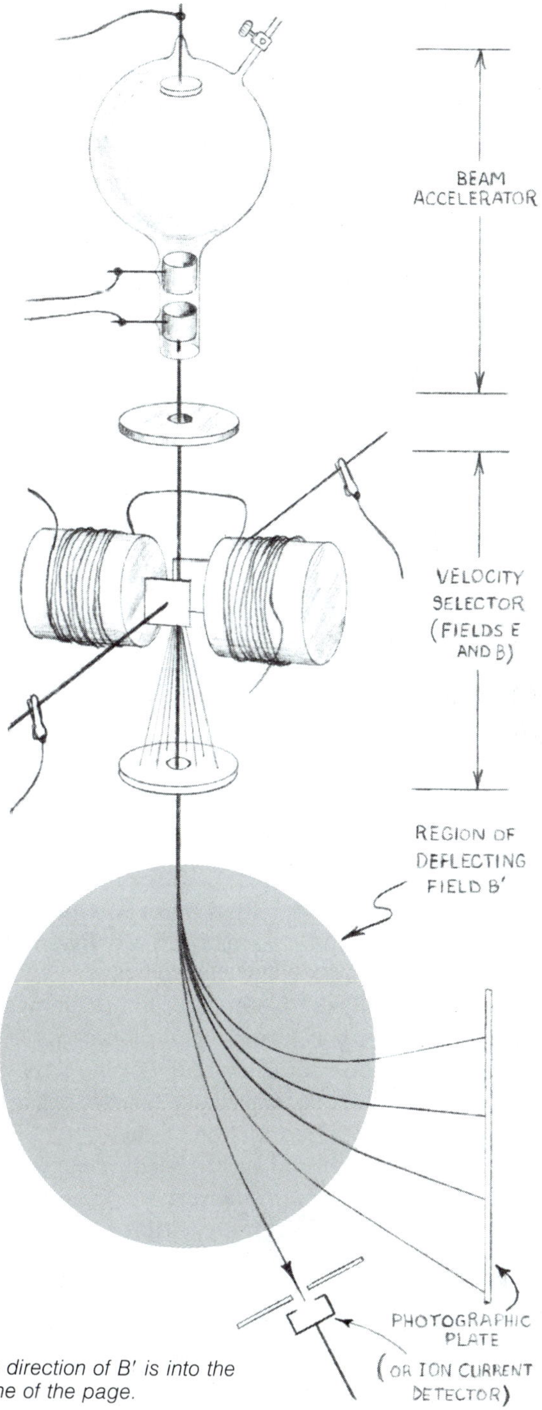

The direction of B′ is into the plane of the page.

The magnetic separation of isotopes begins by electrically charging the atoms of a sample of material, for example, by means of an electric discharge through a sample of gas. The resulting ions are then further accelerated by means of the electric potential difference between the lower pair of electrodes, and a beam emerges.

Before the different isotopes in the beam are separated, there is usually a preliminary stage that allows only those ions with a certain velocity to pass through. In one type, the ion beam initially enters a region of crossed magnetic fields B and E, produced by current in coils and charged plates as shown. There, each ion experiences a magnetic force of magnitude qvB and an electric force of magnitude qE. The magnetic and electric forces act on an ion in opposite directions, and only for ions of a certain speed will the forces be balanced, allowing them to pass straight through the crossed fields and the hole in the diaphragm below them. For each of these ions, $qvB = qE$; so their speed $v = E/B$. Because only ions with this speed in the original direction remain in the beam, this portion of the first part of the apparatus is called a *velocity selector*.

The separation of isotopes in the beam is now accomplished in another magnetic field of strength B'. As the beam enters this field, the magnetic field causes a centripetal force to act on each ion, deflecting it into a circular arc whose radius R depends upon the ion's charge-to-mass ratio. That is, $qvB' = mv^2/R$, and so $q/m = v/B'R$.

The divided beams of ions fall on either a photographic plate (in a mass spectro*graph*) or a sensitive ion current detector (in a mass spectro*meter*), allowing the radii R of their deflections to be calculated from the geometry of the apparatus. Since v, B', and R can be determined from measurements, the charge-to-mass ratio of each beam of ions can be calculated directly.

Because this method uses electric and magnetic fields, it is called the *electromagnetic method of separation of isotopes.*

the gas which had passed through the wall was passed through a porous wall again, with the same process repeated many times. Aston measured the average atomic mass of each fraction of the gas by density measurements and found values of 20.15 amu for the fraction that had passed through the wall many times, and 20.28 units for the fraction that had been left behind in many tries. The difference in average atomic mass indicated that the neon was, indeed, a mixture of isotopes.

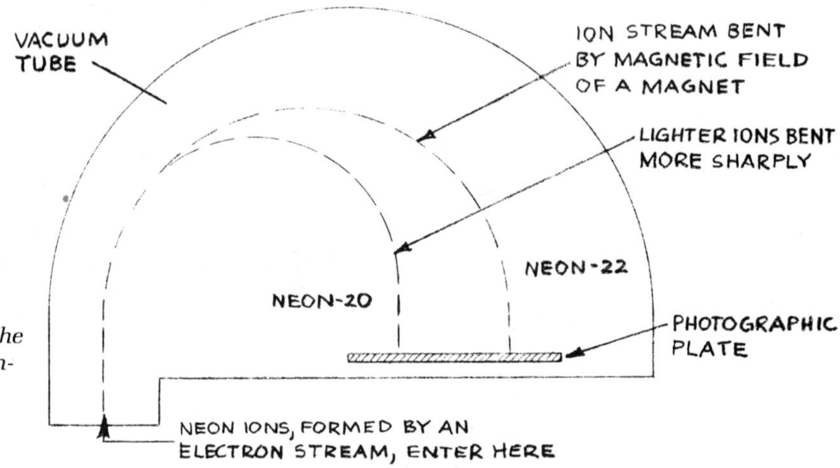

An alternative representation of the mass spectrograph used by Thomson and Aston to measure the atomic weight of neon.

Even more impressive was the change in the relative intensities of the two traces (for atomic masses 20 and 22) in the mass spectrograph. The line corresponding to ions with 22 amu was more prominent in the analysis of the fraction of the gas that had been left behind, showing that this fraction was "enriched" in atoms of mass 22. The optical emission spectrum of the enriched sample was the same as that of the original neon sample, proving that no substance other than neon was present.

These results of separating isotopes at least partially by gas diffusion encouraged Aston to improve the method of determining the atomic masses of the isotopes of many elements other than neon. Today, the number and the atomic masses of virtually all naturally found isotopes of the whole list of elements have been determined. As an example, the figure below shows the mass spectrograph record obtained for the element germanium, indicating that this element has five isotopes. A picture of this kind is called a *mass spectrogram*.

Both the electromagnetic method and the gas-diffusion method of separating isotopes have been modified for large-scale

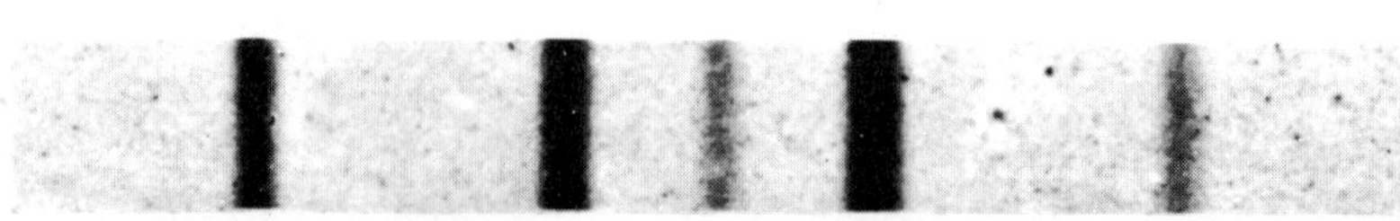

A photographic record of the mass spectrum of germanium, showing the isotopes of mass numbers 70, 72, 73, 74, and 76.

applications. The electromagnetic method shown in principle on page 663 is used by the Department of Energy to provide samples of separated isotopes for research. The gas-diffusion method used by Aston in achieving a small degree of separation of the neon isotopes has been developed on an enormous scale to separate the isotopes of uranium in connection with the manufacture of nuclear bombs and the production of nuclear power.

Although the mass of a *neutral* atom cannot be measured in a mass spectrograph (why not?), it is the custom to compute and list isotopic masses for neutral atoms, based on the measurement of ions.

SG 5

8. What were three experimental results that supported the belief in the existence of two isotopes of neon?

9. Isotopes at a given speed are separated by the electromagnetic method in a mass spectrograph because more massive ions are deflected less than lighter ions going at the same speed. Why are isotopes separated in diffusing through a porous wall?

22.6 | Summary of a useful notation for nuclides; nuclear reactions

It will be useful to summarize and recapitulate some ideas and notations. Because of the existence of isotopes, it was no longer possible to designate an atomic species only by means of the atomic number Z. To distinguish among the isotopes of an element some new symbols were introduced. One is the *mass number, A,* defined as the whole number closest to the measured atomic mass (see Table 22-1). For example, the lighter and heavier isotopes of neon are characterized by the pairs of values: $Z = 10, A = 20$ and $Z = 10, A = 22$. (An element that consists of a single isotope can, of course, also be characterized by its Z and A values.)

These values of Z and A are determined by the properties of the atomic nucleus; according to the Rutherford–Bohr model of the atom, the atomic number Z is the magnitude of the positive charge of the nucleus in elementary charge units. The mass number A is very nearly equal to (but a bit less than) the atomic mass of the nucleus because the total mass of the electrons around the nucleus is very small compared to the mass of the nucleus.

The term *nuclide* is used to denote an atomic species characterized by particular values of Z and A. An *isotope* is then one of a group of two or more nuclides, all having the same atomic number Z but different mass numbers A. A radioactive atomic species is called a radioactive nuclide, or *radionuclide* for short. A nuclide is usually denoted by the chemical symbol with

The current international convention is to write both Z and A values on the *left*: ${}_{Z}^{A}X$. For purposes of clarity in this introductory text, however, the former convention, ${}_{Z}X^{A}$, is used.

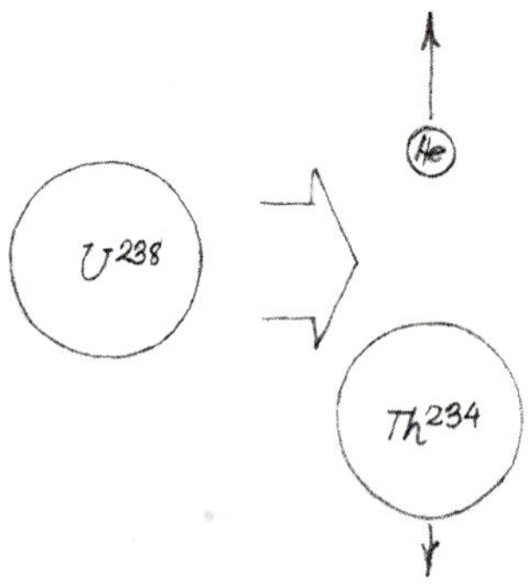

a subscript at the lower left giving the atomic number and a superscript at the upper right giving the mass number. In the symbol ${}_{Z}X^{A}$ for a certain nuclide, Z stands for the atomic number, X stands for the chemical symbol, and A stands for the mass number. For example, ${}_{4}Be^{9}$ is the nuclide beryllium with atomic number 4 and mass number 9; the symbols ${}_{10}Ne^{20}$ and ${}_{10}Ne^{22}$ represent the neon isotopes discussed above. The Z-value is the same for all the isotopes of a given element (X), and so it is often omitted, except when needed for balancing equations (as you will shortly see). Thus, you can write Ne^{20} for ${}_{10}Ne^{20}$, or U^{238} for ${}_{92}U^{238}$

The introduction of the mass number and the symbol for a nuclide makes it possible to represent radioactive nuclides in an easy and consistent way (as was done in Table 21-1). In addition, radioactive decay can be expressed by a simple "equation" representing the changes that occur in the decay process. For example, the first step in the uranium–radium series, namely, the decay of uranium-238 into thorium-234, may be written:

$${}_{92}U^{238} \rightarrow {}_{90}Th^{234} + {}_{2}He^{4}$$

The symbol ${}_{2}He^{4}$ stands for the helium nucleus (α particle); the other two symbols represent the initial and final atomic nuclei, each with the appropriate charge and mass number. The arrow stands for "decays into." The "equation" represents a *nuclear reaction* and is analogous to an equation for a chemical reaction. The atomic numbers on the two sides of the equation must balance because the electric charge of the nucleus must be conserved: 92 = 90 + 2. Also, the mass numbers must balance because of conservation of mass: 238 = 234 + 4.

There is also an antineutrino ($\bar{\nu}$) given off together with the β particle. The neutrino and antineutrino are two particles that will be discussed briefly in Sec. 23.6. Z and A are both zero for neutrinos and gamma rays: ${}_{0}\nu^{0}$; ${}_{0}\gamma^{0}$.

For another example, in the table of the uranium–radium series, ${}_{90}Th^{234}$ (thorium-234) decays by β emission, becoming ${}_{91}Pa^{234}$ (protactinium-234). Since a β particle (electron) has charge $-q_e$ and has an extremely small mass, the symbol ${}_{-1}e^{0}$ is used for it. This β-decay process may then be represented by the equation:

SG 6–9

$${}_{90}Th^{234} \rightarrow {}_{91}Pa^{234} + {}_{-1}e^{0} + {}_{0}\bar{\nu}^{0}$$

?

10. Write the complete symbol for the nuclide with atomic mass 194 and atomic number 78. Of which element is it an isotope?

11. Complete the following equation for α decay. Tell what law or rule you relied on.

$${}_{Z}X^{A} \rightarrow {}_{2}He^{4} + {}_{Z-2}X^{?}$$

12. In the same way, complete the following equation for β decay:

$${}_{Z}X^{A} \rightarrow {}_{-1}e^{0} + {}_{?}X^{A} + {}_{0}\bar{\nu}^{0}$$

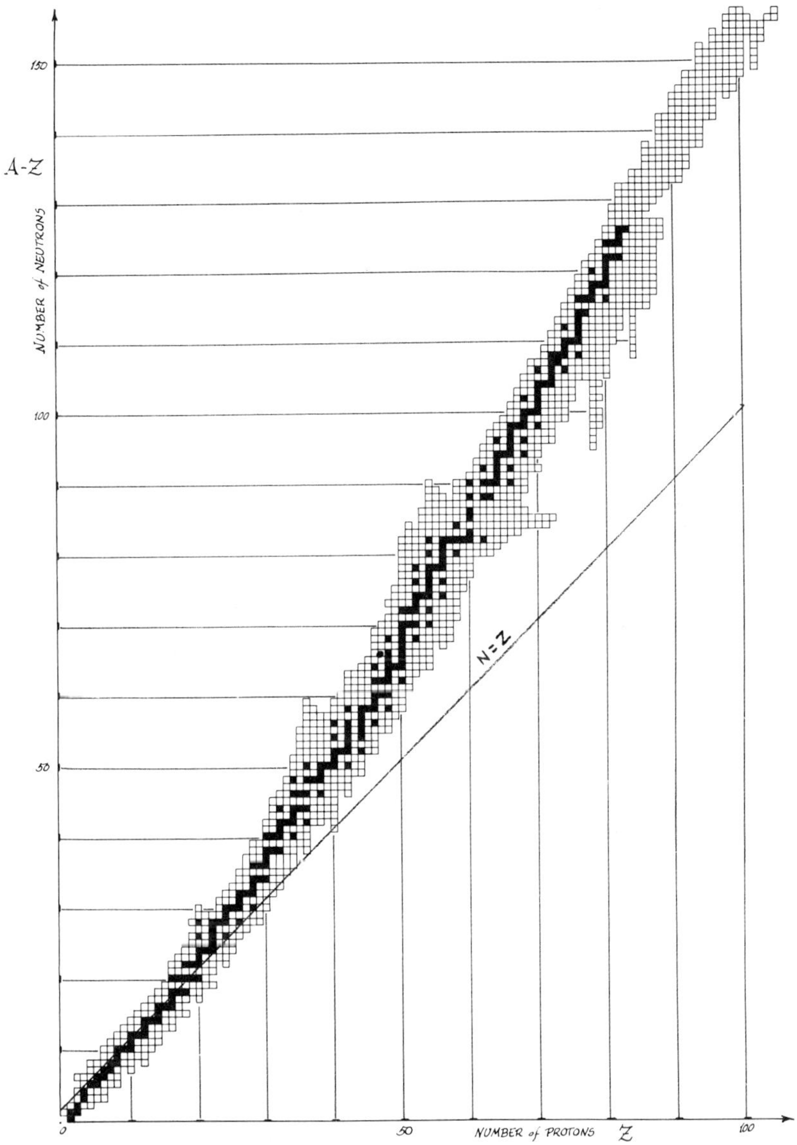

Chart of the known nuclides. Each black square represents a stable, natural nuclide. Each open square represents a known, unstable nuclide, with only a small number of these found naturally, the rest being artificial. Note that all isotopes of a given element are found in a vertical column centered on the element's atomic number Z. (As will be seen in the next chapter, the Z number is the number of protons in the nucleus, and A − Z, the difference between the atomic mass and atomic number, is the number of neutrons.)

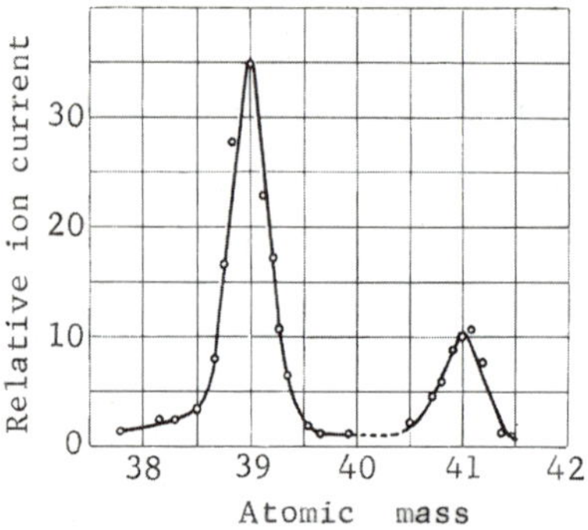

Record of determination of abundance of the isotopes of potassium in a mass spectrometer. In a mass spectrometer, the current due to the ions is detected (see page 663). Comparison of the current due to each isotope permits fairly precise estimates of the relative abundances of the isotopes.

22.7 | The stable isotopes of the elements and their relative abundances

Mass spectra, such as the one of germanium shown on page 664, have now been determined for all the elements that have at least one stable isotope. These are the elements with atomic numbers between 1 (hydrogen) and 83 (bismuth). A few of the results are listed in Table 22-1. The table also includes isotopes of the unstable (radioactive) elements uranium and thorium because they have such long half-lives that they are still present in large quantities in some rocks. Uranium has three naturally occurring isotopes, one of which, U^{235}, has remarkable properties (to be discussed) that have made it important in military and political affairs as well as in science and industry. As can be seen in the table, the relative abundance of U^{235} is very low, and it must first be separated from the far more abundant U^{238} before it can be used in some applications. Such applications and some of their social effects will be discussed in Chapter 24.

Of the elements having atomic numbers between 1 and 83, only about one fourth are single species; the others all have two or more isotopes. As a result, the 83 elements together actually are made up of 284 naturally occurring nuclides. All but 25 of these nuclides are stable. Many elements have only one stable nuclide, others have several, and tin has the greatest number, 10. Carbon and nitrogen each have two, and oxygen has three. Table 22-1 shows that the isotope O^{16} has a very high relative abundance, the isotopes O^{17} and O^{18} being relatively rare.

There are 25 naturally occurring *unstable* nuclides not associated with the decay chains of the heavy radionuclides. They show only a small degree of radioactivity. The most common of these light nuclides is ${}_{19}K^{40}$, an isotope of potassium that has a relative abundance of only 0.012%. This isotope, which emits β particles, has so lengthy a half-life (1.3×10^9 years) that its presence makes it very useful for determining the ages of certain rocks. Such information, coupled with information on the decay of U^{238}, can be used to estimate the age of the earth.

Hydrogen, the lightest element, has two stable isotopes, of which the heavier one, with atomic mass number 2, has a relative abundance of only 0.02%. The hydrogen isotopes are exceptional in that the rare isotope has an atomic mass twice that of the common isotope. As a result, the differences between the properties of the two isotopes are more marked than in any other pair of isotopes. The hydrogen isotope of atomic mass number 2 has therefore been given its own name, *deuterium*, with the symbol D; sometimes it is called "heavy hydrogen." It occurs in so-called "heavy water" or "deuterium oxide," for which the formula is written $({}_1H^2)_2O$ or D_2O.

The American chemist Harold C. Urey received the 1934 Nobel Prize in chemistry for his discovery of "heavy" hydrogen.

TABLE 22-1. RELATIVE NATURAL ABUNDANCES AND MASSES OF SOME NUCLIDES

The masses are given in atomic mass units (amu) based on ${}_6C^{12}$ = *12.000000.*

Element	*Chemical Symbol*	*Atomic Number (Z)*	*Mass Number (A)*	*Relative Abundance (%)*	*Mass of Neutral Atom (amu)*
Hydrogen	H	1	1	99.98	1.007825
		1	2	0.02	2.014102
Helium	He	2	4	100.00	4.002604
Lithium	Li	3	6	7.42	6.015126
		3	7	92.58	7.016005
Beryllium	Be	4	9	100.00	9.012186
Carbon	C	6	12	98.89	12.000000
		6	13	1.11	13.003354
Nitrogen	N	7	14	99.63	14.003074
		7	15	0.37	15.000108
Oxygen	O	8	16	99.76	15.994915
		8	17	0.04	16.999134
		8	18	0.20	17.999160
Neon	Ne	10	20	90.92	19.992440
		10	21	0.26	20.993849
		10	22	8.82	21.991385
Aluminum	Al	13	27	100.00	26.981535
Chlorine	Cl	17	35	75.53	34.968855
		17	37	24.47	36.965896
Platinum	Pt	78	190	0.01	189.9600
		78	192	0.78	191.9614
		78	194	32.90	193.9628
		78	195	33.80	194.9648
		78	196	25.30	195.9650
		78	198	7.21	197.9675
Gold	Au	79	197	100.00	196.9666
Lead	Pb	82	204	1.50	203.9731
		82	206	23.60	205.9745
		82	207	22.60	206.9759
		82	208	52.30	207.9766
Thorium	Th	90	232	100.00	232.0382
Uranium	U	92	234	0.006	234.0409
		92	235	0.720	235.0439
		92	238	99.274	238.0508

Mass of bare nucleus of hydorgen: 1.00727 amu
Mass of electron: 0.000549 amu

Heavy water differs from ordinary water in some important respects. Its density is 1.11 g/cm^3 as compared with 1.00 for ordinary water; at one atmosphere pressure, it freezes at 3.82°C and boils at 101.42°C (the corresponding temperatures for ordinary water being 0°C and 100°C). Naturally occurring water contains only about one atom of H^2 per 7,000 atoms of H^1, but methods have been developed for enriching the fraction of H^2 and also for producing nearly pure D_2O in large amounts. Heavy water is important in some types of devices for the controlled release of nuclear energy, as will be explained in Chapter 24.

Interesting and important regularities have been discovered among the natural abundances, and some are still sources of puzzles. The number of nuclides with the various combinations of even and odd values of Z and A are listed in Table 22-2. It is evident that naturally occurring nuclides with even Z and even A are much more numerous than those with any other combination. Elements with even Z have, on the average, more isotopes per element than do those with odd Z. Every theory of the nucleus has to try to account for these regularities, which are related to the stability of atomic nuclei. Information of this kind is analogous to observations of the positions of planets, to data on chemical compounds, and to atomic spectra. All of these provide material for the building of theories and models.

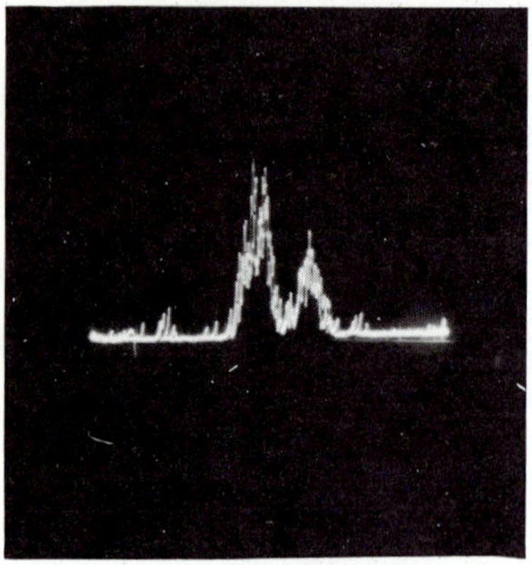

This is a photograph of the oscilloscope display of a high-resolution mass spectrometer when both hydrogen and helium are present. The high peak, on the left, indicates the He^3 isotope of mass 3.016030 amu. The other peak indicates H^3, the extra-heavy hydrogen isotope, otherwise known as tritium, whose mass is 3.016049 amu. The mass difference between the two nuclides is therefore about two parts in 300,000. This difference is easily observable.

TABLE 22-2. SOME INTERESTING DATA CONCERNING NATURALLY OCCURRING NUCLIDES

	Number of Stable Elements	Number of Nuclides Odd A	Even A	Total	Average No. of Isotopes per Element
Odd Z	40	53	8	61	1.5
Even Z	43	57	166	223	5.2
Total	83	110	174	284	3.4

?

13. What is deuterium?

14. What is "heavy water"?

15. Neon actually has three isotopes (see Table 22-1). Why did Thomson and Aston find evidence for only two isotopes?

22.8 | Atomic masses

The masses of most of the stable nuclides have been determined, and the results are of fundamental importance in quantitative work in nuclear physics and its applications. The standard of mass adopted by physicists for expressing the atomic mass of any nuclide was slightly different from that used by chemists for the chemical atomic weights. The chemists' scale was defined by assigning the value 16.0000 amu to ordinary oxygen. But, as can be seen in Table 22-1, oxygen is a mixture of three isotopes, two of which, O^{17} and O^{18}, have very small abundances. For isotopic mass measurements, the value 16.0000 was assigned to the most abundant isotope, O^{16}, and this mass was used as the standard by physicists. For some years, up to 1960, the atomic mass unit, 1 amu, was defined as 1/16 of the mass of a neutral O^{16} atom. Since 1960, O^{16} has been replaced by C^{12} as the standard, and the atomic mass unit is now defined by both physicists and chemists as 1/12 of the mass of the neutral C^{12} atom. The main reason for the choice of carbon is that mass-spectrographic measurements of atomic masses are much more accurate than

SG 10–12

the older chemical methods. Carbon forms an exceptional variety of compounds, from light to very heavy, which can be used as comparison standards in the mass spectrograph.

The results that have been obtained for the atomic masses of some elements of special interest are listed in Table 22-1. Atomic masses can be determined with great accuracy, and, when expressed in atomic mass units, they all turn out to be very close to integers. For each nuclide, the measured mass differs from an integer by less than 0.06 amu. This result is known as Aston's *whole-number rule,* and provides the justification for using the mass number A in the symbol $_ZX^A$ for a nuclide or atom. As you will see in the next chapter, the explanation or physical basis for this rule is connected with the structure of the nucleus.

? *16. What nuclide is the current standard for atomic mass? Why has it been chosen?*

study guide

1. The *Project Physics* learning materials particularly appropriate for Chapter 22 include the following *Transparencies:*

Radioactivity Displacement Rules
Mass Spectrometer
Chart of Nuclides
Nuclear Equations

2. Soddy's proposal of the existence of isotopes meant that not all atoms of the same element are identical. Explain why this proposal does *not* require that the atoms of a given element show differences in chemical behavior.

3. After Soddy's proposal of the existence of isotopes, how could one go about determining whether an apparently new element was really new and should be given a separate place in the periodic table, or was simply an isotope of an already known element?

4. At the National Bureau of Standards, in 1932, 3.8 L of liquid hydrogen was evaporated slowly until only about 1 g remained. This residue allowed the first experimental check on the existence of the "heavy" hydrogen isotope H^2.

(a) With the help of the kinetic theory of matter, explain why the evaporation should leave in the residue an increased concentration of the isotope of greater atomic mass.

(b) Why should the evaporation method be especially effective with hydrogen?

5. A mass spectrograph similar to that sketched on page 672 causes singly charged ions of chlorine-37 to travel a semi-circular path and strike a photographic plate (in the magnetic field at the right). Use the equation on page 663 to answer the following questions:

(a) Show that the path radius is proportional to the ion mass.

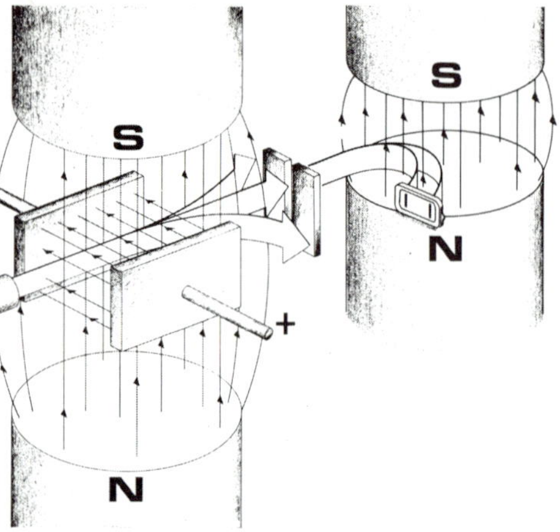

(b) If the path diameter for chlorine ions is about 1.0 m, how far apart will the traces of Cl^{37} and Cl^{35} be on the photographic plate?

(c) What would be the diameter of the orbit of lead-208 ions if the same electric and magnetic field intensities were used to analyze a sample of lead?

(d) The problems of maintaining a uniform magnetic field over surfaces larger than 1 m^2 are considerable. What separation between lead-207 and lead-208 would be achieved if the diameter of the orbit of lead-208 were held to 1.000 m?

6. Supply the missing data indicated by these transformation "equations":

(a) ${}_{?}Pb^{212} \rightarrow Bi^{212} + ?$

(b) ${}_{?}Bi^{212} \rightarrow \quad + {}_{-1}e^{0}$

(c) $? \rightarrow Pb^{208} + {}_{2}He^{4}$

7. A radioactive series, originally called the actinium series, is now known to start with the uranium isotope ${}_{92}U^{235}$. This parent member undergoes transmutations by emitting in succession the following particles: α, β, α, β, α, α, α, α, β, α, β. The last of these disintegrations yields ${}_{82}Pb^{207}$, which is stable. From this information, and by consulting the periodic table, determine the complete symbol for each member of the series. List the members of the series in a column, and beside each member give its mode of decay (similar to what was done in Table 21-1).

8. In the following diagram of the thorium series, which begins with ${}_{90}Th^{232}$, the symbols used are those that were originally assigned to the members of the sequence:
Supply the missing data; then, by consulting the periodic table, replace the old symbols with the present ones. Indicate where alternative possibilities exist in the series.

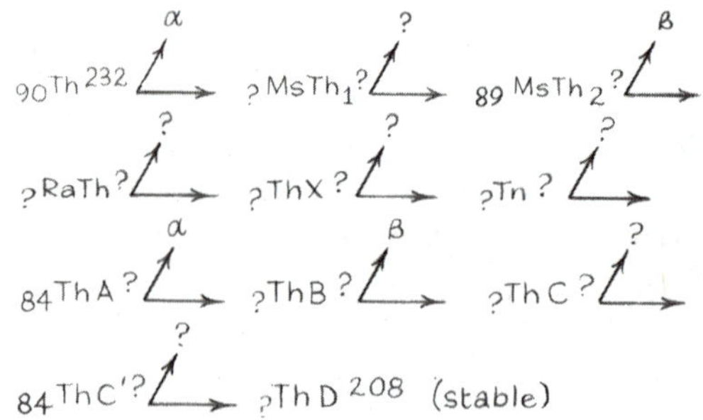

9. From ${}_{94}PU^{241}$, an isotope of plutonium produced artificially by bombarding uranium in a nuclear reactor, a radioactive series has been traced for which the first seven members are ${}_{94}Pu^{241}$, ${}_{95}Am^{241}$, ${}_{93}Np^{237}$, ${}_{91}Pa^{233}$, ${}_{92}U^{233}$, ${}_{90}Th^{229}$, and ${}_{88}Ra^{225}$. Outline the disintegration series for these first seven members, showing the modes of decay as in the preceding question.

10. A trace of radioactivity found in natural carbon makes it possible to estimate the age of materials that were once living. The radioactivity of the carbon is due to the presence of a small amount of the unstable isotope, carbon-14. This isotope is created mainly in the upper atmosphere by transformation (induced by cosmic rays) of the stable isotope carbon-13 to carbon-14. The rate of production of carbon-14 from carbon-13 matches the rate of β decay of carbon-14 into nitrogen-14, so the percentage of total carbon in the atmosphere consisting of carbon-14 is relatively constant. When carbon dioxide is used by plants in photosynthesis, the cell incorporates the isotopes of carbon in the same proportions as exist in the atmosphere. The average activity of the carbon at that point is 15.3 β emissions per minute per gram of carbon. When the interaction of the living plant with the atmosphere stops, for example, when a branch is broken off a living tree for use as a tool, the radioactivity begins to decrease at a rate characteristic of carbon-14. This rate has been measured, and the half-life of carbon is known to be 5,760 years. So if the activity is measured at some later time and if the half-life of carbon-14 is known, then one can use the decay curve given on page 646 to determine the time elapsed since the branch was taken from the tree. For example, suppose the activity was found to have dropped from the normal rate of 15.3 to 9.2 β emissions per minute per gram of carbon. Knowing the half-life, determine the time elapsed.

Use the same procedure to calculate the age of charcoal found in an ancient Indian fire pit if the activity of the carbon in the charcoal is now found to be 1.0 β emission per minute per gram of carbon. What assumption are you making in this part of the problem?

11. **(a)** Find the average atomic mass of carbon by calculating the "weighted average" of the atomic masses of the two natural carbon isotopes. (Use the data of Table 22-1.)

(b) Find the average atomic mass of lithium.

(c) Find the average atomic mass of ordinary lead.

12. The mass of a neutral helium atom is 4.00260 amu, and that of an electron is 0.00055 amu. From these data find the mass of the α particle in atomic mass units.

13. The age of a rock containing uranium can be estimated by measuring the relative amount of U^{238} and Pb^{206} in a sample of the rock. Consider a rock sample that is found to contain three times as many U^{238} atoms as Pb^{206} atoms.

(a) What fraction of the U^{238} contained in the sample when it was formed has decayed, assuming there was no Pb^{206} in the rock initially?

(b) Refer to the graph on page 646 and estimate the fraction of a half-life needed for that fraction of the U^{238} to decay.

(c) How old is the rock?

(d) For this simple method to work, it is necessary to assume that each U^{238} atom that decays appears as a Pb^{206} atom, in other words, that the half-lives of all the intermediate substances in the uranium chain are negligible compared to that of U^{238}. Is this assumption a valid one?

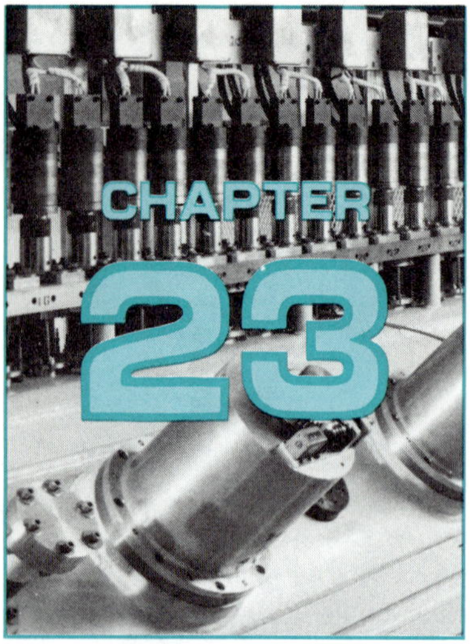

Probing the Nucleus

23.1 The problem of the structure of the atomic nucleus

SG 1 The discoveries of radioactivity and isotopes raised new questions about the structure of atoms, questions that involved the atomic nucleus. You saw in Sec. 22.2 that the transformation rules of radioactivity could be understood in terms of the Rutherford–Bohr model of the atom. But that model said nothing about the nucleus other than that it is small, has charge and mass, and may emit an α or a β particle. This implies that the nucleus has a structure that changes when a radioactive process occurs. The question arose: Can a theory or model of the atomic nucleus be developed that will explain the facts of radioactivity and the existence of isotopes?

The answer to this question makes up much of *nuclear physics*. The problem of nuclear structure can be broken down

Ernest O. Lawrence (left) and M. S. Livingston (right) are shown standing beside the magnet for one of the earliest cyclotrons. Lawrence and Livingston invented the cyclotron in 1931, thereby initiating the development of high-energy physics in the United States.

into two question: (1) what are the building blocks of which the nucleus is made, and (2) how are the nuclear building blocks put together? Answers to the first question are considered in this chapter. The next chapter will take up the question of how the nucleus is held together. The attempt to solve the problem of nuclear structure, although not yet completed, has led to many new basic discoveries and to large-scale practical applications. It has also had important social and political consequences, stretching far beyond physics into the life of society in general. Some of these consequences will be discussed in Chapter 24.

The *Project Physics* Supplemental Unit A, entitled "Elementary Particles," goes one step further, into the nature and structure of the subatomic particles themselves.

23.2 | The proton–electron hypothesis of nuclear structure

The emission of α and β particles by radioactive nuclei suggested that a model of the nucleus might be constructed by starting with α and β particles as building blocks. Such a model would make it easy to see, for example, how a number of α particles could be emitted, in succession, in a radioactive series. But not all nuclei are radioactive, nor do all nuclei have masses that are

SG 2

"Proton" comes from the Greek "protos" (first). It is not known who suggested the name originally; it is found in the literature as far back as 1908. In 1920, Rutherford's formal proposal of the name proton was accepted by the British Association for the Advancement of Science.

multiples of the α-particle mass. For example, the nucleus of an atom of the lightest element, hydrogen, with an atomic mass of one unit (two units in the case of the heavy isotope), is too light to contain an α particle; so is the light isotope of helium, ${}_2He^3$.

A positively charged particle with mass of one unit would seem to be more satisfactory as a nuclear building block. Such a particle does indeed exist: the nucleus of the common isotope of hydrogen. This particle has been named the *proton*. According to the Rutherford–Bohr theory of atomic structure, the hydrogen atom consists of a proton with a single electron revolving around it.

In the preceding chapter (Sec. 22.4), Aston's *whole-number rule*, which expressed the experimental result that the atomic masses of the nuclides are very close to whole numbers, was discussed. This rule, together with the properties of the proton (for example, its single positive charge) made it appear possible that all atomic nuclei are made up of protons. Could a nucleus of mass number A consist of A protons? If this were the case, the charge of the nucleus would be A units, but, except for hydrogen, the nuclear charge Z is found to be always less than A, usually less than ½ A. To get around this difficulty, it was assumed that in addition to the protons, atomic nuclei contain just enough electrons to cancel the charge of the extra protons; that is, they were supposed to contain $A-Z$ electrons. These electrons would contribute only a small amount to the mass of the nucleus, but together with the protons they would make the net charge equal to $+Z$ units, as required. It seemed plausible to consider the atom as consisting of a nucleus made up of A protons and $A-Z$ electrons, with Z additional electrons outside the nucleus to make the entire atom electrically neutral. For example, an atom of ${}_8O^{16}$ would have a nucleus with 16 protons and eight electrons, with eight additional electrons outside the nucleus. This model of the nucleus is known as the *proton–electron hypothesis* of nuclear composition.

SG 3

The proton–electron hypothesis seemed to be consistent with the emission of α and β particles by atoms of radioactive substances. Since it was assumed that the nucleus contained electrons, explanation of β decay was no problem. When the nucleus is in an appropriate state, it may simply eject one of its electrons. It also seemed reasonable that an α particle could be formed, in the nucleus, by the combination of four protons and two electrons. (An α particle might exist already formed in the nucleus, or it might be formed at the instant of emission.)

Careful inspection of the modern values of nuclide masses (page 669) shows that nuclides can not be considered as simple conglomerates of hydrogen and electrons.

The proton–electron hypothesis is similar to an earlier idea suggested by the English physician William Prout in 1815. On the basis of the small number of atomic masses then known, Prout proposed that all atomic masses are multiples of the atomic mass of hydrogen and that therefore all the elements might be built up of hydrogen. Prout's hypothesis was discarded when, later in the nineteenth century, the atomic masses of some elements were found to be fractional, in particular, those of chlorine (35.46 units) and copper (63.54 units). With the discovery of isotopes, however, it was realized that the fractional atomic masses of chlorine and copper, like that of neon, arse because these elements are *mixtures* of isotopes, with each separate isotope having an atomic mass close to a whole number.

Although the proton–electron hypothesis was satisfactory in some respects (for example, in accounting for the whole-number rule for isotope masses and in being consistent with the emission of α and β particles by radioactive nuclides) it led to serious difficulties and had to be given up.

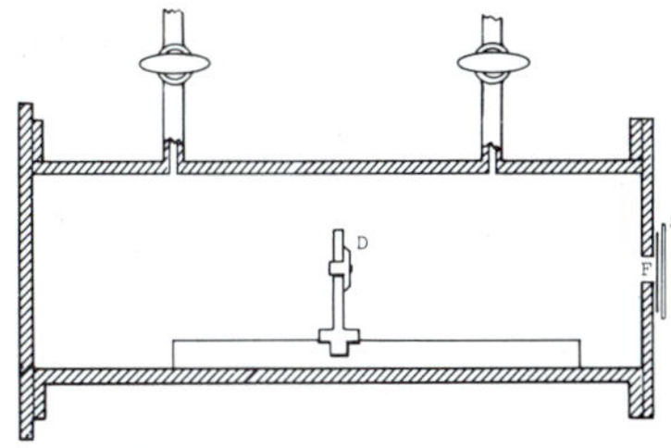

Diagram of Rutherford's apparatus used to detect the protons from disintegrations produced by particles. The α source was on a movable stand, D. Nitrogen nuclei in the nitrogen gas that filled the box are transmuted by the α particles. At the end of the box was a piece of silver foil, F, thick enough to stop particles but not protons. Behind the foil was a lead sulfide screen, S, which would show flashes of light when struck by protons. To see the flashes, the screen had to be watched through a microscope with a dark-adapted eye.

?

1. *Why was the idea of hydrogen atoms being a basic building block of all atoms given up in the nineteenth century?*
2. *On the basis of the proton–electron hypothesis, what would a nucleus of $_6C^{12}$ contain?*
3. *Does the proton–electron hypothesis work out for, say, $_2He^4$?*

23.3 | The discovery of artificial transmutation

A path that led to a better understanding of nuclear composition was opened, almost by accident, in 1919. In that year, Rutherford found that when nitrogen gas was bombarded with α particles from bismuth-214, swift particles were produced that could travel farther in the gas than did the α particles themselves. When these particles struck a scintillation screen, they produced flashes of light fainter than those produced by α particles, about the intensity that would be expected for positive hydrogen ions (protons). Measurements of the effect of a magnetic field on the paths of the particles suggested that they were indeed protons. Rutherford ruled out, by means of careful experiments, the possibility that the protons came from hydrogen present as an impurity in the nitrogen. Since the nitrogen atoms in the gas were the only possible source of protons, Rutherford concluded that an α particle, in colliding with a nitrogen nucleus, can occasionally knock a small particle (a proton) out of the nitrogen nucleus. In other words, Rutherford deduced that an α particle can cause the artificial disintegration of a nitrogen nucleus, with one of the products of the disintegration being a proton. But this process does not happen easily. The experimental results showed that only one proton was produced for about 1 million α particles passing through the gas.

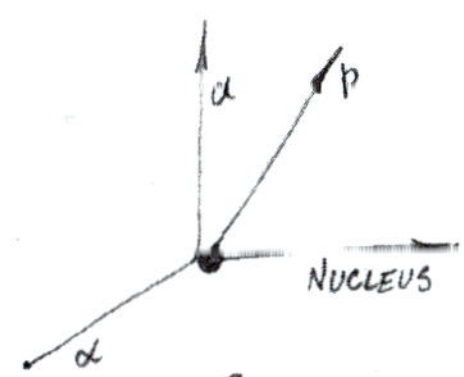

a

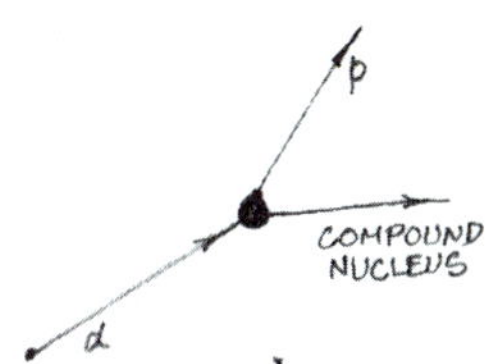

b

Between 1921 and 1924, Rutherford and Chadwick extended the work on nitrogen to other elements and found evidence for

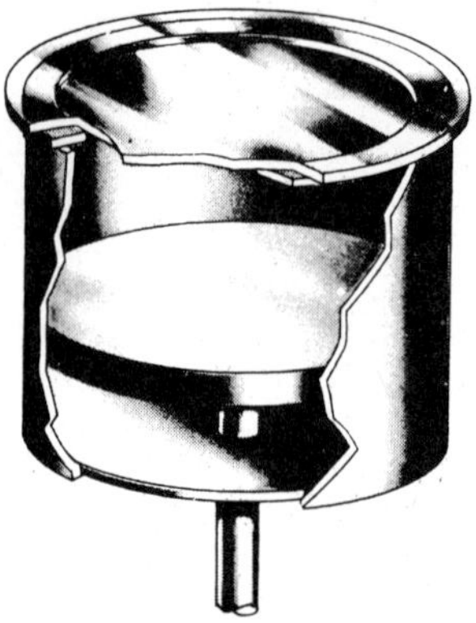

The Wilson cloud chamber. When the piston is moved down rapidly, the gas in the cylinder cools and becomes supersaturated with water vapor. The water vapor will condense on the ions created along the path of a high-energy charged particle, thereby making the track. For his invention of the cloud chamber, C. T. R. Wilson (1869–1959) of Scotland shared the 1927 Nobel Prize in physics with Arthur H. Compton. (See also page 692 margin.)

the artificial disintegration of all the light elements from boron to potassium, with the exception of carbon and oxygen. (These elements were later shown also to undergo artificial disintegration.)

The next step was to determine the nature of the nuclear process leading to the emission of the proton. Two hypotheses were suggested for this process: (a) the nucleus of the bombarded atom loses a proton, "chipped off" as the result of a collision with a swift α particle; or (b) the α particle is *captured* by the nucleus of the atom it hits, forming a new nucleus that, a moment later, emits a proton. It was possible to distinguish experimentally between these two possible cases by using a device called a "cloud chamber," which reveals the path or track of an individual charged particle. The cloud chamber was invented by C. T. R. Wilson and perfected by him over a period of years. In 1911, it became a major scientific instrument; a simplified diagram is shown at the left. If hypothesis (a) holds, the chipped-off proton should create four tracks in a photograph of a disintegration event: the track of an α particle before the collision, the track of the same α particle after collision, and the tracks of both the proton and the recoiling nucleus after collision. In case (b), on the other hand, the α particle should disappear in the collision, and only three tracks would be seen: that of the α particle before collision and those of the proton and recoil nucleus after the collision. The choice between the two possibilities was settled in 1925 when P. M. S. Blackett studied the tracks produced when particles passed through nitrogen gas in a cloud chamber. He found, as shown in the photograph below,

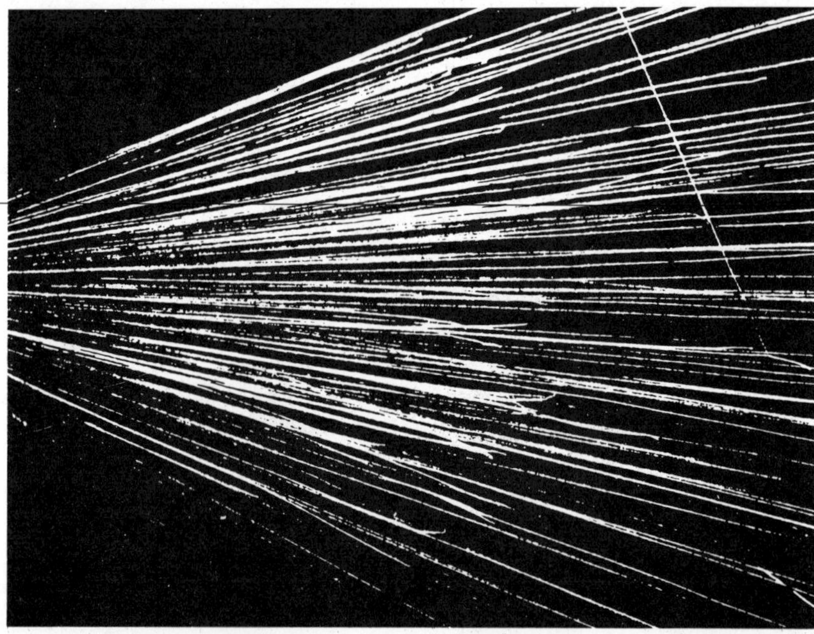

Alpha-particle tracks from a source at the left, in a cloud chamber filled with nitrogen gas. At the far right, one α particle has hit a nitrogen nucleus; a proton is ejected upward toward the left, and the resulting oxygen nucleus recoils downward to the right. (From P. M. S. Blackett, 1925.)

that the only tracks in which artificial disintegration could be seen were those of the incident α particle, a proton, and the recoil nucleus. The absence of a track corresponding to the presence of an α particle after the collision proved that the α particle disappeared completely and that case (b) is the correct interpretation of artificial disintegration.

The process in which an α particle is absorbed by a nitrogen nucleus and a proton is emitted may be represented by an "equation" that is analogous to the representation used near the end of Sec. 22.6 to represent radioactive decay. The equation expresses the fact that the total mass number is the same before and after the collision (that is, there is conservation of mass number) and the fact that the total charge is the same before and after the collision (there is conservation of charge). The atomic number, the mass number, and the nuclear charge are known for the target nucleus ${}_7N^{14}$, for the incident α particle ${}_2He^4$, and for the proton ${}_1H^1$. The product nucleus will therefore have the atomic number $7 + 2 - 1 = 8$ (which is the atomic number for oxygen) and will have the mass number $14 + 4 - 1 = 17$. Therefore, the product nucleus must be ${}_8O^{17}$, an isotope of oxygen. The disintegration process may therefore be represented by the nuclear reaction:

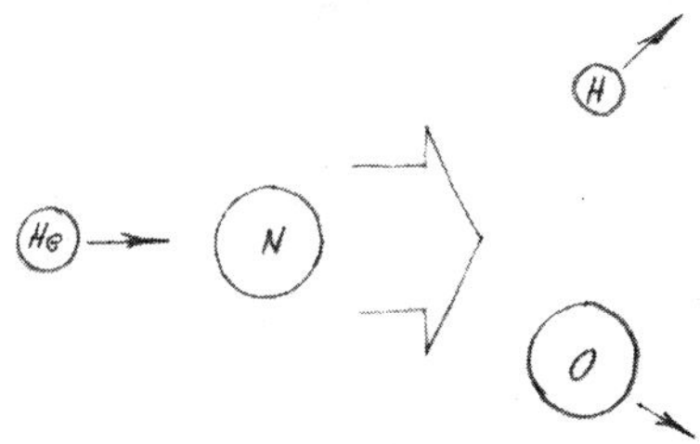

$$ {}_2He^4 + {}_7N^{14} \rightarrow {}_8O^{17} + {}_1H^1 $$

This reaction shows that a transmutation of an atom of one chemical element into an atom of another chemical element has taken place. The transmutation did not occur spontaneously, as it does inthe case of natural radioactivity; it was produced by exposing target atoms (nuclei) to projectiles emitted from a radioactive nuclide. In the paper in which he reported this first artificially produced nuclear reaction, Rutherford said:

SG 4

> The results as a whole suggest that, if α particles—or similar projectiles—of still greater energy were available for experiment, we might expect to break down the nuclear structure of many of the lighter atoms.

This call for greater energies of "projectiles" was soon answered by the construction of accelerators. (See Sec. 23.7.)

The further study of reactions involving light nuclei led (as you will see in the next section) to the discovery of a new particle, the *neutron,* and to a better theory of the constitution of the nucleus. Many types of reactions have been observed with nuclei of all masses, from the lightest to the heaviest, and the possibilities indicated by Rutherford have been realized to an extent far beyond what he would have imagined in 1919.

?

4. What evidence showed that the bombarding α particle was temporarily absorbed *by the nitrogen nucleus rather than simply broken up and bounced off?*

23.4 | The discovery of the neutron

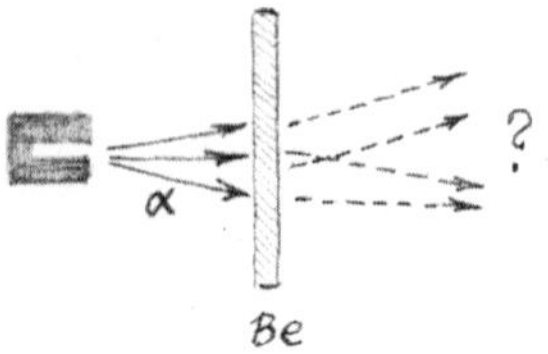

In 1920, Rutherford suggested that a proton inside the nucleus might have an electron tied to it so closely as to form a neutral particle. Rutherford even suggested the name *neutron* for this hypothetical particle. Physicists looked for neutrons, but the search presented at least two difficulties: (1) they could find no naturally occurring neutron-emitting materials; and (2) the methods used for detecting atomic particles all depended on effects of the electric charge of the particles and so could not be applied directly to neutral particles. Until 1932, the search for neutrons was unsuccessful.

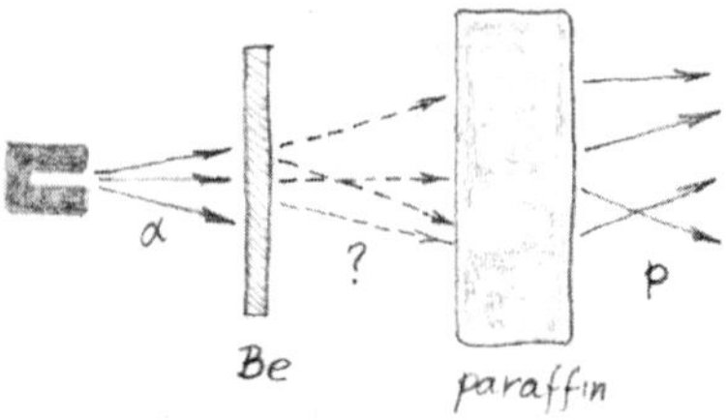

The proof of the existence of neutrons came in 1932 as the climax of a series of experiments on nuclear reactions made by physicists in different countries. The discovery of the neutron is a good example of how physicists operate, how they think about problems and arrive at solutions; it is an excellent "case history" in experimental science. Working in Germany in 1930, W. G. Bothe and H. Becker found that when samples of boron or of beryllium were bombarded with α particles, they emitted radiations that appeared to be of the same kind as γ rays, at least insofar as the γ rays had no electric charge. Beryllium gave a particularly marked effect of this kind. Observations by physicists in Germany, France, and Great Britain showed that the radiation from the beryllium penetrated farther (through lead, for example) than any γ radiation found up to that time and had an energy of about 10 MeV. The radiation was thus much more energetic than the γ rays (that is, high-energy photons) previously observed and, as a result, aroused much interest.

James Chadwick (born 1891) received the Nobel Prize in physics in 1935 for his discovery of the neutron.

Among those who investigated this radiation were the French physicists Frédéric Joliot and his wife Irène Curie, a daughter of the discoverers of radium. They studied the absorption of the radiation in paraffin, a material rich in hydrogen. In the course of their experiments, Joliot and Curie found that the radiation from beryllium, when it fell on paraffin, ejected large numbers of hydrogen nuclei (protons) from the paraffin. The energies of these protons were found to be about 5 MeV. Using the principles of conservation of momentum and energy, they calculated the energy a γ ray would need if it were to transfer 5 MeV to a proton in a collision. The result was about 50 MeV, a value much greater than the 10 MeV that had been measured for the radiation. In addition, the number of protons produced was found to be much greater than that predicted on the assumption that the radiation consisted of γ rays.

These discrepancies (between the results of two sets of experiments and between theory and experiment) left physicists in a dilemma. Either they could conclude that the conservation principles of momentum and of energy did not apply to the collisions between the radiation and the protons in the paraffin, or they could adopt another hypothesis about the nature of the radiation. Now, if there is any one thing physicists do not want to do it is to give up the principles of conservation of momentum and of energy. These principles are so basic to scientific thought and have proven so useful that physicists tried very hard to find an alternative to giving them up.

The English physicist James Chadwick found similarly perplexing results for recoiling nuclei from several other light elements, including helium, lithium, carbon, nitrogen, and argon. In 1932, Chadwick proposed a successful alternative hypothesis about the nature of the radiation. Chadwick's first published report of his hypothesis is reproduced on the next page. In a later, more complete paper, "The Existence of a Neutron," he said:

> If we suppose that the radiation is not a quantum radiation [γ ray], but consists of particles of mass very nearly equal to that of the proton, all the difficulties connected with the collisions disappear, both with regard to their frequency and to the energy transfers to different masses. In order to explain the great penetrating power of the radiation, we must further assume that the particle has no net charge. We must suppose it to consist of a proton and electron in close combination, the 'neutron' discussed by Rutherford in his Bakerian Lecture of 1920.

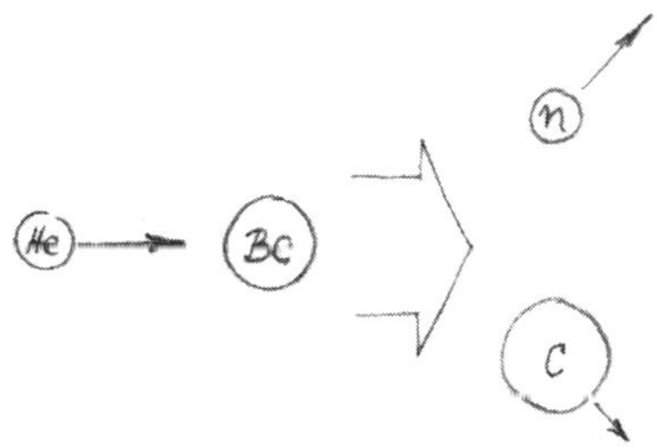

Thus, according to Chadwick's hypothesis, when an element such as beryllium is bombarded with α particles, a nuclear reaction can take place that produces neutrons:

$${}_2He^4 + {}_4Be^9 \rightarrow {}_6C^{12} + {}_0n^1$$

SG 5, 6

Here, the symbol ${}_0n^1$ represents the neutron postulated by Chadwick, with zero charge and mass number equal to 1. Such neutrons, because they have no electric charge, could penetrate bricks of a material as dense as lead without giving up their energy. When neutrons go through paraffin, there would occasionally be head-on collisions with hydrogen nuclei (protons). The recoiling protons could then be observed because of the ionization they produce. Thus, Chadwick's chargeless particle hypothesis could account in a qualitative way for the observed effects of the mysteriously penetrating radiation.

Chadwick's estimate that the particle's mass must be nearly equal to the mass of a proton was made by applying the laws of conservation of momentum and energy to the case of perfectly elastic collisions, that is, simply applying the laws that worked well for the case of interacting billiard balls and other objects treated in "classical" physics. In a perfectly elastic head-on collision between two bodies, as you saw in Chapter 9, almost all

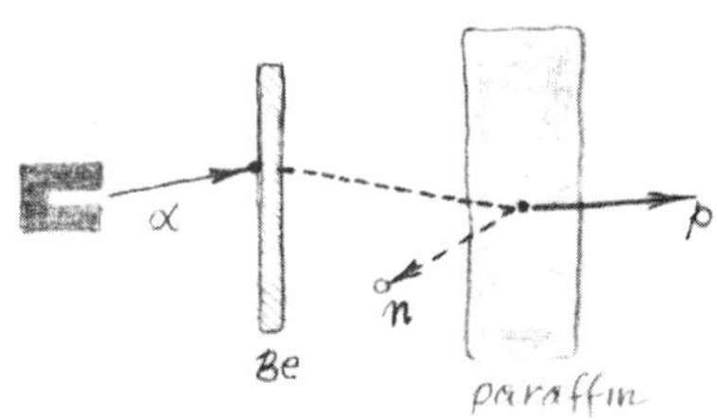

Paraffin wax contains 14 hydrocarbon compounds ranging from $C_{18}H_{38}$ to $C_{32}H_{66}$.

Letters to the Editor

[*The Editor does not hold himself responsible for opinions expressed by his correspondents. Neither can he undertake to return, nor to correspond with the writers of, rejected manuscripts intended for this or any other part of* NATURE. *No notice is taken of anonymous communications.*]

Possible Existence of a Neutron

IT has been shown by Bothe and others that beryllium when bombarded by α-particles of polonium emits a radiation of great penetrating power, which has an absorption coefficient in lead of about 0·3 (cm.)$^{-1}$. Recently Mme. Curie-Joliot and M. Joliot found, when measuring the ionisation produced by this beryllium radiation in a vessel with a thin window, that the ionisation increased when matter containing hydrogen was placed in front of the window. The effect appeared to be due to the ejection of protons with velocities up to a maximum of nearly 3×10^9 cm. per sec. They suggested that the transference of energy to the proton was by a process similar to the Compton effect, and estimated that the beryllium radiation had a quantum energy of 50×10^6 electron volts.

I have made some experiments using the valve counter to examine the properties of this radiation excited in beryllium. The valve counter consists of a small ionisation chamber connected to an amplifier, and the sudden production of ions by the entry of a particle, such as a proton or α-particle, is recorded by the deflexion of an oscillograph. These experiments have shown that the radiation ejects particles from hydrogen, helium, lithium, beryllium, carbon, air, and argon. The particles ejected from hydrogen behave, as regards range and ionising power, like protons with speeds up to about $3{\cdot}2 \times 10^9$ cm. per sec. The particles from the other elements have a large ionising power, and appear to be in each case recoil atoms of the elements.

If we ascribe the ejection of the proton to a Compton recoil from a quantum of 52×10^6 electron volts, then the nitrogen recoil atom arising by a similar process should have an energy not greater than about 400,000 volts, should produce not more than about 10,000 ions, and have a range in air at N.T.P. of about 1·3 mm. Actually, some of the recoil atoms in nitrogen produce at least 30,000 ions. In collaboration with Dr. Feather, I have observed the recoil atoms in an expansion chamber, and their range, estimated visually, was sometimes as much as 3 mm. at N.T.P.

These results, and others I have obtained in the course of the work, are very difficult to explain on the assumption that the radiation from beryllium is a quantum radiation, if energy and momentum are to be conserved in the collisions. The difficulties disappear, however, if it be assumed that the radiation consists of particles of mass 1 and charge 0, or neutrons. The capture of the α-particle by the Be^9 nucleus may be supposed to result in the formation of a C^{12} nucleus and the emission of the neutron. From the energy relations of this process the velocity of the neutron emitted in the forward direction may well be about 3×10^9 cm. per sec. The collisions of this neutron with the atoms through which it passes give rise to the recoil atoms, and the observed energies of the recoil atoms are in fair agreement with this view. Moreover, I have observed that the protons ejected from hydrogen by the radiation emitted in the opposite direction to that of the exciting α-particle appear to have a much smaller range than those ejected by the forward radiation. This again receives a simple explanation on the neutron hypothesis.

If it be supposed that the radiation consists of quanta, then the capture of the α-particle by the Be^9 nucleus will form a C^{13} nucleus. The mass defect of C^{13} is known with sufficient accuracy to show that the energy of the quantum emitted in this process cannot be greater than about 14×10^6 volts. It is difficult to make such a quantum responsible for the effects observed.

It is to be expected that many of the effects of a neutron in passing through matter should resemble those of a quantum of high energy, and it is not easy to reach the final decision between the two hypotheses. Up to the present, all the evidence is in favour of the neutron, while the quantum hypothesis can only be upheld if the conservation of energy and momentum be relinquished at some point.

J. CHADWICK.

Cavendish Laboratory,
Cambridge, Feb. 17.

Chadwick's first publication of the "neutron hypothesis" to explain the Joliot–Curie experimental results.

of the kinetic energy of the initially moving body will be transferred to the initially stationary body only if the bodies have approximately equal masses. In collisions that are not precisely head-on, less kinetic energy will be transferred. Therefore, on the *average*, a kinetic energy of about 5 MeV for the recoiling protons would be about right for collisions produced by neutrons with energies about 10 MeV, if the neutron and proton masses were approximately equal.

As explained in Sec. 14.8, the electron volt (eV) is a unit of energy.

$$1\text{keV} = 10^3\text{eV}$$
$$1\text{MeV} = 10^6\text{eV}$$
$$1\text{BeV} = 10^9\text{eV}$$

Chadwick was able to make a more precise calculation of the neutron's mass by applying the conservation laws to data on collisions with nuclei of different masses; the details of the derivation are shown on page 686. Chadwick found the mass of the neutron to be 1.16 amu. The difficulties of measuring the kinetic energies of the recoiling nuclei made this only an approximate value, but it was good enough to show that the neutron has a mass very close to that of the proton; thus, Chadwick's hypothesis did indeed offer a satisfactory solution to the problem of the "radiation" emitted when beryllium or boron was bombarded with particles.

SG 7, 8

Much research has been done since 1932 on the properties of neutrons and on the interactions between neutrons and atoms. An entire branch of study called *neutron physics* has been developed. Neutron physics deals with the production of neutrons, their detection, and their interaction with atomic nuclei and with matter in bulk. This research has led, among other things, to the discovery of nuclear fission, to be discussed in Chapter 24.

The best methods now available for determining the neutron mass give 1.008665 amu (based on the scale C^{12} = 12 exactly).

?

5. Why could the penetrating radiation from bombarded beryllium not be considered γ rays?

6. Why did the mass of a neutron have to be found by measurements on protons ejected by the neutrons in collision?

7. How could the principles of conservation discussed in Unit 3 be used to find the mass of the neutron?

23.5 | The proton–neutron theory of the composition of atomic nuclei

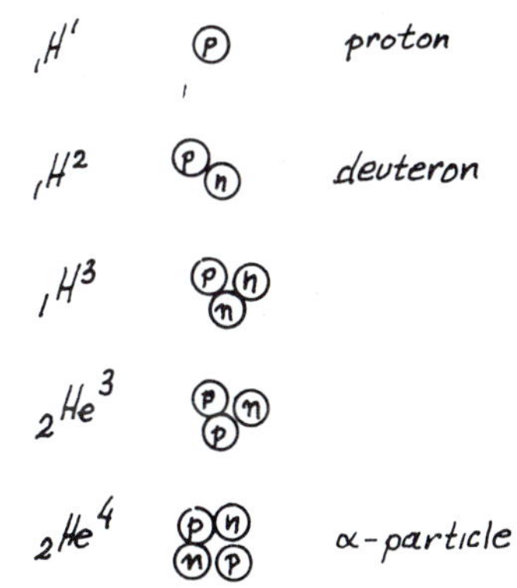

The discovery of the neutron, with an atomic mass close to one unit and with no electric charge, confirmed Rutherford's suggestion that the atomic nucleus is made up of protons and neutrons. This hypothesis was soon used as the basis of a detailed theory of the nucleus by Heisenberg in 1932 and is still the basis of attempts to describe the properties and structure of the nucleus. According to the proton–neutron hypothesis, the nucleus of an atom having atomic number Z and mass number A consists of Z protons and $A-Z$ neutrons. The nuclei of the isotopes of a given element differ only in the number of neutrons they contain. Thus, the nucleus of the hydrogen isotope of mass number 1 contains one proton; the nucleus of the hydrogen isotope of mass number 2 contains one proton and one neutron. (That nucleus is called a deuteron.) The nucleus of the neon isotope Ne^{20} contains 10 protons and 10 neutrons, while that of Ne^{22} contains 10 protons and 12 neutrons. The atomic number Z identified with the charge on the nucleus, is the number of protons in the nucleus. The mass number A is the total number of protons and neutrons. If the term *nucleons* refers to both kinds of nuclear particles, then A is simply the number of nucleons.

SG 9, 10

Is the proton–neutron hypothesis for the structure of nuclei fully consistent with the facts of radioactivity, such as α and β emission and the transformation rules? If two protons and two neutrons could combine, the resulting particle would have $Z = 2$ and $A = 4$, just the properties of the α particle. The emission of two protons and two neutrons (in the combined form of an α particle) would be consistent with the first transformation rule of radioactivity. (The α particle might exist as such in the nucleus, or it might be formed at the instant of emission; the latter possibility is now considered more likely.) But if the nucleus consists of protons and neutrons, where could a β particle come from? This question is more difficult to answer than that of the origin of an α particle. The second transformation rule of radioactivity provides a clue: When a nucleus emits a β particle, its charge Z increases by one unit while its mass number A remains unchanged. This would happen if a neutron were to change into a proton and a β particle.

This idea was not a return to the proton–electron hypothesis discussed in Sec. 23.2. Physicists had already come to the conclusion that electrons are not present in the nucleus, so β decay was not considered to be a simple separation of a proton and electron; it would have to be a *transformation* of a neutron that *created* a proton and electron. However, there were additional experimental data that raised difficulties for such a simple transformation idea.

?

8. According to the proton–neutron theory of the nucleus, what is in the nucleus of $_7N^{14}$?

9. Describe an ordinary helium atom in terms of the three elementary particles: the proton, the neutron, and (outside the nucleus) the electron.

10. If nuclei do not contain β particles, how can β emission be explained?

23.6 | The neutrino

The description of β decay in terms of the transformation of a neutron in the nucleus is part of one of the most fascinating stories in modern physics: the prediction and eventual discovery of the particles called the *neutrino* and the *antineutrino*. Quantitative studies of the energy relations in β decay during the 1920's and 1930's raised a difficult and serious question. Methods were devised for determining the energy change in a nucleus during β decay. According to the principle of conservation of energy, the energy lost by the nucleus should be equal to the energy carried off by the β particle, but the kinetic energy of the β particles had a whole range of measured values, all smaller than the amount of energy lost by the nucleus. Some of the energy lost by the nucleus seemed to have disappeared. Measurements made on a large number of β-emitters indicated that on the average about two thirds of the energy lost by the β-decaying nuclei seemed to disappear. Attempts to find the missing energy failed. For example, some physicists thought that the missing energy might be carried off by γ rays, but no such γ rays could be detected experimentally. The principle of conservation of energy seemed to be violated in β decay. Similar discrepancies were found in measurements of the momentum of the emitted electron and the recoiling nucleus.

As in the case of the experiments that led to the discovery of the neutron, physicists tried very hard to find an alternative to accepting the failure of the principles of conservation of energy and momentum. The Austrian physicist Wolfgang Pauli suggested in 1933 that another, hitherto unnoticed, particle is emitted in β decay along with the electron and that this particle carries off the missing energy and momentum. This hypothetical particle could have no electric charge, because the positive charge of the proton and the negative charge of the β particle together are equal to the zero charge of the original neutron. The mass–energy balance in the decay of the neutron indicated that the rest mass of the hypothetical particle should be very small, much smaller than the mass of an electron and possibly even zero. The combination of zero electric charge and zero or nearly zero mass would make the particle extremely hard to detect.

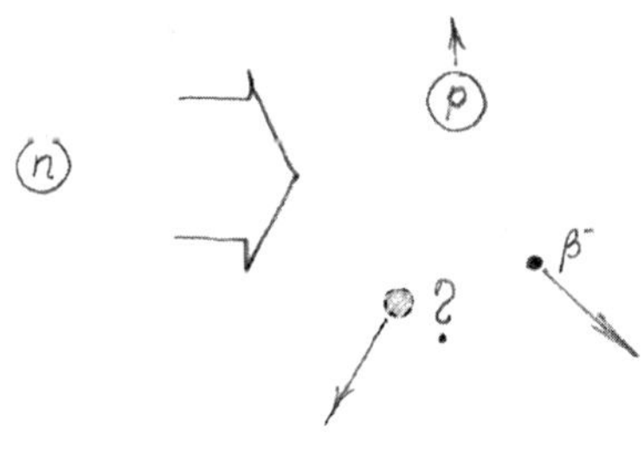

The Italian physicist Enrico Fermi called the suggested particle the *neutrino* ("little neutral one" in Italian). In 1934, Fermi constructed a theory of β decay based on Pauli's suggestion. This theory has been successful in describing the known facts of β decay. From 1934 on, the neutrino was accepted as a "real" particle for two reasons, both theoretical: It saved the principle of conservation of energy in β decay, and it could be used successfully both to describe the results of experiments in β decay and to predict the results of new experiments.

It is now known that a *free* neutron, that is, a neutron separated from an atom, sooner or later decays into a proton, an electron, and a neutrino. (The half-life of a beam of free neutrons has been measured to be 12 min.)

Close Up

Determining the Neutron's Mass

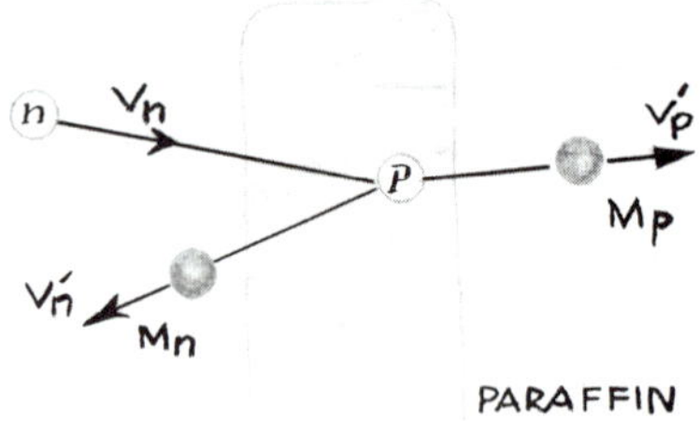

(a) The sketch above represents an elastic collision of a neutron (n) and a proton (p). If it were a head-on collision, the neutron would rebound straight back and the proton would be seen to emerge along the *same* line. To determine the mass of the neutron, m_n, you may use the principles of conservation of kinetic energy and conservation of momentum, which provide two algebraic equations that must both hold. The case is particularly simple if you consider a perfectly elastic head-on collision. As shown at the right, an expression for the proton's recoil speed v_p' can be derived by combining the equations algebraically (solving the momentum equation for v_n, substituting the resulting expression for v_n' in the energy equation, expanding, collecting terms, and solving for v_p'). However, this expression includes the term v_n, the neutron's initial speed, which cannot be measured directly. You can eliminate v_n from the equation by analyzing another collision and combining the results with what you already have.

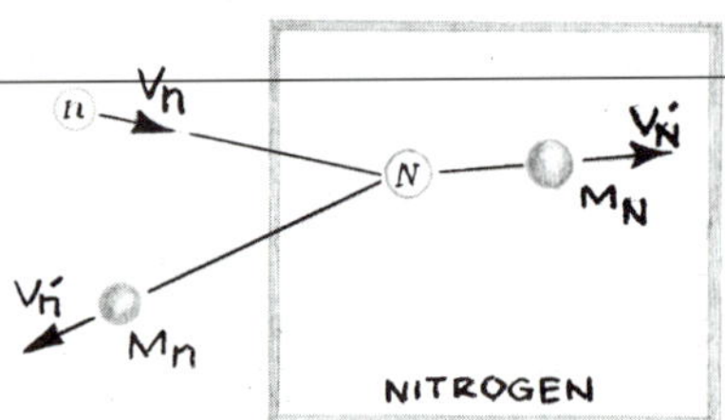

(b) The sketch above represents a perfectly elastic collision between a neutron (n) and a nitrogen nucleus (N). When the collision is head-on, you can write energy and momentum equations similar to what you wrote before, but this time *leading to an expression for the recoil speed of the nitrogen nucleus, v_N'*. This expression also includes the unmeasurable quantity v_n.

CONSERVATION OF ENERGY

$$\tfrac{1}{2}M_nV_n^2 = \tfrac{1}{2}M_nV_n'^2 + \tfrac{1}{2}M_pV_p'^2$$

CONSERVATION OF MOMENTUM

$$M_nV_n = M_nV_n' + M_pV_p'$$

$$V_n' = \frac{M_nV_n - M_pV_p'}{M_n}$$

$$\tfrac{1}{2}M_nV_n^2 = \tfrac{1}{2}M_n\left(\frac{M_nV_n - M_pV_p'}{M_n}\right)^2 + \tfrac{1}{2}M_pV_p'^2$$

$$M_nV_n^2 = \frac{M_n^2V_n^2 - 2M_nM_pV_nV_p' + M_p^2V_p'^2}{M_n} + M_pV_p'^2$$

$$M_n^2V_n^2 = M_n^2V_n^2 - 2M_nM_pV_nV_p' + M_p^2V_p'^2 + M_nM_pV_p'^2$$

$$M_p^2V_p'^2 + M_nM_pV_p'^2 = 2M_nM_pV_nV_p'$$

$$M_pV_p' + M_nV_p' = 2M_nV_n$$

$$V_p' = \frac{2M_nV_n}{M_p + M_n}$$

CONSERVATION OF ENERGY

$$\tfrac{1}{2}M_nV_n^2 = \tfrac{1}{2}M_nV_n'^2 + \tfrac{1}{2}M_NV_N'^2$$

CONSERVATION OF MOMENTUM

$$M_nV_n = M_nV_n' + M_NV_N'$$

$$V_N' = \frac{2M_nV_n}{M_N + M_n}$$

$$\frac{V_p'}{V_N'} = \frac{M_N + M_n}{M_p + M_n}$$

$$M_n = \frac{M_NV_N' - M_pV_p'}{V_p' - V_N'}$$

(c) The v_p equation and v_N' equation are then combined algebraically (eliminating v_n), and solved for m_n. The expression for m_n now contains only terms that can be measured, so the mass of the neutron, m_n, can be calculated. Note that only the ideas developed for ordinary elastic collisions are used here. (See SG 7 and 8.)

Many unsuccessful attempts were made to capture neutrinos over a period of 25 years. Finally, in 1956, neutrinos were detected in an experiment using the extremely large flow of neutrinos that comes out of a nuclear reactor (see Chapter 24). The detection of neutrinos is an indirect process that involves detecting the products of a reaction *provoked* by a neutrino. The reaction used was a reverse β decay, the production of a proton from a neutron. Because the proper meeting of a proton, an electron, and a neutrino at the same place and same time is an exceedingly unlikely event, and the resulting neutron difficult to detect, "catching" the neutrinos required a very elaborate and sensitive trap. (See photo at the right.) Again the faith of physicists in the principle of conservation of energy was justified.

There is one more complication. It is now known that there are several kinds of neutrinos. The one involved in β decay (as discussed so far) is now referred to as an *antineutrino* and is denoted by the symbol $\bar{\nu}$. The transformation of a neutron during β emission is then written:

$$_0n^1 \rightarrow {}_1p^1 + {}_{-1}e^0 + \bar{\nu}$$

The first detection of neutrinos was in this tank. Reactions provoked by neutrinos cause flashes of light in the liquid with which the tank is filled. The flashes are detected by the photoelectric tubes that stud the tank wall. This work was done by two American physicists, Clyde Cowan and Frederick Reines.

Chapter 4 of *Project Physics* Supplemental Unit B, "Discoveries in Physics," discusses the story of the neutrino in much more detail.

? *11. Why was an almost undetectable particle invented to patch up the theory of β decay?*

23.7 | The need for particle accelerators

Up to 1932, the study of nuclear reactions was limited by the kind of projectile that could be used to bombard nuclei. Only α particles from the naturally radioactive nuclides could bring about reactions. Progress was limited because α particles could be obtained only in beams of low intensity and with fairly low kinetic energies. These relatively low-energy particles could produce transmutations only in light elements. When heavier elements are bombarded with α particles, the repulsive electric force exerted by the greater charge of the heavy nucleus on an α particle makes it difficult for the α particle to reach the nucleus. The probability of a nuclear reaction taking place becomes very small or zero. Because the interest in nuclear reactions was great, physicists in many countries sought methods of increasing the energy of charged particles to be used as projectiles.

The proton accelerator at Fermilab, where the 9.6-BeV "Upsilon" particle was discovered.

There were advantages to be gained in working with particles like the proton or the deuteron (the nucleus of the deuterium or heavy hydrogen atom) that have only one positive charge. Having only a single charge, these particles would experience smaller repulsive electric forces than would α particles in the neighborhood of a nucleus and thus would be more successful

A Van de Graaff generator, built on a vertical axis.

in getting close enough to produce transmutations, even of *heavy* (and therefore high-charge) target nuclei. Protons or deuterons could be obtained from positive-ray tubes, but their energies were rather low. Some device was needed to accelerate these particles to higher energies, as Rutherford was among the first to say (see page 686). Such devices might also offer other advantages. The speed (and energy) of the bombarding particles could be controlled by the experimenter, and very intense projectile beams might be obtained. It would then be possible to find how nuclear reactions depend on the energy of the bombarding particles.

Since 1930, many devices for accelerating charged particles have been invented and developed. In each case, the particles used (electrons, protons, deuterons, α particles, or heavy ions) are accelerated by an electric field. In some cases, a magnetic field is used to control the path of particles, that is, to steer them. The simplest type has a single high-voltage step. These machines cannot be practically operated above about 10 million volts, so they cannot be used to increase electron or proton energies above about 10 MeV.

Another type has a long series of low-voltage steps applied as the particle travels in a straight line. Some of these machines produce electron energies up to 20 GeV (1 GeV = 10^4 eV, and used to be written also as 1 BeV). A third general type uses magnetic fields to hold the particles in a circular path, returning
SG 16 them over and over to the same low-voltage accelerating fields. The first machine of this type was the cyclotron (see the photograph on page 675). Other circular types are illustrated on pages 690 and 691. Some of these accelerators produce 7 GeV electrons or 500 GeV protons. Accelerators producing 1,000 GeV (1 TeV) will soon be in operation. Accelerators have become basic tools for research in nuclear and high-energy physics; the way they operate and the way a typical experiment was actually done are the subject of the two *Project Physics* films, "Synchrotron," and "People and Particles." Accelerators also are used in the production of radioactive isotopes and serve as radiation sources for medical and industrial purposes.

Table 22-3 summarizes the major types of particle accelerators now being used or planned. One of the most powerful is a 500-GeV particle accelerator now in operation at the National Accelerator Laboratory (Fermilab) in Batavia, Illinois. Such "machines" are among the most complex and grandiose structures ever built. Indeed, they are monuments to human imagination and ingenuity, the ability to reason and to collaborate in groups on peaceful projects that further the understanding of nature. Basically, the "machines" are tools to help physicists find out as much as they can about the structure of nuclear particles and the forces holding them together.

TABLE 22-3. MAJOR TYPES OF PARTICLE ACCELERATORS

Type	Principle of Operation	Maximum Energy	Particles	Notes and Examples of Use
Once-Through Acceleration				
Cockcroft–Walton	direct high-voltage potential	≈ 4 MeV	various	commercially available
Van de Graaff generator	high voltage by transport of charges on moving belt	≈ 3 MeV ≈ 14 MeV	electrons protons	commercially available
Linear accelerator	successive application of high-frequency voltages	≈ 10 MeV per particle	heavy ions	Lawrence Radiation Laboratory and Yale University
Linear accelerator	pulsed high-frequency wave	≈ 20GeV	electrons	Stanford University, 3.2 km
Cyclic Acceleration				
Betatron	magnetic induction (electrons in an evacuated tube accelerated by variable field of electromagnet)	≈300 MeV	electrons	largest machine is at University of Illinois
Cyclotron	voltage of constant frequency applied to particles in fixed magnetic field	≈ 12 MeV ≈ 24 MeV ≈ 48 MeV	protons deuterons He nuclei	commercially available; numerous installations, including some built by students at high schools
Synchro-cyclotron	voltage of variable frequency applied to particles in fixed magnetic field	≈ 750 MeV	protons	460-cm unit at Lawrence Radiation Laboratory, Berkeley
Electron synchrotron	voltage of constant frequency applied to particles orbiting in variable magnetic field	≈ 7 GeV	electrons	Hamburg, Germany (7.5 GeV); Cambridge Electron Accelerator (6 GeV) operated by Harvard and M.I.T. 1962–1973
Proton synchrotron	synchronized voltage of high frequency applied to particles orbiting in variable magnetic field	≈ 12 GeV	protons	6.2 BeV "Bevatron" at Lawrence Radiation Laboratory; 3 BeV Cosmotron at Brookhaven; 3 BeV at Princeton; and 12.5 BeV synchrotron at Argonne National Laboratory
Alternating gradient synchrotron	same as synchrotron except successive segments of magnetic field have opposite curvature	≈ 30 GeV	protons	Brookhaven National Laboratory (Long Island), and CERN, Switzerland (where a 500-GeV accelerator is also in operation)
		≈ 76 GeV	protons	Serpukhov, U.S.S.R. (also a planned 1,000-GeV accelerator)
Strong-focusing synchrotron	same as alternating gradient synchrotron	≈ 500 GeV	protons	Enrico Fermi National Accelerator Laboratory, Batavia, Illinois (Fermilab)
Energy doubler/saver (Tevatron)	uses super-conducting magnets beneath main ring	≈ 1,000 GeV (1 TeV)	protons	Fermilab

With the discovery of the neutron in 1932, it was believed that three "elementary" particles act as the building blocks of matter: the proton, the neutron, and the electron. The existence of new particles, such as neutrinos and antineutrinos, has been mentioned. As high-energy accelerators became available, additional "elementary" particles were discovered, one after

Close Up

Accelerators

Research into the nature of matter has disclosed the structure of the atom and the atomic nucleus. Current research is focused on the particles that make up the nucleus. Matter responds to four different types of force: (1) the strong force, (2) the electromagnetic force, (3) the weak force, and (4) the gravitational force. By observing how particles react when influenced by these forces, scientists have discovered the existence of many new and bizarre particles with properties like "charm" and "color." The discovery of these particles has only become possible through the use of particle accelerators of increasingly higher energy. The first accelerator was built by E. O. Lawrence in 1930. Particle physics research in the United States is carried on at Brookhaven National Laboratory, the Stanford Linear Accelerator Center, Argonne National Laboratory, and the Fermi National Accelerator Laboratory. In Europe, the most powerful accelerators are at CERN in Geneva, Switzerland, and near Serpukhov in the Soviet Union. Probing the nature of matter is an international endeavor. For example, scientists from over 20 countries participate in research at Fermilab.

The photos on these two pages show facilities at Brookhaven (shown below) and Fermilab (opposite, bottom).

RAB60

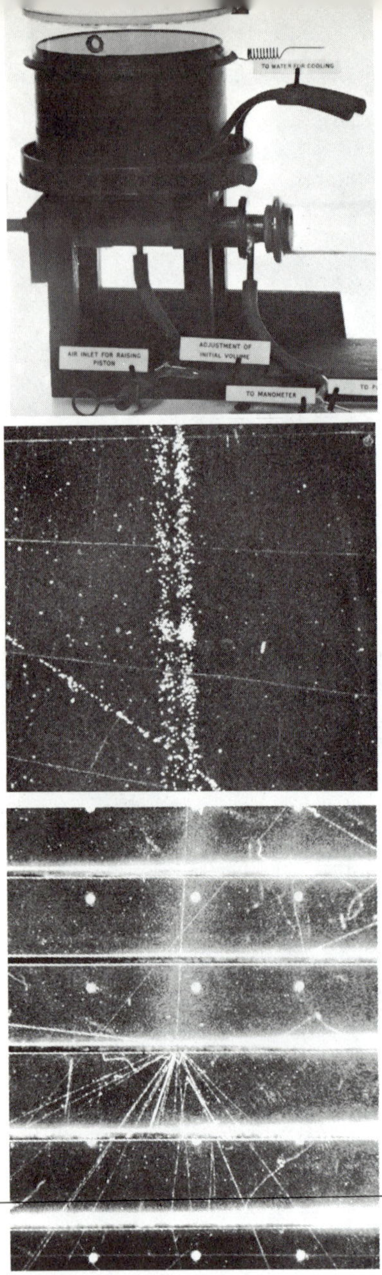

The top photograph shows C. T. R. Wilson's cloud chamber. (See also page 678.) The middle photograph shows particle tracks in a cloud chamber. (The positively and negatively charged ions had separated before the cloud was formed, so the track shows up as two vertical streaks.) In the bottom photograph, high voltages between the plates in a spark chamber cause sparks to jump along the ionized trails left by high-energy charged particles.

another. These particles are grouped into "families" according to their properties. Most of these particles exist only briefly; typical lifetimes are of the order of 10^{-8} sec or less. A whole new field, high-energy physics, has evolved, and the aim of the high-energy physicist of today is to discern the order and structure behind the large number of "elementary" particles that have been discovered.

How do physicists detect these particles? A number of methods by which physicists can observe and measure radioactive emissions have already been mentioned. They include the electroscope and the electrometer employed since the early days of radioactivity, the Geiger counter (see Sec. 19.3), and the Wilson cloud chamber. In addition, various types of ionization chambers, scintillation counters, photographic emulsions, semiconductor devices, spark chambers, and bubble chambers (some of which are displayed on pages 694 and 695) are also in use. One of the supplemental units in the *Project Physics* course, entitled "Elementary Particles," describes in detail the devices and the discoveries made with them.

?

12. Why can low-energy α particles cause transmutations only in nuclei of relatively small atomic number?

13. Why are protons more effective projectiles for producing nuclear reactions than are α particles or heavy ions?

14. What are some of the devices for producing high-energy particles to be used as projectiles? What are some devices for detecting nuclear reactions induced by such projectiles?

23.8 | Nuclear reactions

The development of the cyclotron and other particle accelerators led to great advances in the study of nuclear reactions. Nearly all of the stable nuclides have now been bombarded with protons, deuterons, α particles, neutrons, and γ rays, and hundreds of nuclear reactions have been examined. Examples of reactions induced by α particles and protons have already been discussed.

Since the first known alchemical writings during the third or fourth centuries A.D., and throughout the historical development of chemistry, the dream of transmuting materials (usually into gold) has always haunted some people. In most nuclear reactions, one element is indeed changed into another; in a sense the ancient dream of the alchemist has come true, but it is unlikely to make a fortune for anyone. It is possible to transmute various elements into gold, but such transformations are, of course, completely different, both in method and purpose, from the attempts of the ancient alchemists. (Moreover, they are all entirely uneconomical methods for "making gold.")

Gold has only one stable isotope found in nature, ${}_{79}Au^{197}$; other gold isotopes can be made, but are radioactive. Two types of nuclear reactions induced by deuterons both result in the stable isotope of gold:

$$ {}_{1}H^{2} + {}_{80}Hg^{199} \rightarrow {}_{79}Au^{197} + {}_{2}He^{4} $$
$$ {}_{1}H^{2} + {}_{78}Pt^{196} \rightarrow {}_{79}Au^{197} + {}_{0}n^{1} $$

In both cases, an accelerator is needed to produce high-energy deuterons; in bombarding a mercury isotope, α particles are produced in addition to the desired gold. In bombarding platinum, neutrons are produced in addition to the gold.

The last reaction, in which a *neutron* was produced, is an example of reactions that have become especially important because of the usefulness of the neutrons. Neutrons can be produced when nuclei are bombarded with protons, deuterons, or α particles, as in the reactions:

$$ {}_{28}Ni^{58} + {}_{1}H^{1} \rightarrow {}_{29}Cu^{58} + {}_{0}n^{1} $$
$$ {}_{6}C^{12} + {}_{1}H^{2} \rightarrow {}_{7}N^{13} + {}_{0}n^{1} $$
$$ {}_{4}Be^{9} + {}_{2}He^{4} \rightarrow {}_{6}C^{12} + {}_{0}n^{1} $$

The neutrons produced by such bombardment can, in turn, be used to induce other nuclear reactions. As noted before, neutrons are especially effective as "bullets," because they have no electric charge. They are not subject to repulsive electrostatic forces in the neighborhood of a positively charged nucleus and are therefore more likely to penetrate nuclei than are protons, deuterons, or α particles.

Because of the neutron's lack of electrical charge, many more reactions have been induced by neutrons than by any other kind of particle. Enrico Fermi was the first to undertake a systematic program of research involving the use of neutrons as projectiles in nuclear reactions. Starting in 1934, Fermi and his group bombarded many elements, from the lightest to the heaviest, with neutrons, and studied the properties of the nuclides produced. The research described in the Prologue to Unit 1 was undertaken as part of this program.

A typical neutron-induced reaction, again one resulting in gold, is:

$$ {}_{0}n^{1} + {}_{80}Hg^{198} \rightarrow {}_{79}Au^{197} + {}_{1}H^{2} $$

In another very common type of neutron-induced reaction, the neutron is captured, and a γ ray is emitted, as in the following example:

$$ {}_{0}n^{1} + {}_{78}Pt^{196} \rightarrow {}_{78}Pt^{197} + \gamma $$

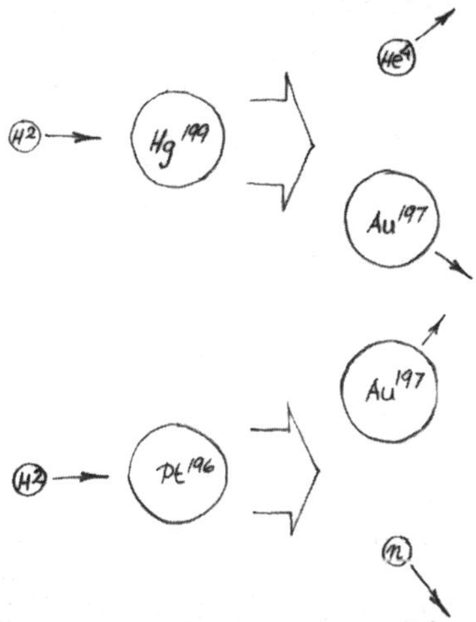

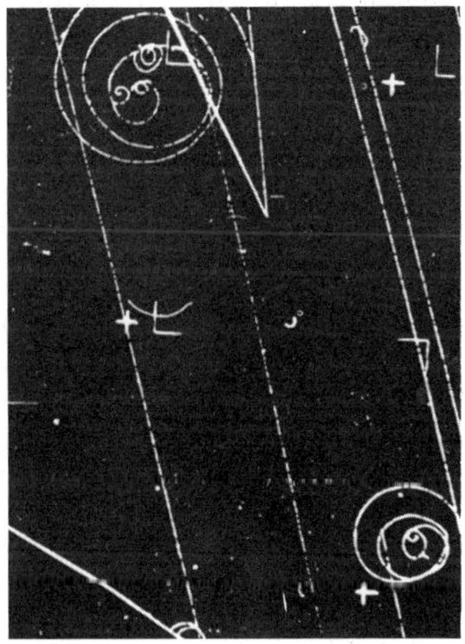

In this bubble-chamber picture, a neutron is produced at the bottom, at the end of the 1.25-cm long track, near the + mark. This neutron in turn causes a reaction near the center of the plate. (Neutral particles do not leave tracks in bubble chambers.)

The transmutation into gold is discussed only as an example of a nuclear reaction; a more useful reaction is the transmutation of gold into something else, for example:

$$ {}_{79}Au^{197} + {}_{0}n^{1} \rightarrow {}_{80}Hg^{198} + {}_{-1}e^{0} $$

This reaction can be used to obtain pure samples of a single mercury isotope (alchemy turned upside down.)

Close Up

Bubble Chambers

Below left: The tiny bubble chamber (only 3 cm long) invented by D. A. Glaser in 1952. (Note the particle track.) Glaser was 26 at the time of his invention of the bubble chamber and was later awarded the Nobel Prize. Below right: The 4.5-m bubble chamber at the Fermi National Accelerator Laboratory. High-energy particles from the accelerator enter the chamber from the right. The particles interact with liquid hydrogen in the chamber to form tracks, which are photographed by seven cameras in the top of the chamber.

The bubble chamber photo on the opposite page was taken in a 25-cm liquid-hydrogen bubble chamber at the Lawrence Radiation Laboratory of the University of California. The chamber is also shown on the opposite page (below, left) with the liquid-nitrogen shield removed. The accompanying diagram (below, right) gives some of the details of the bubble chamber and its auxiliary equipment.

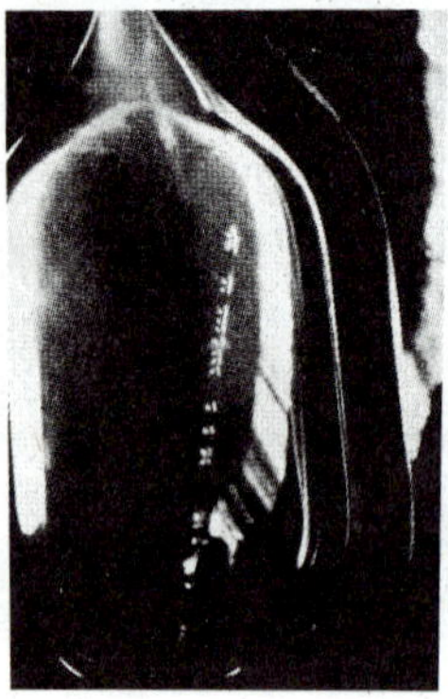

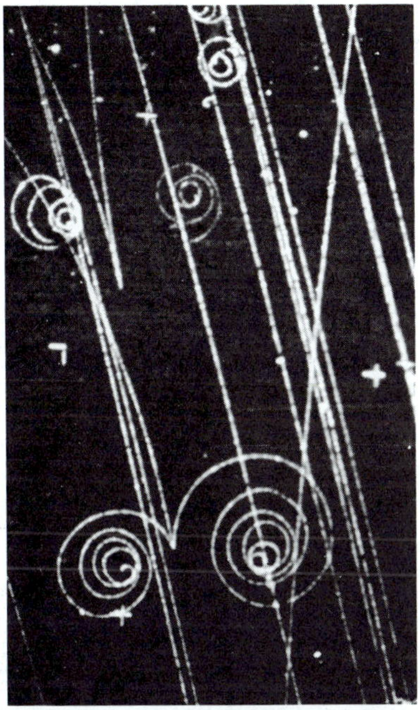

The bubble chamber photo at the left illustrates one of the major discoveries of modern physics, the interconversion of energy and matter (discussed in Chapter 20). The diagram at the right shows the significant tracks recorded in the photo. In the upper left of the diagram, an electron–positron pair is formed by a gamma ray (not visible in bubble chamber pictures) interacting with a hydrogen nucleus. (The discovery of positrons is described briefly on page 697.) An applied magnetic field causes the electron and the positron to be deflected in opposite directions. (In what direction was the magnetic field?) In the lower left of the diagram, a gamma ray forms another electron–positron pair; the additional electron (third track, upward) was knocked out of a hydrogen atom during this process.

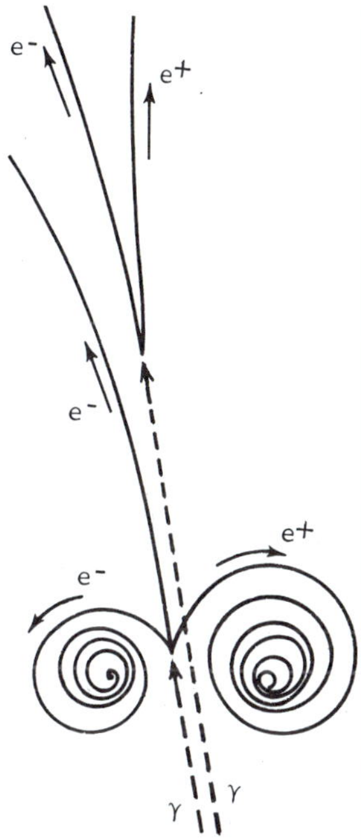

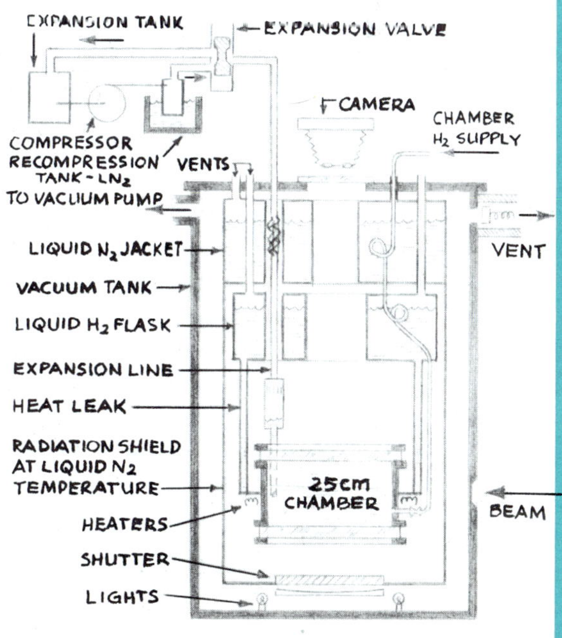

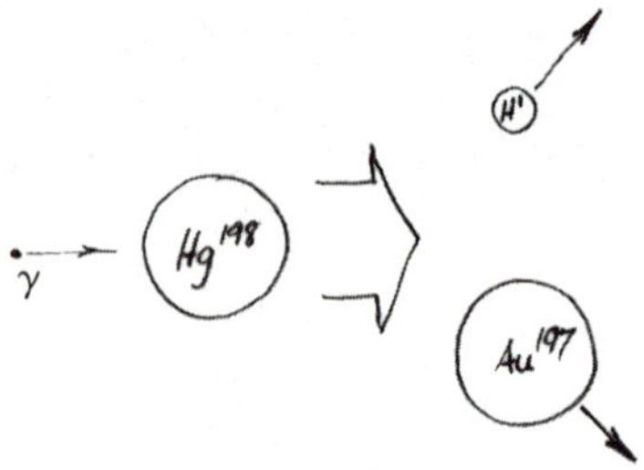

Note that since there is no change in the atomic number, the element here remains the same. An isotope of the target nucleus is produced with a mass number greater by one unit than that of the target nucleus. The new nucleus so produced has more energy than it needs to be stable and is said to be produced in an "excited state." The nucleus returns to its lowest energy state by emitting one or more γ rays.

Some nuclei can also undergo reactions when bombarded with γ rays; an example (for illustration's sake, once again resulting in gold) is the reaction:

$$\gamma + {}_{80}\mathrm{Hg}^{198} \rightarrow {}_{79}\mathrm{Au}^{197} + {}_{1}\mathrm{H}^{1}$$

In this case, the energy of the γ ray excites the mercury target nucleus, which becomes unstable, ejects a proton, and thereby becomes a gold nucleus.

SG 11 The amount of gold that can be produced by the above reaction is very small; these reactions are simply illustrations of
SG 12 some typical artificial transmutations. The examples given barely suggest the rich variety of such reactions that have been observed. The products of these reactions may change as the energy of the bombarding particles changes. Nuclear reactions are important, not only because they indicate an ability to produce new nuclides, but also because they provide important data about nuclear structure. A model of nuclear structure, to be successful, must enable physicists to predict the results of these nuclear reactions, just as a successful model of atomic structure must allow chemists to predict the results of chemical reactions.

?

15. What property of neutrons makes them particularly useful for producing nuclear reactions?

16. Complete the following equation for a nuclear reaction:

$${}_{13}Al^{27} + {}_{1}H^{2} \rightarrow {}_{0}n^{1} + {}_{?}Si^{?}$$

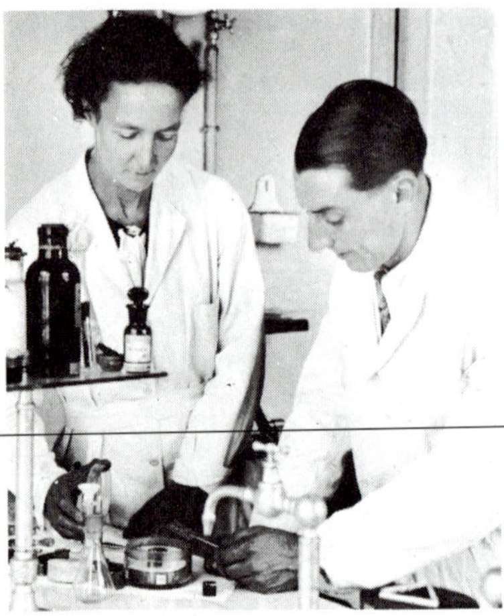

Irène Curie and F. Joliot in their laboratory. They were married in 1926. In 1935, they shared the Nobel Prize for chemistry.

23.9 | Artificially induced radioactivity

The discussion of nuclear reactions has hinted at another interesting discovery. In the last section, you saw that the capture of a neutron by platinum-196 results in platinum-197 and the emission of a γ ray. As listed in Table 22-1, six different isotopes of platinum are found in nature, but platinum-197 is not among these. The question arises: Is platinum-197, produced by neutron capture, stable? The answer is no; it is radioactive and decays by the emission of a β particle to gold-197 (the only stable gold isotope):

$$_{78}Pt^{197} \rightarrow {}_{79}Au^{197} + {}_{-1}e^{0} + \bar{\nu}$$

The half-life of platinum-197 is 20 hr.

The production of radioactive platinum-197 in a nuclear reaction is an example of *artificially induced radioactivity*. This phenomenon was discovered in 1934 by Irène Curie and F. Joliot. They were studying the effects of α particles on the nuclei of light elements. When they bombarded boron, magnesium, and aluminum with α particles from polonium, they observed the immediate ejection of protons and neutrons from the bombarded nuclei, as expected. But, in addition to these particles, *positive electrons*, or *positrons*, also were observed to be emitted. The positron is a particle whose mass is the same as that of the electron, and whose charge has the same magnitude but opposite sign to that of the electron.

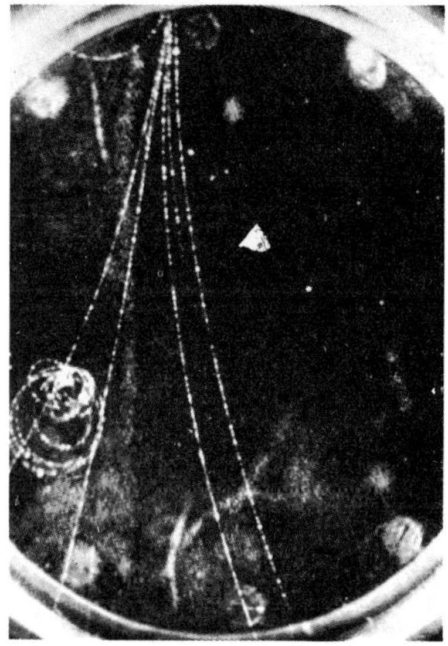

One of the earliest records of a "shower" of electrons and positrons; it shows their tracks curving in opposite directions in a strong magnetic field. The shower was caused by cosmic rays and was recorded in a Wilson cloud chamber taken to an altitude of 4.3 km.

The positron had been discovered by the American physicist C. D. Anderson in 1932 while studying cosmic rays. (Cosmic rays are highly penetrating radiations which originate outside the earth and consist of protons, electrons, neutrons, photons, and other particles.) Employing a cloud chamber situated in a magnetic field, Anderson observed some tracks which, judged by the density of ionization along the track, could have been produced only by high-speed particles having the same mass and *magnitude* of charge as an electron; but the curvature was opposite in direction to that of the high-speed electron tracks. Anderson concluded that the particles producing them must have been positively charged electrons, to which the name positron was given (symbol β^{+}, or $_{1}e^{0}$).

In the Joliot–Curie experiment, the production of positrons along with neutrons as a result of the bombardment of a light element with a α particles seemed to indicate that a new type of nuclear reaction was occurring. Further experiments by this couple showed that the light-element targets *continued to emit positrons*, even after the source of the α particles had been removed. When the rate of emission of the positrons was plotted against time elapsed since removal of the α-particle source, curves for each target were obtained that were similar to the curves obtained in natural β radioactivity. (The half-life of the emitter was found to be 2.5 min.) The results seemed to show that an initially stable nuclide had been changed into a radioactive one. In the case of the bombardment of $_{13}Al^{27}$ by α particles, which produced neutrons as well as a new radioactive material, a nuclear reaction would produce a nuclide of mass number 30 (= 27 + 4 − 1) and atomic number 15 (= 13 + 2 − 0), an isotope of phosphorus. The reaction would be:

$$_{13}Al^{27} + {}_{2}He^{4} \rightarrow {}_{0}n^{1} + {}_{15}P^{30}$$

Curie and Joliot ran chemical separations similar to those made in the study of the naturally radioactive elements and

Spectrum of γ rays emitted from a human hair after it had been irradiated by neutrons. The peaks in intensity are created by activated metals normally present in hair in minute amounts. Because everyone has a slightly different hair spectrum, activation analysis is a useful identification tool. By the same technique, the composition of most other objects can be determined.

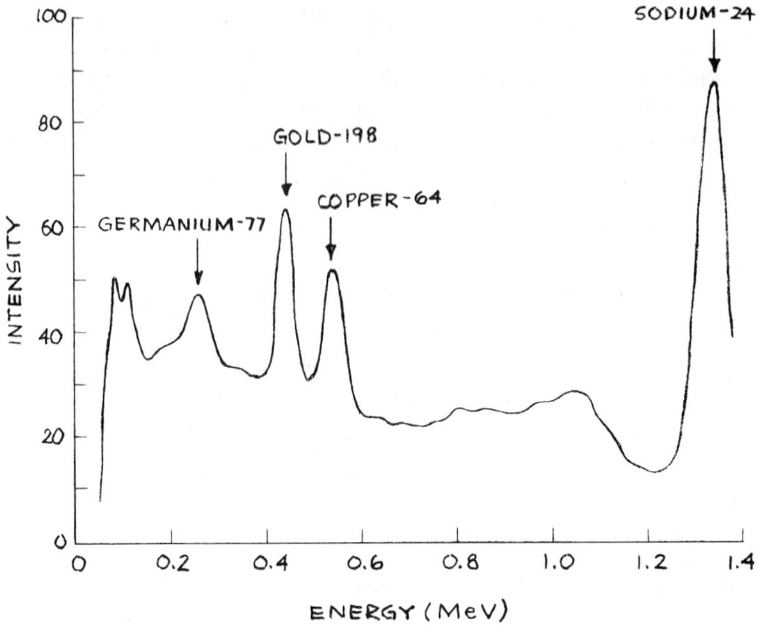

showed that the target, after bombardment, indeed contained a small amount of phosphorus and an isotope that was radioactive. Phosphorus occurs in nature only as $_{15}P^{31}$; no isotope of phosphorus with mass number 30 had ever been found to occur naturally. It was reasonable to suppose that if P^{30} were made in a nuclear reaction, it would not be stable but radioactive. If it decayed by emission of a positron, that reaction would be expressed in the following manner:

$$_{15}P^{30} \rightarrow {_{14}Si^{30}} + {_{1}e^{0}} + \nu$$

where $_{14}Si^{30}$ is a known isotope of silicon, $_{1}e^{0}$ represents a positron, and ν is a neutrino.

This kind of decay implies that a proton in the nucleus may be transformed into a neutron that remains in the nucleus, a positron that is emitted, and a neutrino:

$$_{1}p^{1} \rightarrow {_{0}n^{1}} + {_{1}e^{0}} + \nu \text{ (within the nucleus)}$$

In sum, after the discovery that the bombardment of light nuclides by α particles could lead to radioactive products, it was found that nuclear reactions induced by protons, deuterons, neutrons, and photons could also result in radioactive products. As in the case of the natural radionuclides, an artificial radionuclide can be characterized by its half-life and the type of radiation it emits. When the products of nuclear reactions are radioactive, they can be traced in chemical separations by means of their characteristic half-lives or decay products. (They cannot be traced chemically because very small amounts are involved, often less than 1 millionth of a gram.) The special branch of

Among the various modes of decay of artificial radioactive nuclides are α, β^-, β^+, γ emissions and capture of an orbital electron by the nucleus.

chemistry that deals with the separation and identification of the radioactive products of nuclear reactions is called *radiochemistry* and has become an important part of nuclear science. The breadth of this field is indicated by the fact that since 1935 about 1,200 artificially radioactive nuclides have been made and identified, many of which are in use in research and industry.

SG 13–16

? *17. Complete the following equation for a positive β decay:*

$${}_7N^{13} \rightarrow {}_1e^0 + {}_??^?$$

How many neutrons and protons were there in the nitrogen nucleus before decay? How many in the resulting product nucleus afterward?

study guide

1. The *Project Physics* learning materials particularly appropriate for Chapter 23 include the following:

Activity
Neutron Detection Problem Analogue

Film Loop
Collisions With an Object of Unknown Mass

Films
People and Particles
Synchrotron

2. Why would it be difficult to explain the nucleus of ${}_{92}U^{235}$ as a mixture of alpha particles and electrons?

3. On the basis of the proton–electron hypothesis of nuclear composition, how many protons would you expect to find in the ${}_{92}U^{235}$ nucleus? How many electrons?

4. Complete the following nuclear equations:

(a) ${}_5B^{10} + {}_2He^4 \rightarrow (\quad) + {}_1H^1$

(b) ${}_{11}Na^{23} + {}_2He^4 \rightarrow (\quad) + {}_1H^1$

(c) ${}_{13}Al^{27} + {}_2He^4 \rightarrow (\quad) + {}_1H^1$

(d) $(\quad) + {}_2He^4 \rightarrow {}_{17}Cl^{35} + {}_1H^1$

(e) $(\quad) + {}_2He^4 \rightarrow {}_{20}Ca^{42} + {}_1H^1$

5. Complete the following nuclear equations:

(a) ${}_3Li^6 + {}_1H^1 \rightarrow {}_2He^4 + (\quad)$

(b) ${}_4Be^9 + {}_1H^1 \rightarrow {}_2He^4 + (\quad)$

(c) ${}_4Be^9 + {}_1H^1 \rightarrow (\quad) + {}_1H^2$

(d) ${}_5B^{11} + {}_2He^4 \rightarrow {}_7N^{14} + (\quad)$

6. Complete the following nuclear equations (consult the periodic table of elements for the atomic numbers of indicated nuclides):

(a) $Al^{27} + {}_0n^1 \rightarrow Al^{28} + (\quad)$

(b) $Al^{27} + {}_1H^2 \rightarrow {}_1H^1 + (\quad)$

(c) $Al^{27} + {}_1H^1 \rightarrow {}_2He^4 + (\quad)$

(d) $Al^{27} + {}_1H^2 \rightarrow {}_2He^4 + (\quad)$

What aspect of nuclear reactions do equations (**b**) and (**d**) illustrate?

7. (a) Explain briefly why the maximum speed gained by nitrogen nuclei in collisions with neutrons is roughly 10 times less than that gained by hydrogen nuclei in collisions with neutrons.

(b) Where in this course was the physics needed for this problem first developed?

8. One major disadvantage of indirect methods of measurement is that the experimental uncertainty is often larger. If Chadwick had measured a maximum speed of 3.4×10^9 cm/sec for hydrogen nuclei (a change of only 3%) and 4.7×10^8 cm/sec for nitrogen nuclei (no change), what would be the calculated mass of the neutron? By what percentage would the calculated mass of the neutron change because of the 3% shift in the speed measurement?

9. Copy the following table in your notebook and indicate the mass number A, the atomic number Z, the number of protons, and the number of neutrons for each of the nuclei.

	A	Z	*Number of Protons*	*Number of Neutrons*
H^1				
H^2				
He^4				
Li^7				
C^{13}				
U^{238}				
Th^{234}				
Th^{230}				
Pb^{214}				
Pb^{206}				

10. How many electrons are there in a neutral atom of

(a) platinum-196?

(b) gold-198?

(c) mercury-198?

(d) mercury-199?

11. Complete the following nuclear equations:

(a) ${}_{11}Na^{23} + {}_1H^2 \rightarrow {}_1H^1 + (\quad)$

(b) ${}_{11}Na^{23} + {}_0n^1 \rightarrow \gamma + (\quad)$

(c) ${}_{12}Mg^{24} + {}_0n^1 \rightarrow {}_1H^1 + (\quad)$

(d) ${}_{12}Mg^{26} + {}_1H^2 \rightarrow {}_2He^4 + (\quad)$

What aspect of nuclear reactions do these equations illustrate?

12. Describe the following reactions in *words*:

$${}_{13}Al^{27} + {}_0n^1 \rightarrow {}_{12}Mg^{27} + {}_1H^1$$
$${}_{12}Mg^{27} \rightarrow {}_{13}Al^{27} + {}_{-1}e^0 + \bar{\nu} + \gamma\ (T_{1/2} = 9.5 \text{ min})$$

13. It is often necessary to infer information in the absence of direct evidence. Thus, when a hiker following the tracks of a rabbit in the snow finds that the tracks suddenly stop with no evidence of other tracks or of hiding places, the hiker may infer something about the possible presence of owls or eagles.

The bubble chamber photograph below shows, among other things, the tracks of two nuclear particles that originate or terminate at a point in the lower center. Describe interactions that might occur at that point in terms of your knowledge of the law of conservation of momentum.

14. How may the discovery of artificially radioactive nuclides have helped the development of theories of nuclear structure?

15. If you have seen one or more of the films "Synchrotron," "People and Particles," and "The World of Enrico Fermi," write an essay on either:

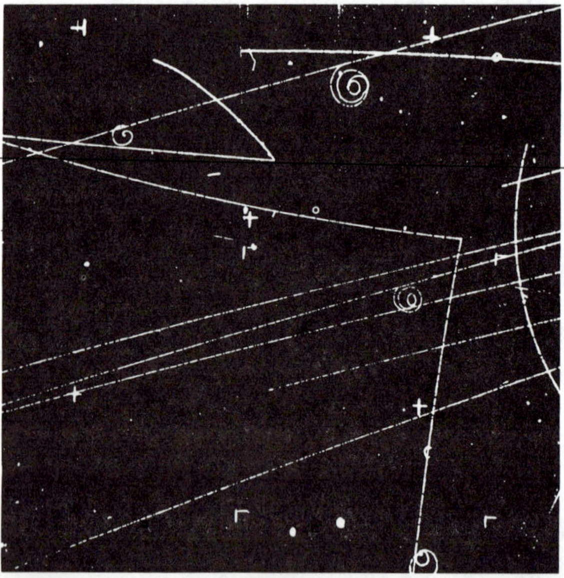

(**a**) the way research teams work together in modern high-energy physics;

(**b**) the reasons why some parts of modern experimental physics require large "machines" to do research; or

(**c**) why in many major countries millions of dollars of public money are appropriated to build and run these machines.

16. Compare the mass of a neutral helium atom with the sum of the masses of four hydrogen nuclei plus two electrons outside (to get a neutral helium atom). What conclusions do you draw?

17. Activity: Neutron Detection Problem Analogue (Chadwick's Problem). It is impossible to determine both the mass and the velocity of a neutron from measurements of the mass and the final velocity of a target particle that the neutron has hit. To help you understand this, try the following:

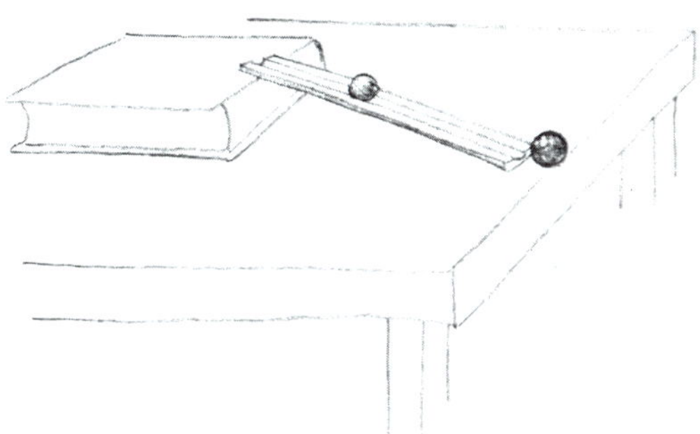

Set up an inclined groove on a table as shown in the diagram. Let a small ball bearing roll part way down the groove, hitting the larger target ball and knocking it off the table. Note the point where the target ball strikes the floor. Now use another smaller ball as the projectile. Can you adjust the point of release until the target ball strikes the *same spot* on the floor as it did when you used the large projectile? If so, then two different combinations of mass and velocity for the projectile cause the same velocity of the target ball. Are there more combinations of mass and velocity of the target ball? Are there more combinations of mass and velocity of the "neutron" that will give the same result?

Now repeat the experiment, but this time have the same projectile collide in turn with two different target balls of different masses, and measure the velocities of the targets.

Use these velocity values to calculate the mass of the incoming neutron. (Hint: Refer to Sec. 23.4. You need only the ratio of the final velocities achieved by the different targets; therefore, you can use the ratio of the two distances measured along the floor from directly below the edge of the table, since they are directly proportional to the velocities.) See also *Film Loop 48*.

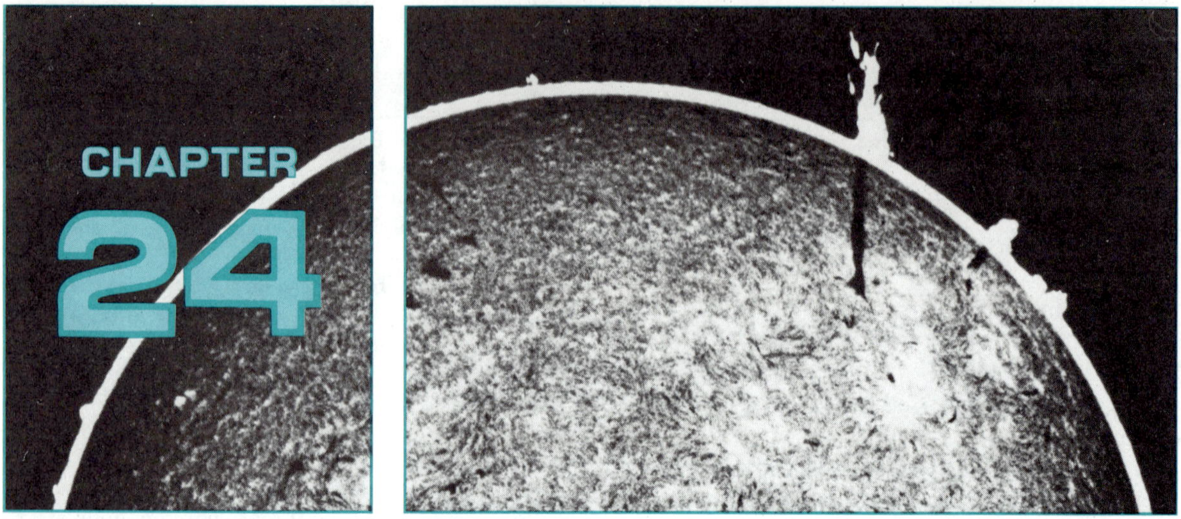

CHAPTER 24

Nuclear Energy; Nuclear Forces

24.1 | Conservation of energy in nuclear reactions

SG 1

In the discussion of nuclear reactions in the last chapter, the emphasis was on the transformations of nuclei and on the properties of the nuclides formed. But there is another property of these reactions that is important, that is, the absorption or release of energy.

In both kinds of chemical reactions, the small amount of energy that may be required to trigger the reaction is neglected.

You know that in some chemical reactions energy must be supplied from the outside to keep the reaction going, while in others energy is liberated. The formation of water from oxygen and hydrogen is an example of a reaction in which energy is liberated; the reaction between these two gases is usually violent, and heat is given off. Therefore, the water that is formed has less energy than did the substances of which the water is made. On the other hand, when water is decomposed by electrolysis, electrical energy must be supplied by passing a current through the water, and the products of the reaction, the oxygen and hydrogen liberated, have more energy than the water.

Nuclear reactions, too, may absorb energy or liberate energy. One main reason for the interest in nuclear reactions is the fact that the amount of energy absorbed or liberated per nucleus involved can be greater by a factor of 1 million or more than the amount involved per atom in a chemical reaction. Nuclear fission and nuclear fusion (discussed later in this chapter) are two special kinds of nuclear reactions in which the energy release is exceptionally large; these types of reactions are therefore important in industrial and military applications.

It would be a good idea to reread Sec. 20.1 in Unit 5, to review the relativistic relationship of mass and energy. Two important ideas for this chapter are: (a) the mass of a moving body is greater than the rest mass by KE/c^2, and (b) a body at rest has an energy of m_0c^2.

Since there is an equivalence between mass and energy, a large release of energy in a nuclear reaction will be accompanied by corresponding changes in the total rest mass of the interacting nuclei. Therefore, the relation $E = mc^2$ plays an important part in interpreting nuclear reactions.

In this chapter, you will examine the mass and energy relations in nuclear reactions. This study will show how some of the ideas and experimental information of the last three chapters are linked together.

?

1. *Is energy always liberated in a nuclear reaction?*

24.2 | The energy of nuclear binding

The concepts of atomic and nuclear structure, that is, that an atom consists of a nucleus surrounded by electrons and that the nucleus is made up of protons and neutrons, led to a fundamental question: Is the mass of a neutral atom equal to the sum of the masses of the protons, neutrons, and electrons that make up the neutral atom? This question can be answered precisely because the masses of the proton, the neutron, and the electron are known, as are the masses of nearly all the atomic species. A survey of the known atomic masses shows that, for each kind of atom, the atomic mass is always *less* than the sum of the masses of the constituent particles in their free states. The simplest atom containing at least one proton, one neutron, and one electron is deuterium, $_1H^2$; in this case, the masses are:

As early as 1927, Aston concluded from his measurements with a mass spectrograph that when two light nuclei combine to form a heavier one, the new nucleus weighs less than the sum of the original ones.

rest mass of one proton	=	1.007276 amu
rest mass of one neutron	=	1.008665
rest mass of one (orbiting) electron	=	0.000549

total rest mass of constituent particles in free state	= 2.016490 amu
rest mass of deuterium atom	= 2.014102 amu
difference (Δm)	= 0.002388 amu

The energy equivalent of 1 atomic mass unit:

$$1 \text{ amu} = 1.66 \times 10^{-27} \text{kg}$$
$$\Delta E = \Delta mc^2$$
$$= (1.66 \times 10^{-27} \text{kg}) \times (3 \times 10^{8} \text{m/sec})^2$$
$$= 14.9 \times 10^{-11} \text{ J}$$

But $1 \text{ MeV} = 1.60 \times 10^{-12}$ J

$$\Delta E = \frac{14.9 \times 10^{-11} \text{ J}}{1.6 \times 10^{-13} \text{ J/MeV}}$$
$$= 931 \text{ MeV}$$

Although the difference in rest mass, Δm, may appear small, it corresponds to a significant energy difference, because of the factor c^2 in the relation $E = mc^2$. The difference Δm in mass corresponds to the difference ΔE in energy according to the relation: $\Delta E = \Delta mc^2$. A convenient conversion factor from atomic mass (expressed in atomic mass units) to energy (expressed in million electron volts) is, as shown in the margin, 1 amu = 931 MeV. If you therefore consider the formation of a deuterium atom when a proton and a neutron combine (and are joined by a tiny electron), then an amount of mass 0.002388 amu will have to be "lost" in the process. This means that an amount of energy equal to 0.002388 amu × 931 MeV/amu = 2.22 MeV has to be radiated away from this system of combining particles before they settle down as a deuterium atom.

SG 2

The expected energy loss calculated from the difference in rest mass can be compared with the result of a direct experiment. When hydrogen is bombarded with neutrons, a neutron can be captured in the reaction:

$${}_0n^1 + {}_1H^1 \rightarrow {}_1H^2 + \gamma$$

This reaction produces no particle fragments having large kinetic energy, so the mass of 0.002388 amu by which ${}_1H^2$ is lighter than ${}_0n^1 + {}_1H^1$ must be carried away by the γ ray. The energy of the γ ray has been determined experimentally and found to be 2.22 MeV, just as predicted! The inverse reaction, in which deuterium is bombarded with γ rays, has also been studied:

$${}_1H^2 + \gamma \rightarrow {}_1H^1 + {}_0n^1$$

When the energy of the γ rays is less than 2.22 MeV, this reaction cannot occur. But if γ rays of energy 2.22 MeV or greater are used, the reaction does occur; the proton and neutron separate and can be detected.

Following the "capture" of a neutron by the nucleus ${}_1H^1$, energy is liberated in a γ ray. This energy (2.22 MeV) is called the *binding energy* of the deuteron. It can be thought of as the energy released when a proton and neutron combine to form a nucleus. To get the inverse reaction (when ${}_1H^2$ is bombarded with γ rays), energy must be absorbed. So you can think of the binding energy as the amount of energy *needed* to break the nucleus up into its constituent nuclear particles.

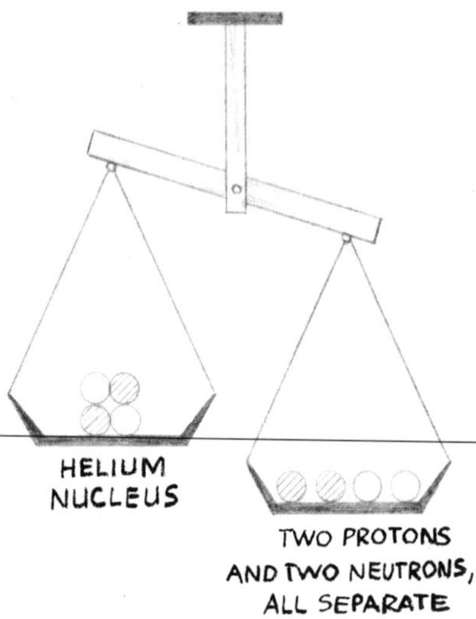

A case where the whole seems to be not equal to the sum of its parts. Two protons and two neutrons, measured separately, are distinctly heavier than a helium nucleus, which consists of the same particles, but are close together. The energy associated with the separate particles explains the difference.

The concept of binding energy, of course, applies to all situations in which simple parts are bound together by some force to form a complex system. For example, the earth is held in orbit around the sun and would need to be given a certain additional amount of kinetic energy to escape from the sun, which binds it by gravitational attraction. In a hydrogen atom, the electron needs 13 eV before it can escape from the nucleus that binds it by an electric attraction. Conversely, when a bare $_1H^1$ nucleus captures an electron and becomes a stable, ordinary neutral atom of hydrogen, the system must give up an amount of energy equal to 13 eV by radiation, exactly the observed energy of the photon emitted in this process of electron capture. However, only the nuclear binding energies are relatively large enough to represent measureable mass differences.

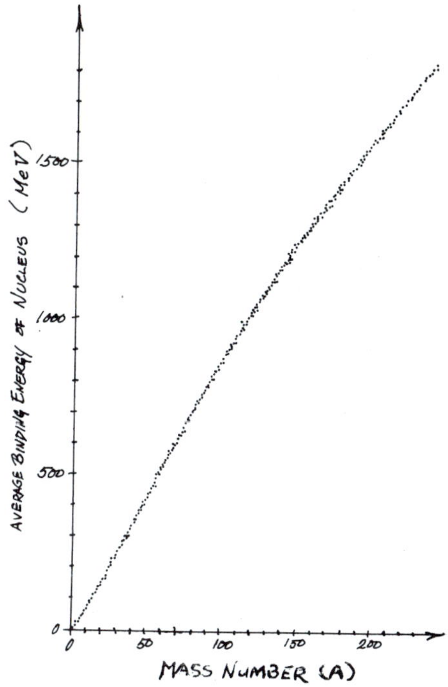

Nuclear binding energy *as a function of the number of particles in the nucleus.*

?

2. When energy is "liberated" during a nuclear reaction, what becomes of it?

3. What is the definition of binding energy for the case of the deuteron nucleus?

24.3 | Nuclear binding energy and stability

The calculation of nuclear binding energy made for deuterium can be extended to all other nuclear species, but it is first necessary to explain a convention. In practice, physicists make such calculations for neutral atoms rather than for bare atomic nuclei. (Experimental values of masses found from mass-spectrographic measurements are for atoms that are missing only one or two electrons.) Since an atom contains electrons orbiting around the nucleus as well as the protons and neutrons inside the nucleus, the mass of one electron outside the nucleus must be included for every proton inside the nucleus in the calculations.

The following example illustrates the calculations done to find the nuclear binding energy of an atom. Compare the actual mass of a carbon-12 atom with the sum of the masses of its component particles:

rest mass of six hydrogen atoms (includes six protons and electrons)	6×1.007825	$= 6.04695$ amu
rest mass of six neutrons	6×1.008665	$= 6.05199$
total rest mass of particles		$= 12.09894$
rest mass of carbon-12 atom		$= 12.00000$
difference in rest mass (Δm)		$= 0.09894$

corresponding energy =

$$0.09894 \text{ amu} \times 931 \text{ MeV/amu} = 92.1 \text{ MeV}$$

In the same manner, you can calculate the nuclear binding energy of any stable atom. The figure in the margin shows in

Notice the unusually high position (above the curve) of the dot near 7.1 MeV, compared to its neighbors. The point is for He^4. The relatively high value of the binding energy of their nucleus is related to its unusually great stability.

graphic form how the nuclear binding energy for stable nuclides actually increases with increasing atomic mass, as more particles are added to form the nucleus. The term *nucleons* refers to both protons and neutrons; therefore, the binding energy of the nucleus increases with the number of nucleons. But, as you see, the result is not a straight line. Such experimental data have important implications.

The implications can be seen more clearly if the *average binding energy per particle* is calculated. In the case of the carbon-12 example, the total binding energy is 92.1 MeV. Since there are 12 particles inside the nucleus (six protons and six neutrons), the average binding energy per particle is 92.1 MeV/12 or 7.68 MeV. In the graph below, the values of average binding energy per particle (MeV) are plotted against the number of particles in the nucleus (mass number, *A*). The significance of the graph lies in its striking shape.

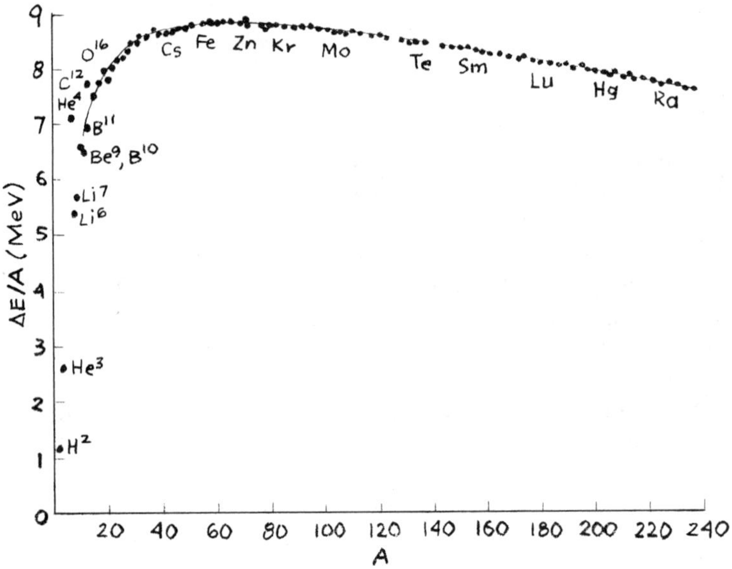

The average binding energy per nucleon *for stable nuclei as a function of the number of particles in the nucleus.*

Note that the binding energy per particle starts with a low value for deuterium (the first point) and then increases rapidly. Some nuclei in the early part of the curve, for example, He^4, C^{12}, and O^{16}, have exceptionally high values as compared with their neighbors. More energy would have to be supplied to remove a particle from one of them than from one of their neighbors. You would therefore expect He^4, C^{12}, and O^{16} to be exceptionally stable. There is evidence in favor of this conclusion, for example, the fact that the four particles making up the He^4 nucleus are emitted as a single unit, the α particle, in radioactivity. The curve has a broad maximum, extending from approximately $A = 50$ to $A = 90$, and then drops off for the heavy elements. Thus, ${}_{29}Cu^{63}$ near the maximum is found to have a binding energy per particle

SG 3

Remember: High binding energy per particle means a great deal of energy needed per particle to take the nucleus apart into its constituent nucleons.

of about 8.75 MeV, while $_{92}U^{235}$, near the high-A end of the curve, has a value of 7.61 MeV. It follows that the nuclei in the neighborhood of the maximum of the curve, like those of copper, should be more difficult to break up than those of uranium.

The idea of binding energy should now make it clear why atomic masses, when precisely measured, are not exactly whole-number multiples of the mass of a hydrogen atom, even though nuclei are just collections of identical protons and neutrons. When those particles combined to make a nucleus, their total rest mass was reduced by an amount corresponding to the binding energy, and the average binding energy varies from nuclide to nuclide, as shown in the graph on page 706.

With the information you now have about the nuclear binding energy, you will be able to calculate and predict the energy needed for or released in nuclear reactions. (The average binding energy curve has other important implications which will be mentioned later.)

? 4. *Which would be more stable, a nuclide with a high* total *binding energy or a nuclide with a high* average *binding energy per nucleon?*

24.4 | The mass–energy balance in nuclear reactions

In the previous section, a very simple nuclear reaction was used to introduce the concept of binding energy. In this section, a more complicated reaction will show an important relation between the binding energy and the energy liberated in a nuclear reaction.

The mass–energy balance in the reaction of a proton with lithium-7 will be analyzed here:

$$_1H^1 + {}_3Li^7 \rightarrow {}_2He^4 + {}_2He^4$$

This reaction has historical interest; it was the first case of a nuclear disintegration brought about by artificially accelerated particles. The analysis of the reaction provided one of the earliest quantitative tests of Einstein's mass–energy relation. The reaction was a good one to analyze because the masses of the proton, the α particle, and the lithium atom were known, and the kinetic energies of the incoming proton and the two resulting α particles could be measured accurately (by their ionizing effects).

rest mass of Li^7 atom = 7.016005 amu
rest mass of H^1 atom = 1.007825 amu
rest mass of He^4 atom = 4.002604 amu

REST MASSES

Before		*After*	
Li^7	7.016005 amu	He^4	4.002604 amu
H^1	1.007825 amu	He^4	4.002604 amu
	8.023830 amu		8.005208 amu

Difference

$$\begin{aligned} & 8.023830 \text{ amu} \\ & -\ 8.005208 \text{ amu} \\ \Delta m &= 0.018622 \text{ amu} \end{aligned}$$

$$0.018622 \text{ amu} \times 931 \text{ MeV/amu} = 17.3 \text{ MeV}$$

The energy to be released in the reaction may be calculated by finding the difference in rest masses before and after the nuclear reaction takes place. The rest mass of the products is less by 0.018622 amu than the rest mass of the initial atoms, corresponding to a deficit of 17.3 MeV. The corresponding deficit in energy, 17.3 MeV, appears as the kinetic energy of the two α particles emitted. (In fact, the incident proton also has kinetic energy, so that the 17.3 MeV represents the *difference* between the kinetic energies of the two emitted α particles and the kinetic energy of the incident proton.)

When the experiment is made, full agreement is found between the expected kinetic energy deficit calculated from the data for the rest masses and the experimental value found for the kinetic energies. This agreement shows that the mass–energy relation is valid. There is a release of energy when the lithium atom is broken up, and this release shows up at the expense of some of the rest mass of its fragments. This experiment was first done in 1932; since then, hundreds of nuclear transformations have been studied, and the results have invariably agreed with the mass–energy relationships calculated by means of the equation $\Delta E = \Delta mc^2$.

As you have seen, then, the kinetic energies of the products of nuclear reactions can be related to the difference in the total rest masses of the products and of the reactants. The kinetic energies can also be related to the binding energies of the nuclei involved. (Experimentally, of course, the binding energies are not measured independently but are derived from the mass and kinetic energy data.) For example, consider the production of two α particles from the reaction of a proton and a Li^7 nucleus. The binding energy of lithium-7 is 39.2 MeV. (Note that in the graph on page 706, the binding energy per nucleon in the case of

SG 4

Li^7 is given as 5.6 MeV, and there are seven particles in the nucleus of Li^7.) The incident proton has no binding energy. The nuclear binding energy of each resulting α particle is 28.3 MeV, making a total of 56.6 MeV for the two α particles. The difference between the total binding energies shows how much more

SG 5–7

strongly the nucleons are bound in an α particle than in a Li^7 nucleus: 56.6 MeV $-$ 39.2 MeV $=$ 17.4 MeV. It is just this difference in the binding energies that must be released somehow by the reaction, and here it is released as the kinetic energy of the two α particles.

The conservation of energy clearly requires that *when the total binding energy of the products exceeds that of the reactants, energy is liberated in the reaction;* otherwise energy is required for the reaction to proceed. The analyses of many nuclear reactions have verified this result. These findings can be expressed in another way: When the average binding energy per particle of the products exceeds that of the reactants, energy is liberated. To express it graphically, energy will be liberated when the products lie higher on the average binding energy curve than the reactants do.

The shape of the average binding energy curve, which drops off at both ends, indicates, therefore, that there are two general nuclear reaction processes by which one can hope to release energy from nuclei: (1) combining light nuclei into a more massive nucleus, or (2) splitting up heavy nuclei into nuclei of medium mass. In either process, the products would have greater average binding energy, so energy would be released. A process in which two nuclei join together to form a heavier nucleus is called *nuclear fusion*. A process in which a heavy nucleus splits into fragments of intermediate mass is called *nuclear fission*. Both fusion and fission have been shown to occur, and the technology of fission has been simplified and exploited in many countries. Fission reactions can be made to take place slowly (as in a nuclear power plant) or very rapidly (as in a nuclear explosion).

?

5. Would breaking up a heavy nucleus into many light nuclei result in the liberation of energy?

24.5 | Nuclear fission: discovery

The discovery of nuclear fission is an example of an unexpected result of great practical importance, obtained during the course of research carried on for reasons having nothing to do with the possible usefulness of the discovery. It is also an excellent example of the combined use of physical and chemical methods in nuclear research and of the effectiveness of teamwork. After Joliot and Curie showed that some products of nuclear reactions are radioactive (Sec. 23.9), Fermi and his colleagues in Italy undertook a systematic study of nuclear reactions induced by neutrons. One of the purposes of this research was to produce new nuclides. As a result, many new radioactive nuclides were made and their half-lives determined. One nuclear reaction used successfully in this study was the capture of a neutron followed at once by the emission of a γ ray. For example, when aluminum is bombarded with neutrons, the following reaction occurs: ${}_0n^1 + {}_{13}Al^{27} \rightarrow {}_{13}Al^{28} + \gamma$. Aluminum-28 is radioactive, with a half-life of 2.3 min, decaying by β emission into silicon: ${}_{13}Al^{28} \rightarrow {}_{14}Si^{28} + {}_{-1}e^0 + \bar{\nu}$. As a result of these two reactions, a nuclide (${}_{14}Si^{28}$) is produced with values of Z and A each greater by one unit than those of the initial nucleus. Fermi thought that if neutrons bombarded uranium, the atomic species having the largest value

A few of the problems encountered by Fermi in his work on these reactions were described in the Prologue to Unit 1. Chapter 3 of the supplemental *Project Physics* Unit B. "Discoveries in Physics" goes into more detail on the discovery of fission.

Enrico Fermi.

of Z known then, an entirely *new* element might be formed by the β decay of the heavier uranium isotope:

$$_{0}n^{1} + {_{92}U^{238}} \rightarrow {_{92}U^{239}} + \gamma$$

$$_{92}U^{239} \rightarrow {_{93}(?)^{239}} - {_{-1}e^{0}} + \bar{\nu}$$

He also speculated that the new nuclide denoted by $_{93}(?)^{239}$ in turn might also undergo β decay:

$$_{93}(?)^{239} \rightarrow {_{94}(?)^{239}} + {_{-1}e^{0}} + \bar{\nu}$$

In this way, two new elements might be produced (one with Z = 93, one with Z = 94). If these reactions could really be made to occur, the result would be the artificial production of an element, or elements, not previously known to exist: *transuranium elements*.

Fermi found in 1934 that the bombardment of uranium with neutrons actually produced new radioactive elements in the
SG 8 target as shown by the emission of rays and a decay activity that defined new, relatively short half-lives. The new elements were at first assumed to be the hypothesized transuranium elements.

The results aroused much interest, and in the next 5 years a number of workers experimented with the neutron bombardment of uranium. Many different radioactive half-lives were found for the radiation from the target, but attempts to identify these half-lives with particular elements led to great confusion. The methods used were similar to those used in the study of the natural radioactive elements (Sec. 21.7). But the difficulty of identification was even greater because a radioactive nuclide formed in a nuclear reaction is usually present in the target area only in an extremely small amount, possibly as little as 10^{-12} g; special techniques to separate these small quantities had to be developed.

The reason for the confusion was found early in 1939 when Otto Hahn and Fritz Strassmann, two German chemists, showed definitely that one of the supposed transuranium elements was actually an isotope of *barium* ($_{56}Ba^{139}$), identified by its half-life of 86 min and its chemical behavior. Another nuclide resulting from the neutron bombardment of uranium was identified as lanthanum ($_{57}La^{140}$), with a half-life of 40 hr.

Lise Meitner and Otto Hahn. Lise Meitner, born in Austria, joined Otto Hahn in 1908 in a research collaboration that lasted 30 years. In 1938, Meitner was forced to leave Germany by the Hitler regime. She was in Sweden when she published the first report on fission with her nephew, O. R. Frisch.

The production of the nuclides $_{56}Ba^{139}$ and $_{57}La^{140}$ from uranium, a nuclide with the atomic number 92 and an atomic mass of nearly 240, required an unknown kind of nuclear reaction, one in which the heavy nucleus is split almost in half. Nothing like it had been known to exist before. If such a process really occurred, it would also have to be possible to find "the other half," that is, to find nuclides with mass between 90 and 100 and atomic numbers of about 35. Indeed, Hahn and Strassmann were able to find in the target material a radioactive isotope of strontium (Z = 38) and one of yttrium (Z = 39) which fulfilled these conditions, as well as isotopes of krypton (Z = 36) and xenon (Z = 54). It was clear from the chemical evidence that the uranium nucleus, when bombarded with neutrons, can indeed split into two nuclei of intermediate atomic mass.

Although Hahn and Strassmann showed that isotopes of intermediate mass did appear, they hesitated to state the conclusion that the uranium nucleus could be split. In their historic report, dated January 9, 1939, they said:

> On the basis of these briefly presented experiments, we must, as chemists, really rename the previously offered scheme and set the symbols Ba, La, Ce in place of Ra, Ac, Th. As "nuclear chemists" with close ties to physics, we cannot decide to make a step so contrary to all existing experience of nuclear physics. After all, a series of strange coincidences may, perhaps, have led to these results.

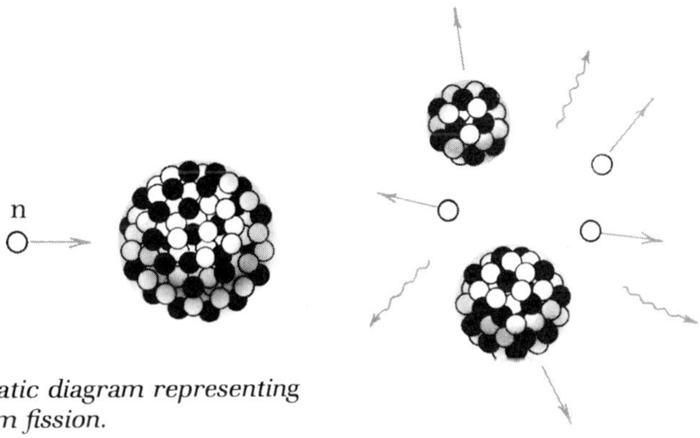

Schematic diagram representing uranium fission.

Starting about 6 years after Fermi's speculation of 1934, it was found possible, by a variety of methods, to create transuranium elements. The new elements up to Z = 106 are listed below. A tiny sample of one of them, curium-244, dissolved in a test tube of water, is shown in the 5-min exposure above (by light produced when the radiation interacts with the surrounding matter).

$_{92}U$	*uranium*
$_{93}Np$	*neptunium*
$_{94}Pu$	*plutonium*
$_{95}Am$	*americium*
$_{96}Cm$	*curium*
$_{97}Bk$	*berkelium*
$_{98}Cf$	*californium*
$_{99}Es$	*eisteinium*
$_{100}Fm$	*fermium*
$_{101}Md$	*mendelevium*
$_{102}No$	*nobelium*
$_{103}Lr$	*lawrencium*
$_{104}Rf$	*rutherfordium*
$_{105}Ha$	*hahnium*
$_{106}$	

The step which Hahn and Strassmann could not bring themselves to take was taken on January 16, 1939 by two Austrian physicists, Lise Meitner and Otto R. Frisch. They suggested that the neutron provoked a disintegration of the uranium nucleus into "two nuclei of roughly equal size," a process they called "nuclear fission" by analogy to the biological division, or fission, of a living cell into two parts. On the basis of comparison of the low average binding energy per nucleon of uranium with the higher average binding energy per nucleon of the products, they predicted that the fragments would have high kinetic energy. This was soon verified experimentally. Shortly afterward, it was found that transuranium elements may, after all, *also* be formed when uranium is bombarded with neutrons. In other words, the capture of a neutron by uranium sometimes leads to fission and sometimes leads to β decay. The β decay results in the formation of isotopes of elements of atomic number 93 and 94, later named *neptunium* and *plutonium*. The presence of both types of reaction, fission and neutron capture followed by β decay, was responsible for the difficulty and confusion in the analysis of the effects of neutrons on the uranium target. Now, the interpretation of the experiments opened two new fields of scientific endeavor: the physics and chemistry of the transuranium elements, and the study of nuclear fission.

Otto R. Frisch.

The discovery of nuclear fission inspired research workers all over the world, and much new information was obtained within a short time. It was found that the uranium nucleus, after capturing a neutron, can split into one of more than 40 different pairs of fragments. Radiochemical analysis showed that nuclides resulting from fission have atomic numbers between 30 and 63 and mass numbers between 72 and 158.

Yet nuclides of medium mass are not the only fission products. Neutrons also are emitted in fission; the average number of neutrons emitted is usually between two and three. The following reaction indicates only one of the many ways in which a uranium nucleus can split:

SG 9, 10

$$_0n^1 + {}_{92}U^{235} \rightarrow {}_{56}Ba^{141} + {}_{36}Kr^{92} + 3\,{}_0n^1$$

Similarly, $_{36}Kr^{92}$ is transformed into $_{40}Zr^{92}$ by four successive β decays. See SG 11.

$_{56}Ba^{141}$ and $_{36}Kr^{92}$ are not "natural" nuclides and are not stable; they are radioactive and decay by β emission. For example, $_{56}Ba^{141}$ can decay into $_{59}Pr^{141}$ by successive emission of three β particles, as shown by the following scheme (the numbers in parentheses are the half-lives):

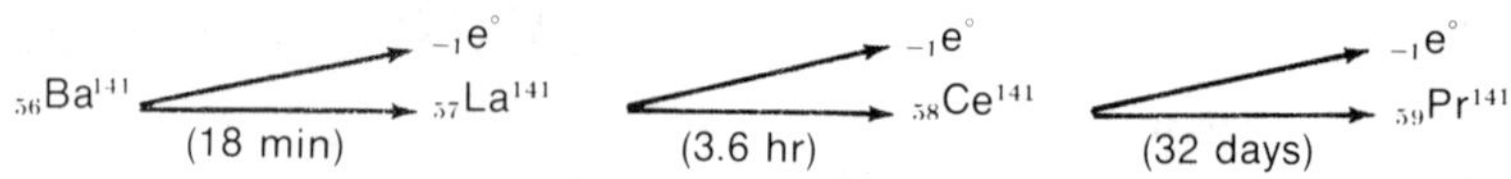

It has been found that only certain nuclides can undergo fission. For those that can, the probability that a nucleus will split when bombarded depends on the energy of the neutrons used in the bombardment. The nuclides $_{92}U^{235}$ and $_{94}Pu^{239}$ can undergo fission when bombarded with neutrons of *any* energy, even 0.01 eV or less. On the other hand, U^{238} and Th^{232} undergo fission only when bombarded with neutrons having kinetic energies of 1 MeV or more.

Plutonium 239 ($_{94}Pu^{239}$) is produced by the capture of a neutron by $_{92}U^{238}$ and the subsequent emission of two β particles, as was discussed on page 710.

The energy released in the fission of a nucleus is about 200 MeV. This value can be calculated either by comparing atomic rest masses of reactants and products or from the average binding energy curve of the graph on page 706. The energy release in fission is more than 20 times larger than in the more common nuclear reactions, where it is usually less than 10 MeV, and more than a million times larger than in chemical reactions.

Under appropriate conditions the neutrons released in fission can, in turn, cause fission in neighboring uranium atoms, and a process known as a *chain reaction* can develop in a sample of uranium. The combination of the large energy release in fission and the possibility of a chain reaction is the basis of the large-scale use of nuclear energy.

?

6. What two successive reactions can result in the appearance of a transuranium element?

7. What product of the fission process makes a chain reaction possible?

24.6 | Nuclear fission: controlling chain reactions

For a chain reaction in a sample of uranium to continue at an even rate, there must be a favorable balance between the net *production* of neutrons by fissions and the *loss* of neutrons due to the following three processes:

1. capture of neutrons by uranium without fission resulting;
2. capture of neutrons by other materials in the sample or in the structure containing the sample;

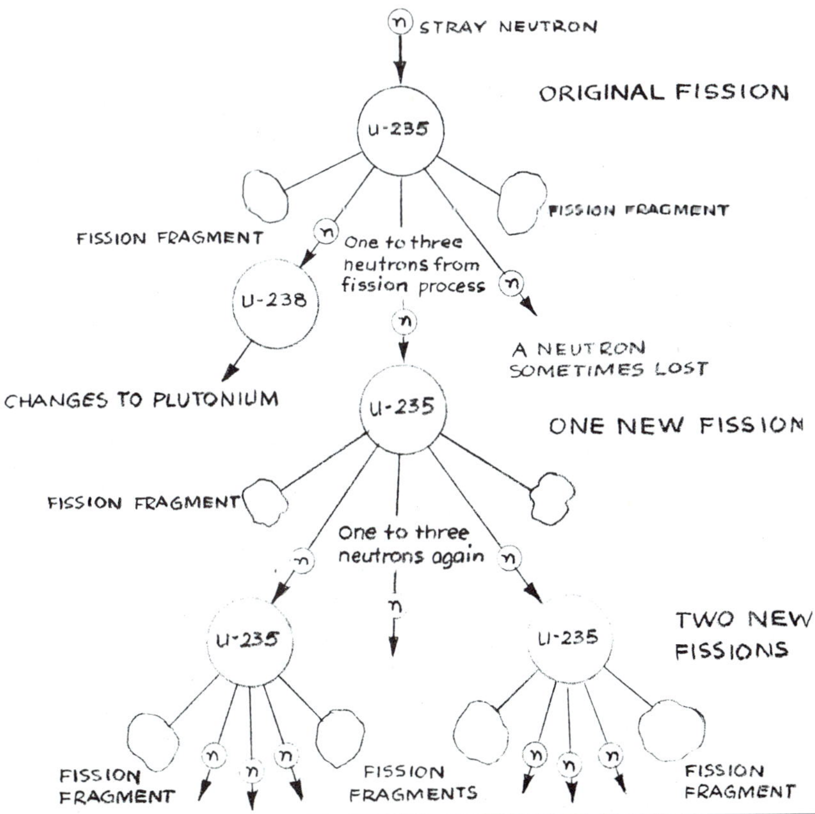

This diagram indicates what happens in a chain reaction resulting from the fission of uranium-235 atoms. (Not shown are other emissions such as α, β, and γ rays.)

3. escape of neutrons from the sample without being captured. If too *many* neutrons escape from or are absorbed in the structure or assembly (called a *reactor*), there will not be enough to sustain the chain reaction. If too *few* neutrons escape or are absorbed, the reaction will continue to build up more and more. The design of nuclear reactors as energy sources involves finding proper sizes, shapes, and materials to maintain or control a balance between neutron production and neutron loss.

Since the nucleus occupies only a tiny fraction of an atom's volume, the chance of a neutron colliding with a uranium nucleus is small, and a neutron can go past the nuclei of billions of uranium (or other) atoms while moving a few centimeters. If the reactor assembly is small, a significant percentage of the fission neutrons can escape from the assembly without causing further fissions. The "leakage" of neutrons can be so large that a chain reaction cannot be sustained. The number of neutrons produced is proportional to the *volume*, but the number of neutrons that escape is proportional to the *surface area*. As the linear size L of the assembly is increased, the volume and area increase in proportion to L^3 and L^2, respectively, so that neutron *production* increases with size more rapidly than neutron *escape* does. For a given combination of materials (uranium and other structural materials that may be needed), there is a size of the reactor, called the critical size, for which the net production of neutrons by fission is just equal to the loss of neutrons by nonfission capture and escape. If the size of the reactor assembly is smaller than this critical size, a chain reaction cannot be sustained. The design of a reactor of reasonable dimensions, with given materials, which will correspond to critical size is an important part of research in the field of *nuclear engineering*.

SG12

Although nuclear reactors can be built in which the fissions are induced by fast neutrons, it has been easier to build reactors with materials in which the fissions are induced by slow neutrons.

Another important consideration in the design of nuclear reactors is the fact that fission is much more probable when U^{235} is bombarded with *slow* neutrons than when it is bombarded with fast neutrons. The neutrons released in fission generally come out at very high speeds, having kinetic energies from about 0.01 MeV to nearly 20 MeV, with an average kinetic energy of about 2 MeV. The fast neutrons can be slowed down in the reactor by the addition of material to which the neutrons can lose energy in collisions. The material should be relatively low in atomic mass so that the neutrons will transfer a significant fraction of their energy in elastic collision; but the material should not also capture and absorb many neutrons. Pure carbon in the form of graphite and also water and beryllium meet these requirements. These substances are called *moderators* because they slow down, or moderate, the newly produced neutrons to lower speeds at which the probability of causing additional fission is high.

Hydrogen atoms in water are very effective in slowing down

neutrons because the mass of a hydrogen nucleus is nearly the same as that of a neutron and because the number of hydrogen atoms per unit volume is high. A neutron can lose a large fraction of its energy in a collision with a hydrogen nucleus. Only about 20 collisions are needed, on the average, to slow down the fast neutron to energies under 1 eV. However, neutrons can also be captured by the hydrogen nucleus in the reaction:

$$_1H^1 + {_0n^1} \rightarrow {_1H^2} + \gamma$$

The probability of this reaction occurring instead of an elastic collision is high enough so that it has been found impossible to achieve a chain reaction with natural uranium and ordinary water.

Sec. 23.4 describes how neutrons lose nearly all their kinetic energy in a head-on collision with a hydrogen nucleus, but most collisions will not be head-on.

Heavy water: $(H^2)_2O$, or D_2O.

$$_1H^2 + {_0n^1} \xrightarrow{\text{(low probability)}} {_1H^3} + \gamma$$

There are also other ways to make reactors. For example, the absorption of a neutron by a deuterium nucleus, such as the nucleus of the heavy isotope of hydrogen found in heavy water, has an extremely small probability. Neutrons do not lose as much energy per collision with H^2 nuclei, but this disadvantage is compensated for by the much lower absorption rate. Therefore, a chain reaction can be achieved easily with natural uranium and heavy water. Reactors with natural uranium as the fuel and heavy water as the moderator have been built in the United States, Canada, France, Sweden, Norway, and other countries.

The contrast between the nuclear properties of hydrogen $_1H^1$ and deuterium ($_1H^2$ or $_1D^2$) has important implications for the development of nuclear reactors. Heavy water is much more expensive than ordinary water, but when it is used with natural uranium (mostly U^{238}), a chain reaction can be achieved efficiently. Ordinary water can be used, if uranium enriched in the isotope U^{235} is used instead of natural uranium. Many reactors "fueled" with enriched uranium and moderated with ordinary water have been built in the United States. In fact, this general reactor type has been used in nearly all the large nuclear power plants built so far and in the reactors used in nuclear-powered ships.

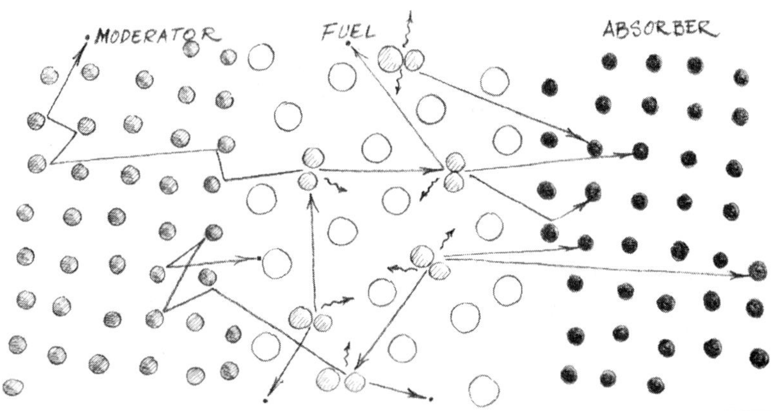

Schematic diagram of three types of functions fulfilled by parts of a nuclear reactor.

The west wall of the football stands of Stagg Field, University of Chicago. Squash courts under these stands were used as the construction site of the first nuclear reactor.

Carbon in the form of graphite has been used as a moderator in many reactors, including the earliest ones. It is not as good a slowing-down agent as water or heavy water; about 120 collisions with carbon atoms are needed to slow down a fast neutron with an initial energy of 2 MeV to the desired energy of about 0.025 eV; in heavy water only about 25 collisions are needed. Although carbon in the form of graphite is not the best moderator and absorbs some neutrons, it does permit a chain reaction to occur when lumps of natural uranium (cylindrical rods, for example) are arranged in a large mass of graphite. The determination of just how this could be done was one of the main problems that had to be solved before the world's first chain reaction was achieved by a team working under Enrico Fermi in December 1942 at the University of Chicago. (It was a crucial experiment because until its success, it was by no means certain that a chain reaction was really possible.) Many graphite-moderated reactors are now in operation throughout the world. Their chief purpose will be discussed in the next section.

The control of a reactor is relatively simple. If fission is occurring too frequently, a few "control" rods are inserted into the reactor. The rods consist of a material (such as cadmium or boron) that absorbs slow neutrons, thereby reducing the number of neutrons in the moderator. Removal of the control rods will allow the rate of the reactor to go up. The sketch at the bottom of page 715 illustrates the basic reactions that occur in a nuclear reactor in which uranium is the fissionable material.

?

8. What is a moderator?

9. What is an advantage and a disadvantage of using water as a moderator in nuclear reactors?

10. How can the rate of reaction be controlled in a reactor?

24.7 | Nuclear fission: large-scale energy release and some of its consequences

The large-scale use of nuclear energy in chain reactions was accomplished in the United States between 1939 and 1945. The work was done under the pressure of World War II, as a result of the cooperative efforts of large numbers of scientists and engineers. The workers in the United States included Americans, Britons, and European refugees from fascist-controlled countries. They were working to obtain a nuclear weapon before the Germans, who were also working on one.

The aim was to produce a so-called *atomic* (more properly, *nuclear*) bomb, essentially an uncontrolled nuclear reactor in

Scale model of the CP-1 (Chicago Pile No. 1) used by Enrico Fermi and his associates when they first achieved a self-sustaining nuclear reaction on December 2, 1942. Alternate layers of graphite, containing uranium metal and/or uranium oxide, were separated by layers of solid graphite blocks. Graphite was used as a moderator, to slow down neutrons in order to increase the likelihood of fissions.

The "pile reactions" to produce Pu-239.

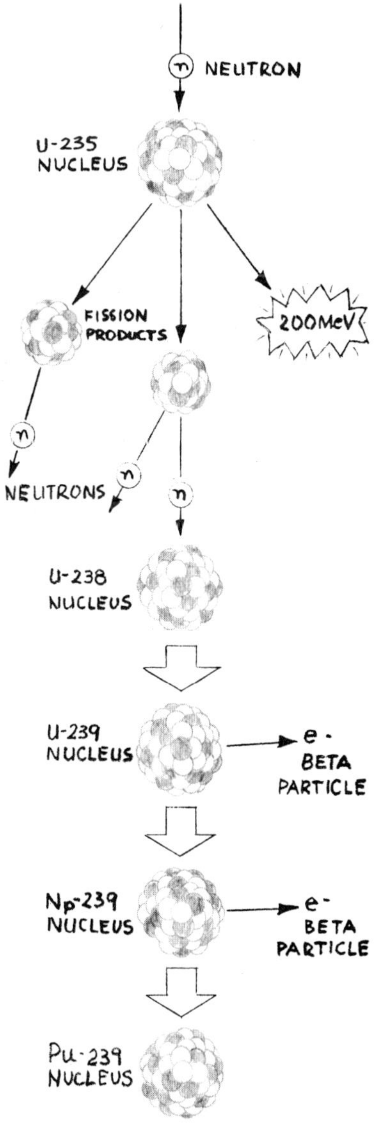

which an extensive chain reaction occurs throughout the material in a few millionths of a second. This reaction differs therefore from the controlled nuclear reactor, in which the operating conditions are so arranged that the energy from fission is released at a much slower and essentially constant rate. In the controlled reactor, the fissionable material is mixed with other materials in such a way that, on the average, only *one* of the neutrons emitted in fission causes the fission of another nucleus; in this way, the chain reaction just sustains itself. In a nuclear bomb, the fissionable material is pure, that is, not mixed with a moderator, and the device is designed so that nearly all of the neutrons emitted in each fission can cause fissions in other nuclei.

Nuclear reactors were used during World War II to produce raw materials for one kind of nuclear bomb, namely to manufacture Pu^{239} from U^{238}. These reactors were designed in such a way that some of the neutrons from the fission of U^{235} were slowed down sufficiently *not* to cause fission in U^{238} atoms. (In natural uranium, only about 0.75% of the atoms are U^{235}.) Instead, the neutrons were absorbed by U^{238} nuclei to form Pu^{239} through the reactions described in the previous section.

Recall that fission of U^{235} can occur with neutrons of any speed, but fission of U^{238} requires high-speed neutrons.

Pu^{239} acts like U^{235}; both materials can sustain a rapid, uncontrolled chain reaction. Nuclear bombs have been made of both materials; a single nuclear bomb, using U^{235}, destroyed the city of Hiroshima, Japan, on August 6, 1945; another bomb, using ${}_{94}Pu^{239}$, destroyed the city of Nagasaki three days later. Since the end of World War II in 1945, the technology of fission has been further developed in two different directions. One direction has been military. Other countries besides the United States have made nuclear weapons, including the United Kingdom, the Soviet Union, France, India, and China. The enormous death-dealing capability of these weapons, and the ever-larger numbers of bombs of many varieties that have been accumulating all over the globe, have increased and made more dangerous the tensions existing throughout the world and have emphasized critically the need for the peaceful settlement of international disputes.

From the beginning, scientists have been prominently involved in activities to alert their government and fellow citizens to the moral and practical problems raised by the nuclear weapons race.

Genetic effects of radiation: effects producing changes in cells that will affect offspring of exposed individual.
Somatic effects: all effects caused by radiation to an exposed individual during the individual's lifetime.

One incidental problem has been that of the radioactive *fallout* from bomb tests. In the explosion of a nuclear bomb, large amounts of radioactive fission products are scattered. These materials can be blown by winds from one part of the world to another and carried down from the atmosphere by rain or snow. Some of the radioactive materials are long-lived; they may be absorbed in growing foodstuffs and eaten by animals and people. It is known that such radioactive materials can cause harmful genetic effects as well as somatic effects. One of the most abundant and long-lived products of the fission of either U^{235} or Pu^{239} is strontium-90 (${}_{38}Sr^{90}$). This isotope of strontium is similar to ${}_{20}Ca^{40}$ in its chemical properties. Therefore, when Sr^{90} from radioactive fallout enters the body, it finds its way into bone material. It decays by emission of 0.54-MeV β particles (half-life = 28 years), which can injure cells and cause leukemia, bone tumor, and possibly other forms of damage, particularly in growing children.

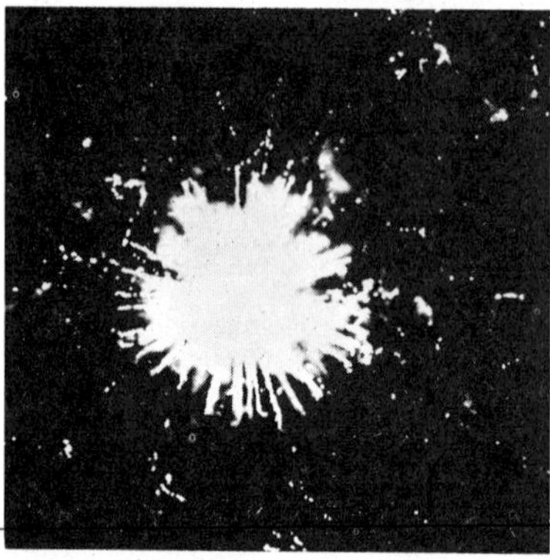

A grain of radioactive dust from the atmosphere caused these α-particle tracks in a photographic emulsion (enlarged 2000 times).

There has been much research and discussion concerning possible harm to present and future generations. Partly as the result of petitions and protests organized by scientists, the United States, the United Kingdom, the Soviet Union, and most other nations (but not France and China) agreed in 1963 to a moratorium on further bomb tests in the atmosphere. Though it allowed continuation of tests underground, the atmospheric test ban treaty was rightly considered a great step forward in simultaneously curbing radioactive pollution and increasing somewhat the chances for further arms control treaties. For example, it is said to have helped pave the way to the treaty, in effect since 1970, by which most nations agreed not to disseminate nuclear weapons to "non-nuclear" nations and set the stage for arms limitation talks that have been going on with some success since 1970.

The second direction in which the use of nuclear energy has been pushed on a large scale has been that of the production of electrical energy from the energy released in fission. In almost all present systems of nuclear-power production, the reactor is the source of heat for running steam turbines; the turbines drive electrical generators just as they do in coal-fired or oil-fired power stations. The fissionable material replaces the coal or oil used in a conventional power plant and so provides a new source of energy in the form of electricity.

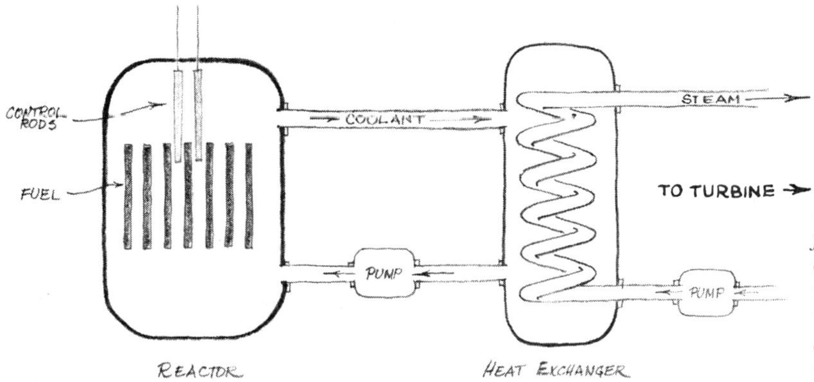

Heat produced in a reactor (by the flying fission fragments) does not directly turn water into steam. As this simplified diagram indicates, the water is heated in a "heat exchanger" by a fluid that circulates through the reactor core.

The ever-increasing use of electrical energy is an important aspect of modern life. The amount of electricity used in the United States, an advanced industrial country, increased by a factor of about 40 between 1920 and 1970. This means that during this 50-year period, the amount doubled approximately every 10 years. Although large supplies of coal and oil still exist, it has become evident that even in the greater efforts to use energy more carefully and frugally, additional sources of energy will be required; nuclear energy from fission (and, in the long run, from fusion) can help fill this need.

The need for new sources of energy was sharply emphasized by the "Energy Crisis," which hit the United States, western European countries, and Japan in 1973–1974. This shortage was made severe because the oil-producing countries in the Mid-East cut back shipments of oil to some highly industrialized countries. These events focused attention on alternative methods of energy, from more pollution-free uses of coal, to solar energy, to the role of the nuclear-power industry in our economy.

The development of nuclear power in the United States has been slower than was expected at the end of World War II. For a variety of reasons, some administrative and some technical, but mostly connected with the "Cold War" with the Soviet Union, the U.S. Atomic Energy Commission (AEC), now replaced by the Department of Energy, did not emphasize applied research on nuclear electric power systems until President Eisenhower so directed in 1953. Nuclear electric power became economically competitive with hydroelectricity and electricity from coal and oil during the 1960's. By the beginning of 1978, 65 nuclear reactors were operating with over 47 million kilowatts capacity, about 9% of the nation's total electric power production. With nearly 90 or more reactors under construction, the nuclear portion of U.S. electricity output was expected to be about 17% by 1980 and about 28% by 1985. In the rest of the world, there were, at the beginning of 1978, about 130 nuclear power reactors in operation with about 50 million kilowatts capacity; about 325 reactors are expected to exist by 1995.

The increasing use of nuclear power has raised questions that are now receiving much public attention. Proponents say that in the United States the cost of electricity from nuclear energy is, on the average, about 20% less than the cost of that from fossil fuels (coal and oil). They see the nuclear electricity industry as viable and economic, and as helping to reduce the dependence of the United States on imported oil. Opponents argue that nuclear power plants are liable to major accidents. These might result in the release of large amounts of radioactivity and might cause the deaths of many people. There might be a catastrophe of a magnitude never before experienced by any industry. Nuclear plants are designed to minimize the chance of such a major accident. Although less serious accidents have occurred, the safety record of the nuclear industry has been remarkably good so far. Opponents of nuclear power argue that even though a major accident is not expected to occur often, there is still a possibility that one will occur. They are also worried about unauthorized diversion of fissionable materials, and the long-range, safe disposal of radioactive wastes. With more and more nuclear power reactors being built, these problems will also increase. It has therefore been suggested by well-qualified experts that the construction of nuclear power plants be stopped until their safety and security has been demonstrated beyond any question. The U.S. government, relying also on expert advice, has not agreed to stop the growth of nuclear power; more reactors are being built, and more research is being done to improve reactor design, to reduce further the possibility of a major accident or diversion, and to find safe methods of waste disposal. The debate is continuing, and citizens are participating, as they should.

In March 1979, five nuclear power plants in various parts of the United States were closed by the Nuclear Regulatory Agency. These plants were designed and built according to a computer formula; an error in the formula created the possibility that the plants would not be safe in the event of an earthquake.

Waste heat, radioactive emissions, and radioactive wastes from nuclear plants constitute possible environmental hazards that are also subjects of lively public discussion. Most plants that produce electricity by means of steam, whether nuclear or fossil-fueled, have an efficiency of between 30 and 40%. This means that for every three units of heat formed to power the generator, one goes to produce electricity and approximately two units are discharged as waste. The power plants using fossil fuels (coal, oil,

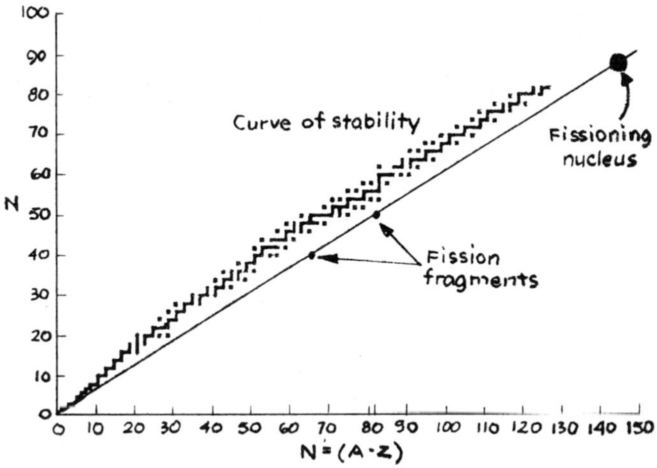

This diagram illustrates that fission fragments tend to have relatively too many neutrons.

and gas) discharge some of their waste heat into the air, while all the waste heat from nuclear power plants goes into the cooling water. Although thermal waste is common to both nuclear and fossil-fueled plants, nuclear plants use their heat less efficiently and produce more waste heat. For the same amount of electricity produced, a fossil-fueled plant discharges about 40% less heat than a nuclear plant of the kind now generally in use. This waste heat has to be disposed of; if dumped into a river or lake, it may have harmful effects on aquatic life and is therefore referred to as "thermal pollution." Environmentalists have attacked the construction of nuclear power plants because of their greater contribution to thermal pollution. Answers to the problem of thermal pollution are available in several directions: increased thermal efficiency of nuclear plants, the use of cooling towers or cooling ponds to decrease the temperature of the waste heat, search for practical uses of the waste heat.

The rate of release of radioactivity from nuclear power plants under normal operating conditions must meet strict standards set by biologists and medical scientists. It can be held to insignificant levels by means of careful design and rigorous operating procedures, except if there should be a major accident of the kind previously discussed.

A problem that is now receiving much attention by engineers as well as by the public is, as mentioned, that of the disposal of the radioactive wastes, or "ashes," resulting from nuclear fission. The immense number of atomic nuclei that must undergo fission in a reactor to produce the desired electrical power results in the formation of an immense number of radioactive fission products. Some of these radioactive products have very long half-lives, up to tens of thousands of years, and even more. A highly complex technology has been developed for separating these products from the still useful fissionable material in partially used fuel assemblies. The products must be handled in special ways while they lose the part of their radioactivity that is due to the shorter-lived fission products. Eventually, the radioactive wastes must be stored where they will do no harm. Present plans for the ultimate disposal of these wastes call for burial deep underground. However, this process presents many problems. For example, it requires the development of long-lasting containers that can be monitored for leakage or the conversion of the wastes into a solid, insoluble, form. Appropriate geological locations must be found where the wastes cause no contamination of underground water or oil, should some leak out after all. The problems are complicated by the fact that these wastes will still have significant radioactivity 10,000 years from now. These wastes may outlast both their containers and even the present form of society in which dangerous materials are not wantonly allowed to be abused by a very few.

The problem of the proliferation of Pu^{239}, which may be used in weapons, is of both national and international interest. In a nuclear power reactor, some U^{238} is converted into Pu^{239}. In operation, this Pu^{239} also undergoes fission, leading to the production of heat energy and electricity. This process makes it possible to increase the "lifetime" of the nuclear fuel elements, that is, the length of time that the fuel elements can be in the reactor before they must be replaced. It is possible, however, to remove the fuel elements before the Pu^{239} is "burned up"; this is what is done in a Pu^{239}-production reactor. The Pu^{239} can be separated by means of a complicated and difficult chemical processing procedure. The Pu^{239} can then, with appropriate physical and metallurgical techniques, be made into nuclear bombs. Thus, a nuclear reactor can be used to produce electricity or to produce Pu^{239} as a by-product, perhaps as material for weapons.

Up to now, only the United States, the Soviet Union, the United Kingdom, France, and China have made and exploded nuclear weapons. In 1974, the government of India exploded a "nuclear device" as a test of peaceful applications, according to that government. This explosion showed that countries other than the five "nuclear powers" could, if they so desired, produce Pu^{239} and, therefore, nuclear weapons. This capability is the so-called "proliferation problem."

Most, but not all, of the industrial countries of the world signed a "Nonproliferation Treaty" in 1970 and renewed it in 1975. President Carter of the United States proposed limitations on the use of Pu^{239} in certain types of reactors to reduce the probability of proliferation. But many other countries that need electric power and have little or no coal or oil did not respond favorably to Carter's proposal. The possibility of proliferation remains an international political and economic problem.

From the examples cited of numerous problems raised by the expected increased use of nuclear energy, it should be clear why there is justifiably so much public discussion of the advantages and disadvantages of nuclear energy. On the one hand, nuclear energy offers a means of dealing with the fuel shortage in which the United States and other countries find themselves. On the other hand, the social costs of the nuclear energy revolution have already been very high in human lives, in money, and in the anxiety of life under the threat of nuclear war. In some ways, these problems are analogous to the human price of industrialization after the development of the steam engine (Unit 3). At the same time, the potential benefit to humanity is great. As in the past, the decisions that will be necessary in the future development of nuclear power cannot be made on the basis of physics alone. Science can help to illuminate alternatives on which essentially political decisions can be based, but it cannot and should not be used by itself to choose among them. Responsible scientific opinion must be supplemented by political insight and a broad humanistic view of society. At the very least, responsible citizens must have some understanding of the scientific principles that will underlie the alternatives among which they must choose.

Among the many problems for public policy created by developments in nuclear power was the Plowshare Program. The crater shown at the left was part of Plowshare's research into the possibility of creating lakes, harbors, and sea-level canals between oceans by exploding nuclear devices. The problems raised included pollution and the dangers of diversion for war purposes. This kind of research has now ceased because the U.S., the U.S.S.R., and the U.K. have agreed to stop above-ground nuclear explosions.

Close Up

Some Developments in Nuclear Science and Technology

1896 H. Becquerel discovers unstable (radioactive) atoms.

1899 Isolation of radium by Curies.

1905 Einstein's statement of equivalence of mass and energy.

1911 Rutherford discovers nucleus.

1919 Rutherford achieves transmutation of one stable chemical element into another.

1920–1925 Improved mass spectrographs show that changes in mass per nuclear particle accompanying nuclear reactions account for energy released by nucleus.

1931 E. O. Lawrence and M. S. Livingston construct first cyclotron.

1932 Chadwick identifies neutrons.

Joliot–Curies in Paris discover artificial radioactivity.

1934 Fermi's group in Rome finds radioactivity induced by neutrons.

1939 Evidence of uranium fission by Hahn and Strassmann, identification of fission products by Meitner and Frisch.

1940 Discovery of neptunium and plutonium at the University of California.

1942 Achievement of first self-sustaining nuclear reaction at the University of Chicago.

1945 First test of a nuclear device, at Alamagordo, New Mexico, followed by the dropping of nuclear bombs on Hiroshima and Nagasaki, at the end of World War II.

1946 President Truman signs the bill creating the U.S. Atomic Energy Commission.

First shipment of radioactive isotopes from Oak Ridge to hospital in St. Louis, Mo.

1951 First significant amount of electricity (100 KW) produced from nuclear energy at testing station in Idaho.

1952 First detonation of a hydrogen bomb at Eniwetok Atoll, Pacific Ocean.

1953 President Eisenhower announces U.S. Atoms-for-Peace program and proposes establishment of an international atomic energy agency.

1954 First nuclear-powered submarine, *Nautilus*, commissioned.

1955 First United Nations International Conference on Peaceful Uses of Atomic Energy held in Geneva, Switzerland.

1956 First commercial power plant begins operation at Calder Hall, England.

1957 Shippingport Atomic Power Plant in Pennsylvania reaches full power of 60,000 KW.

International Atomic Energy Agency formally established.

1959 First nuclear-powered merchant ship, the *Savannah,* launched at Camden, New Jersey.

1961 A radioactive isotope-powered electric generator placed in orbit, the first use of nuclear power in space.

1963 President Kennedy signs the Limited Test Ban Treaty for the United States

1964 President Johnson signs law permitting private ownership of certain nuclear materials.

1966 Beginning of the rapid development of nuclear power plants in the U.S.

1968 "Nonproliferation" agreement, signed by the United States, the Soviet Union, and other countries, limiting the number of countries possessing nuclear weapons.

1970 "Nonproliferation" agreement ratified.

1970–present Nuclear power begins to constitute a significant fraction of the electrical power used in the U.S. and there is widespread discussion of its future.

24.8 | Nuclear fusion

Fusion reactions have been produced in the laboratory by bombarding appropriate light target materials with, for example, high-energy deuterons from a particle accelerator. In these reactions, nuclei result that are heavier than the nuclei of either the "projectiles" or the targets; there are usually also additional particles released and energy. Some typical examples of fusion reactions, together with the energy liberated in each reaction, are:

$$_1H^2 + {_1H^2} \rightarrow {_1H^3} + {_1H^1} + 4 \text{ MeV}$$
$$_1H^2 + {_1H^2} \rightarrow {_2He^3} + {_0n^1} + 3.2 \text{ MeV}$$
$$_1H^2 + {_1H^3} \rightarrow {_2He^4} + {_0n^1} + 17.6 \text{ MeV}$$
$$_1H^2 + {_2He^3} \rightarrow {_2He^4} + {_1H^1} + 18.3 \text{ MeV}$$

In the first of the above equations, the heavier product nucleus is an isotope of hydrogen, called *tritium*, with mass number A = 3; it has been found in small traces in nature, is radioactive with a half-life of about 12 years, and decays by β emission into $_2He^3$, an isotope of helium. When a target containing tritium is bombarded with deuterons, $_2He^4$ can be formed, as in the third equation above, liberating 17.6 MeV of energy. Of this energy, 14.1 MeV appears as kinetic energy of the neutron and 3.5 MeV as kinetic energy of the product nucleus.

Although the energy liberated in a single fusion is less than that in a single fission, the energy released *per unit mass* is much greater. The mass of about 50 helium atoms is approximately equal to the mass of one uranium atom; 50 × 17.6 MeV = 1040 MeV, compared to 200 MeV for a typical fission.

The fusion of tritium and deuterium offers the possibility of providing large sources of energy, for example, in electric power plants. Deuterium occurs in water with an abundance of about one part in seven thousand hydrogen atoms and can be separated from the lighter isotope. Four liters of water contain about 0.13 g of deuterium, which can now be separated at a cost of about $0.08. If this small amount of deuterium could be made to react under appropriate conditions with tritium (perhaps produced by the reaction discussed above), the energy output would be equivalent to that from about 1,140 L of gasoline. The total amount of deuterium in the oceans is estimated to be about 10^{17} kg, and its energy content would be about 10^{20} kw-years. If deuterium and tritium could be used to produce energy, they would provide an enormous source of energy.

There are, however, some difficult problems to be solved before fusion reactions are likely to be useful as steady sources of energy; some of these should be discussed at least briefly. The nuclei which react in the fusion processes are positively charged and repel one another because of the repulsive electric force. The nuclei must, therefore, be made to collide with a high relative speed to overcome the repulsive force tending to keep them apart. Experiments have shown that this can occur when

SG 13, 14

the particles have kinetic energies of about 0.1 MeV or more. The nuclei must also be confined in a region where they can undergo many collisions without escaping, or being absorbed by the walls bounding the region, or losing energy by collisions with too many "cooler" (less energetic) molecules. There must be enough collisions per unit time so that fusion can occur at a rate that will yield more energy than that needed to cause the collisions. The combination of these requirements means that the nuclei must be contained at a temperature of the order of 100 million degrees.

A *plasma* is an ionized gas in which positively and negatively charged particles move about freely.

At the temperature required for fusion, the atoms have been stripped of their electrons, and the resulting nuclei and separated electrons are said to form a *plasma*. No wall made of ordinary material can contain a hot plasma at 10^8 °K (the wall would be vaporized instantly!). But the charged particles of a plasma can, in theory, be contained in an appropriately designed magnetic field. The first problem to be solved, therefore, is to contain the plasma of deuterium and tritium nuclei in a magnetic field, while accelerating the nuclei by means of an electric field to the required kinetic energy (or temperature). The behavior of the charged particles in a plasma is complicated; there are many kinds of instabilities that make the plasma difficult to contain properly and long enough. These problems of the release of energy to form a *controlled* and sustained fusion reaction have not yet been solved on a practical scale, but research on them is being carried on in many countries. Significant advances have been made during the last few years in containment of the plasma and in reaching high temperatures, as high as 5×10^7 °K. There are still difficult technological problems to be overcome, and it may be a generation before electric power will be produced by fusion at costs that will

At Lawrence Livermore Laboratory, scientists are trying to initiate a controlled fusion reaction by imploding a fuel pellet using energy supplied by laser beams.

compete with electricity from coal or uranium. There is considerable international cooperation in this research, including visits of research teams among the United States, Britain, France, and the Soviet Union. Although the effort and expenses are great, the possible payoff in terms of future power resources is enormous.

?

11. Why are very high temperatures required to cause fusion reactions?

12. How could extremely hot gases be kept from contacting the wall of a container?

Pen drawing by Vincent van Gogh.

24.9 | Fusion reactions in stars

One of the most fascinating aspects of nuclear physics is the study of the sources of the energy of different types of stars. The sun is an example. In the sun, the fusion process results in the production of a helium nucleus from four protons. The net results of the reactions can be written as:

$$4\ {}_{1}H^{1} \rightarrow {}_{2}He^{4} + 2\ {}_{+1}e^{0} + 26\ \text{MeV}$$

SG 15

The reaction does not take place in a single step but can proceed through different sets of reactions whose net results are summarized in the above equation; in each case, the overall amount of energy released is 26 MeV.

The fusion of four protons into a helium nucleus is the main source of the energy of the sun. Chemical reactions cannot

For details, see SG 16, 17, and 18.

provide energy at large enough rates (or for long enough duration!) to account for energy production in the sun, but nuclear fusion reactions can. Hydrogen and helium together make up about 99% of the sun's mass, with approximately twice as much H as He. There is enough hydrogen to supply the sun's energy for many millions of years to come.

By which of the several possible sets of reactions does the transformation of hydrogen into helium take place? The direct process of four protons colliding to form a helium nucleus has been ruled out because the probability for such a reaction under solar conditions is too low. It may happen, but not often enough for the amount of energy released. A more likely set of reactions is the process represented in the sketch below. When the temperature is about 10^7 °K, the kinetic energies are large enough to overcome the electric repulsion between protons, and fusion of two protons ($_1H^1$) takes place. The nuclear reaction results in a deuteron ($_1H^2$), a positron ($_{+1}e^0$), and a neutrino. As soon as a deuteron is formed, it reacts with another proton, resulting in helium-3 ($_2He^3$) and a γ ray. The helium-3 nuclei fuse with each other, forming α particles and two protons. In each of these reactions, energy is released, resulting in 26 MeV for the complete cycle of four protons forming a helium nucleus.

One form of the proton–proton fusion chain that releases energy in stars (● protons, ○ neutrons, • positrons, ⇝ γ rays).

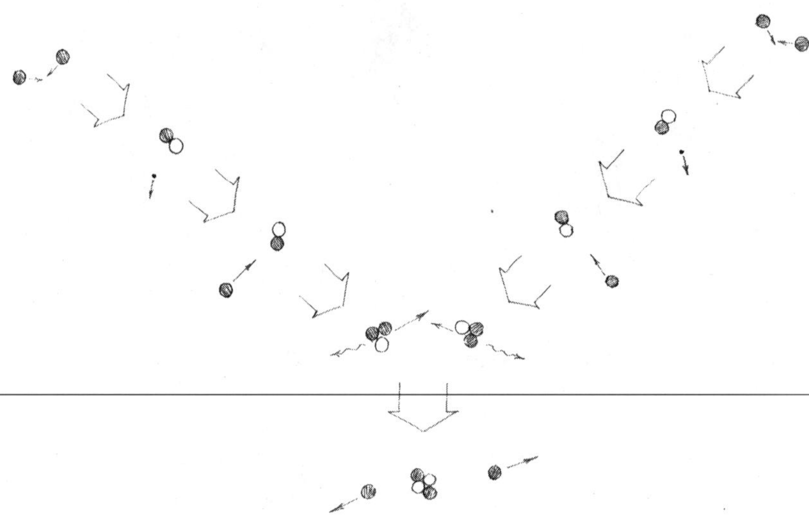

The rates of the reaction depend on the number of nuclei per unit volume and on the temperature; the higher the temperature, the faster the thermal motion of the particles and the more frequent and energetic the collisions. At the temperature of the sun's interior, which has been estimated to be 10–20 million degrees, the kinetic energies resulting from the thermal motion are in the neighborhood of 1 keV.

The release of large amounts of energy by means of fusion processes *on earth* has so far been possible only in thermonuclear explosions, such as hydrogen bombs. A hydrogen bomb consists of a mixture of light elements with a fission bomb. The high particle energies produced by the fission reaction serve to initiate the fusion reaction. The explosion of a fission bomb produces a temperature of about 5×10^7 °K, which is sufficiently high to make fusion possible. The fusion reactions then release additional large amounts of energy. The total energy release is much greater than would be liberated by the fission bomb alone. Moreover, while there is a sort of upper limit beyond which fission bombs become not much more destructive (because they disperse the extra fissionable material before it can undergo fission), there seems to be no such upper limit to the size, and therefore the destructive power, of fusion weapons.

The lack of an upper limit on the destructiveness of fusion bombs is one of the reasons why scientists such as Oppenheimer, Fermi, and Rabi advised against making such weapons, at least as long as there was any reasonable hope for international arms control agreements.

? *13. Is the ratio of the amount of hydrogen to the amount of helium in the sun increasing or decreasing?*

24.10 | The strength of nuclear forces

The large energies involved in nuclear reactions, a million or more times larger than the energies involved in chemical (molecular) reactions, indicate that the forces holding the nucleus together are very much stronger than the forces that hold molecules together. Another clue to the magnitude of nuclear forces is the density of a typical nucleus. The work of Rutherford and his colleagues on the scattering of α particles showed that atomic nuclei have radii in the neighborhood of 10^{-13} cm to 10^{-12} cm; this means that the volume of an atomic nucleus may be as small as 10^{-39} to 10^{-36} cm^3. The mass of one of the lighter atoms is of the order of 10^{-24} g, and this mass is almost all concentrated in the nucleus, with the result that the density of the nucleus may be as high as 10^{12} to 10^{14} g/cm^3. Densities of such magnitude are thousands of billions of times beyond the limits of ordinary experience, since the greatest densities of ordinary material are in the neighborhood of 20 g/cm^3 (uranium, gold, lead). It is evident that the forces that hold the atomic nucleus together must be very different from any forces considered so far. The search for understanding of these forces is one of the most important problems of modern physics. Although a good deal has been learned about nuclear forces, the problem is far from solved.

Information about nuclear forces has been obtained in several ways. It is possible to deduce some of the properties of nuclear forces from the known properties of atomic nuclei, for example, from the binding-energy curve of the graph on page 706. That curve shows that the average binding energy per nucleon has nearly the same value for all but the lightest nuclei, about 8 MeV per nucleon. In other words, the *total* binding energy of a nucleus is roughly proportional to the number of nucleons. If every particle in the nucleus were to exert a force on every other particle, it would be expected that the energy of the interactions, and therefore the binding energy, would be approximately proportional to the number of interacting pairs. But the number of pairs of nucleons goes up nearly in proportion to the square of the number of nucleons. Therefore, the binding energy calculated by assuming such interacting pairs is very different from the experimental results. To deal with this contradiction, it is necessary to assume that a nuclear particle does not interact with all other nuclear particles, but only with a limited number of them, that is, only with its nearest neighbors. For this to be the case, the nuclear forces must have a *short range*; the nuclear forces must fall off very rapidly as the distance between two nucleons increases. This decrease must be more rapid than the $1/r^2$ decrease of the gravitational force between two particles, or the $1/r^2$ decrease of the Coulomb electric force between two charges.

The chief problem studied by the team of physicists in the documentary film *People and Particles* is whether the electric force between charged particles of very small distances varies inversely as the square of the distance. (It does.)

The presence of protons in the nucleus also tells something about nuclear forces. Since there are only positively charged and neutral particles in the nucleus, the electric forces must be repulsive. The nucleus is very small, of the order of 10^{-12} cm in diameter; therefore, these repulsive forces must be enormous. Why then do the pieces that make up the nucleus not fly apart? It seems reasonable to assume that the electric repulsion is overcome at very small distances by very strong attractive forces between the nuclear particles. Information about such specifically nuclear forces can be obtained by studying the scattering of protons or neutrons by materials containing protons. Scattering experiments and the theory needed to account for their results form an important branch of nuclear physics. These experiments show that such attractive nuclear forces do indeed exist. Many of the properties of these forces are now known. But the problems of nuclear forces and how they hold the nucleus together lie at the frontier of nuclear research.

In the absence of a complete theory of nuclear forces and structure, models of the nucleus have been developed. Several models are in use, each for a specific aspect of nuclear phenomena, because no one model adequately describes the whole wide range of phenomena, from particle emission in radioactive decay to nuclear reactions and fission. Two of the most prominent of these models are described briefly in the next two sections: the liquid-drop model and the shell model.

?

14. Why is it assumed that there are special nuclear forces to hold the nucleus together?

15. Why is it assumed that the nuclear force is very short-range?

24.11 | The liquid-drop nuclear model

In the liquid-drop model, the nucleus is regarded as analogous to a charged drop of liquid. This model was suggested because the molecules in a liquid drop are also held together by short-range forces. According to this model, the particles in the nucleus, like the molecules in a liquid drop, are in continual random motion; the nucleus retains its integrity because of forces analogous to the surface tension of the liquid drop. The model also suggests an analogy between the evaporation of molecules from the surface of a drop and the escape of α particles from the nucleus (the actual mechanisms for the two processes are, however, quite different).

This model has been especially useful in describing nuclear reactions. A particle may enter the nucleus from outside and impart enough additional kinetic energy to the protons and neutrons to permit the escape of a proton or a neutron, or a combination such as a deuteron or an α particle. A detailed quantitative theory of nuclear reactions based on this idea has been developed.

The usefulness of the liquid-drop model is well shown in its ability to account for fission. As you know, when a sample of U^{235} is bombarded with slow neutrons, that is, neutrons whose kinetic energy is very small, a U^{235} nucleus may capture a neutron to form a U^{236} nucleus. The energy made available inside the nucleus by the captured neutron can be calculated:

mass of U^{235} nucleus	=	235.04393 amu
mass of neutron	=	1.00867 amu
total mass	=	236.05260 amu
mass of (unexcited) U^{236} nucleus	=	236.04573 amu
difference in mass	=	0.00687 amu

corresponding excess energy $= 0.00687 \text{ amu} \times 931 \text{ MeV/amu}$
$= 6.4 \text{ MeV}$

Therefore, at the instant when the neutron is captured, the U^{236} nucleus formed has this additional energy, 6.4 MeV, which is called the *excitation energy* due to the neutron capture. This energy is several million electron volts, even though the kinetic energy of the neutron (less than 1 eV) is relatively so small that it can be neglected in this calculation.

What happens to the excited U^{236} nucleus? This problem was studied theoretically in 1939 by Niels Bohr, who had come to the United States, and John A. Wheeler, an American physicist. They showed that, according to the liquid-drop model, the U^{236} should be able to act like a drop of water when the latter is "excited" by being given mechanical energy. The nucleus can be deformed into an elongated or dumbbell-like shape whose two (charged) parts may be beyond the range of the nuclear forces of attraction. The electric force of repulsion between the two parts of the deformed nucleus can overcome the short-range attractive forces, causing the nucleus to split, that is, to undergo fission, and causing the fragments to separate with high speeds. Each of the fragments will then quickly assume a spherical (or nearly spherical) form because within it the attractive nuclear forces again predominate. A schematic picture of a possible sequence of stages is sketched below.

The liquid-drop model gives a simple answer to the question: Why do some nuclides (U^{235} and Pu^{239}) undergo fission with slow neutrons while others (Th^{232} and U^{238}) undergo fission only with fast neutrons? The answer is that a certain minimum amount of energy must be available to a nucleus to deform it enough so that the repulsive electric forces can overcome the attractive nuclear forces. This amount, called the *activation energy*, can be calculated with the aid of the mathematical theory of the liquid-drop model. When U^{235} captures a neutron to make U^{236}, the excitation energy of the U^{236} nucleus is greater than the energy required for fission, even if the exciting neutron has very low kinetic energy. This calculation was made by Bohr and Wheeler in 1939; they found that their model predicted, correctly, that U^{235} undergoes fission with slow neutrons. The theory also predicted that when U^{238} captures a slow neutron to form U^{239}, the excitation energy is *smaller* than the activation energy by 0.9 MeV. Therefore, U^{238} should not undergo fission unless bombarded with neutrons with kinetic energies of 0.9 MeV or more. The accuracy of this prediction was verified by experiment.

Fission sometimes occurs spontaneously, but so rarely that it can be neglected for this treatment.

SG 19, 20

16. According to the liquid-drop model, what kind of force is responsible for fission of a nucleus?

17. Why does U^{238} require fast *neutrons to provoke fission? Why does fission occur in U^{239} with* slow *neutrons?*

24.12 | The shell model

Another nuclear model is required to account for other properties of the nucleus that could not be accounted for by the liquid-drop model. In Sec. 22.7, you saw that nuclides with even numbers of neutrons and protons are more stable than nuclides that contain odd numbers of either protons or neutrons. Detailed experimental studies of nuclear stability have shown that nuclei having 2, 8, 20, 50, or 82 protons, or 2, 8, 20, 50, 82, or 126 neutrons are unusually numerous and stable. These nuclei have greater binding energies than do closely similar nuclei. When the exceptional properties of nuclei with these numbers of protons and neutrons became clear, in 1948, no available theory or model of the nucleus could account for this situation. The numbers 2, 8, 20, 50, 82, and 126 were referred to as "magic numbers."

It was known from the study of chemical properties that atoms with atomic numbers 2, 10, 18, 36, 54, and 86 (gases helium to radon) have special chemical stability. This property was explained in the Bohr–Rutherford model of the atom by the idea that the electrons around each nucleus tend to arrange themselves in concentric shells, with each shell able to contain only a certain maximum number of electrons: two for the innermost shell, eight for the next, and so on. An especially stable atom is one with a full electron shell on the outside. Although the Bohr–Rutherford model has been replaced by a more successful one based on quantum mechanics, the idea of shells still provides a useful picture, and a nuclear model called the *nuclear shell model* has been developed to deal with the observation that some nuclei are particularly stable.

In the nuclear shell model, it is assumed that protons can, in a rough way of speaking, arrange themselves in shells, and that neutrons can, independently, do likewise; in the "magic-number" nuclei the shells are filled. The model has been worked out in great detail on the basis of quantum mechanics and has been successful in correlating the properties of nuclides that emit α or β particles and γ photons and in describing the electric and magnetic fields around nuclei. But the nuclear shell model does not help explain fission, and there are fundamental differences between this model and the liquid-drop model. For example, the shell model emphasizes definite patterns in which nucleons are arranged, while the liquid-drop model pictures the nuclear material in random motion. Each model is successful in accounting for some nuclear phenomena, but fails for others.

When two seemingly contradictory theories or models must be used in a field of physics, a strong effort is made to develop a more general viewpoint, or theory, which can include the two as special cases. Such a nuclear theory is being developed; it is called the *collective model*, and one of the physicists who has

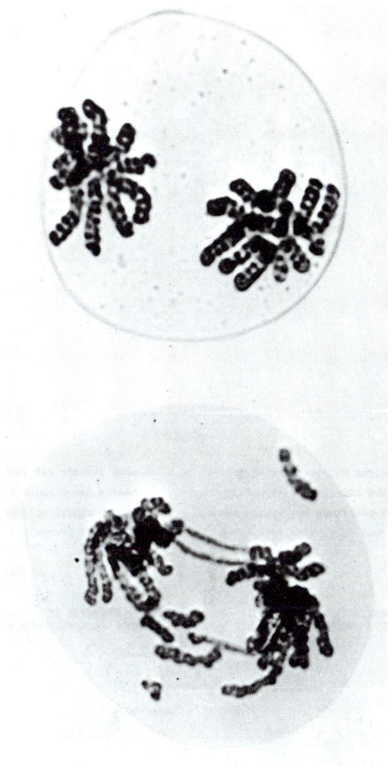

The upper portion of this photo shows normal plant cell chromosomes divided into two groups. Below, the same cell is shown after X-ray exposure; fragments and bridges between groups are typical radiation-induced abnormalities.

worked on this model is the Danish physicist Aage Bohr, the son of Niels Bohr. This model represents an advance beyond the shell and liquid-drop models in correlating nuclear data. It also has limits; thus, it does not answer fundamental questions about nuclear forces, which are still among the chief problems in modern physics.

?

18. According to the shell model, what gives nuclei having a "magic number" of protons and neutrons their special properties?

19. Which is more accurate, the liquid-drop or the shell model of the nucleus?

An autoradiograph of a fern frond made after the plant had absorbed a solution containing radioactive sulfur ($_{16}S^{35}$).

24.13 | Biological and medical applications of nuclear physics

Section 24.7 discussed military applications of nuclear energy and the use of nuclear energy as a source of electric power for cities, industries, and agriculture. There are many other applications that may, in the long run, turn out to be more important than some of those already mentioned. These may be included under the general headings of *radiation biology* and *radiation medicine*. The fields of science indicated by these names are broad. This section will indicate, by means of a few examples, some of the problems that are being worked on. In this work, radiations are used in the study of biological phenomena, in the diagnosis and treatment of disease, and in the improvement of agriculture.

The physical and chemical effects of various kinds of radiations on biological materials are being studied to find out, for example, how radiation produces genetic changes. Since it has been discovered that many of the key chemical processes in cells are organized by single chains of molecules, it is clear that a single particle of radiation can, by breaking a chemical bond in such a chain, cause a permanent and perhaps disastrous change in the cell.

The metabolism of plants and animals is being studied with the aid of extremely small amounts of radioactive nuclides called *isotopic tracers*, or "tagged atoms." A radioactive isotope (for example, C^{14}) acts chemically (and therefore physiologically) like a stable isotope (C^{12}). Thus, by following a radioactive tracer with counters, the behavior of a chemical material can be followed as it goes through various metabolic processes. The role of micronutrients (elements that are essential, in extremely small amounts, for the well-being of plants and animals) can be studied in this way. Agricultural experiments with fertilizers

Damaged trees surround a radioactive cesium-137 capsule. The capsule had been kept in place for nearly 6 months in an experiment to study the effects of ionizing radiation on biological systems.

An aerial view of Brookhaven National Laboratory, where the experiment shown above was performed.

containing radioactive isotopes have shown at what point in the growth of a plant the fertilizer is essential. In chemistry, radioactive isotopes help in the determination of the details of chemical reactions and of the structure of complex molecules, such as proteins, vitamins, and enzymes.

Perhaps the most rewarding uses of radioisotopes have been in medical research, diagnosis, and therapy. For example, tracers can help to determine the rate of flow of blood through the heart and to the limbs, thus aiding in the diagnosis of abnormal conditions. Intense doses of radiation can do serious damage to all living cells, but diseased cells are often more easily damaged than normal cells. Radiation can, therefore, be used to treat some diseases, such as cancer. Some parts of the body take up particular elements perferentially. For example, the thyroid gland absorbs iodine easily. Specially prepared radioisotopes of such elements can be administered to the victims of certain diseases, thus supplying desired radiation right at the site of the disease. This method has been used in the treatment of cancer of the thyroid gland, blood diseases, and brain tumors and in the diagnosis of thyroid, liver, and kidney ailments.

SOME TYPICAL ISOTOPE APPLICATIONS

SG 21

Isotope	*Half-Life*	*Important Uses*
${}_{1}H^{3}$	11 years	Used as a tag in organic substances.
${}_{6}C^{14}$	4,700 years	Used as a tag in studying the synthesis of many organic substances. When ${}_{6}C^{14}$ is incorporated in food material, its presence can be traced in the metabolic products.
${}_{11}Na^{24}$	15 hours	Useful in a wide variety of biochemical investigations because of its solubility and chemical properties.
${}_{15}P^{32}$	14 days	For the study of bone metabolism, the treatment of blood diseases and the diagnosis of tumors.
${}_{16}S^{35}$	87 days	Has numerous chemical and industrial applications.
${}_{27}Co^{60}$	5.3 years	Because of its intense γ emission, may be used as a low-cost substitute for radium in radiography and therapy.
${}_{53}I^{131}$	8 days	For the study of thyroid metabolism and the treatment of thyroid diseases.

The increased use of radioisotopes has been closely related to advances in the chemistry of radioactive pharmaceuticals. Together with advances in electronics and nuclear instrumentation, these advances have led to the emergence during the 1960's and 1970's of an important medical specialty called "nuclear medicine." The promise that nuclear physics held for the future of biology and medicine is being realized. This realization symbolizes the meaning of science at its best: *Research in science lays open to our understanding the secrets of nature, and from the application of this knowledge to human needs, all people can benefit.*

study guide

1. The *Project Physics* learning materials particularly appropriate for Chapter 24 include:

Activity
Two Models of a Chain Reaction

Film
The World of Enrico Fermi

Transparency
Binding Energy Curves

2. Suppose that a nucleus of ${}_6C^{13}$ is formed by adding a neutron to a ${}_6C^{12}$ atom. Neglecting any kinetic energy the neutron may have, calculate the energy that becomes available to the nucleus because of the absorption of that neutron to make ${}_6C^{13}$. The atomic masses of C^{12} and C^{13} (in an unexcited state) are 12.000000 and 13.003354 amu.

3. The atomic mass of He^4 is 4.00260 amu; what is the average binding energy per particle?

4. Suppose that a proton with relatively small kinetic energy induces the following reaction:

$${}_3Li^7 + {}_1H^1 \rightarrow {}_2He^4 + {}_2He^4$$

If the lithium nucleus were initially at rest, what would be the relative directions of the two α particles? What would be the kinetic energy of each α particle?

5. The first nuclear transmutation (obtained by Rutherford in 1919) was the reaction:

$${}_7N^{14} + {}_2He^4 \rightarrow {}_8O^{17} + {}_1H^1$$

The atomic masses involved are:

N^{14}: 14.003074 amu
O^{17}: 16.999134 amu
He^4: 4.002604 amu
H^1: 1.007825 amu

Is energy absorbed or released in this reaction? How much energy (MeV) is absorbed or released?

6. In an experiment on the reaction given in SG 5, the α particles used had a kinetic energy of 7.68 MeV, and the energy of the protons was 5.93 MeV. What was the energy of the "recoiling" O^{17} nucleus?

7. Calculate the amount of energy (MeV) liberated in the following nuclear reaction:

$${}_7N^{14} + {}_1H^2 \rightarrow {}_7N^{15} + {}_1H^1$$

The atomic masses are:

N^{14}: 14.003074 amu
H^2: 2.014102 amu
N^{15}: 15.000108 amu
H^1: 1.007825 amu

8. Appreciable amounts of the uranium isotope ${}_{92}U^{233}$ do not occur outside the laboratory; ${}_{92}U^{233}$ is formed after the thorium nucleus ${}_{90}Th^{232}$ has captured a neutron. Give the probable steps leading from ${}_{90}Th^{232}$ to ${}_{92}U^{233}$.

9. Use the graph at the top right-hand corner of page 705 to find the binding energies for U^{235}, Ba^{141}, and Kr^{92}. Use these values to show that the energy released in the fission of U^{235} is approximately 200 MeV.

10. Possible end products of U^{235} fission, when provoked by capture of slow neutrons, are ${}_{57}La^{139}$ and ${}_{42}Mo^{95}$. This reaction may be described by the equation:

$${}_{92}U^{235} + {}_0n^1 \rightarrow {}_{57}La^{139} + {}_{42}Mo^{95} + 2{}_0n^1 + 7({}_{-1}e^0)$$

The mass of ${}_{57}La^{139}$ is 138.91 amu; that of ${}_{42}Mo^{95}$ is 94.9057 amu. How much energy is released per atom in this particular fission? (The mass of the seven electrons may be neglected.)

11. Write a set of equations that describe the decay of the fission product ${}_{336}Kr^{92}$ into ${}_{40}Zr^{92}$.

12. Loss of neutrons from a structure containing fissionable material depends on its shape as well as its size. For some shapes, it is impossible to reach a critical size because the neutron loss through the surface is too great. With what shape would a mass of fissionable material suffer the *least* loss of neutrons by passage through the surface? the *most*?

13. Why are the high temperatures produced by the explosion of a fission bomb necessary to initiate fusion in a thermonuclear device?

14. It is generally agreed that stars are formed when vast clouds of hydrogen gas collapse under the mutual gravitational attraction of their particles. How might this process lead to fusion reactions beginning in such stars? (Hint: The cloud has gravitational potential energy.)

15. One of the energy sources in the sun is the production of helium nuclei by four protons as described in Sec. 24.9: $4\,_1H^1 \rightarrow {}_2He^4 + 2\,_{+1}e^0$. Show that about 27 MeV of energy are released in each cycle.

16. Fusion reactions in the sun convert a vast amount of hydrogen into radiant energy each second.

(**a**) Knowing that the energy output of the sun is 3.90×10^{26} J/sec, calculate the rate at which the sun is losing mass.

(**b**) Convert the value 3.90×10^{26} J/sec to horsepower. (Recall that 1 horsepower is equivalent to 746 W.)

17. A source of energy in the sun may be the "carbon cycle," proposed by Hans Bethe, which is outlined below.

(**a**) Complete the six steps of the cycle.

(**b**) After a cycle has been completed, which nuclides used in the cycle have been changed (and in what ways), and which have come out the same as they entered the cycle?

$${}_6C^{12} + {}_1H^1 \rightarrow (\quad\quad) + \gamma$$
$$(\quad\quad) \rightarrow {}_6C^{13} + {}_{+1}e^0 + \nu$$
$${}_6C^{13} + {}_1H^1 \rightarrow (\quad\quad) + \gamma$$
$$(\quad\quad) + {}_1H^1 \rightarrow {}_8O^{15} + \gamma$$
$${}_8O^{15} \rightarrow (\quad\quad) + {}_{+1}e^0 + \nu$$
$$(\quad\quad) + {}_1H^1 \rightarrow {}_6C^{12} + {}_2He^4$$

18. Another reaction that may take place in the sun is:

$$He^3 + He^4 \rightarrow Be^7 + \gamma$$

The atomic mass of He^3 is 3.016030 amu, and that of Be^7 is 7.016929. Is energy absorbed or released? How much energy?

19. The atomic masses of ${}_{92}U^{233}$ and ${}_{92}U^{234}$ are 233.039498 and 234.040900 amu. The activation energy for the fission of the nucleus ${}_{92}U^{234}$ is 4.6 MeV. Is U^{233} fissionable by slow neutrons?

20. Bombardment of ${}_{94}Pu^{241}$ with slow neutrons sometimes leads to the reaction:

$${}_{94}Pu^{241} + {}_0n^1 \rightarrow {}_{94}Pu^{242} + \gamma$$

The atomic masses of Pu^{241} and Pu^{242} are 241.056711 amu and 242.058710 amu. The activation energy of Pu^{241} is 5.0 MeV. Is Pu^{241} fissionable with slow neutrons?

21. The chemical structural formula for the energy-carrying adenosine triphosphate (ATP) molecule in living cell is

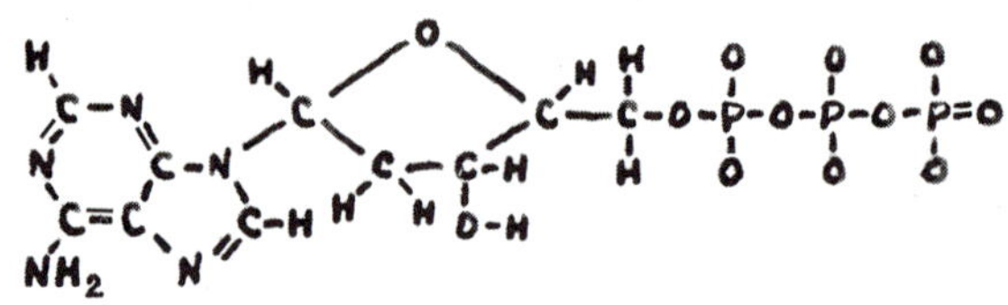

Energy is provided to some other molecule when the end phosphate group 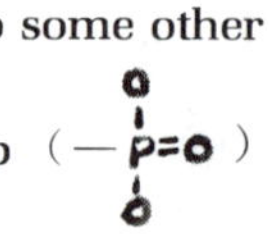 is transferred to it, changing the ATP to adenosine *di*phosphate (ADP). Energy from the oxidation of food is used to attach new phosphate groups to the ADP, changing it once again to ATP. Suggest a procedure by which you could determine the rate at which new molecules of ATP are formed.

22. Write an essay on one of the following topics:

(**a**) The various ways a citizen can help assure that technological innovations will be made and used in a manner benefiting society as a whole.

(**b**) The differences between technology and basic science.

(**c**) The responsibilities of scientists to society.

(**d**) The responsibilities of society to further science.

(**e**) The fields of physics or related sciences in which you may want to do further study.

23. Activity: Two Models of a Chain Reaction.

Mousetraps

Carefully put six or more set mousetraps in a large cardboard box. Place two small corks on each trap in such a position that they will be thrown about violently when the trap is sprung. Place a sheet of clear plastic over the top. Then drop one cork in through the corner before you slide the cover on completely. Can you imagine the situation with trillions of tiny mousetraps and corks in a much smaller space?

What in the nucleus is represented by the potential energy of the mousetrap spring? What do the corks represent? Does the model have a critical size? Describe the effect of the box cover.

Match Heads

Break off the heads of a dozen wooden matches about 0.3 cm below the match head. Arrange the match heads as shown in the drawing. Place wads of wet paper at certain points. Light a match and place it at point A. Observe what happens to the right and left sides of the arrangement. What component of a nuclear reactor is represented by the wet paper? How could you modify this model to demonstrate the function of a moderator?

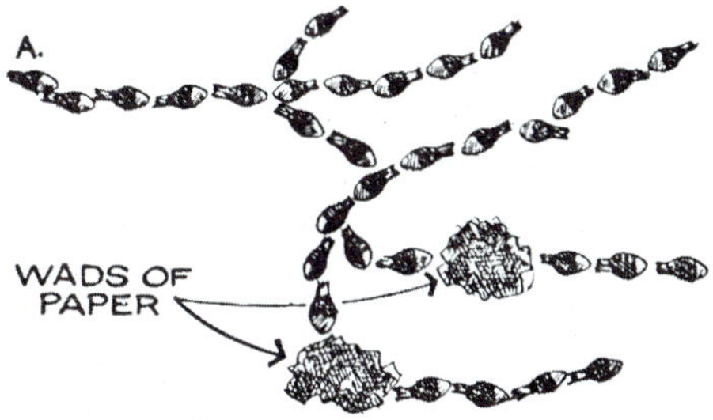

Comment on how good an analogue this is of a nuclear chain reaction.

EPILOGUE

This unit has traced the development of nuclear physics from the discovery of radioactivity to current work in nuclear fission and fusion. Radioactivity provided the starting place and tools to work with. Radioactivity revealed the naturally occurring transmutation of elements and so led to the achievement of artificial transmutations. The naturally occurring radioactive series pointed to the existence of isotopes, both radioactive and stable. Artificial transmutation has increased by many hundreds the number of nuclear species available for study and use.

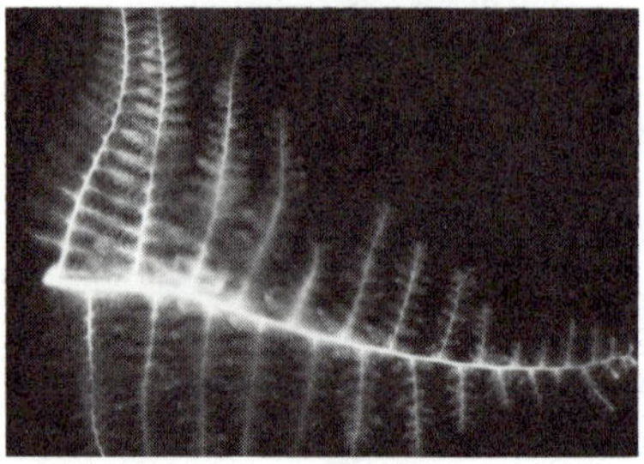

Nuclear physicists and chemists study the reactions of the stable and radioactive nuclides. The collection and correlation of a vast body of experimental data now available are the result of the work of the nineteenth-century chemists and spectroscopists. Nuclear models are built, changed, and replaced by newer and, perhaps, better models. But the detailed nature of nuclear forces is still the subject of much research, especially in the field of high-energy physics.

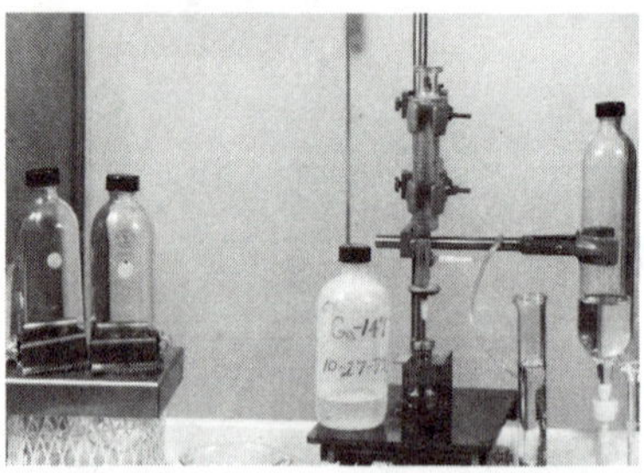

Yet that is only one of the fields that remains to be explored. The nucleus also has magnetic properties that affect the behavior of atoms. Sometimes it helps to study these properties when the atoms of matter are at very low temperatures, as close to absolute zero as possible. Nuclear physics overlaps with solid-state physics and with low-temperature physics; at low temperatures extraordinary things happen, and quanta again help to explain them.

The study of light through the development of devices such as the laser attracts many physicists. These devices are made possible by, and contribute to, the increasing understanding of how complex atomic systems jump from one energy to another, and how they can be made to change where and when they are needed.

The properties of liquids are still only imperfectly understood. Thales of Miletus was perhaps the first person on record to make a large-scale scientific speculation when he proposed, over 26 centuries ago, that maybe everything in the world is basically made of water in combinations of its various states. Thales was wrong, but even today scientists are trying to develop an adequate theory of the behavior of water molecules.

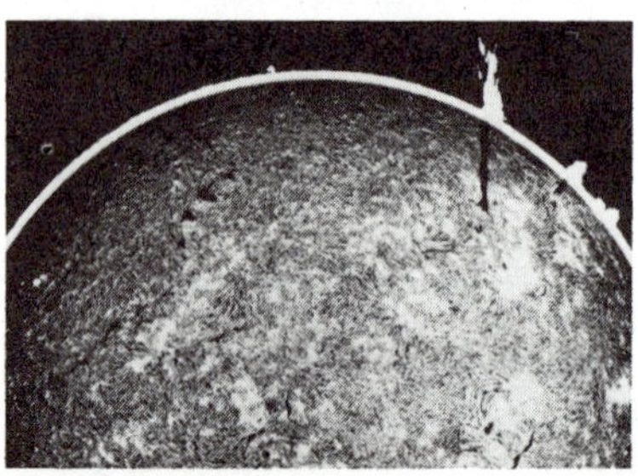

All the subjects mentioned touch on engineering, where physics and other disciplines are put to use to fashion the "artificial world." All of the engineering fields involve physics. Nuclear engineering and space engineering are the most recent and, at the moment, perhaps the most glamorous. But today the chemical engineer, the mechanical engineer, and the metallurgist all use the physicist's way of understanding the properties of atoms and atomic nuclei, because it is no longer enough to know only the properties of matter in bulk.

The radiations talked about (α, β, and γ rays) are tools for industry, biology, and medicine. They help to cure, preserve, study, understand. Neutrons are not only constituents of the nucleus, they are also probes for studies in science and in industry.

So the study of atoms and nuclei, indeed the whole course, has been an introduction not only to physics but also to the many fields with which physics is closely linked. It has been an introduction to an ever-expanding world in which much is known and understood; where much more, and perhaps the most wonderful part, is waiting to be discovered.

Acknowledgments

The authors and publisher have made every effort to trace the ownership of all selections found in this book and to make full acknowledgment for their use. Many of the selections are in the public domain.

Grateful acknowledgment is hereby made to the following authors, publishers, agents, and individuals for use of their copyrighted material.

Unit 1

Pp. 2–4 Fermi, Laura, *Atoms in the Family,* pp. 83–100 not inclusive, copyright 1954 by University of Chicago Press. **Pp. 43–45, 46–48, 53–54, 55–59** Galilei, Galileo, *Two New Sciences,* translated by Crew and DeSalvio, pp. 62–243 not inclusive, copyright 1914 by Dover Publications, Inc. **P. 60** Aristotle, *De Caelo,* translated by J. L. Stokes, Book I, Chapter 6, Oxford University Press, p. 273b. **P. 80** Newton, Sir Isaac, *The Principia,* Vol. I, Motte's translation revised by Florian Cajori, pp. 13–14, copyright © 1961, by University of California Press **P. 88** Ibid., pp. XIII–XV. **P. 117** Childe, V. Gordon, "Rotary Motion," *A History of Technology,* Vol. I, Oxford University Press, p. 187. **Pp. 126–127** Whitehead, A. N., *Science and the Modern World,* a Mentor Book published by The New American Library, pp. 46–47.

Unit 2

Pp. 148–149 Ptolemy, "The Almagest," translated by Taliaferro, R. Catesby, *Great Books of the Western World,* Vol. 16, pp. 5–12 not inclusive, copyright 1952 by Encyclopaedia Britannica, Inc. **P. 157** Rosen, Edward, translator of "The Commentariolus of Copernicus," *Three Copernican Treatises,* Columbia University Press, pp. 57–58. **Pp. 157, 159** Copernicus, "De Revolutionibus" ("On the Revolutions of Heavenly Spheres"), *Great Books of the Western World,* Vol. 16, pp. 514 and 508, copyright 1939 by Encyclopaedia Britannica, Inc. **Pp. 157–159** Rosen, op. cit., pp. 58–59. **P. 165** Knedler, John W., Jr., *Masterworks of Science,* Doubleday and Company, Inc., pp. 67–69. **P. 166** Copernicus, *Great Books of the Western World,* op. cit., pp. 514, 515, copyright 1939 by Encyclopaedia Britannica, Inc. **P. 168** Butterfield, H., *Origins of Modern Science, 1300–1800,* p. 41, copyright © 1957 by The Macmillan Company. **P. 189** Shapley, Harlow, and Howarth, Helen E., *A Source Book in Astronomy,* copyright © 1956 by Harvard University Press. **P. 194** Drake, Stillman, *Discoveries and Opinions of Galileo,* pp. 29, 31, 51, copyright © 1957 by Doubleday and Company, Inc. **Pp. 197–198** Galilei, Galileo, *Dialogue Concerning the Two Chief World Systems,* translated by Stillman Drake, pp. 123–124, copyright © 1967 by University of California Press. **P. 198** Von Gebler, Karl, *Galileo Galilei and the Roman Curia,* translated by Mrs. George Sturge, C. Kegan Paul and Company, p. 26. **P. 209** Needham, Joseph, and Pagel, Walter, eds., *Background to Modern Science,* The Macmillan Company, pp. 73–74. **P. 209** Stukeley, William, *Memoirs of Sir Isaac Newton's Life,* Taylor and Francis, p. 21, written in 1752. **P. 210** Anthony, H. D., *Sir Isaac Newton,* p. 96, copyright © 1960 by Abelard-Schuman Ltd. **Pp. 211–212** Newton, Sir Isaac, *The Principia,* Vol. I, Motte's translation revised by Cajori, pp. xvii–xviii, copyright © 1961 by University of California Press. **Pp. 211–212, 222** Ibid., Vol II, pp. 398–408 and p. 547, copyright © 1962 by University of California Press. **P. 212** Pope, Alexander, *Epitaph Intended for Sir Isaac Newton* (1732). **P. 243** Wilson, James, *The Works of the Honourable James Wilson,* published under direction of Wilson, Bird, Bronson, and Chauncey, pp. 411–412, not inclusive. **P. 245** Darwin, Erasmus, quoted by Charles Darwin in his preface to Krause, *Erasmus Darwin,* D. Appleton and Company, p. 44.

Unit 3

Pp. 246–247 Desagulier, J. T., *The Newtonian System of the World, The Best Model of Government, an Allegorical Poem.* **P. 247** Descartes, René, *The Philosophical Works of Descartes,* translated by Haldane, Elizabeth S. and Ross, G. R. T., Vol. 1, p. 299, copyright 1931 by Cambridge University Press. **P. 248** "A Free Inquiry into the received Notion of Nature," *The Works of Robert Boyle,* Vol. IV, p. 362. **P. 248** Newton, Sir Isaac, *The Principia,* Motte's translation revised by Florian Cajori, preface to first edition (1687), p. xviii, copyright © 1961 by University of California Press. **P. 249** Lucretius, *On the Nature of Things,* translated by H. Munro, George Bell & Sons, Book II, p. 51. **Pp. 249–250** Bacon, Francis, *Complete Essays Including the New Atlantis and Novum Organum,* edited by Finch, Henry L., Washington Square Press. **P. 250** Lavoisier, Antoine, *Elements of Chemistry* (1790), a facsimile of an original translation by Robert Kerr with new introduction by Professor Douglas McKie, Dover Publications, Inc. **P. 267** Gerhardt, *Die Philosophische Schriften von Gottfried Wilhelm Leibniz, 1875–90,* Vol. IV, p. 286, English translation in Dugas, R., *Mechanics in the Seventeenth Century,* Central Book Co., p. 407. **P. 285** Joule, James Prescott, *The Scientific Papers of James Prescott Joule,* an illustrated facsimile of the 1884 edition, copyright © 1963 by Dawson of Pall Mall. **P. 289** Savery, Thomas, in Hart, I. B., *James Watt & the History of Steam Power,* copyright © 1961 by Crowell Collier-Macmillan & Co. **P. 305** Joule, op. cit. **P. 307** Helmholtz, H. L. F., *Popular Scientific Lectures,* edited by Morris Kline, copyright 1873, 1881, © 1962 by Dover Publications, Inc. **P. 313** Poincaré, Henri, La Science et l'Hypothèse, E. Flammarion, excerpt translated by McClaim, John. **P. 331** Lord Kelvin, "The Size of Atoms," *Nature,* Vol. I, p. 551. **P. 338** Frost, Robert, from "West-Running Brook," *The Poetry of Robert Frost,* edited by Edward Connery Lathem, p. 259, copyright © 1969 by Holt, Rinehart and Winston, Publishers, copyright © 1956 by Robert Frost. Reprinted by permission of Holt, Rinehart and Winston, Publishers. **P. 344** Lord Kelvin, "The Kinetic Theory of the Dissipation of Energy," *Proceedings of the Royal Society of Edinburgh,* Vol. 8, p. 325. **P. 345** Nietzsche, Friedrich, "Der Wille zur Macht," *Nietzsche: An Anthology of His Works,* ed. Manthey-Zorn, Otto, Washington Square Press, p. 90. **P. 345** Shelley, Percy Bysshe, *Hellas,* ed. Wise, Thomas J., Reeves & Turner, pp. 51–52 not inclusive. **P. 346** Poincaré, Henri, *Kinetic Theory,* ed. Stephen Brush, Vol, 2, p. 206, copyright © 1966 by Pergamon Press. **P. 373** Huygens, Christian, *Treatise on Light,* first published in Paris (1690), translated by Thompson, Silvanus P., copyright 1912 by Dover Publications, Inc. **P. 390** Power, Henry, *Experimental Philosophy* (1664), reprinted 1966 by Johnson Reprint Corp., p. 192. **P. 390** Donne, John, "The First Anniversary," *Donne's Poetical Works,* ed. Grierson, Herbert J., Oxford University Press (Clarendon Press imprint), Vol. 1, p. 237.

Unit 4

P. 394 Newton, Sir Isaac, *Newton's Philosophy of Nature,* ed., H. S. Thayer, Haffner Publishing Co., N.Y., 1953, pp. 68–81. **P. 396** Blackmore, Richard, *Creation, A Philosophical Poem,*

Robert Johnson, p. 91. **P. 405** Young, Thomas, *Course of Lectures on Natural Philosophy and the Mechanical Arts,* Cox. **Pp. 407, 409–410** Newton, Sir Isaac, *Newton's Philosophy of Nature,* ed., H. S. Thayer, Haffner Publishing Co., N.Y., 1953, pp. 68–81. **P. 410** Thompson, James, *The Poetical Works of James Thompson,* William Tegg and Co., pp. 145–146 and p. 10. **P. 411** Goethe, J. W., *Goethe As A Scientist,* Magnus, Rudolf, translated by Heinz Norden, Henry Schuman, pp. 184–185, copyright 1949 by Heinz Norden. **P. 417** Young, Thomas, *Miscellaneous Works* (London, 1855) Vol. I, p. 415. **Pp. 417–418** Galilei, Galileo, *Dialogues Concerning Two New Sciences,* translated by Crew H. and de Salvio, A., copyright 1952 by Dover Publications, pp. 42–43. **Pp. 423–424** Gilbert, William, *De Magnete,* translated by Mottelay, P. Fleury, copyright 1958 by Dover Publications, pp. 1–14 and p. 121. **P. 439** von Guericke, Otto, Magie, Wm. F., *A Sourcebook in Physics,* McGraw-Hill, 1935, p. 393. **Pp. 442–443** Volta, Alessandro, *On the Electricity Excited by the Mere Contact of Conducting Substances of Different Kinds,* Philosophical Transactions of the Royal Society of London, 1800, p. 403. **Pp. 453–454** Ampère, André-Marie, Whittaker, E. T., *A History of the Theories of Aether and Electricity,* Harper & Brothers, N.Y., 1960, p. 84. **P. 468** Faraday, Michael, MacDonald, D. K. C., *Faraday, Maxwell & Kelvin,* Science Studies Series, Doubleday, Garden City, 1964, p. 11. **P. 470** Faraday, Michael, *Faraday's Diary,* G. Bell & Sons, London, 1932, Vol. I, pp. 375–376. **P. 501** Maxwell, James Clerk, *The Scientific Papers of James Clerk Maxwell,* ed., Niven, W. D., Dover Publications, N.Y., 1965, p. 158. **Pp. 505–506** Maxwell, James Clerk, *The Scientific Papers of James Clerk Maxwell,* ed., Niven, W. D., Dover Publications, N.Y., 1965, p. 500. **P. 517** Maxwell, James Clerk, *The Scientific Papers of James Clerk Maxwell,* ed., Niven, W. D., Dover Publications, N.Y., 1965, p. 775 of Vol. II. **P. 517** Maxwell, James Clerk, *The Scientific Papers of James Clerk Maxwell,* ed., Niven, W. D., Dover Publications, N.Y., 1965, p. 763 of Vol. II. **Pp. 522–523** Feynman, R. P., *The Feynman Lectures on Physics,* Feynman, R. P., Leighton, R. B., and Sands, M., Vol. 2, Addison-Wesley, 1964, pp. 9-10.

Much of the historical information in Chapter 15, in particular the discussion of the ac-dc controversy, is based on the book by Harold L. Sharlin, *The Making of the Electrical Age* (Abelard-Schuman, New York, 1963). We are grateful to Professor Sharlin for giving his permission to use this material.

Unit 5

P. 528 From "The First Chapter of Aristotle's 'Foundations of Scientific Thought' (Metaphysica, Liber A)," translated by Daniel E. Gershenson and Daniel A. Greenburg, in *The Natural Philosopher,* Vol. II, copyright © 1963 by the Blaisdell Publishing Company, pp. 14–15. **P. 529** From *The Life of the Honorable Henry Cavendish,* by George Wilson, printed for the Cavendish Society, 1851, pp. 186–187. **P. 530** From "Elements of Chemistry" by Antoine Laurent Lavoisier, translated by Robert Kerr, in *Great Books of the Western World,* Vol. 45, copyright 1952 by Encyclopaedia Britannica, Inc., pp. 3–4. **Pp. 551–552** Einstein, Albert, translated by Professor Irving Kaplan, Massachusetts Institute of Technology. **P. 556** Roentgen, W. K. **P. 565** From "Opticks" by Isaac Newton from *Great Books of the Western World,* Vol. 34, copyright 1952 by Encyclopaedia Britannica, Inc., pp. 525–531. **Pp. 574–575** From *Background to Modern Science,* Needham, Joseph and Pagel, Walter, eds., copyright 1938 by The Macmillan Company, pp. 68–69. **P. 595** Letter from Rutherford to Bohr, March 1913. **P. 619** From *Atomic Physics* by Max Born, copyright 1952 by Blackie & Son, Ltd., p. 95. **P. 620** Letter from Albert Einstein to Max Born, 1926. **P. 624** From *A Philosophical Essay on Possibilities* by Pierre Simon Laplace, translated by Frederick W. Truscott and Frederick L. Emory, copyright 1951 by Dover Publications, Inc., p. 4.

Unit 6

P. 631 Becquerel, Henri, in Magie, Williams Francis, *A Source Book in Physics,* copyright © 1964 by Harvard University Press, pp. 610–611. **P. 634** Curie, Pierre and Marie, "Sur une substance nouvelle radio-active, contenue dans la pechblende," *Comptes Rendus Des Séances De L'Académie Des Sciences,* Vol. 127, pp. 175 and 1215. **P. 635** Curie, Marie, in *Madame Curie, A Biography by Eve Curie,* translated by Vincent Sheean, copyright 1938 by Doubleday and Co., Inc., p. 170. **P. 679** Rutherford, Ernest, "Collision of Alpha Particles with Light Atoms. IV. An Anomalous Effect in Nitrogen," *Philosophical Magazine,* Series VI, Vol. 37, Apr. 1919, p. 587. **P. 681** Chadwick, James, "The Existence of a Neutron," *Proceedings of the Royal Society of London,* Vol. CXXXVI, May 1932, p. 697. **P. 682** Chadwick, James, in "Letters to the Editor," *Nature,* Vol. 129, No. 3552, Feb. 27, 1932, p. 312. Reprinted by permission. **P. 711** Hahn, Otto and Strassman, F., "On the Determination and Behavior of the Alkaline Earth Metals Resulting from Uranium Irradiated with Neutrons." *Die Naturwissenschaften,* Vol. 27, Jan. 1939, p. 15.

Picture Credits

Unit 1 opener, Dr. Harold E. Edgerton, M.I.T.; Prologue **p. 2** courtesy, Univ. of Chicago; **p. 5** Fermi Lab; **p. 6** (left) Mt. Wilson and Palomar Observatories, (right) courtesy, Prof. Erwin W. Mueller, Pennsylvania University; **p. 7** (left) Museum of Comparative Zoology, Harvard University, (right) Brookhaven National Laboratory. Chapter 1 opener (L.L.) Rue/Monkmeyer; **p. 9** Yale University Art Gallery, Collection Société Anonyme; **p. 14** Vannucci/Leo de Wys, Inc.; **p. 15** Bill Hurter/Leo de Wys, Inc; **p. 21** (top) W. O. Roberts, Director, High Altitude Observatory, Climax, Colorado, (center) Dr. Leland Ernest, Dept. of Biology, Eastern Nazarene College, (bottom) from the filmstrip "Investigating a Glacier" © 1966, Encyclopaedia Britannica Educational Corporation, Chicago; **p. 26** (top left) Bayerisches Nationalmuseum, Munich, (center left and bottom) George Eastman House, Rochester, N.Y., (center right) Berndt/Stock/Boston; **p. 27** (all) Dr. Harold E. Edgerton, M.I.T.; **p. 30** Frank Siteman/Stock/Boston; **p. 35** International Museum of Photography at George Eastman House, Rochester, N.Y. Chapter 2 opener, Yerkes Observatory, University of Chicago, William Bay, Wisconsin; **p. 37** Vatican Museum, Rome; **p. 43** (top) Burndy Library, (bottom) DeGolyer Collection, University of Oklahoma Libraries; **p. 44** DeGolyer Collection, University of Oklahoma Libraries; **p. 45** courtesy of Educational Development Center, Newton, Mass.; **p. 46** Cabinet des Dessins, Louvre Museum; **p. 53** Alinari/EPA. Chapter 3 opener, Falk/Monkmeyer; **p. 67** Wide World Photos; **p. 70** Herwig/Stock/Boston; **p. 71** (top) UPI, (bottom) Hamlin/Stock/Boston; **p. 74** C. T. Polumbaum, © Time Inc.; **p. 80** Fundamental Photographs, N.Y.; **p. 83** U.S. Bureau of Standards; **p. 85** The Granger Collection; **p. 86** NASA; **p. 87** Wide World Photos; **p. 88** Dr. Harold E. Edgerton, M.I.T.; **p. 94** The Museum of Modern Art, New York; **p. 97** U.S. Air Force. Chapter 4 opener, NASA; **p. 101** Verne, Jules, "De la terre à la lune," Paris, 1866; **p. 102** NASA; **p. 104** NASA; **p. 105** (top) Houghton Library, Harvard University, (bottom) from PSSC Physics, D. C. Heath & Co., Boston, 1965; **p. 106** (top) S. Wayman/Life Magazine, © Time Inc., (bottom) National Film Board of Canada; **p. 114** A. Rickerby/Life Magazine, © Time Inc.; **p. 118** (bottom) NASA; **p. 119** Collection, The Museum of Modern Art, New York; **p. 121** (top) A. Gregory, Jr., (center) Mt. Wilson and Palomar Observatories, (bottom) T. Duffy/Focus on Sports; **p. 125** A. E. Clark/Life Magazine © Time Inc.; **p. 126,** photos that appear have been credited previously.

Unit 2 opener, O. Bessim/Monkmeyer; Prologue **p. 129** (top) The American Museum of Natural History, New York, (bottom) courtesy of the Trustees of The British Museum; **p. 130** Stephen Perrin. Chapter 5 opener, NASA; **p. 134** E. Schultess/Black Star Publishing Co.; **p. 136** (left) J. Stofan/Amphoto, (right) J. Bufkin, Macon, Missouri, Feb., 1964; **p. 137** © The Anasazi Project, Inc., Karl Kernberger; **p. 138** Mt. Wilson and Palomar Observatories; **p. 145** DeGolyer Collection, University of Oklahoma Libraries. Chapter 6 opener, F. Baldwin/De Wys Inc.; **p. 155** Muzeum Okregowe in Torun, courtesy of Dr. Owen Gingerich; **p. 157 p. 166** Jagiellonian Library, University of Krakow, Poland; **p. 167** Keuffel & Esser Co.; **p. 172** (top left) from Atlas Major, Vol, I, Jan Blaeu, 1664, (center) Danish Information Service, (bottom left) The Mansell Collection, London, (bottom right) Danish Information Center; **p. 173** Smithsonian Astrophysical Observatory, courtesy of Dr. Owen Gingerich; **p. 174** (top) Smithsonian Institution, (bottom) from Brahe, Tycho, "Astronomiae Instauratae Mechanica," Philip von Ohr, Wandsbeck, 1598; **p. 176** (left) Lick Observatory, (bottom) Smithsonian Astrophysical Observatory, (right) Cerro Tololo Inter American Observatory. Chapter 7 opener, NASA, The Bettmann Archive; **p. 182** Kepler, Johannes, "Mysterium cosmographicum," Linz, 1596; **p. 193** Instituto e Museo di Storia della Scienza, Florence, Italy; DeGolyer Collection, University of Oklahoma Libraries; **p. 195** (top) Griffith Observatory, Los Angeles, California; **p. 196** (top) Lowell Observatory; **p. 200** Alinari/Art Reference; **p. 201** Bill Bridges; **p. 203** NASA. Chapter 8 opener, M. Forsyth/Monkmeyer; **p. 206** Yerkes Observatory; **p. 208** (left) Harvard Observatory, (right) engraved by Bt. Reading from a painting by Sir Peter Lely, Trinity College, Cambridge; **p. 221** Descartes, René, "Principia Philosophiae," 1664; **p. 223** Berry/Magnum Photos; **p. 228** NASA; **p. 231** Cavendish, Henry, "Philosophical Transactions of the Royal Society," vol. 88, 198; **p. 241,** photos that appear have been credited previously. **p. 242** Print Collection of the Federal Institute of Technology, Zurich.

Unit 3 opener, Hobart/Monkmeyer; Prologue, **p. 247** Swiss Federation of Watch Manufacturers; **p. 248** Whitworth Art Gallery, University of Manchester. Chapter 9 opener, courtesy of Brunswick Division, Brunswick Corporation; **p. 249** (top) Joel Gordon © 1978, (bottom) Dean Loomis/© Time Inc.; **p. 250** (top) Colgate University; **p. 251** T. Hugh/Photo Researchers; **p. 252** (left) painted by Jacques Louis David, Courtesy of the Rockefeller University; **p. 256** Boston Patriots, Football Club; **p. 258** National Film Board of Canada; **p. 261** The Boeing Company; **p. 263** (top) Mt. Wilson and Palomar Observatory, (bottom) H. Groskwsky/Life Magazine, © Time Inc.; **p. 265** Royal Netherlands Academy of Sciences and Letters, Amsterdam; **p. 266 p. 267** Burndy Library, Norwalk, Conn.; **p. 272** "Le Petit Prince" by Antoine de Saint-Exupéry, © 1943 by Harcourt Brace & World, Inc. Chapter 10 opener, U. S. Forest Service, **p. 273** Dept. of Energy; **p. 275** J. Berndt/Stock Boston; **p. 277** D. Strickler/Monkmeyer; **278** G. Bellerose/Stock/Boston; **p. 278** Albert B. Gregory Jr.; **p. 280** Dr. Harold E. Edgerton, M.I.T.; **p. 281** Dimitri Kessel/Life Magazine, © Time Inc.; **p. 283** (top) B. Suess/Focus on Sports, (bottom) T. Duffy/Focus on Sports; **p. 284** The Granger Collection; **p. 285** American Institute of Physics; **p. 287** (top to bottom) C. W. Kirland/Life Magazine © Time Inc., Life Magazine © Time Inc., Agricola, Georgius, "De Re Metallica," 1556, Houghton Library, Harvard University, M. G. Walker/Life Magazine © Time Inc.; **p. 288** Science Museum, London; **p. 289** Newcomen Society, London; **p. 291** (top) Boston & Maine Corporation, (bottom) Lewis Hine/George Eastman House Collection; **p. 293** (all except top right) Science Museum, London; **p. 294** Franco Tosi/Westinghouse; **p. 295** American Institute of Physics; **p. 298** Professor Keith R. Porter, Dept. of Biology, Harvard University; **p. 304** Courtesy of The Metropolitan Museum of Art, bequest of Samuel A. Lewisohn, 1951; **p. 306** (top) The Granger Collection, (bottom) Harvard Observatory, Photo; **p. 307** Koenigsberger, Leo, "Hermann von Helmholtz," Braunschweig Druck und Verlag, von Friedrich Vieweg und Sohn; **p. 312** AIP Niels Bohr Library/Lande Collection; **p. 315** H. Redi, Life Magazine, © Time Inc.; **p. 317** Courtesy of Kenyon & Eckhardt, Inc. and Fugazy Travel Bureau, Inc. Chapter 11 opener, The Bettmann Archive; **p. 323** U. S. Air Force; **p. 326 p. 327** The Granger Collection; **p. 332** De Wys Inc.; **p. 339** Educational Development Center; **p. 341** (bottom) P. E. Genereux, E. Lynn; **p. 343** PSSC Physics, D. C. Heath & Co., 1965; **p. 344** (top to bottom) a, b, d, Dr. Harold E. Edgerton, M.I.T., Colgate University, Ansel Adams, 1944; **p. 345** Thompson, Silvanus P., "Life of William Thompson, Vol. 1, Macmillan & Co., Ltd, London; **p. 348** Greek National Tourist Office, New York City. Chapter 12 opener, B. Adams/Monkmeyer; **p. 354** W. Bischof/Magnum Photos; **p. 358** D. Plowden/Photo Researchers; **p. 364** W. Bischof/Magnum Photos; **p. 368** (bottom) Museum of Fine Arts Boston; **p. 372** National Film Board of Canada; **p. 374** courtesy of College of Engineering University of California, Berkeley; **p. 375 p. 377** (except top) courtesy, Film Studio Educational Development Center; **p. 379** (right) U. S. Navy; **p. 383** (top left) Schaefer and Seawell; **p. 385** (left) Bell Telephone Laboratories, (right) Hedrich-Blessing; **p. 390,** photos that appear have been credited previously.

Unit 4 opener, General Electric; Prologue, **p. 393** Euler, Leonhard, "Theoria Motuum Planetarum et Cometarum," Berlin, 1774; **p. 394** (left) Descartes, René, "Principia Philosophiae," 1664, (right) Maxwell, James Clerk, "A Treatise on Electricity and Magnetism, Vol. II, Oxford Press, 1892; **p. 395** National Radio Astronomy Observatory, Greenbank, West Virginia. Chapter 13 opener, General Motors; **p. 396** Minor White; **p. 398** (top right) Greek National Tourist Office, (bottom left) Bausch & Lomb Optical Co., (bottom right) Air France; **p. 399** Frisius, Reiner Gemma, "De Radio Astronomico et Geometrico Liber," 1545; **p. 402** (top) Bausch & Lomb Optical Co., (bottom) General Electric Co.; **p. 404** Bausch & Lomb Optical Co.; **p. 405** engraved by G. R. Ward, from a picture by Sir Thomas Lawrence; **p. 406** (bottom) Cabinet des Estampes Bibliothèque Nationale, Paris; **p. 407** © 1962, Springer-Verlag OHG, Berlin; **p. 408** (top, d) National Radio Astronomy Observatory, Greenbank, West Virginia, (center and bottom) © 1962, Springer-Verlag OHG, Berlin; **p. 416** Collection, The Museum of Modern Art, Lillie P. Bliss Collection. Chapter 14 opener, Fundamental Photographs, N.Y.; **p. 422** courtesy of the Lawrence Radiation Laboratory, University of California; **p. 425** courtesy of Burndy Library; **p. 427** Musée de Versailles; **p. 428** from Coulomb's memoire to the French Academy of Sciences, 1785; **p. 440** American Academy of Arts & Sciences; **p. 441** (bottom) courtesy of The Burndy Library; **p. 442** The Science Museum, London; **p. 446** courtesy of Stanford University; **p. 452** (left) Nationalhistoriske Museum, Frederiksborg; **p. 453** (top) from "Photographs of Physical Phenomena," © Kodansha, Tokyo, 1968, (bottom) Bibliothèque de l'Académie des Sciences; **p. 454** Crown copyright, Science Museum, London; **p. 458** Dr. V. P. Hessler; **p. 460** Radio Corporation of America. Chapter 15 opener, S. Falk/Monkmeyer; **p. 465** Freelance Photographers Guild; **p. 466** Deutsches Museum, Munich; **p. 468** The Royal Institute of Great Britain; **p. 475** E. Herwig/Stock/Boston; **p. 476** Tennessee Valley Authority; **p. 477** (top and bottom) Figuier, Louis, "Les Nouvelles Conquêtes de la Science," Paris, 1884; **p. 478** (top) Tannahill Research Library/Henry Ford Museum, Dearborn, Michigan, (right) Thomas Alva Edison Foundation; **p. 479** U.S. Patent Office Picture Collection; **p. 480** N. Y. Herald, Dec. 21, 1879; **p. 484** (left) Tennessee Valley Authority; **p. 484** (right) Power Authority of the State of N.Y.; **p. 486** (left) Westinghouse Electric Corp., (bottom) Edison Electric Inst., N.Y.C.; **p. 487** (left) Empire State Building Observatory, (right) Wide World Photos; **p. 489** Power Authority of the State of New York; **p. 490** Institute of Energy Conversion, University of Delaware; **p. 492** (left) Institute of Energy Conversion, (right) NASA; **p. 495** D. Brody/Stock/Boston. Chapter 16 opener, Rogers/Monkmeyer; **p. 499** Dr. V. P. Hessler; **p. 501** Courtesy of Trinity College Library, Cambridge; **p. 508** HRW Photo/Russell Dian; **p. 513** courtesy of Eastman Kodak Co.; **p. 514** (top) National Center for Atmospheric Research, (center) Kitt Peak National Observatory, (bottom) Lick Observatory, University of California; **p. 515** (top left) U. S. Air Force Photo, courtesy of News Photo Branch, Defense Information Service; **p. 515** (top right and center right) M.I.T. Lincoln Laboratory; **p. 515** (bottom left) American Science & Engineering, Inc.; **p. 516** (top) courtesy of G. E. X-ray Dept., (center) Eastman Kodak Co., (bottom) Brookhaven National Lab; **p. 517** American Institute of Physics, Niels Bohr Library; **p. 521** John Tenniel drawings; **p. 522,** photos that appear have been credited previously.

Unit 5, Prologue, **p. 525** courtesy, Museum of Fine Arts Boston, Henry Lillie Pierce Residuary Fund; **p. 528** Fisher Scientific Co., Medford, Mass. Chapter 17 opener, H. Rogers/Monkmeyer; **p. 533** © Hawthorn Books, Inc., N.Y.C.; **p. 540** Burndy Library, Norwalk, Conn.; **p. 541** Courtesy Bell Laboratories; **p. 543** Burndy Library, courtesy of AIP Niels Bohr Library; **p. 547** California Institute of Technology; **p. 552** HRW/Russell Dian; **p. 553** (top right) courtesy of New York Times, (top center) American Institute of Physics, (top left) courtesy of Europa Verlag, Zurich, (bottom) courtesy of California Institute of Technology Archives; **p. 554** courtesy of the Niels Bohr Library, American Institute of Physics; **p. 555** (top left) Dr. Max F. Millikan, (top right) Harper Library, University of Chicago, (bottom) The Smithsonian Institution; **p. 558** Eastman Kodak Co.; **p. 560** (right) High Voltage Engineering Corporation, (left) American Institute of Radiology; **p. 561** (left) Eastman Kodak Co., (right) S. Feldman/Photo Researchers, (bottom) UPI. Chapter 19 opener, Hale Observatories; **p. 566** Science Museum, London, lent by Sir Lawrence Bragg, F. R. S.; **p. 572** courtesy of Dr. Owen J. Gingerich, Smithsonian Astrophysical Observatory; **p. 575** UPI; **p. 584** (left and top right) American Institute of Physics, (bottom) courtesy of Niels Bohr Library, American Institute of Physics; **p. 591** Dept. of Energy, Lawrence Lab. Chapter 20 opener, The Ealing Press, Cambridge, Mass.; **p. 601** PSSC film "Matter Waves," Production #68, Education Development Center; **p. 607** American Institute of Physics; **p. 608** Professor Harry Meiners, Rennselaer Polytechnic Institute; **p. 609** Académie des Sciences, Paris; **p. 611** American Institute of Physics; **p. 612** (top and bottom) American Institute of Physics; **p. 613** Professor W. K. Heisenberg; **p. 614** (left) Perkin-Elmer Corp., (right) Professor E. W. Mueller; **p. 618** © 1961 by John Wiley & Sons, Inc.; **p. 621** Hawthorn Books Inc.; **p. 625,** photos that appear have been credited previously.

Unit 6 opener, NCAR Photo; Prologue, **p. 628** Yankee Atomic Electric, Rowe, Mass. Chapter 21 opener, Brookhaven National Lab; **p. 630** (bottom) Burndy Library, Norwalk, Conn.; **p. 636** (left) The Smithsonian Institute, (top right) Institut de Physique Nucléaire, Paris, (center) courtesy of Mrs.

Eve Curie Labouisse. Photo from Laboratoire Curie, Institut de Physique Nucléaire, (bottom) Culver Pictures; **p. 637** Brown Brothers; **p. 639** Tracerlab, Waltham, Mass.; **p. 643** Atomic Energy Commission, Savannah River Plant. Chapter 22 opener, Oak Ridge National Laboratory; **p. 655** Professor Kenneth Bainbridge, Harvard University; **p. 657** Brown Brothers; **p. 660** Cavendish Laboratory, Cambridge University, Cambridge, England; **p. 661** Professor Alfred O. C. Nier, University of Minnesota; **p. 662** (top) Burndy Library, Norwalk, Conn., (bottom) British Crown Copyright, Science Museum, London; **p. 668** News Office, Columbia University; **p. 670** Dr. Parker E. Moreland, Jr., Chemistry Division, Argonne National Laboratory. Chapter 23 opener, Brookhaven National Lab; **p. 675** Professor M. S. Livingston, Dept. of Physics, M.I.T., and Director, Cambridge Electron Accelerator, Harvard University; **p. 678** (bottom) Sir P. M. S. Blackett, London; **p. 680** Nobel Foundation, Stockholm; **p. 687** (top) courtesy of the Los Alamos Scientific Laboratory, (bottom) FermiLab; **p. 688** Museum of Science and Hayden Planetarium; **p. 690** (top) Fermilab, (bottom) Brookhaven National Lab; **p. 691** (top left) Fermilab, (top right) Brookhaven National Lab, (bottom) FermiLab; **p. 692** (top) Cavendish Laboratory, University of Cambridge, (center) Professor W. E. Hazen, University of Michigan, (bottom) Professor W. B. Fretter, Berkeley; **p. 693** The Ealing Press, Cambridge, Mass.; **p. 694** (left) Brookhaven National Lab, (right) FermiLab; **p. 695** (top left, bottom left) The Ealing Press, Cambridge, Mass.; **p. 696** French Embassy Press and Information Division, N. Y. C.; **p. 697** Professor Carl Anderson, California Institute of Technology; **p. 700** The Ealing Press, Cambridge, Mass. Chapter 24 opener, NCAR Photo; **p. 702** Lester V. Bergman & Assoc., Inc.; **p. 710** (top) The University of Chicago, (bottom) American Institute of Physics; **p. 711** AEC, Savannah River Plant; **p. 712** American Institute of Physics; **p. 716** Office of Public Relations, The University of Chicago; **p. 717** University of Chicago; **p. 718** Air Force Cambridge Research Laboratories; **p. 723** Lawrence Radiation Laboratory, Livermore; **p. 726** University of California, Lawrence Livermore Lab; **p. 727** "Les Saintes-Maries," Oskar Reinhart Collection, Winterthur; **p. 733 p. 734 p. 735** Brookhaven National Lab; **p. 740,** photos that appear have been credited previously.

Index